160.00

M & T CHEMICALS INC.
RESEARCH LABORATORY
LIBRARY
BOX 1104, RAHWAY, N. J.

D1496200

The chemistry of the
metal—carbon bond
Volume 3

THE CHEMISTRY OF FUNCTIONAL GROUPS

A series of advanced treatises under the general editorship of Professor Saul Patai

The chemistry of alkenes (2 volumes)
The chemistry of the carbonyl group (2 volumes)
The chemistry of the ether linkage
The chemistry of the amino group
The chemistry of the nitro and nitroso groups (2 parts)
The chemistry of carboxylic acids and esters
The chemistry of the carbon—nitrogen double bond
The chemistry of amides
The chemistry of the cyano group
The chemistry of the hydroxyl group (2 parts)
The chemistry of the azido group
The chemistry of the acyl halides
The chemistry of the carbon—halogen bond (2 parts)
The chemistry of the quinonoid compounds (2 parts)
The chemistry of the thiol group (2 parts)
The chemistry of the hydrazo, azo and azoxy groups (2 parts)
The chemistry of the amidines and imidates
The chemistry of cyanates and their thio derivatives (2 parts)
The chemistry of diazonium and diazo groups (2 parts)
The chemistry of the carbon—carbon triple bond (2 parts)
Supplement A: The chemistry of double-bonded functional groups (2 parts)
Supplement B: The chemistry of acid derivatives (2 parts)
The chemistry of ketenes, allenes and related compounds (2 parts)
The chemistry of the sulphonium group (2 parts)
Supplement E: The chemistry of ethers, crown ethers, hydroxyl groups and their sulphur analogues
Supplement F: The chemistry of amino, nitroso and nitro compounds and their derivatives
Supplement C: The chemistry of triple-bonded functional groups (2 parts)
The chemistry of the metal—carbon bond Volume 1
Supplement D: The chemistry of halides pseudo-halides and azides
The chemistry of peroxides
The chemistry of the metal—carbon bond volume 2
The chemistry of organic selenium and tellurium compounds

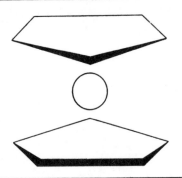

The chemistry of the metal—carbon bond
Volume 3
Carbon—carbon bond formation using organometallic compounds

Edited by

FRANK R. HARTLEY

*The Royal Military College of Science
Shrivenham, England*

and

SAUL PATAI

*The Hebrew University
Jerusalem, Israel*

1985

JOHN WILEY & SONS

CHICHESTER – NEW YORK – BRISBANE – TORONTO – SINGAPORE

An Interscience ® Publication

Copyright © 1985 by John Wiley & Sons Ltd.

All rights reserved.

No part of this book may be reproduced by any means, or transmitted, or translated into a machine language without the written permission of the publisher.

Library of Congress Cataloging in Publication Data:
Main entry under title:

Carbon–carbon bond formation using organometallic compounds.

  (The Chemistry of the metal–carbon bond; v. 3)
(The Chemistry of functional groups)
  'An Interscience publication.'
  Includes indexes.
  1. Organometallic compounds. 2. Chemical bonds.
I. Hartley, F. R. II. Patai, Saul. III. Series.
IV. Series: Chemistry of functional groups.
QD410.C43   1984 vol. 3   547'.05 s   84-13078
[QD411]   [547'.05]

ISBN 0 471 90557 7

British Library Cataloguing in Publication Data:

The Chemistry of the metal–carbon bond.—
  (The chemistry of functional groups)
  Vol. 3: Carbon–carbon bond formation using
organometallic compounds
  1. Organometallic compounds
  I. Hartley, Frank   II. Patai, Saul
  547'.05   QD411

ISBN 0 471 90557 7

Photoset by Thomson Press (India) Ltd., and printed and bound in Great Britain at The Bath Press, Avon

Volume 3—Contributing authors

Gordon K. Anderson	Department of Chemistry, University of Missouri–St. Louis, St. Louis, Missouri 63121, USA
G. Paolo Chiusoli	Istituto di Chimica Organica dell'Università, Via M. D'Azeglio 85, 43100 Parma, Italy
Chit–Kay Chu	Chemistry Department, University of Malaya, Kuala Lumpur 22-11, Malaysia
V. G. Kumar Das	Chemistry Department, University of Malaya, Kuala Lumpur 22-11, Malaysia
Julian A. Davies	Department of Chemistry, College of Arts and Sciences, University of Toledo, Toledo, Ohio 43606, USA
Donald M. Fenton	Science and Technology Division, Union Oil Company of California, P.O. Box 76, Brea, California 92621, USA
G. Henrici-Olivé	Department of Chemistry, University of California at San Diego, San Diego, California, USA
Jon E. Johansen	Division of Applied Chemistry, SINTEF, N-7034 Trondheim-NTH, Norway
Léone Miginiac	Laboratoire de Synthèse Organique, Université de Poitiers, 40 Avenue du Recteur Pineau, 86022 Poitiers, France
Eric L. Moorehead	Science and Technology Division, Union Oil Company of California, P.O. Box 76, Brea, California 92621, USA
S. Olivé	Department of Chemistry, University of California at San Diego, San Diego, California, USA
Olav-T. Onsager	Laboratory of Industrial Chemistry, Norwegian Institute of Technology, University of Trondheim, N-7034 Trondheim-NTH, Norway
Giuseppe Salerno	Dipartimento di Chimica, Università della Calabria, Arcavacata di Rende, Cosenza, Italy
Fumie Sato	Department of Chemical Engineering, Tokyo Institute of Technology, Meguro, Tokyo 152, Japan

Volume 3—contributing authors

Randy J. Shaver	Department of Chemistry, College of Arts and Sciences, University of Toledo, Toledo, Ohio 43606, USA
Jiro Tsuji	Department of Chemical Engineering, Tokyo Institute of Technology, Meguro, Tokyo 152, Japan
Mark J. Winter	Department of Chemistry, The University, Sheffield S3 7HF, UK

Foreword

The Chemistry of the Metal—Carbon Bond is a multi-volume work within the well established series of books covering *The Chemistry of Functional Groups*. It aims to cover the chemistry of the metal—carbon bond as a whole, but lays emphasis on the carbon end. It should therefore be of particular interest to the organic chemist. The general plan of the material will be the same as in previous books in the series with the exception that, because of the large amount of material involved, this will be a multi-volume work.

The first volume was concerned with:
 (a) The structure and thermochemistry of organometallic compounds.
 (b) The preparation of organometallic compounds.
 (c) The analysis and spectroscopic characterization of organometallic compounds.

The second volume was concerned with cleavage of the metal—carbon bond, insertions into metal—carbon bonds, nucleophilic and electrophilic attack of metal—carbon bonds, oxidative addition, and reductive elimination. It also included a chapter on the structure and bonding of Main Group organometallic compounds.

This volume is concerned with the use of organometallic compounds to create carbon—carbon bonds. Chapters commissioned on the use of alkali metal, alkaline earth metal and Group III organometallic compounds as well as Ziegler–Natta polymerization olefin methathesis, and the use of metal—arene compounds in carbon—carbon bond formation have not yet been completed and will be included in the next volume. The remaining volume in the series will cover the use of organometallic compounds in the synthesis of carbon—hydrogen and other carbon—element bonds, as well as the preparation and use of Main Group organometallic compounds in organic synthesis.

In classifying organometallic compounds we have used Cotton's hapto nomenclature (η-) to indicate the number of carbon atoms directly linked to a single metal atom.

In common with other volumes in *The Chemistry of the Functional Groups* series, the emphasis is laid on the functional group treated and on the effects which it exerts on the chemical and physical properties, primarily in the immediate vicinity of the group in question, and secondarily on the behaviour of the whole molecule. The coverage is restricted in that material included in easily and generally available secondary or tertiary sources, such as Chemical Reviews and various 'Advances' and 'Progress' series as well as textbooks (i.e. in books which are usually found in the chemical libraries of universities and research institutes) is not, as a rule, repeated in detail, unless it is necessary for the balanced treatment of the subject. Therefore, each of the authors has been asked *not* to give an encyclopaedic coverage of his subject, but to concentrate on the most important recent developments and mainly on material that has not been adequately covered by reviews or other secondary sources by the time of writing of the chapter, and to address himself to a reader who is assumed to be at a fairly advanced postgraduate level. With these restrictions, it is realised that no plan can be devised for a volume that would give a *complete* coverage of the subject with *no* overlap between the chapters, while at the same time preserving the readability of the text. The Editors set themselves the goal of attaining *reasonable* coverage with *moderate* overlap, with a minimum of cross-references between the chapters of each volume. In this manner sufficient freedom is given to each author to produce readable quasi-monographic chapters. Such a plan necessarily means that the breadth, depth and thought-provoking nature of each chapter will differ with the views and inclinations of the author.

The publication of the Functional Group Series would never have started without the support of many people. Foremost among these is the late Dr Arnold Weissberger, whose reassurance and trust encouraged the start of the task. This volume would never have reached fruition without and Mrs Baylis's and Mrs Hunt's help with typing and the efficient and patient cooperation of several staff members of the Publisher, whose code of ethics does not allow us to thank them by name. Many of our colleagues in England, Israel and elsewhere gave help in solving many problems, especially Professor Z. Rappoport. Finally, that the project ever reached completion is due to the essential support and partnership of our wives and families.

Shrivenham, England   FRANK HARTLEY
Jerusalem, Israel   SAUL PATAI

Contents

1. Carbon—carbon bond formation using tin and lead organometallics   1
   V. G. Kumar Das and C.-K. Chu

2. Carbon—carbon bond formation using organometallic compounds of zinc, cadmium, and mercury   99
   L. Miginiac

3. Carbon—carbon bond formation using η^3-allyl complexes   143

   Part 1. η^3-Allylnickel complexes   143
   G.P. Chiusoli and G. Salerno

   Part 2. η^3-Allylpalladium complexes   163
   J. Tsuji

   Part 3. Other η^3-allyl transition metal complexes   200
   F. Sato

4. Olefin oligomerization   205
   O.-T. Onsager and J. E. Johansen

5. Alkyne oligomerization   259
   M. J. Winter

6. Transition metal carbonyls in organic synthesis   295
   J. A. Davies and R. J. Shaver

7. Olefin and alcohol carbonylation   335
   G. K. Anderson and J. A. Davies

8. Olefin hydroformylation   361
   J. A. Davies

9. The Fischer–Tropsch synthesis   391
   G. Henrici-Olivé and S. Olivé

10. Olefin Carbonylation   435
    D. M. Fenton and E. L. Moorehead

Author index   447
Subject index   479

List of Abbreviations Used

In order to economize on printing expenses, whenever possible abbreviations are used instead of full names of radicals and instead of explicitly drawn structures. The following list, arranged alphabetically, contains most of these abbreviations.

A	appearance potential
Ac	acetyl (CH_3CO)
acac	acetylacetone
ac	acrylonitrile
acacen	bis(acetylacetonate)ethylenediamine
Ad	adamantyl
aibn	azobisisobutyronitrile
all	allyl
An	actinide metal
ap	antiplanar
appe	$Ph_2AsCH_2CH_2PPh_2$
Ar	aryl
bae	bis(acetylacetonate)ethylenediamine
bipy	2,2′-bipyridyl
Bz	benzyl
Bu	butyl (also t-Bu or Bu^t)
cd	circular dichroism
cdt	(E,E,E)-cyclododeca-1,5,9-triene
cht	cycloheptatriene
CI	chemical ionization
CIDNP	chemically induced dynamic nuclear polarization
CNDO	complete neglect of differential overlap
coct	cyclooctene
1,5-cod	cycloocta-1,5-diene
cot	cyclooctatetraene
Cp	η^5-cyclopentadienyl
Cp*	η^5-pentamethylcyclopentadienyl
C.P.	cross-polarization
Cy	cyclohexyl
dba	dibenzylideneacetone
dbn	1,5-diazabicyclo[5.4.0]non-5-ene

List of abbreviations used

dbu	1,8-diazabicyclo[5.4.0]undec-7-ene
dccd	dicylohexylcarbodiimide
def	diethyl fumarate
diars	*o*-bis(dimethylarsino)benzene
dibah	diisobutylaluminium hydride
diop	2,3-*o*-isopropylidene-2,3-dihydroxy-1,4-bis(diphenylphosphino)butane
dme	1,2-dimethoxyethane
dmfm	dimethyl fumarate
dmm	dimethyl maleate
dmpe	bis(1,2-dimethylphosphino)ethane
dotnH	bis(diacetylmonoxime)propylene-1,3-diamine
dpm	dipivaloylmethanato
dppb	bis(1,4-diphenylphosphino)butane
dppe	bis(1,2-diphenylphosphino)ethane
dppm	bis(1,1-diphenylphosphino)methane
dppp	bis(1,3-diphenylphosphino)propane
dmso	dimethyl sulphoxide
$E_{1/2}$	half-wave potential
ece	electron transfer–chemical step–further electron transfer
ee	enantiomeric excess
EI	electron impact
E_p	peak potential
ESCA	electron spectroscopy for chemical analysis
Et	ethyl
eV	electronvolt
Fc	ferrocene
FD	field desorption
FI	field ionization
fmn	fumaronitrile
fod	$F_3C(CF_2)_2COCH=C(O)C(CH_3)_3$
Fp	$Fe(\eta^5\text{-}C_5H_5)(CO)_2$
Fp*	$Fe(\eta^5\text{-}C_5H_5)(CO)(PPh_3)$
FT	Fourier transform
Fu	furyl (OC_4H_5)
Hex	hexyl (C_6H_{13})
hfac	hexafluoroacetone
hfacac	hexafluoroacetylacetonato
hmdb	hexamethyl(Dewar)benzene
hmpa	hexamethylphosphoramide
hmpt	hexamethylphosphorotriamide
HOMO	highest occupied molecular orbital
I	ionization potential
ICR	ion cyclotron resonance
I_D	ionization potential
INDOR	inter-nuclear double resonance
LCAO	linear combination of atomic orbitals

lda	lithium diisopropylamide
Ln	lanthanide metal
LUMO	lowest unoccupied molecular orbitals
M	metal
M	parent molecule
ma	maleic anhydride
MAS	magic angle spinning
m-cpba	*m*-chloroperbenzoic acid
Me	methyl
MNDO	modified neglect of diatomic overlap
mnp	2-methyl-2-nitrosopropane
Naph	naphthyl
nbd	norbornadiene
nbs	*N*-bromosuccinimide
nmp	*N*-methylpyrrolidone
oA	*o*-allylphenyldimethylarsine
Pe	pentenyl
Pen	pentyl (C_5H_{11}-)
phen	*o*-phenanthroline
Pip	piperidyl ($C_5H_{10}N$-)
Ph	phenyl
pmdeta	pentamethyldiethylenetriamine
ppm	parts per million
Pr	propyl (also *i*-Pr or Pr^i)
PRDDO	partial retention of diatomic differential overlap
pyr	pyridyl (C_5H_4N)
pz	pyrazolyl
Quin	quinolyl (NC_9H_6)
R	any radical
RT	room temperature
salen	bis(salicylaldehyde)ethylenediamine
salophen	bis(salicylaldehyde)-*o*-phenylenediamine
sce	standard calomel electrode
set	single electron transfer
SOMO	singly occupied molecular orbital
sp	synplanar
SPT	selective population transfer
tba	tribenzylideneacetylacetone
tcne	tetracyanoethylene
thf	tetrahydrofuran
Thi	thienyl (SC_4H_3)
tmed	tetramethylethylenediamine
tms	tetramethylsilane (only used in this context as a free standing symbol)
tms	trimethylsilyl (only used in this context either with a dash after it or adjacent to a chemical symbol)

Tol	tolyl ($CH_3C_6H_4$)
tond	1,3,5,7-tetramethyl-2,6,9-trioxobicyclo[3.3.1]nona-3,7-diene
tos	tosyl
trityl	triphenylmethyl (Ph_3C)
ttfa	thallium tris(trifluoroacetate)
un	olefin or acetylene
X	halide
Xyl	xylyl ($Me_2C_6H_3$)

In addition, entries in the 'List of Radical Names' in *IUPAC Nomenclature of Organic Chemistry*, 1979 Edition, Pergamon Press, Oxford, 1979, pp. 305–322) will also be used in their unabbreviated forms, both in the text and in structures.

We are sorry for any inconvenience to our readers. However, the rapidly rising costs of production make it absolutely necessary to use every means to reduce expenses— otherwise the whole existence of our Series would be in jeopardy.

The Chemistry of the Metal—Carbon Bond, Vol. 3
Edited by F. R. Hartley and S. Patai
© 1985 John Wiley & Sons Ltd.

CHAPTER **1**

Carbon—carbon bond formation using tin and lead organometallics

V. G. KUMAR DAS and CHIT-KAY CHU

Chemistry Department, University of Malaya, Kuala Lumpur 22–11, Malaysia

I. INTRODUCTION	2
II. REACTIONS OF α-FUNCTIONALLY SUBSTITUTED ORGANO-TIN AND -LEAD COMPOUNDS	3
A. α-Elimination Leading to Carbene Formation	3
B. C—C Bond Formation in Direct Reactions with Carbon Electrophiles	5
C. C—C Bond Formation via Lithium Derivatives	8
1. Halogen–lithium and hydrogen–lithium exchange	8
2. Transmetallation	12
III. REACTIONS OF β- AND γ-FUNCTIONALLY SUBSTITUTED ORGANO-TIN AND -LEAD COMPOUNDS	17
A. C—C Bond Formation via β-Substituted Derivatives	17
B. C—C Bond Formation via γ-Substituted Derivatives	21
IV. REACTIONS OF α,β- AND β,γ-UNSATURATED ORGANO-TIN AND -LEAD COMPOUNDS	24
A. Vinyl- and Alkynyl-Metal Derivatives	24
1. Alkynyl transfer reactions	25
2. Other C—C bond-forming reactions of alkynylmetals	27
3. Vinyl transfer reactions	30
4. Diels–Alder reactions of alkenylstannanes	34
B. Allylmetal Derivatives	36
1. Allyl transfer reactions	37
a. Reactions with organic halides, -acetates, and acid halides .	37
b. Carbonyl addition reactions	41
2. Reactions with electrophilic olefins and alkynes	50
C. Aryl- and Benzyl-Metal Derivatives	52
V. REACTIONS OF ORGANOTIN COMPOUNDS CONTAINING SATURATED OR REMOTELY UNSATURATED CARBON CHAINS	55

	A. Tetraalkyl Derivatives.	55
	B. Derivatives with Remote Side-chain Unsaturation.	58
VI.	REACTIONS OF HETEROATOM-BONDED ORGANO-TIN AND -LEAD COMPOUNDS	61
	A. O-Stannyl and O-Plumbyl Derivatives.	61
	1. Enolstannanes	61
	2. Alkoxides and other derivatives	66
	B. N-Stannyl and N-Plumbyl Derivatives.	67
	C. Organotin Hydrides	70
VII.	MISCELLANEOUS REACTIONS	77
	A. Oxidative Addition–Reductive Elimination	77
	B. C—C Coupling via Stannylenes	78
	C. Reactions of Lead Tetraacetate and Aryllead Tricarboxylates	78
	D. Catalytic Reactions.	83
	1. Ziegler-type polymerization.	83
	2. Olefin metathesis	84
	3. Other	88
VIII.	ACKNOWLEDGEMENT	89
IX.	REFERENCES	89

I. INTRODUCTION

Organometallic compounds have found increasing use in recent years for the stoichiometric and catalytic synthesis of a wide range of organic compounds. Although the art remains largely empirical, the measure of success already achieved in respect of chemo-, regio-, and stereo-selective synthesis[1] of several cyclic and open-chain systems using Main Group and transition metal organometallics has created a surge of interest in research directed at their application potential. An aspect of organic synthesis of paramount concern to us in this series is that of C—C bond formation, and in this Chapter we focus attention on the use of tin and lead organometallics. A perusal of the literature quickly reveals that among Main Group metals the entry of organotin and, to a very much lesser extent, organolead compounds into the important and diverse field of organic synthesis is relatively recent. The majority of useful documented reactions of organotin compounds, particularly the hydrides, alkoxides, and amines, are concerned with the synthesis of bonds linking carbon to hydrogen and other heteroatoms. Excellent reviews bearing on the above may be found in several recent monographs[2-4] and articles[5-18] on the subject, and in the volumes edited by Sawyer[19]. Work particularly originating in the last 5 years strongly indicates that organotin compounds can find utility for the creation of new C—C bonds, thus extending their overall synthetic potential.

In this chapter we attempt to examine the several organo-tin and -lead initiated coupling reactions that result in C—C bond formation within the mechanistic framework of the known reactivities[2,3,20-24] of these organometals towards electrophiles, nucleophiles, and free radicals. In as much as a large number of such reactions reported in the literature appear as isolated accounts, little purpose is served in approaching this review purely from the standpoint of the projected utility of the reactions in organic synthesis. We attempt instead to place the results in perspective by giving a comprehensive coverage of these reactions based on metal atom functionality.

A unified discussion of both organo-tin and -lead compounds is presented, despite some obvious differences in their chemistries and the relative infancy of investigations directed to the use of organoleads in organic synthesis. Literature coverage for this chapter extends up to the close of 1982, but several relevant publications accessible to us after this date have also been included.

II. REACTIONS OF α-FUNCTIONALLY SUBSTITUTED ORGANO-TIN AND -LEAD COMPOUNDS

A. α-Elimination Leading to Carbene Formation

C—C bond formation via the generation of carbenes from organo-tin and -lead compounds has attracted much attention since 1960, when Clark and Willis[25] found that the pyrolysis of trimethyl(trifluoromethyl)tin with or without tetrafluoroethylene gave perfluorocyclopropane.

$$3Me_3SnCF_3 \xrightarrow[\text{(sealed tube)}]{150\ °C,\ 20\ h} 3Me_3SnF + \text{cyclo-}C_3F_6 \quad (1)$$
$$95\%$$

Seyferth et al.[26] reported that the harsh reaction conditions for the generation of the $:CF_2$ species could be avoided by using NaI in dme as solvent, making this method a more useful route for synthesizing *gem*-difluorocyclopropanes (equation 2). Although the generation

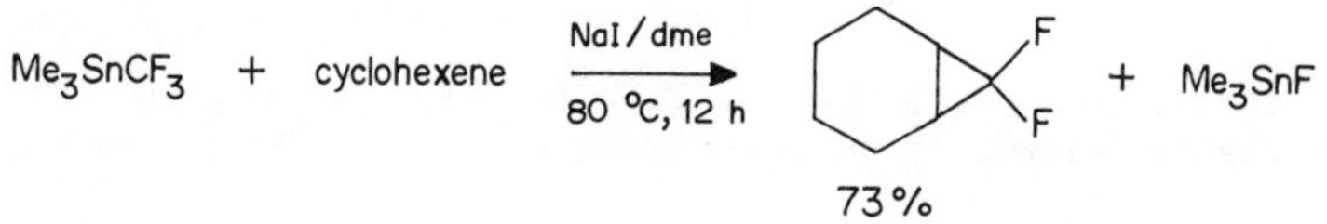

$$Me_3SnCF_3 + \text{cyclohexene} \xrightarrow[80\ °C,\ 12\ h]{NaI/dme} \text{[bicyclic CF}_2\text{]} + Me_3SnF \quad (2)$$
$$73\%$$

of other halocarbenes from trialkyl(trihalomethyl)tins has since been demonstrated[27–29] (for example, equation 3), the hydrolytic instability of the organotin substrates precludes their routine use in synthesis.

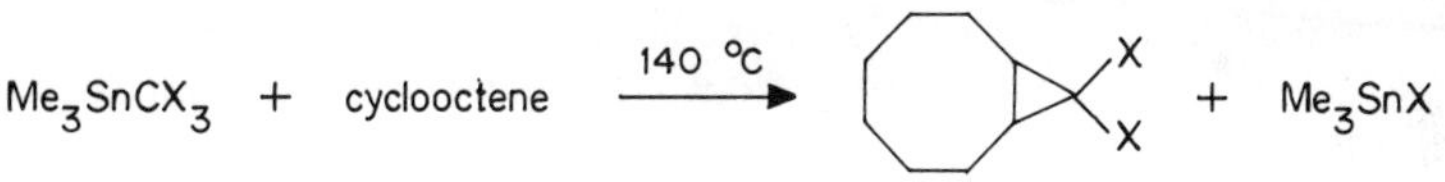

$$Me_3SnCX_3 + \text{cyclooctene} \xrightarrow{140\ °C} \text{[bicyclic CX}_2\text{]} + Me_3SnX \quad (3)$$

$$X = Cl\ (94\%)\ ;\ Br\ (66\%)$$

Triphenyl(trihalomethyl)plumbanes[30] have also been found to be good divalent carbon transfer agents, particularly at temperatures between 120 and 150 °C. Ph_3PbCCl_3, for example, has been used to form cyclopropane derivatives in reactions with olefinic compounds[31,32]. Other divalent carbon transfer agents such as $(R_3M)_2CX_2$ and R_3MCHX_2 (M = Sn or Pb) have also been investigated[28,33] but, in comparison with $PhHgCCl_3$[4], they appear to be less useful in that the reaction conditions of higher temperatures and longer reaction times tend to impinge on the thermal stability of the products and hence the yields.

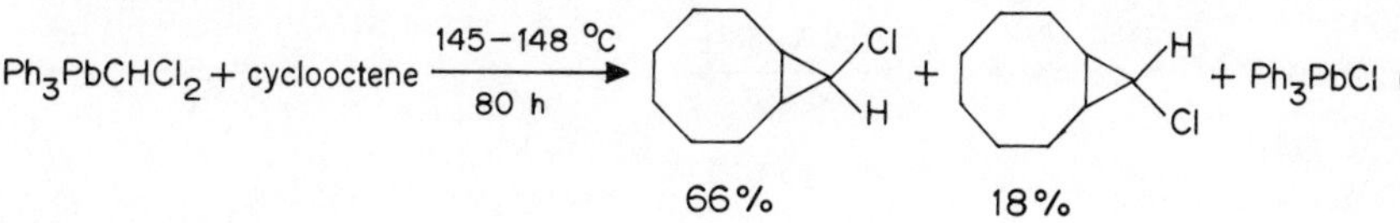

$$Ph_3PbCHCl_2 + \text{cyclooctene} \xrightarrow[80\ h]{145-148\ °C} \text{[bicyclic CHCl]} + \text{[bicyclic CHCl]} + Ph_3PbCl \quad (4)$$
$$66\% \qquad 18\%$$

7-Bromo-7-norcaranyltriphenyltin, **1**, has been shown to yield the spiro-derivative **3** on refluxing in cyclooctene. A cyclopropylidene intermediate, **2**, has been speculated[33] for this reaction.

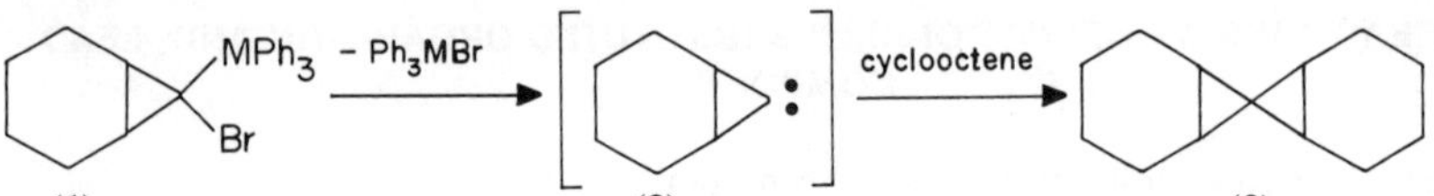

$$\text{(1)} \xrightarrow[-Ph_3MBr]{MPh_3} [\text{(2)}] \xrightarrow{\text{cyclooctene}} \text{(3)} \quad (5)$$

M = Sn, Pb

Lotts and coworkers[35] have recently reported that Me_3SnCH_2Cl undergoes metallation in cyclohexene to yield exclusively *anti*-7-trimethylstannylnorcarane, via the trimethylstannyl carbene **4**.

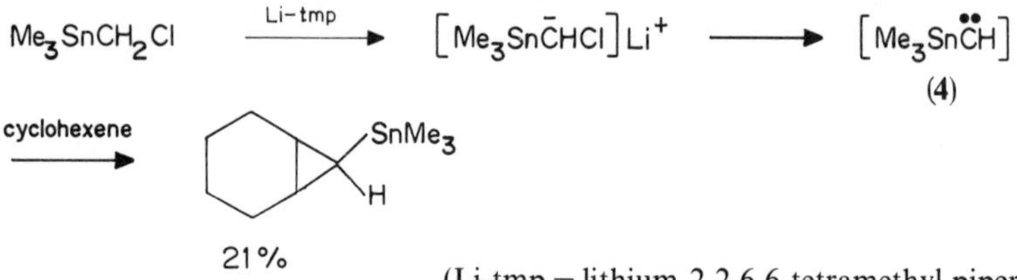

$$Me_3SnCH_2Cl \xrightarrow{Li\text{-}tmp} [Me_3Sn\bar{C}HCl]Li^+ \longrightarrow [Me_3Sn\ddot{C}H] \quad (6)$$
$$\text{(4)}$$

$\xrightarrow{\text{cyclohexene}}$ norcarane-SnMe$_3$ (H) 21%

(Li-tmp = lithium 2,2,6,6-tetramethyl piperidine)

This reaction contrasts with that reported[36] for Ph_3SnCH_2I (equation 7), wherein transmetallation followed by methylene transfer appear clearly to be implicated. In

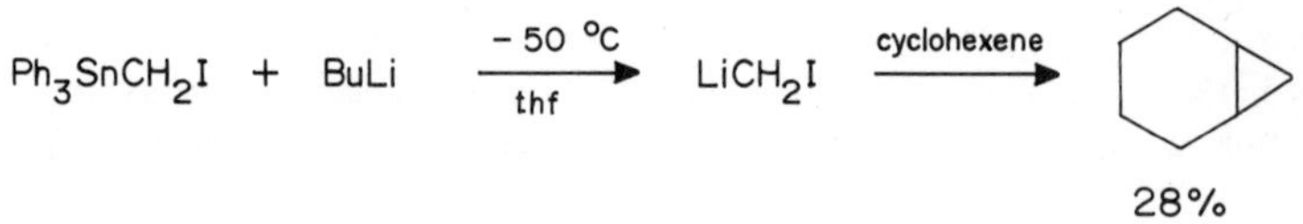

$$Ph_3SnCH_2I + BuLi \xrightarrow[\text{thf}]{-50\,°C} LiCH_2I \xrightarrow{\text{cyclohexene}} \text{norcarane} \quad (7)$$
28%

another circumstance, however, the carbethoxy-stabilized triphenylstannyl carbene, $Ph_3Sn\ddot{C}CO_2Et$[37], has been generated from ethyl diazo(triphenylstannyl)acetate and trapped using isobutene.

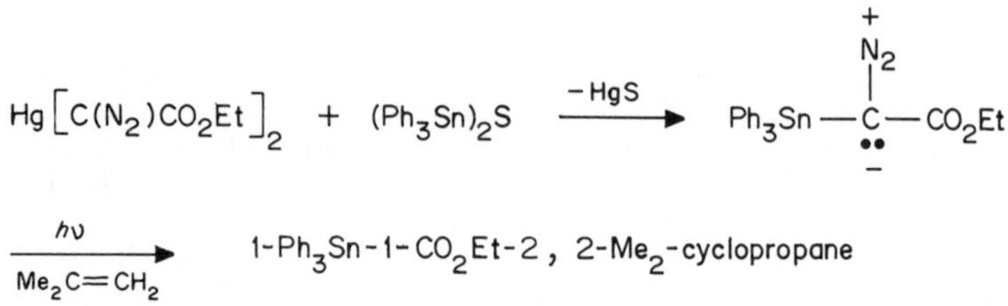

$$Hg[C(N_2)CO_2Et]_2 + (Ph_3Sn)_2S \xrightarrow{-HgS} Ph_3Sn-\underset{-}{\overset{N_2^+}{\underset{\cdot\cdot}{C}}}-CO_2Et$$

$$\xrightarrow[Me_2C=CH_2]{h\nu} 1\text{-}Ph_3Sn\text{-}1\text{-}CO_2Et\text{-}2,2\text{-}Me_2\text{-cyclopropane} \quad (8)$$

The reaction of $Bu_3SnCHCl(OEt)$, **5**, with excess of cyclohexene has been shown to lead to ethoxynorcaranes[38] (equation 9). The formation of the carbene, :CHOEt, via α-elimination from **5** has been inferred for this reaction.

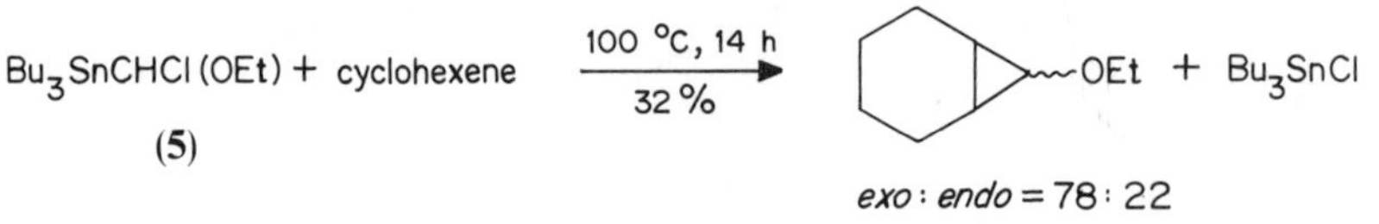

$$Bu_3SnCHCl(OEt) + \text{cyclohexene} \xrightarrow[32\%]{100\,°C,\,14\,h} \text{norcarane-OEt} + Bu_3SnCl \quad (9)$$
(5)

exo : *endo* = 78 : 22

Interestingly, stannyl- and plumbyl-diazoalkanes do not undergo α-elimination to yield the corresponding carbenes. With dicarboxylate acetylenes, these reagents have been shown[39,40] to undergo 1,3-dipolar cycloaddition reactions to give unstable organometallic substituted isopyrazoles (6), which immediately form the more stable pyrazoles (7) by metallotropic rearrangement as indicated in equation 10.

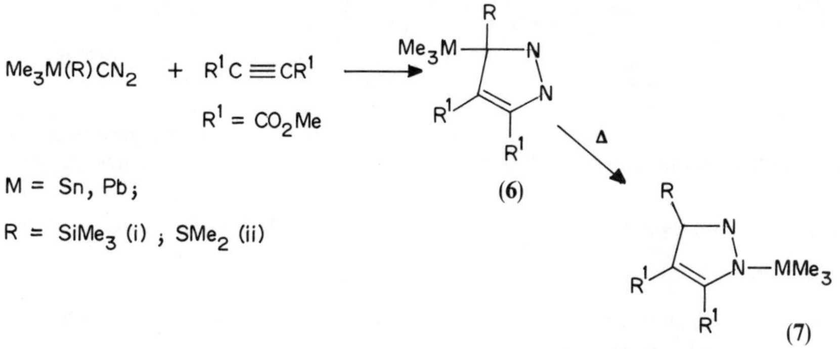

$$Me_3M(R)CN_2 + R^1C\equiv CR^1 \longrightarrow \text{(6)} \xrightarrow{\Delta} \text{(7)} \tag{10}$$

$R^1 = CO_2Me$

M = Sn, Pb;
R = SiMe$_3$ (i) ; SMe$_2$ (ii)

M = Sn; 69% (i) ; 82% (ii)
M = Pb; 82% (i) ; 71% (ii)

B. C—C Bond Formation in Direct Reactions with Carbon Electrophiles

The tendency of α-functionally substituted organotin compounds to undergo electrophilic attack at the Sn—C bond may be particularly enhanced when substituents with a negative mesomeric effect such as CN or CO_2R reside on the α-carbon[41]. Thus with aldehydes and ketones as electrophiles, addition reactions result (equation 11). The

$$R_3Sn-\underset{Y}{\overset{|}{C}}- + R^1COR^2 \longrightarrow R_3SnO-\underset{R^2}{\overset{R^1}{\underset{|}{C}}}-\underset{Y}{\overset{|}{C}}- \tag{11}$$

Y = CN or CO_2R

additions are similarly facilitated by electron-withdrawing groups on the attacking electrophile. In polar solvents such as hmpt, even organic halides can cleave the Sn—C bond[42] (equation 12).

$$Bu_3SnCH(Pr^i)Y + RBr \xrightarrow{hmpt} RCHY(Pr^i) + Bu_3SnBr \tag{12}$$

R = allyl or benzyl ; Y = CN, CO_2Et

The addition of tributyl(trihalomethyl)tin compounds to aldehydes has been described[43] as proceeding via nucleophilic attack of the CX_3 group at the carbonyl centre

$$Bu_3SnCX_3 + RCHO \xrightarrow[20\ h]{80\ °C} RCH(CX_3)OSnBu_3 \tag{13}$$

R = Me, Et, Pr^i, Ph, PhCH=CH; X = Cl, Br

(equation 13). With α-chiral aldehydes such as 2-phenylpropanal[43], the reaction leads to two diastereomeric products:

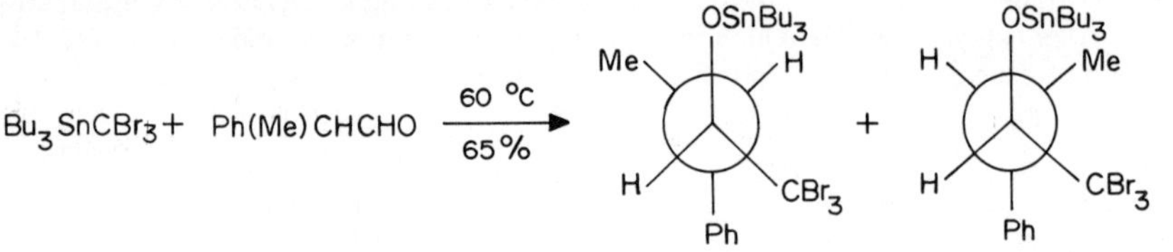

$$Bu_3SnCBr_3 + Ph(Me)CHCHO \xrightarrow[65\%]{60\ °C}$$ [Newman projection with OSnBu₃, Me, H, CBr₃, Ph] + [Newman projection with OSnBu₃, Me, H, CBr₃, Ph] (14)

Stannyl ethers are also obtained from the thermal reaction of α-stannylated nitrosamines with aromatic aldehydes[44] (for example, equation 15). Similarly, phosphono-

$$Bu^tN(NO)CH_2SnMe_3 + PhCH=CHCHO \xrightarrow[2.\ H_3O^+]{1.\ 70\ °C,\ 10\ h} Bu^tN(NO)CH_2CH(OH)CH=CHPh$$
$$54\%$$
(15)

methyl triorganostannanes have been reported to undergo C—C bond-forming reactions with aldehydes and isocyanates[45]:

$$Et_3SnCH_2P(=O)(OEt)R + R^1CHO \longrightarrow Et_3SnOCHR^1CH_2P(=O)(OEt)R$$

R = Ph, OEt ; R¹ = Pr, CCl₃ (16)

$$Et_3SnCH_2P(=O)(OEt)R + PhNCO \longrightarrow Et_3SnNPhCOCH_2P(=O)(OEt)R$$
(17)

R = OEt, Ph

As noted for α-substituted nitriles and esters (equation 11), trialkylstannylacetones also readily undergo exothermic additions[46,47] with aldehydes and ketones as in reactions 18 and 19.

$$Et_3SnCH_2COMe + PhCHO \xrightarrow[1\ h]{20\ °C} Et_3SnOCH(Ph)CH_2COMe$$
(8)
$$\downarrow (COOH)_2$$ (18)
$$PhCH(OH)CH_2COMe$$
$$67\%$$

$$8 + \text{(cyclohexanone)} \xrightarrow[48\ h]{25\ °C} \text{(cyclohexyl with OSnEt_3, CH_2COMe)} \xrightarrow{(COOH)_2} \text{(cyclohexyl with OH, CH_2COMe)}$$ (19)

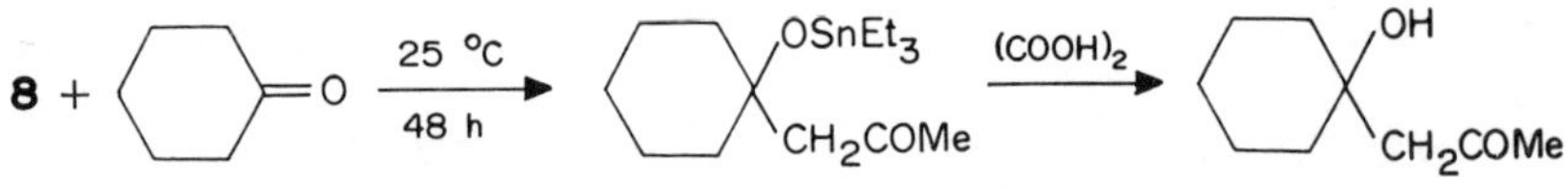

The ease of cleavage of the Sn—O bond in the resulting stannyl ethers with organic acids such as oxalic or malonic acid makes this method an attractive route for the

1. Carbon—carbon bond formation

preparation of β-hydroxy-substituted ketones. With several other stannyl ketones such as $Bu_3SnCH(Me)COEt$, the O-metallotropic form (enolstannane) predominates[48,49] (70%).

$$Bu_3SnCH(Me)COEt \rightleftarrows (E)-Bu_3SnOC(Et)=CHMe + (Z)-Bu_3SnOC(Et)=CHMe$$
$$ 75\% 25\% \quad (20)$$

The addition reactions of these types are considered in more detail in Section VI.A.

Kinetic studies[50] on the carbon-functional derivatives R_3SnCH_2Y show that the rate of carbonyl addition increases with increasing ability of the Y group to stabilise the α-carbon centre. This effect facilitates an S_E1 mechanism in which the ionization of the organotin compound in the rate-determining step results in the formation of an ion-pair (Scheme 1).

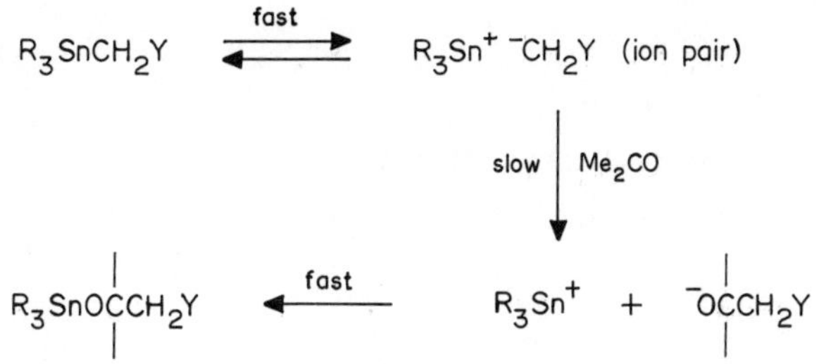

$$R_3SnCH_2Y \underset{}{\overset{fast}{\rightleftarrows}} R_3Sn^+ \; ^-CH_2Y \text{ (ion pair)}$$

$$\downarrow \text{slow} \; | \; Me_2CO$$

$$R_3SnOCCH_2Y \underset{}{\overset{fast}{\longleftarrow}} R_3Sn^+ \; + \; ^-OCCH_2Y$$

SCHEME 1

Electron-withdrawing substituents at the carbonyl group accelerate the addition as in reactions 21–23[41,50,51]. This Reformatsky-type reaction offers an attractive route to β-hydroxy-nitriles, -esters, -ketones, and -amides.

$$Et_3SnCH_2CO_2Et + C_6F_5CHO \xrightarrow{RT} Et_3SnOCH(C_6F_5)CH_2CO_2Et \quad (21)$$

$$Et_3SnCH_2CO_2Et + C_6H_5CHO \xrightarrow[ZnCl_2]{60\,°C,/8\,h} Et_3SnOCH(C_6H_5)CH_2CO_2Et \quad (22)$$

$$Et_3SnCH_2COMe + nCCl_3CHO \xrightarrow{20\,°C} Et_3Sn[-O-CH(CCl_3)-]_n CH_2COMe \quad (23)$$

Bis-α-functionally substituted organotin compounds have been shown to react in an analogous manner with ketones[41] (equation 24).

$$Et_2Sn(CH_2CO_2Me)_2 + 2CF_3COPh \xrightarrow[20\,h]{65-75\,°C} Et_2Sn[O-C(Ph)(CF_3)CH_2CO_2Me]_2$$
$$ 75\% \quad (24)$$

With α,β-unsaturated ketones, stannylacetones yield 1,2-addition products[41] (equation 25). β-Diketones and β-ketoesters which contain active hydrogens, on the other hand,

$$\textbf{8} + PhCH=CHCHO \longrightarrow PhCH=CHCH(CH_2COMe)OSnEt_3 \quad (25)$$

$$R_3SnCH_2COMe + MeCOCH_2COR^1 \longrightarrow R_3SnOC(Me){=}CHCOR^1 + Me_2CO \quad (26)$$

R^1 = alkyl or alkoxy

follow a different course of reaction (reaction 26)[52]. Bu_3SnCH_2COMe reacts with Et_3SiCH_2COCl, as with alkyl halides, to eliminate Bu_3SnCl but the β-diketo product that is formed suffers rearrangement[53]:

$$Bu_3SnCH_2COMe + Et_3SiCH_2COCl \longrightarrow Et_3SiCH_2COCH_2COMe + Bu_3SnCl \quad (27)$$

$$Et_3SiCH_2COCH_2COMe \xrightarrow{\Delta} Et_3SiCH_2COOC({=}CH_2)Me$$

C. C—C Bond Formation via Lithium Derivatives

Apart from their own intrinsic reactivity, α-hetero-substituted organo-tin and -lead compounds yield metallated derivatives which are useful precursors for hetero-substituted carbanions[54].

1. Halogen–lithium and hydrogen–lithium exchange

A convenient route to α-stannyl-organolithium compounds is via halogen/lithium exchange reactions (reaction 28)[55]. These reactions are exothermic and generally proceed in very high yields[56,57] as illustrated in equation 29. Addition of Ph_3SnCH_2Li (**9**) to carbonyls readily affords β-hydroxystannanes (equation 30). On heating to about 110–

$$R_3SnCBr_3 + RLi \longrightarrow R_3SnCBr_2Li \quad (28)$$

$$R_3SnCH_2I + BuLi \underset{}{\overset{\text{ether}}{\rightleftharpoons}} R_3SnCH_2Li + BuI \quad (29)$$

R = Ph, Me, Bu — Ph, 98–100%; Me, 92%; Bu, 98–100%

$$Ph_3SnCH_2Li\ (\mathbf{9}) + R^1R^2CO \longrightarrow Ph_3SnCH_2C(OLi)R^1R^2 \xrightarrow{H_2O} Ph_3SnCH_2C(OH)R^1R^2 \xrightarrow{\Delta} R^1R^2C{=}CH_2 + Ph_3SnOH \quad (30)$$

$R^1 = H$, $R^2 = Ph$, 65%

$R^1 = Me$, $R^2 = Ph$, 41%

175 °C, the β-hydroxystannanes decompose to yield the corresponding alkenes[58], $R^1R^2C{=}CH_2$. Interestingly, **9** itself undergoes oxidative coupling when heated in the presence of $CuCl_2$ (reaction 31).

1. Carbon—carbon bond formation

$$9 \xrightarrow{CuCl_2} Ph_3SnCH_2CH_2SnPh_3 \quad (31)$$
$$(10) \quad 60\%$$

The reaction of **9** with allyl bromide in the presence of copper(I) chloride leads to a moderate yield of the homoallylstannane **11**. Presumably a transmetallated product Ph_3SnCH_2Cu is initially formed, which then adds on to the allyl bromide to yield **11** via an oxidative addition/reductive elimination mechanism (Scheme 2).

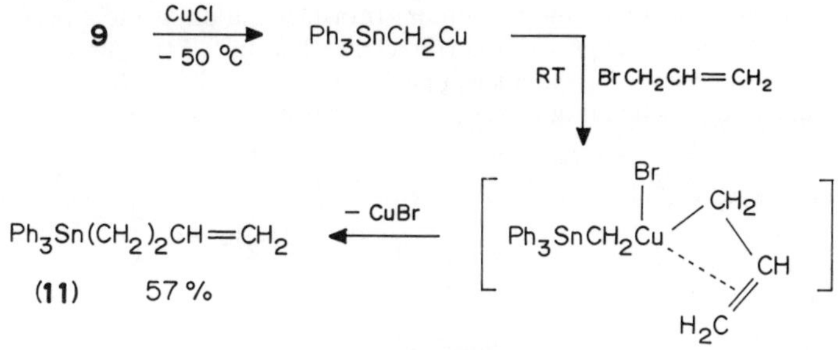

SCHEME 2

There are few reports in the literature on the direct metallation (H/Li exchange) of α-functionally substituted organotin compounds to yield α-stannylorganolithium. An example of this has already been cited in equation 6. Another is equation 32. Compound **12** undergoes Michael-type 1,4-additions with cyclic enones[59] to give products in good yields (Scheme 3).

$$(MeS)_2CHLi + R_3SnCl \xrightarrow{thf/hmpt} (MeS)_2CHSnR_3 \xrightarrow{Pr^i_2NLi} (MeS)_2C(SnR_3)Li \quad (32)$$
$$(12)$$

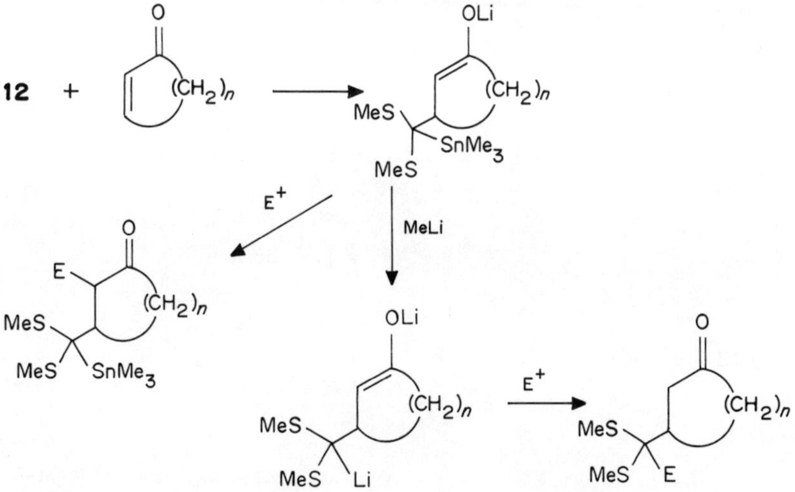

$n = 2$: $R = Me$; $E = H$; 70%
$R = Me$; $E = n\text{-}C_5H_{11}$; 65%

SCHEME 3

In contrast, 1,3-dithiane derivatives apparently undergo only 1,2-additions[60] with cyclic enones (reaction 33).

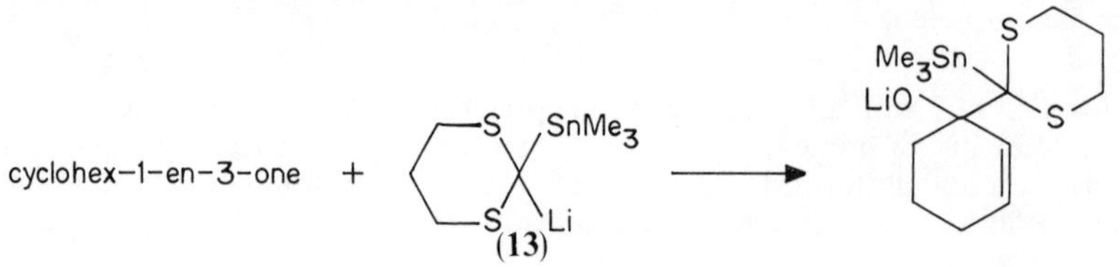

cyclohex-1-en-3-one + (13) ⟶ [product] (33)

It may be noted in passing that tin/lithium transmetallation at the 2-position of the dithiane is a faster and milder process than direct metallation[60]. This is important because it facilitates the generation of the nucleophilic 2-dithiane centre (14) which would not otherwise survive the metallation process under the usual conditions (Scheme 4, step b).

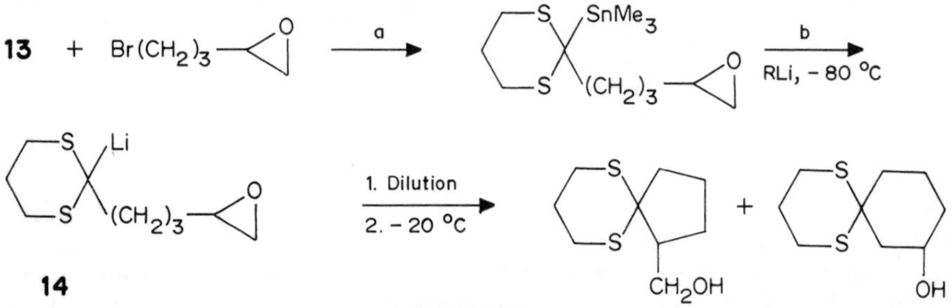

13 + Br(CH$_2$)$_3$-epoxide $\xrightarrow{a}$ [SnMe$_3$ intermediate] $\xrightarrow[RLi,\ -80\ °C]{b}$

14 $\xrightarrow[\text{2. } -20\ °C]{\text{1. Dilution}}$ [spiro product, CH$_2$OH] + [spiro product, OH]

SCHEME 4

α-Substituted sulphur and selenium carbanions[61] have in recent years been shown to be versatile reagents for engendering C—C bond formation. Equations 34–36 exemplify this.

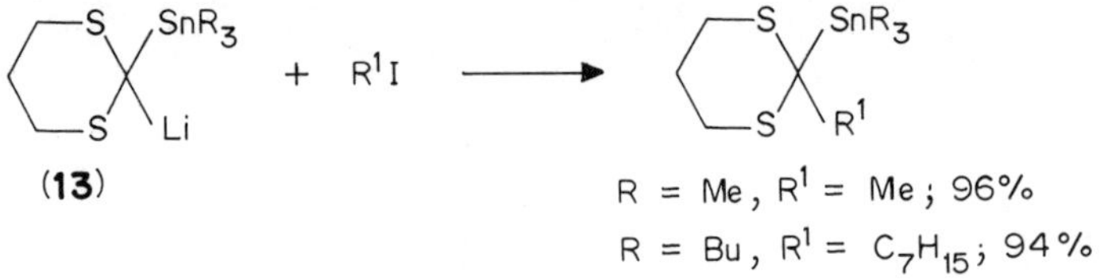

(13) + R^{1}I ⟶ [product] (34)

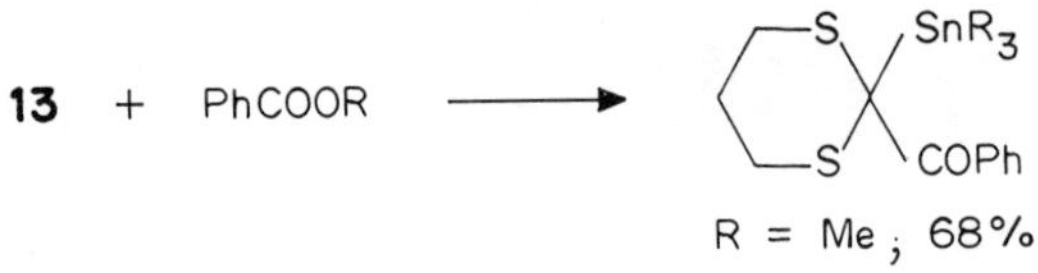

R = Me, R^1 = Me; 96%
R = Bu, R^1 = C$_7$H$_{15}$; 94%

13 + PhCOOR ⟶ [product with COPh] (35)

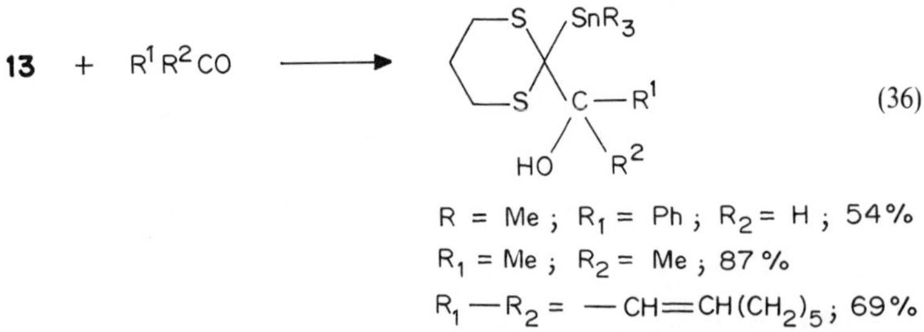

R = Me; 68%

13 + R^{1}R^{2}CO ⟶ [product with C—R^1, HO, R^2] (36)

R = Me; R$_1$ = Ph; R$_2$ = H; 54%
R$_1$ = Me; R$_2$ = Me; 87%
R$_1$—R$_2$ = —CH=CH(CH$_2$)$_5$; 69%

1. Carbon—carbon bond formation

The high polarizability of the sulphur and selenium atoms in derivatives such as **13** renders them good nucleophilic reagents and also facilitates the cleavage of the Sn—C bond. A point of obvious synthetic appeal is that desulphurization of the products formed can be readily achieved by reduction with Raney nickel[62] or hydrazine[63], or by the use of lithium tetrahydroaluminate with $CuCl_2/ZnCl_2^{64}$. Thus benzaldehyde can be transformed alternatively to the cis-(phenylthio) styrene (85%) or the trans-isomer (84%) from (phenylthio)(triphenylstannyl) methyllithium or (phenylthio)(triphenylplumbyl) methyllithium[65] (Scheme 5).

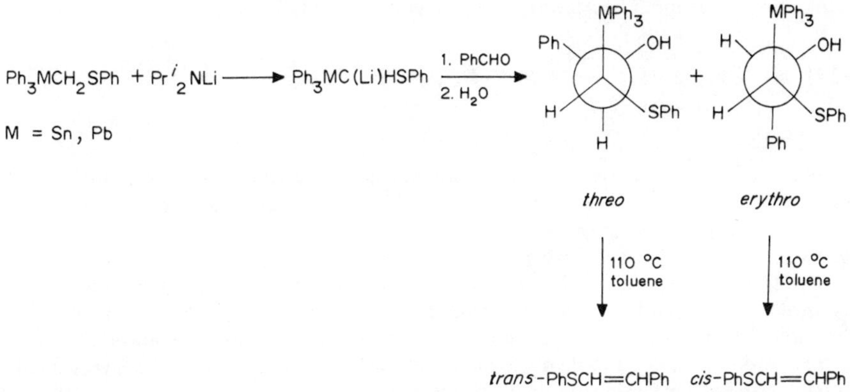

$Ph_3MCH_2SPh + Pr^i_2NLi \longrightarrow Ph_3MC(Li)HSPh \xrightarrow[2.\ H_2O]{1.\ PhCHO}$

M = Sn, Pb

threo → 110 °C toluene → trans-PhSCH=CHPh

erythro → 110 °C toluene → cis-PhSCH=CHPh

SCHEME 5

1,3-Dienes and styrene derivatives can be prepared in good yields from allyl 2-pyridyl sulphide and allyl phenyl sulphone derivatives[66]. Although mechanistic details of the reactions have not been elucidated, the stability of allyl carbanions as well as the regioselectivity towards α-alkylation of allyl anions are apparently significant factors in the synthesis of 1,3-dienes.

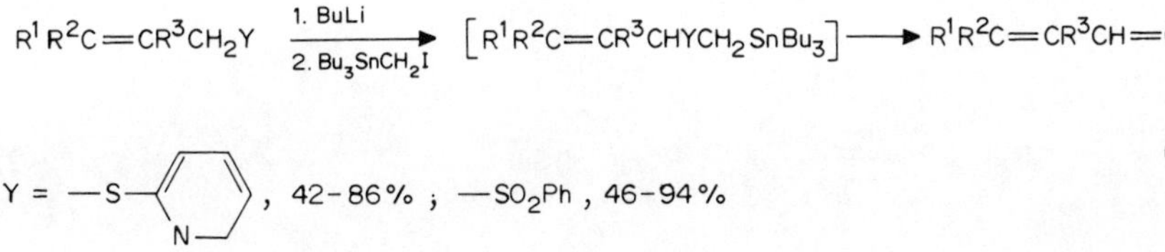

$$R^1R^2C=CR^3CH_2Y \xrightarrow[2.\ Bu_3SnCH_2I]{1.\ BuLi} [R^1R^2C=CR^3CHYCH_2SnBu_3] \longrightarrow R^1R^2C=CR^3CH=CH_2$$

(37)

Y = —S—(2-pyridyl), 42–86% ; —SO_2Ph, 46–94%

The successive addition of two equivalents of BuLi to Bu_3SnCH_2OH followed by an electrophile such as an alkyl halide, aldehyde, or ketone gives the corresponding hydroxymethyl derivative[67], probably via $LiCH_2OLi$ formed as intermediate. This route provides the first hydroxymethylation of a carbonyl compound (equation 38).

$$Bu_3SnCH_2OH + 2BuLi + (Me_3C)_2CO \longrightarrow (Me_3C)_2C(OH)CH_2OH \quad (38)$$

41%

2. Transmetallation

In these reactions, the C—MR$_3$ function serves as a protected carbanion which is unmasked to provide synthetically useful carbon nucleophiles. The substitution of the stannyl or plumbyl group is conveniently achieved using organolithium reagents so that much of the derived chemistry is largely that of the transmetallated product.[†] For this reason and also because studies on the regiochemical control potential of the C—MR$_3$ metalloids *per se* have been little described in the literature, only selected reactions exemplifying the synthetic scope of this protection strategy are discussed here.

Gröbel and Seebach[68] employed the stannyl group to facilitate lithiation of sulphur-containing nucleophilic acylating reagents (equation 39).

$$R^1R^2C=C(SR^3)SnMe_3 \xrightarrow{BuLi} [R^1R^2C=CSR^3]^- Li^+ \xrightarrow{E^+} R^1R^2C=C(SR^3)E$$

(15)   (39)

The reagent **15** is a useful synthetic equivalent of the enolate of α-thiolated acetaldehyde[69,70]. Thus, when $R^1 = H$ and $R^2 = OC_2H_5$, the equivalent enolate species is (OHC$\bar{C}$HSR^3). In principle, it is also possible to substitute SOR^3 for SR^3 (ref. 71) and allenyl $(R^1R^2C=C=CSR^3)^-$ for vinyl derivatives[72].

The transmetallation of dialkoxymethyltributyltin (**16**), a masked aldehyde anionic equivalent, with BuLi gives a new organolithium reagent which can react with benzyl bromide, benzaldehyde and cyclohexenone as shown in equations 40 and 41[16].

The addition of organotin nucleophiles to carbonyl compounds gives synthetically useful carbinyl carbanion equivalents (reaction 42)[73]. Still[57] developed an efficient

$$Bu_3SnCH(OEt)_2 + BuLi \longrightarrow LiCH(OEt)_2 \xrightarrow[\text{2. } H_2O]{\text{1. BzBr}} BzCHO \quad (40)$$
(**16**)   78%

$$\mathbf{16} + BuLi \xrightarrow[\substack{\text{2. cyclohexanone} \\ \text{3. } H_2O}]{\text{1. CuI}} \text{(EtO)}_2\text{CH—cyclohexanone=O} \quad (41)$$

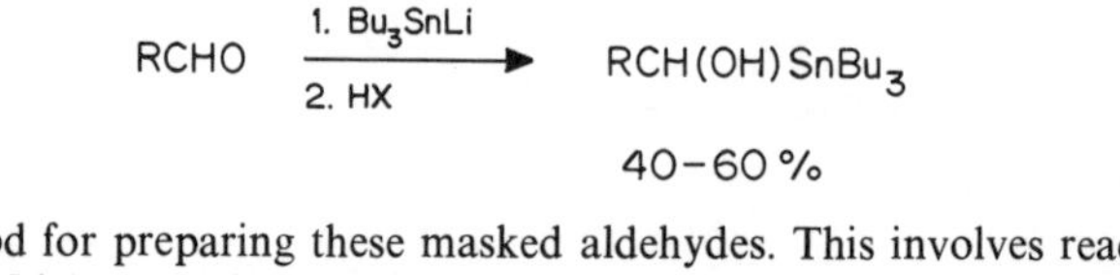

$$RCHO \xrightarrow[\text{2. HX}]{\text{1. Bu}_3\text{SnLi}} RCH(OH)SnBu_3 \quad (42)$$
40–60%

method for preparing these masked aldehydes. This involves reacting a thf solution of Bu$_3$SnLi (prepared *in situ* from deprotonation of Bu$_3$SnH with (Pr^i)$_2$NLi at 0 °C) with aldehydes at −78 °C. High yields of the carbinols are thus obtained. Since the carbinols are generally labile the hydroxy group is protected via conversion to the ethoxyethyl derivative (**17**). The protected carbinol, (*E*)-1-methoxylmethyl but-2-enyl tributylstannane, derived from crotonaldehyde, for example, has found synthetic utility as a useful precursor for the stereoselective synthesis of *trans*-3,4-disubstituted 4-butanolides[73a]

[†] Tin/lithium exchange reactions are especially facile when the organolithium reagent to be generated is stabilized either by unsaturation or by a heteroatom.

$$Bu_3SnCHR(OH) \xrightarrow{EtOCH(Me)Cl} RCH(SnBu_3)OCH(Me)OEt \quad (43)$$
$$90\% \quad (\mathbf{17})$$

(Scheme 6). A useful general reaction of protected carbinols such as **17** is that they rapidly undergo tin/lithium exchange with BuLi to form α-alkoxyl organolithium reagents, RCH(OR^1)Li, which readily react *in situ* with carbonyl compounds. Equation 44 is illustrative.

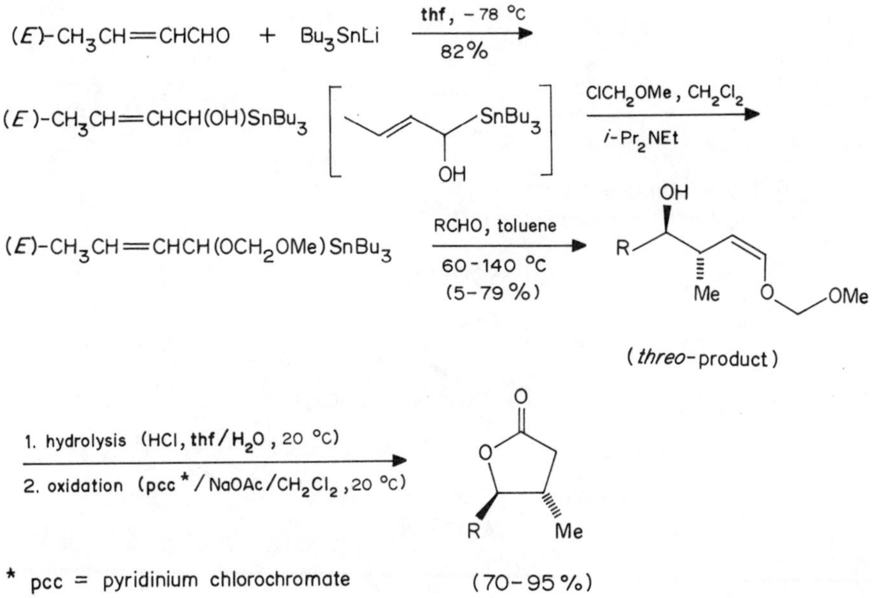

(E)-CH$_3$CH=CHCHO + Bu$_3$SnLi $\xrightarrow[82\%]{\text{thf, }-78\ °C}$

(E)-CH$_3$CH=CHCH(OH)SnBu$_3$ [structure with SnBu$_3$ and OH] $\xrightarrow{\text{ClCH}_2\text{OMe, CH}_2\text{Cl}_2}_{i\text{-Pr}_2\text{NEt}}$

(E)-CH$_3$CH=CHCH(OCH$_2$OMe)SnBu$_3$ $\xrightarrow[\substack{60-140\ °C \\ (5-79\%)}]{RCHO,\ toluene}$ [threo-product structure with R, OH, Me, OCH$_2$OMe]

(*threo*-product)

$\xrightarrow[\text{2. oxidation (pcc}^*/\text{NaOAc/CH}_2\text{Cl}_2,\ 20\ °C)}]{\text{1. hydrolysis (HCl, thf/H}_2\text{O, 20 °C)}}$ [lactone with R and Me]

(70–95%)

* pcc = pyridinium chlorochromate

SCHEME 6

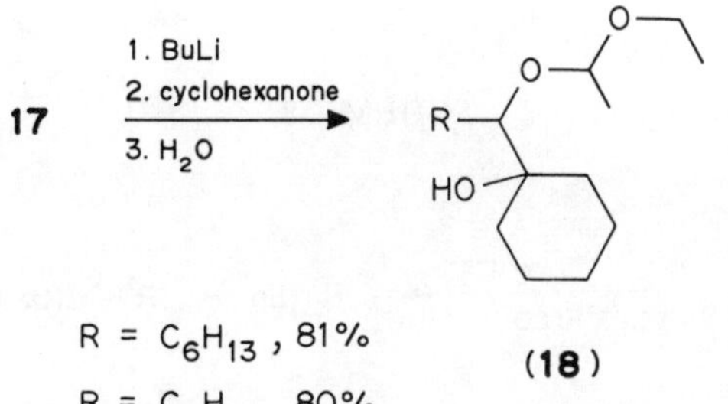

17 $\xrightarrow[\text{3. H}_2\text{O}]{\substack{\text{1. BuLi} \\ \text{2. cyclohexanone}}}$ [structure **(18)**] (44)

R = C$_6$H$_{13}$, 81%
R = C$_6$H$_{11}$, 80%

Significantly, no 1-butylcyclohexanol could be detected. This implies that the tin/lithium exchange goes virtually to completion. The synthetic potential of these α-alkoxyl organolithium reagents, RCH(OR^1)Li, in natural product synthesis is illustrated by the simple synthesis of dendrolasin and (±)-9-hydroxydendrolasin from furan-3-carboxaldehyde[57] (Scheme 7). Noteworthy features of the reaction are the relatively mild

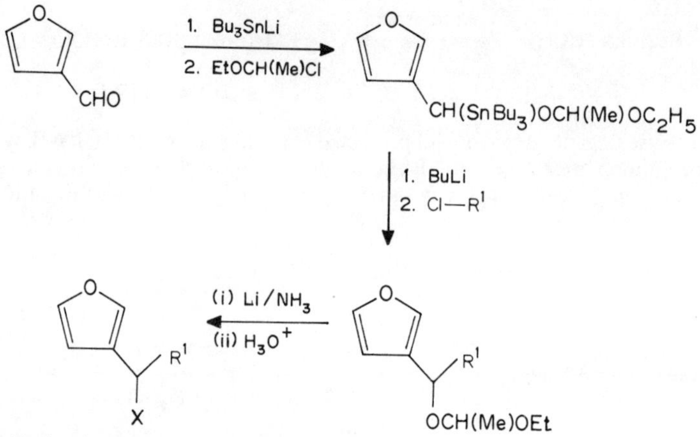
1. Bu3SnLi
2. EtOCH(Me)Cl
CHO
CH(SnBu3)OCH(Me)OC2H5
1. BuLi
2. Cl—R1
(i) Li/NH3
(ii) H3O+
R1
X
OCH(Me)OEt

(i) dendrolasin (X = H); 92%
(ii) 9-hydroxydendrolasin (X = OH); 96%

SCHEME 7

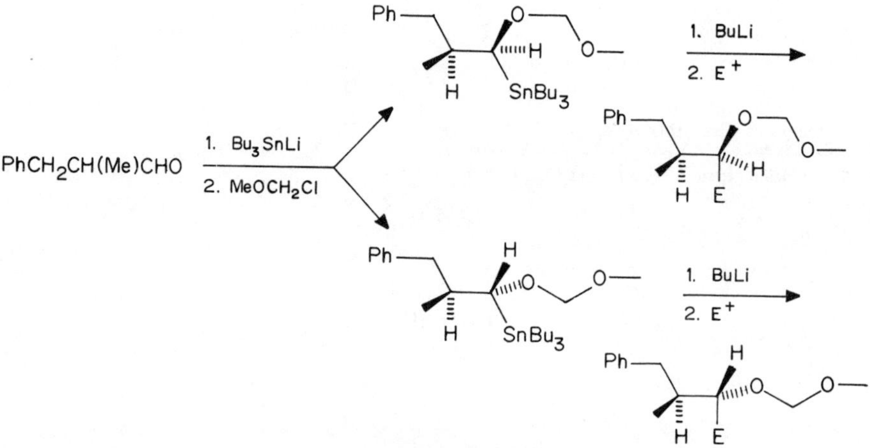
PhCH2CH(Me)CHO
1. Bu3SnLi
2. MeOCH2Cl
Ph
H
SnBu3
1. BuLi
2. E+
E

SCHEME 8

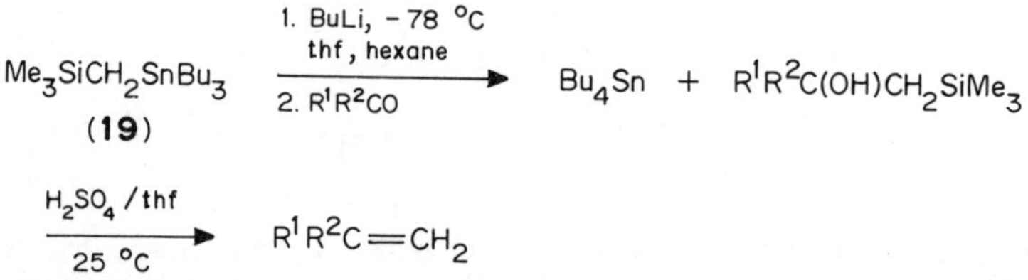
Me3SiCH2SnBu3
(19)
1. BuLi, −78 °C
thf, hexane
2. R1R2CO
Bu4Sn + R1R2C(OH)CH2SiMe3
H2SO4 /thf
25 °C
R1R2C=CH2
(45)

Product: p-$ClC_6H_4CH{=}CH_2$, 72%; 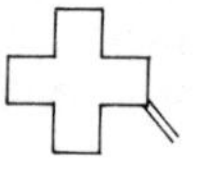, 90%

1. Carbon—carbon bond formation

reaction conditions and the high stereospecificity[74] (Scheme 8). The α-trimethylsilylstannane **19**, prepared in near quantitative yield from Me_3SiCH_2Cl and Bu_3SnLi at 0 °C in a mixture of thf and hexane[75], undergoes ready transmetallation in the same medium to afford Me_3SiCH_2Li, a versatile reagent for methylenation of aldehydes and ketones (equation 45). The Pb—C bond of $(Ph_3Pb)_3CH$ can be similarly cleaved with PhLi and benzaldehyde, giving the styrylplumbane (reaction 46)[76].

$$(Ph_3Pb)_3CH \xrightarrow[98\%]{PhLi} (Ph_3Pb)_2CHLi \xrightarrow[81\%]{PhCHO} (Ph_3Pb)_2CHCH(OH)Ph \qquad (46)$$

$$\xrightarrow[76\%]{SiO_2} (E)\text{-}Ph_3PbCH{=}CHPh$$

As noted earlier, α-stannylcarbinols are unstable, but these may be converted to the more stable α-stannylalkyl halides by reaction with Ph_3P and CBr_4 in CH_2Cl_2 at room temperature[77]. Interestingly, the halides do not yield the expected α-halocarbanions with BuLi. α-Stannylalkyl bromide, for instance, gives Bu_4Sn (60%) and a mixture of E/Z (79:21) olefinic coupling products[77].

$$PhCH_2CH_2CH(X)SnBu_3 \xrightarrow[\substack{-78\,°C\,(Br,Cl) \\ 0\,°C\,(I)}]{BuLi} PhCH_2CH_2CH{=}CHCH_2CH_2Ph \qquad (47)$$

X = Cl, Br, I

X = Cl, 65%; % E/Z, 77:23
X = Br, 65%; % E/Z, 79:21
X = I, 60%; % E/Z, 50:50

$$RCH(X)SnBu_3 \underset{}{\overset{BuLi}{\rightleftharpoons}} Li^+RCHX^- + Bu_4Sn \qquad (47a)$$

(20)   **(21)**

fast ↘ 20

$$LiX + Bu_3SnCHRCHRX \xrightarrow{BuLi} RCH{=}CHR + LiX$$

(22)   (E/Z)

It has been found that the reaction of the lithiocarbanion **21** (generated by tin/lithium exchange) with α-stannylalkyl halide proceeds much faster than nucleophilic attack by BuLi[78]. This results in the formation of the coupled product **22**. Destannylation with BuLi yields the olefinic coupling products indicated in equation 47a. Cross-coupling has also been effected using a mixture of α-stannylalkyl iodide and chloride with BuLi at −78 °C (equation 48)[78]. The reaction yielded 43% cross-coupling product (E/Z = 76:24) and 5%

$$Me(CH_2)_8CH(I)SnBu_3 + PhCH_2CH_2CH(Cl)SnBu_3 \longrightarrow$$
$$Ph(CH_2)_2CH{=}CH(CH_2)_8CH_3 + Ph(CH_2)_2CH{=}CH(CH_2)_2Ph \qquad (48)$$

homocoupling product. The method thus affords an important route for the conversion of aldehydes to olefinic coupling products.

A high stereoselective [2,3]-sigmatropic Wittig rearrangement has been successfully used for the synthesis of stereospecific Z-trisubstituted homoallylic alcohols[79] (equation 49). The method holds considerable promise since most trisubstituted olefin syntheses

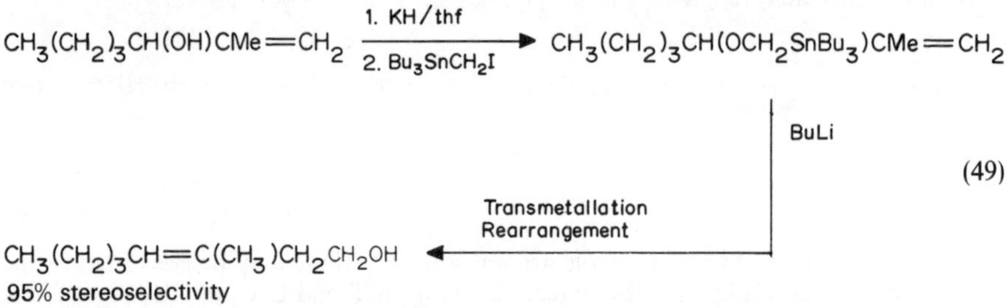

$$CH_3(CH_2)_3CH(OH)CMe=CH_2 \xrightarrow[2.\ Bu_3SnCH_2I]{1.\ KH/thf} CH_3(CH_2)_3CH(OCH_2SnBu_3)CMe=CH_2$$

$$\downarrow BuLi$$

$$\text{Transmetallation Rearrangement}$$

$$CH_3(CH_2)_3CH=C(CH_3)CH_2CH_2OH$$
95% stereoselectivity

(49)

produce the E-isomers as the major product. Trienyl acetate, an active component in the sex attractant of a major citrus pest, has been prepared from ethyl ester of β,β-dimethylacrylic acid by this method (equation 50). A convenient synthesis of the

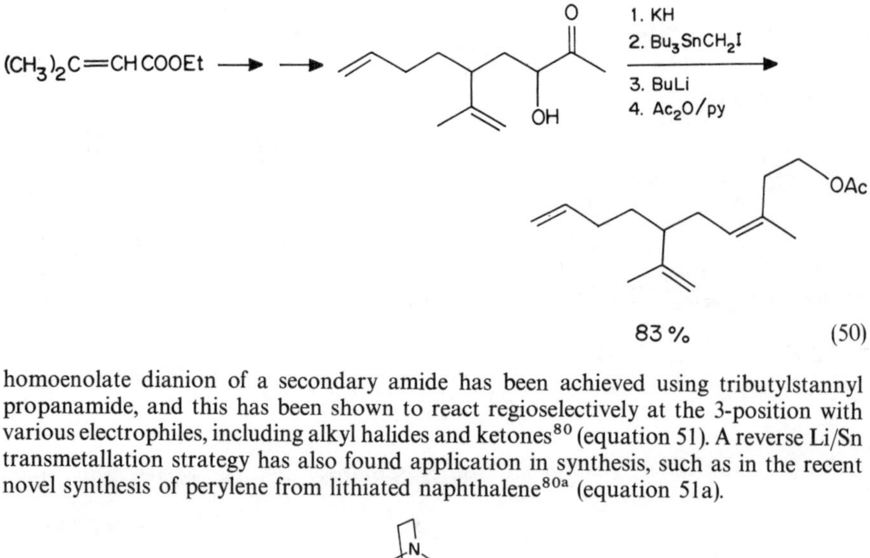

$$(CH_3)_2C=CHCOOEt \longrightarrow \longrightarrow$$

1. KH
2. Bu_3SnCH_2I
3. BuLi
4. Ac_2O/py

83%   (50)

homoenolate dianion of a secondary amide has been achieved using tributylstannyl propanamide, and this has been shown to react regioselectively at the 3-position with various electrophiles, including alkyl halides and ketones[80] (equation 51). A reverse Li/Sn transmetallation strategy has also found application in synthesis, such as in the recent novel synthesis of perylene from lithiated naphthalene[80a] (equation 51a).

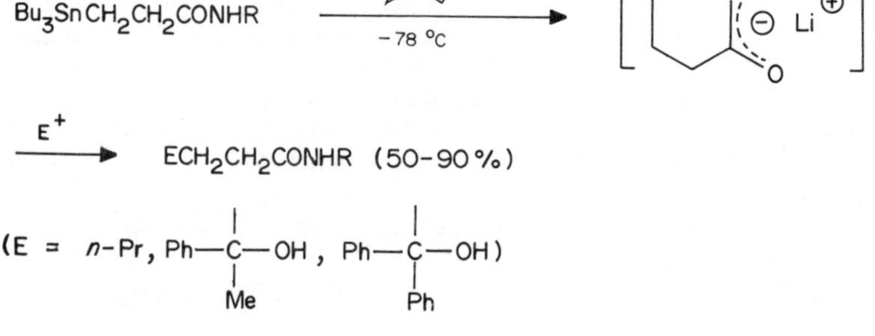

$$Bu_3SnCH_2CH_2CONHR \xrightarrow[-78\ ^\circ C]{BuLi/\ \text{(TMEDA)}\ /\ thf} \left[\begin{array}{c} Li \leftarrow N-R \\ \ominus\ Li^\oplus \\ O \end{array} \right]$$

$$\xrightarrow{E^+} ECH_2CH_2CONHR\ (50-90\%)$$

(51)

$$(E = n\text{-Pr},\ Ph-\underset{Me}{\overset{|}{C}}-OH,\ Ph-\underset{Ph}{\overset{|}{C}}-OH)$$

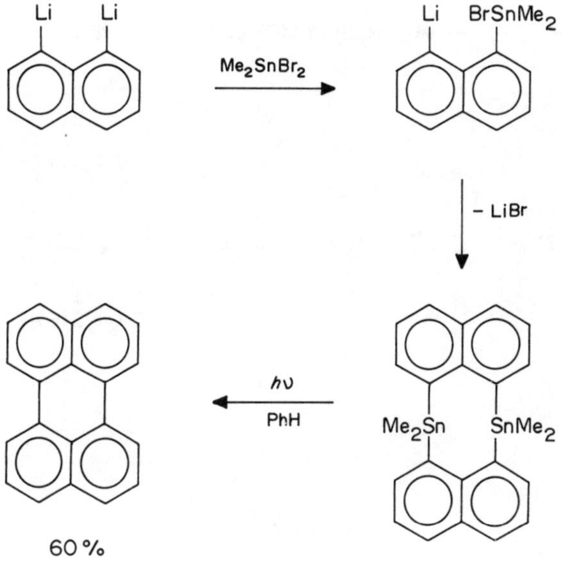

1. Carbon—carbon bond formation

(51a)

60%

III. REACTIONS OF β- AND γ-FUNCTIONALLY SUBSTITUTED ORGANO-TIN AND -LEAD COMPOUNDS

A. C—C Bond Formation via β-Substituted Derivatives

A widely observed reaction of β-substituted organo-tin and -lead compounds is 1,2-elimination[81-83]. The reaction proceeds with relative ease when the β-carbon carries a heterofunctional group and a kinetic study[83] has indicated that for β-stannyl hydroxides (equation 52) the reaction is first order in the organometallic reagent. For thermal or spontaneous reactions (for example, equation 30), *syn*-elimination appears probable by analogy to the silyl case[84], whereas *anti*-elimination has been deduced in the presence of added electrophiles (equation 53).

$$Ph_3MCR_2CR_2OH \xrightarrow[\text{or MeOH/H}_2\text{O}]{\text{HClO}_4 \; \text{AcOH/H}_2\text{O}} Ph_3MCR_2CR_2\overset{+}{O}H_2 \longrightarrow CR_2{=\!\!=}CR_2 \quad (52)$$

M = Sn or Pb

95–100%

$$Ph_3SnCH(Me)CH(Me)OH \xrightarrow[\text{AcOH/H}_2\text{O}]{\text{HClO}_4} MeCH{=\!\!=}CHMe \quad (53)$$

100% Z-isomer

β-Sulphides[81,85], $R_3Sn(CH_3)_2SR^1$, like the hydroxides, also undergo preferential alkene elimination (Scheme 9, step b) over Sn—C bond cleavage of the unsubstituted side-chain (step a), particularly in the presence of Br_2, ArSCl, or MeI (EY). Step a, on the other hand,

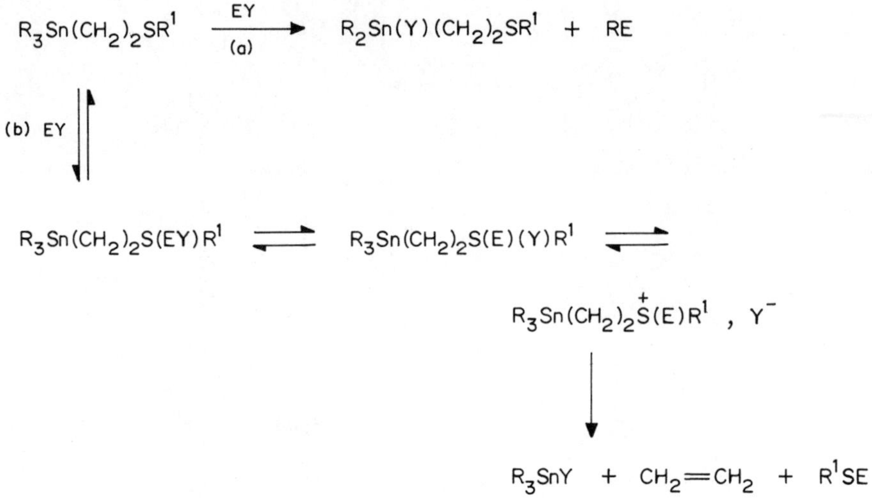

$$R_3Sn(CH_2)_2SR^1 \xrightarrow[(a)]{EY} R_2Sn(Y)(CH_2)_2SR^1 + RE$$

(b) EY ↕

$$R_3Sn(CH_2)_2S(EY)R^1 \rightleftarrows R_3Sn(CH_2)_2S(E)(Y)R^1 \rightleftarrows$$

$$R_3Sn(CH_2)_2\overset{+}{S}(E)R^1 , Y^-$$

↓

$$R_3SnY + CH_2{=}CH_2 + R^1SE$$

SCHEME 9[85]

occurs exclusively with electrophiles such as $HgCl_2$ or I_2, which do not favour the formation of $R_3Sn(CH_2)_2S(R^1)(E)$, and more generally[86] when carbon functional substituents such as CN or CO_2R are attached to the β- or γ-carbon (equation 54).

$$R_3Sn(CH_2)_nCO_2Me + 2Br_2 \xrightarrow{PhH} RSn(Br)_2(CH_2)_nCO_2Me + 2RBr \quad (54)$$

n = 2 or 3

Wardell[85] has noted that, notwithstanding the possibility of a concerted pathway, for alkene elimination to occur a positive character to the β-carbon might have to be formed during the reaction. However, the intervention of a β-carbocation intermediate is thought to be unlikely, as evidenced by the high stereospecificity of the elimination encountered for β-stannyl hydroxides[82] (equation 53).

β-Stannylalkyl radicals and cations are, of course, known and interpretation of their stability in terms of hyperconjugation appears to be generally accepted[2,87]. The reactivity of the β-carbon extends even in the absence of functional groups, as exemplified below by the ready insertions of $:CCl_2$ into the C_β—H bond[88] (equations 55 and 56). β-Trimethylstannylethylidenetriphenylphosphonium iodide (23), obtained from equation

$$Me_3SnCD(Me)CH_2CH_3 \xrightarrow[PhH]{PhHgCCl_2Br} Me_3SnCD(Me)CH(CCl_2)CH_3 \quad (55)$$

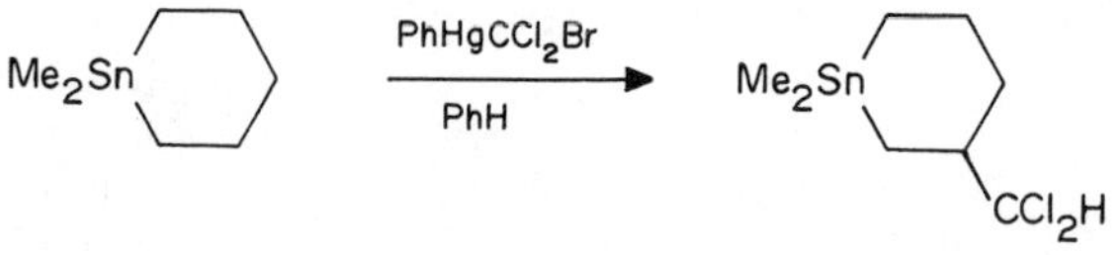

$$Me_2Sn{-}\hspace{-0.2em}\bigcirc \xrightarrow[PhH]{PhHgCCl_2Br} Me_2Sn{-}\hspace{-0.2em}\bigcirc{-}CCl_2H \quad (56)$$

78%

57, can be readily converted to β-trimethylstannylethylidenetriphenylphosphorane[89] (**24**). The ylide **24**, which is unstable, undergoes Wittig addition with aldehydes and ketones to yield allylic organotin products[90,91]. β-Tributylstannylpropionaldehyde (**25**), which can

$$Ph_3P=CH_2 + Me_2SnCH_2I \longrightarrow [Ph_3PCH_2CH_2SnMe_3]^+, I^- \quad (57)$$
$$89\% \quad (\mathbf{23})$$

$$\mathbf{23} + Pr^i_2NLi \longrightarrow Ph_3P=CHCH_2SnMe_3 \xrightarrow[thf, -93°C]{cyclohexanone}$$
$$69\% \quad (\mathbf{24})$$

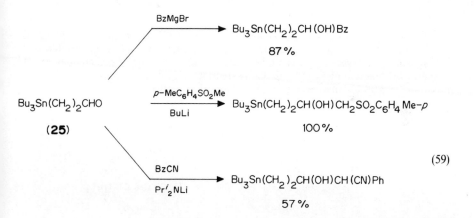

be obtained by oxidation of $Bu_3Sn(CH_2)_3OH$ which either N-chlorosuccinimide[92] (81%) or $CrO_3 \cdot PyHCl^{93}$ (55%), undergoes facile conversion to γ-hydroxypropylstannanes[94], hence making it a synthetically useful reagent (equation 59).

β-Stannyl ketones, which may be prepared *in situ* from the 1,4-addition of trialkylstannyllithium to α-enones[95], can be cleanly alkylated with reactive alkyl halides, in particular alkyl iodides[96]. The synthesis of **26** from cyclopent-1-en-2-one[95] illustrates this. **26** can be readily converted to dihydrojasmone (**27**) in 71% overall yield[95] via the reaction sequence shown in equation 61. Taken together, equations 60 and 61 offer an efficient dialkylative transposition route involving the *in situ* generation of β-stannyl ketones. The stannyl β-keto phosphonate ester (**28**) has been used to generate uniquely the 1,4-dicarbanion (**29**), a useful homoenolate dianion synthon[97]. **29** with various carbon electrophiles reacts exclusively at the δ carbon (equation 62).

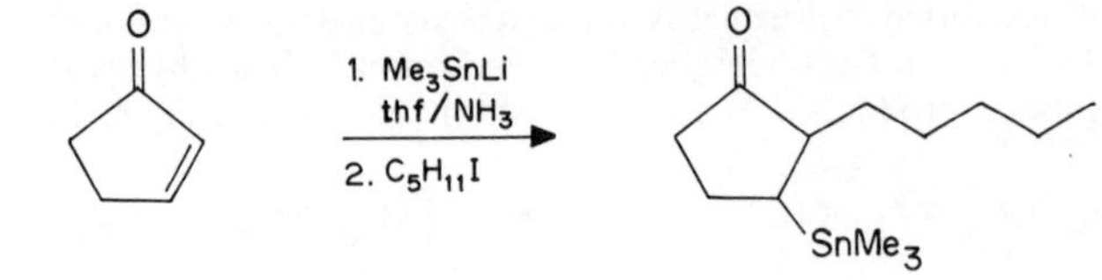

O
1. Me_3SnLi
thf/NH_3
2. $C_5H_{11}I$
O
$SnMe_3$

(26)

(60)

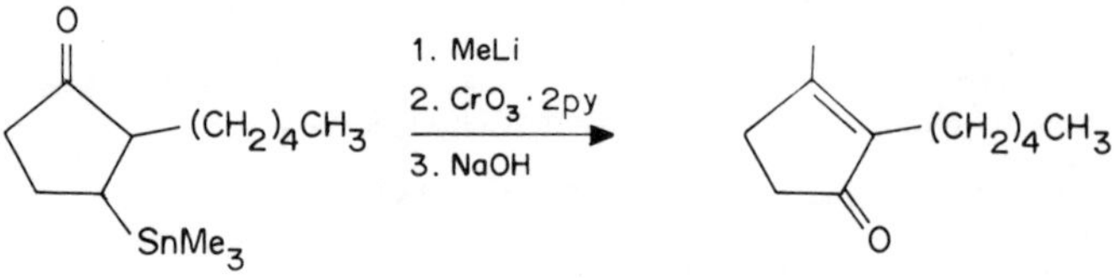

O
$(CH_2)_4CH_3$
$SnMe_3$
1. MeLi
2. $CrO_3 \cdot 2py$
3. NaOH
$(CH_2)_4CH_3$
O

(27)

(61)

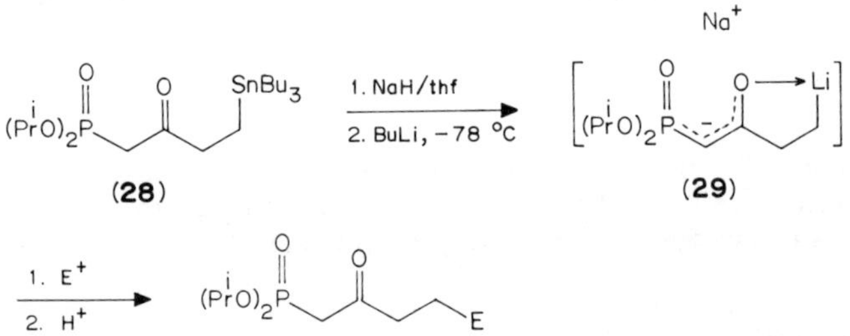

O
O
$SnBu_3$
$(Pr^iO)_2P$
1. NaH/thf
2. BuLi, −78 °C
Na^+
O
O
Li
$(Pr^iO)_2P$
1. E^+
2. H^+
O
O
$(Pr^iO)_2P$
E

(28) (29)

(62)

Lithiation of 2-(tributylstannyl)propionitrile[98,99] with Pr^i_2NLi yields **30**, which reacts readily with organic halides such as MeI, BzBr, 1-chloromethylnaphthalene, 3-chloro-2-methylprop-1-ene, *trans*-cinnamyl bromide, and 2-bromoethylphenyl ether (equation 63).

Apart from functioning as a C—C chain extension reagent, **31** has also been used as an important reagent for the generation of cyclopropane derivatives[99].

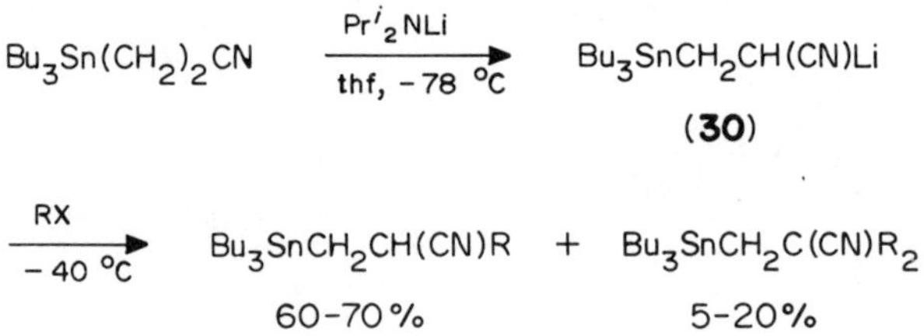

$Bu_3Sn(CH_2)_2CN$
Pr^i_2NLi
thf, −78 °C
$Bu_3SnCH_2CH(CN)Li$
RX
−40 °C
$Bu_3SnCH_2CH(CN)R$
+
$Bu_3SnCH_2C(CN)R_2$
60–70%
5–20%

(30)

(31)

(63)

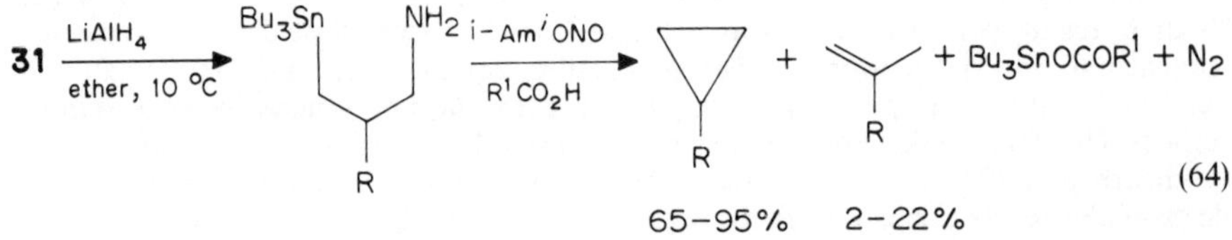

31
$LiAlH_4$
ether, 10 °C
Bu_3Sn
NH_2
R
i-Am^iONO
R^1CO_2H
R
+
R
+ $Bu_3SnOCOR^1$ + N_2
65–95%
2–22%

(64)

B. C—C Bond Formation via γ-Substituted Derivatives

A reaction of some consequence with γ-substituted organotin compounds is 1,3-elimination resulting in cyclopropane ring formation[100]. The mechanism of reaction 65 has not been clarified. It may involve a concerted reaction pathway (for example, equation 66), or a carbocation intermediate (equation 66a). The concerted pathway has been

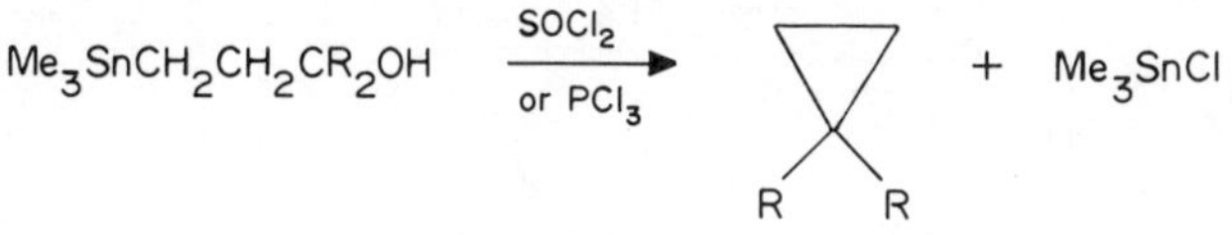

$$Me_3SnCH_2CH_2CR_2OH \xrightarrow{SOCl_2 \text{ or } PCl_3} \underset{R \quad R}{\triangle} + Me_3SnCl \quad (65)$$

$$R = H, 90\% \; ; \; R = Me, 80-85\%$$

$$Me_3Sn\!\!\smallsetminus\!\!\curvearrowright\!\!CR_2\text{—OX} \longrightarrow \underset{R \quad R}{\triangle} + Me_3Sn^+ \quad (66)$$

$$X = PCl_2, SOCl$$

$$Me_3Sn\!\!\smallsetminus\!\!\curvearrowright\!\!\overset{+}{\underset{R}{\diagdown}}{\!\!}^R \longrightarrow \underset{R}{\triangle}\!\!\!{\!\!}^R + Me_3Sn^+ \quad (66a)$$

advanced for the mesylate[100a] (32) for which the elimination in acetic acid apparently does not involve ion-pair formation (i.e. solvolysis). However, deoxymetallations of γ-hydroxypropylstannanes have been envisaged[94] to follow the mechanistic scheme indicated in equation 68. Whereas simple secondary alcohols with γ-stannyl substituent tend to decompose with no noticeable cyclopropane formation, the cyclization appears to be particularly facile with the corresponding γ-stannyl tertiary and benzyl alcohols[101]. It is

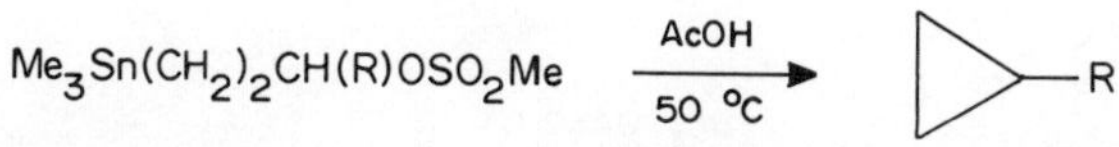

$$Me_3Sn(CH_2)_2CH(R)OSO_2Me \xrightarrow[50\ °C]{AcOH} \triangle\!\!-\!\!R \quad (67)$$

$$R = H, Me \qquad\qquad\qquad >95\%$$

$$(\mathbf{32})$$

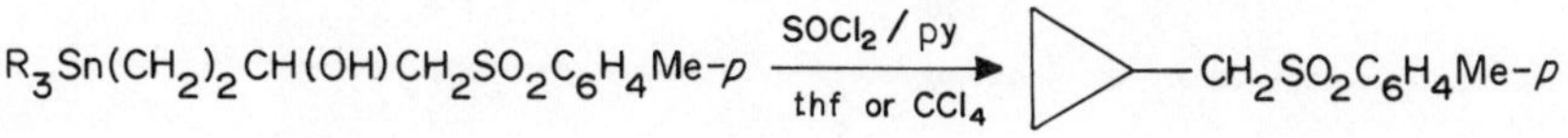

$$R_3Sn(CH_2)_2CH(OH)CH_2SO_2C_6H_4Me\text{-}p \xrightarrow[\text{thf or } CCl_4]{SOCl_2/py} \triangle\!\!-\!\!CH_2SO_2C_6H_4Me\text{-}p \quad (68)$$

$$RT \qquad\qquad\qquad 81\%$$

instructive that despite the highly favourable opportunity for phenyl shift with formation of a tertiary cation, no such rearrangement is observed for the γ-stannyl alcohols depicted in equation 69. Evidence that the above ring closure proceeds with high stereospecificity with inversion of configuration at the hydroxyl-bearing carbon is illustrated in the set of reactions in Scheme 10[101].

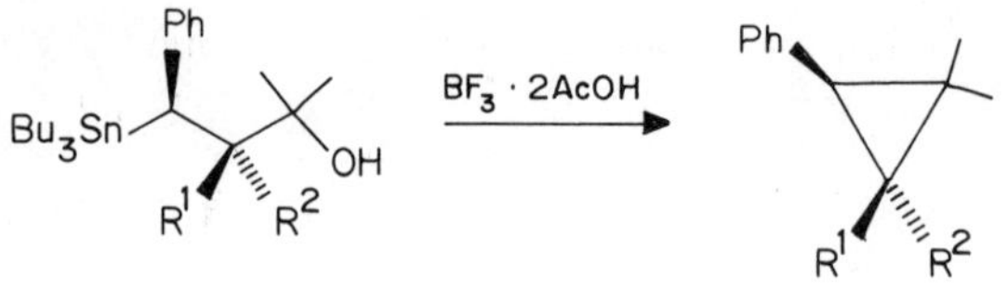

(69)

R^1 = Me, R^2 = Ph ; 96%

R^1 = Ph, R^2 = Me ; 73%

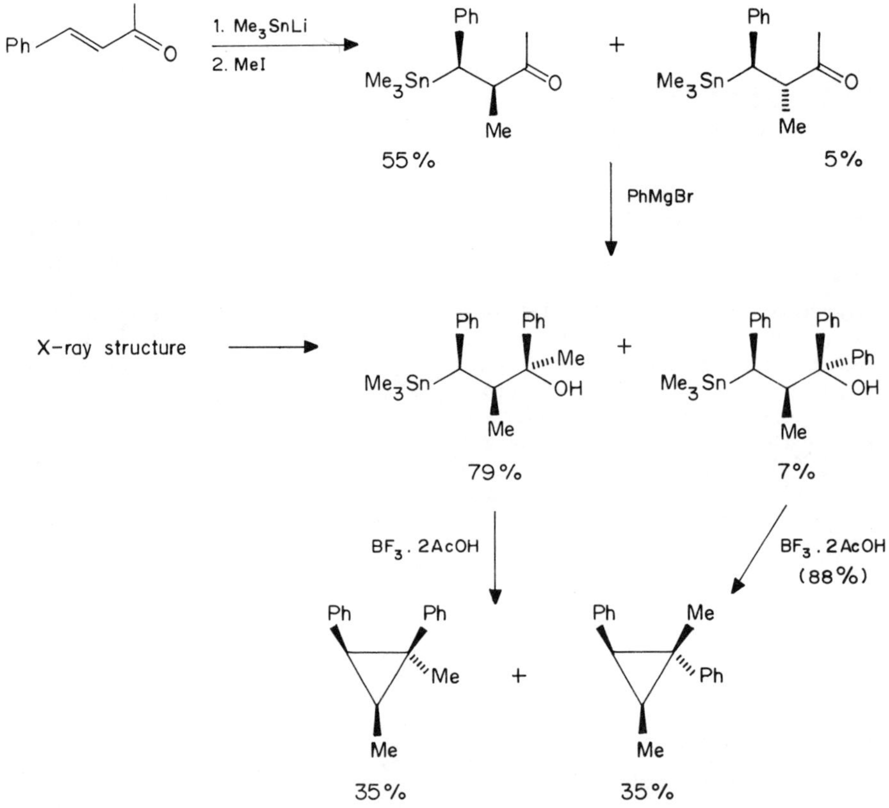

SCHEME 10

1. Carbon—carbon bond formation

A non-classical ion (or a rapidly equilibrating classical ion) has been postulated[100] to account for the formation of norbornene from *anti*-7-trimethylstannyl-2-*exo*-norbornanol (33).

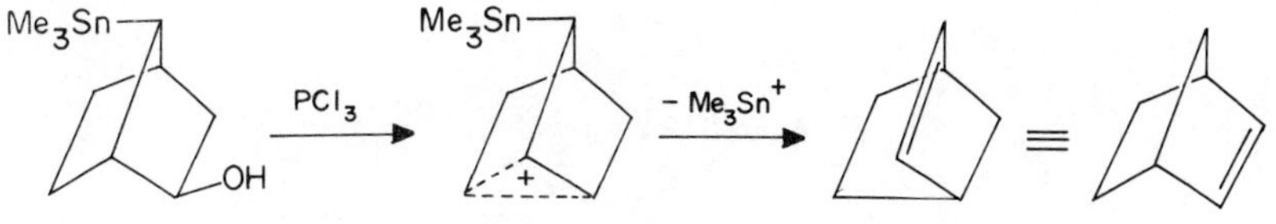

(70)

(33)   (34)

In this case, the particular geometry of 33 is unfavourable for a concerted 1,3-cyclization, but the placement of positive charge, in part, at the β-carbon in the intermediate 34 facilitates rapid elimination leading to C=C formation. A similar stabilized ion (35) has been proposed by Hartman and Traylor[102] for the formation of retroene in equation 71, but an alternative cyclic transition state 36 has since been suggested[103] as being the more probable for this reaction.

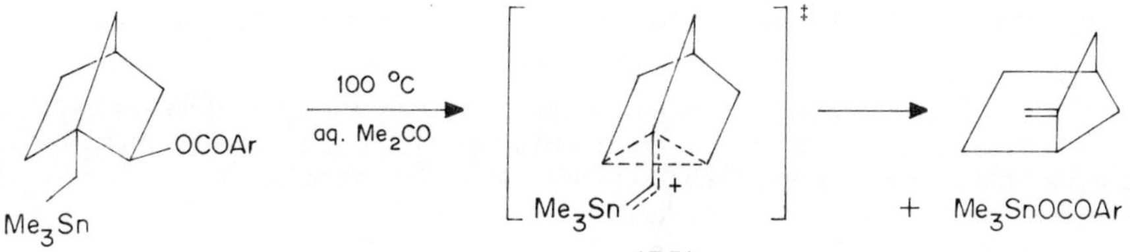

(35)

or   (71)

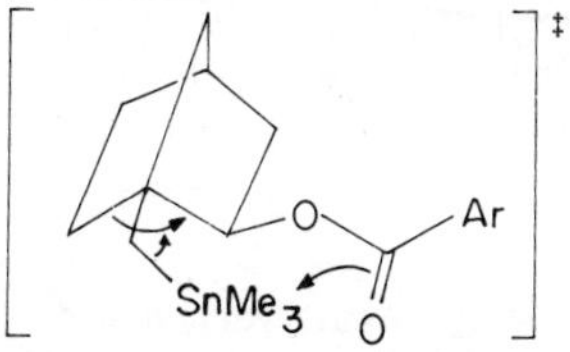

(36)

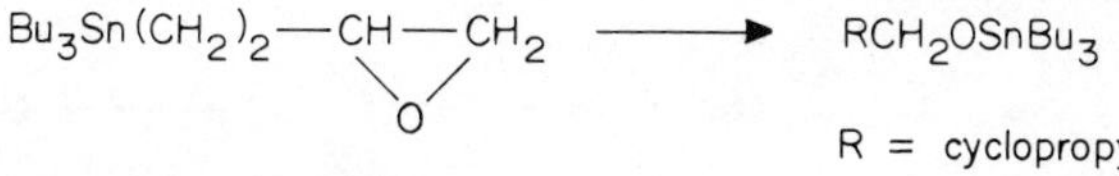

$Bu_3Sn(CH_2)_2-CH-CH_2 \longrightarrow RCH_2OSnBu_3$
       \\ O /
       R = cyclopropyl   (72)[104]

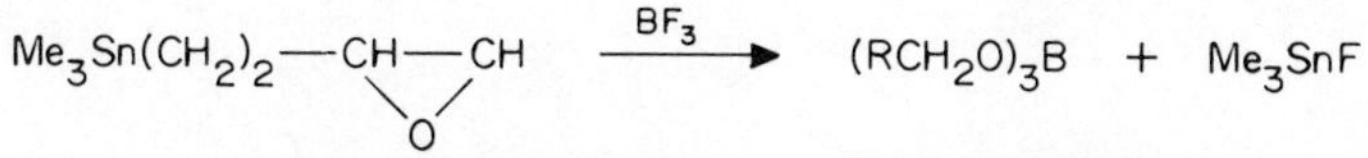

$Me_3Sn(CH_2)_2-CH-CH \xrightarrow{BF_3} (RCH_2O)_3B + Me_3SnF$
       \\ O /
       R = cyclopropyl   (73)[105]

$Me_3Sn(CH_2)_3OTos \xrightarrow{125\ °C}$ cyclopropane + $Me_3SnOTos$ (74)[106]

Intramolecular eliminations resulting in cyclizations have also been observed in a number of other reactions, for example, equations 72–74. A recent report[106a] has described a convenient synthesis for (Z)-γ-hydroxyvinylstannanes and their conversion to α,β-unsaturated γ-lactones (equation 74a).

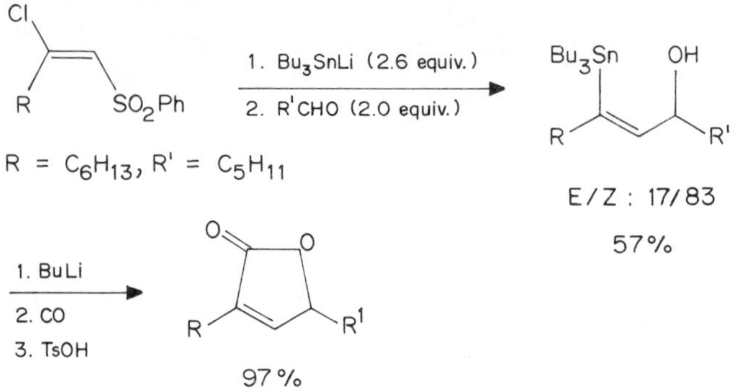

$$\begin{array}{c}\text{Cl}\\ \diagup\\ R\quad SO_2Ph\end{array} \xrightarrow[\text{2. R'CHO (2.0 equiv.)}]{\text{1. Bu}_3\text{SnLi (2.6 equiv.)}} \begin{array}{c}Bu_3Sn\quad OH\\ \diagup\quad\diagup\\ R\qquad R'\end{array}$$

$R = C_6H_{13}, R' = C_5H_{11}$

E/Z : 17/83

57%

(74a)

$$\xrightarrow[\begin{array}{c}\text{1. BuLi}\\\text{2. CO}\\\text{3. TsOH}\end{array}]{} \begin{array}{c}O\\\parallel\\ O \diagup \diagdown O\\ R\diagdown\quad\diagup R^1\end{array}$$

97%

IV. REACTIONS OF α, β- AND β,γ-UNSATURATED ORGANO-TIN AND -LEAD COMPOUNDS

The focus of discussion here will be on reactions in which the organometal substrates act as carbon nucleophiles. In the presence of reactive carbon electrophiles as partners, C—C bond formation is engendered. The reactions are to be delineated from those in which the C—MR$_3$ function serves as a masked carbanionic equivalent. For reasons previously cited, only minimal reference is made to these latter reactions in this section.

In respect of the electrophilic substitution reactions of the first category, depending on the substrate and reaction conditions, both homolytic and heterolytic or concerted reaction pathways are encountered. Whereas from a synthetic point of view such reactions resulting in unsaturated organic group transfer from the metal are of special interest, C—C bond-forming reactions are also known to occur at the sites of unsaturation in several cases, particularly with alkynylstannanes. These latter reactions, in which the metal atom presumably exerts little influence, are included to the extent that they lead to products which are either unique or possess high synthetic potential.

A. Vinyl- and Alkynyl-metal Derivatives

Mechanistic studies of the reactions of these substrates with a variety of electrophiles have revealed that the reactions proceed via an intermediate β-stannyl (or plumbyl) carbocationic species which derives stabilization through σ-π conjugation (hyperconjugation) (reactions 75 and 76)[2,22,23]. The hyperconjugation effect not only facilitates

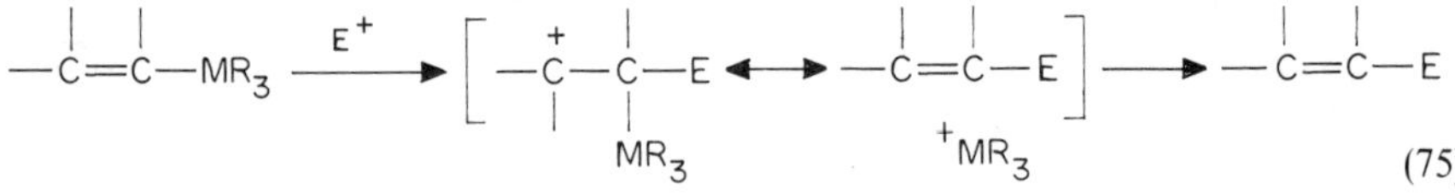

$$-\overset{|}{\underset{|}{C}}=\overset{|}{\underset{|}{C}}-MR_3 \xrightarrow{E^+} \left[-\overset{+}{\underset{|}{C}}-\overset{|}{\underset{MR_3}{C}}-E \longleftrightarrow -\overset{|}{\underset{|}{C}}=\overset{|}{\underset{^+MR_3}{C}}-E \right] \longrightarrow -\overset{|}{\underset{|}{C}}=\overset{|}{\underset{|}{C}}-E \qquad (75)$$

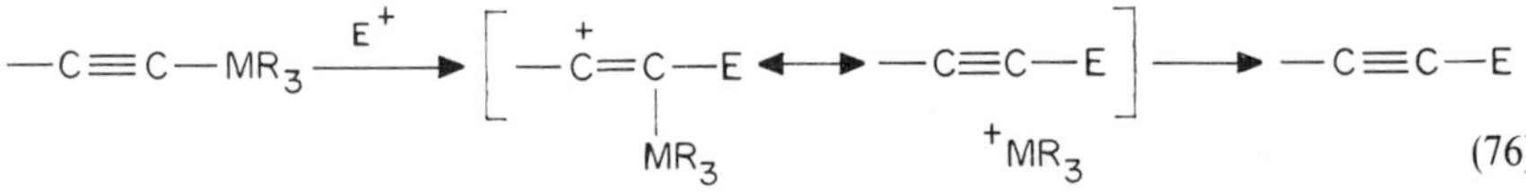

$$-C\equiv C-MR_3 \xrightarrow{E^+} \left[-\overset{+}{\underset{MR_3}{C}}=C-E \longleftrightarrow -C\equiv C-E \atop {}^+MR_3 \right] \longrightarrow -C\equiv C-E \qquad (76)$$

1. Carbon—carbon bond formation

interaction between the electrophile (E^+) and the π-system, but also cleavage of the M—C bond. Frequently, the carbon electrophiles (e.g. organic halides and carbonyl compounds) require catalytic activation (see below) for the reactions to occur.

1. Alkynyl transfer reactions

Acid chlorides have been shown to react with both alkynyl-lead and -tin compounds to give alkynyl ketones (equations 77 and 78). The Et_3SnCl formed in reaction 77 also exerts a catalytic influence on the reaction and, indeed, appears necessary to promote the reaction with PhCOCl[107]. For acetylenic tins, the coupling reaction is particularly facilitated by the

$$Et_3SnC{\equiv}CR' + RCOCl \xrightarrow[5-6\ h]{120-150\ ^\circ C} RCOC{\equiv}CR' + R'C{\equiv}CH + Et_3SnCl$$

R' = Bu, Ph, CH_2Cl ; R = Me, Ph   R = R' = Ph ; 59%   (77)[107]

$$Et_3PbC{\equiv}CR + MeCOCl \longrightarrow MeCOC{\equiv}CR + Et_3PbCl \quad (78)^{108}$$

R = Bu, Ph   R = Bu; 84%

$$Me_3SnC{\equiv}CPr + PhCOCl \xrightarrow[CHCl_3,\ 65\ ^\circ C,\ 23\ h]{[BzPdCl(PPh_3)_2]} PhCOC{\equiv}CPr \quad (79)$$

87%

$$Bu_3SnC{\equiv}CPh + Pr'COCl \xrightarrow[84\ ^\circ C,\ 2\ h]{[Pd(PPh_3)_4]\ (CH_2Cl)_2} Pr'COC{\equiv}CPh \quad (79a)$$

71%

use of palladium catalysts[108a,b] (equations 79, 79a). Organic halides are generally unreactive towards alkynylstannanes, but Beletskaya and co-workers[109] were able to obtain high yields of the desired coupling products with aryl halides using palladium catalysts (reaction 80). Stannylated ynamines react with acyl and imidolyl chlorides[110]

$$Me_3SnC{\equiv}CPh + ArX \xrightarrow[(CH_2Cl)_2]{[ArPdI(PPh_3)_2]} ArC{\equiv}CPh + Me_3SnX$$

(80)

Ar = Ph, $4\text{-}NO_2C_6H_4$, $2,4\text{-}(NO_2)_2C_6H_3$, $2,4,6\text{-}(NO_2)_3C_6H_2$; X = Cl, Br, I

and with ketenes[111] to give substituted ynamines in good yields. The alkynyl transfer in these reactions is achieved via an addition–elimination sequence (Scheme 11). A mechanistically analogous scheme probably also prevails for the reactions of alkoxyethynyltins[112] with acid chlorides and $(Me_3C)(CN)C{=}C{=}O$.

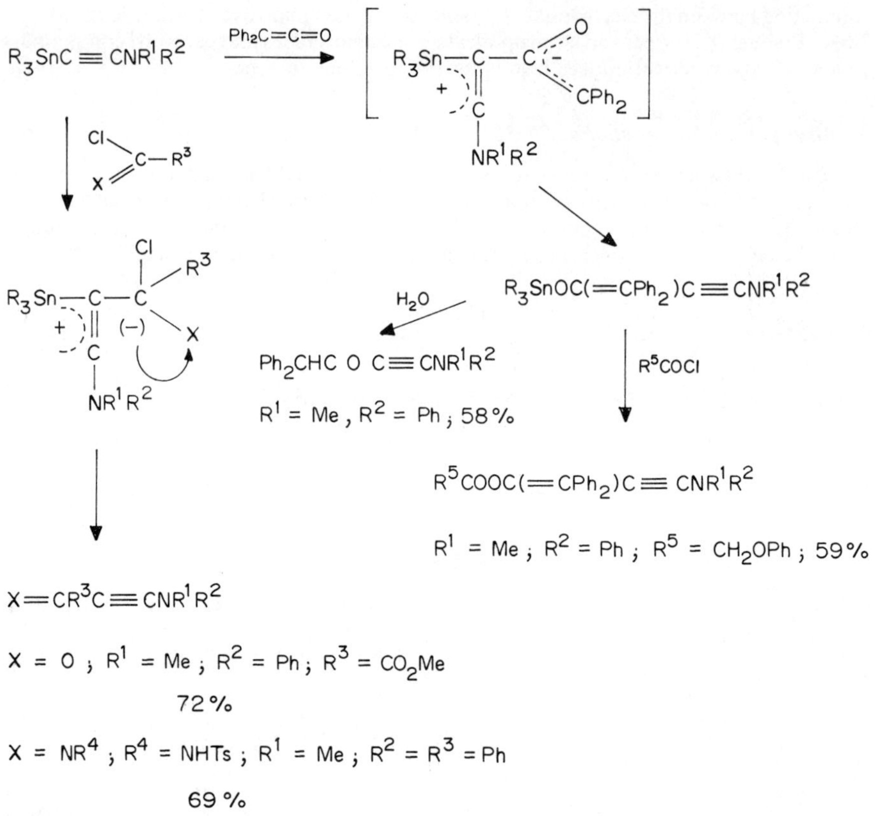

$R_3SnC≡CNR^1R^2$ →[$Ph_2C=C=O$] [$R_3Sn—C—C(=O)/CPh_2$ with NR^1R^2]

↓ $Cl\!\!\diagdown C—R^3 / X$

$R_3Sn—C—C(Cl)(R^3) / X$ with NR^1R^2

→[H_2O] $Ph_2CHC\,O\,C≡CNR^1R^2$
$R^1 = Me, R^2 = Ph; 58\%$

$R_3SnOC(=CPh_2)C≡CNR^1R^2$

↓ R^5COCl

$R^5COOC(=CPh_2)C≡CNR^1R^2$
$R^1 = Me; R^2 = Ph; R^5 = CH_2OPh; 59\%$

↓

$X=CR^3C≡CNR^1R^2$

$X = O; R^1 = Me; R^2 = Ph; R^3 = CO_2Me$
72%

$X = NR^4; R^4 = NHTs; R^1 = Me; R^2 = R^3 = Ph$
69%

SCHEME 11

$R^1_3SnC≡COR^2 + RCOCl \longrightarrow RCOC≡COR^2 + R^1_3SnCl$ (81)

27–60%

$R^1 = Me, Ph; R^2 = Me, Et; R = CHPhOAc, Bz, CH_2OPh, CH_2Br$

$Ph_3SnC≡COMe + (Me_3C)(CN)C=C=O$

$\longrightarrow (Me_3C)(CN)C=C(OSnPh_3)C≡COMe \xrightarrow{H_2O}$

$(Me_3C)(CN)CHC(O)C≡COMe$ (81a)

$Et_3SnC≡CPh \xrightarrow[\text{2. }H_2O]{\text{1. }Cl_3CCHO} HOCH(CCl_3)C≡CPh$ (82)

61%

1. Carbon—carbon bond formation

A particularly reactive electrophile towards alkynyltins is chloral[113] (equation 82). The phenyl substituent on the ethynyltin (equation 82) disfavours addition to less reactive aldehydes and ketones[113], but the use of catalysts has been proposed[47] (equation 83). With propargyltins 37, alkynyl transfer is not observed; instead allenic alcohols are obtained on aqueous work-up (reaction 84)[114]. The isomeric allenyltins (38), on the other hand, yield propargyl alcohols (reaction 85)[114]. The results were explained on the basis of a

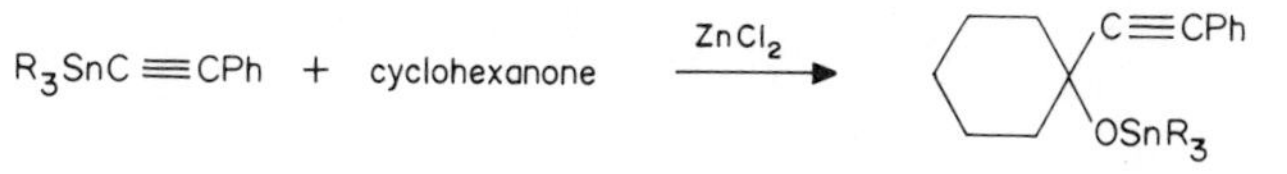

$$R_3SnC \equiv CPh + \text{cyclohexanone} \xrightarrow{ZnCl_2} \text{[cyclohexane ring with } C \equiv CPh \text{ and } OSnR_3] \quad (83)$$

$$\underset{(37)}{Me_3SnCH_2C \equiv CMe} \xrightarrow[2.\ H_2O]{1.\ Cl_3CCHO,\ 20\ °C} \underset{90\%}{Cl_3CCH(OH)C(Me)=C=CH_2} \quad (84)$$

$$\underset{(38)}{Me_3SnCH=C=CHMe} \xrightarrow[2.\ H_2O]{1.\ Cl_3CCHO,\ 60\ °C} \underset{80\%}{Cl_3CCH(OH)CH(Me)C \equiv CH} \quad (85)$$

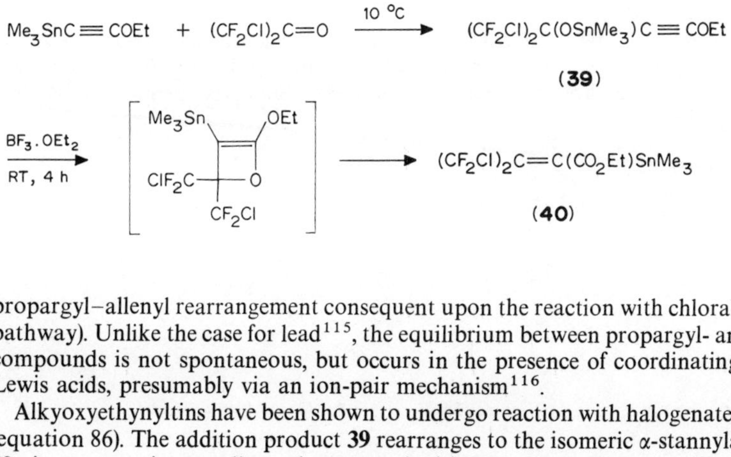

$$Me_3SnC \equiv COEt + (CF_2Cl)_2C=O \xrightarrow{10\ °C} \underset{(39)}{(CF_2Cl)_2C(OSnMe_3)C \equiv COEt}$$

$$\xrightarrow[RT,\ 4\ h]{BF_3 \cdot OEt_2} \begin{bmatrix} Me_3Sn \diagdown \diagup OEt \\ \Box \\ ClF_2C \diagup \diagdown O \\ CF_2Cl \end{bmatrix} \longrightarrow \underset{(40)}{(CF_2Cl)_2C=C(CO_2Et)SnMe_3} \quad (86)$$

propargyl–allenyl rearrangement consequent upon the reaction with chloral (S_E1' or S_E2' pathway). Unlike the case for lead[115], the equilibrium between propargyl- and allenyl-tin compounds is not spontaneous, but occurs in the presence of coordinating solvents or Lewis acids, presumably via an ion-pair mechanism[116].

Alkyoxyethynyltins have been shown to undergo reaction with halogenated ketones[117] (equation 86). The addition product 39 rearranges to the isomeric α-stannylated acrylate 40 via an oxete intermediate when treated with $BF_3 \cdot OEt_2$. The oxete intermediate has been isolated for the analogous reaction with alkoxyethynylsilane[117].

2. Other C—C bond-forming reactions of alkynylmetals

A category of reactions which has been widely observed is that of addition across the triple bond in alkynylstannanes. Thus, diazomethane affords 3(5)-substituted pyrazoles[118] (equation 87). The reaction is envisaged to proceed by attack of the nucleophilic centre of the diazo group ($\bar{C}H_2-\overset{+}{N}\equiv N: \leftrightarrow \bar{C}H_2-\ddot{N}=N^+:$) on the β-stannyl carbocationic site on the alkyne. With $Et_3SnC \equiv CC \equiv CH$, the 1,3-dipolar addition of CH_2N_2 takes place at

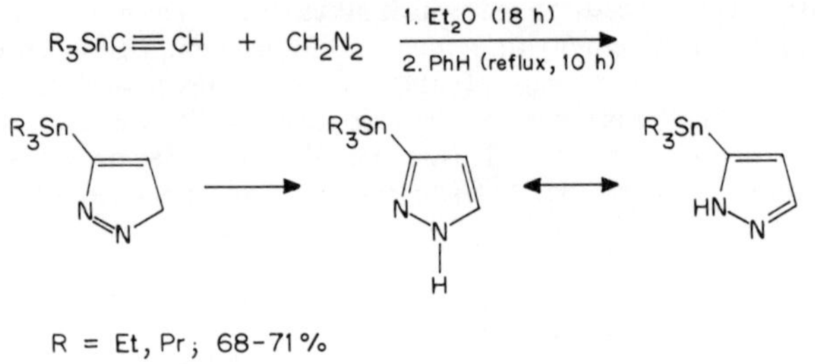

$$R_3SnC \equiv CH + CH_2N_2 \xrightarrow[\text{2. PhH (reflux, 10 h)}]{\text{1. Et}_2\text{O (18 h)}}$$

(87)

R = Et, Pr; 68-71%

the terminal alkyne group to yield the ethynylpyrazole derivative[119] **41**. Destannylation of the product was not attempted but the use of mild acids such as Ac_2H_2, $AcCH_2CO_2Et$,

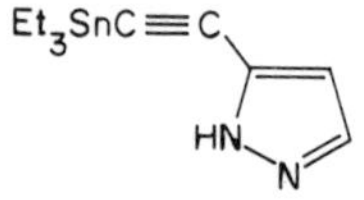

81%
(**41**)

$BzCH_2NO_2$, Dimedone, or Barbituric acid, has recently been proposed[120] for this purpose.

Replacement of CH_2N_2 with $(CF_3)_2CN_2$ in the reaction with alkynylstannanes leads to the formation of cyclopropene, *albeit* in low yield[121]. The reaction presumably involves addition of $(CF_3)_2C$: to the triple bond (reaction 88). Similarly, perfluoroalkynyltin

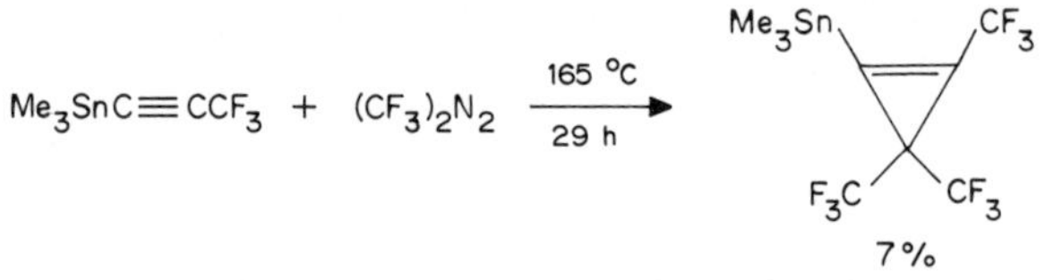

$$Me_3SnC \equiv CCF_3 + (CF_3)_2N_2 \xrightarrow[\text{29 h}]{\text{165 °C}}$$

7%

(88)

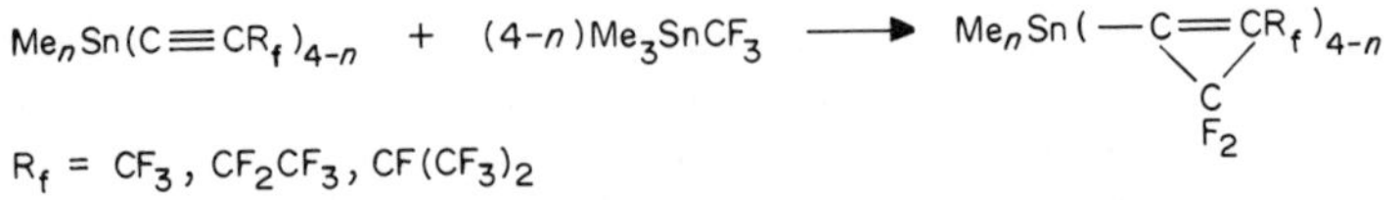

$$Me_nSn(C \equiv CR_f)_{4-n} + (4-n)Me_3SnCF_3 \longrightarrow Me_nSn(-\underset{\underset{F_2}{C}}{C} = CR_f)_{4-n}$$

(89)

$R_f = CF_3, CF_2CF_3, CF(CF_3)_2$

derivatives react with Me_3SnCF_3 at 150 °C to give perfluorocyclopropyltins[122] (equation 89). Insertion and cyclization steps appear to be implicated in the reactions with arylisocyanates[108] which yield 4-benzylidene-1,3-diarylhydantoin (**42**) (Scheme 12). C—C bond formation is also engendered in the several addition reactions of organoboranes studied by Wrackmeyer and coworkers[123], which yield products whose nature depends on the ratio of reactants, the substituents present on the reactants, and the solvent used. The

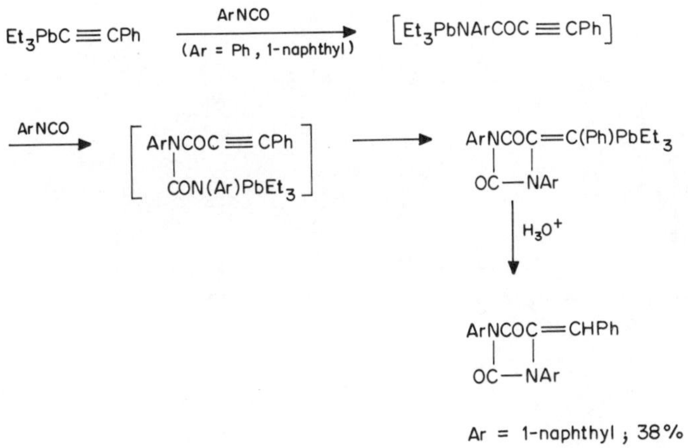
Et3PbC≡CPh
ArNCO
(Ar = Ph, 1-naphthyl)
[Et3PbNArCOC≡CPh]
ArNCO
ArNCOC≡CPh
CON(Ar)PbEt3
ArNCOC=C(Ph)PbEt3
OC—NAr
H3O+
ArNCOC=CHPh
OC—NAr

Ar = 1-naphthyl ; 38%

(**42**)

SCHEME 12

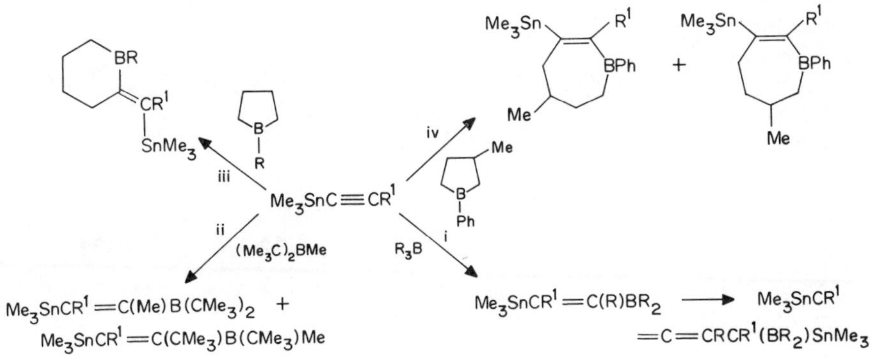
BR
CR1
SnMe3
B
R
iii
Me3SnC≡CR1
ii
(Me3C)2BMe
R3B
i
iv
Me
B
Ph
Me3Sn
R1
BPh
Me
+
Me3Sn
R1
BPh
Me
Me3SnCR1=C(Me)B(CMe3)2 +
Me3SnCR1=C(CMe3)B(CMe3)Me
Me3SnCR1=C(R)BR2
Me3SnCR1
=C=CRCR1(BR2)SnMe3

i. $R^1 = Me_3Sn$; R = Me, Et
ii. $R^1 = H$
iii. $R^1 = H$, Me, CMe_3, $SiMe_3$, $SnMe_3$; R = Me, Ph, 2-thienyl
iv. $R^1 = Me_3Sn$

SCHEME 13

products, as illustrated in Scheme 13, include several with high synthetic potential.

With bis(alkynyl)stannanes[124], five- and six-membered heterocyclic compounds are formed (equation 90).

Alkynyltins have also been shown to participate in Diels–Alder reactions (reactions 91 and 92). The 2-stannanorboranane adduct **43** formed with perchlorocyclopentadiene (equation 91) rearranges on photolysis in ether to give the substituted quadricyclane **44** in approximately 45% yield[125]. Interestingly, the presence of $SnCl_2$ or its phosphine adduct has been reported to promote the facile cycloreversion of quadricyclane to norbornadiene[127]. With α-pyrones, alkynyltins form substituted benzenes[126] (equation 92).

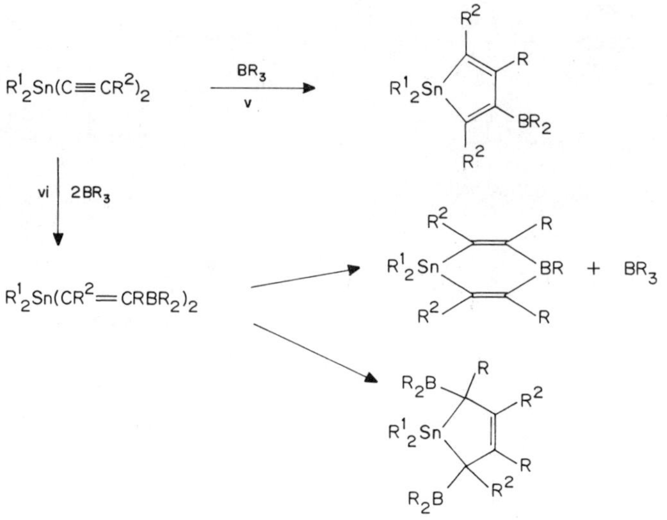

$$R^1_2Sn(C\equiv CR^2)_2 \xrightarrow{BR_3}{v} \text{[stannole product]}$$

(90)

v. R^1 = Me, Et; R^2 = H, CMe$_3$, SiMe$_3$; R = Et, HCMe$_2$, CMe$_3$
vi. R^2 = Me

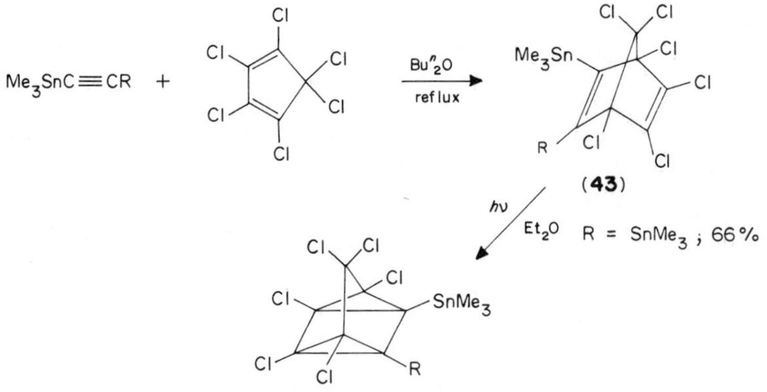

$$Me_3SnC\equiv CR + \text{[hexachlorocyclopentadiene]} \xrightarrow[\text{reflux}]{Bu^n_2O} \textbf{(43)}$$

(91)[125]

$h\nu$ / Et$_2$O   R = SnMe$_3$; 66%

(44)

$$\text{[cyclohexadienone]} + Me_3SnC\equiv CSnMe_3 \xrightarrow[155\,°C,\,27\,h]{PhBr} \text{[bis-stannyl arene]} + CO_2$$

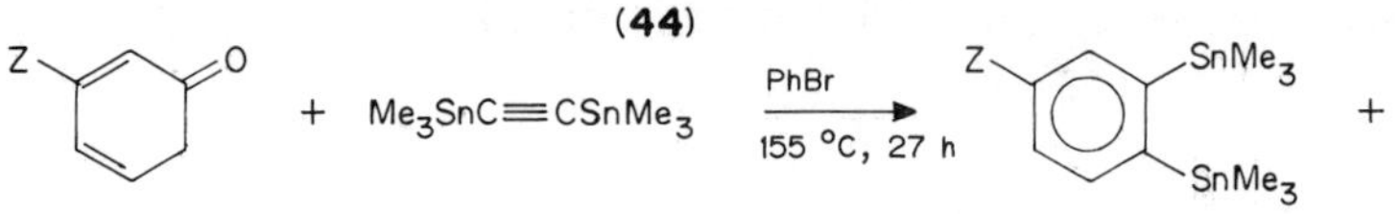

(92)[126]

Z = H, 50% ; CO$_2$Me, 50-65% ; Me, 50%

3. Vinyl transfer reactions

In organic synthesis, vinylstannanes have proved valuable precursors for vinyl group transfer. Palladium catalysts have been shown to be especially effective in this respect. Thus, with acylhalides, α-enone products have been obtained in excellent yields (reactions 93 and 94)[128]. Moderate yields of the products have also been obtained using AlCl$_3$ as

$$Bu_3SnCH\!=\!CH_2 \xrightarrow[\text{PhH, 40 °C}]{\text{PhCOCl, }[Pd(PPh_3)_4]} PhCOCH\!=\!CH_2 \quad (93)$$
$$87\%$$

$$Bu_3SnCH\!=\!CH_2 \xrightarrow[\text{[BzPdCl(PPh}_3)_2]]{\text{MeO}_2C(CH_2)_3COCl} MeO_2C(CH_2)_3COCH\!=\!CH_2 \quad (94)$$
$$92\%$$

$$Bu_3SnCH\!=\!CHMe \xrightarrow[\text{AlCl}_3,\ -10\ °C]{\text{MeCOCl, CH}_2Cl_2} (Z)\text{-MeCOCH}\!=\!CHMe \quad (95)$$
$$64\%\ Z + 36\%\ E \qquad\qquad 65\%$$

$$Bu_3SnCH\!=\!CHPh \xrightarrow[\text{AlCl}_3,\ -10\ °C]{\text{MeCOCl, CH}_2Cl_2} (Z)\text{-MeCOCH}\!=\!CHPh \quad (96)$$
$$86\%\ E + 14\%\ Z \qquad\qquad 49\%$$

catalyst[129] from both acyl halides and acyl anhydrides. A high stereoselectivity was noted for these reactions (reactions 95 and 96). It appears likely here that the elimination of the tin electrofuge from the intermediate, $[RCHCH(E)SnBu_3]$, proceeds so that the least steric interaction occurs along the newly formed C=C bond.

Vinyl transfers to organic halides proceed in poor yields (20–40%) in the absence of catalysts (reaction 97)[129]. Greater success however, has been achieved in the vinylation of benzyl and aromatic halides[130], that is, halides not containing β-H, using mono- or tetra-vinyltins and palladium catalysts. The cross-coupling products are obtained in high yield, as in reaction 98. The mechanism of this reaction is discussed in Section V (Scheme 22).

Allyl acetates[131] have also been successfully reacted with vinyltins using $[Pd(PPh_3)_4]$ as catalyst (reaction 99).

$$Bu_3SnCH\!=\!CH_2 + XCH_2CO_2Et \xrightarrow{120-150\ °C} Bu_3SnX + CH_2\!=\!CHCH_2CO_2Et \quad (97)$$
$$X = Br, I$$

$$BzBr + Bu_3SnCH\!=\!CH_2 \xrightarrow[\text{hmpt, 65 °C}]{[BzPdCl(PPh_3)_2]} BzCH\!=\!CH_2 + Bu_3SnBr \quad (98)$$
$$100\%$$

$$PhCH\!=\!CHSnMe_3 + CH_2\!=\!CHCH_2OAc \xrightarrow[\text{hmpt, 20 °C, 2.5 h}]{[Pd(PPh_3)_4]\ (5\%)} (E)\text{-PhCH}\!=\!CHCH_2CH\!=\!CH_2$$
$$E\text{-isomer} \qquad\qquad\qquad\qquad\qquad\qquad\qquad\qquad 90\%$$
$$(99)$$

Transfer of the vinylstannyl group to organic halides and α-enones[132] (Scheme 14) has been achieved in high yield using (E)-1,2-bis(tributylstannyl)ethene (**45**).

The particular value in organic synthesis of these reactions is that the vinylstannyl group functions as a precursor in the establishment of an angular ethynyl group in fused

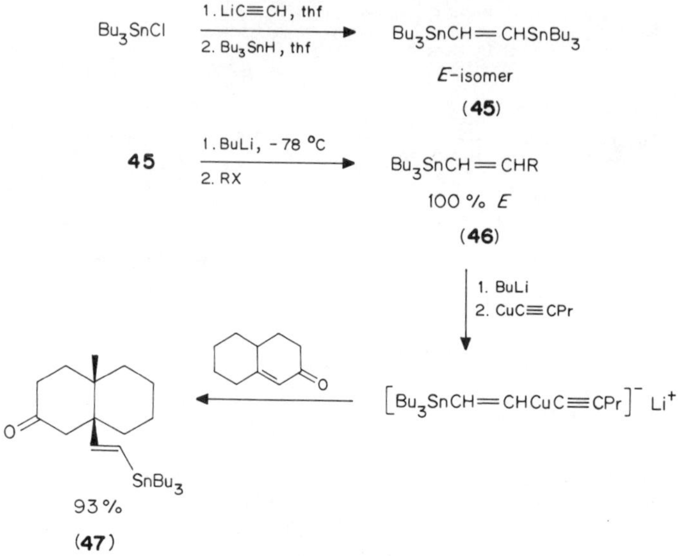

SCHEME 14 (reactions shown):

$Bu_3SnCl \xrightarrow[\text{2. Bu}_3\text{SnH, thf}]{\text{1. LiC}\equiv\text{CH, thf}} Bu_3SnCH=CHSnBu_3$

E-isomer (**45**)

45 $\xrightarrow[\text{2. RX}]{\text{1. BuLi, }-78\,°C} Bu_3SnCH=CHR$

100% *E* (**46**)

1. BuLi
2. CuC≡CPr

$[Bu_3SnCH=CHCuC\equiv CPr]^- Li^+$

(**47**) 93%

SCHEME 14

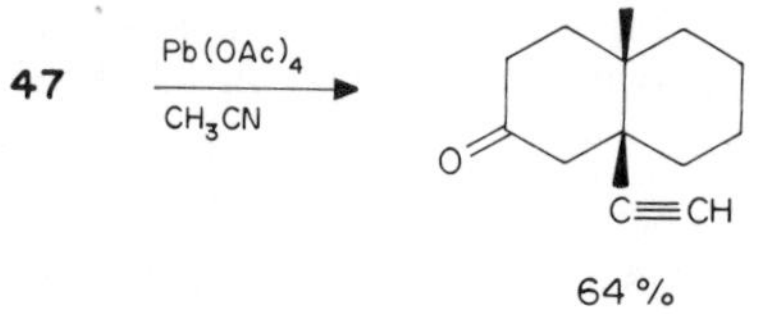

47 $\xrightarrow[\text{CH}_3\text{CN}]{\text{Pb(OAc)}_4}$ [decalone with C≡CH] 64%   (100)

cyclic structures[132] (equation 100). In the reaction of **46** with Pb(OAc)$_4$ (equation 101), the intermediate formation of an organolead species has been proposed (equation 102)[132]. The steps outlined in Scheme 14, namely lithiation of vinylstannane, conversion to the mixed cuprate, and conjugate addition to enone, have been similarly employed in the

46 $\xrightarrow[\text{CH}_3\text{CN}]{\text{Pb(OAc)}_4}$ HC≡CR   (101)

$\begin{bmatrix} Bu_3SnCH\overset{+}{C}HR \\ | \\ Pb(OAc)_3 \end{bmatrix} \cdot OAc^- \longrightarrow [(AcO)_3PbCH=CHR] + Bu_3SnOAc$

$\qquad\qquad\qquad\qquad\qquad\qquad\downarrow$
$\qquad\qquad\qquad\qquad\qquad\; HC\equiv CR + Pb(OAc)_2 + AcOH$ (102)

synthesis of *dl*-PGE$_2$ and certain 15-deoxy-16-hydroxyprostaglandins[133]. Thus the vinylstannane **48**, obtained in 87% yield from the silylether of oct-1-yn-3-ol, was used as the precursor to yield 42% of *dl*-PGE$_2$ and *dl*-15-epi-PGE$_2$ in the approximate ratio 40:60 (Scheme 15).

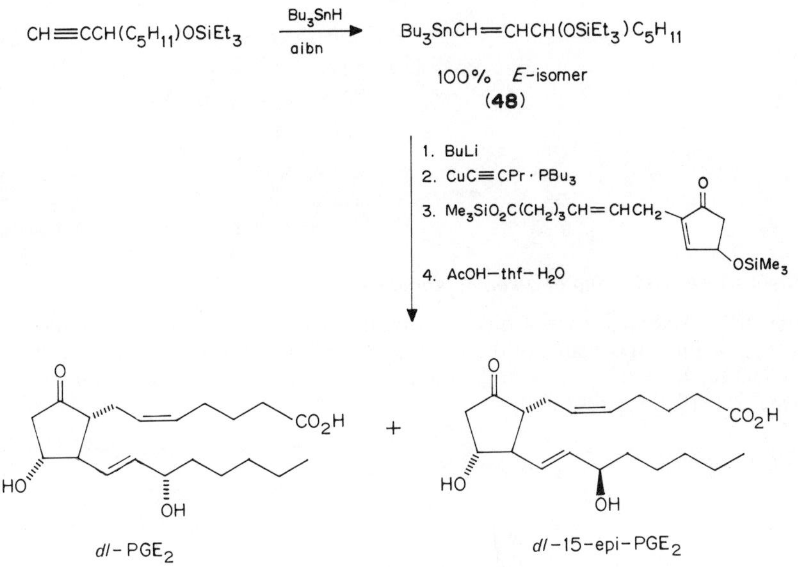

1. Carbon—carbon bond formation

$$CH\equiv CCH(C_5H_{11})OSiEt_3 \xrightarrow[aibn]{Bu_3SnH} Bu_3SnCH=CHCH(OSiEt_3)C_5H_{11}$$

100% E-isomer
(48)

1. BuLi
2. CuC≡CPr·PBu$_3$
3. Me$_3$SiO$_2$C(CH$_2$)$_3$CH=CHCH$_2$—[cyclopentenone-OSiMe$_3$]
4. AcOH–thf–H$_2$O

dl-PGE$_2$ + dl-15-epi-PGE$_2$

SCHEME 15

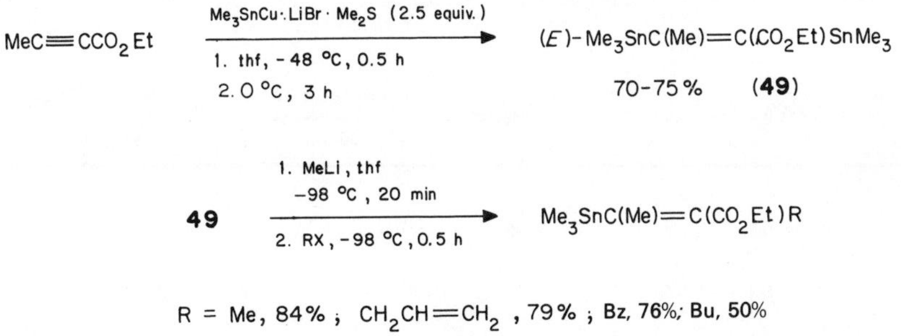

$$MeC\equiv CCO_2Et \xrightarrow[\substack{1.\ thf,\ -48\ °C,\ 0.5\ h \\ 2.\ 0\ °C,\ 3\ h}]{Me_3SnCu\cdot LiBr\cdot Me_2S\ (2.5\ equiv.)} (E)\text{-}Me_3SnC(Me)=C(CO_2Et)SnMe_3$$

70-75% **(49)**

$$\mathbf{49} \xrightarrow[\substack{1.\ MeLi,\ thf \\ -98\ °C,\ 20\ min \\ 2.\ RX,\ -98\ °C,\ 0.5\ h}]{} Me_3SnC(Me)=C(CO_2Et)R$$

R = Me, 84%; CH$_2$CH=CH$_2$, 79%; Bz, 76%; Bu, 50%

SCHEME 16

Destannylation of **48**, which is conveniently achieved by using Br$_2$ in CCl$_4$ at $-20\,°C$, yields functionalized vinyl halides[133] which find use in synthesis. Similar in scope to **48** is the compound **49**, which reacts with alkyl halides (Scheme 16) with retention of configuration with respect to the olefinic linkage[134]. The reactions in both Schemes 15 and 16 constitute recent examples from the literature which illustrate the synthetic potential of the protected carbanionic moiety in C—SnR$_3$.

Unlike the case for alkyltins, the thermal or catalytic reactions of vinyltins with aldehydes and ketones have not been reported, although several examples are known where this interaction is achieved employing the metalloid protection strategy. Equation 103, depicting the use of (E)-2-ethoxyvinyllithium as a nucleophilic acetaldehyde equivalent[135], is illustrative of these.

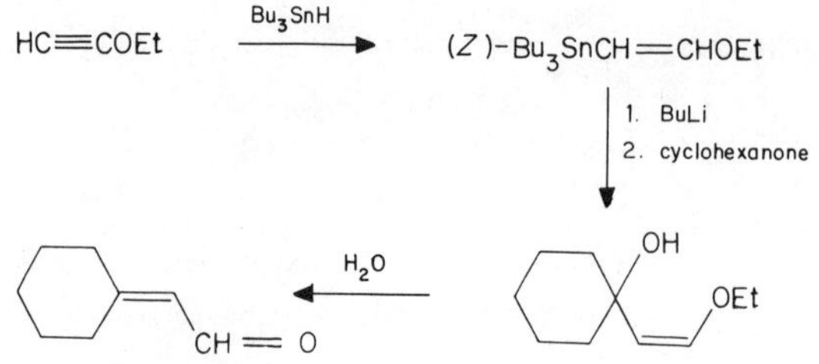

$$HC{\equiv}COEt \xrightarrow{Bu_3SnH} (Z)\text{-}Bu_3SnCH{=}CHOEt \quad (103)$$

$$\downarrow \begin{array}{l}1.\ BuLi\\ 2.\ cyclohexanone\end{array}$$

[cyclohexylidene-CH=O] ← H_2O — [1-(cyclohexyl)(CH=CHOEt)OH]

4. Diels–Alder reactions of alkenylstannanes

The Diels–Alder reactions of vinyltins appear to have been only limitedly studied. The most thoroughly investigated of these are the reactions of $RR^1C{=}CHSnMe_3$ (R = Me, Ph, CO_2Me; R^1 = H or D) with 2,3-dimethylbuta-1,3-diene and hexachlorocyclopentadiene[136] (equations 104–106).

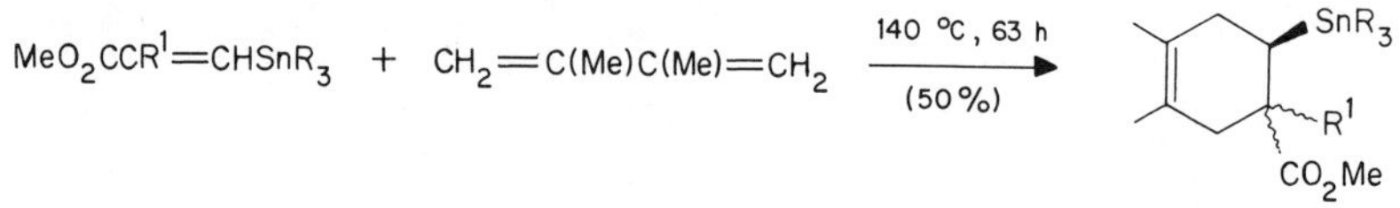

$$MeO_2CCR^1{=}CHSnR_3 + CH_2{=}C(Me)C(Me){=}CH_2 \xrightarrow[(50\%)]{140\ °C,\ 63\ h} \text{[cyclohexene with SnR}_3,\ R^1,\ CO_2Me\text{]} \quad (104)$$

(a)   92% E + 8% Z (R^1 = H; R = Me)   94 : 6

(b)   25% E + 75% Z (R^1 = H; R = Me)   17 : 83

(c)   35% E + 65% Z (R^1 = D; R = Bu)   37 : 63

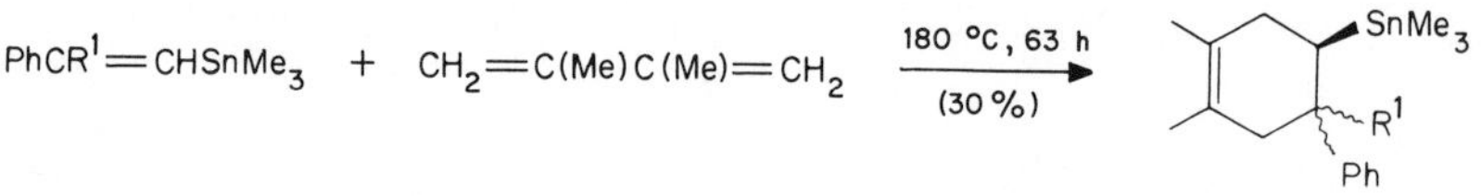

$$PhCR^1{=}CHSnMe_3 + CH_2{=}C(Me)C(Me){=}CH_2 \xrightarrow[(30\%)]{180\ °C,\ 63\ h} \text{[cyclohexene with SnMe}_3,\ R^1,\ Ph\text{]} \quad (105)$$

(a)   75% E + 25% Z (R^1 = D)   80 : 20

(b)   30% E + 70% Z (R^1 = H)   81 : 19

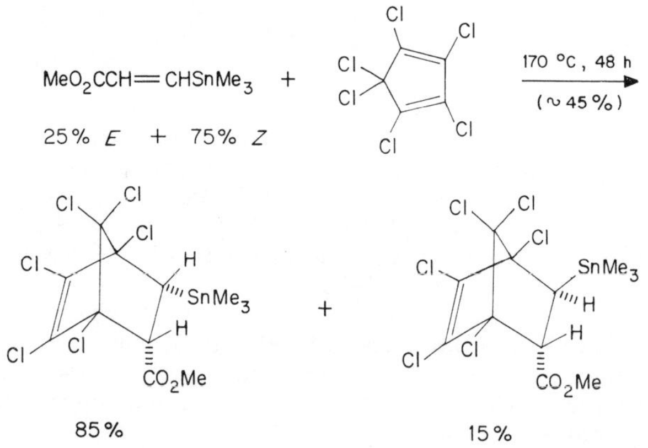

$$MeO_2CCH{=}CHSnMe_3 + \text{[hexachlorocyclopentadiene]} \xrightarrow{170\ °C,\ 48\ h} (\sim 45\%)$$

25% E + 75% Z

[norbornene adduct with SnMe₃ endo, CO₂Me] + [norbornene adduct with SnMe₃ exo, CO₂Me]   (106)

85%   15%

Equation 104 reveals a nearly stereospecific cycloaddition. The stereochemistry of the diasteromeric products obtained (equation 104) was established from a single ^{119}Sn n.m.r. spectrum using chemical shift values and ^{3}J(SnD) coupling constants. This is illustrated in Figure 1, which depicts the more highly populated of the half-chair conformations possible for the major and minor products.

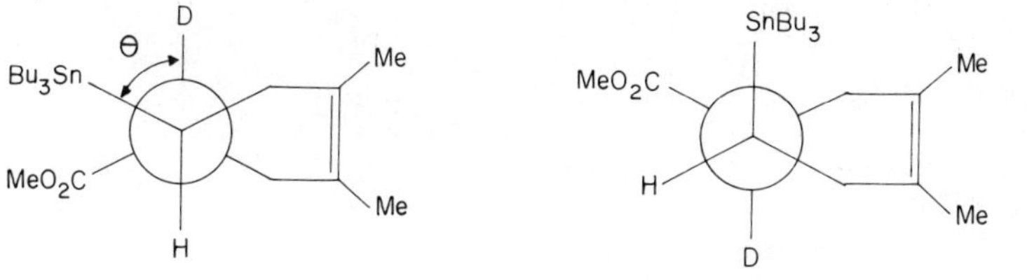
D
θ
Bu3Sn
Me
MeO2C
Me
H
SnBu3
MeO2C
Me
H
Me
D

	E-isomer	*Z*-isomer
δ_{Sn}:	− 15.5 ppm (t)	− 13.02 ppm (t)
[^{3}J(SnD)]	< 3 Hz	15.2 Hz
Sn–D dihedral angle (θ) =	60°	180°

FIGURE 1

The dominant isomer obtained for case c in equation 104 corresponded to the *Z*-configurational isomer. For the case of styryltins (equation 105), non-stereospecificity was ascribed to extensive $Z \rightarrow E$ isomerization of the alkenyltin at the temperature of the reaction. For reaction 106, ^{1}H n.m.r. was used to establish the identity of the cycloadducts. The major adduct (*Z*-isomer) is that which is formed stereospecifically with *endo*-orientation of the CO_2Me group and *exo*-orientation of both H atoms. Other examples of Diels–Alder additions are illustrated in equations 107–109.

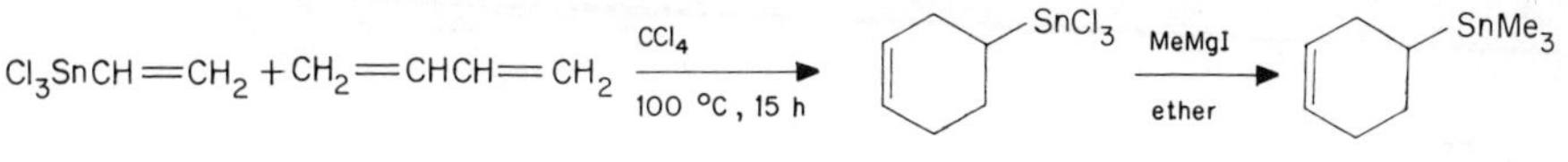
Cl3SnCH=CH2 + CH2=CHCH=CH2
CCl4
100 °C, 15 h
SnCl3
MeMgI
ether
SnMe3
31 %

(107)[137]

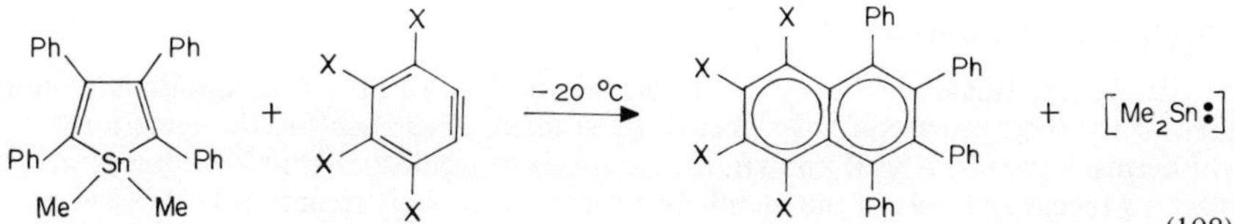
Ph
Ph
Ph
Sn
Ph
Me
Me
+
X
X
X
X
− 20 °C
X
Ph
X
Ph
X
Ph
X
Ph
+
[Me2Sn:]

(108)[138]

(**50**) X = H or F X = H ; 65 %

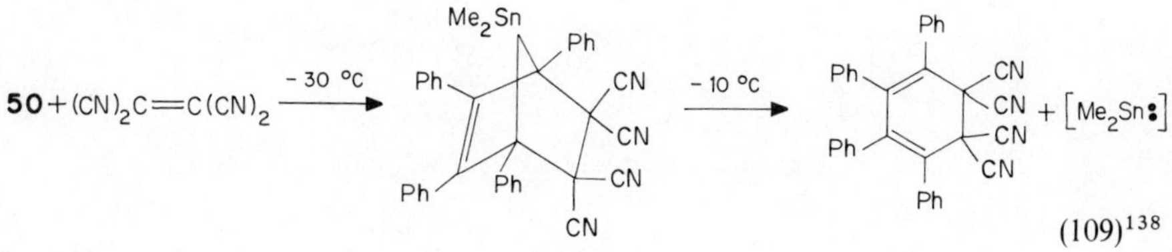
50 + (CN)2C=C(CN)2
− 30 °C
Me2Sn
Ph
Ph
CN
CN
CN
Ph
Ph
CN
− 10 °C
Ph
Ph
CN
CN
CN
CN
Ph
Ph
+ [Me2Sn:]

(109)[138]

For the case of the stannol reagent **50**, the expected bicyclic adduct was observed with tetracyanoethylene, but not with benzyne[138] (equations 108 and 109). On treatment with bromine **50** yields the acyclic product **51**, the oxidative pyrolysis[139] of which in benzene has been shown to proceed through the intermediate tetraphenylbutadiene, trapped as the $NiBr_2$ complex **52** (equation 110).

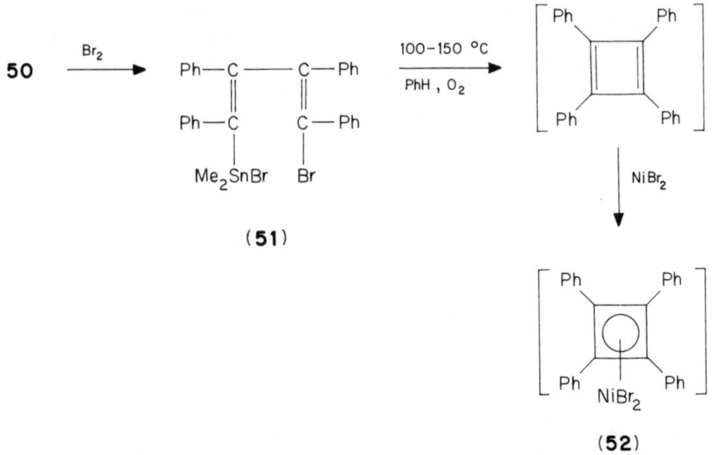

$$ \text{50} \xrightarrow{Br_2} \text{51} \xrightarrow[PhH, O_2]{100-150\ °C} [\text{tetraphenylcyclobutadiene}] \xrightarrow{NiBr_2} \text{52} \quad (110) $$

A retro-Diels–Alder reaction has been reported by Considine and coworkers[140] in the thermolysis of both *exo*- and *endo*-norborn-2-en-5-yltrimethyltin (equation 111).

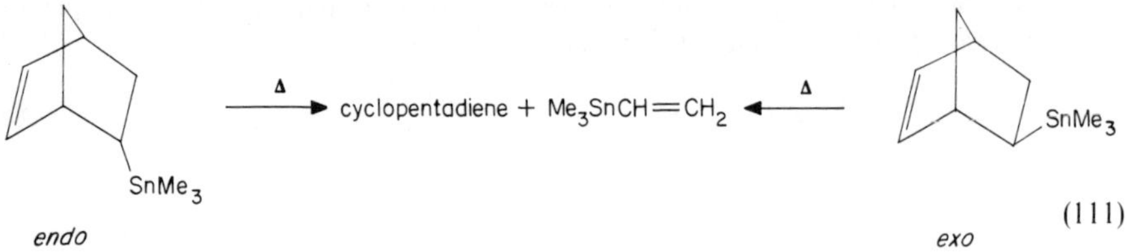

$$ endo\text{-norbornenyl-SnMe}_3 \xrightarrow{\Delta} \text{cyclopentadiene} + Me_3SnCH{=}CH_2 \xleftarrow{\Delta} exo\text{-norbornenyl-SnMe}_3 \quad (111) $$

B. Allylmetal Derivatives

Allyllead compounds are the most unstable thermally of all Pb—C compounds[141] and consequently their reactions have been little studied. Their use for the generation of allyllithiums, however, is well known; the synthesis of γ-chloroallyl anion equivalent[142] serves as a recent example of the metalloid protection strategy (reaction 112). By way of

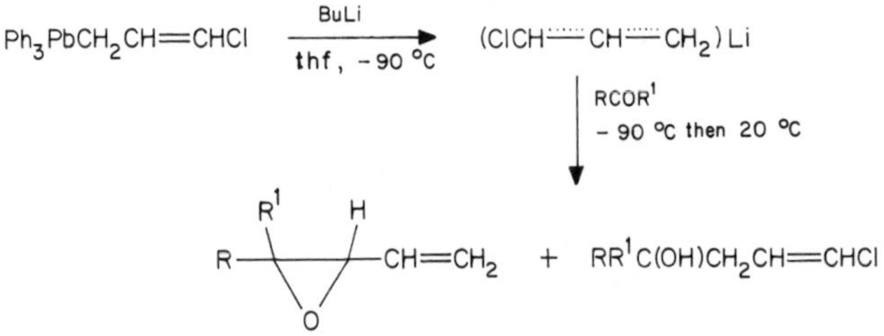

$$ Ph_3PbCH_2CH{=}CHCl \xrightarrow[thf,\ -90\ °C]{BuLi} (ClCH{\cdots}CH{\cdots}CH_2)Li \xrightarrow[-90\ °C\ then\ 20\ °C]{RCOR^1} \underset{O}{R\overset{R^1\ \ H}{\triangle}}CH{=}CH_2 + RR^1C(OH)CH_2CH{=}CHCl \quad (112) $$

1. Carbon—carbon bond formation

contrast, the reactive allyltin compounds have proved to be among the most useful organotin reagents for engendering C—C bond formation.

A comparison of electrophilic reactivity of allyltrimethyltin with vinyltrimethyltin showed that the former is up to 100 times more reactive towards HCl in methanol containing 4-5% water[143]. The generally marked reactivity of allyltins has been ascribed to hyperconjugative stabilization effects (equation 113)[2,22,23,144]. Radical pathways are

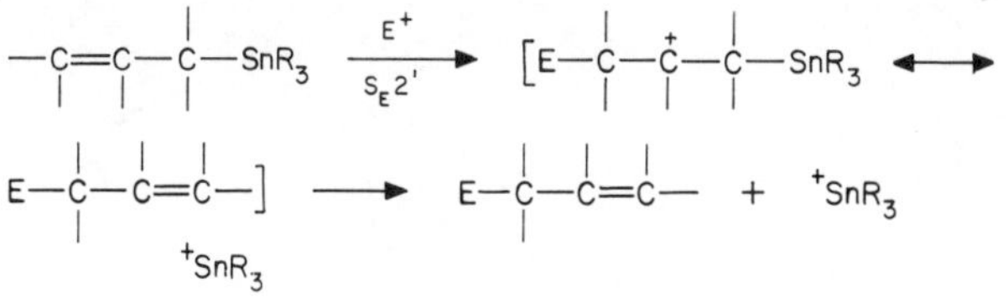

$$\begin{aligned}-\overset{|}{C}=\overset{|}{C}-\overset{|}{C}-SnR_3 \xrightarrow{E^+}_{S_E2'} \left[E-\overset{|}{C}-\overset{|}{C}-\overset{+}{\overset{|}{C}}-SnR_3 \longleftrightarrow \right. \\ \left. E-\overset{|}{\underset{{}^+SnR_3}{C}}-\overset{|}{C}=\overset{|}{C}- \right] \longrightarrow E-\overset{|}{C}-\overset{|}{C}=\overset{|}{C}- + {}^+SnR_3 \end{aligned} \quad (113)$$

also accessible to allyltins, as has been evidenced in the several thermal- and light-induced reactions with organic halides (R^1X). An S_H2' mechanism (reaction 114) has been generally postulated[145-150].

$$R^1\bullet + CH_2=CHCH_2SnR_3 \longrightarrow R^1CH_2CH=CH_2 + R_3Sn\bullet$$
$$R_3Sn\bullet + R^1X \longrightarrow R_3SnX + R^1\bullet \quad (114)$$

In comparison with allylsilanes, which already find established use in organic synthesis[151], allyltins have evoked similar interest only in the last few years. Of potential significance is that not only are allyltins more nucleophilic than their silicon analogues[152], but they appear to be more readily accessible preparatively from a wide variety of allyl—X units via several methods[153], including the use of organostannylanionoids[154] (R_3SnLi).

1. Allyl transfer reactions

a. Reactions with organic halides, -acetates and acid halides

Grignon et al.[145] in their thorough investigations on the thermal reactions of allyltins with alkyl and aromatic halides, reported that these uncatalysed reactions proceed by a radical pathway and tolerate certain functional groups which would not have survived in more conventional reactions employing allyl-Grignard or allyllithium reagents (reactions 115–117). These radical reactions are known[150a] to tolerate a variety of functional groups in the

$$Bu_3SnCH_2CH=CH_2 + (\pm)MeCH(Br)CO_2Me \xrightarrow[10\ h]{100\ ^\circ C} (\pm)CH_2=CHCH_2CH(Me)CO_2Me + Bu_3SnBr$$
(**53**)   70%   (115)

53 + Z—C$_6$H$_4$—Br $\xrightarrow{200\ ^\circ C}$ Z—C$_6$H$_4$—CH$_2$CH=CH$_2$ + Bu$_3$SnBr   (116)

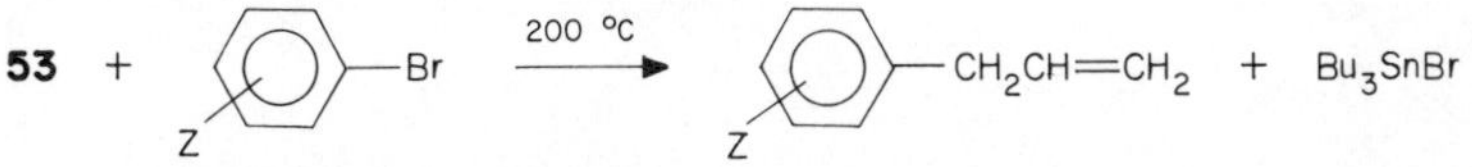

Z = p-OMe, p-Me, p-F, m-CF$_3$; log k_Z/k_H = 0.87 σ

53 + CH$_2$=CH(CH$_2$)$_4$I $\xrightarrow[70\ h]{190\ ^\circ C}$ cyclopentyl-CH$_2$CH$_2$CH=CH$_2$ + cyclohexyl-CH$_2$CH=CH$_2$ + Bu$_3$SnI

                                    76%   4%   (117)

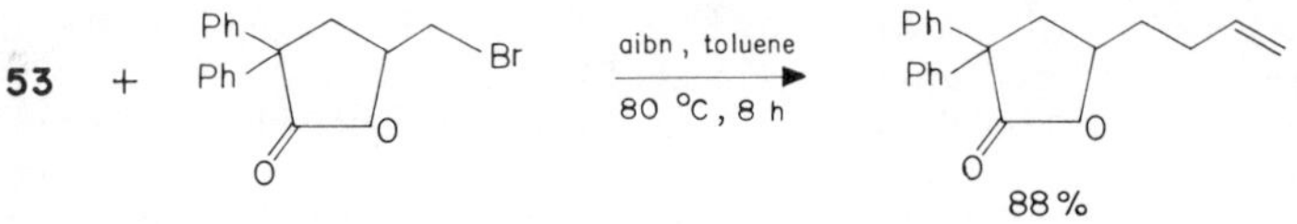

[Reaction 118: compound **53** + (Ph,Ph-substituted γ-butyrolactone with CH₂Br) → aibn, toluene, 80 °C, 8 h → allyl-substituted product, 88%] (118)

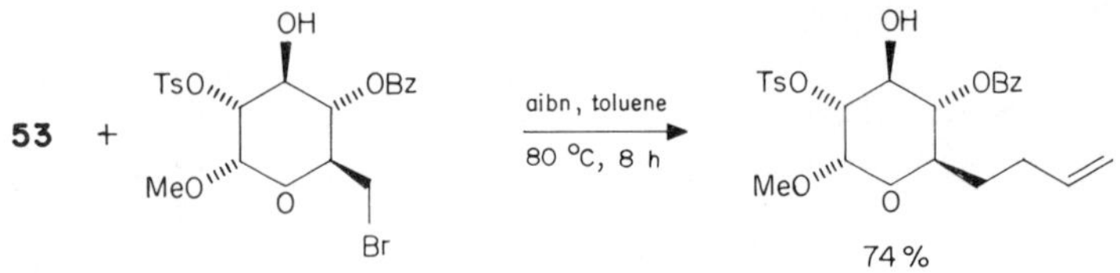

[Reaction 118a: **53** + (cyclohexane with OH, TsO, OBz, MeO, O, CH₂Br substituents) → aibn, toluene, 80 °C, 8 h → allyl-substituted product, 74%] (118a)

organic halide and to proceed with a high degree of chemoselectivity as illustrated by equations 118 and 118a. With substituted allyltins the reaction is accompained by allylic rearrangement as in reactions 119 and 120[145,146]. In the case of the reaction with chloral, however, a concerted mechanism appears to be more likely, with the product arising from an intermediary alkenoxy compound[146] (equation 120a).

$$Bu_3SnCH_2CH{=}CHMe + CCl_4 \xrightarrow[36\ h]{u.v.} CH_2{=}CHCH(Me)CCl_3 + Bu_3SnCl \quad (119)$$

$$55\%$$

$$Bu_3SnCH_2CH{=}CHMe + Cl_3CCHO \xrightarrow[5\ h]{200\ °C} CH_2{=}CHCH(Me)CCl_2CHO + Bu_3SnCl \quad (120)$$

$$70\%$$

$$Bu_3SnCH_2CH{=}CHMe + Cl_3CCHO \xrightarrow[20\ h]{100\ °C} Cl_3CCH(OSnBu_3)CH(CH_3)CH{=}CH_2 \quad (120a)$$

Aromatic halides containing substituents which exert a negative mesomeric effect do not enter into thermal reactions with allyltins, except in the presence of catalysts such as [Pd(PPh₃)₄][155]. Aryl bromides were found to be the best substrates of the aryl halides. Although the mechanism of the reaction was not investigated, it has been postulated[155] that ArPdX species may be involved as intermediates.

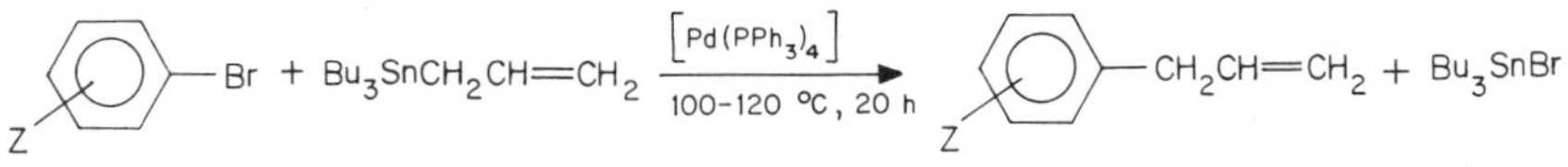

$$Z{-}C_6H_4{-}Br + Bu_3SnCH_2CH{=}CH_2 \xrightarrow[100-120\ °C,\ 20\ h]{[Pd(PPh_3)_4]} Z{-}C_6H_4{-}CH_2CH{=}CH_2 + Bu_3SnBr \quad (121)$$

Z = H, 96% ; Cl, 100% ; Me, 96% ; MeCO, 98% ; NO_2, 72%

Palladium catalysts have also featured in the unsymmetrical allyl–allyl coupling between allyltins and allyl bromides[156] and acetates[157]. Homocoupling products are not formed in these reactions and allylic rearrangement is manifested only in the tin partner. The reaction offers promise as a general procedure for the synthesis of compounds containing the polyisoprenoid structure (reactions 122–126). With $ZnCl_2$ in place of the palladium(II) catalyst, an equally efficacious coupling is achieved which proceeds with no

$Me_2C=CHCH_2Br + (CH_2=CHCH_2)_4Sn \xrightarrow[\text{thf, 60 °C, 48 h}]{[BzPdCl(PPh_3)_2]} Me_2C=CH(CH_2)_2CH=CH_2$
81%
(122)

$CH_2=CHCH_2Br + Me_2C=CHCH_2SnBu_3 \xrightarrow[\text{CHCl}_3, 60 °C, 48 h]{[BzPdCl(PPh_3)_2]} Me_2C=CH(CH_2)_2CH=CH_2 +$
3%

$CH_2=CHCMe_2CH_2CH=CH_2$
48%
(123)

$PhCH=CHCH_2OAc + Me_2C=CHCH_2SnBu_3 \xrightarrow[\text{thf, 6 h}]{[Pd(PPh_3)_4]} PhCH=CHCH_2CMe_2CH=CH_2$
(54)
4%
(124)

$54 + (CH_2=CHCH_2)_4Sn \xrightarrow[\text{thf, 6 h}]{[Pd(PPh_3)_4]} PhCH=CH(CH_2)_2CH=CH_2$
71%
(125)

$54 + MeCH=CHCH_2SnBu_3 \xrightarrow[\text{thf, 6 h}]{[Pd(PPh_3)_4]} PhCH=CHCH_2CH(Me)CH=CH_2$ (126)
32%

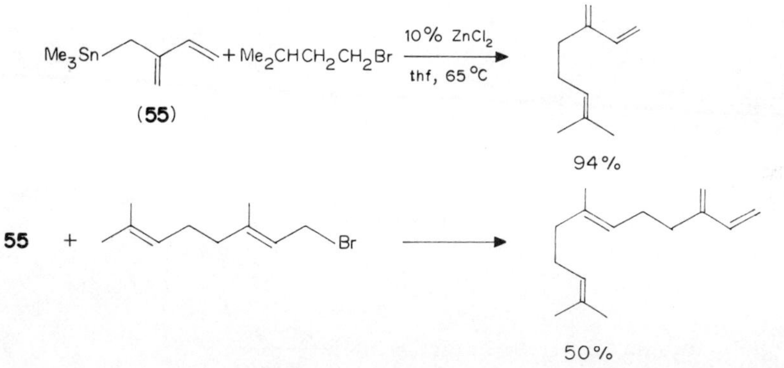

$Me_3Sn\!\!\diagup\!\!\diagdown + Me_2CHCH_2CH_2Br \xrightarrow[\text{thf, 65 °C}]{10\% ZnCl_2}$
(55)
94%

$55 + \text{(geranyl bromide)} \longrightarrow$ (126a)
50%

allylic transposition[156a]. Myrcene and β-farnesene have been synthesized by this scheme (equation 126a).

The low yield of product obtained in the reaction of tributylprenyltin with cinnamyl acetate (equation 124) in the presence of palladium(0) catalyst has been ascribed to steric hindrance at the tertiary carbon centre of the prenyl group. Interestingly, no butyl transfer occurs in this reaction, whereas with triphenylprenyltin the phenyl group is transferred.

The mechanism envisaged[157] for the above reactions featuring either palladium(II) or palladium(0) as catalyst favours the initial formation of a π-allylpalladium complex, which then engages in electrophilic attack at the terminal vinyl carbon of the allyltin reagent (reaction 127). This pathway is preferred to the formation of a palladium(0) complex from

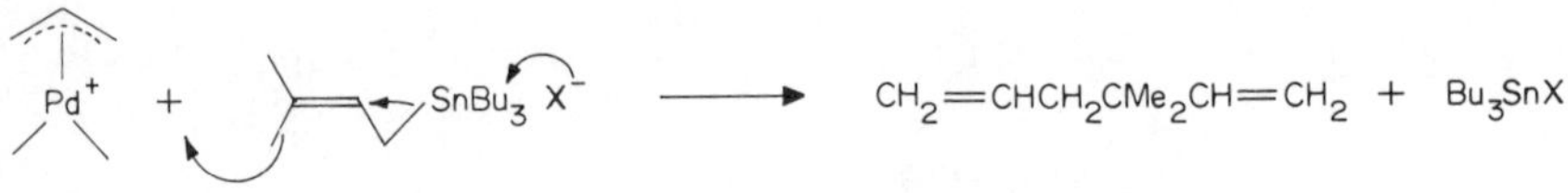

$$Pd^+ + \text{allylSnBu}_3 \; X^- \longrightarrow CH_2=CHCH_2CMe_2CH=CH_2 + Bu_3SnX$$

X = Br or OAc   (127)

palladium(II) via metathesis followed by reductive elimination, since no allylic transposition was observed in the allyl halide or acetate and Me_4Sn was found to be without effect on prenyl bromide.

When both the allyltin and allyl acetate partners reside in the same molecule such that the allyl acetate exists in an Z-configurational arrangement, intramolecular cyclization may be induced by the use of suitable catalysts. An extremely selective synthesis of limonene has been reported by Oshima and coworkers[158] using this approach (equation 128). The reaction of carboxylic acid halides with allyltins appears similarly facile, particularly in the presence of transition metal catalysts. Rhodium catalysts have aroused considerable interest in this respect since it has been reported[159] that no allylic transposition results with these catalysts. Reactions 129 and 130 afford a useful synthetic

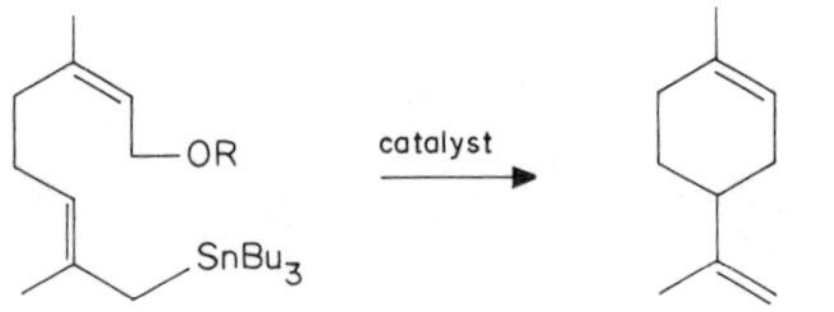

(128)

R = H; catalyst, $TiCl_4$–$PhNHMe/CH_2Cl_2$; 73%
R = COMe; catalyst, $MeAl(OCOCF_3)_2$/hexane; 51%

$$PhCOCl + Bu_3SnCH_2CH=CH_2 \xrightarrow[PhH, 80\,°C, 5\,h]{[RhCl(PPh_3)_3]} PhCOCH_2CH=CH_2 \quad (129)$$
86%

$$EtCOCl + Bu_3SnCH_2CH=CHMe \xrightarrow[PhH, 80\,°C, 12\,h]{[RhCl(PPh_3)_3]} EtCOCH_2CH=CHMe \quad (129a)$$
64%

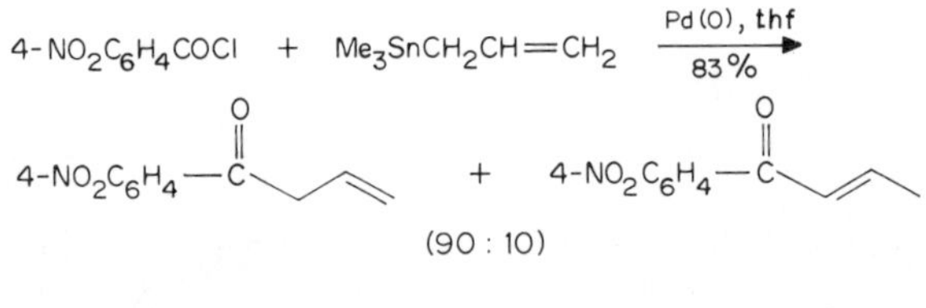

$$4\text{-}NO_2C_6H_4COCl + Me_3SnCH_2CH=CH_2 \xrightarrow[83\%]{Pd(0),\,thf}$$

$$4\text{-}NO_2C_6H_4-\overset{O}{\underset{\|}{C}}-CH_2CH=CH_2 + 4\text{-}NO_2C_6H_4-\overset{O}{\underset{\|}{C}}-CH=CHCH_3 \quad (130)$$

(90 : 10)

route to allyl ketones under mild conditions, since these compounds are generally difficult to isolate from acidic or basic reaction mixtures without causing isomerization to occur to the corresponding conjugated enones. An allylic rhodium(I) intermediate has been postulated as the actual allylating agent. Selective allyl transfer has also been successfully demonstrated using $[Pd(PPh_3)_4]$ as catalyst[108a]. Allylation of acyl halides under mild conditions in the absence of catalysts has recently been reported using the more reactive

1. Carbon—carbon bond formation

allylchlorotin reagents[159a]. Dibutylcrotyltin chloride, for example, yields methylallyl ketone (equation 130a) in high yield from both acetyl and pivaloyl chlorides, indicating that the reaction proceeds with allylic rearrangement.

$$Bu_2Sn(CH(Me)=CHCH_2)Cl + RCOCl \xrightarrow[24-30\ h]{25\ °C} RCOCH(Me)CH=CH_2 \quad (130a)$$

R = Me, Me$_3$C ; 80–90%

b. Carbonyl addition reactions

Allyltins have been intensively studied in the last few years in respect of their addition to carbonyl compounds. With aldehydes and ketones, homoallylic alcohols result on aqueous work-up (reaction 131). The reaction of crotyltins with aldehydes has been

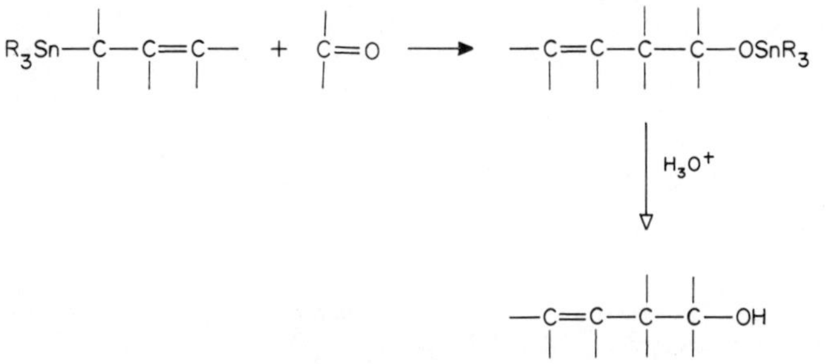

$$R_3Sn-\overset{|}{C}-\overset{|}{C}=\overset{|}{C}- + \overset{|}{C}=O \longrightarrow -\overset{|}{C}=\overset{|}{C}-\overset{|}{C}-\overset{|}{C}-OSnR_3$$

$$\Big\downarrow H_3O^+ \quad (131)$$

$$-\overset{|}{C}=\overset{|}{C}-\overset{|}{C}-\overset{|}{C}-OH$$

recently reviewed by Hoffmann[160]. Both allyl- and crotyl-tin compounds react with aromatic aldehydes only on heating[†] to temperatures in the range 100–200 °C (reactions 132 and 133)[47,146]. Ambient temperatures, on the other hand, sufficed for the more reactive chloral[146,161,162] and perhalogenoacetones[163] (reactions 134–136). With crotyltins the reactions were always accompanied by allylic rearrangement. Further, in the reaction with chloral[162] (equation 134), investigation of the stereochemistry of addition

$$Bu_3SnCH_2CH=CHMe + p\text{-}Cl\text{-}C_6H_4CHO \xrightarrow[10\ h]{200\ °C} p\text{-}Cl\text{-}C_6H_4CH(OSnBu_3)CH(Me)CH=CH_2$$
$$70\% \quad (132)$$

$$Et_3SnCH_2CH=CH_2 + p\text{-}NO_2\text{-}C_6H_4CHO \xrightarrow[2\ h]{100\ °C} p\text{-}NO_2\text{-}C_6H_4CH(OSnEt_3)CH_2CH=CH_2$$
$$80\% \quad (133)$$

$$Bu_3SnCH_2CH=CHMe + Cl_3CCHO \xrightarrow[10\ h]{20\ °C} Cl_3CCH(OH)CH(Me)CH=CH_2 \quad (134)$$

$$Me_3SnCH_2CH=CHMe + (CF_3)_2CO \xrightarrow[10\ h]{25\ °C} (F_3C)_2C(OSnMe_3)CH(Me)CH=CH_2$$
$$69\% \quad (135)$$

[†]A recent report [Y. Yamamoto, K. Maruyama and K. Matsumoto, *J. Chem. Soc., Chem. Commun.*, 489 (1983)], however, indicates that facile allylation with allylic stannanes may be accomplished at room temperature under conditions of high pressure (10 Kbar).

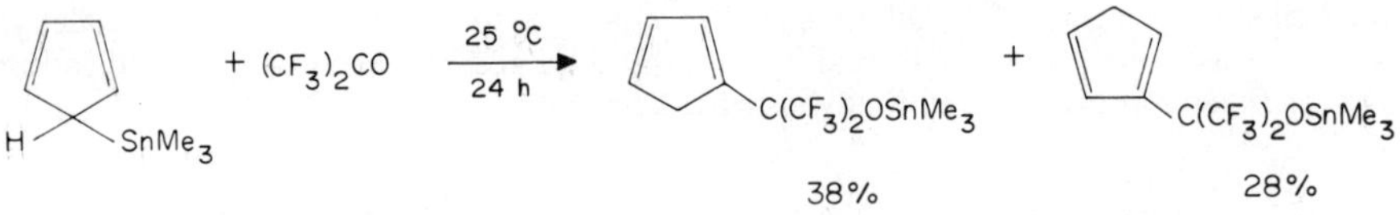

$$\text{(cyclopentadienyl-CH}_2\text{SnMe}_3) + (CF_3)_2CO \xrightarrow[24\ h]{25\ °C} \text{(cyclopentenyl-CH}_2\text{C(CF}_3)_2\text{OSnMe}_3) + \text{(cyclopentadienyl-C(CF}_3)_2\text{OSnMe}_3)$$
38%   28%

(136)

under conditions of kinetic control showed that (Z)-crotyltin yielded diastereospecifically the *erythro* (*syn*) product, while (E)-crotyltin afforded the *threo* (*anti*) product (Scheme 17).

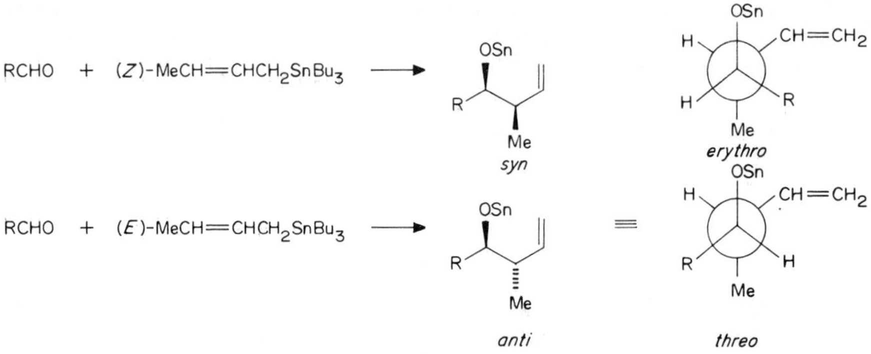

$$RCHO + (Z)\text{-MeCH=CHCH}_2\text{SnBu}_3 \longrightarrow$$

syn / erythro

$$RCHO + (E)\text{-MeCH=CHCH}_2\text{SnBu}_3 \longrightarrow$$

anti / threo

SCHEME 17

At this point, a word on the stereo-nomenclature of the diastereomeric products is in order. The term *erythro* is used for the product for which the Newman projection leads to the same group sequences around the two chiral carbons as given by the Cahn–Ingold–Prelog priority rule. The other configuration is termed *threo*. Thus in the example above (Scheme 17), the *erythro* isomer is the *RR* enantiomer, and the *threo* isomer is the *SR* enantiomer. An alternative nomenclature (among others) uses the terms *syn* and *anti*. This has the advantage that it allows of easy visualization of structures in most cases. The *syn* compound (corresponding to *erythro*) is that for which in the 'sawhorse' formalism the groups of interest, namely $OSnR_3$ and Me, are located on the same side of the newly formed C—C bond. In this chapter, the terms *erythro* and *threo* will be used since these currently enjoy wide acceptance in the literature on aldol reactions.

Returning to equation 134, it was additionally observed that when the temperature of the reaction was raised to 100 °C only the *threo* isomer prevailed. This was explained[162] in terms of rapid $Z \rightarrow E$ isomerization at this temperature. High diastereoselectivity has also been reported in the reaction of aldehydes with trihaloallyltin compounds, prepared *in situ* from allyl iodides (equations 137 to 139). Qualitative comparisons of reactivity of allyltins

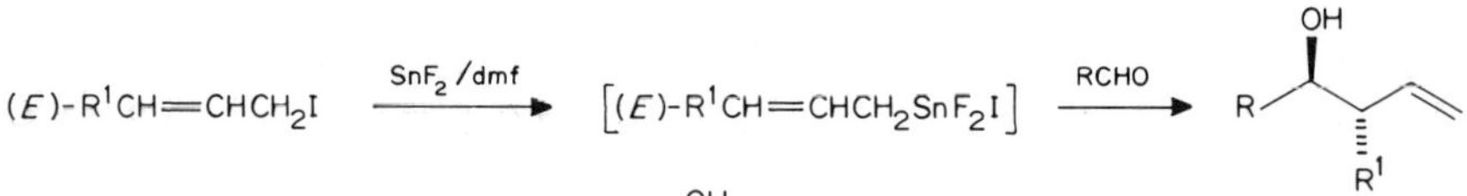

$$(E)\text{-}R^1CH=CHCH_2I \xrightarrow{SnF_2/dmf} [(E)\text{-}R^1CH=CHCH_2SnF_2I] \xrightarrow{RCHO} R\text{-CH(OH)-CH(R}^1)\text{-CH=CH}_2$$

(137)[164]

$$(Z)\text{-}R^1CH=CHCH_2I \xrightarrow[\text{2. RCHO}]{\text{1. SnF}_2/dmf} R\text{-CH(OH)-CH(R}^1)\text{-CH=CH}_2$$

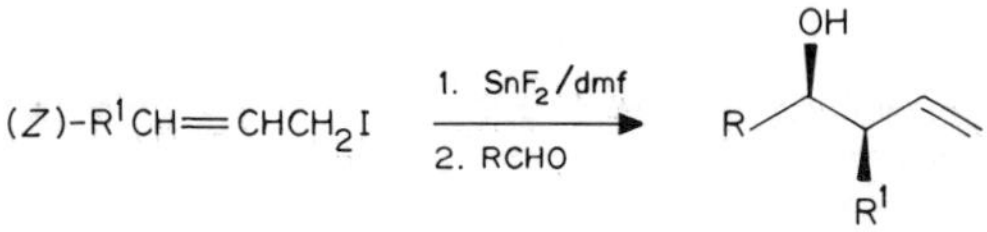

(138)[164]

1. Carbon—carbon bond formation

$$\text{(139)}^{165}$$

in aldol additions have been made[165-168]. In general, crotyltins appear to be less reactive than allyltins towards aldehydes and ketones[166]. Halogen[165,166] or additional allyl moieties[168] on tin enhance the reactivity. Thus, compared with tributylallyltin, both tetraallyltin and dibutyl(crotyl)tin chloride (57) react exothermically with aldehydes at room temperature and with non-activated ketones at moderate temperatures.

$$\text{Bu}_2\text{SnClCH}_2\text{CH}=\text{CHMe} + \text{RCHO} \xrightarrow[\text{2. aq. NH}_4\text{Cl}]{\text{1. 25 °C}} \text{RCH(OH)CH(Me)CH}=\text{CH}_2 \quad (140)$$

(57) 75-98%

R = Me, Et, CHMe$_2$, CHMeEt, Ph

In the reaction depicted in equation 140, an *erythro/threo* product ratio of 1:2 was obtained, independent of the *E/Z* ratio of the crotyl residues[169]. From this, the *threo* isomer formation rate was inferred to be twice that of the *erythro* isomer. With bulky R groups on the aldehyde, the reaction proceeded with a high degree of stereoselectivity. A pericyclic transition state has been proposed by Tagliavini and coworkers[169] to describe the features of allylic rearrangement and stereoselectivity. As shown in Scheme 18, two stereochemically different transition states lead to two diastereomers, *threo-* and *erythro-α*-methylallylcarbinol. The *E*-configuration transition state leads to the *threo*-isomer (*RS* and *SR* enantiomers), and is more favourable than the *Z*-state, which is influenced by the steric hindrance arising from the two opposed R and Me groups.

A recent investigation[170] on the preference for *Z*- or *E*-geometry in the ethylenic part of the homoallylic alcohols derived from Bu$_2$SnClCH(Me)CH=CH$_2$ (58) clearly shows that the *Z*-geometry is obtained exclusively. Under the conditions of the reaction (equation 141), formation of 'branched' alcohols via the isomerization of 58 to 57 was found to be marginal. A six-membered cyclic transition-state structure has been postulated with the Me group occupying an axial position in the chair conformation of the ring (Figure 2). In this conformation, the Me group is in a *gauche* conformation with respect to one R group and to one oxygen atom, whereas in the alternative equatorial case (Figure 2b), the Me group is in an unfavourable eclipsed conformation with respect to the Cl atom. The preference for the first-mentioned transition state is retained even with increasing steric size of the R groups on tin, no alcohols with *E*-geometry being detected. In the above example, the tin atom is shown in the trigonal bipyramidal configuration with the Cl and O atoms in apical positions as is found for most pentacoordinated triorganotin(IV) structures[20,23,171].

In both Scheme 18 and Figure 2, six-membered cyclic transition-state structures were considered involving some degree of coordination of tin to the carbonyl moiety. This appears reasonable in view of the known Lewis acidities of triorganotin halides. Hoffmann[160] has pointed out that on steric grounds boat-like transition states may even be preferred to chair forms in many cases, particularly when energy differences between the two transition states become comparable to that commonly encountered in Claisen

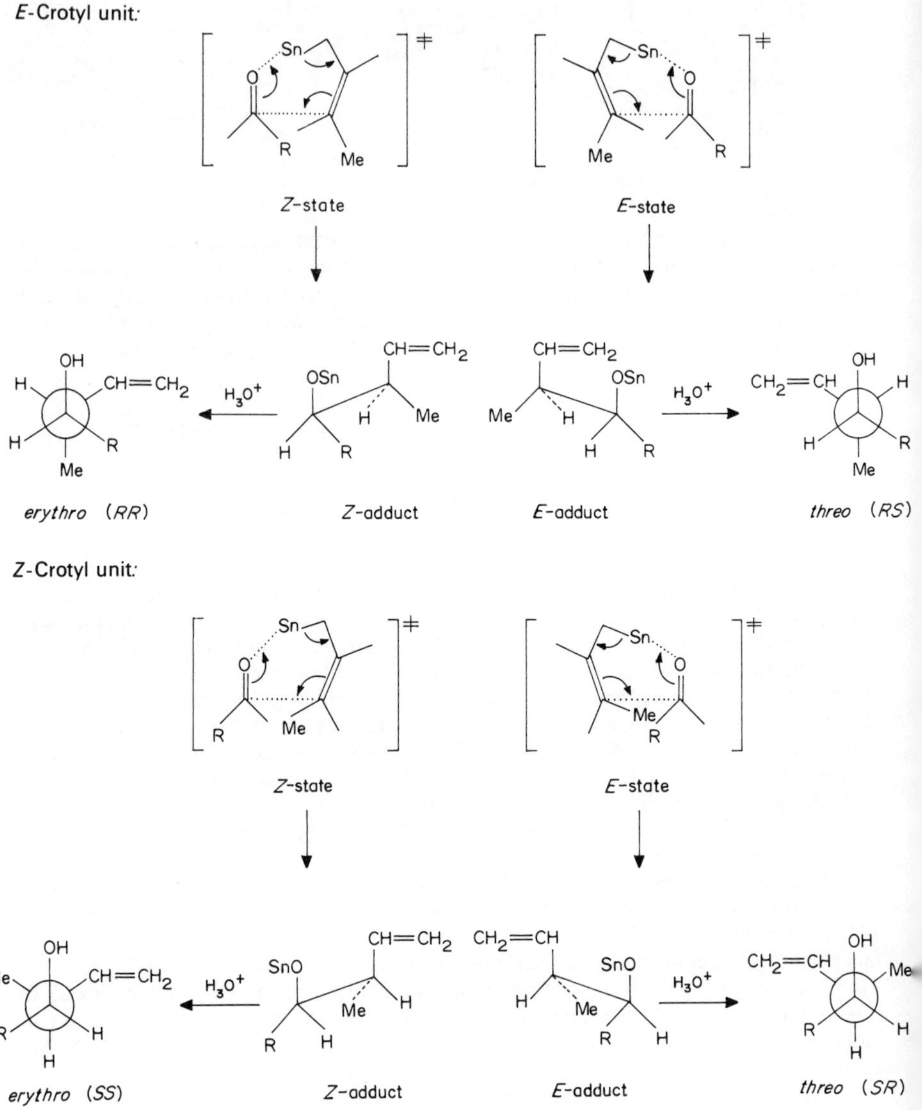

E-Crotyl unit:

Z-state   *E*-state

erythro (*RR*)   *Z*-adduct   *E*-adduct   threo (*RS*)

Z-Crotyl unit:

Z-state   *E*-state

erythro (*SS*)   *Z*-adduct   *E*-adduct   threo (*SR*)

SCHEME 18 (adapted from ref. 169)

$$R_2SnClCH(Me)CH=CH_2 + R^1CHO \xrightarrow[\text{2. aq. NH}_4\text{Cl}]{\text{1. 25 °C}} R^1CH(OH)CH_2CH=CHMe \quad (141)$$

(58)   80–100%

R = Me, Bu, Ph ; R^1 = Et, *s*-Bu, *i*-Pr, *t*-Bu, Ph

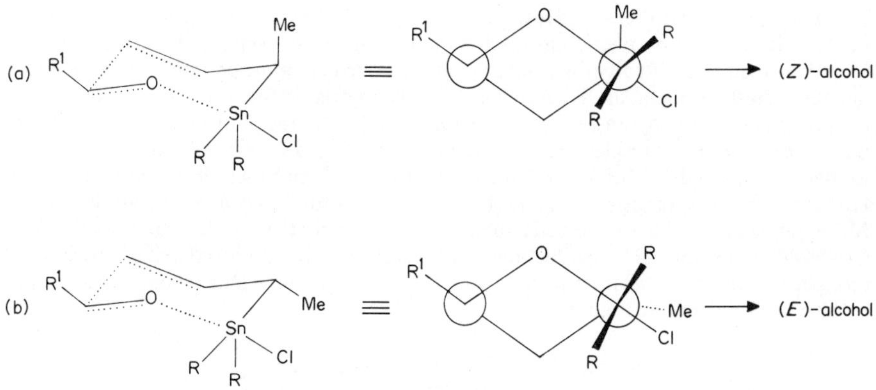

1. Carbon—carbon bond formation

FIGURE 2

rearrangements[172]. However, on account of the poor acceptor properties of tetraorganotins[20,23] and the weak donor characteristics of most aldehydes and ketones towards tin[173], open-chain transition states may provide a favourable alternative to cyclic transition-state structures in the case of trialkylallyltins. This would also appear to be the case when Lewis acid catalysts (e.g. BF_3) are added which complex with the carbonyl group of the aldehyde or ketone. Open-chain transition states, on the other hand, may admit the possibility of reaction without allylic rearrangement (i.e. S_E2 instead of S_E2') and of stereoconvergence [i.e. dominant formation of only one diastereomeric product from (E)- and (Z)-allyltins] (Figure 3). Both these possibilities have been evidenced, for example,

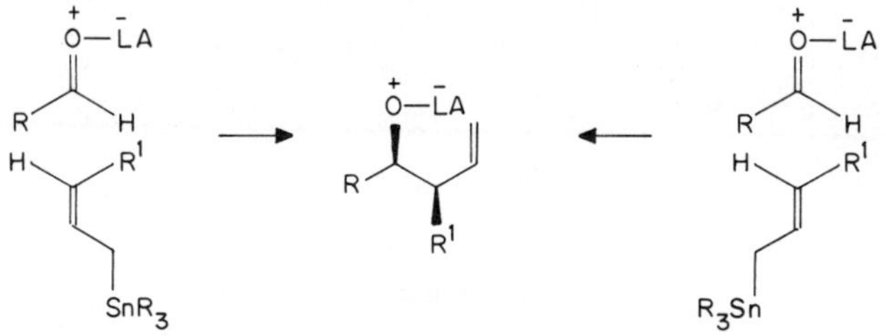

FIGURE 3

in the BF_3-catalysed addition of crotyl- and prenyl-tin compounds to several substituted quinones[174] which proceed without allylic rearrangement (see below) and in the stereoconvergent reaction of tributylcrotyltin with aldehydes (equation 142) reported by

$Bu_3SnCH_2CH{=}CHMe$ + RCHO $\xrightarrow[\text{2. } H_2O]{\text{1. } BF_3\cdot OEt_2,\ CH_2Cl_2,\ -78 \text{ to } 0\ °C,\ N_2}$ (142)

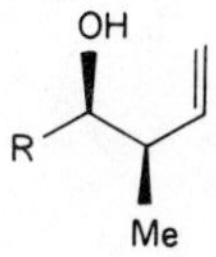

R = Me; 91% erythro : 9% threo
i-Pr; 95% erythro : 5% threo
Et_2CH, Ph; 98% erthro : 2% threo
i-Bu; 90% erythro : 10% threo

>90%

Maruyama and coworkers[175]. However, when the alkyl or alkenyl substituent at the γ-position is replaced by phenyl group, the *threo*-product is formed preferentially[176].

In their reaction with allyltins, α-chiral aldehydes may *a priori* be expected to yield two different diastereomers as a result of 1,2-asymmetric induction. The products are designated as 'Cram' or 'anti-Cram', depending on their mode of formation via attack on the *si*-side or *re*-side, respectively, of the aldehyde. Based on Cram's rules, the 'Cram' isomer is expected to be formed preferentially[177] (cf. product 56 in equation 139). However, in the synthesis of the Prelog–Djerassi lactone (59) using tributylcrotyltin, Maruyama *et al.*[178] have observed a high diastereoselectivity in the formation of the *anti*-Cram *erythro* product. The stereoconvergent reaction (143) was considered to proceed via an eight-membered cyclic intermediate of crown conformation which favours attack at the '*re*-face' as indicated in Figure 4.

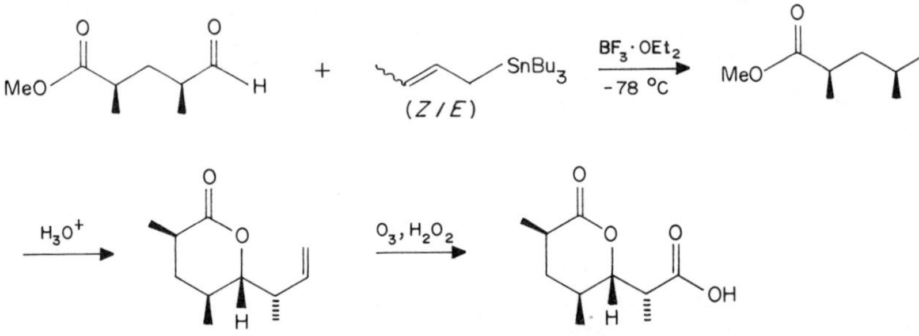

(143)

94%, *erythro*, anti-Cram

(59)

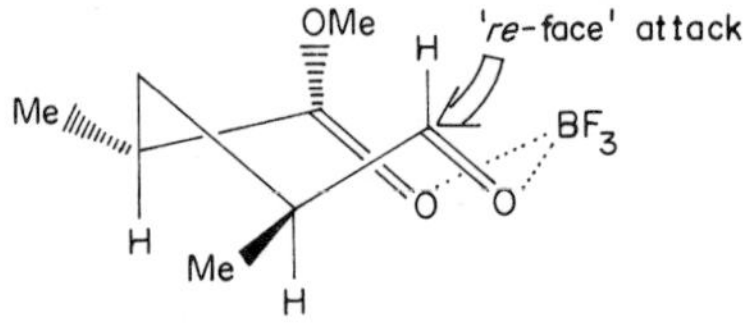

FIGURE 4

Notwithstanding the *erythro*-selectivity of crotyltin reactions such as depicted in equations 142 and 143, the generation of *erythro*-homoallylic alcohols has also been particularly enhanced by the recent stereoselective synthesis of (Z)-2-alkenyltins[161] (Scheme 19). Of added significance is that these *erythro*-alcohols can be almost

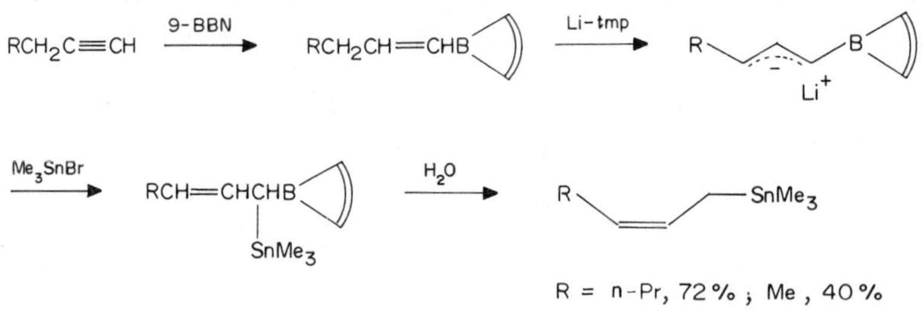

R = n-Pr, 72%; Me, 40%

SCHEME 19

1. Carbon—carbon bond formation

quantitatively converted via Wacker oxidation (equation 144) to erythro-β-hydroxycarbonyl compounds[161] (60), which form a basic synthetic block for macrolide, polyether or ansamycin antibiotics. An alternative synthesis for 60 from enol stannanes via aldol condensation reactions with aldehydes is discussed in Section VI.A.

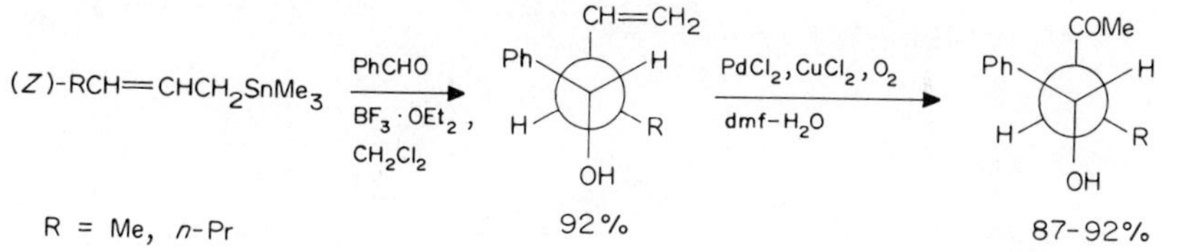

$$(Z)\text{-RCH}=\text{CHCH}_2\text{SnMe}_3 \xrightarrow[\text{CH}_2\text{Cl}_2]{\text{PhCHO}, \text{BF}_3 \cdot \text{OEt}_2} \text{[Ph, H, R, OH Newman projection]} \quad 92\% \xrightarrow[\text{dmf-H}_2\text{O}]{\text{PdCl}_2, \text{CuCl}_2, \text{O}_2} \text{[Ph, COMe, H, R, OH Newman projection]} \quad 87-92\%$$

R = Me, n-Pr

(60) (144)

Hydroxymethylation of allyltins with trioxan–$BF_3 \cdot OEt_2$ has also been recently explored as a means of synthesizing homoallylic alcohols[153]. (±)-Lavandalol (61) and related homoallylic alcohols were prepared by this method.

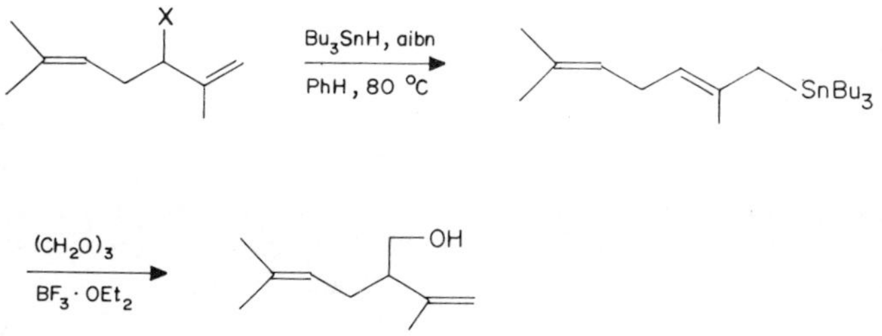

$$\xrightarrow[\text{PhH, 80 °C}]{\text{Bu}_3\text{SnH, aibn}} \xrightarrow[\text{BF}_3 \cdot \text{OEt}_2]{(\text{CH}_2\text{O})_3} \quad 57\%$$

(61)

X = tos, SCOSMe

The high stereoselectivity of aldol additions in the presence of $BF_3 \cdot OEt_2$ has already been noted. Of equal significance in organic synthesis is that such reactions also proceed with high chemoselectivity. The reactions[179] shown in equations 146–149 are illustrative.

$$R^1COR^2 + R^3{}_3SnCH_2CH=CR^4{}_2 \xrightarrow[\text{2. H}_2\text{O}]{\text{1. BF}_3 \cdot \text{OEt}_2, -78 \text{ °C, CH}_2\text{Cl}_2} R^1R^2C(OH)C(R^4)_2CH=CH_2 \quad (146)$$

>90%

R^1 = alkyl, aryl; R^2 = H, alkyl; R^3 = Me, Bu; R^4 = H, Me

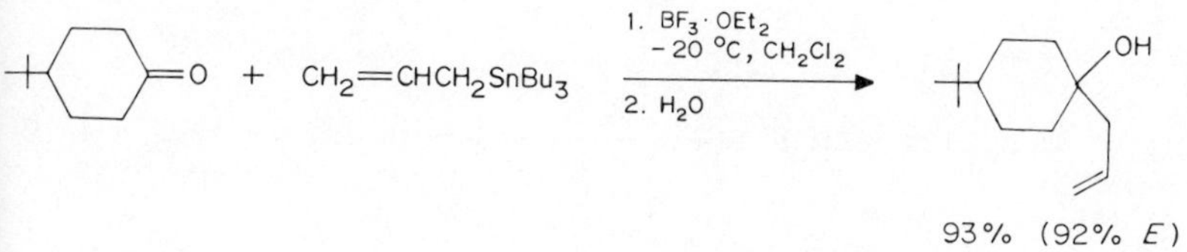

$$\text{[4-t-Bu-cyclohexanone]} + CH_2=CHCH_2SnBu_3 \xrightarrow[\text{2. H}_2\text{O}]{\text{1. BF}_3 \cdot \text{OEt}_2, -20 \text{ °C, CH}_2\text{Cl}_2} \text{[4-t-Bu-1-allyl-cyclohexanol]} \quad (147)$$

93% (92% E)

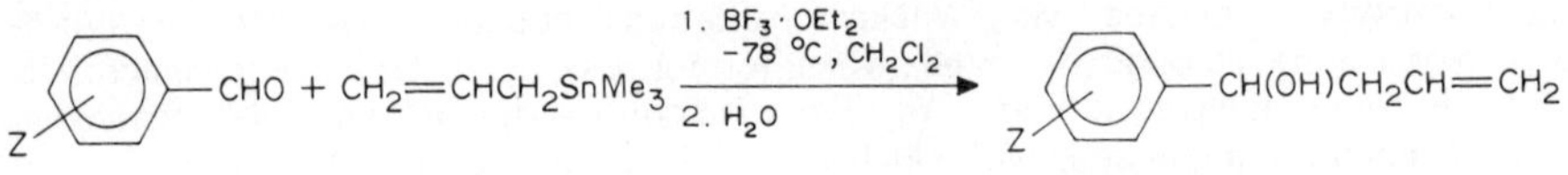

$$\underset{Z}{\underset{}{\bigcirc}}-CHO + CH_2=CHCH_2SnMe_3 \xrightarrow[2.\ H_2O]{1.\ BF_3\cdot OEt_2 \atop -78\ ^\circ C,\ CH_2Cl_2} \underset{Z}{\underset{}{\bigcirc}}-CH(OH)CH_2CH=CH_2$$

66-99%

Z = o-Br, o-OH, m-NO$_2$, p-Me, p-CN, p-OMe   (148)

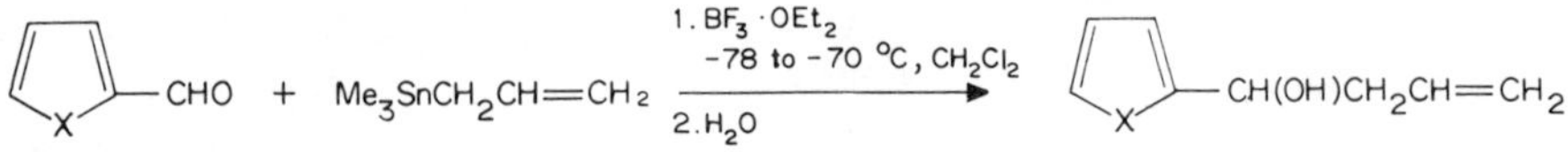

$$\underset{X}{\underset{}{\bigcirc}}-CHO + Me_3SnCH_2CH=CH_2 \xrightarrow[2.H_2O]{1.\ BF_3\cdot OEt_2 \atop -78\ to\ -70\ ^\circ C,\ CH_2Cl_2} \underset{X}{\underset{}{\bigcirc}}-CH(OH)CH_2CH=CH_2$$

X = S, O   88-99%   (149)

$$CH_3(CH_2)_4COCH_3 + (CH_3CH_2CH_2)_2CO \xrightarrow[BF_3\cdot OEt_2]{Bu_3SnCH_2CH=CH_2} CH_3(CH_2)_4CMe(OH)CH_2CH=CH_2$$

                                  (53%)   78%

$+ (CH_3CH_2CH_2)_2C(OH)CH_2CH=CH_2$

22%   (150)

Aldehydes were found to be more reactive than ketones and terminal ketones were attacked more easily than inner ones (equation 150). Conjugate addition occurs with α,β-unsaturated ketones[180]. Interestingly, the palladium-catalysed coupling of allylstannanes (as well as acetonyltins) with carbonyl compounds containing α-, β-, or γ-halogens give oxiranes, oxetanes, and tetrahydrofurans, respectively[181]. The cyclization reactions are clean, with very few side reactions. However, the oxiranes tend to undergo rearrangement and dehydration reactions at prolonged reaction times at higher temperatures.

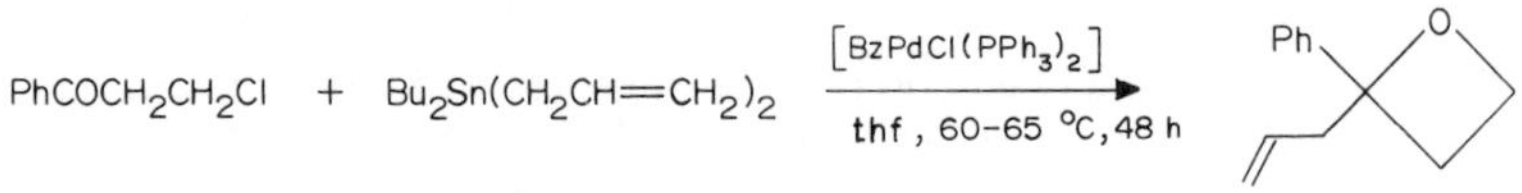

$$PhCOCH_2CH_2Cl + Bu_2Sn(CH_2CH=CH_2)_2 \xrightarrow[thf,\ 60-65\ ^\circ C,\ 48\ h]{[BzPdCl(PPh_3)_2]}$$ [oxetane product]   (151)

85%

$$PhCOCH_2Br + Bu_2Sn(CH_2CH=CH_2)_2 \xrightarrow{5\ h}$$ [oxirane product]   (151a)

75%

$$RCOCHR^1Cl + Bu_3Sn(CH_2CH=CH_2) \xrightarrow[20\ h]{100\ ^\circ C}$$ [oxirane product]   (151b)

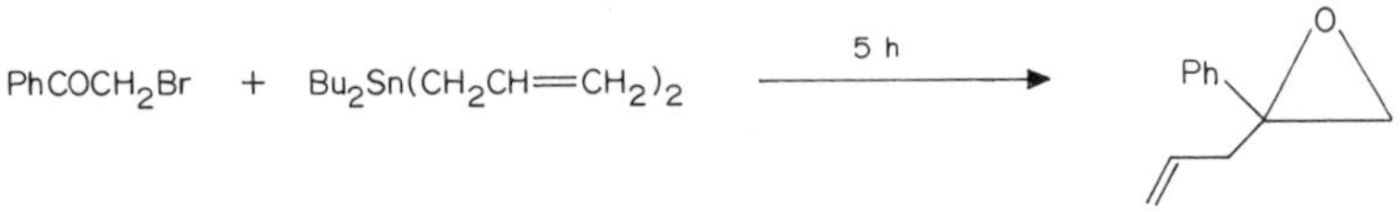

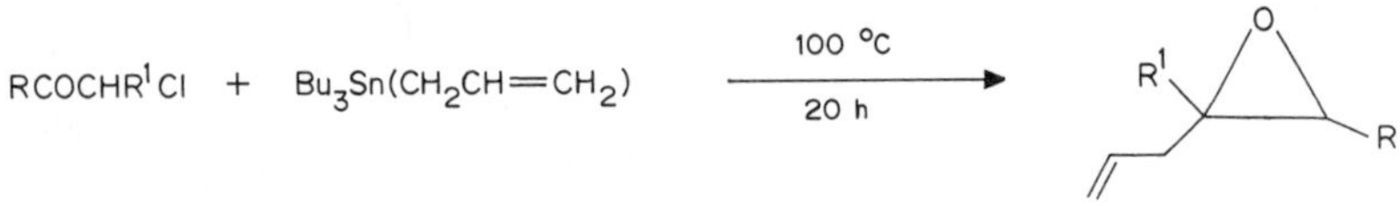

Surprisingly, with α-chloroacetophenone addition to the carbonyl group rather than cyclization was observed, yielding halohydrins in good yields. However, the use of [Pd(Ph$_3$P)$_4$] as catalyst has been reported elsewhere[181a] to yield oxiranes with α-chloroketones. In contrast, replacement of halogen in α-haloketones by allyl was observed in the presence of aibn[182].

The reaction of quinones with allyltin compounds in the presence of BF$_3$·OEt$_2$ yields allylhydroquinones with high regioselectivity (reaction 152)[174,183]. The reaction proceeds

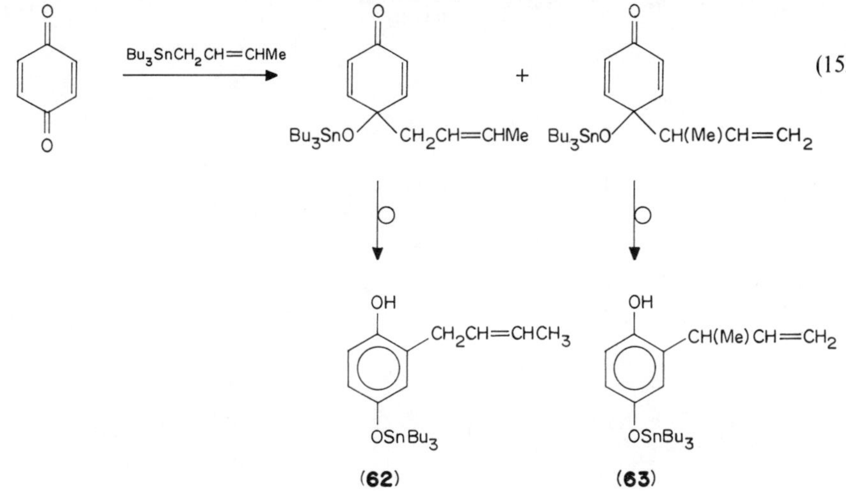

via 1,2-addition of allylstannane to quinone, followed by dienone–phenol rearrangement. The boron trifluoride etherate, besides activating the carbonyl group of quinone, also catalyses the rearrangement. With tributylcrotyltin, the γ-adduct (63) is the dominant product (95%). The α-adduct, involving no allylic transposition in the tin reagent, however, predominates for steric reasons with 2,5(6)-dimethylquinones, and also in prenyltin reactions (equation 153).

Several naturally occurring isoprenylquinones (e.g. ubiquinones, plastoquinones, menaquinones, and phylloquinones) have become synthetically accessible via such BF$_3$-

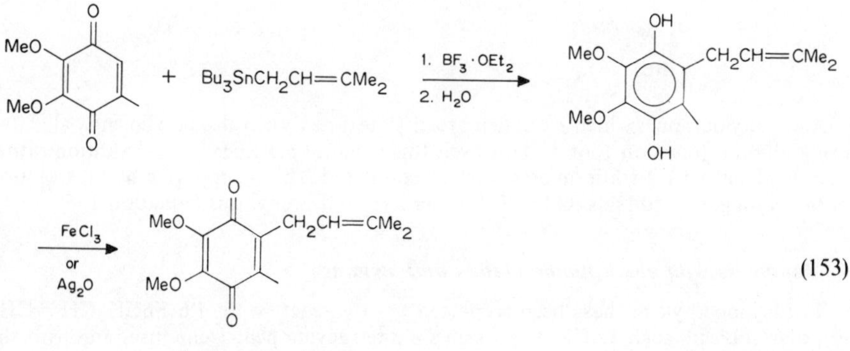

catalysed allyl transfer reactions. Vitamin K_1 (**65**), for example, has been successfully synthesized using polyprenyltin (**64**) without loss of stereochemistry in the phytyl side-chain (96%E + 4%Z) (equation 155). o-Benzoquinones (**66**) and 1,2-naphthoquinones (**67**) (R=H) have also been shown[184] to react with allylstannanes in dichloromethane in the presence of $BF_3 \cdot OEt_2$ to yield 62–93% and 65–78% of the respective diols, which were oxidized to the allylquinones (**66**) and (**67**) (R = allyl).

$$(E)\text{-ClCH}_2\text{CH}=\text{CMeR} + \text{Me}_3\text{SnLi} \xrightarrow[\text{thf, } N_2]{-60 \,°\text{to } 25\,°C} \text{Me}_3\text{SnCH}_2\text{CH}=\text{CMeR} \quad 154)$$

$$96\% \, E \,, 4\% \, Z$$
$$(\mathbf{64})$$

$$R = -\!\!\left[\text{CH}_2\text{CH}_2\text{CH}_2\text{CH(Me)CH}_2\right]_3\!\!-\!\text{H}$$

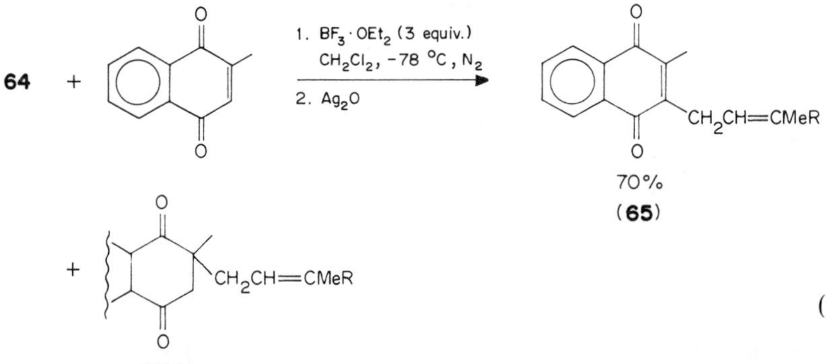

(155)

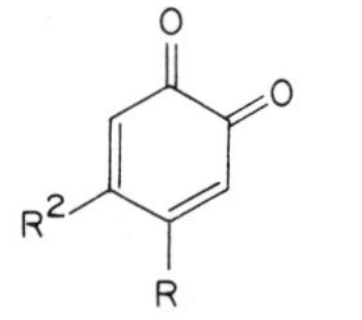

R^2 = H, Me, CMe_3

(**66**)

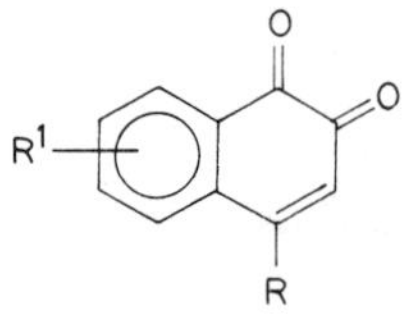

R^1 = H, 6-Br, 3-OMe

(**67**)

Alkenonylquinones which are difficult to allylate directly can also be efficiently allylated using allyltins (reaction 156).[185] However, this reaction proceeds by 1,4-addition rather than by the usual 1,2-addition previously encountered. This is supported by the isolation of the conjugate γ-adducts (**68**) with both crotyl- and prenyl-tins (equation 156).

2. Reactions with electrophilic olefins and alkynes

Tetracyanoethylene has been reported[186] to react with $Ph_3SnCH_2CH=CH_2$ in polar solvents such as CH_3CN to give 4,5-tetracyanopent-1-ene in addition to the cyclo-adduct. A zwitterionic intermediate has been inferred for this reaction (Scheme 20).

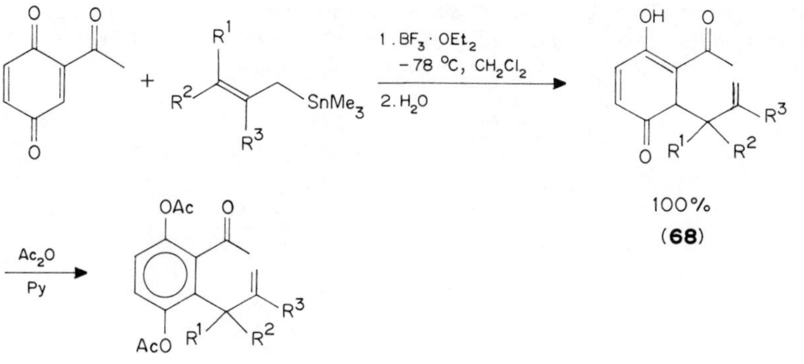

(156)

$R^1 = R^3 = H$, $R^2 = Me$; $R^1 = R^2 = Me$, $R^3 = H$

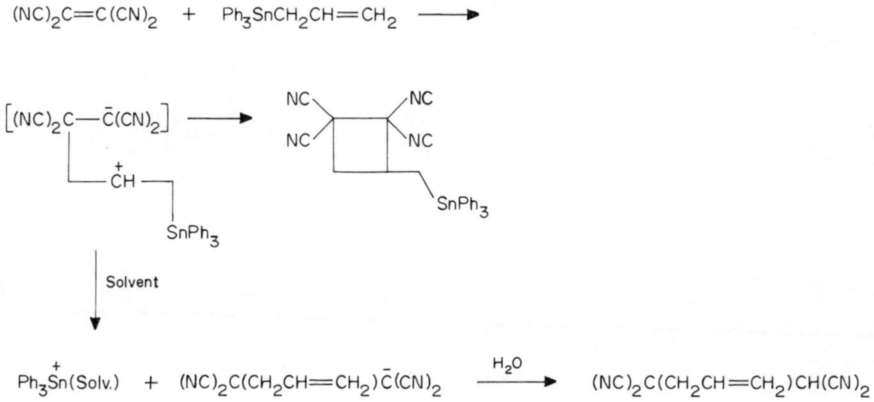

SCHEME 20

The thermal Diels–Alder reactions of mono- and bis-trimethylstannylcyclopentadienes with hexafluorobut-2-yne yield the appropriate bicyclic adducts in moderate yields[187] (equation 157). The bis-substituted compound, however, isomerizes on forming the adduct.

The use of the trimethylstannylisoprene (**69**) as a five-carbon isoprene synthon has been

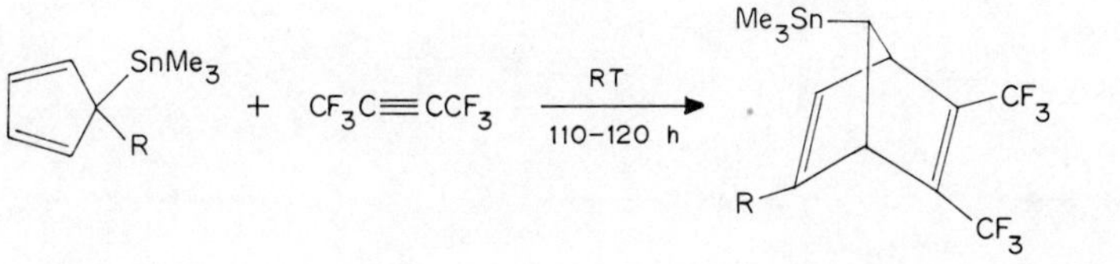

(157)

R = H, $SnMe_3$; 52–53%

recently described by Wilson et al.[188] **69** is obtained uniquely from cyclobut-1-enylmethyllithium (Scheme 21) and readily participates in a Diels-Alder reaction with methyl acrylate leading to the synthesis of δ-terpinol (**70**) as shown.

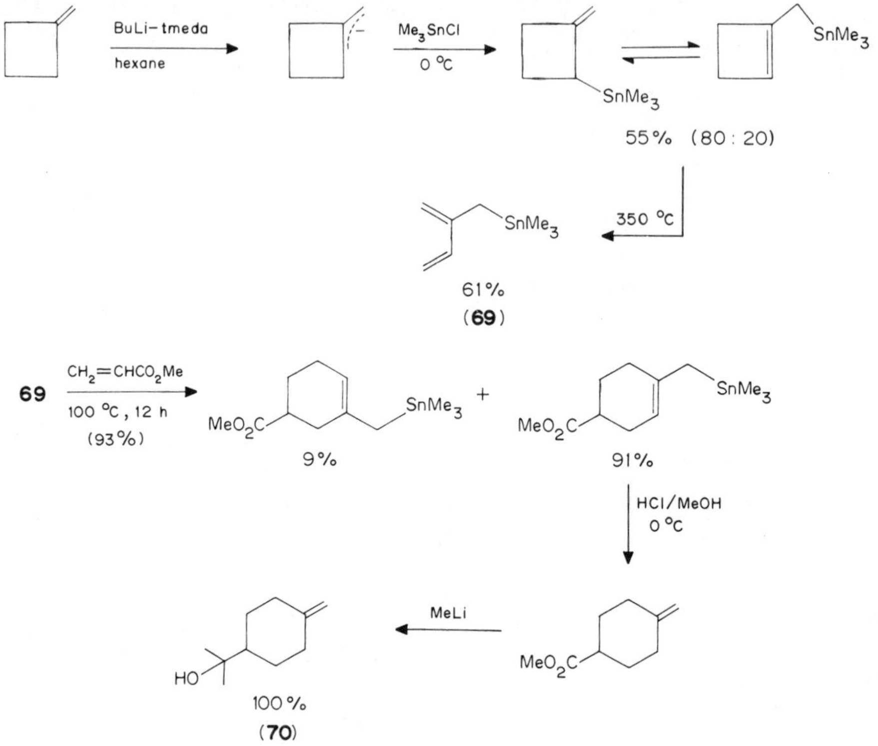

SCHEME 21

C. Aryl- and Benzyl-metal Derivatives

The interaction of carbon electrophiles with aryl- and benzyl-stannanes and -plumbanes via heterolytic pathways may be represented by the schemes[2,21] shown in equations 158 and 159.

The transfer of a phenyl group over prenyl in triphenylprenyltin in its reaction with cinnamyl acetate[157] to yield 1,3-diphenylpropene (19%) as the only coupling product has

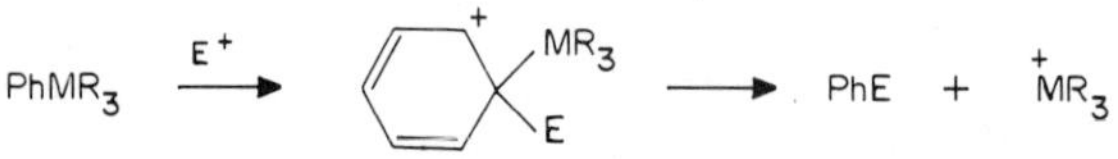

(158)

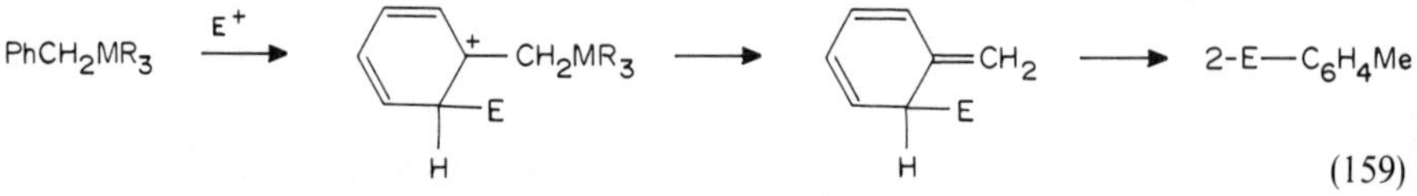

(159)

already been referred to. Although this result contradicts bond-energy arguments as a factor in the rationale of the coupling reaction, the specific cleavage of the aryl and benzyl groups in ArSnR$_3$ and BzSnR$_3$ compounds by various electrophilic reagents has been well documented. Some typical recent examples are shown in reactions 160–169. In equation

$$Ph_4Sn + PhCOCl \xrightarrow[\text{PhH, 140 °C, 5 h}]{[Pd(PPh_3)_4]} PhCOPh \quad 85\% \tag{160)[128]}$$

$$BzSnBu_3 + MeCOCl \xrightarrow[\text{PhH, 80 °C, 12 h}]{[RhCl(PPh_3)_3]} BzCOMe \quad 69\% \tag{161)[159]}$$

$$Ph_3SnMe + BzBr \xrightarrow[\text{hmpt, 65 °C}]{[BzPdCl(PPh_3)_2]} BzPh \quad 95\% \tag{162)[128]}$$

$$Me_3SnBz + (NC)_2C=C(CN)_2 \xrightarrow[\text{CH}_2\text{Cl}_2]{20\text{ °C}} Bz(CN)C=C(CN)_2 \tag{163)[189]}$$

$$RSnMe_3 + (4-NO_2C_6H_4)_3CBF_4 \xrightarrow[\text{RT}]{CH_2Cl_2} RC(C_6H_4NO_2\text{-}4)_3 + (C_6H_4NO_2\text{-}4)_3C^{\bullet} \quad 100\%$$

R = 9-fluorenyl, indenyl $\tag{164)[190]}$

$$RSnMe_3 \xrightarrow[\substack{[(\eta^3\text{-}C_3H_5)PdCl]_2 \\ \text{hmpt, 20 °C}}]{R^1COCl} RCOR^1 \quad 65\text{-}97\%$$

R = Bz, Ph, 4-NO$_2$C$_6$H$_4$; R^1 = Me, Ph, 4-NO$_2$C$_6$H$_4$, 2-FC$_6$H$_4$, Bz, PhCH=CH $\tag{165)[191]}$

$$RSnMe_3 \begin{cases} \xrightarrow[\text{dmf}]{R^1X,\ [(\eta^3\text{-}C_3H_5)PdCl]_2} RR^1 \quad 70\text{-}95\% & (166)[192] \\ \xrightarrow[\text{(CH}_2\text{Cl)}_2]{ArX,\ [ArPdI(PPh_3)_2]} RAr & (167)[109] \end{cases}$$

R = Ph, 4-MeOC$_6$H$_4$, 4-MeC$_6$H$_4$, 4-ClC$_6$H$_4$, 9-fluorenyl, 9-cyanofluorenyl, indenyl, 3-methylindenyl;
R^{1}X = MeI, CH$_2$=CHCH$_2$Br;
Ar = Ph, 4-NO$_2$C$_6$H$_4$, 2,4-(NO$_2$)$_2$C$_6$H$_3$, 2,4,6-(NO$_2$)$_3$C$_6$H$_2$;
X = Cl, Br, I

$$ArSnMe_3 + 4\text{-}NO_2C_6H_4I + CO \xrightarrow[\text{hmpt, 20 °C}]{[(\eta^3\text{-}C_3H_5)PdCl]_2} ArCOC_6H_4NO_2\text{-}4 \quad 58\text{-}98\% \tag{168)[193]}$$

Ar = Ph, 4-MeC$_6$H$_4$, 4-ClC$_6$H$_4$, 4-NO$_2$C$_6$H$_4$

$$4\text{-}RC_6H_4SnMe_3 \;+\; CH_2=CHCH_2OAc \xrightarrow[\text{hmpt, 20 °C, 6 h}]{[Pd(PPh_3)_4],\, 5\%} \underset{100\%}{4\text{-}RC_6H_4CH_2CH=CH_2} \quad (169)^{194}$$

169, the use of thf as the solvent leads to the additional formation of biaryls (2.5–4 h; 24–31%).

Heteroarylstannanes have, in general, received sparse attention in the literature, and few workers have examined the feasibility of engendering C—C bond formation using these substrates. 1-methyl-2-(trimethylstannyl)pyrrole[195] has been shown to react with ethyl chloroformate to give the 2-carbethoxy derivative in low yield. With benzaldehyde, however, the ketone **71** is unexpectedly formed in about 57% yield (equation 170), presumably via oxidation of the stannyl ether intermediate formed under the reaction conditions. 2-Trimethylstannylthiophene was found to be unreactive towards PhCHO but gave 2-thienyl phenyl ketone with PhCOCl (reaction 171)[195]. A unique intramolecular

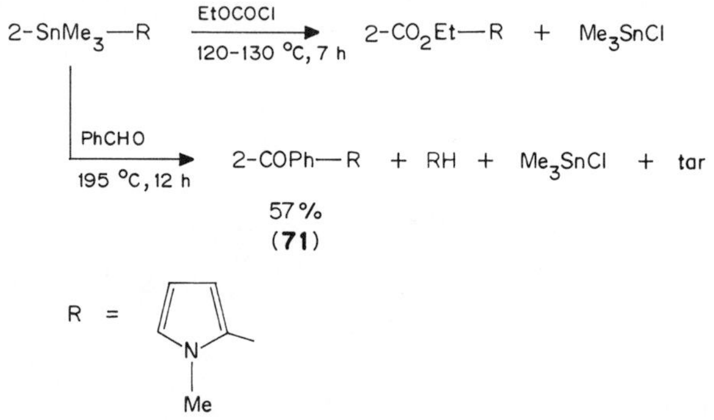

$$2\text{-}SnMe_3\text{—}R \xrightarrow[120\text{-}130\,°C,\,7\,h]{EtOCOCl} 2\text{-}CO_2Et\text{—}R \;+\; Me_3SnCl$$

$$\downarrow \text{PhCHO},\; 195\,°C,\, 12\,h$$

$$2\text{-}COPh\text{—}R \;+\; RH \;+\; Me_3SnCl \;+\; \text{tar}$$
$$\underset{(\mathbf{71})}{57\%}$$

$$R \;=\; \text{1-methylpyrrol-2-yl}$$

(170)

$$RSnMe_3 \;+\; PhCOCl \xrightarrow[6\,h]{172\,°C} \underset{45\%}{RCOPh} \;+\; Me_3SnCl$$

$$R \;=\; \text{thien-2-yl}$$

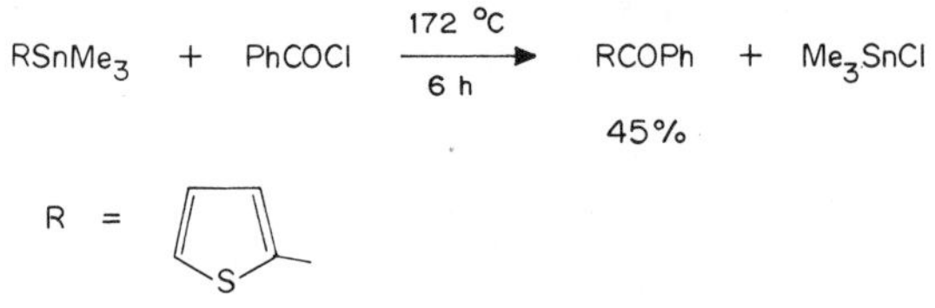

(171)

cyclization of the α-stannylated indole ring on to the proximate olefinic bond has been recently reported by Trost and Fortunak[196] for **72**. The cyclization proceeds in the presence of a palladium catalyst to give **73** in moderately high yield.

3- and 4-trimethylstannyl and -tributylplumbyl derivatives of pyridine have been shown to yield the tetrahydrobipyridyls **75** and **76** when reacted with $(Me_3Si)_2Hg^{197}$. The

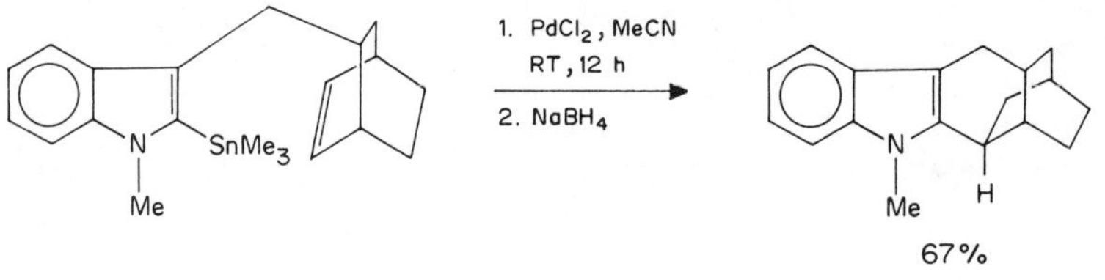

$$\mathbf{(72)} \xrightarrow[\text{2. NaBH}_4]{\text{1. PdCl}_2,\, MeCN,\, RT,\, 12\,h} \underset{67\%}{\mathbf{(73)}}$$

(172)

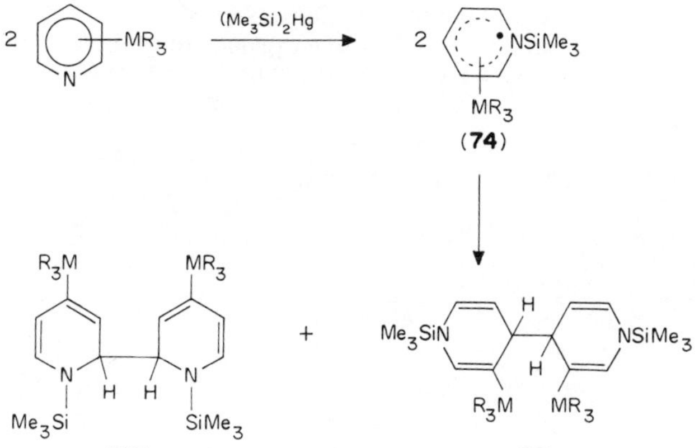

1. Carbon—carbon bond formation

(173)

intermediacy of the pyridyl free radical **74** has been implicated in these reactions. 1-(Trimethylstannyl)-2-iodotetraphenylbenzene (**77**) appears to be a useful precursor for the synthesis of highly hindered biphenyls[198]. The reaction proceeds via bimolecular condensation (equation 174), there being no evidence for benzyne formation.

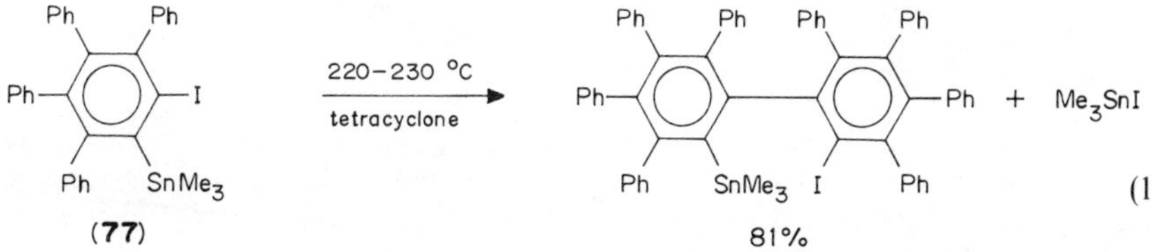

(174)

V. REACTIONS OF ORGANOTIN COMPOUNDS CONTAINING SATURATED OR REMOTELY UNSATURATED CARBON CHAINS

A. Tetraalkyl Derivatives

Tetraorganotin compounds containing saturated alkyl groups on tin are generally inert in their reactivities to external carbon electrophiles. The highly reactive electrophile $Ph_3C^+BF_4^-$ reacts with $EtSnMe_3$ to eliminate ethene via hydride abstraction from the β-carbon atom of the alkyl group attached to tin[199]. With other $RSnMe_3$ compounds (R = Me, allyl, 9-fluorenyl, indenyl, PhC≡C—), Ph_3CR is formed quantitatively at room temperature in CH_2Cl_2[190]. The latter reaction is considered as an electrophilic substitution process, but for R = Ph (or with the use of $(p\text{-}NO_2C_6H_4)_3C^+BF_4^-$ in place of $Ph_3C^+BF_4^-$) a redox process obtains (cf. equation 164).

The less reactive alkyl halides and acid chlorides do not react except in the presence of catalysts. The synthetic potential of such C—C coupling reactions has only recently been recognized[108a,128,130,191]. With palladium(II) catalysts in hmpt[†], high yields of the cross-

---

[†]The use of chloroform gives somewhat lower yields although facilitating workup[108a].

coupling products have been obtained (equations 175 and 176) with benzyl and aryl halides[130] and acid chlorides[108a,130]. Only one organic group is transferred from the tetraorganotin reagent; the second leaves about 100 times more slowly from R_3SnX[108a].

$$R^1Br + R_4Sn \xrightarrow[\text{hmpt, 62 °C, 20 h, argon}]{[BzPdCl(PPh_3)_2]} RR^1 + R_3SnBr \qquad (175)$$

$$R^1COCl + R_4Sn \xrightarrow[\text{hmpt, 65 °C}]{[BzPdCl(PPh_3)_2]} RCOR^1 + R_3SnCl \qquad (176)$$

$$p\text{-}Z\text{-}C_6H_4COCl + Me_4Sn \xrightarrow{Pd(II)} p\text{-}Z\text{-}C_6H_4COMe \qquad (177)$$
$$86\text{-}98\%$$

$$Z = CHO, NO_2, CN, OMe, CO_2R, Cl$$

$$RCOCl + Me_4Sn \xrightarrow{Pd(II)} RCOMe \qquad (178)$$
$$91\%$$

$$R = \text{(2-methylfuryl)}$$

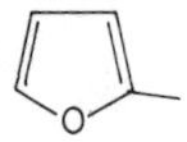

Reaction 176, in particular, tolerates a variety of functional groups on the acid chloride, as shown in reactions 177 and 178. Tributylvinyltin and triphenylmethytin transfer exclusively the vinyl and phenyl groups, respectively, with the vinyl transfer proceeding at a much faster rate.

The mechanism shown in Scheme 22 has been proposed[130] for palladium(II) catalysis in respect of the reaction represented by equation 176. In the transmetallation step a, Bz

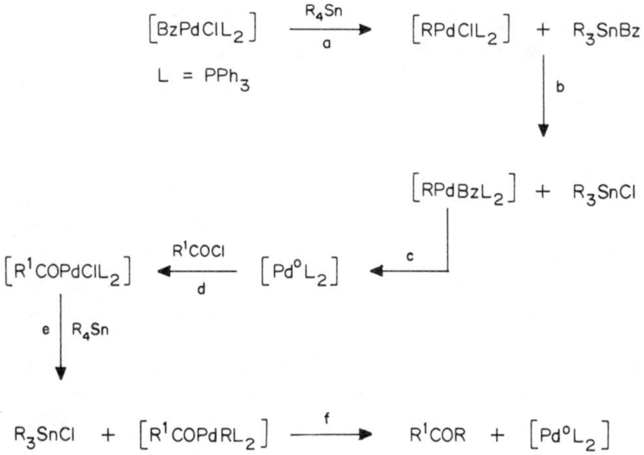

$$[BzPdClL_2] \xrightarrow[a]{R_4Sn} [RPdClL_2] + R_3SnBz$$
$$L = PPh_3$$
$$\downarrow b$$
$$[RPdBzL_2] + R_3SnCl$$
$$\downarrow c$$
$$[R^1COPdClL_2] \xleftarrow[d]{R^1COCl} [Pd^0L_2]$$
$$\downarrow e \; R_4Sn$$
$$R_3SnCl + [R^1COPdRL_2] \xrightarrow{f} R^1COR + [Pd^0L_2]$$

SCHEME 22

rather than Cl is replaced. In step b, R_3SnBz rather than R_4Sn participates on account of the greater reactivity of the former to generate a diorganopalladium complex from which the active catalyst, $[Pd^0(PPh_3)_2]$ is formed by reductive elimination (step c). Oxidative addition of the acid halide (step d), followed by transmetallation with R_4Sn (step e) and reductive elimination (step f) yields the product ketone[†].

For the catalysis of the coupling reaction with benzyl and aryl halides (equation 175), however, the reductive elimination has been proposed[130] to take place preferentially from a palladium(IV) intermediate. Using optically active complexes, the elimination process was shown to proceed with retention of configuration at carbon (Scheme 23).

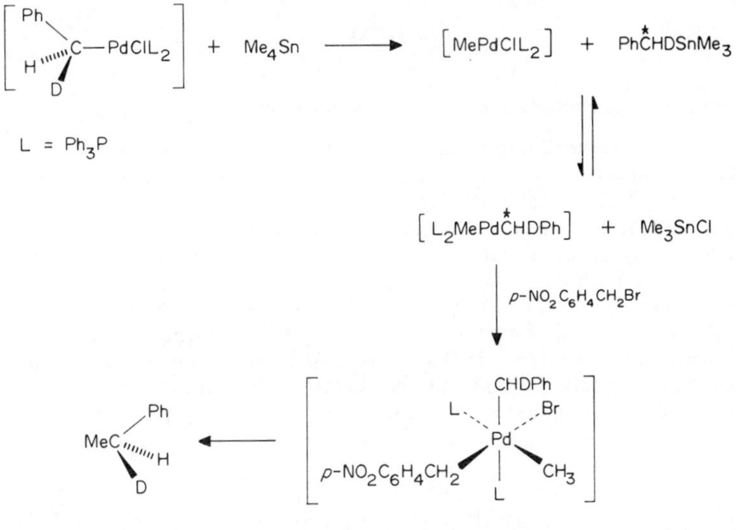

SCHEME 23

Tetraorgano-tin and -lead compounds have also previously featured in the *in situ* preparation of organopalladium catalysts such as reported by Heck[199a] for the alkylation/arylation of olefins (see for example, equation 179).

$$YCH=CH_2 + R_4M \xrightarrow[\text{RT, 24 h}]{Li_2[PdCl_4], MeOH} (E)\text{-}YCH=CHR \qquad (179)$$

$$80\text{-}100\%$$

Y = Ph, CO_2Me ; R = Me, Ph ;
M = Sn, Pb

$$R^1X + R_4Sn + CO \xrightarrow[\text{120 °C}]{\text{autoclave}} R^1COR + R_3SnX \qquad (179a)$$
$$\text{(30 atm)} \quad [PhPdI(Ph_3P)_2] \quad \text{(a) 73\% ; (b) 62\%}$$

(a) R^1X = PhI ; R = Bu   (b) R^1X = (E)-PhCH=CHBr ; R = Me

[†]The transmetallation step (e), slow relative to steps (d) and (f), has recently been shown to involve electrophilic cleavage of the C–Sn bond of R_4Sn, with the acylchloropalladium(II) species acting as the electrophile [J. W. Labadie and J. K. Stille, *J. Am. Chem. Soc.*, **105**, 6129 (1983)].

Unsymmetrical ketones have also been directly synthesized in recent years from organic halides, carbon monoxide, and tetralkyltins in the presence of palladium or nickel complex catalysts[200] (equation 179a). The mechanism in Scheme 24 has been envisaged.

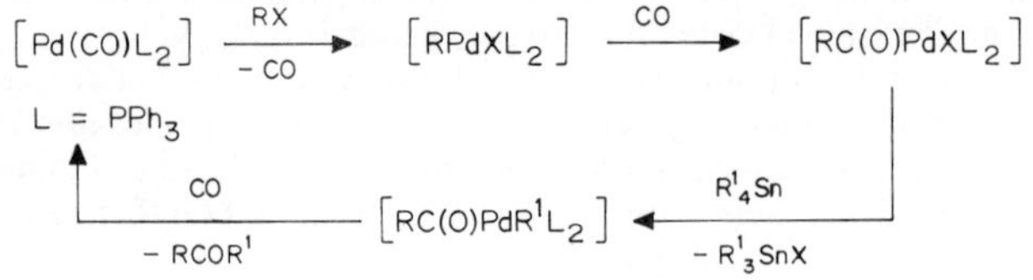

SCHEME 24

B. Derivatives with Remote Side-chain Unsaturation

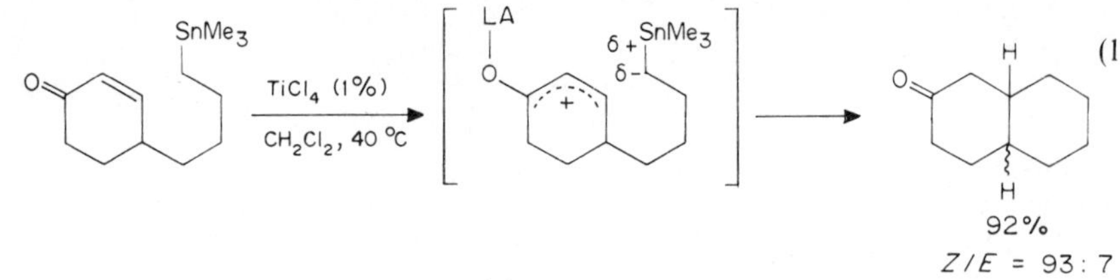

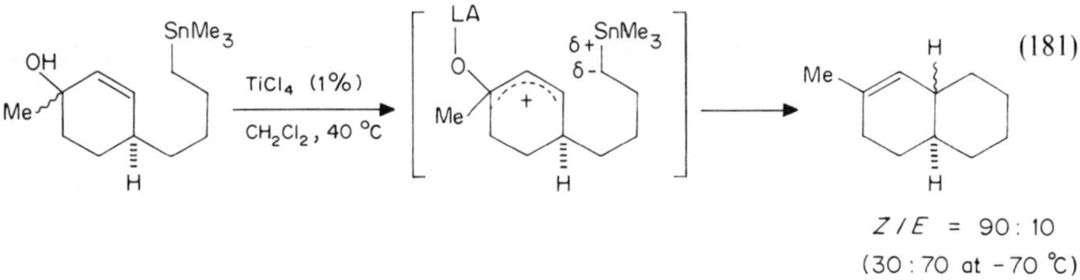

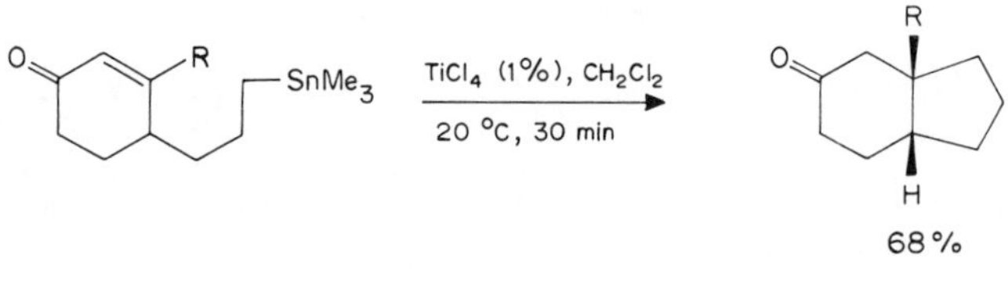

Among tetraorganotin compounds with remote unsaturation in the alkyl side-chains, the most promising from a synthetic point of view appears to be the (trimethylstannyl)alkyl-substituted cyclohexenones[201] and cyclohexenols[202] in which the carbonyl and hydroxy functions, respectively, are located adjacent to the C=C double bond. The weakly polarized nature of the C—Sn σ-bond ensures compatibility of the tetraalkyltin with both the α-enone and the allylic alcohol functions in these molecules until electrophilic activation of the functional groups transpires. As shown below for the case of the 4-substituted compounds, activation of the α,β-enone and allylic alcohol moieties with a Lewis acid (e.g. $TiCl_4$) generates a β-electrophilic site whose interaction with the stereoproximate C—Sn bond leads to intramolecular cyclization (equations 180 and 181). The carbocyclization leading to five-membered rings appears qualitatively to be a kinetically faster process and to be less sensitive to steric factors than six-membered rings (equations 182 and 183). However, when the site adjacent to the (4^1-trimethylstannyl)butyl side-chain in the 4-cyclohexenyl cation is tertiary, an alternative reaction occurs. As shown in equations 184 and 185, β-hydride transfer products are exclusively formed. With 3-substituted cyclohexenones, spirocyclic compounds are obtained in good yield (equation 186). Although a very high selectivity for 1,4-addition of the carbanionic nucleophile to the

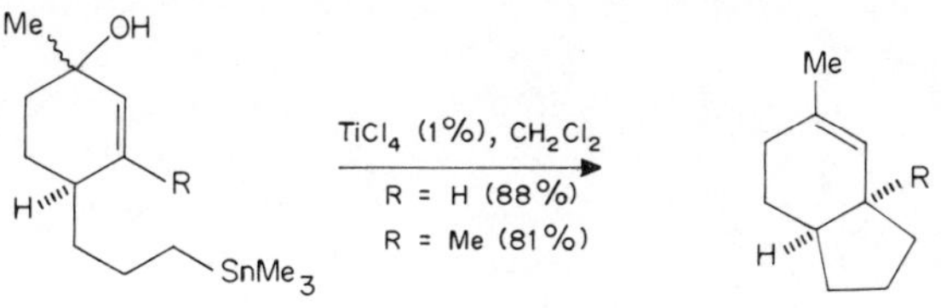

(183)

(184)

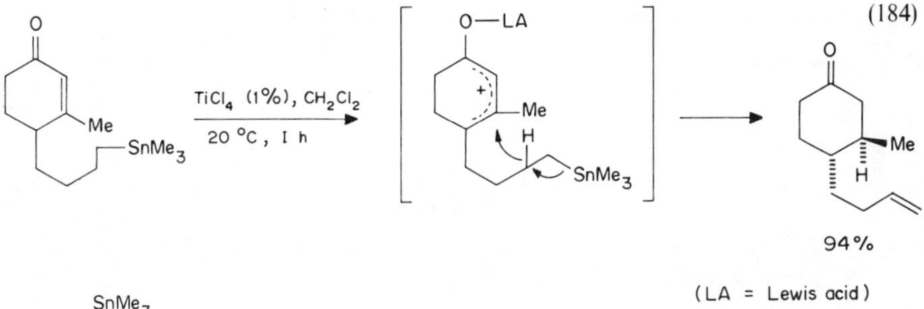

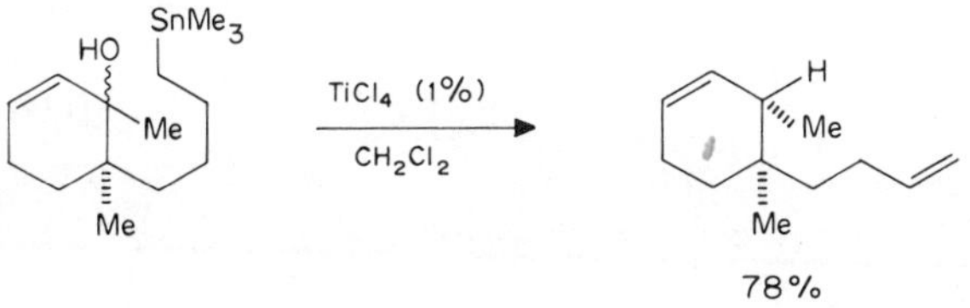

(LA = Lewis acid)

(185)

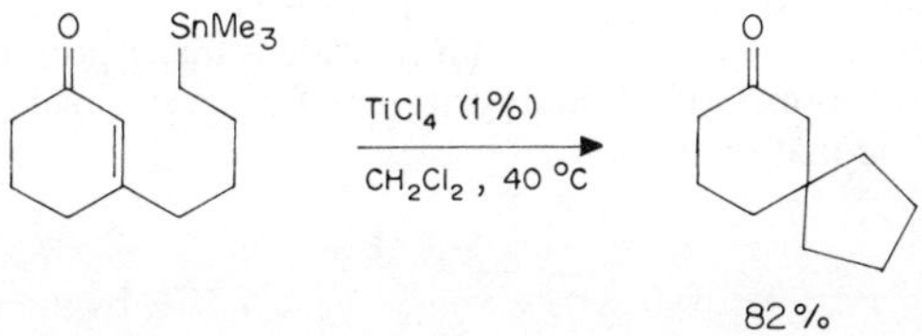

(186)

enone moiety is seen in the above examples, the intrinsically less preferred 1,2-addition can be made to predominate when entropy factors of the cyclization substrate inhibit the conjugate addition pathway. A case in point involves 6-substituted cyclohexenone (equation 187).

The above carbocyclization schemes employing the C—Sn σ-bond as a latent carbanionic nucleophile with the *in situ* generation of an allyl carbocation offer considerable scope for the synthesis of medium-sized carbocyclic rings which are often synthetically inaccessible through direct annular processes.

Functionalized homoallyltin compounds, accessible from the reaction of β-tributylstannylpropionaldehyde with appropriate Wittig reagents (equation 188), are

(187)

$$Bu_3Sn(CH_2)_2CHO + Ph_3P{=}CHR \xrightarrow[\text{or thf, RT}]{\text{PhH, 80 °C}} Bu_3Sn(CH_2)_2CH{=}CHR \quad (188)$$

80–93%

R = COMe, COPh, CO_2Me, p-$NO_2C_6H_4$

(189)

R = Me or Ph ; 90%

readily destannylated on treatment with CF_3COOH in thf or CH_2Cl_2 at room temperature to yield cyclopropanes[203] (equation 189) in good yields. Cyclopropane formation is also facilitated in the presence of halogens[104], ArSCl[104], and tosOH[204] (equation 190). Noteworthy is the absence of β-hydride transfer products[199] formed as in equation 191 in this reaction. Cyclodestannylation to form a strained ring system has also been encountered[104] (equation 192).

(190)

e.g. EY = 4-ClC_6H_4SCl ; yield = 80% (CH_2Cl_2, 0 °C)

(191)

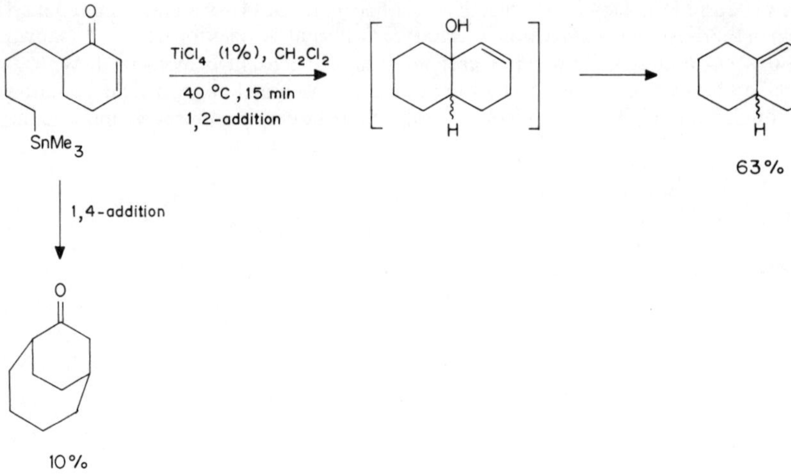

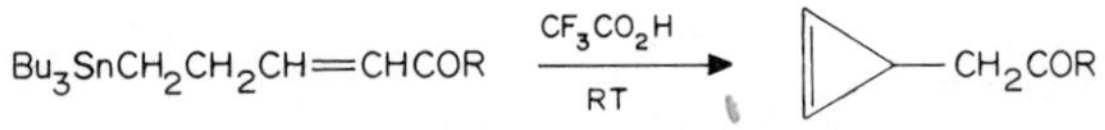

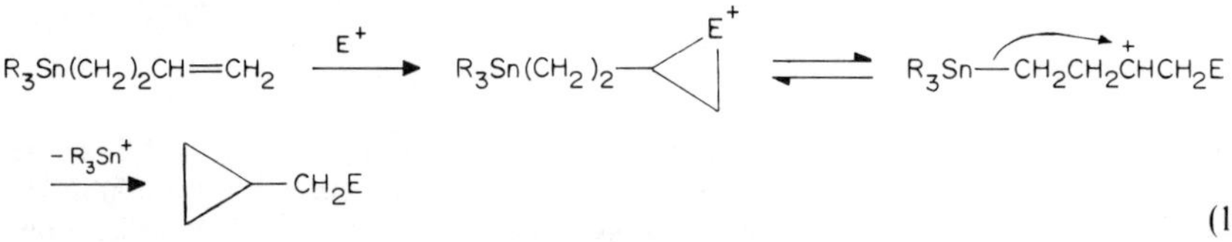

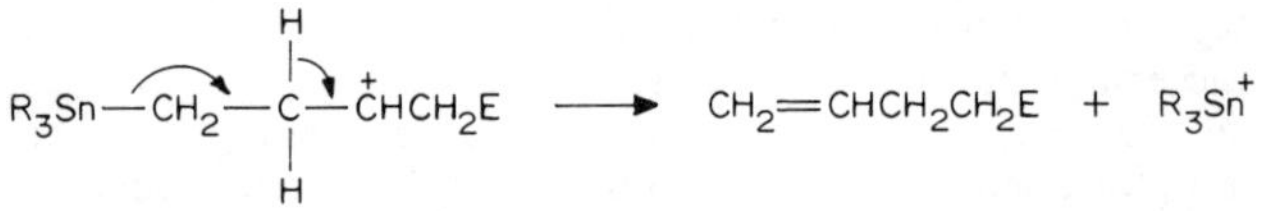

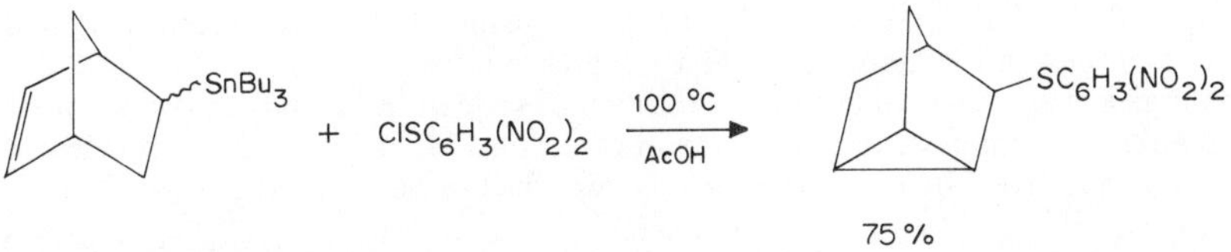

1. Carbon—carbon bond formation

$$\text{(norbornenyl)SnBu}_3 + \text{ClSC}_6\text{H}_3(\text{NO}_2)_2 \xrightarrow[\text{AcOH}]{100\,°\text{C}} \text{(norbornyl)SC}_6\text{H}_3(\text{NO}_2)_2 \quad 75\% \tag{192}$$

An equally versatile cyclodestannylation reaction is the oxyselenation of the C=C double bond in homoallylstannanes with N-phenylselenonaphthalimide[205] (N-psp) (equation 193).

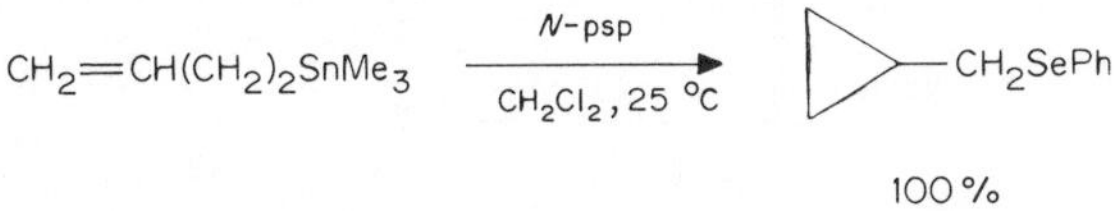

$$\text{CH}_2\!\!=\!\!\text{CH}(\text{CH}_2)_2\text{SnMe}_3 \xrightarrow[\text{CH}_2\text{Cl}_2,\,25\,°\text{C}]{N\text{-psp}} \text{(cyclopropyl)}\!-\!\text{CH}_2\text{SePh} \quad 100\% \tag{193}$$

VI. REACTIONS OF HETEROATOM-BONDED ORGANO-TIN AND -LEAD COMPOUNDS

A. O-Stannyl and O-Plumbyl Derivatives

1. Enolstannanes

These have featured strongly in recent years in enolate chemistry on account of their high regio-, chemo-, and stereo-selectivity. Enolstannanes appear to be excellent nucleophiles and afford several useful C—C bond-forming reactions with alkyl halides, allyl acetates, and aldehydes as electrophilic partners. Enolstannanes exist as two metallotropic (C— and O—) forms in equilibrium. The C-isomer (95%) is the most stable form for $\text{Bu}_3\text{SnCH}_2\text{COMe}$, but for $\text{Bu}_3\text{SnCH}(\text{Me})\text{COEt}$ the O-isomer (70%) is dominant[48,49] (see equation 20).

For $\text{Bu}_3\text{SnCH}(\text{Me})\text{COMe}$, on the other hand, it has been reported[48] that the O-isomer constitutes only 23% of the equilibrium mixture but exists exclusively in the E-geometry. Cyclopentenyl- and cyclohexenyl-tin enolates exist also in the stereochemically pure E-geometry, but allow of regioisomerism when substituents are present in the ring (equation 194). The 'kinetic' enolate, **78**, may be obtained from the corresponding lithium enolate at low temperatures (equation 195), but equilibration to the 'thermodynamic' enolate **79**, with consequent variation in the regioselectivity of the reaction, would attend the interaction of the tin enolate with 'sluggish' electrophiles. Alkylation of enolstannanes

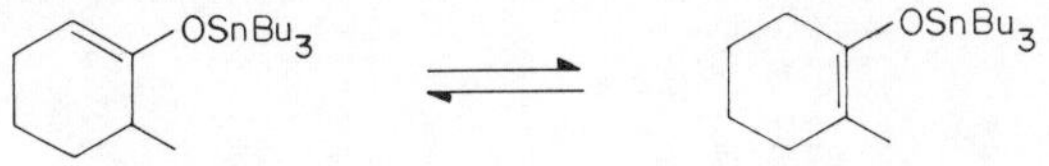

$$\text{(78)} \rightleftharpoons \text{(79)} \tag{194}$$

$$\text{(2-methylcyclohexanone)} \xrightarrow[\text{thf, }-78\,°\text{C}]{\text{Pr}^i_2\text{NLi}} \text{(OLi enolate)} \xrightarrow[-78\,°\text{C}]{\text{Bu}_3\text{SnCl}} \text{(OSnBu}_3\text{ enolate)} \tag{195}$$

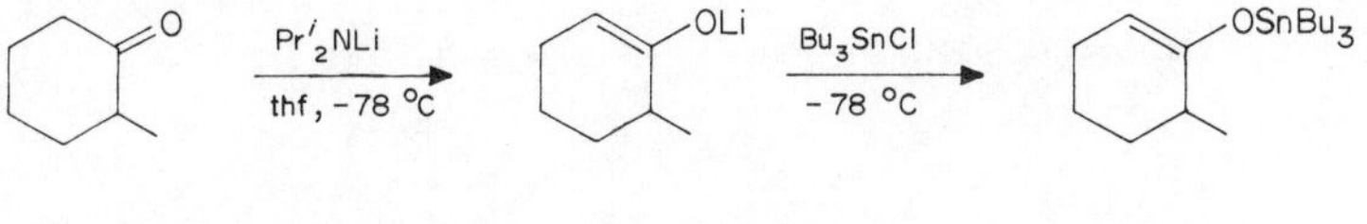

with various alkyl halides[48,206] proceeds clearly to give monoalkylated products in good yield. The examples in reactions 196–199 are representative.

The poor yield due to steric hindrance of the halide in equation 197 may be overcome by using the mixed solvent system[207] depicted in equation 199, or by using a Grignard or organolithium additive (equation 200) such as described in Section II.C.

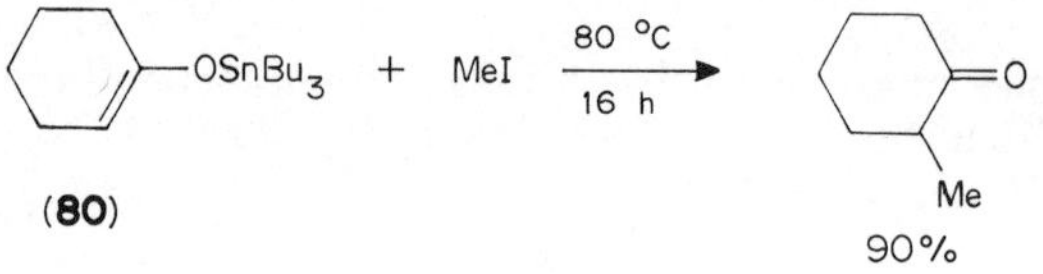

$$\text{(80)} \quad \text{cyclohexenyl-OSnBu}_3 + \text{MeI} \xrightarrow[16\text{ h}]{80\ °C} \text{2-methylcyclohexanone} \quad 90\% \tag{196}$$

$$\mathbf{80} + \text{Pr}^i\text{I} \xrightarrow[16\text{ h}]{80\ °C} \text{2-isopropylcyclohexanone} \quad 5\% \tag{197}$$

$$\mathbf{80} + \text{MeI} \xrightarrow[0\ °C,\ 1\text{ h}]{\text{dme}} \text{2-methylcyclohexanone} \quad 93\% \tag{198}$$

$$\mathbf{80} + \text{BuI} \xrightarrow[\text{RT, 20 h}]{\text{dme/hmpt (10:1, v/v)}} \text{2-butylcyclohexanone} \quad 63\% \tag{199}$$

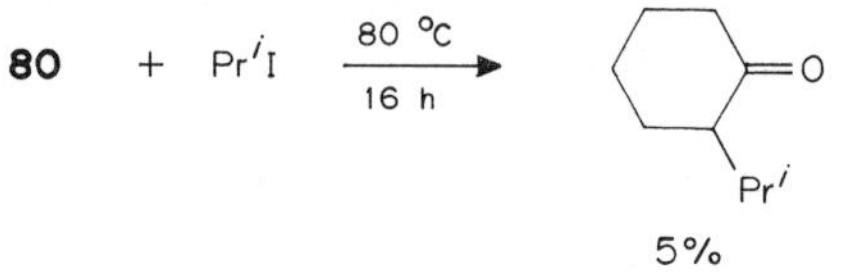

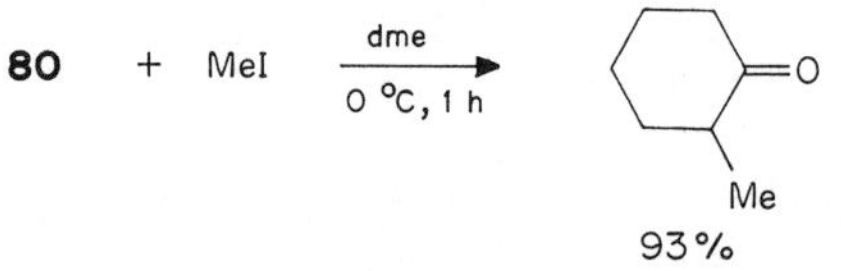

$$\text{(79)} \quad \text{2-methylcyclohexenyl-OSnBu}_3 \xrightarrow[\text{PhLi}]{\text{dme, MeI}} \text{2,2-dimethylcyclohexanone (73\%)} + \text{2,6,6-trimethylcyclohexanone (4\%)} \tag{200}$$

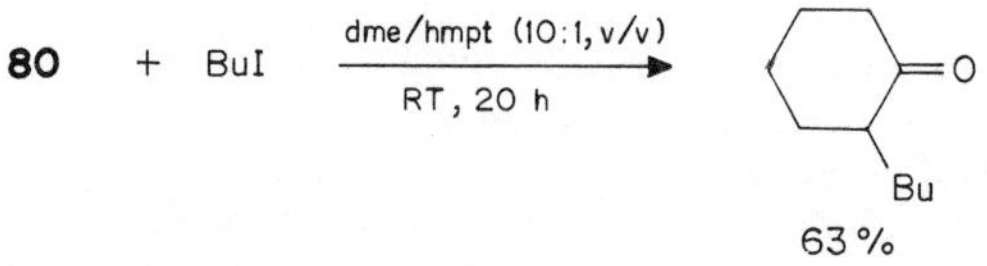

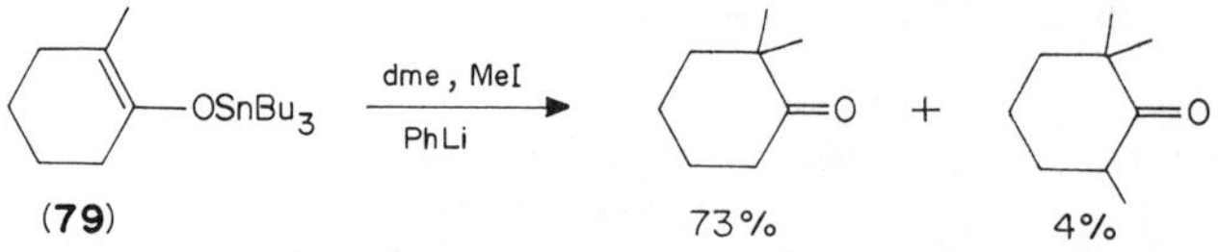

The regiospecificity of the reaction is clearly apparent in equation 201, while equation 202 reveals the beneficial influence of steric hindrance in the product in preventing side reactions involving addition of the aldehydic function to the tin enolate. In no case does O-alkylation occur except for the reaction of $\text{Bu}_3\text{SnCH}_2\text{COMe}$ (5% O-isomer) with

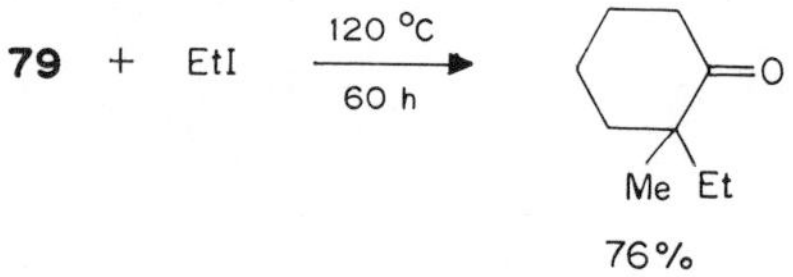

$$\mathbf{79} + \text{EtI} \xrightarrow[60\text{ h}]{120\ °C} \text{2-methyl-2-ethylcyclohexanone} \quad 76\% \tag{201}$$

$$\text{Me}_2\text{C}{=}\text{CHOSnBu}_3 + \text{MeI} \xrightarrow[14\text{ h}]{90\ °C} \text{Bu}^t\text{CHO} \quad 86\% \tag{202}$$

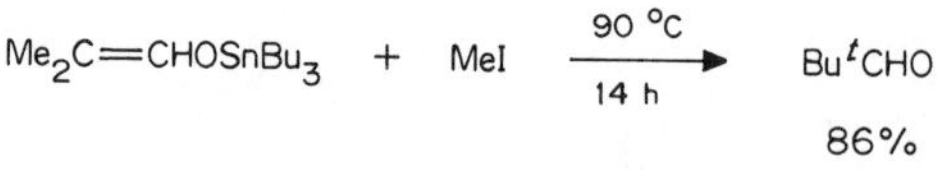

1. Carbon—carbon bond formation

$$Bu_3SnCH_2COMe + BrCH_2COOEt \xrightarrow[14\ h]{140\ °C} MeCOCH_2CH_2COOEt \quad (203)$$
$$50\%$$

ClCH$_2$OMe (equation 204). With allyl acetates, a facile reaction is obtained with enolstannanes of cyclohexanone in thf at room temperature in the presence of [Pd(PPh$_3$)$_4$] (5 mol%) as catalyst[208]. The reaction exhibits a high regioselectivity for alkylation at the less substituted end of the allyl moiety with formation of the E-isomer (equation 205).

$$Bu_3SnCH_2COMe + ClCH_2OMe \xrightarrow[14\ h]{100\ °C} CH_2{=}C(Me)OCH_2OMe \quad (204)$$
$$63\%$$

$$\mathbf{79} + Me_2C{=}C(Me)CH(Me)OAc \xrightarrow[thf]{[Pd(PPh_3)_4]} \text{(cyclohexanone-CH(Me)C(Me){=}CMe}_2) \quad (205)$$

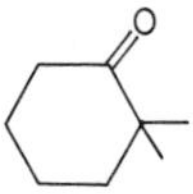

Of special interest from the point of view of organic synthesis is that allylic alkylation proceeds with a high degree of chemoselectivity[208] (equations 206–207a). A π-allylpalladium complex is considered to be initially formed, which undergoes attack by the

$$\mathbf{79} + Br(CH_2)_5CH(OAc)C(Me){=}CH_2 \xrightarrow[RT,\ 19\ h]{Pd(0)} RCH_2C(Me){=}CH(CH_2)_5Br \quad (206)$$
$$89\%$$

$$\mathbf{79} + MeCO(CH_2)_4CH(OAc)CH{=}CH_2 \xrightarrow[RT,\ 3\ h]{Pd(0)} RCH_2CH{=}CH(CH_2)_4COMe \quad (206a)$$
$$91\%$$

$$\mathbf{79} + CH_2{=}C(OEt)CH_2OAc \xrightarrow[RT,\ 41\ h]{Pd(0)} RCH_2C(OEt){=}CH_2 \quad (207)$$
$$72\%$$

$$\mathbf{79} + EtO_2CCH{=}C(Me)CH{=}CHCH{=}C(Me)CH_2OAc \xrightarrow[RT,\ 22\ h]{Pd(0)} RCH_2C(Me){=}CHCH{=}$$
$$CHC(Me){=}CHCO_2Et \quad 77\%$$

R = (2-methylcyclohexanone-2-yl) (207a)

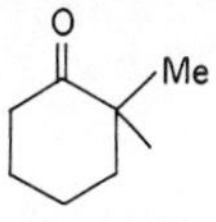

enolstannane. The reaction of enolstannanes with halo ketones and halo aldehydes yields cyclic ethers[181] instead of the expected 1,4-diketones. As previously noted for the allylstannanes (equations 151, 151a), the cyclic ethers are obtained in good yields under relatively mild conditions in the presence of a palladium catalyst. However, acetonyl-oxiranes undergo rearrangement with dehydration to yield 2-methylfurans either on prolonged heating or catalysis by acid. The furans may also be obtained directly in the absence of catalysts[208a].

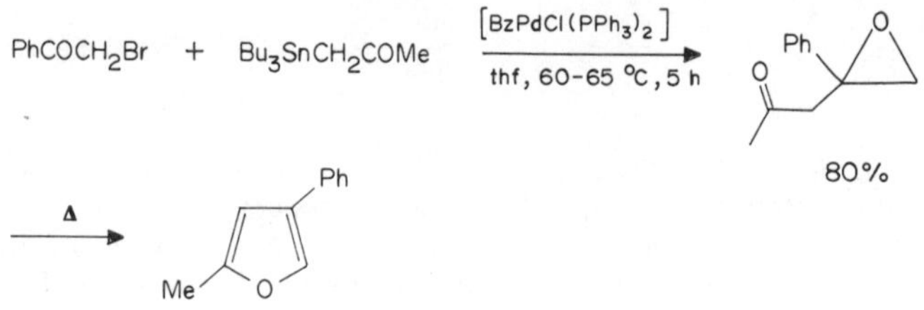

$$\text{PhCOCH}_2\text{Br} + \text{Bu}_3\text{SnCH}_2\text{COMe} \xrightarrow[\text{thf, 60-65 °C, 5 h}]{[\text{BzPdCl(PPh}_3)_2]} \quad \text{(208)}$$

80%

$$\xrightarrow{\Delta}$$

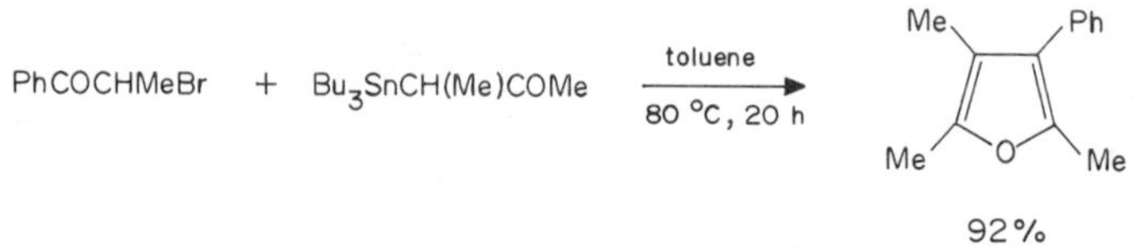

$$\text{PhCOCHMeBr} + \text{Bu}_3\text{SnCH(Me)COMe} \xrightarrow[\text{80 °C, 20 h}]{\text{toluene}} \quad \text{(209)}$$

92%

Enolstannanes have also been shown to undergo rapid aldol condensation with aldehydes in the absence of Lewis acid and transition metal catalysts to give products stereoselectively[209,210] and in good yields (80–90%). Based on studies to date, a *threo*-selectivity seems indicated for the aldol formed from trialkyltin enolates with *E*-geometry. The results are in consonance with studies on other metal enolates[2,211], where it was demonstrated that for kinetically controlled reactions, the *erythro*-isomer is the favoured product from *Z*-metal enolates (M = Li, Mg, Zn) while the *threo*-isomer ordinarily predominates from *E*-enolates. Under thermodynamic control, the *threo*-product is favoured, irrespective of the geometry of the starting enolates. A cyclic transition state has been envisaged to account for the stereoselection in which the steric bulk of the substituent R^2 at the α-position appears to be an important factor (Scheme 25). Only chair conformations were considered although, as pointed out in Section IV.B. 1.b, boat-like transition states are also conceivable. For trialkyltin enolates of *E*-geometry derived from pentan-3-one and propiophenone, *threo* adducts appear to be favoured [209,210] in reactions with aldehydes. With triphenyltin enolates, however, the results appear to be 'anomalous' in indicating a preferred *erythro*-selectivity irrespective of the geometry of the starting enolate (equation 210).

Even with cyclohexenyl and cyclopentenyl enolates with fixed *E*-geometry, the *erythro* products were predominant when the tin atom carried phenyl groups[210], normal *threo* adducts being obtained[209,210] with alkyl groups on tin under conditions of kinetic control. A boat transition state has been postulated to explain this anomaly for the case of cyclohexanone triphenylstannyl enol ether by Shenvi and Stille[209] (Figure 5).

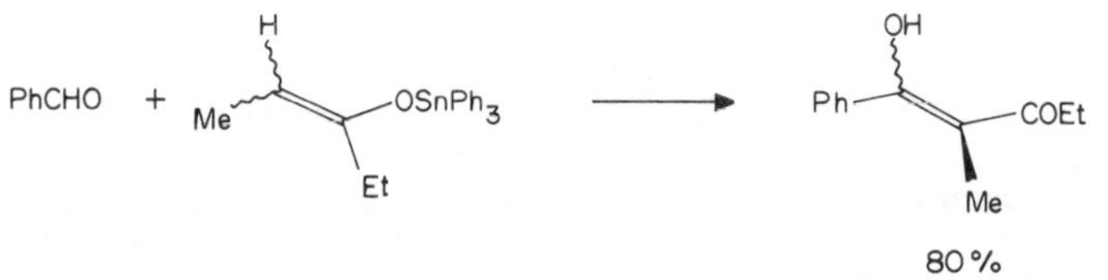

$$\text{PhCHO} + \quad \longrightarrow \quad \text{(210)}$$

80%

E/Z = 8 : 92
86 : 14

threo/erythro = 18 : 82
26 : 74

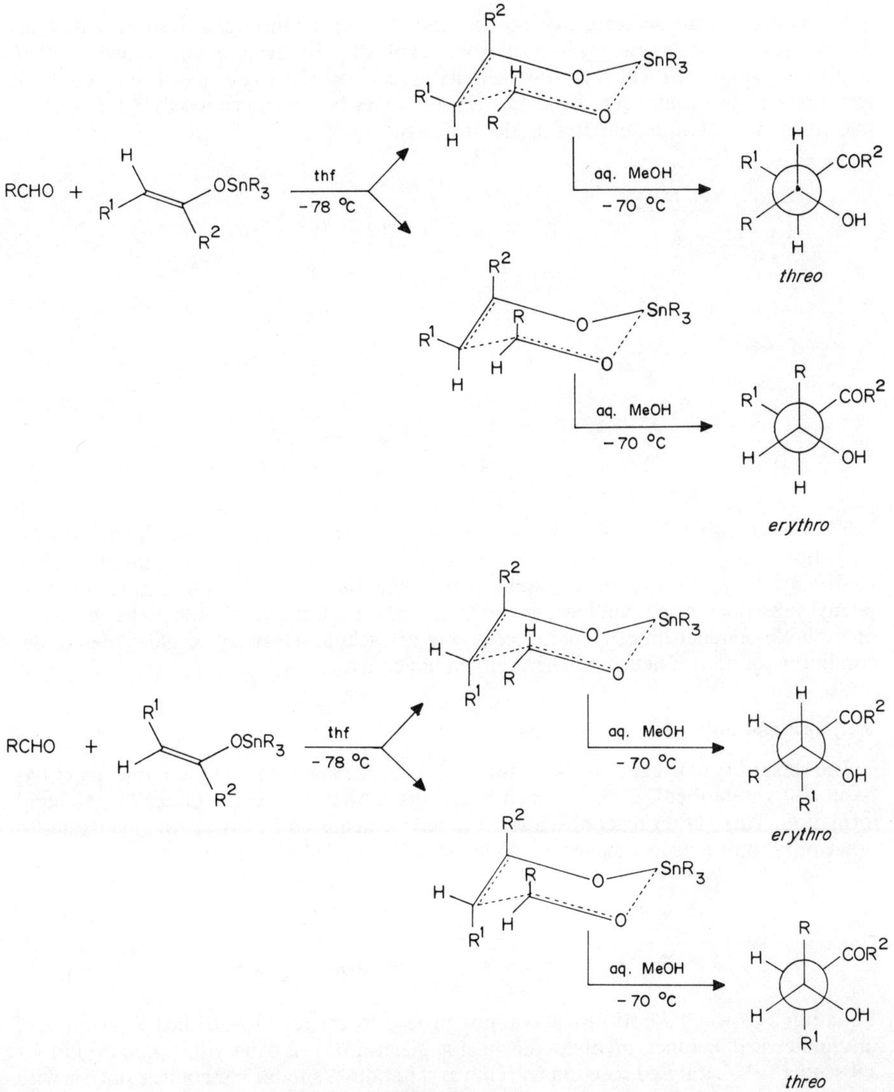

1. Carbon—carbon bond formation

SCHEME 25

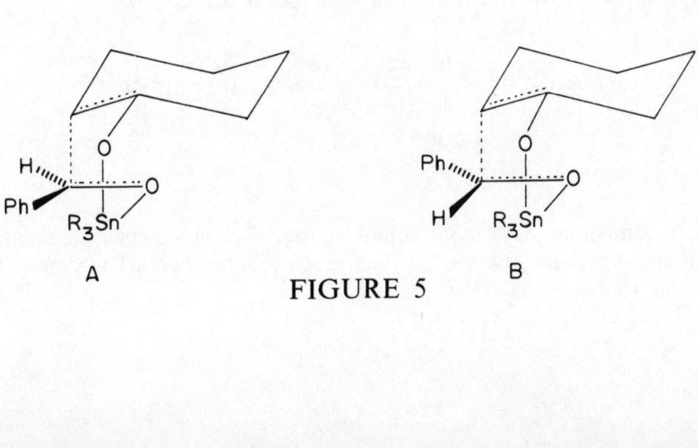

FIGURE 5

With bulky groups on tin, the *erythro* product is favoured through transition state B; in other cases the observed *threo*-selectivity is explicable in terms of the transition state structure A (Figure 5). The *erythro*-selectivity of reaction 210 may also be explained [210] in terms of an open-chain transition state (Figure 6), as has been proposed [212] for *erythro*-selectivity of zirconium enolates in aldol condensations.

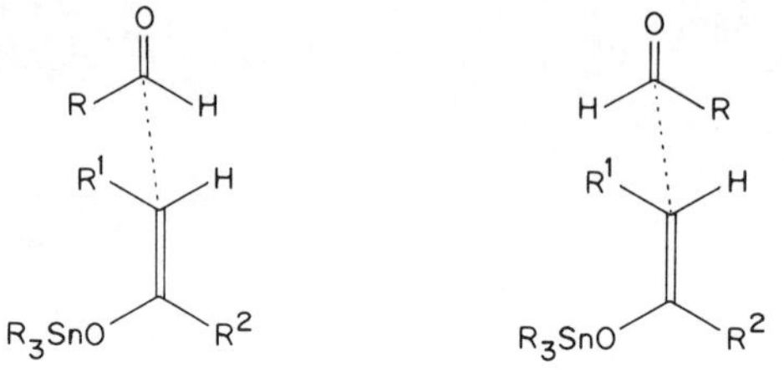

syn C   *anti* D

FIGURE 6

Thus, with either Z- or E-enolates, the transition state D is sterically favoured over transition state C; this leads to a dominant *erythro* product. However, more detailed studies are required in order to establish the generality of the variation introduced by phenyl substituents on tin in these enolate reactions. Such studies should preferably focus on isolable stereochemically pure enolates and include a variety of aldehydes under conditions of both kinetic and thermodynamic control.

2. Alkoxides and other derivatives

The versatility of organotin alkoxides[5,6,15-18,48] as reagents in synthetic chemistry has been well established[†]. None of their reactions, however, lead to direct C—C bond formation. These latter reactions, however, may be achieved by converting the oxides to other organotin reagents as, for example, in reactions 211 and 211a.

$$R_3SnOR + HC{\equiv}CR^1 \longrightarrow R_3SnC{\equiv}CR^1 + ROH \qquad (211)^{213}$$

$$R_3SnOR + CH_2{=}C{=}O \longrightarrow R_3SnCH_2COOR \qquad (211a)^{214}$$

Equation 212 which constitutes a 'one-pot' procedure for regio-controlled α-arylation of unsymmetrical ketones involves the *in-situ* generation of tributyltin enolate and its subsequent Pd-catalysed reaction with the aryl halide. A similar bimetallic catalysis with tin enolate intermediacy has been reported [214b] in the regioselective allylation of enol acetates (to allyl ketones) by allylic carbonates (equation 212a).

$$R^1C(OAc){=}CR^2R^3 \xrightarrow[\substack{[PdCl_2(o\text{-}tolyl)_3P] \\ 100\ °C,\ 5\ h}]{Bu_3SnOMe,\ PhBr} R^1COC(Ph)R^2R^3 \qquad (212)^{214a}$$

[†]Recently, cyclic stannoxanes have been found to be efficient covalent templates in the self-condensation of β-lactones to macrocyclic polylactones, e.g. enterobactin [A. Shanzer and J. Libman, *J. Chem. Soc., Chem. Commun.*, 846 (1983)].

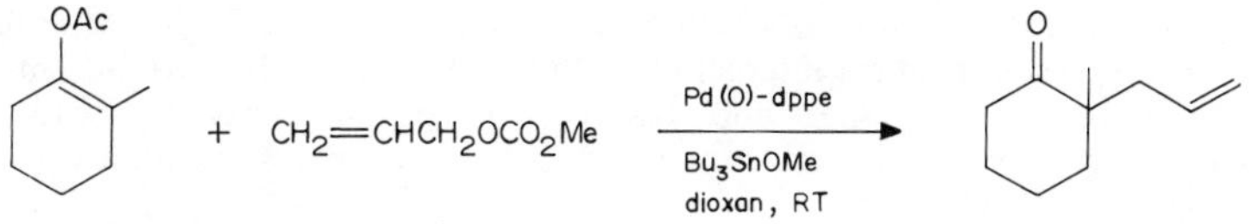

1. Carbon—carbon bond formation

$$\text{(cyclohexenyl OAc with methyl)} + CH_2=CHCH_2OCO_2Me \xrightarrow[\text{Bu}_3\text{SnOMe}]{\text{Pd(0)-dppe}} \text{(2-allyl-2-methylcyclohexanone)} \quad (212a)$$
dioxan, RT

Organolead alkoxides, because of their greater reactivity over organotin alkoxides, have found application in catalysing the addition reactions of active methylene compounds to acrylonitrile and isocyanates[215] (equations 213 and 214).

$$CH_2(CO_2Et)_2 + CH_2=CHCN \xrightarrow{Et_3PbOMe} (EtO_2C)_2CHCH_2CH_2CN + \quad (213)$$

$$(EtO_2C)_2C(CH_2CH_2CN)_2$$

$$MeCOCH_2CO_2Et + PhNCO \xrightarrow{Et_3PbOMe} PhNHCOCH(CO_2Et)(COMe) \quad (214)$$

The products are obtained in good yield. A limitation noted for these reactions is that with less reactive substrates the lead alkoxide may induce polymerization of the acrylonitrile and isocyanate acceptors.

The stannyl ester $Bu_3SnOCOCCl_3$ has been shown to undergo an interesting reaction with cyclopentadiene in the presence of Ph_3P, leading to the bicyclic product **81**[216]. The organotin reagent clearly serves as a precursor for $Cl_2C=C=O$, which presumably arises from the intermediate $[Bu_3SnOC(\overset{+}{O}PPh_3)=CCl_2\cdot Cl^-]$.

Unsaturated triorganostannyl esters[217-220] function as monomers in several polymerization and copolymerization reactions, e.g. equation 216. The resulting linear organotin

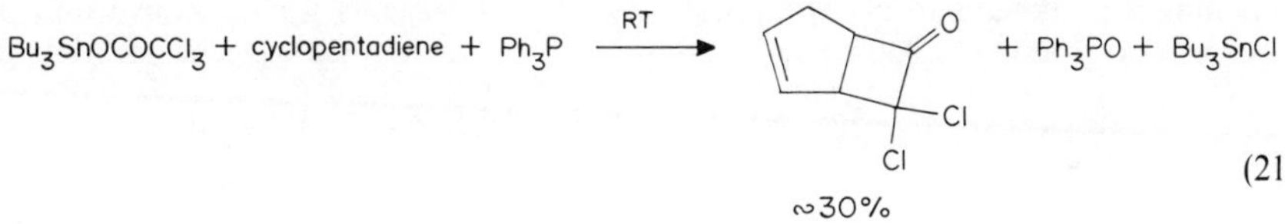

$$Bu_3SnOCOCCl_3 + \text{cyclopentadiene} + Ph_3P \xrightarrow{RT} \text{(bicyclic product)} + Ph_3PO + Bu_3SnCl \quad (215)$$

≈30%
(**81**)

$$n/2\ CH_2=CHCl + n/2\ CH_2=CR^1CO_2SnR_3 \longrightarrow [-CH_2CHClCH_2C(CO_2SnR_3)R^1-]_n$$

R^1 = H or Me (216)

polymers (R = Bu or Ph) have been formulated into useful slow-release, marine antifouling coatings.

Organolead esters, in particular aryllead tricarboxylates, have featured strongly in recent years in C—C bond formation reactions. These are discussed together with reactions of $Pb(OAc)_4$ [which in many instances proceed through monoorganolead(IV) species] in Section VII.C.

B. *N*-Stannyl and *N*-Plumbyl Derivatives

Unlike organotin alkoxides and enolates, preparative difficulties impinge on the synthetic utility of organotin amines. The amines have generally proved to be more reactive than the alkoxides in both substitution and addition processes[4,5,48]. Organotin enamines (see below) result from the interaction of the organotin halides with lithium or magnesium derivatives of enamines, and also on treating distannazanes with tin enolates[221] (equation 217). Like the tin enolates, the tin enamines are metallotropic.

$$(Bu_3Sn)_2NEt + EtCH=CHOSnBu_3 \longrightarrow \underset{\text{major}}{EtCH=CHN(Et)SnBu_3} + \underset{\text{minor}}{EtCH(SnBu_3)CH=NEt} \quad (217)$$

C—C bond formation reactions of organotin amines have been little explored except through the circumstance of their further conversion[222] to vinyl, alkynyl, and α-functionally substituted organotins, as in reactions 218 and 219.

$$R_3SnNR^1_2 + CH_2=CHCN \longrightarrow R_3SnCH(CN)CH_2NR^1_2 \quad (218)$$

$$Me_3SnNMe_2 + PhC\equiv CCl \longrightarrow Ph(Me_3Sn)C=C(NMe_2)Cl \quad (219)$$

With organotin amines containing a dimethylamino group, interaction with simple ketones such as acetone results in an aldol condensation reaction (equation 220). This has been ascribed to the basic properties of the nitrogen atom[223]. With other alkylamino groups, organotin enamines or organotin enolates are usually formed[4].

$$2Me_3SnNMe_2 + 2MeCOMe \longrightarrow Me_2C=CHCOMe + 2Me_2NH + (Me_3Sn)_2O \quad (220)$$

Me_3SnNMe_2 has also been used as a precursor in an interesting set of reactions leading to the formation of the dihydropyrazole derivative[4] **82** (equation 221). Organotin amines containing a dialkylamino group appear to be useful reagents for the conversion of oxiranes into 2-amino alkanols in high yields[223a] (reaction 221a). The reaction is

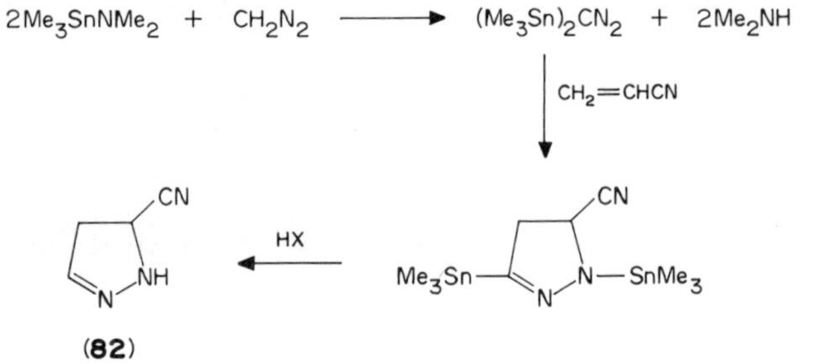

$$2Me_3SnNMe_2 + CH_2N_2 \longrightarrow (Me_3Sn)_2CN_2 + 2Me_2NH \quad (221)$$

$$\downarrow CH_2=CHCN$$

(82) ← HX — Me₃Sn-N,N-SnMe₃ pyrazoline with CN

$$H\text{-oxirane-Me} + Me_3SnNR_2 \xrightarrow[\text{10 h}]{CH_2Cl_2,\text{ reflux}} \underset{NR_2}{MeCHCH_2OSnMe_3} \quad 70-80\%$$

$$(R_2 = Et_2, (CH_2)_5)$$

$$\xrightarrow[\text{ether}]{\text{malonic acid}} \underset{NR_2}{MeCHCH_2OH} \quad 98\% \quad (221a)$$

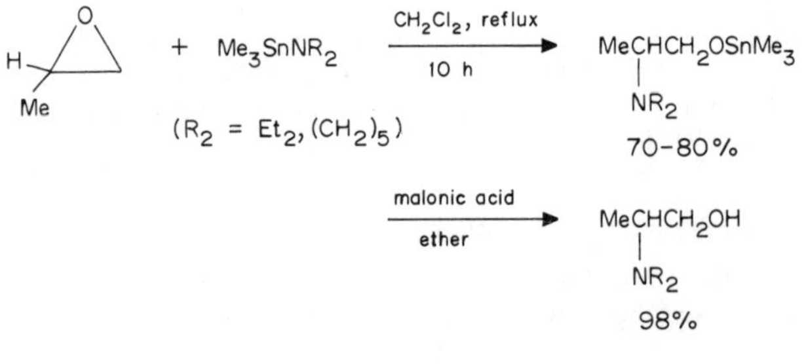

1. Carbon—carbon bond formation

regiospecific, probably due to the strong affinity of the tin atom to the oxygen atom of the oxirane ring, and is independent of the types of organotin compound (organotin alkoxides and halides are just as effective) and oxirane used.

Tributylstanylpyrrole has been shown[224] to react with benzyl, allyl and acyl halides leading to alkylation of the heterocycle in the 2-position (reaction 222). With BzBr (120 °C, 20 h), the 2-substituted product was obtained in 33% yield, with a small amount (17.8%) of the 3-isomer. MeCOCl yielded 43.3% of the 2-acyl product (25 °C, 2 h) and 2% of the N-acyl derivative.

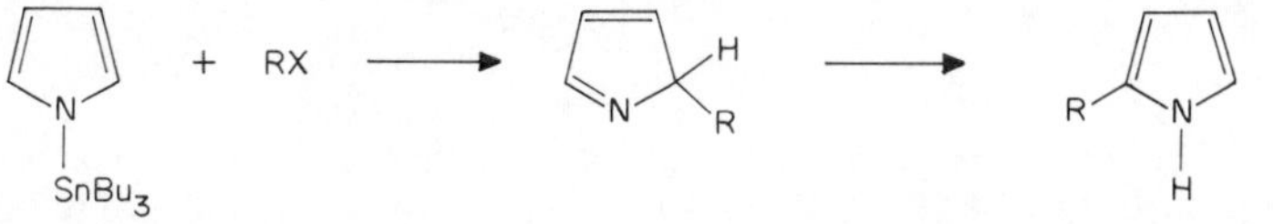

Trimethyltin(diphenylmethylene)amine, $Ph_2C=NSnMe_3$, has been shown[225] to react readily at room temperature with the diketone **83** to give the 1:1 adduct **84** in quantitative yield. Treatment of **84** with ethane thiol caused rapid cyclization to occur to yield the heterocyclic derivative **85** (equation 223). N-Plumbyl ketone imines[226] (**86**) yield C-alkylated and C-acetylated compounds in good yields in exothermic reactions with allyl bromide and benzoyl chloride, respectively (Scheme 26). The reactions probably proceed by 1,2-addition to C=C double bond with subsequent β-elimination of Bu_3PbX.

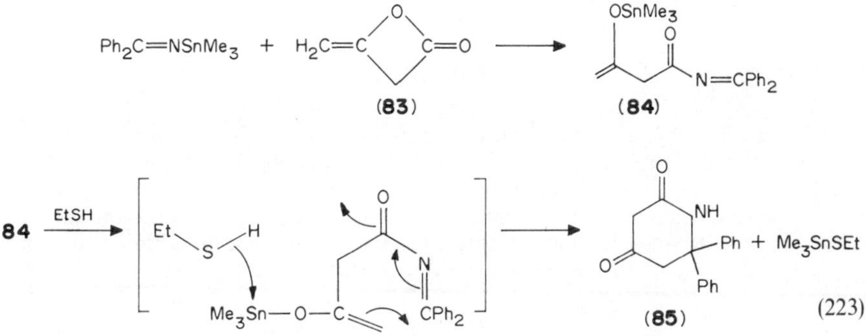

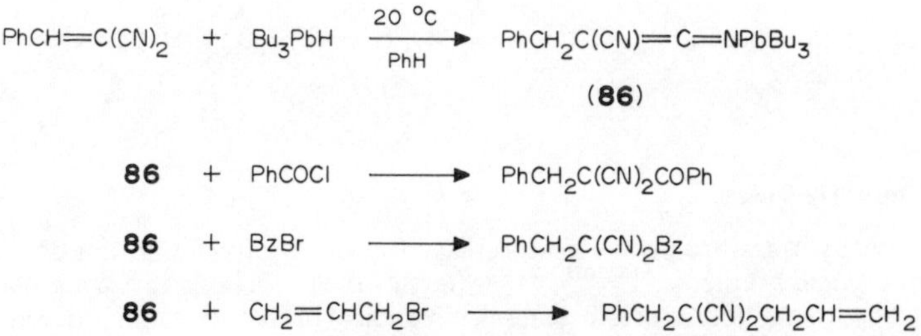

$$PhCH=C(CN)_2 + Bu_3PbH \xrightarrow[PhH]{20\,°C} PhCH_2C(CN)=C=NPbBu_3$$
$$(86)$$

$$86 + PhCOCl \longrightarrow PhCH_2C(CN)_2COPh$$
$$86 + BzBr \longrightarrow PhCH_2C(CN)_2Bz$$
$$86 + CH_2=CHCH_2Br \longrightarrow PhCH_2C(CN)_2CH_2CH=CH_2$$

SCHEME 26

N-Aryl-N^1-tributylstannyl hydrazines[227], obtained from the reaction of N-arylhydrazine with Bu_3SnNEt_2, give biphenyl derivatives on oxidation with a variety of reagents. Whereas HgO in benzene yielded only 35% biphenyl, the use of chloranil increased the yield to 93%[227] (equation 224).

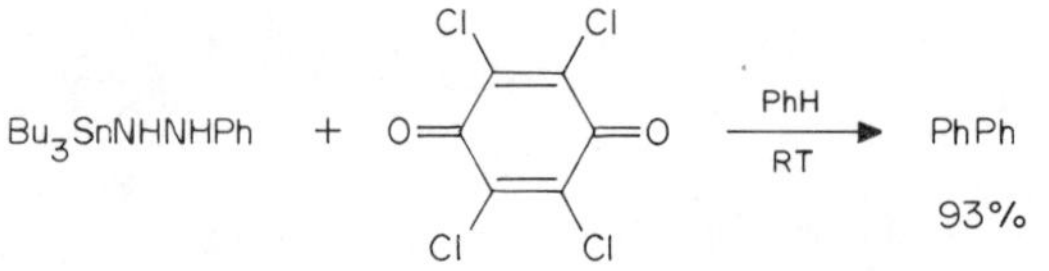

$$Bu_3SnNHNHPh \ + \ \text{chloranil} \ \xrightarrow[RT]{PhH} \ PhPh \quad 93\% \quad (224)$$

In pyridine or anisole as solvent, a mixture of phenylpyridines (o-, 59%; m-, 41%) and methoxybiphenyls (o-, 64%; m-, 22.5%; p-, 13%), respectively, are produced. With $Bu_3SnNHNHC_6H_4Me$-p, the dominant product (93%) in benzene medium is 4-methylbiphenyl, with 4,4'-dimethylbiphenyl (6%) and the unsubstituted biphenyl (0.5%) constituting the other products. A radical mechanism with the intermediate formation of the stannyldiimine $Bu_3SnN=NPh$ has been postulated[227] for these reactions (Scheme 27).

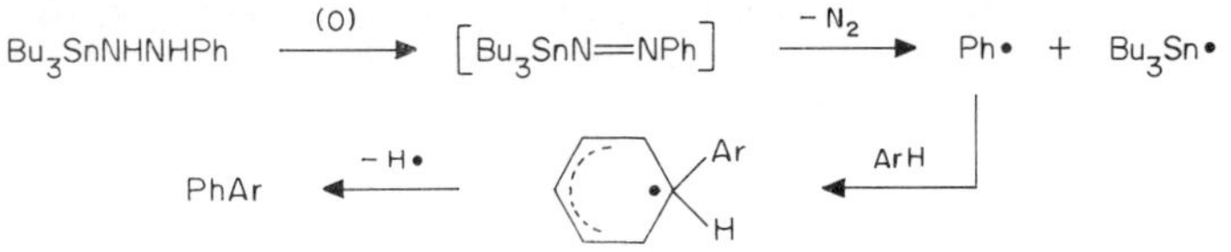

$$Bu_3SnNHNHPh \ \xrightarrow{(O)} \ [Bu_3SnN=NPh] \ \xrightarrow{-N_2} \ Ph\bullet \ + \ Bu_3Sn\bullet$$

$$PhAr \ \xleftarrow{-H\bullet} \ \text{(cyclohexadienyl)} \ \xleftarrow{ArH} \ $$

SCHEME 27

Only a preliminary account of the use of organotin enamines in synthesis has appeared in the literature[221], relating to the preparation of functionally substituted carbonyl compounds. Among the conditions identified for the success of the reaction (equation 225) are: (1) the alkyl chain bonded to N in the enamine must begin by a CH_2; (2) the ethylenic part of the molecule must be monosubstituted on the C atom in the β-position with respect to N; and (3) the alkene must be strongly electrophilic.

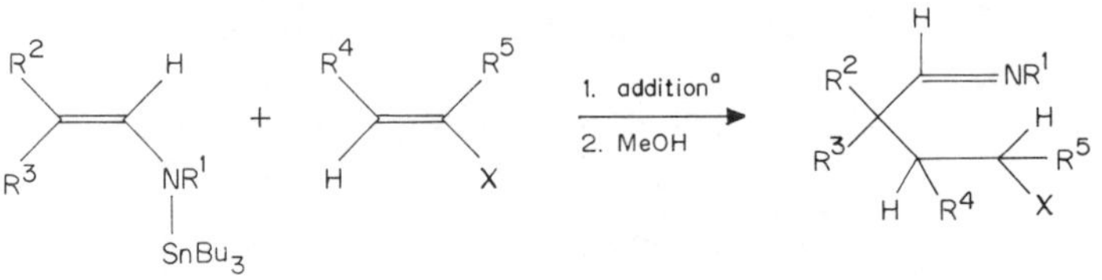

$$\underset{\underset{SnBu_3}{|}}{\overset{R^2}{\underset{R^3}{\diagdown}}C=C\overset{H}{\diagup}_{NR^1}} \ + \ \overset{R^4}{\underset{H}{\diagdown}}C=C\overset{R^5}{\diagup}_{X} \ \xrightarrow[2.\ MeOH]{1.\ addition^a} \ \overset{R^2}{\underset{R^3}{\diagdown}}C\text{—}C\overset{H}{\diagup}=NR^1 \ \ldots \quad (225)$$

^{a} E.g. 4 h at 60 °C for $R^1 = i$-Bu, $R^2 = Me$, $R^3 = R^4 = R^5 = H$, $X = CO_2Me$; yield 90%

C. Organotin Hydrides

Organotin hydrides are very versatile reagents which have found important applications in organic synthesis[5-14,20,23,228]. Many of their reactions involving hydrostannation and hydrostannolysis are, in general, free radical in character and proceed through $R_3Sn\cdot$ intermediates[20,23,229]. It may be expected, therefore, that with suitable organic substrates the intermediate carbon-centred radical that is generated could enter into intramolecular processes, leading to C—C bond formation, provided that these rates are favourably faster than hydrogen transfer rates. An account up to 1970 of the several known reactions wherein this expectation is strongly featured was published by Kuivila[230]. An up-

1. Carbon—carbon bond formation

to-date review of reactions of this category is beyond the scope of this chapter. Only selected examples from the literature illustrating the synthetic scope of these reactions will be considered here, since detailed treatments of the subject can be found in periodic reviews of advances in free-radical chemistry.

Typically, intramolecular cyclization reactions have been observed with radicals possessing unsaturation in the 5- or 6-positions. The Bu_3SnH reductions of alkenyl bromides[231–234,236,237] are exemplified in Schemes 28–31 and that of alkynyl bromides[238] in Scheme 32. The reductions are often conveniently carried out in benzene and due attention is given to dilution effects in respect of the hydride so as to favour the homolytic carbocyclizations.

Scheme 28 illustrates the cyclization pathways for 1-bromo-hex-5-ene substituted in the

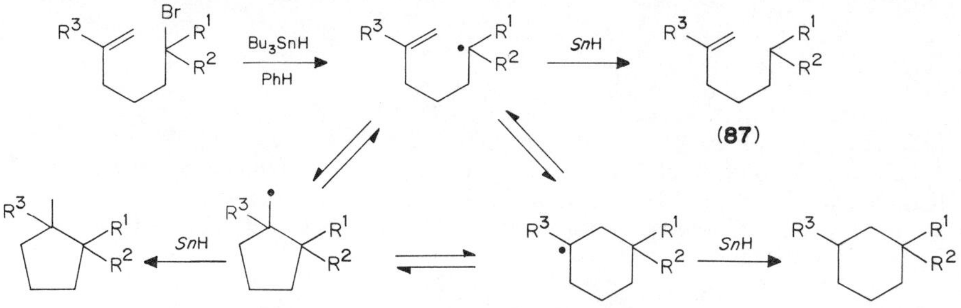

$R^1 = R^2 = R^3 = Me$; % products (**87, 90, 91**) = 8.9, 47.1, 44.0; $T = 40\,°C$ (ref. 232)

$R^1 = R^2 = R^3 = H$; % products = 7, 78 trace; $T = 40\,°C$ (ref. 231)

$R^1 = Ph, R^2 = R^3 = H$; % products = 0.6, 32.3, 77.1; $T = 100\,°C$ (ref. 233)

SCHEME 28

1- and 5-positions. At a given stannane concentration, both the yields of the cyclized products and the ratios of six-membered to five-membered ring products increase with temperature (40–100 °C) and with 1-substitution, while 5-substitution decreases this ratio. Confirmation of the interconversion of the radicals **88** and **89** in the above scheme has been secured in separate experiments on (*E*)-2-phenylcyclopentylmethyl bromide for which the product ratio **91:90** was noted to be significantly higher. A small amount of the indene derivative **92** also arises in the reaction from intramolecular radical attack on the aromatic ring.

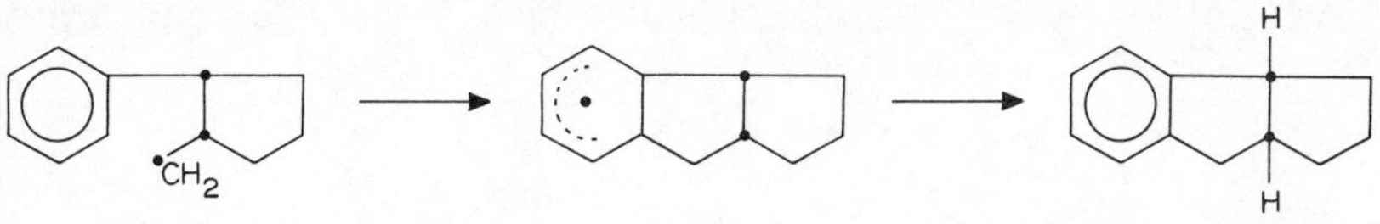

(226)

(**92**)

More recent studies by Beckwith an Lawrence[234] have indicated the dominant role of stereoelectronic effects in determining the direction of intramolecular addition in hexenyl radicals. Thus, the *exo*-mode of cyclization of the radical **93** (R = H), yielding the cyclopentyl derivative **94**, was found to be 100 times faster than the *endo* ring closure, which gives the cyclohexyl derivative (Scheme 29).

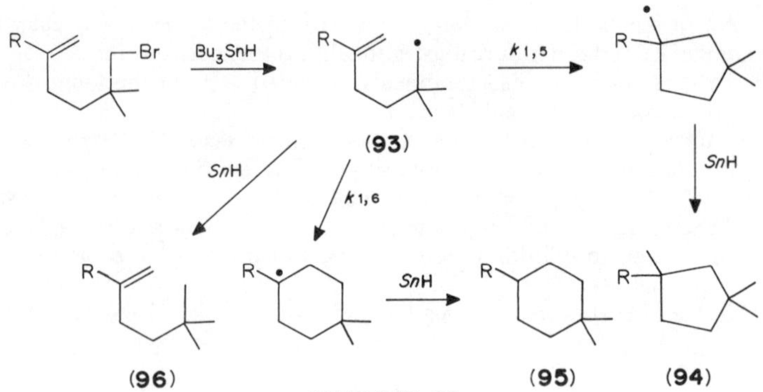

SCHEME 29

For R = H, **94** was formed in 90% yield and **95** in 10% yield at 80 °C, indicating high regiospecificity, whereas for R = Me, the product distribution was 57% **96**, 30% **94**, and 14% **95**. Also, for the radical **93** (R = H), $k_{1,5}$ for ring closure was found to be 10 times larger than that for $CH_2=CH(CH_2)_3CH_2^\bullet$. Both high regiospecificity and stereoselectivity have been observed in the homolytic cyclization of the bromoacetal[235] **97** induced by Bu_3SnH. Thus, (Z)-γ-butyrolactone (**98**) was predominantly obtained (equation 227) via this

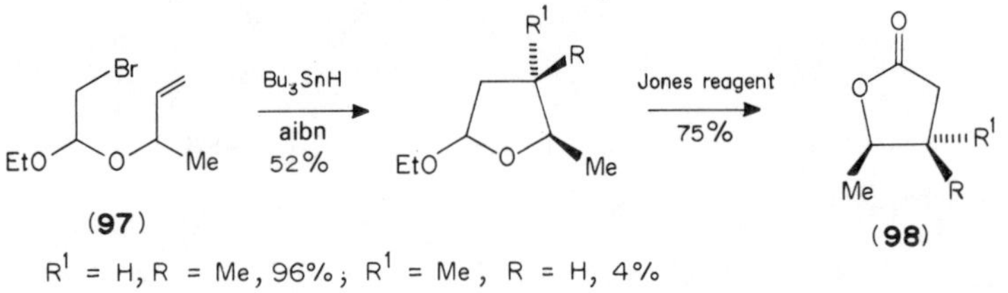

$$\tag{227}$$

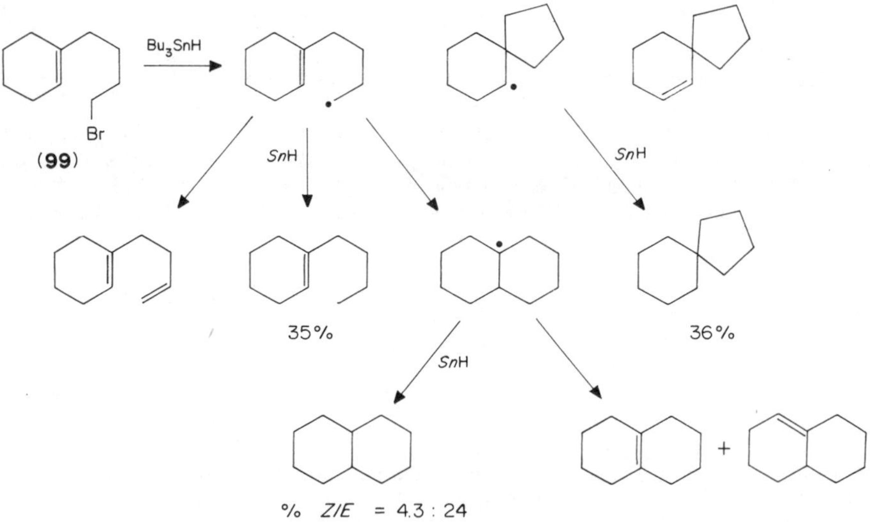

R^1 = H, R = Me, 96%; R^1 = Me, R = H, 4%

% Z/E = 4.3 : 24

SCHEME 30

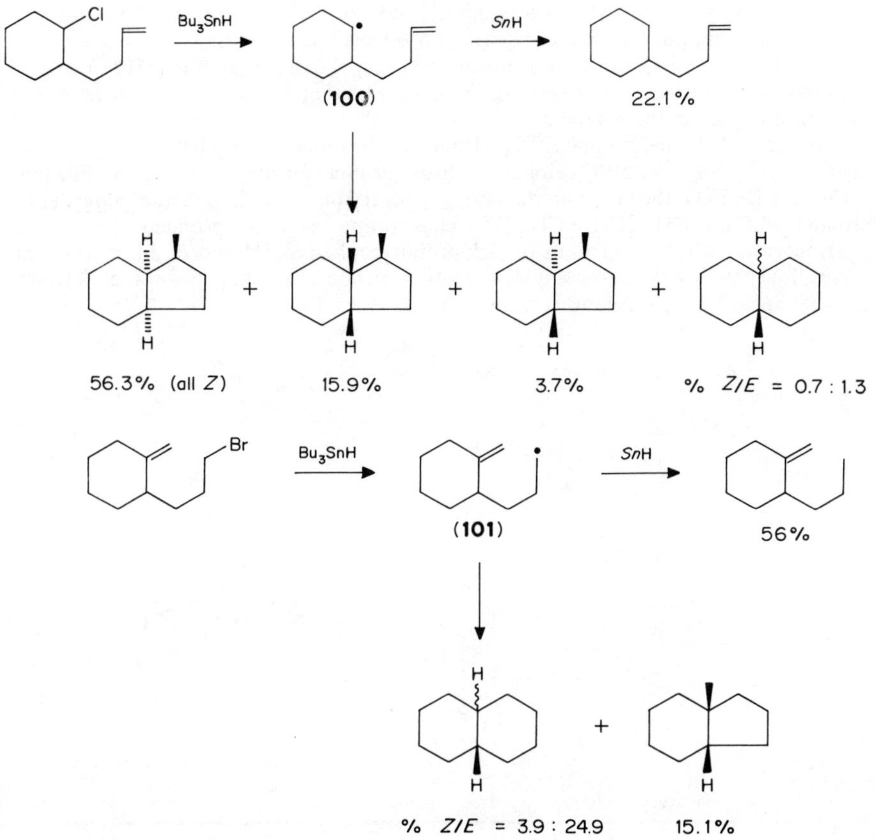

1. Carbon—carbon bond formation

SCHEME 31

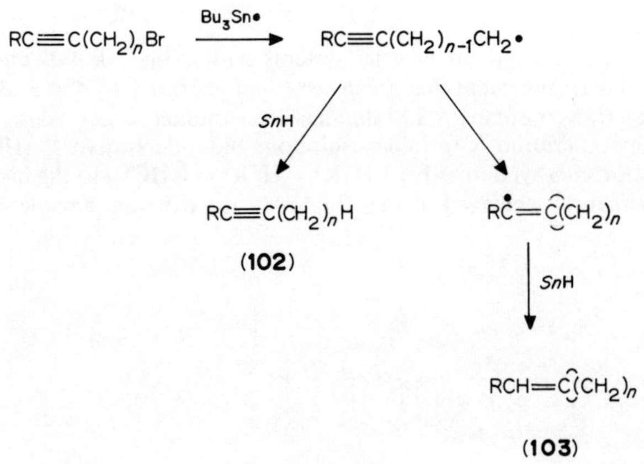

SCHEME 32

method. Homolytic carbocyclization studies[236] on 4-(cyclohex-1-enyl)butyl bromide (**99**) indicate the formation of a variety of cyclic products (Scheme 30).

Related studies on 1,2- and 4,5-disubstituted hexenyl radicals (**100**) and (**101**) have also been recently reported[237] (Scheme 31), and it appears that 5-substitution disfavours 1,5-ring closure (*exo*) in these systems.

In the case of alkynyl bromides[238] (Scheme 32), the reduction products are obtained in high yields (70–90%), with **102** being the exclusive product for $BuC\equiv C(CH_2)_2Br$. For $PhC\equiv C(CH_2)_4Br$, **103** is the major product (93%), with **102** (4%) and the dehydrohalogenation product, $PhC\equiv C(CH_2)_2CH=CH_2$ (3%), constituting the minor products.

Tributyltin hydride reductions of γ-chlorobutyrophenone[239] and of 5-hexenoyl and citronelloyl chloride[240] follow similar homolytic carbocyclization pathways. In the latter cases (Scheme 33), ketones rather than aldehydes were formed.

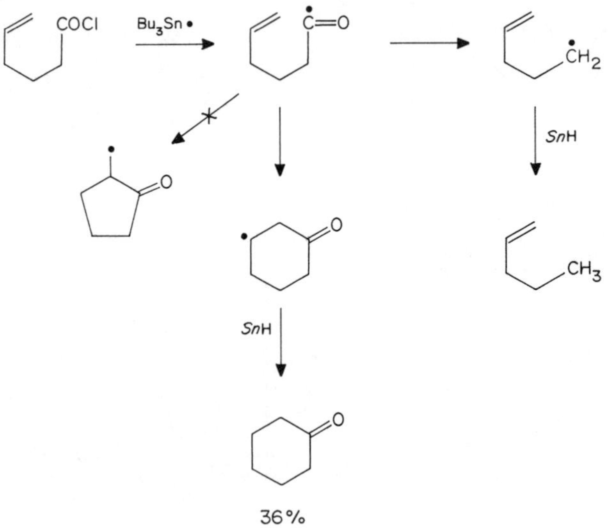

SCHEME 33

In reductions carried out on bicyclic systems with both tributyl- and triphenyl-tin hydrides, radical rearrangements have also been demonstrated[241–243]. Radical rearrangement involving a transannular 1,4-aryl shift has been uniquely observed in the reduction of the α-halomethylpiperidino-*N*-(*p*-toluenesulphonamide) derivative[244] (**104**). The reduction of the dichloromethyl derivative **104** ($R^1 = H$, $R^2 = CHCl_2$) to the methyl derivative may be conveniently achieved using Bu_3SnH in refluxing anisole. However, the

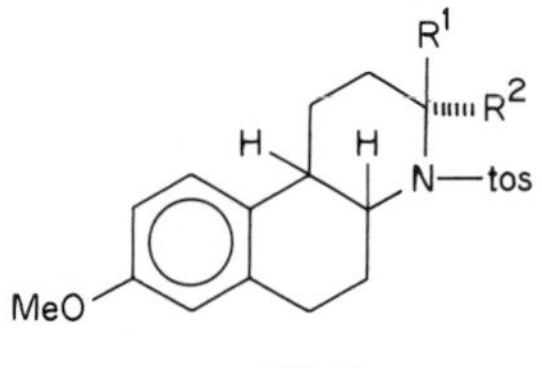

(**104**)

1. Carbon—carbon bond formation

chloromethyl derivative **105** ($R^1 = CH_2Cl$, $R^2 = H$) under identical conditions afforded a 50% yield of **106** (X = Cl) instead of the expected product ($R^1 = Me$, $R^2 = H$). A nearly quantitative yield of **106** (X = Br) was obtained with the corresponding bromomethyl derivative. In the absence of solvent, only 30% yields of the rearranged products were obtained. These results can be rationalized in terms of Scheme 34[244].

Ph$_3$SnH reduction of **107** similarly produced the unexpected products **108** and **109** rather than **110** (Scheme 35)[245]. Free-radical annelation in bicyclic β-lactams[246] has also

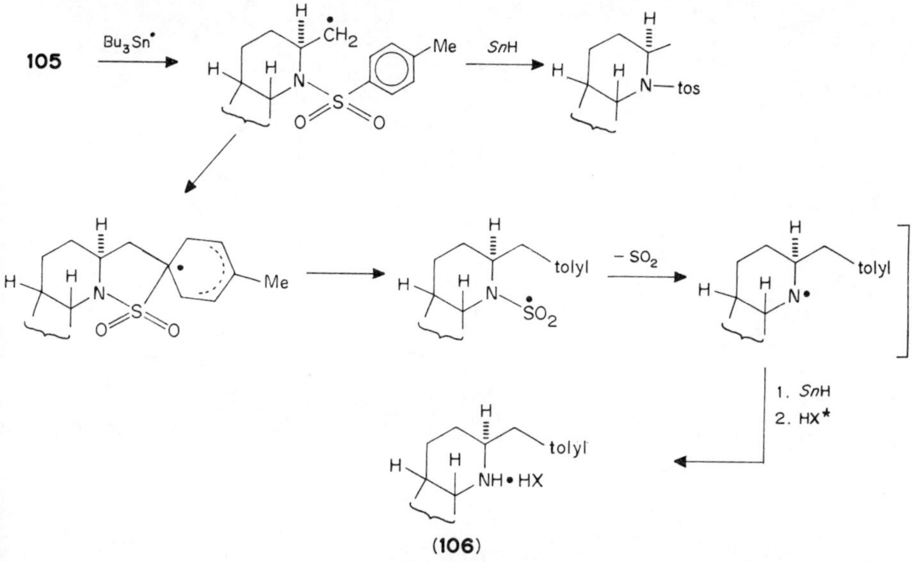

SCHEME 34

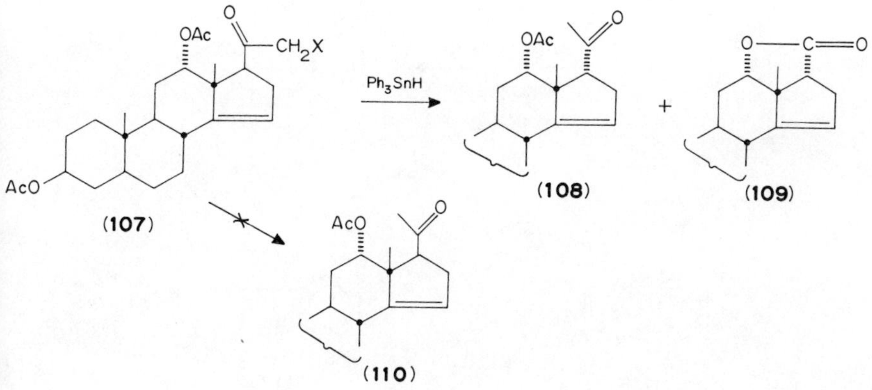

SCHEME 35

been initiated using Bu_3SnH (Scheme 36). The hydrogenated products derived from the radicals **111–114** are formed in the amounts 16, 4, 52.2 and 16.8%, respectively.

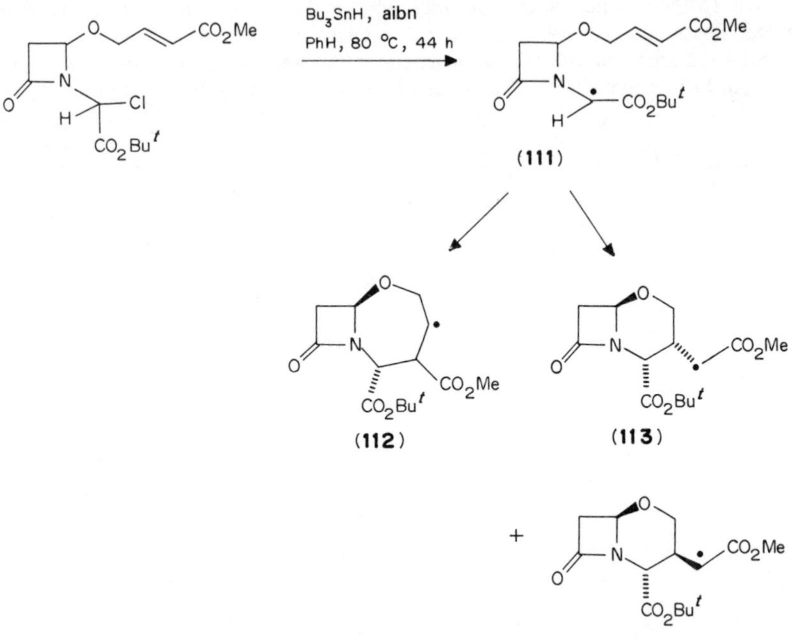

SCHEME 36

Recently, Ueno et al.[247] utilized Bu_3SnH for the reduction of compounds **115** (X = NH or O) to dihydroindole or dihydrobenzofuran. An intramolecular S_H' process has been suggested for the cyclization (Scheme 37). For concentrations of Bu_3SnH around 0.02M,

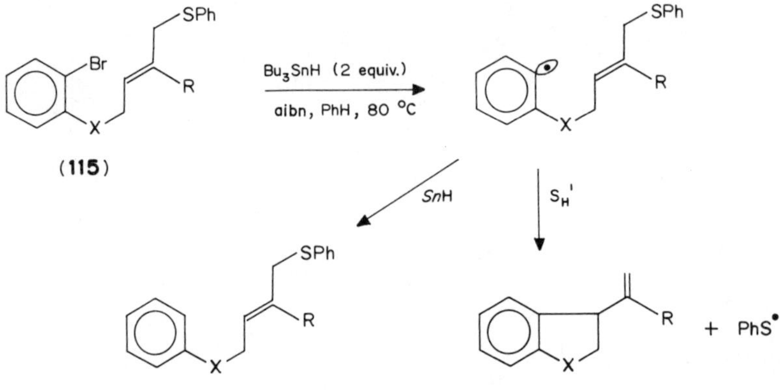

SCHEME 37

compound **117** was obtained exclusively in near quantitative yields (R = H; X = NH or O), but at 0.5M concentration of the tin reagent, approximately equal amounts of **116** and **117** were obtained in both cases. The driving force for the carbocyclization is undoubtedly the formation of the phenylthiyl radical which yields the thiostannane Bu_3SnSPh^{248} with the tin hydride. In a more recent extension of the work, β-vinyl γ-butyroketones have been synthesized[235] in 95% yield starting from the bromoacetal **118** (equation 228).

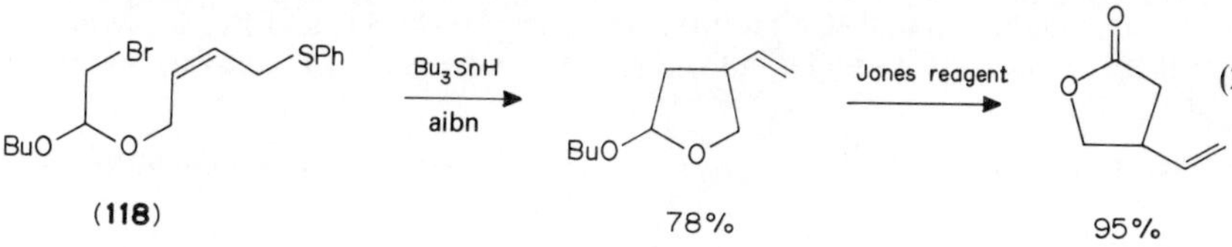

$$\text{BuO-CH(Br)-O-CH}_2\text{-CH=CH-CH}_2\text{-SPh} \xrightarrow[\text{aibn}]{Bu_3SnH} \text{(furan with vinyl, BuO)} \xrightarrow{\text{Jones reagent}} \text{(γ-butyrolactone with vinyl)} \quad (228)$$

(**118**)   78%   95%

VII. MISCELLANEOUS REACTIONS

A. Oxidative Addition–Reductive Elimination

With the exception of bis(cyclopentadienyl)tin derivatives, most dialkylstannanes exist as cyclic oligomers[249]. They are generally very susceptible to oxidative reaction with halogens or methyl iodide and there is evidence that the reaction involves a radical chain mechanism[250,251] (equation 229). Bis(pentamethylcyclopentadienyl)tin(II), $(pcp)_2Sn(II)$,

$$(Cp)_2Sn + X\bullet \longrightarrow (Cp)_2\overset{\bullet}{Sn}X \xrightarrow{XY} (Cp)_2SnXY + X\bullet \quad (229)$$

for example, undergoes oxidative addition with halogens to give $(pcp)_2Sn(IV)X_2$, which reductively eliminates bis[(1,2,3,4,5-pentamethyl)cyclopenta-2,4-dienyl], pcp—pcp, in the presence of a nucleophilic solvent such as pyridine (reaction 230)[252].

$$(pcp)_2Sn(II) \xrightarrow[PhH]{I_2} (pcp)_2Sn(IV)I_2 \xrightarrow{py} pcp\text{—}pcp + SnI_2\cdot py \quad (230)$$
$$\qquad\qquad\qquad\qquad\qquad\qquad\qquad\qquad 18\%$$

Cross-coupling of cyclopentadienyl and alkyl groups has also been reported[253] in reactions of $(Cp)_2Sn(II)$ with Ph_3CBr, BzBr, BzCl, and $CH_2=CHCH_2Br$.

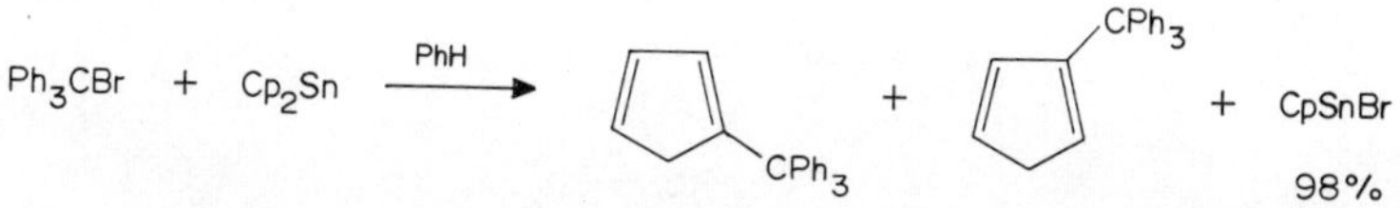

$$Ph_3CBr + Cp_2Sn \xrightarrow{PhH} \text{(Cp-CPh}_3\text{)} + \text{(Cp with CPh}_3\text{)} + CpSnBr \quad (231)$$
$$\qquad\qquad\qquad\qquad\qquad\qquad\qquad\qquad\qquad 98\%$$

The lead analogue, Cp_2Pb, undergoes similar oxidative addition reaction with MeI to yield $Cp_2PbMeI^{254,255}$, which very readily undergoes reductive elimination to give MeCp and CpPbI (reaction 232). Although copious amounts of the elimination products,

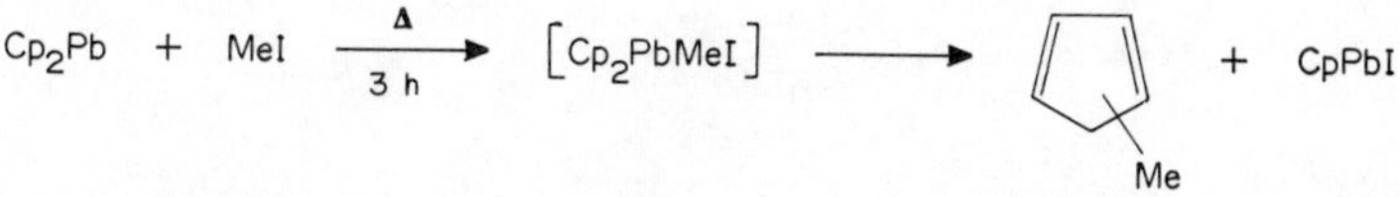

$$Cp_2Pb + MeI \xrightarrow[3\text{ h}]{\Delta} [Cp_2PbMeI] \longrightarrow \text{(MeCp)} + CpPbI \quad (232)$$

CpMX, generally result in these reactions, the yields of cross-coupling products remain low. The reaction of Cp_2Pb with MeCOCl in Et_2O at $-80\,°C$, for example, has been reported[254] to give a high yield of CpPbCl and, although the reaction is presumably one of oxidative addition, no acetylcyclopentadiene has been isolated. The relatively insoluble nature of the CpMX derivatives has been considered to provide the driving force for the reductive elimination step in reactions such as 231 and 232.

B. C—C Coupling via Stannylenes

C—C coupling of aldehydes has been recently achieved in a highly stereoselective manner using suitable stannylene precursors[256] such as Me_8Sn_3 and Bu_6Sn_2 (reaction 233). The percentage of D,L-isomer in the glycol derivative is influenced by the nature of

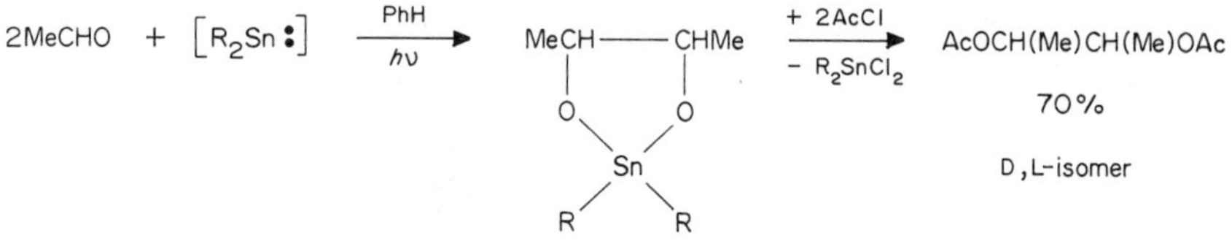

$$2MeCHO + [R_2Sn:] \xrightarrow[h\nu]{PhH} \underset{(119)}{MeCH\underset{O}{|}\underset{\underset{R}{Sn}\underset{R}{\diagdown}}{\diagup}CHMe} \xrightarrow[-R_2SnCl_2]{+2AcCl} \underset{\text{D,L-isomer}}{\underset{70\%}{AcOCH(Me)CH(Me)OAc}} \quad (233)$$

the R group in the distannane in the order Me (89%) < Et (96%) < Bu (100%). Apparently a high steric requirement in the reactive intermediate **119** is a condition for the stereoselectivity of the C—C coupling.

Aldehydes such as EtCHO, PrCHO, and PhCHO have been shown to give higher yields of C—C coupling products than ketones under the same reaction conditions. This may be attributed to the lower carbonyl activity and higher steric hindrance of the ketones in the formation of the reactive intermediate **119**. Evidence has also been presented to show that the reaction proceeds via a stannylenoid mechanism starting from a relatively long-lived triplet state.

C. Reactions of Lead Tetraacetate and Aryllead Tricarboxylates

Lead tetraacetate, LTA, in the presence of copper(II) ions has been employed for oxidative decarboxylation of primary acids to give good yields of terminal olefins and esters[257,258].

$$RCH_2CH_2CO_2H \xrightarrow[Cu^{2+}]{LTA} RCH_2CH_2CO_2Pb(OAc)_3 \xrightarrow{-CO_2} RCH_2CH_2^{\bullet} \quad (234)$$

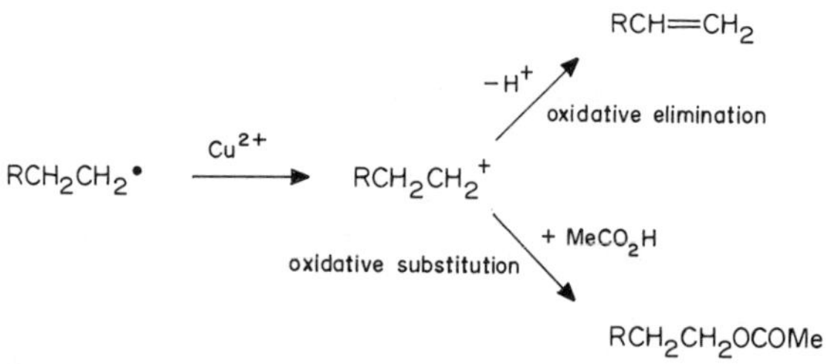

$$RCH_2CH_2^{\bullet} \xrightarrow{Cu^{2+}} RCH_2CH_2^{+} \begin{array}{c} \xrightarrow{-H^+} RCH=CH_2 \\ \text{oxidative elimination} \\ \xrightarrow[\text{oxidative substitution}]{+MeCO_2H} RCH_2CH_2OCOMe \end{array} \quad (235)$$

There is strong evidence that these reactions proceed by a free-radical chain mechanism[259]. Thus, the homolytic alkylation of benzene with **120** or **121** decarboxylated by LTA in the presence of catalytic amount of pyridine and Cu^{2+} ions gave, among other

1. Carbon—carbon bond formation

acyclic olefins, only the (*E*)-cyclopropyl isomer (**122**). The absence of stereospecificity in the decarboxylation can be explained on the basis of the mobile configuration of the cyclopropyl radical intermediate.

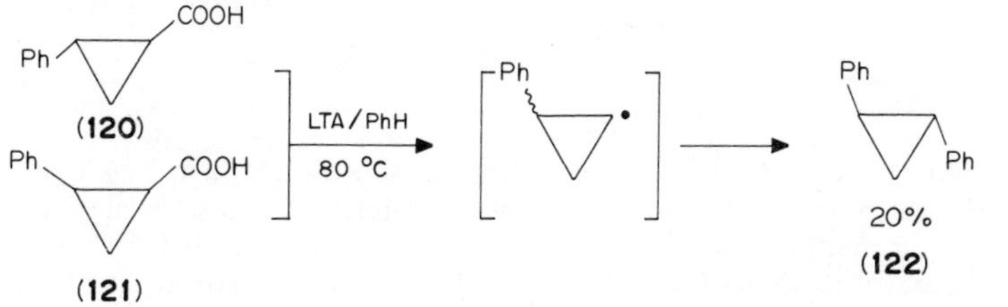

$$\begin{array}{c} \text{Ph} \\ \text{(120)} \\ \text{Ph} \\ \text{(121)} \end{array} \xrightarrow[80\ ^\circ\text{C}]{\text{LTA/PhH}} \left[\begin{array}{c} \text{Ph} \\ \bullet \end{array} \right] \longrightarrow \begin{array}{c} \text{Ph} \\ \text{Ph} \\ 20\% \\ \text{(122)} \end{array} \quad (236)$$

Benzenoid compounds may also be oxidised by LTA in trifluoro- or trichloro-acetic acid. With electron-rich benzenoids such as polymethylated benzenes and anisoles, LTA yields, in addition to acyloxation and plumbylation[260] products, both biaryls and diarylmethanes through oxidative dimerization. The proportions of the products vary markedly with the aromatic compounds and the conditions of reaction. The factors governing the formation of biaryls and diarylmethanes have been investigated by Norman and coworkers[261-264], Wolters and coworkers[265], and Sternhell and coworkers[266-270]. Benzene, for example, gives phenol in almost 80% yield with LTA in CF_3COOH, whereas anisole gives mainly the dimethoxybiphenyls[263]. Also, anisole gives about 50% of bis(*p*-methoxyphenyl)lead bis(trichloroacetate) (**123**) in CCl_3COOH, which is believed to mediate in the formation of dimethoxybiphenyl. This may be achieved in two ways, depending on the conditions of reaction. First, it can act as a source of *p*-methoxyphenyl cation (**124**), which can effect substitution in 123 (equation 238a) or in anisole (equation 238b). Secondly, it may act as a source of Pb(IV) for the oxidation of anisole (equation 239).

$$(p\text{-MeOC}_6\text{H}_4)_2\text{Pb}(\text{OCOCCl}_3)_2 \xrightarrow{CF_3CO_2H} (p\text{-MeOC}_6\text{H}_4)\text{Pb}(\text{OCOCCl}_3)_2(\text{OCOCF}_3)$$
(**123**)

(237)

$$(p\text{-MeOC}_6\text{H}_4)^+$$
(**124**)

124 $\xrightarrow{+\ 123}$ 2,4'- and 3,4'-dimethoxybiphenyls (238a)

$\xrightarrow{+\ \text{MeOPh}}$ 2,4'-, 3,4', and 4,4'-dimethoxybiphenyls (238b)

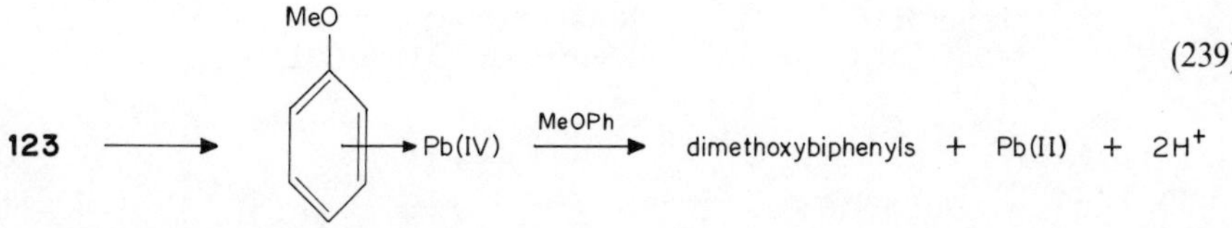

123 $\longrightarrow$ MeO—C$_6$H$_4$—Pb(IV) $\xrightarrow{\text{MeOPh}}$ dimethoxybiphenyls + Pb(II) + 2H$^+$ (239)

In addition, the methoxyphenyl cation can also account for the formation of dimethoxyphenylmethanes in the presence of adventitious water found in the trihaloacetic acid (reactions 240 and 241).

$$MeOPh + (p\text{-}MeOC_6H_4)^+ \longrightarrow PhOCH_2^+ + MeOPh \qquad (240)$$

$$PhOCH_2^+ + H_2O \xrightarrow{-H^+} PhOCH_2OH \xrightarrow{-PhOH} CH_2O \xrightarrow[-H_2O]{2MeOPh} (p\text{-}MeOC_6H_4)_2CH_2 \qquad (241)$$

On the other hand, the oxidation of methyl-substituted benzenes by LTA in CF_3COOH, which yields mainly biaryls and diarylmethanes, has been proposed to proceed via the formation of an aromatic radical cation as in reaction 242[264]. The radical cation can react with another benzenoid molecule to yield an intermediate which carries an unpaired electron on one ring and a positive charge on the other (equation 243). This

$$ArMe \xrightarrow{-e} ArCH_3^{+\bullet} \qquad (242)$$

$$ArCH_3^{+\bullet} + Ar'H \longrightarrow CH_3Ar^+\text{-}\dot{A}r'H + CH_3\dot{A}r\text{---}Ar^+H \qquad (243)$$

finally leads to the formation of biaryls. Alternatively, the radical cation may undergo further oxidation to form a benzylic cation from which diarylmethanes are derived (reactions 244–246).

$$ArCH_3^{+\bullet} \xrightarrow{-e} ArCH_3^{2+} \xrightarrow{-H^+} ArCH_2^+ \qquad (244)$$

or

$$ArCH_3^{+\bullet} \xrightarrow{-H^+} ArCH_2^{\bullet} \xrightarrow{-e} ArCH_2^+ \qquad (245)$$

$$ArCH_2^+ + Ar'H \xrightarrow{-H^+} ArCH_2Ar' \qquad (246)$$

Another interesting use of the LTA/CF_3COOH oxidant solution is in the alkylation and arylation of fused ring systems such as adamantane[271]. The mechanism of reaction 247, however, has not been clarified.

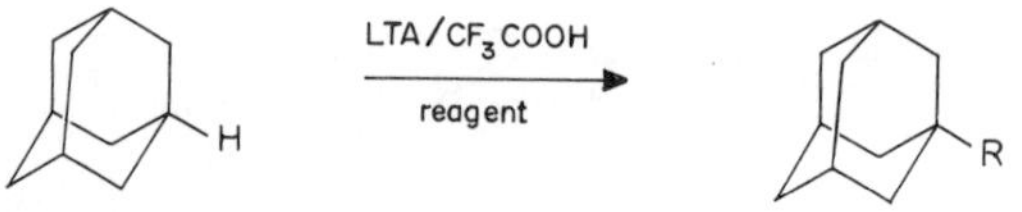

$$\text{Adamantane-H} \xrightarrow[\text{reagent}]{LTA/CF_3COOH} \text{Adamantane-R} \qquad (247)$$

Reagent
anisole/H^+
phenol/H^+
$MeCOCH_2CO_2Et/H^+$

Product
$R = p$-anisyl, 84%
$R = p$-hydroxyphenyl, 81%
$R = -CH(COMe)CO_2Et$ 87%

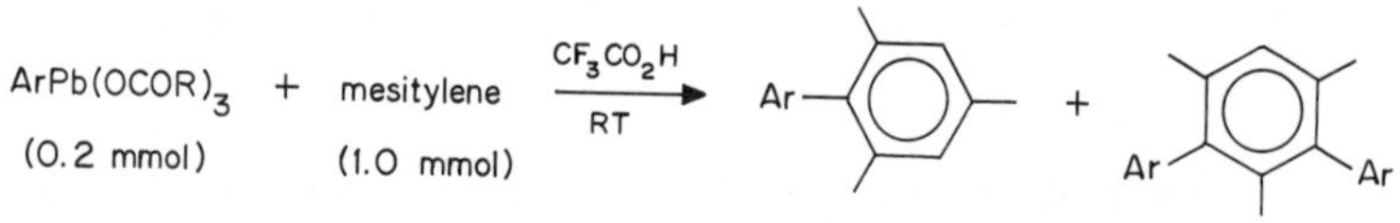

$$ArPb(OCOR)_3 + \text{mesitylene} \xrightarrow[RT]{CF_3CO_2H} Ar\text{-mesityl} + Ar\text{-mesityl-}Ar \qquad (248)$$

(0.2 mmol) (1.0 mmol)

$R = Me, CF_3$

$Ar = p\text{-}FC_6H_4$; 88%   ($<19\%$)
$p\text{-}MeC_6H_4$; 72%

Aryllead tricarboxylate has also been used directly as an arylating agent in trihaloacetic acids[266-270]. Aromatics which are more reactive than toluene generally give good yields of unsymmetrical biaryls, for example equation 248. It has been observed that arylleadtriacetates bearing electron-releasing or electron-withdrawing groups react with electron-rich mesitylene to give more than 70% yields of biaryls. Benzene and toluene, on the other hand, give relatively lower yields (< 10%) of unsymmetricals biaryls with $ArPb(OAc)_3$. The yield, however, may be considerably increased by using $AlCl_3$, $Al(OCOCF_3)_3$, or $Al(OCOC_2F_5)_3$ in place of CF_3COOH. The reaction products generally exhibit high stereospecificity, which is consistent with products being formed via an electrophilic substitution mechanism[260] (equations 249 and 250) and not via a free radical pathway.

$$ArPb(OCOCF_3)_3 \xrightarrow{CF_3CO_2H} ArPb^+(OCOCF_3)_2 \rightleftharpoons Ar^+ + Pb(OCOCF_3)_2 \quad (249)$$

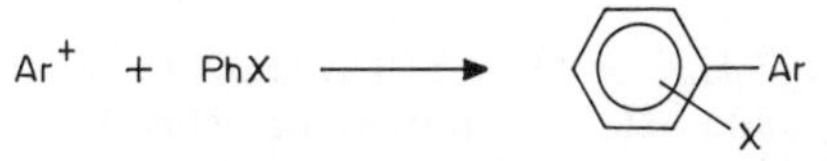

$$Ar^+ + PhX \longrightarrow \underset{X}{\underset{|}{\bigcirc}}-Ar \quad (250)$$

Further, it has been observed that the rate of arylation of polymethylbenzenes with p-$FC_6H_4Pb(OCOMe)_3$ increases in the order p-xylene < hemimellitene, mesitylene < durene, which is also the order of increasing π-donor ability. This result strongly suggests that the rate-determining step in the mechanism involves the formation of a π-

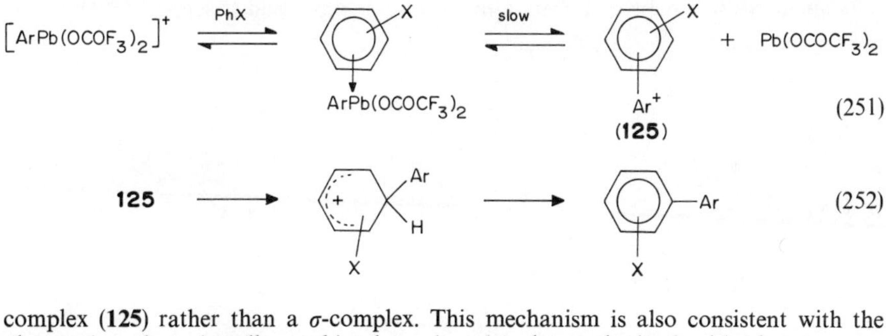

$$[ArPb(OCOF_3)_2]^+ \underset{}{\overset{PhX}{\rightleftharpoons}} \underset{ArPb(OCOCF_3)_2}{\underset{|}{\bigcirc}}^X \underset{}{\overset{slow}{\rightleftharpoons}} \underset{Ar^+}{\underset{|}{\bigcirc}}^X + Pb(OCOCF_3)_2 \quad (251)$$
$$(125)$$

$$125 \longrightarrow \underset{X}{\underset{|}{\bigcirc}}\overset{Ar}{\underset{H}{\diagup}} \longrightarrow \underset{X}{\underset{|}{\bigcirc}}-Ar \quad (252)$$

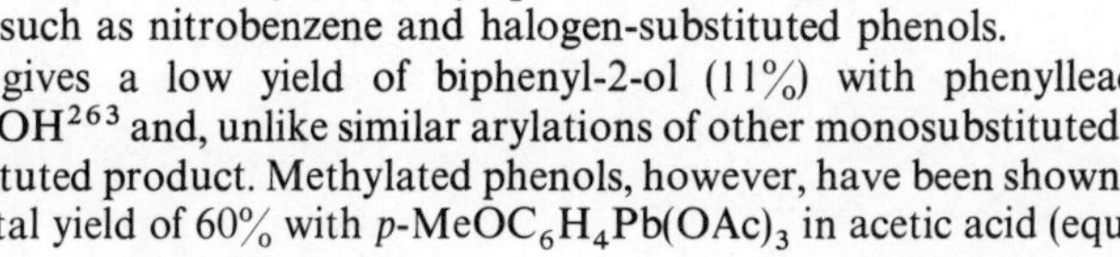

complex (125) rather than a σ-complex. This mechanism is also consistent with the observation that virtually no biaryl product has been obtained with electron-poor aromatics such as nitrobenzene and halogen-substituted phenols.

Phenol gives a low yield of biphenyl-2-ol (11%) with phenyllead triacetate in $CHCl_2COOH$[263] and, unlike similar arylations of other monosubstituted benzenes, gives no p-substituted product. Methylated phenols, however, have been shown to give 126 and 127 in a total yield of 60% with p-$MeOC_6H_4Pb(OAc)_3$ in acetic acid (equation 253). The

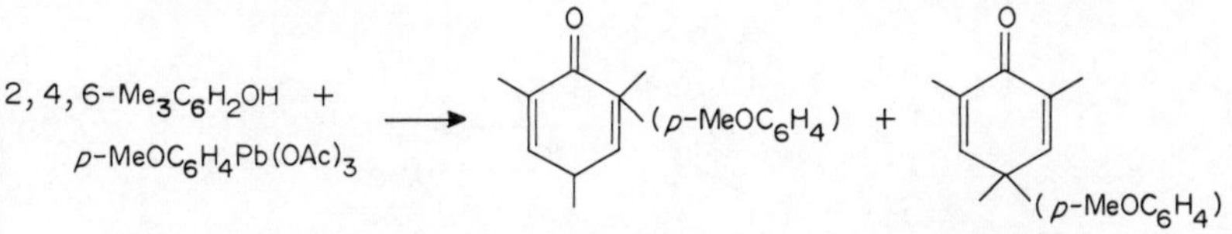

$$2,4,6\text{-}Me_3C_6H_2OH + p\text{-}MeOC_6H_4Pb(OAc)_3 \longrightarrow \underset{(126)}{\text{(}p\text{-}MeOC_6H_4\text{)}} + \underset{(127)}{\text{(}p\text{-}MeOC_6H_4\text{)}} \quad (253)$$

rate of arylation of phenols seems to depend on both the number and positions of the methyl substituents; hence high yields have been obtained for *ortho*-substituted phenols.

Bulky substituents at the *ortho*- and *para*-positions of phenol inhibit the rate of arylation. Whereas 2,4,6-tri-*tert*-butylphenol does not yield C-arylated products, 2,6-di-*tert*-butyl-4-methylphenol reacts very slowly to give a 4-aryldienone (equation 254). Increased acidic

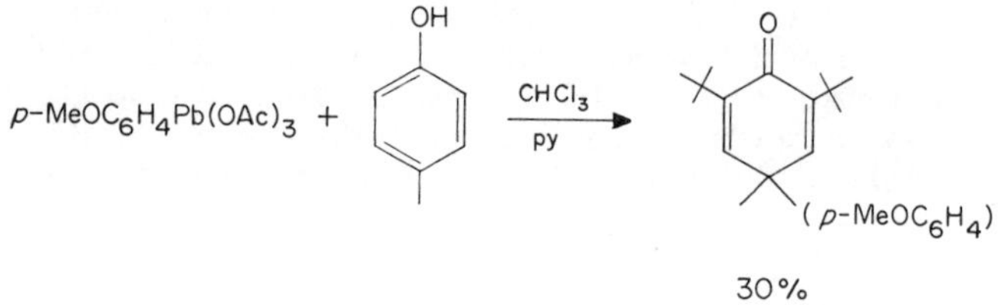

$$p\text{-MeOC}_6\text{H}_4\text{Pb(OAc)}_3 + \text{(p-cresol)} \xrightarrow[\text{py}]{\text{CHCl}_3} \text{(4-aryldienone, } p\text{-MeOC}_6\text{H}_4\text{)} \quad 30\% \tag{254}$$

reaction conditions obtained e.g. by replacing CH_3COOH with CF_3COOH reduced the yield, whereas neutral solvents such as $CHCl_3$ and CH_2Cl_2 improved the yields. Addition of pyridine to the reaction system has been observed to catalyse the arylation reaction presumably by complexing with the $p\text{-MeOC}_6\text{H}_4\text{Pb(OAc)}_3$.

Similarly, the reaction of β-diketones such as acetylacetone gives, with a 1:2 ratio of diketone to $ArPb(OAc)_3$, the monoarylated compounds as the major products (equation 255)[269]. Cyclic diketones such as dimedone, on the other hand, give diarylated products (equation 255a). Similar conditions have also been employed to prepare the anti-inflammatory drug, naproxen, from 6-methoxy-2-naphthyllead triacetate[269a] (equation 256).

$$\text{MeCOCH}_2\text{COMe} + \text{RPb(OAc)}_3 \xrightarrow[\text{py}]{\text{CHCl}_3} \text{MeCOCHRCOMe} + \text{MeCOCR}_2\text{COMe} \tag{255}$$

R = $p\text{-MeOC}_6\text{H}_4$:   44%   2.4%
R = MeC_6H_4 :   32%   17%

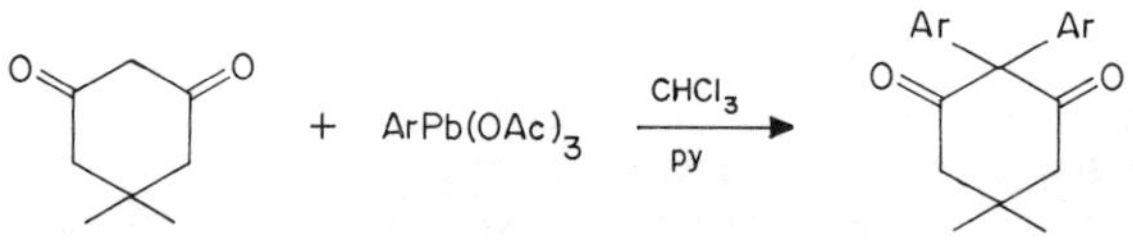

$$\text{(dimedone)} + \text{ArPb(OAc)}_3 \xrightarrow[\text{py}]{\text{CHCl}_3} \text{(diarylated dimedone)} \tag{255a}$$

Ar = $p\text{-MeOC}_6\text{H}_4$: 76%
Ar = $p\text{-MeC}_6\text{H}_4$: 82%

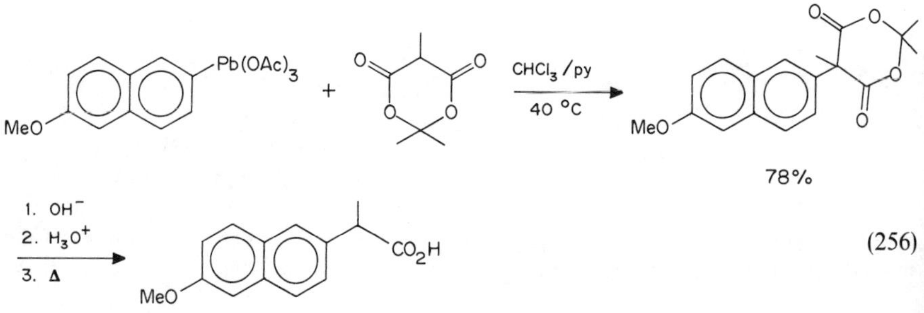

$$\text{MeO-naphthyl-Pb(OAc)}_3 + \text{(Meldrum's acid)} \xrightarrow{\text{CHCl}_3/\text{py}}_{40\,^\circ\text{C}} \text{product} \quad 78\%$$

1. OH^-
2. H_3O^+
3. Δ

$\longrightarrow$ MeO-naphthyl-CH(Me)-CO$_2$H   (256)

Ketones, on conversion to β-ketoesters **(128)** and subsequent reaction with aryllead triacetate, give α-arylated ketones in good yields[269b] (equation 256a).

$$R^1COCH_2R^2 + O{=}C(OMe)_2 \xrightarrow{NaH} R^1COCH(R^2)CO_2Me$$

$$\textbf{(128)}$$

$$\xrightarrow[\substack{CHCl_3, \text{ py} \\ 40\,°C, 1-48\text{ h}}]{ArPb(OAc)_3} R^1COC(R^2)(Ar)CO_2Me \xrightarrow{NaCl}_{dmso} R^1COCH(R^2)Ar \qquad (256a)$$

D. Catalytic Reactions

The rapid development of organometallic chemistry owes a great deal to the industrial application of organometallics as synthetic intermediates such as those in the Wacker process, hydroformylation, and the Ziegler–Natta polymerization of olefins. Although organo-tin and -lead compounds are relative 'newcomers' in catalytic processes and as yet commercially unexploited, they have nevertheless received much interest in the last decade. The potential of these compounds as initiators and co-catalysts in Ziegler-type polymerizations and metathesis is well documented in the patent literature. A comprehensive review on this subject is outside the scope of this chapter. However, a brief survey of some of the more recent investigations in this area is given in this section.

1. Ziegler-type polymerization

When used with transition metal compounds, organo-tins and -leads may generate complex coordinated catalyst structures or active intermediates. Thus, the organo-tin or -lead may function as a substitute for the aluminium alkyl in a Ziegler-type catalyst and produce alkylated transition metal atoms of lower valence which subsequently become the active sites for olefin polymerization. The reactions between organotins and $TiCl_4$, for example, have provided evidence for the formation of organotitanium compounds[272] and for the presence of Ti—H bonds[273].

A particular advantage with catalyst compositions consisting of R_4M (R = Et, Pr, Bu, Ph; M = Sn, Pb) and a transition metal halide or hydroxide is that an aqueous medium may be used[274]. Typically, the components may be mixed in a 1:1 or 4:1 ratio and constitute 0.1–2% by weight in the aqueous medium. A wide range of temperatures (20–150 °C) and pressures (10–25 atm) may be used, the best results being obtained on exclusion of air and with efficient stirring.

Ternary catalyst systems are also commonly used, consisting of a mixture of transition metal compound, aluminium halide, and an organo-tin or -lead compound; variations often lead to specific properties of the polymer products. The polymerization of ethylene, for example, can be effected by catalysts such as $[Co(acac)_2]/AlCl_3/R_4Sn$[275], $VCl_4/EtAlCl_2/Ph_4Sn$[276], and $VOCl_3/AlBr_3/Ph_4Sn$[277] and also in systems where R_4Sn may be substituted by organotin oxides[278] or chromates[279].

The function of the tin compound has been identified in such systems as one of alkylating or arylating the aluminium halide[276]. In the case of Et_4Pb, the aluminium halide component of the ternary system may be replaced by CCl_4 or a metal such as Mg, Al, or Zn[280].

Both two- and three-component catalyst systems, in general, have been noted to yield highly crystalline, high-melting, tough polymers characteristic of products prepared by Ziegler-type catalysts. Itoi and Nishida[281,282] obtained high molecular weight polyethylene of increased bulk density and softening point from the polymerization of ethylene using $TiCl_3/AlCl_3$ catalyst activated by R_2SnH_2 or R_3SnH (R = Et, Pr, Bu, or Ph), with heptane as solvent. Catalytic mixtures containing $AlCl_3/VCl_5/Ph_4Sn$ in hexane have been reported[283] to polymerize ethylene even at vanadium concentrations as low as 1 ppm. The polyethylene obtained is linear, possesses a narrow range of molecular weight distribution and contains less than one unsaturated group for every 10 000 carbon atoms.

Increased catalytic activity and improved yields of the products have been engendered in several cases by the addition of organo-tin and -lead compounds. Thus, the yield of polypropylene was increased from 0.9 to 21.6% on treating the catalyst $Cr_2O_3/SiO_2/CO$ with Et_4Sn^{284}. Likewise, poly(vinyl fluoride) has been prepared at relatively low temperatures (ca. 20 °C) and pressures when the catalyst containing $AgNO_3$ (or acetate) was activated with Et_4Pb^{285}. It is interesting that alkyltins formed in situ from dialkylmagnesium and tin halides have been found to increase productivity of polyolefins[286]. Further, the polyolefins obtained have been shown to be less soluble in heptane. The catalyst $Mg/n-C_5H_{11}Cl/TiCl_4$ activated with an organoaluminium compound gave 1820 g of polypropylene per gram of catalyst per hour, whereas the same catalyst previously milled with $SnCl_4$ gave an improved yield of $3110 g^{287}$.

The polymerization of lactones to form polyesters is also facilitated by the use of an organo-tin and -lead compound as a component catalyst. R_4Sn, R_2SnX_2, R_2SnO, or an organic salt of lead (salicylate, benzoate) are particularly important catalysts because of their ability to promote the formation of virtually colourless polyesters and in a very short reaction time, in certain cases from weeks to a few hours[278]. This method is also useful for copolymerizing lactones with vinyl polymers to yield polymers represented by formula 129. Other copolymerizations using organotin compounds include those between

$$-[CH_2CH(CH_2CH_2)_n]-$$
$$|$$
$$OH$$

(129)

ethylene and styrene[288], buta-1,3-diene[289], and isoprene[290]. Polypropylene, for example, has been graft polymerized to butyl methacrylate or methacrylic acid in the presence of dibenzyltin oxide[291] and polyethylene has been modified with divinylbenzene using tert-butyltin oxide as catalyst. This is useful because modified polyolefins have improved impact strength and tensile strength. It has also been noted that these graft polymerizations do not occur in the absence of tetravalent organotin compounds.

2. Olefin metathesis†

There has been much controversy as to whether olefin metathesis reactions proceed via a pairwise or non-pairwise mechanism, and it is now generally accepted that the latter is the more probable mechanism[292]. This non-pairwise mechanism involves the formation of a carbene—metal species (130), which then interacts with the olefin to form a metallocyclobutane intermediate, 131, as depicted in Scheme 38^{293}.

Tetraalkyltin compounds appear to be instrumental in the initiation step leading to the formation of the carbenoid species[294]. Grübbs and Hoppin[295] investigated the metathesis reaction of deca-2,8-diene (132) using Me_4Sn/WCl_6 as catalyst and found that the intial product was propene followed by the usual metathesis reaction products (equation 257). The use of $(CD_3)_4Sn$ yielded perdeuteriated methane and ethylene, which was explained by the authors in terms of the formation of a carbenoid species, as shown in equation 258. Careful deuterium labelling studies using $[1,1,1,10,10,10-d_6]$deca-2,8-diene led them

† For a recent review see J. C. Mol, J. Mol. Catal., 15, 35 (1982)

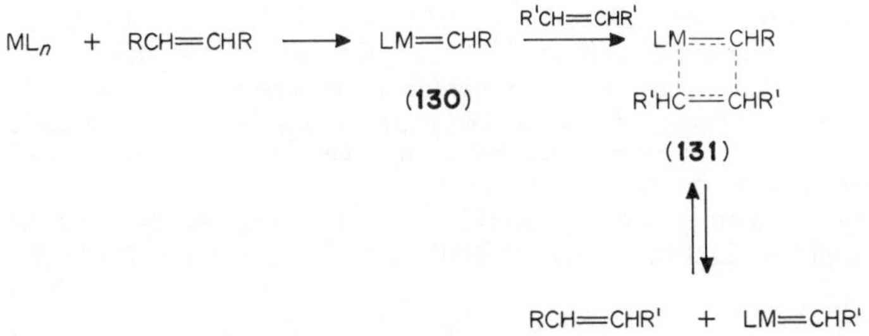

1. Carbon—carbon bond formation

$$ML_n + RCH{=}CHR \longrightarrow LM{=}CHR \xrightarrow{R'CH=CHR'} \begin{array}{c} LM\text{---}CHR \\ | \quad\quad | \\ R'HC\text{---}CHR' \end{array}$$

$$(130) \quad\quad\quad (131)$$

$$\updownarrow$$

$$RCH{=}CHR' + LM{=}CHR'$$

SCHEME 38

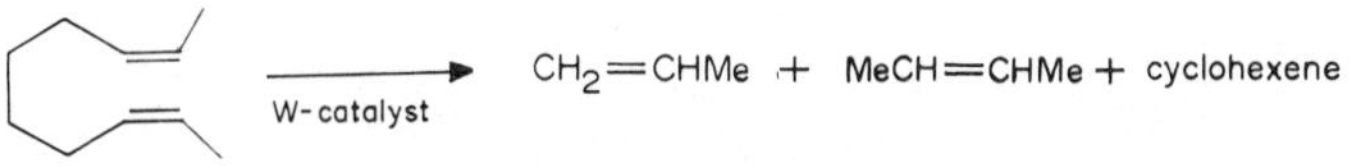

$$\text{(cyclooctadiene)} \xrightarrow{\text{W-catalyst}} CH_2{=}CHMe + MeCH{=}CHMe + \text{cyclohexene} \quad (257)$$

(132)

$$WCl_6 + (CD_3)_4Sn \longrightarrow [Cl_nW{=}CD_2] \longrightarrow CD_4 + CD_2{=}CD_2 + \text{W-catalyst} \quad (258)$$

further to propose that initiation of the metathesis proceeded via the set of reactions indicated in Scheme 39.

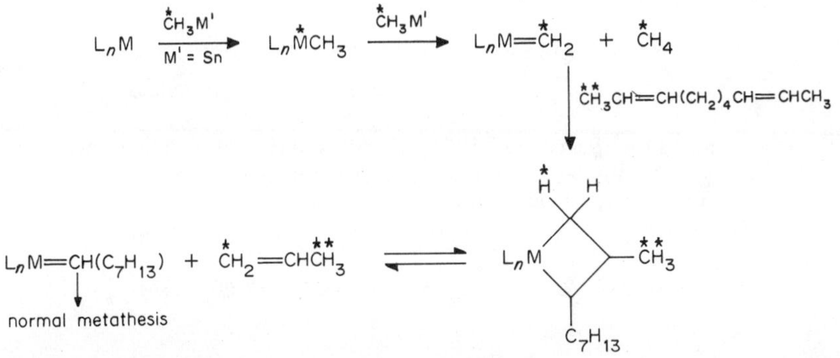

$$L_nM \xrightarrow[M'=Sn]{\overset{*}{C}H_3M'} L_n\overset{*}{M}CH_3 \xrightarrow{\overset{*}{C}H_3M'} L_nM{=}\overset{*}{C}H_2 + \overset{*}{C}H_4$$

$$\downarrow \overset{**}{C}H_3CH{=}CH(CH_2)_4CH{=}CHCH_3$$

$$L_nM{=}CH(C_7H_{13}) + \overset{*}{C}H_2{=}CH\overset{**}{C}H_3 \rightleftarrows L_nM\begin{array}{c}\overset{*}{H}\quad H\\ \square \\ C_7H_{13}\end{array}\overset{**}{C}H_3$$

↓ normal metathesis

SCHEME 39

The stereospecificity of metathesis products is dependent on several factors, such as the nature of ligands attached to the transition metal catalyst [as in-$W(CO)_5$, WF_6, etc.], steric influences of the reacting olefins, and in some cases the polarity of the solvent system. Basset et al.[296] noted that diverse catalyst systems did not bear any correlation with the stereospecificity of the Z- and E-products and as such the ligand composition about the transition metal was not significant. On the other hand, Dall'Asta[297] found that the reaction of (Z)-pent-2-ene led selectively to the formation of E-olefinic products on modifying the ligands on the transition metal catalyst. Yet (Z)-pent-2-ene afforded butenes and hexenes having 95% Z-structure when $[(CO)_5W{=}CPh_2]$ catalyst was used[298]. It therefore appears that, whereas stereospecificity is dependent on steric factors, the specific catalyst effects cannot be ignored. Solvent influences on polymer configuration have also been observed. Thus, Masuda et al.[299] and Hasegawa[300] in their studies on the polymerization of phenylpropyne have noted that a decrease in solvent polarity leads to an increase in polymer having Z content. The same has also been observed upon replacing WCl_6 with $MoCl_5$[301]; attempts to polymerize phenylpropyne in the presence of $MoCl_5$ or WCl_6 without organotin compounds failed.

Organotin compounds are important co-catalysts in self-metathesis reactions (equations 259 and 260), and also in cross-metathesis reactions (equation 261) of olefins.

$$R_1\text{CH=CH}R_2 \rightleftharpoons R_1\text{CH=CH}R_1 + R_2\text{CH=CH}R_2 \quad (259)$$

$$CH_2\text{=CH}(CH_2)_nY \xrightarrow{WCl_6/R_4Sn} Y(CH_2)_n CH\text{=}CH(CH_2)_nY \quad (260)$$
(major product)

$$\text{1-octene} + Z\text{-oct-2-ene} \xrightarrow[PhCl]{WCl_6/Ph_4Sn^{302}} \text{non-2-ene} + \text{6-tridecene} + \text{7-tetradecene} \quad (261)$$
(or E) (major product)

A number of workers[303–307] have reported the use of WCl_6/Me_4Sn homogeneous catalyst for cross-metathesis reactions of aliphatic unsaturated acids, esters, and polymers. Thus, methyl oleate reacts with hex-3-ene at 70 °C to give equal amounts of octadec-9-ene and dimethyl ester (equation 262).

$$Me(CH_2)_7CH\text{=}CH(CH_2)_7CO_2Me \longrightarrow Me(CH_2)_7CH\text{=}CH(CH_2)_7Me + MeO_2C(CH_2)_7CH\text{=}CH(CH_2)_7CO_2Me \quad (262)$$

It has been noted that the catalyst WCl_6/Me_4Sn is unique in the homo-metathesis of fatty acid esters in that similar systems using other tetraalkyltins are inactive[303]. The co-metathesis of cycloocta-1,5-diene with dihydromuconic methyl ester using this catalyst gave an α,ω-dimethylcarboxylate–polybuta-1,4-diene-like polyene (equation 263)[307]. The

$$C_8H_{12} + MeO_2CCH_2CH\text{=}CHCH_2CO_2Me \xrightarrow[60\ °C]{PhCl} MeO_2CCH_2CH[\text{=}CHCH_2CH_2CH]_{n-1}CHCH_2CO_2Me \quad (263)$$

observed distribution of the lower metathesis product does not correspond to the statistical one evaluated by considering dihydromuconic methyl ester as a transfer site and cyclooctadiene as two propagation sites. Hence the formation of the trienic molecule ($n = 3$) is much more favoured than that of the dienic molecule ($n = 2$) and also the metathesis involves preferentially the double bonds in the neighbourhood of the ester group.

The effects of substituents on the olefins undergoing metathesis have been investigated for both homogeneous and heterogeneous catalyst systems[308–310]. The catalyst WCl_6/Me_4Sn ($Sn/W = 2$), which gives the best results for cross-metathesis of hex-2-ene with methyl oleate, has featured strongly in such investigations[308]. It has been observed that electron-withdrawing substituents close to the olefinic double bond inhibit metathesis cleavage of the W=C bond. Hence no cross-metathesis products have been observed for hex-2-ene and $RCH=CH(CH_2)_nCO_2R'$ when $n = 0$[311]. The same has been observed for the corresponding halides $RCH=CH(CH_2)_nX$, while this inhibition effect becomes even more pronounced in the cyano compound $RCH=CH(CH_2)_nCN$, where no metathesis product has been observed for $n < 4$[312].

Ring-opening polymerization of cyclopentene to yield a polypentenamer[313] as product can also be promoted by organotin compounds at the initiating step, which requires the disproportionation of the transition metal–alkyl complex[314] (equation 264). The ring-

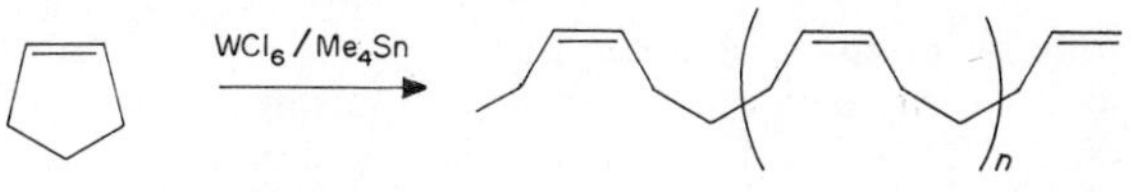

$$\text{cyclopentene} \xrightarrow{WCl_6/Me_4Sn} \text{polypentenamer} \qquad (264)$$

opening step is well demonstrated by the metathesis of partially fluorinated bicyclo[2.2.1]hept-2-enes and bycyclo[2.2.1]hepta-2,5-dienes in the presence of Ph_4Sn/WCl_6 catalyst (Sn/W = 2)[315].

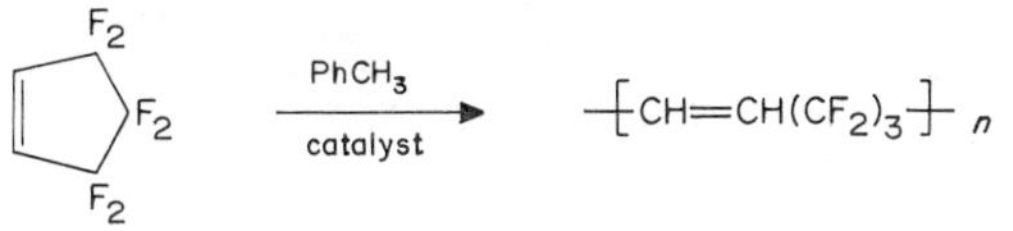

$$\xrightarrow{PhCH_3, \text{catalyst}} \{CH=CH(CF_2)_3\}_n \qquad (265)$$

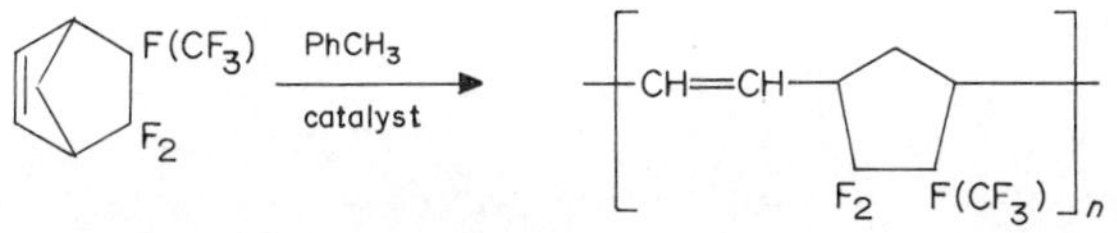

$$\xrightarrow{PhCH_3, \text{catalyst}} \{CH=CH-\text{ring}(F_2, F(CF_3))\}_n \qquad (266)$$

Cyclic olefins such as cyclopentenes and cyclooctenes can react with acyclic olefins in the presence of Bu_4Sn or Bu_4Pb co-catalyst to give telomers of the cyclic olefins containing terminal groups of the acyclic olefins[293] (equation 267). The acyclic olefins act as chain

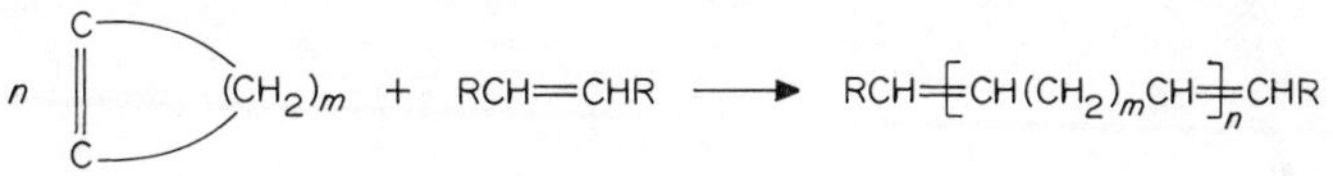

$$n \text{ cycloolefin}(CH_2)_m + RCH=CHR \longrightarrow RCH\{CH(CH_2)_mCH\}_n CHR \qquad (267)$$

scission agents in cycloolefin polymerization and hence are used as molecular weight regulators. But-1-ene, for example, has been used as a regulator in the formation of polypentenamer rubber which contains 92% Z-double bonds[316].

A naturally occurring lactone, ambrettolide (**133**), has been found to undergo ring-opening metathesis with WCl_6/Me_4Sn as catalyst (Sn/W = 5) to give a high molecular weight, unsaturated polyester[317] (equation 268). Equations 268 and 269 represent a new

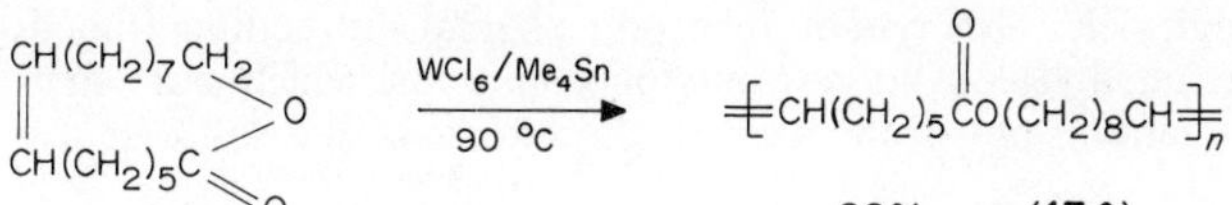

$$\underset{(\textbf{133})}{\underset{CH(CH_2)_5C}{CH(CH_2)_7CH_2}}\!\!>\!\!O \xrightarrow[90\,°C]{WCl_6/Me_4Sn} \{CH(CH_2)_5\overset{O}{\overset{\|}{C}}O(CH_2)_8CH\}_n \qquad (268)$$

$$80\% \quad (\textbf{134})$$

$$\textbf{133} + \text{cyclooct-1,5-diene} \xrightarrow[90\,°C]{WCl_6/Me_4Sn}$$

$$\{CH(CH_2)_2CH=CH(CH_2)_2CH=CH(CH_2)_5CO_2CH_2(CH_2)_7CH\}_n \qquad (269)$$

$$(\textbf{135})$$

synthetic route for the production of industrially useful linear vulcanizable polyester rubber, **135**. In addition, the lactone **133** may be inserted into a polyalkene such as (Z)-polybuta-1,4-diene to yield polyester-modified polyalkenylenes (equation 270).

$$\mathbf{133} + \text{Me}(CH_2)_2CH=CH(CH_2)_2\text{Me} \rightleftharpoons \text{Me}(CH_2)_2CH=CH(CH_2)_5CO_2(CH_2)_8CH=CH(CH_2)_2\text{Me} \quad (270)$$

Some novel metathesis reactions using Me_4Sn/WCl_6 as catalyst have also been reported for the synthesis of relatively high molecular weight polyalkylenes containing pendant ester groups (equations 271–273)[318]. The molecular weight and percentage conversion appear to increase with increasing strain in the monomer, and it is noteworthy that these ring-opening polymerization reactions leave the cyclopropane ring intact.

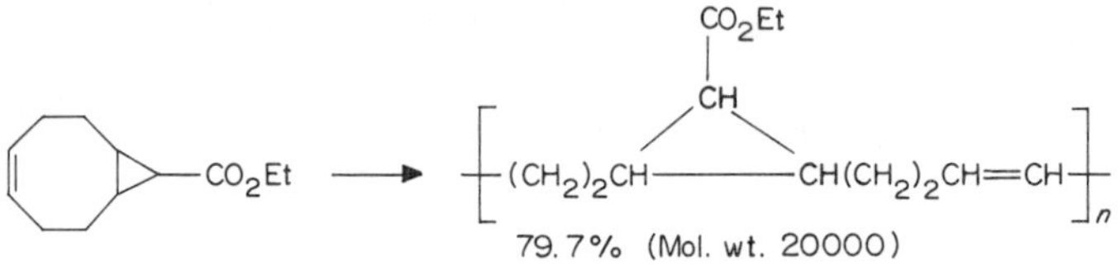

$$\longrightarrow [-(CH_2)_2CH-CH(CH_2)_2CH=CH-]_n \quad (271)$$
79.7% (Mol. wt. 20000)

$$\longrightarrow [-(CH_2)_2CH=CH(CH_2)_2CH-CH(CH_2)_2CH=CH-]_n \quad (272)$$
35.4% (Mol. wt. 5000)

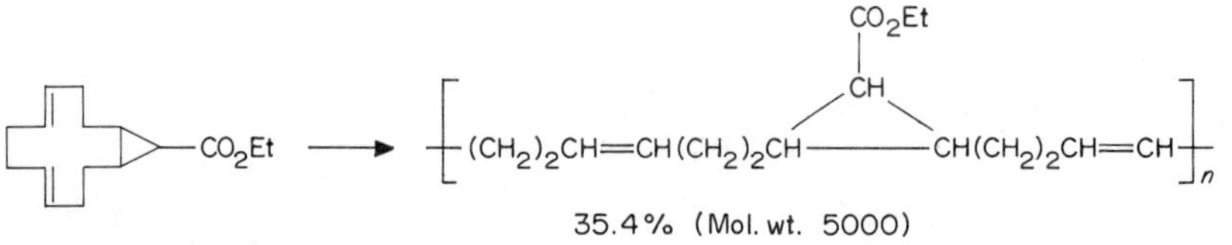

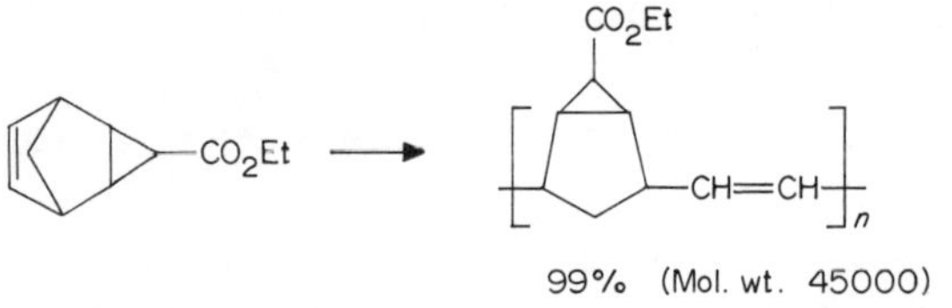

$$\longrightarrow [-CH=CH-]_n \quad (273)$$
99% (Mol. wt. 45000)

In non-catalytic roles, organotin compounds are also known to improve the processing characteristics of the polymer such as in organolithium-catalysed polymerizations[319]. A new role for the tin atom has recently been devised by Schumann et al.[320] in catalysts of the type $[(CO)_3Ni(PBu^t_2SnClMe_2)]$, which readily cyclotrimerize alkynes into benzene derivatives in high yield. The organotinphosphino ligand with sodium trimethylsilanolate gives dimethyl(trimethylsiloxy)(di-*tert*-butylphosphino)tin, which is the first model substance for an organotin phosphine fixed on the surface of aerosol.

3. Other

Examples of non-polymerization reactions where organo-tin and -lead compounds are used catalytically to promote C—C bond formation are sparse. The application of organotin alkoxides in catalysing the addition reactions of active methylene compounds to acrylonitrile and isocyanates has already been mentioned (Section VI.C). A further example appears in the lead tetraalkyl-catalysed condensation of unsaturated organic compounds of formula $R_2C=CHX$ with alkylaromatic hydrocarbons, $ArCR_2H$, containing at least one hydrogen atom directly bonded to the α-carbon (reaction 274)[321]. The

1. Carbon—carbon bond formation

$$ArCHR^1R^2 + XCH=CR^3R^4 \longrightarrow ArCR^1R^2CR^3R^4CH_2X \text{ or } ArCR^1R^2CHXCHR^3R^4 \quad (274)$$

reactants and catalyst (preferably supported on charcoal or granular silica or alumina) are contacted in an autoclave and reasonable yields of the products may be obtained by operating at temperatures in the range 50–400 °C and pressures of 1–200 atm. Equations 275 and 276 illustrate the scope of such side-chain alkylation reactions. A free-radical

$$p\text{-R}\text{—}C_6H_4CMe_2H + CH_2\text{=}CHCN \longrightarrow p\text{-R}\text{—}C_6H_4CMe_2(CH_2)_2CN \quad (275)$$

R = C_9 or C_{12} group

$$RCHMe_2 + CH_2\text{=}CHNO_2 \longrightarrow RC(Me)_2CH_2CH_2NO_2 \quad (276)$$

R = (2-thienyl)

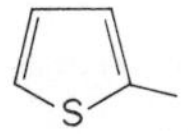

pathway initiated by alkyl radicals from R_4Pb has been implicated for the above reactions, and also for the following cyclization reaction, which proceeds only in the presence of Et_4Pb^{322}.

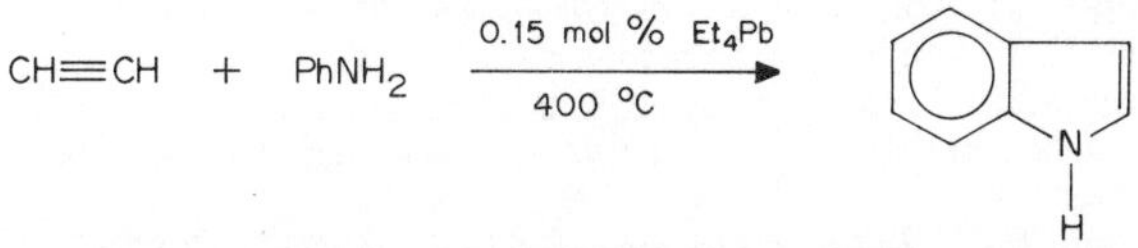

$$CH\equiv CH + PhNH_2 \xrightarrow[400\ °C]{0.15\ mol\ \%\ Et_4Pb} \text{indole} \quad (277)$$

VIII. ACKNOWLEDGEMENT

The authors thank The Institute of Advanced Studies, University of Malaya, for facilities and support in connection with the preparation of this chapter.

IX. REFERENCES

1. For a discussion of the definitions and the terminology used in this chapter, see Y. Izumi and A. Tsai, *Stereodifferentiating Reactions*, Academic Press, New York, 1977; V. Gold, *Pure Appl. Chem.*, **51**, 1725 (1979).
2. E. Negishi, *Organometallics in Organic Synthesis*, Vol. 1, Wiley, New York, 1980.
3. J. J. Zuckerman (Ed.), *Organotin Compounds: New Chemistry and Applications*, Advances in Chemistry Series, No. 157, American Chemical Society, Washington, DC. 1976.
4. B. J. Aylett, *Organometallic Compounds, Vol. 1, Part 2, Groups IV and V*, 4th ed., Chapman and Hall, London, 1979.
5. M. Pereyre and J. C. Pommier, *J. Organomet. Chem. Libr.*, **1**, 161 (1976).
6. M. Pereyre and J. P. Quintard, *Pure Appl. Chem.*, **53**, 2401 (1981).
7. N. Ono, H. Miyake, R. Tamura, and A. Kaji, *Tetrahedron Lett.*, **22**, 1705 (1981); N. Ono, H. Miyake, A. Kamimura, N. Tsukui, and A. Kaji, *Tetrahedron Lett.*, **29**, 2957 (1982).
8. Y. Ueno, H. Sano, and M. Okawara, *Tetrahedron Lett.*, **21**, 1767 (1980); Y. Ueno, H. Sano, S. Aoki, and M. Okawara, *Tetrahedron Lett.*, **22**, 2675 (1981).
9. E. Keinan and P. A. Gleize, *Tetrahedron Lett.*, **23**, 477 (1982).
10. M. Seikina, A. Kune, and T. Hatta, *J. Chem. Soc., Chem. Commun.*, 969 (1981).
11. D. H. R. Barton, W. Hartwig, R. S. H. Motherwell, W. B. Motherwell, and A. Strange, *Tetrahedron Lett.*, **23**, 2019 (1982).
12. P. Four and F. Guibe, *J. Org. Chem.*, **46**, 4439 (1981); *Tetrahedron Lett.*, **23**, 1825 (1982).
13. M. E. Jung and L. A. Light, *Tetrahedron Lett.*, **23**, 3851 (1982).
14. S. David and A. Thieffry, *J. Chem. Soc., Perkin Trans. 1*, 1568 (1979).
15. J. M. Reuter and R. G. Saloman, *Tetrahedron Lett.*, **35**, 3199 (1978).

16. T. Ogawa, K. Katano, and M. Matsui, *Carbohydr. Res.*, **60**, C13 (1978); M. A. Nashed, *Carbohydr. Res.*, **60**, 200 (1978).
17. P. Didier and J. C. Pommier, *J. Organomet. Chem.*, **150**, 203 (1978).
18. T. L. Su, R. S. Klein, and J. J. Fox, *J. Org. Chem.*, **47**, 1506 (1982).
19. A. K. Sawyer (Ed.), *Organotin Compounds*, Vols. 1–3, Marcel Dekker, New York, 1971 and 1972.
20. R. C. Poller, *The Chemistry of Organotin Compounds*, Logos Press, London, 1970; *Rev. Silicon, Germanium, Tin, Lead Cmpd.*, **3**, 243 (1978).
21. C. Eaborn, *J. Organomet. Chem.*, **100**, 43 (1975).
22. M. H. Abraham, Electrophilic Substitution at a Saturated Carbon Atom in *Comprehensive Chemical Kinetics* (Eds. C. H. Bamford and C. F. H. Tipper), Vol. 12, pp. 1–243 Elsevier, Amsterdam, 1973.
23. A. G. Davies and P. J. Smith, in *Comprehensive Organometallic Chemistry* (Ed. G. Wilkinson), Pergamon Press Oxford, 1982, Chapter 11, pp. 519–627; *Adv. Inorg. Chem. Radiochem.*, **23**, 1 (1980).
24. H. Shapiro and F W. Frey, *The Organic Compounds of Lead*, Wiley–Interscience, New York, 1968.
25. H. C. Clark and C. J. Willis, *J. Am. Chem. Soc.*, **82**, 1888 (1960).
26. D. Seyferth, J. Y. P. Mui, M. E. Gordon, and J. M. Burlitch, *J. Am. Chem. Soc.*, **87**, 681 (1965).
27. D. Seyferth, F. M. Armbrecht, B. Prokai, and R. J. Cross, *J. Organomet. Chem.*, **6**, 573 (1966).
28. D. Seyferth and F. M. Armbrecht, *J. Am. Chem. Soc.*, **91**, 2616 (1969).
29. D. Seyferth and R. L. Lambert, *J. Organomet. Chem.*, **91**, 31 (1975).
30. L. C. Willemsens and G. J. M. van der Kerk, *J. Organomet. Chem.*, **23**, 471 (1970).
31. D. Seyferth, G. J. Murphy, R. L. Lambert, and R. E. Mammarella, *J. Organomet. Chem.*, **90**, 173 (1975).
32. C. M. Werner and J. G. Noltes, *J. Organomet. Chem.*, **24**, C4 (1970).
33. D. Seyferth, F. M. Armbrecht, and B. Schneider, *J. Am. Chem. Soc.*, **91**, 1954 (1969).
34. D. Seyferth and D. C. Mueller, *J. Organomet. Chem.*, **25**, 293 (1970).
35. R. A. Olofson, D. H. Hoskin, and K. D. Lotts, *Tetrahedron Lett.*, **19**, 1677 (1978).
36. R. Kriegesmann, *Dissertation*, Univ. Münster (1980).
37. U. Schollkopf and N. Rieber, *Angew Chem., Int. Ed. Engl.*, **6**, 884 (1967).
38. J. P. Quintard, B. Elissondo, and M. Pereyre, *J. Organomet. Chem.*, **212**, C31 (1981).
39. M. Birkhahn, R. Hohlfeld, W. Massa, R. Schmidt, and J. Lorberth, *J. Organomet. Chem.*, **192**, 47 (1980).
40. R. Gruning and J. Lorberth, *J. Organomet. Chem.*, **129**, 55 (1977).
41. J. G. Noltes, F. Verbeek, and H. M. J. C. Creemers, *Organomet. Chem. Synth.*, **1**, 57 (1970/71).
42. M. Pereyre, G. Colin, and J. Valade, *C. R. Acad. Sci., Ser. C*, **264**, 1204 (1967).
43. C. Furet, C. Servens, and M. Pereyre, *J. Organomet. Chem.*, **102**, 423 (1975).
44. B. Renger, H. Hugel, W. Wykypiel, and D. Seebach, *Chem. Ber.*, **111**, 2630 (1978).
45. H. Weichmann, B. Ochsler, I. Duchek, and A. Tzschach, *J. Organomet. Chem.*, **182**, 465 (1979).
46. S. V. Ponomarev and I. F. Lutsenko, *Zh. Obsch. Khim.*, **34**, 3450 (1964).
47. K. Konig and W. P. Neumann, *Tetrahedron Lett.*, 495 (1967).
48. J. C. Pommier and M. Pereyre, in *Organotin Compounds: New Chemistry and Applications* (Ed. J. J. Zuckerman), Advances in Chemistry Series, No. 157, American Chemical Society, Washington, DC, 1976, Ch. 6, pp. 82–112.
49. M. Pereyre, D. Bellegarde, J. Mendelsohn, and J. Valade, *J. Organomet. Chem.*, **11**, 97 (1968).
50. A. J. Leusink, H. A. Budding, and W. Drenth, *J. Organomet. Chem.*, **13**, 163 (1968).
51. J. G. Noltes, H. M. J. C. Creemers, and G. J. M. van der Kerk, *J. Organomet. chem.*, **11**, P21 (1968).
52. S. V. Ponamarev, E. V. Machigin, and I. F. Lutsenko, *Zh. Obsch. Khim.*, **36**, 548 (1966).
53. N. I. Savel'eva, A. S. Kostyuk, Y. I. Baukov, and I. F. Lutsenko, *Zh. Obsch. Khim.*, **41**, 2339 (1971).
54. D. J. Peterson, *Organomet. Chem. Rev., Sect. A*, **7**, 295 (1972).
55. D. Seyferth, F. M. Armbrecht and E. M. Hansen, *J. Organomet. Chem.*, **44**, 299 (1973).

56. D. Seyferth and S. B. Andrews, *J. Organomet. Chem.*, **30**, 151 (1971).
57. W. C. Still, *J. Am. Chem. Soc.*, **100**, 1481 (1978).
58. T. Kauffmann, R. Kriegesmann, B. Altpeter, and F. Steinseifer, *Chem. Ber.*, **115**, 1810 (1982); T. Kauffmann, *Angew. Chem.*, **89**, 900 (1977).
59. D. Seebach and R. Burstinghaus, *Angew. Chem., Int. Ed. Engl.*, **14**, 57 (1975).
60. B. T. Grobel and D. Seebach, *Synthesis*, 357 (1977).
61. T. Kauffmann, K. J. Echsler, A. Hamsen, R. Kriegesmann, F. Steinseifer, and A. Vahrenhorst, *Tetrahedron Lett.*, **4391** (1978).
62. G. R. Petit and E. E. van Tamelen, *Org. React.*, **12**, 356 (1962).
63. V. Georgian, R. Harrison, and N. Grubisch, *J. Am. Chem. Soc.*, **81**, 5834 (1959); W. H. Baarschers and T. L. Loh, *Tetrahedron Lett.*, 3483 (1971).
64. T. Mukaiyama, K. Narasaka, M. Maekawa, and M. Furusato, *Bull. Chem. Soc. Jpn.*, **44**, 2285 (1971).
65. T. Kauffmann, R. Kriegesmann, and A. Hamsen, *Chem. Ber.*, **115**, 1818 (1982).
66. M. Ochiai, S Tada, K. Sumi, and E. Fujita, *Tetrahedron Lett.*, **23**, 2205 (1982).
67. D. Seebach and N. Meyer, *Angew. Chem., Int. Ed. Engl.*, **15**, 438 (1976).
68. B. T. Gröbel and D. Seebach, *Chem. Ber.*, **110**, 867 (1977).
69. K. Oshima, K. Shimoji, H. Takahashi, H. Yamamoto, and H. Nozaki, *J. Am. Chem. Soc.*, **95**, 2694 (1973).
70. E. Shaumann and W. Walter, *Chem. Ber.*, **107**, 3562 (1974).
71. P. Vermeer, J. Meijer, and C. Eylander, *Recl. Trav. Chim. Pays-Bas*, **93**, 240 (1974).
72. R M. Carlson, R. W. Jones, and A. S. Hatcher, *Tetrahedron Lett.*, 1741 (1975).
73. J. C. Lahournere and J. Valade, *C. R. Acad. Sci., Ser. C*, 270 (1970).
73a. A. J. Pratt and E. J. Thomas, *J. Chem. Soc., Chem. Commun.*, 1115 (1982).
74. W. C. Still and C. Sreekumar, *J. Am. Chem. Soc.*, **102**, 1201 (1980).
75. D. E. Seitz and A. Zapata, *Tetrahedron Lett.*, **21**, 3451 (1980).
76. A. Rensing, K.-J. Echsler, and T. Kauffman, *Tetrahedron Lett.*, **21**, 2807 (1980).
77. R. C. Weiss and E. I. Snyder, *J. Org. Chem.*, **36**, 403 (1971).
78. Y. Torisawa, M. Shibasaki, and S. Ikegami, *Tetrahedron Lett.*, **22**, 2397 (1981).
79. W. C. Still, J. H. McDonald, D. B. Collum, and A. Mitra, *Tetrahedron Lett.*, 593 (1979); W. C. Still and A. Mitra, *J. Am. Chem. Soc.*, **100**, 1927 (1978).
80. R. Goswami and D. E. Corcoran, *Tetrahedron Lett.*, **23**, 1463 (1982).
80a. J Meinwald, S. Knapp, T. Tatsuoka, J. Finar, and J. Clardy, *Tetrahedron Lett.*, 2247 (1977).
81. R. D. Taylor and J. L. Wardell, *J. Organomet. Chem.*, **94**, 15 (1975).
82. G J. M. van der Kerk, J. G. Noltes, and J. G. A. Luijten, *J. Appl. Chem.*, **7**, 356 (1957).
83. D. D. Davis and C. E. Gray, *J. Org. Chem.*, **35**, 1303 (1970).
84. A. W. P. Jarvie, *Organomet. Chem. Rev., Sect. A*, **6**, 153 (1970).
85. J. L. Wardell in *Organotin Compounds: New Chemistry and Applications* (Ed. J. J. Zuckerman), Advances in Chemistry Series, No. 157, American Chemical Society, Washington, DC, 1976, Ch. 7, pp. 113–122.
86. G. J. M. van der Kerk and J. G. Noltes, *J. Appl. Chem.*, **9**, 179 (1959).
87. A. G. Davies, in *Organotin Compounds: New Chemistry and Applications* (Ed. J. J. Zuckerman), Advances in Chemistry Series, No. 157, American Chemical Society, Washington, DC, 1976, Ch. 2, pp. 26–40.
88. D. Seyferth, S. S. Washburne, C. J. Attridge, and K. Yamamoto, *J. Am. Chem. Soc.*, **92**, 4405 (1970).
89. D. Seyferth, K. R. Wursthorn, and R. E. Mammarella, *J. Organomet. Chem.*, **25**, 179 (1979).
90. D. Seyferth and K. R. Wursthorn, *J. Organomet. Chem.*, **182**, 455 (1979).
91. S. J. Hannon and T. G. Traylor, *J. Chem. Soc., Chem. Commun.*, 631 (1975).
92. E. J. Corey and C. U. Kim, *J. Am. Chem. Soc.*, **94**, 7586 (1972).
93. E. J. Corey and J. W. Suggs, *Tetrahedron Lett.*, 2647 (1975).
94. Y. Ueno, M. Ohta, and M. Okawara, *Tetrahedron Lett.*, **23**, 2577 (1982).
95. W. C. Still, *J. Am. Chem. Soc.*, **99**, 4836 (1977).
96. J. W. Patterson, Jr., and J. H. Fried, *J. Org. Chem.*, **39**, 2506 (1974).
97. R. Goswami, *J. Am. Chem. Soc.*, **102**, 5973 (1980).
98. S. Teratake, *Chem. Lett.*, 1123 (1974).
99. S. Teratake and S. Morikawa, *Chem. Lett.*, 1333 (1975).

100. D. D. Davis, R. L. Chambers, and H. T. Johnson, *J. Organomet. Chem.*, **25**, C13 (1970).
100a. D. D. Davis and R. J. Black, *J. Organomet. Chem.*, **82**, C30 (1974).
101. I. Fleming and C. J. Urch, *Tetrahedron Lett.*, **24**, 4591 (1983).
102. G. D. Hartman and T. G. Traylor, *J. Am. Chem. Soc.*, **97**, 6147 (1975).
103. D. Farcasiu, *Tetrahedron Lett.*, 595 (1977).
104. D. J. Peterson, M. D. Robbins, and J. R. Hansen, *J. Organomet. Chem.*, **73**, 237 (1974); D. J. Peterson and M. D. Robbins, *U. S. Pat.*, 3959 324 (1976).
105. H. G. Kuivila and N. M. Scarpa, *J. Am. Chem. Soc.*, **92**, 6990 (1970).
106. J. C. Pommier and H. G. Kuivila, *J. Organomet. Chem.*, **74**, 67 (1974).
106a. M. Ochiai, T. Ukita and E. Fujita, *Tetrahedron Lett.*, **24**, 4025 (1983).
107. M. F. Shostakovskii, N. P. Ivanova, and R. Mirskov, *Khim. Atsetilina Tekhnol. Karbida Kal'tsiya*, 141 (1972); *Chem. Abstr.*, **79**, 115693b (1973).
108. A. G. Davies and R. J. Puddephatt, *J. Chem. soc. C*, 317 (1968).
108a. J. W. Labadie, D. Tueting and J. K. Stille, *J. Org. Chem.*, **48**, 4634 (1983).
108b. M. W. Logue and K. Teng, *J. Org. Chem.*, **47**, 2549 (1982).
109. A. N. Kashin, I. G. Bumagina, N. A. Bumagin, and I. P. Beletskaya, *Zh. Org. Khim.*, **17**, 21 (1981).
110. G. Himbert, *Angew, Chem., Int. Ed. Engl.*, **18**, 405 (1979).
111. G. Himbert, L. Henn, and R. Hoge, *J. Organomet. Chem.*, **184**, 317 (1980).
112. G. Himbert and L. Henn, *Tetrahedron Lett.*, **22**, 2637 (1981).
113. R. G. Mirskov and V. M. Vlasov, *Zh. Obsch. Khim.*, **36**, 352, 1562 (1966).
114. M. Lequan and G. Guillerm, *J. Organomet. Chem.*, **54**, 153 (1973).
115. H. Hartmann and K. Komorniczyk, *Naturwissenschaften*, **51**, 214 (1964).
116. G. Guillerm, F. Maganem, M. Lequan, and K. R. Brower, *J. Organomet. Chem.*, **67**, 43 (1974).
117. L. I. Livantsova, R. A. Bekker, I. A. Savost'yanova, G. I. Oleneva, and I. F. Lutsenko, *Zh. Obsch. Khim.*, **51**, 1297 (1981).
118. L. G. Sharanina, V. S. Zavgorodnii, and A. A. Petrov, *Zh. Obsch. Khim.*, **38**, 1146 (1968).
119. V. S. Zavgorodnii, A. I. Maleeva, and A. A. Petrov, *Zh. Obsch. Khim.*, **41**, 2230 (1971).
120. N. V. Komarov, A. A. Andreev, E. A. Kovtun, and V. S. Senicher, *Zh. Obsch. Khim.*, **50**, 1427 (1980).
121. N. R. Cullen and M. C. Waldman, *Can. J. Chem.*, **48**, 1885 (1970).
122. N. R. Cullen and M. C. Waldman, *J. Fluorine Chem.*, **1**, 151 (1971).
123. B. Wrackmeyer and H. Noth, *J. Organomet. Chem.*, **108**, C21 (1976); B. Wrackmeyer, *J. Organomet. Chem.*, **205**, 1 (1981); B. Wrackmeyer and R. Zentgraf, *J. Chem. Soc., Chem. Commun.*, 402 (1978); B. Wrackmeyer, *Z. Naturforsch., Teil B*, **32**, 140 (1977); **33**, 385 (1978).
124. A. Schmidt and B. Wrackmeyer, *Z. Naturforsch., Teil B*, **33**, 855 (1978); L. Killian and B. Wrackmeyer, *J. Organomet. Chem.*, **132**, 213 (1977); **153**, 153 (1978).
125. D. Seyferth and A. B. Evnin, *J. Am. Chem. Soc.*, **89**, 1468 (1967).
126. A. B. Evnin and D. Seyferth, *J. Am. Chem. Soc.*, **89**, 952 (1967); D. Seyferth and D. L. White, *J. Organomet. Chem.*, **34**, 119 (1972).
127. M. E. Landis, D. Gremaud, and T. B. Patrick, *Tetrahedron Lett.*, **23**, 375 (1982).
128. D. Milstein and J. K. Stille, *J. Am. Chem. Soc.*, **100**, 3636 (1978); *J. Org. Chem.*, **44**, 1613 (1979).
129. M. L. Saihi and M. Pereyre, *Bull. Soc. Chim. Fr.*, 1251 (1977).
130. D. Milstein and J. K. Stille, *J. Am. Chem. Soc.*, **101**, 4981 (1979).
131. I. P. Beletskaya, A. N. Kasatkin, S. A. Lebedev, and N. A. Bumagin, *Izv. Akad. Nauk SSSR*, **30**, 2414 (1981).
132. E. J. Corey and R. H. Wollenberg, *J. Am. Chem. Soc.*, **96**, 5581 (1974).
133. S. L. Chen, R. E. Schaub, and C. V. Grudzinskas, *J. Org. Chem.*, **43**, 3450 (1978).
134. E. Piers and J. M. Chong, *J. Org. Chem.*, **47**, 1602 (1982).
135. R. H. Wollenberg, K. F. Albizati, and R. Peries, *J. Am. Chem. Soc.*, **99**, 7365 (1977).
136. A. Rahm and M. D. Castaing, *Synth. React. Inorg. Met.-Org. Chem.*, **12**, 243 (1982); M. Pereyre, J. P. Quintard, and A. Rahm, *Pure Appl. Chem.*, **54**, 29 (1982); J. P. Quintard, M. D. Castaing, G. Dumarten, A. Rahm, and M. Pereyre, *J. Chem. Soc., Chem. Commun.*, 1004 (1980).
137. C. Minot, A. Laporterie, and J. Dubac, *Tetrahedron*, **32**, 1523 (1976).
138. C. Grugel, W. P. Neumann, and M. Schriewer, *Angew. Chem., Int. Ed. Engl.*, **18**, 543 (1979).

139. H. H. Freedman, *J. Am. Chem. Soc.*, **83**, 2194 (1961); V. R. Sandal and H. H. Freedman, *J. Am. Chem. Soc.*, **90**, 2059 (1968); H. H. Freedman and D. R. Petersen, *J. Am. Chem. Soc.*, **84**, 2837 (1962).
140. J. D. Kennedy, H. G. Kuivila, F. L. Pelczar, R. Y. Tien, and J. L. Considine, *J. Organomet. Chem.*, **61**, 167 (1973).
141. W. P. Neumann and K. Kuhlein, *Adv. Organomet. Chem.*, **7**, 242 (1968).
142. A. Doucoureau, B. Mauze, and L. Miginiac, *J. Organomet. Chem.*, **236**, 139 (1982).
143. J. C. Cochran and H. G. Kuivila, *Organometallics*, **1**, 97 (1982).
144. A. Schweig, U. Weidner, and G. Manuel, *J. Organomet. Chem.*, **54**, 145 (1973); A. Schweig and U. Weidner, *J. Organomet. Chem.*, **67**, C4 (1967).
145. J. Grignon, C. Servens, and M. Pereyre, *J. Organomet. Chem.*, **96**, 225 (1975).
146. C. Servens and M. Pereyre, *J. Organomet. Chem.*, **26**, C4 (1971).
147. J. Grignon and M. Pereyre, *J. Organomet. Chem.*, **61**, C33 (1973).
148. M. Kosugi, K. Kurino, K. Takayama, and T. Migita, *J. Organomet. Chem.*, **56**, C11 (1973).
149. U. Schroer and W. P. Neumann, *J. Organomet. Chem.*, **105**, 183 (1976).
150. M. G. Voronkov, S. Kh. Khangazheev, R. G. Mirskov, and V. I. Rakhlin, *Zh. Obsch. Khim.*, **50**, 1426 (1980).
150a. G. E. Keck and J. B. Yates, *J. Am. Chem. Soc.*, **104**, 5829 (1982).
151. T. H. Chan and I. Fleming, *Synthesis*, 761 (1979); W. E. Colvin, *Chem. Soc. Rev.*, **7**, 15 (1978); H. Sakurai, *Pure Appl. Chem.*, **54**, 1 (1982).
152. N. H. Andersen, D. A. McCrae, D. B. Grotjahn, S. Y. Gabhe, L. J. Theodore, R. M. Ippolito, and T. K. Sarkar, *Tetrahedron*, **37**, 4069 (1981), and references cited therein.
153. E.g. Y. Ueno, H. Sano and M. Okawara, *Tetrahedron Lett.*, 1767 (1980); Y. Ueno, S. Aoki, and M. Okawara, *J. Am. Chem. Soc.*, **101**, 5414 (1979); *J. Chem. Soc., Chem. Commun.*, 683 (1980); Y. Ueno, H. Sano, S. Aoki, and M. Okawara, *Tetrahedron Lett.*, **22**, 2675 (1981).
154. E. Matarasso-Tchiroukhine, and P. Cadiot, *J. Organomet. Chem.*, **121**, 155 and 169 (1976).
155. M. Kosugi, K. Sasazawa, Y. Shimizu, and T. Migita, *Chem. Lett.*, 301 (1977).
156. J. Godschalx and J. K. Stille, *Tetrahedron Lett.*, **21**, 2599 (1980).
156a. J. P. Godschalx and J. K. Stille, *Tetrahedron Lett.*, **24**, 1905 (1983).
157. B. M. Trost and E. Keinan, *Tetrahedron Lett.*, **21**, 2595 (1980).
158. T. Saito, A. Itoh, K. Oshima, and H. Nozaki, *Tetrahedron Lett.*, 3519 (1979).
159. M. Kosugi, Y. Shimizu, and T. Migita, *J. Organomet. Chem.*, **129**, C36 (1977).
159a. A. Gambaro, V. Peruzzo and D. Marton, *J. Organomet. Chem.*, **258**, 291 (1983).
160. R. W. Hoffmann, *Angew. Chem., Int. Ed. Engl.*, **21**, 555 (1982).
161. H. Yatagai, Y. Yamamoto, and K. Maruyama, *J. Am. Chem. Soc.*, **102**, 4548 (1980).
162. C. Servens and M. Pereyre, *J. Organomet. Chem.*, **35**, C20 (1972).
163. E. W. Abel and R. J. Rowley, *J. Organomet. Chem.*, **84**, 199 (1975).
164. T. Mukayama, Private communication cited in ref. 152.
165. H. Nagaoka and Y. Kishi, *Tetrahedron*, **37**, 3873 (1981).
166. V. Peruzzo and G. Tagliavini, *J. Organomet. Chem.*, **162**, 32 (1978); A. Gambaro, V. Peruzzo, G. Plazzogna, and G. Tagliavini, *J. Organomet. Chem.*, **197**, 45 (1980); A Gambaro, D. Marton, V. Peruzzo and G. Tagliavini, *J. Organomet. Chem.*, **204**, 191 (1981).
167. T. Mukaiyama, T. Harada, and S. Shoda, *Chem. Lett.*, 1507 (1980).
168. G. Daude and M. Pereyre, *Organomet. Chem.*, **190**, 43 (1980).
169. A. Gambaro, D. Marton, V. Peruzzo, and G. Tagliavini, *J. Organomet. Chem.*, **226**, 149 (1982).
170. A. Gambaro, P. Ganis, D. Marton, V. Peruzzo, and G. Tagliavini, *J. Organomet. Chem.*, **231**, 307 (1982).
171. B. Y. K. Ho and J. J. Zuckerman, *J. Organomet. Chem.*, **49**, 1 (1973).
172. P. Vittorelli, H.-J. Hansen, and H. Schmid, *Helv. Chim. Acta*, **58**, 1293 (1975).
173. V. G. Kumar Das and S.-W. Ng, *Malaysian J. Sci.*, **5(B)**, 143 (1978); S.-W. Ng, C. L. Barnes, M. B. Hossain, D. van der Helm, J. J. Zuckerman, and V. G. Kumar Das, *J. Am. Chem. Soc.*, **104**, 5359 (1982).
174. K. Maruyama and Y. Naruta, *J. Org. Chem.*, **43**, 3796 (1978); Y. Naruta, *J. Am. Chem. Soc.*, **102**, 3774 (1980); Y. Naruta, *J. Org. Chem.*, **45**, 4097 (1980).
175. Y. Yamomoto, H. Yatagai, Y. Naruta, and K. Maruyama, *J. Am. Chem. Soc.*, **102**, 7107 (1980).
176. M. Koreeda and Y. Tanaka, *Chem. Lett.*, 1299 (1982).

177. D. J. Cram and F. A. Abd. Elhafez, *J. Am. Chem. Soc.*, **74**, 5828 (1952).
178. K. Maruyama, Y. Ishihara, and Y. Yamamoto, *Tetrahedron Lett.*, **22**, 4235 (1981).
179. Y. Naruta, S. Ushida, and K. Maruyama, *Chem. Lett.*, 919 (1979); see also ref. 161.
180. A. Hosomi, H. Iguchi, M. Endo, and H. Sakurai, *Chem. Lett.*, 977 (1979).
181. I. Pri-Bar, P. S. Pearlman and J. K. Stille, *J. Org. Chem.*, **48**, 4629 (1983).
181a. M. Kasugi, H. Arai, A. Yoshimo, and T. Migita, *Chem. Lett.*, 795 (1978).
182. Y. Naruta and K. Maruyama, *Chem. Lett.*, 881 (1979).
183. Y. Naruta H. Uno, and K. Maruyama, *Nippon Kagaku Kaishi*, 831 (1981).
184. K. Maruyama, A. Takuwa, Y. Naruta, K. Satao, and O. Soga, *Chem. Lett.*, **1**, 42 (1981).
185. Y. Naruta, H. Uno, and K. Maruyama, *Tetrahedron Lett.*, **22**, 5221 (1981).
186. G. D. Hartman and T. G. Traylor, *Tetrahedron Lett.*, **939** (1975).
187. A. R. L. Bursics, M. Murray, and F. G. A. Stone, *J. Organomet. Chem.*, **111**, 31 (1976).
188. S. R. Wilson, L. R. Phillips, and K. J. Natalie, Jr., *J. Am. Chem. Soc.*, **101**, 3340 (1979).
189. O. A. Reutov, V. I. Rozenberg, G. V. Gavrilova, and V. A. Nikanorov, *J. Organomet. Chem.*, **177**, 101 (1979).
190. A. N. Kashin, N. A. Bumagin, I. P. Beletskaya, and O. A. Reutov, *J. Organomet. Chem.*, **171**, 321 (1979).
191. A. N. Kashin, I. G. Bumagina, N. A. Bumagin, and I. P. Beletskaya, *Izv. Akad. Nauk. SSSR*, 1433 (1981).
192. N. A. Bumagin, I. G. Bumagina, A. N. Kashin, and I. P. Belatskaya, *Zh. Obsch. Khim.*, **52**, 714 (1982); A. N. Kashin, V. A. Khutoryanskii, O. I. Margorskaya, I. P. Beletskaya, and O. A. Reutov, *Izv. Akad. Nauk. SSSR*, 2151 (1976).
193. A. N. Kashin, I. G. Bumagina, N. A. Bumagin, and I. P. Beletskaya, *Izv. Akad. Nauk. SSSR*, 1675 (1981).
194. I. P. Beletskaya, A. N. Kasatkin, S. A. Lebedev, and N. A. Bumagin, *Izv. Akad. Nauk. SSSR*, 2414 (1981).
195. J. R. Pratt, F. H. Pinkerton, and S. F. Thames, *J. Organomet. Chem.*, **38**, 29 (1972).
196. B. M. Trost and J. M. D. Fortunak, *Organometallics*, **1**, 7 (1982).
197. T. N. Mitchell, *J. Chem. Soc., Perkin Trans.* 2, 1149 (1976).
198. D. Seyferth, C. Sarafidis, and A. B. Evnin, *J. Organomet. Chem.*, **2**, 417 (1964).
199. J. M. Jerkunica and T. G. Traylor, *J. Am. Chem. Soc.*, **93**, 6278 (1971).
199a. R. F. Heck, *J. Am. Chem. Soc.*, **90**, 5518 (1968); **91**, 6707 (1969).
200. M. Tanaka, *Tetrahedron Lett.*, **21**, 2601 (1979); *Chem. Abstr.*, **95**, 6827h (1981); *Synthesis*, 47 (1981).
201. T. L. MacDonald and S. Mahalingam, *J. Am. Chem. Soc.*, **102**, 2113 (1980).
202. T. L. MacDonald and S. Mahalingam, *Tetrahedron Lett.*, **22**, 2077 (1981).
203. Y. Ueno, M. Ohta, and M. Okawara, *Tetrahedron Lett.*, **23**, 2577 (1982).
204. J. C. Pommier and H. G. Kuivila, *J. Organomet. Chem.*, **74**, 67 (1974).
205. K. C. Nicolau, D. A. Claremon, W. E. Barnette, and S. P. Seitz, *J. Am. Chem. Soc.*, **101**, 3704 (1979).
206. Y. Odic and M. Pereyre, *J. Organomet. Chem.*, **55**, 273 (1973).
207. P. A. Tardella, *Tetrahedron Lett.*, 1117 (1969).
208. B. M. Trost and E. Keinan, *Tetrahedron Lett.*, **21**, 2591 (1980).
208a. M. Kosugi, I. Takano, I. Hoshino and T. Migita, *J. Chem. Soc., Chem. Commun.*, 989 (1983).
209. S. Shenvi and J. K. Stille, *Tetrahedron Lett.*, **23**, 627 (1982).
210. Y. Yamamoto, H. Yatagai, and K. Maruyama, *J. Chem. Soc., Chem. Commun.*, 162 (1981).
211. E.g. J. E. Dubois and P. Fellmann, *Tetrahedron*, **34**, 1349 (1978); W. A. Kleschik, C. T Buse, and C. H. Heathcock, *J. Am. Chem. Soc.*, **99**, 247 (1977); C. T. Buse and C. H. Heathcock, *J. Am. Chem. Soc.*, **99**, 8109 (1977).
212. Y. Yamamoto and K. Maruyama, *Tetrahedron Lett.*, **81**, 4607 (1980).
213. W. P. Neumann and F. G. Kleiner, *Tetrahedron Lett.*, 3779 (1964).
214. W. W. Limburg and H. W. Post, *Recl. Trav. Chim. Pays-Bas*, **81**, 430 (1962).
214a. M. Kosugi, I. Hagiwara, T. Sumiya and T. Migita, *J. Chem. Soc., Chem. Commun.*, 343 (1983).
214b. J Tsuji, I. Minami and I. Shimizu, *Tetrahedron Lett.*, **24**, 4713 (1983).
215. A. G. Davies and R. J. Puddephatt, *J. Chem. Soc. C*, 1479 (1968).
216. I. Okada and R. Okawara, *J. Organomet. Chem.*, **42**, 117 (1972).

217. R. V. Subramaniam and B. K. Garg, *Polym. Plast. Technol. Eng.*, **11**, 81 (1978).
218. M. C. Henry and W. Davidsohn, in *Organotin Compounds* (Ed. A. K. Sawyer), Vol. 3, Marcel Dekker, New York, 1972, p. 975.
219. Z. M. Rzaev, D. A. Kochkin, and P. I. Zubov, *Dokl. Akad. Nauk. SSSR,*, **172**, 364 (1967).
220. J. Montemarano and W. J. Dyckman, *J. Paint Technol.*, **47**, 59 (1975).
221. B. De Jeso and J. C. Pommier, *J. Organomet. Chem.*, **122**, Cl (1976); J. M. Brocas, B. De Jeso, and J. C. Pommier, *J. Organomet. Chem.*, **120**, 217 (1976).
222. M. F. Lappert and B. Prokai, *Adv. Organomet. Chem.*, **5**, 225 (1967).
223. J Lorberth, *J. Organomet. Chem.*, **16**, 235 (1969).
223a. M. Fiorenza and A. Ricci, *Synthesis*, 640 (1983).
224. J. C. Pommier and D. Lucas, *J. Organomet. Chem.*, **57**, 139 (1973).
225. Y. Ishii, H. Suzuki, K. Itoh and I. Matsuda, in *Abstracts of Papers, VIth International Conference on Organometallic Chemistry, August 1973*, No. 172. *Amherst, MA* (Eds. M. Rausch and S. A. Gardner).
226. W. P. Neumann and K. Kuhlein, *Tetrahedron Lett.*, 3415 (1966).
227. S. Hashimoto, K. Kano, and H. Okamota, *Bull. Chem. Soc. Jpn.*, **45**, 967 (1972).
228. H. Schumann and I. Schumann, 'Organotin Hydrides', *Gmelin Handbuch der Anorganischen Chemie*, Band 35, Teil 4, Springer-Verlag, Berlin, New York, 1976; V. P. Baillargeon and J. K. Stille, *J. Am. Chem. Soc.*, **105**, 7175 (1983).
229. H. G. Kuivila, *Adv. Organomet. Chem.*, **1**, 47 (1964); see also ref. 87.
230. H. G. Kuivila, *Synthesis*, 499 (1971); *Accounts Chem. Res.*, **1**, 299 (1968).
231. C. Walling, J H. Cooley, A. A. Ponaras, and E. J. Racah, *J. Am. Chem. Soc.*, **88**, 5361 (1966).
232. C. Walling and A. Cioffari, *J. Am. Chem. Soc.*, **94**, 6059 (1972).
233. C. Walling and A. Cioffari, *J. Am. Chem. Soc.*, **94**, 6064 (1972).
234. A. L. J. Beckwith and T. Lawrence, *J. Chem. Soc. Perkin. Trans. 2*, 1535 (1979).
235. Y. Ueno, K. Chino, M. Watanabe, O. Moriya, and M. Okawara, *J. Am. Chem. Soc.*, **104**, 5564 (1982).
236. D. L. Struble, A. L. J. Beckwith, and G. E. Gream, *Tetrahedron Lett.*, 3701 (1968).
237. A. L. J. Beckwith, G. Phillipou, and A. K. Serelis, *Tetrahedron Lett.*, **22**, 2811 (1981).
238. J. K. Grandall and D. J. Keyton, *Tetrahedron Lett.*, 1653 (1969).
239. A. G. Kuivila and L. W. Menapace, *J. Am. Chem. Soc.*, **86**, 3047 (1964).
240. Z. Cekovic, *Tetrahedron Lett.*, 749 (1972).
241. C. R. Warner, R. J. Strunk, and H. G. Kuivila, *J. Org. Chem.*, **31**, 3381 (1966).
242. H. P. Loffler, *Chem. Ber.*, **107**, 2691 (1974).
243. T. G. Burrowes and W. R. Jackson, *Aust. J. Chem.*, **28**, 639 (1975).
244. R. Loven and W. N. Speckamp, *Tetrahedron Lett.*, 1567 (1972).
245. P. G. Harrison, *J. Organomet. Chem.*, **58**, 49 (1973).
246. M. D. Bachi and C. Hoornhaert, *Tetrahedron Lett.*, **22**, 2689 (1981).
247. Y. Ueno, K. Chino, and M. Okawara, *Tetrahedron Lett.*, **23**, 2575 (1982).
248. Y. Ueno, S. Aoki, and M. Okawara, *J. Am. Chem. Soc.*, **101**, 5414 (1979).
249. W P. Neumann, in *The Organic Chemistry of Tin* (Ed. W. P. Neumann), Wiley–Interscience, New York, 1970.
250. J. D. Cotton, P. J. Davidson, and M. F. Lappert, *J. Chem. Soc., Dalton Trans.*, 2275 (1976).
251. O. M. Nefedow, S. P. Kolesnikov, and I. A. Ioffe, *Organomet. Chem. Rev.*, **5**, 181 (1977); A. V. Kramer and J. A. Osborn, *J. Am. Chem. Soc.*, **96**, 7832 (1974).
252. P. Jutzi and F. Kohl, *J. Organomet. Chem.*, **161**, 141 (1979).
253. K. D. Bos, E. J. Bulten, and J. G. Noltes, *J. Organomet. Chem.*, **99**, 397 (1975).
254. A. K. Holliday, P. H. Makin and R. J. Puddephatt, *J. Chem. Soc., Dalton Trans.*, 435 (1976).
255. H. P. Fritz and K. E. Schwarzhans, *Chem. Ber.*, **97**, 1390 (1964).
256. C. Grügel, W. P. Neumann, J. Sauer, and P. Seifert, *Tetrahedron Lett.*, **31**, 2847 (1978).
257. R. A. Sheldon and J. K. Kochi, *Org. React.*, **19**, 279 (1971).
258. R. M. Moriarty, in *Selective Organic Transformations* (Ed. B. S. Thyagarajan), pp. 183–237, Interscience, New York, 1972.
259. T. Aratani, Y. Nakanisi, and H. Nozaki, *Tetrahedron Lett.*, 1809 (1969).
260. L. M. Stock and T. L. Wright, *J. Org. Chem.*, **45**, 4645 (1980).
261. D. R Harvey and R. O. C. Norman, *J. Chem. Soc.*, 4860 (1964).
262. R. O. C. Norman and C. B. Thomas, *J. Chem. Soc. B*, 421 (1970).

263. R. O. C. Norman, C. B. Thomas, and J. S. Willson, *J. Chem. Soc. B*, 518 (1971).
264. R. O. C. Norman, C. B. Thomas, and J. S. Willson, *J. Chem. Soc., Perkin Trans. 1*, 325 (1973).
265. L. C. Willemsens, D. Devos, J. Spierenburg, and J. Wolters, *J. Organomet. Chem.*, **39**, C61 (1972); see also D. Devos, Thesis, University of Leiden (1975).
266. H. C. Bell, J. R. Kalman, J. T. Pinhey, and S. Sternhell, *Aust. J. Chem.*, **32**, 1521 (1979).
267. H. C. Bell, J. R. Kalman, G L. May, J. T. Pinhey, and S. Sternhell, *Aust. J. Chem.*, **32**, 1531 (1979).
268. H. C. Bell, J. T. Pinhey and S. Sternhell, *Aust. J. Chem.*, **32**, 1551 (1979).
269. J. T. Pinhey and B. A. Rowe, *Aust. J. Chem.*, **32**, 1561 (1979).
269a. R. P. Kozyrod and J. T. Pinhey, *Tetrahedron Lett.*, **24**, 1301 (1983).
269b. J. T. Pinhey and B. A. Rowe, *Aust. J. Chem.*, **33**, 113 (1980).
270. J. R. Kalman, J. T. Pinhey, and S. Sternhell, *Tetrahedron Lett.*, 5369 (1972).
271. S. R Jones and J. M. Mellor, *Synthesis*, 32 (1976); *J. Chem. Soc., Perkin Trans. 1*, 2576 (1976).
272. O. A. Osipov and O. E. Kashireninov, *Zh. Obshch. Khim.*, **32**, 1717 (1962); Y. Takami, *Kogyo Kagoku Zasshi*, **65**, 234 (1962); *Chem. Abstr.*, **58**, 1482g (1963).
273. G. V. Sorokin, M. V. Pozdnyakova, N. I. Ter-Asaturova, V. N. Perchenko, and N. S. Nametkin, *Dokl. Akad. Nauk. SSSR*, **174**, 376 (1967).
274. W. L. Loeb, *U. S. Pat.*, 3 166 547 (1965).
275. Japan Synthetic Rubber Co. Ltd., *Br. Pat.*, 1 100 933 (1968); *Chem. Abstr.*, **68**, 60365m (1968).
276. H. J. Meijer, J. W. G. van den Hurk, and G. J. M. van der Kerk, *Recl. Trav. Chim. Pays-Bas*, **85**, 1018 (1966).
277. Y. Nakamura, T. Ouchi, and M. Imoto, *Kobunshi Ronbunshu*, **31**, 676 (1974).
278. D. M. Young, F. Hostettler, and C. F. Horn, *U.S., Pat.* 2 890 208 (1959).
279. T. Pullukat, U.S. Pat., 3 928 304 (1976); *Chem. Abstr.*, **84**, 90802 (1976).
280. H. J. Nienburg, H. Böhm, and R. Herbeck, *Ger. Pat.*, 1 105 167 (1961); *Chem. Abstr.*, **55**, 2654 (1961).
281. K. Itoi, *Jpn. Pat.*, 6 919 865; *Chem. Abstr.*, **71**, 113460j (1969); *Jpn. Pat.*, 6 826 180; *Chem. Abstr.*, **70**, 88429a (1969); *Jpn. Pat.*, 6 826 181; *Chem Abstr.*, **70**, 97388 (1969).
282. K. Itoi and T. Nishida, *Jpn. Pat.*, 6 830 316; *Chem. Abstr.*, **70**, 97384j (1069); *Ger. Offen.*, 1 804 490; *Chem. Abstr.*, **71**, 71710h (1969).
283. W. L. Carrick, J. Karol, G. L. Karapinko, and J. J. Smith, *J. Am. Chem. Soc.*, **82**, 1502 (1960).
284. W. P. Long, *U.S. Pat.*, 3 639 379 (1972); *Chem. Abstr.*, **76**, 141532f (1972).
285. A. Damiel, M. Levy, and D. Vovsi, *Ger. Offen.*, 2 227 914 (1973); *Chem. Abstr.*, **78**, 98285e (1973).
286. H. Sakurai, Y. Katayama, T. Ikegami, and M. Furusato, *Ger. Offen.*, 2 946 562 (1981); *Chem. Abstr.*, **95**, 43967z (1981); *Ger. Offen.*, 3 028 479 (1981); *Chem. Abstr.*, **95**, 8056e (1981).
287. C. M. Selman, *U.S. Pat.*, 4 287 091 (1981); *Chem. Abstr.*, **95**, 170119z (1981).
288. K. Itoi, *Jpn. Pat.*, 6 826 301 (1967); *Chem. Abstr.*, **70**, 78515m (1967).
289. A. Kawasaki and T. Maruyama, *Jpn. Pat.*, 7 302 937 (1974); *Chem. Abstr.*, **80**, 4097m (1974).
290. Z. S. Nurkeeva, N. A. Plate, and V. V. Mal'tsev, *Vysokomol. Soedin., Ser. A*, **14**, 2047 (1972); *Chem. Abstr.*, **78**, 72669h (1973).
291. Asahi Chem. Ind. Co. Ltd., *Fr. Pat.*, 1 633 691; *Chem. Abstr.*, **71**, 40016g (1969).
292. N. Calderon, J. P. Lawrence, and E. A. Ofstead, *Adv. Organomet. Chem.*, **17**, 449 (1979).
293. J. L. Herisson and Y. Chauvin, *Makromol. Chem.*, **141**, 161 (1970).
294. R. R. Schrock, *J. Am. Chem. Soc.*, **97**, 6577 (1975).
295. R. H. Grübbs and C. R. Hoppin, *J. Chem. Soc., Chem. Commun.*, 634 (1977).
296. J. M. Basset, J. L. Bilhou, R. Mutin, and A. Theolier, *J. Am. Chem. Soc.*, **97**, 7376 (1975).
297. G. Dall'Asta, *Proc. Int. Congr. Pure Appl. Chem.*, 24th, **1**, 133 (1973).
298. T. J. Katz and W. H. Hersh, *Tetrahedron Lett.*, 585 (1977).
299. T. Masuda, N. Sasaki, and T Higashimura, *Makromolecules*, **8**, 717 (1975).
300. K. Hasegawa, *Eur. Polym. J.*, **13**, 315 (1977); **13**, 47 (1977).
301. N. Sasaki, T. Masuda, and T. Higashimura, *Makromolecules*, **9**, 664 (1976).
302. A. Uchida, M. Hinenoya, and T. Yamamoto, *J. Chem. Soc., Dalton Trans.*, 1089 (1981).

1. Carbon—carbon bond formation

303. C. Boelhouwer and E. Verkuijlen, *Int. Metathesis Symp., Mainz, 1976*, 28 (1976).
304. P. B. van Dam, M. C. Mittelmeijer, and C. Boelhouwer, *J. Chem. Soc., Chem. Commun.*, 1221 (1972).
305. R. Baker and M. J. Grimmin, *Tetrahedron Lett.*, 1441 (1977); J. C. Mol and E. F. G. Woerlee, *J. Chem. Soc., Chem. Commun.*, 330 (1979).
306. W. Ast., G. Rheinwald, and R. Kerber, *Makromol. Chem.*, **177**, 39 (1976).
307. C. P. Pinazzi, I. Campistron, M. C. Croissandeau, and D. Reyx, *J. Mol. Catal.*, **8**, 325 (1980).
308. J. Otton, Y. Colleuille, and J. Varagnat, *J. Mol. Catal.*, **8**, 313 (1980).
309. E. Verkuijlen, F. Kapteiju, J. C. Mol, and C. Boelhouwer, *J. Chem. Soc., Chem. Commun.*, 198 (1977).
310. P. G. Gassman and T. H. Johnson, *J. Am. Chem. Soc.*, **98**, 6057 (1976).
311. W. Ast, G. Rheinwald and R. Kerber, *Recl. Trav. Chim. Pays-Bas*, **96**, M127 (1977); J. Vevisalles and D. Villemin, *Tetrahedron*, **36**, 3181 (1980).
312. C. Sanchez, R. Kieffer, and A. Kieeneman, *J. Prakt. Chem.*, **230**, 329 (1978).
313. S. R. Wilson and D. E. Schalk, *J. Org. Chem.*, **41**, 3929 (1976).
314. R. R. Schrock and G. W. Parshall, *Chem. Rev.*, **76**, 243, (1976).
315. W. J. Feast and B. Wilson, *J. Mol. Catal.*, **8**, 277 (1980).
316. G. Lehnert, G. Pampus, and D. Maertens, *Ger. Offen.*, 2 163 395 (1973); *Chem. Abstr.*, **79**, 93222t (1973); G. Lehnert, D. Maertens, and J. Witte, *Ger. Offen.*, 2 106 302 (1972); *Chem. Abstr.*, **77**, 152 882s (1972).
317. W. Ast, G. Rheinwald, and R. Kerber, *Makromol. Chem.*, **177**, 1341 (1976).
318. W. Ast, G Rheinwald, and R. Kerber, *Makromol. Chem.*, **177**, 1349 (1976).
319. C. A. Uraneck and G. R Kahle, *Belg. Pat.*, 644 681 (1964); *Chem. Abstr.*, **63**, 13 542d (1965); *Neth. Pat.*, 6 602 265 (1967); *Chem. Abstr.*, **66**, 3523k (1967); *U. S. Pat.*, 3 278 508 (1966); *Chem. Abstr.*, **65**, 20334g (1966).
320. H. Schumann, J. Held, W.-W. Du Mont, G. Rodewald, and B. Wobke, in *Organotin Compounds: New Chemistry and Applications* (Ed. J. J. Zuckerman), Advances in Chemistry Series, No. 157, American Chemical Society, Washington, DC. 1976, pp. 57–69.
321. J. A. Chenicek and H. S. Bloch, *U.S. Pat.*, 2 867 673 (1959).
322. S. Horie, *Nippon Kagaku Zasshi*, **78**, 1795 (1957).

CHAPTER 2

Carbon—carbon bond formation using organometallic compounds of zinc, cadmium, and mercury

LÉONE MIGINIAC

Laboratoire de Synthèse Organique, Université de Poitiers, 40 Avenue du Recteur Pineau, 86022 Poitiers, France

I. INTRODUCTION	101
II. REACTIONS OF ZINC ORGANOMETALLICS	101
A. Addition Reactions	101
1. Reactions of allylic and benzylic organozinc compounds	101
a. Carbonyl derivatives and epoxides	101
b. C=N derivatives and nitriles	102
c. Isolated olefinic double bonds	102
d. Isolated acetylenic triple bonds	103
e. Conjugated enynes	105
2. Reactions of propargylic and allenic organozinc compounds	107
a. Carbonyl derivatives	107
b. Imines	107
c. Isolated acetylenic triple bonds	108
3. Reactions of saturated and miscellaneous organozinc compounds	108
a. Carbonyl derivatives and epoxides	108
b. Imines	109
c. Terminal alkynes	109
d. Conjugated enynes	109
B. Substitution Reactions	110
1. Reactions of allylic organozinc compounds	110
a. Derivatives with a mobile halogen atom	110
b. Derivatives with a reactive alkoxy group	111
2. Reactions of propargylic and allenic organozinc compounds	111
a. Derivatives with a mobile halogen atom	112
b. Derivatives with a reactive alkoxy group	112
3. Reactions of saturated and miscellaneous	

        organozinc compounds. 112
          a. Derivatives with a mobile halogen atom. 112
          b. Palladium-catalysed reaction with aryl,
            alkenyl, and allenic halides 113
          c. Derivatives with a reactive alkoxy group 113
    C. Carbene and Carbenoid Intermediates from Organozinc Compounds 113
      1. Addition and insertion reactions 113
          a. Carbon—carbon double bonds. 113
          b. Carbon—carbon triple bonds 115
          c. C=N derivatives. 115
      2. Wittig alkene synthesis. 115
    D. Reformatsky and Related Reactions 115
      1. Addition reactions 116
          a. Carbonyl derivatives 116
          b. C=N derivatives and nitriles 116
          c. Terminal alkynes. 117
      2. Substitution reactions 118
          a. Derivatives with a mobile halogen atom. 118
          b. Derivatives with an alkoxy or acyloxy group 119
III. REACTIONS OF CADMIUM ORGANOMETALLICS 119
    A. Addition Reactions 119
      1. Reactions of allylic organocadmium compounds
        and analogues 119
      2. Reactions of saturated and phenylic organocadmium
        compounds 120
      3. Reactions of organocadmium reagents prepared from
        α-bromoesters 122
    B. Substitution Reactions 122
      1. Alkylation 122
          a. Alkyl, allylic, and benzylic halides 122
          b. Halides with other functional groups 122
      2. Acylation 123
    C. Carbene and Carbenoid Intermediates from
       Organocadmium Compounds. 124
IV. REACTIONS OF MERCURY ORGANOMETALLICS 124
    A. Alkene and Alkyne Additions and Heck Reaction 124
    B. Substitution Reactions 126
      1. Alkylation 126
      2. Acylation 127
    C. Dimerization 128
    D. Carbene and Carbenoid Intermediates from Organomercury Compounds 129
      1. Addition reactions 129
          a. Carbon—carbon double bonds. 129
          b. Carbon—carbon triple bonds 130
          c. Other double bond systems 130
      2. Insertion reactions 130
          a. C—H bonds 131
          b. B—C bonds 131
          c. Si—C and Ge—C bonds 131
      3. Wittig alkene synthesis. 131
V. REFERENCES. 132

I. INTRODUCTION

Organo-zinc, -cadmium, and -mercury reagents are among the oldest organometallics known (1849–53); they were the first synthetically useful organometallic compounds, but were almost entirely superseded by the more conveniently prepared and versatile organo-magnesium and lithium reagents. However, many recent developments have demonstrated their new utility in organic synthesis:
1. their saturated and phenylic derivatives, being generally much less reactive than organo-magnesium and -lithium compounds, are more selective and react with fewer reactive functional groups, ignoring many other important functional groups present in the same molecule;
2. on the other hand, allylic, benzylic, propargylic, and related organometallic reagents recently prepared from these metals show high reactivity towards carbonyl and other unsaturated compounds;
3. in addition, these organometallics are able to enter into peculiar reactions which are not undergone by organo-magnesium and -lithium compounds.

This chapter is divided into three parts, dealing with the main results obtained recently with zinc, cadmium and mercury organometallics.

II. REACTIONS OF ZINC ORGANOMETALLICS

Several general reviews[1-3] covering the literature up to and including 1972 have been published on this subject.

A. Addition Reactions

We shall consider successively:
1. reactions of allylic and benzylic organozinc compounds;
2. reactions of propargylic and allenic organozinc compounds;
3. reactions of saturated and miscellaneous organozinc compounds.

1. Reactions of allylic and benzylic organozinc compounds

a. Carbonyl derivatives and epoxides

Organozinc compounds prepared from allyl, 1-methylallyl, benzyl, crotyl, and cinnamyl bromide react readily with aldehydes, ketones, esters, acid anhydrides, and carbon dioxide[4,5]; organozinc derivatives of γ-ethylallyl, γ-vinylallyl, and γ-ethynylallyl bromide react with ketones, ethyl formate, N,N-dimethylformamide and carbon dioxide[5-7]. An allylic rearrangement is generally observed (reaction 1). Other examples of the reactions of

$$R^1CH=CHCH_2ZnX + R^2COR^3 \longrightarrow CH_2=CHCH(R^1)C(OH)R^2R^3 \quad (1)$$

allylic organozinc bromides or diallylzinc with carbonyl compounds have been reported[1-3,5,8-17] and the stereochemistry of addition has received attention[13]; 1,4-addition to alkylidene malonates[18-20] and alkylidene cyanoacetates[19] has been observed.

In the reaction of allylic organozinc reagents with carbonyl derivatives, it is possible to observe isomerization of the branched product **1** to the thermodynamically more stable linear product **2** (reaction 2). The ratio of linear to branched-chain alcohol increases with increasing reaction time and depends principally on the steric strain within the carbonyl

$$R^2R^3C(OZnBr)CHR^1CH=CH_2 \longrightarrow R^2R^3C(OZnBr)CH_2CH=CHR^1 \quad (2)$$
$$\qquad\qquad\mathbf{1} \qquad\qquad\qquad\qquad\qquad\qquad \mathbf{2}$$

derivative. These results have been interpreted in terms of reversible condensation[21–23]; this reversibility has been observed in numerous cases of allylic organozinc reagents in their reactions with aldehydes, ketones[11,12,21–29], and esters[30].

Unlike their saturated analogues, allylic organozinc compounds are sufficiently nucleophilic to open the epoxide ring (reactions 3 and 4)[5,31–34]. The stereochemistry of the reaction of diallylzinc with 1-phenyl-1,2-epoxypropane has been studied[35].

$$(R^1CH{=}CHCH_2)_2Zn \;+\; CH_3{-}CH\underset{O}{\diagdown\!\diagup}CH_2 \longrightarrow CH_3CHOHCH_2CHR^1CH{=}CH_2 \quad (3)$$

$$(R^1CH{=}CHCH_2)_2Zn \;+\; Ph{-}CH\underset{O}{\diagdown\!\diagup}CH_2 \longrightarrow PhCH(CH_2OH)CHR^1CH{=}CH_2 \quad (4)$$

b. C=N derivatives and nitriles

Allylic organozinc compounds react readily with a wide variety of imines (reaction 5)[5,36,37]. In many cases, it has been shown that this reaction is reversible and leads to thermodynamically more stable secondary linear amines with longer reaction time (reaction 6)[38–40]. Allylic organozinc reagents also react with many C=N derivatives such

$$R^1CH{=}CHCH_2ZnBr \;+\; R^2R^3C{=}NR^4 \longrightarrow R^4NHCR^2R^3CHR^1CH{=}CH_2 \quad (5)$$

$$R^2R^3C[N(ZnBr)R^4]CHR^1CH{=}CH_2 \longrightarrow R^2R^3C[N(ZnBr)R^4]CH_2CH{=}CHR^1 \quad (6)$$

as iminoethers[41–43], amidines[41], iminocarbonates[44,45], carbodiimides[46], and O-alkyloximes (reactions 7–9)[36,47]. Apparently, they do not react with isoureas[45], gua-

$$2R_1CH{=}CHCH_2ZnBr \;+\; PhN{=}CHOC_2H_5 \longrightarrow PhNHCH(CHR^1CH{=}CH_2)_2 \quad (7)$$

$$2R_1CH{=}CHCH_2ZnBr \;+\; PhN{=}C(OC_2H_5)_2 \longrightarrow PhNHCH(CHR^1CH{=}CH_2)_3 \quad (8)$$

$$R_1CH{=}CHCH_2ZnBr \;+\; R^2N{=}C{=}NR^2 \longrightarrow R^2N{=}C(NHR^2)CR^1{=}CHCH_3 \quad (9)$$

nidines[45], or O-alkylbenzohydroximates[48]. With nitriles, allylic organozinc compounds lead to a mixture of β- and α-ethylenic ketones (reaction 10)[49]. A reaction is also observed with aryl isonitriles (reaction 11)[43].

$$R^1CH{=}CHCH_2ZnBr \;+\; R^2C{\equiv}N \longrightarrow R^2COCHR^1CH{=}CH_2 \;+\; R^2COCR^1{=}CHCH_3 \quad (10)$$

$$3\,CH_2{=}CHCH_2ZnBr \;+\; C_6H_5N{\equiv}C{:} \longrightarrow C_6H_5NHC(CH_2CH{=}CH_2)_3 \quad (11)$$
$$30\%$$

c. Isolated olefinic double bonds

Allylic organozinc reagents undergo addition reactions across isolated olefinic double bonds of various ethylenic alcohols[5,50], ethers[5,50], and amines (reactions 12 and 13)[41,45,46,50–53]. These reactions are often regioselective, but sometimes both terminal

$$CH_2{=}CHCH_2ZnBr \;+\; CH_2{=}CHCH_2OH \longrightarrow CH_3CH(CH_2CH{=}CH_2)CH_2OH \quad (12)$$
$$23\%$$

$$C_2H_5CH=CHCH_2ZnBr + CH_2=CHCH_2CH_2N(C_2H_5)_2 \longrightarrow \qquad (13)$$

$$CH_2=CHCH(C_2H_5)CH_2CH_2CH_2CH_2N(C_2H_5)_2$$
$$70\%$$

and non-terminal addition reactions occur (reaction 14)[50–52]. Also, allylic organozinc compounds add on either end of the ethylenic bond of 2-vinylpyridine to give a mixture of two isomeric 2-alkylpyridines[54].

$$CH_2=CHCH_2ZnBr + CH_2=CH(CH_2)_nNR^1R^2 \longrightarrow \begin{cases} CH_2=CHCH_2CH_2CH_2(CH_2)_nNR^1R^2 \\ CH_3CH(CH_2CH=CH_2)(CH_2)_nNR^1R^2 \end{cases}$$
$$(14)$$

By means of the same processes of addition, dimerization of allylic organozinc derivatives has been observed (reaction 15)[55,56]. Under more drastic conditions, allylic organozinc reagents add to the double bond of alkenes such as ethylene or styrene

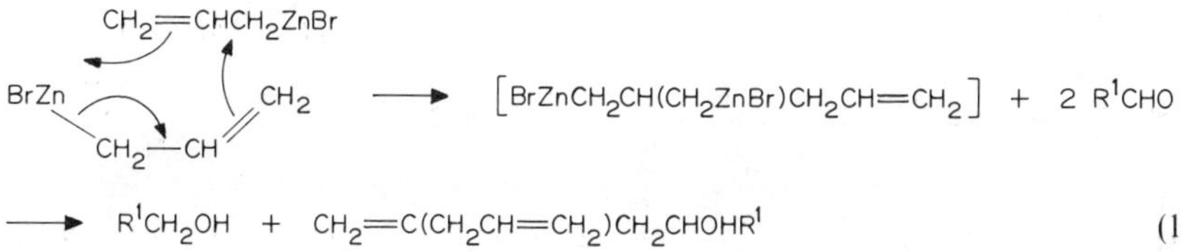

$$\longrightarrow R^1CH_2OH + CH_2=C(CH_2CH=CH_2)CH_2CHOHR^1 \qquad (15)$$

$$CH_2=CR^1CR^2R^3CH_2CH_3 \qquad (16)$$

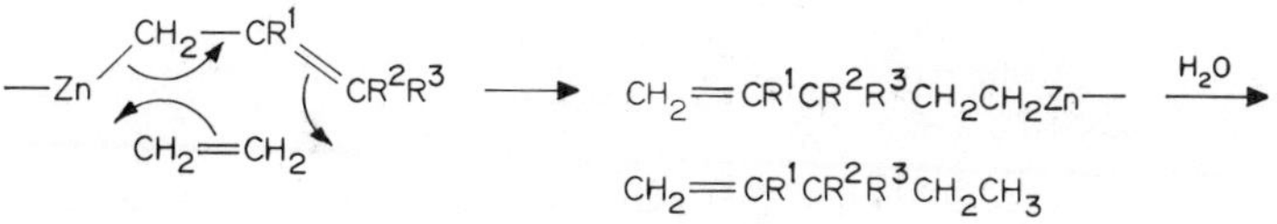

(reaction 16)[57–59], and the regioselectivity of the addition of bis(2-methallyl)zinc to *para*- and *meta*-substituted styrenes has been investigated[60].

d. Isolated acetylenic triple bonds

Allylic organozinc compounds are able to add to a wide variety of alkynes. Other organometallic reagents of the same type (lithium, magnesium, boron, aluminium, or copper) react with difficulty or do not react at all with alkynes; thus, organozinc derivatives must be considered as complementary to the other organometallic reagents in their addition reactions[61].

(i) *Addition to non-functionalized alkynes*

Allylic organozinc derivatives react with acetylene and monosubstituted alkynes in tetrahydrofuran (reaction 17)[5,52,62–65]. Under certain experimental conditions[52,63,65], the reaction may proceed further by addition of another equivalent of the allylic organozinc derivative, to lead after hydrolysis to compounds such as $CH_3C(R^2)(CHR^1CH=CH_2)_2$. The regiochemistry of the addition is always of the Markovnikoff type, whatever the R^2 group of the alkyne. The addition of allylic

$R_1CH=CHCH_2ZnBr$ + $HC\equiv CR^2$ ⟶ $BrZnC\equiv CR^2$

$\xrightarrow{R_1CH=CHCH_2ZnBr}$ $(BrZn)_2C=CR^2CHR^1CH=CH_2$ $\xrightarrow{H_2O}$ $H_2C=CR^2CHR^1CH=CH_2$ (17)

R^1 = H, alkyl, phenyl, vinyl; R^2 = H, alkyl, phenyl

organozinc halides to alkynes is a reversible process[64,66]. When the allylic organozinc is substituted, allylic rearrangement occurs widely and often exclusively but, since the addition is reversible, the amount of the minor linear isomer, thermodynamically more stable than the branched isomer, can be increased by longer reaction times (reaction 18).

$R^1CH=CHCH_2ZnBr$ + $HC\equiv CR^2$ ⟶ $(BrZn)_2C=CR^2CR^1CH=CH_2$ +

$(BrZn)_2C=CR^2CH_2CH=CHR^1$ (18)

Disubstituted non-functionalized alkynes do not react with allylic organozinc derivatives[61,62]; however, one example of intramolecular cyclisation has been observed (reaction 19)[67].

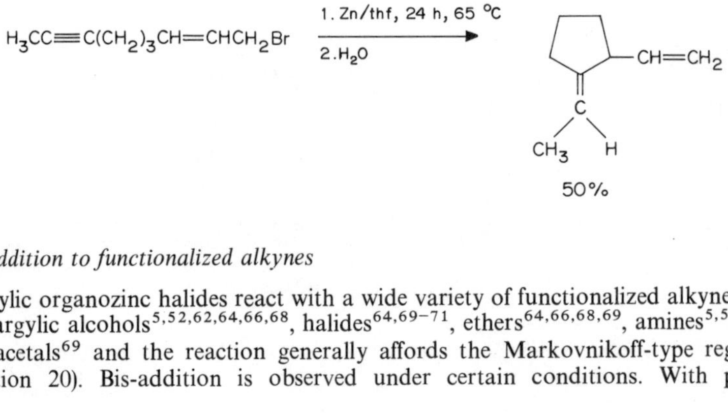

$H_3CC\equiv C(CH_2)_3CH=CHCH_2Br$ $\xrightarrow[2.H_2O]{1.\,Zn/thf,\,24\,h,\,65\,°C}$ [cyclopentane structure with =C(CH_3)H exocyclic and -CH=CH_2 substituent] (19)

50%

(ii) Addition to functionalized alkynes

Allylic organozinc halides react with a wide variety of functionalized alkynes such as propargylic alcohols[5,52,62,64,66,68], halides[64,69-71], ethers[64,66,68,69], amines[5,53,64,66,72], and acetals[69] and the reaction generally affords the Markovnikoff-type regioisomer (reaction 20). Bis-addition is observed under certain conditions. With propargyl

3 $CH_3CH=CHCH_2ZnBr$ + $HC\equiv CCH_2OH$ $\xrightarrow{3\,h,\,20\,°C}$ $CH_2=C[CH(CH_3)CH=CH_2]CH_2OH$

93%

(20)

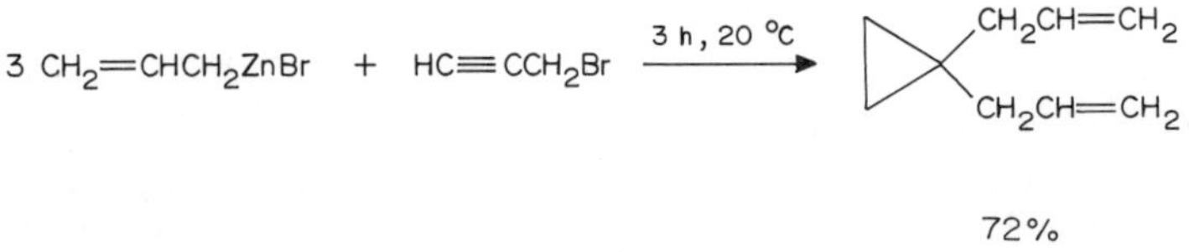

3 $CH_2=CHCH_2ZnBr$ + $HC\equiv CCH_2Br$ $\xrightarrow{3\,h,\,20\,°C}$ [cyclopropane with two $CH_2CH=CH_2$ groups] (21)

72%

bromide[64,70] and propargyl ethers[69,71], the bis-addition products lead to cyclopropanes by 1,3-elimination (reaction 21). Allylic organozincs also add to the allenic bond of α- or β-allenic amines (reaction 22)[5,53,73].

$$R^1CH=CHCH_2ZnBr + CH_2=C=C(CH_3)CHR^2NR^3R^4 \longrightarrow$$

$$CH_2=CHCHR^1CH_2CH=C(CH_3)CHR^2NR^3R^4 \quad (22)$$

R^2 = H, alkyl, phenyl ; R^3, R^4 = H, alkyl   40-72%

e. Conjugated enynes

A recent review by the present author[74] summarizes the results obtained in addition reactions between common organometallic compounds and conjugated enynes.

(i) Conjugated enynes with a terminal triple bond

Allylic organozinc halides add readily to the triple bond of simple enynes[53,62,64,66,74–77] to produce allylic conjugated dienes; with some enynes and when a large excess of organozinc compound is used, a diallylic alkene resulting from a double addition may also be produced (reaction 23). With a substituted allylic organozinc

$$CH_2=CHCH_2ZnBr + HC\equiv CCH=CHBu^n \longrightarrow CH_2=C(CH_2CH=CH_2)CH=CHBu^n +$$
$$43\%$$
$$CH_3-C(CH_2CH=CH_2)_2CH=CHBu^n$$
$$22\% \quad (23)$$

$$R^1CH=CHCH_2ZnBr + HC\equiv CCR^2=CR^3R^4 \xrightarrow{H_2O}$$
$$CH_2=C(CHR^1CH=CH_2)CR^2=CR^3R^4 + CH_2=C(CH_2CH=CHR^1)CR^2=CR^3R^4 \quad (24)$$
$$\mathbf{3} \qquad\qquad \mathbf{4}$$

compound, the reaction leads to a mixture of two products, **3** and **4** (reaction 24). The ratio **3:4** varies according to the experimental conditions; this fact can be explained by the reversible character of the reaction[62,64,66,76,77]. Allylic organozinc derivatives add to functional conjugated enynes (alcohols, ethers, amines) in the same manner (reaction 25)[75,76,78,79]. With an allylic substituted organozinc compound, the reaction proceeds with complete allylic rearrangement.

$$CH_2=CHCH_2ZnBr + HC\equiv CCH=CHCH_2OH \longrightarrow CH_2=C(CH_2CH=CH_2)CH=CHCH_2OH +$$
$$63\%$$
$$CH_3C(CH_2CH=CH_2)_2CH=CHCH_2OH$$
$$14\% \quad (25)$$

According to the experimental conditions, the reaction with an enynic bromide[70] leads to products resulting from a mono-addition and a substitution (S_N2 and S_N2') or forms a

$$3\ CH_2=CH-CH_2ZnBr + HC\equiv CCH=CHCH_2Br \longrightarrow$$
$$CH_2=C(CH_2CH=CH_2)CH=CH(CH_2)_2CH=CH_2 +$$
$$CH_2=C(CH_2CH=CH_2)CH(CH_2CH=CH_2)CH=CH_2 \quad (26)$$

vinylcyclopropane resulting from a bis-addition following of an 1,3-elimination (reactions 26 and 27).

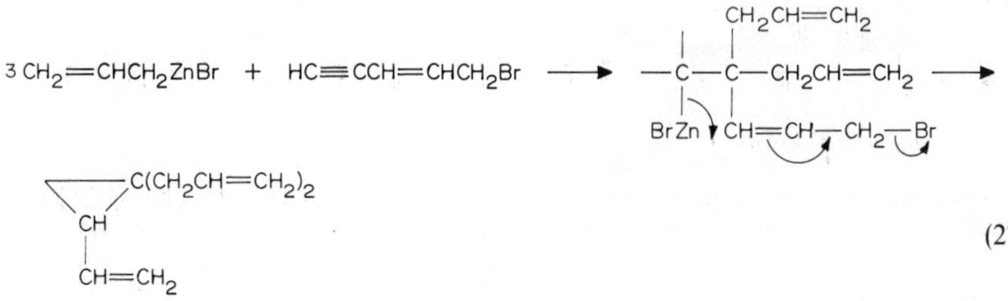

$$3\ CH_2=CHCH_2ZnBr + HC\equiv CCH=CHCH_2Br \longrightarrow$$

[structure with intermediate showing BrZn, CH=CH-CH₂-Br, and -C-C-CH₂CH=CH₂ with CH₂CH=CH₂ branch]

[cyclopropane product structure with C(CH₂CH=CH₂)₂ and CH=CH₂ substituents]

(27)

(ii) *Conjugated enynes with an internal triple bond*

Allylic organozinc compounds do not react with conjugated enynes (hydrocarbons, α-functional enynes) having an internal triple bond[75-77,79,80]. With certain α,α'-difunctional enynes, an addition to the triple bond has been observed[81], but this reaction is often not regioselective (reaction 28). When the group next to the double bond is a good leaving group [OCH_3 instead of $N(CH_3)_2$], addition and substitution occur (reaction 29)[81].

$$HOCH_2C\equiv CCH=CHCH_2N(CH_3)_2 + CH_2=CHCH_2ZnBr \longrightarrow$$

$$HOCH_2CH=C(CH_2CH=CH_2)CH=CHCH_2N(CH_3)_2 \quad (28)$$
$$5\%$$

$$+\ HOCH_2C(CH_2CH=CH_2)=CHCH=CHCH_2N(CH_3)_2$$
$$45\%$$

$$HOCH_2C\equiv CCH=CHCH_2OCH_3 + CH_2=CHCH_2ZnBr \longrightarrow$$

$$HOCH_2C(ZnBr)=C(CH_2CH=CH_2)CH=CHCH_2OCH_3\ +$$

$$\downarrow H_2O$$

$$HOCH_2CH=C(CH_2CH=CH_2)CH=CHCH_2OCH_3 \quad (29)$$
$$15\%$$

$$HOCH_2C(CH_2CH=CH_2)=C(ZnBr)CH=CHCH_2OCH_3$$

$$\begin{array}{l} 1.\ CH_2=CHCH_2ZnBr \\ 2.\ H_2O \end{array}$$

$$HOCH_2C(CH_2CH=CH_2)=CHCH(CH_2CH=CH_2)CH=CH_2$$
$$20\%$$

2. Reactions of propargylic and allenic organozinc compounds

a. Carbonyl derivatives

Organozinc reagents prepared from propargylic halides readily undergo Grignard-type addition reactions with carbonyl and other unsaturated compounds[1-4,82-84]. Organozinc compounds prepared from halides $HC\equiv CCHXR^1$ mainly have an allenic structure and give with ketones mainly the rearranged β-acetylenic alcohols (reaction 30)[4,82,83,85]. It is

$$R^1CH=C=CHZnBr + R^2COR^3 \longrightarrow R^2R^3C(OH)CHR^1C\equiv CH + R^2R^3C(OH)CH=C=CHR^1 \quad (30)$$

currently assumed that the reaction proceeds by means of a six-centred electronic transfer ($S_E i'$ process) or by electrophilic attack of the carbonyl carbon on the organozinc compound ($S_E 2'$ process). The proportion of the allenic isomer increases with increasing solvating strength of solvent and with decreasing electrophilic character of the carbonyl carbon atom[85,86].

Organozinc derivatives prepared from halides $R^1C\equiv CCH_2X$ exist with both allenic and acetylenic structures; subsequent reaction with aldehydes and ketones give mixtures of α-allenic and β-acetylenic alcohols (reaction 31)[4,87]. The relative proportions of isomers

$$R^1C\equiv CCH_2ZnBr \rightleftharpoons R^1C(ZnBr)=C=CH_2 \quad \xrightarrow{\text{1. } R^2R^3CO}_{\text{2. } H_2O} \quad R^2R^3C(OH)CR^1=C=CH_2 + R^2R^3C(OH)CH_2C\equiv CR^1 \quad (31)$$

depend on chiefly the structure of the carbonyl compound; bulkiness in the latter favours the formation of the β-acetylenic alcohol[87]. As in the allylic organozinc series, the reaction between organozinc derivatives prepared from propargyl bromides and ketones is reversible[85,87]. These organozinc compounds give only 1,2-addition products with α-ethylenic ketones[88] or α-acetylenic ketones[8]. On the other hand, their reactions with alkylidene malonates[88,89], alkylidene cyanacetates, and N-disubstituted α-ethylenic amides[90] lead to mixtures of acetylenic and allenic products resulting exclusively from 1,4-addition of the organozinc reagent.

b. Imines

Aldimines react with organozinc compounds with an allenic structure to give a mixture of α-allenic and β-acetylenic secondary amines, the latter being the predominant product (reaction 32)[73,82,91-93]. Allenylzinc bromide reacts similarly with iminoethers to lead to the expected secondary amine (reaction 33)[43]. With organozinc prepared from bromides

$$R^1CH=C=CHZnBr + R^2CH=NR^3 \longrightarrow$$
$$R^2CH(NHR^3)CHR^1C\equiv CH + R^2CH(NHR^3)CH=C=CHR^1 \quad (32)$$

$$CH_2=C=CHZnBr + PhN=CHOC_2H_5 \longrightarrow PhNHCH(CH_2C\equiv CH)_2 \quad (33)$$

$R^1C\equiv CCH_2Br$, the reaction furnishes mixtures of β-acetylenic and α-allenic secondary amines, the latter being the major product[73,93]; the pure allenic amines may be obtained by fractional distillation[73].

The stereochemistry of the reaction between allenic organozinc derivatives and Schiff's bases of the type $PhCH=NR^2$ has been studied[94,95]: the product obtained is always predominantly the *threo* isomer. Reversibility of reactions of these organozinc compounds with imines has been observed[87].

c. Isolated acetylenic triple bonds

Organozincs prepared from propargylic bromides add to the triple bond of simple or functionalized alkynes[61,96,97], to produce the Markovnikoff regioisomer (reactions 34 and 35). In the reaction with propargyl bromide, a second molecule of organozinc

$$CH_3C\equiv CCH_2ZnBr + n\text{-}C_6H_{13}C\equiv CH \xrightarrow[2. H_2O]{1.\ 24\ h,\ 30\ °C} CH_2=C(C_6H_{13})CH_2C\equiv CCH_3 \quad (34)$$
$$70\%$$

$$CH_3C\equiv CCH_2ZnBr + HC\equiv CCH_2OH \xrightarrow[2. H_2O]{1.\ 24\ h,\ 30\ °C} CH_2=C(CH_2OH)CH_2C\equiv CCH_3 \quad (35)$$
$$63\%$$

derivative undergoes addition, followed by an intramolecular displacement of the bromine atom, to lead to the cyclopropane **5**.

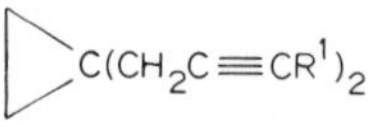

$$\triangleright C(CH_2C\equiv CR^1)_2$$

5

3. Reactions of saturated and miscellaneous organozinc compounds

a. Carbonyl derivatives and epoxides

Saturated organozinc compounds display little reactivity towards aldehydes and generally do not react with ketones and other unsaturated compounds[1-3]; with α,β-unsaturated ketones, they undergo 1,4-addition, without carbonyl addition. However, addition of metal halides increases the reaction rate: the yields of carbinols formed in the reaction of aldehydes and ketones with (a) pure R_2Zn, (b) $R_2Zn + 2MgX_2$, and (c) $2RMgX + ZnX_2$ clearly demonstrate the activing effect of metal halides on the reactivity of organozinc compounds[98]. The stereochemistry of the attack of 4-*tert*-butylcyclohexanone by saturated organozinc derivatives has been studied[99-101].

Normal dialkylzincs do not react with epoxides. However, the *in situ* reagents $2RMgX + ZnX_2$ readily give secondary alcohols without primary alcohols (reaction

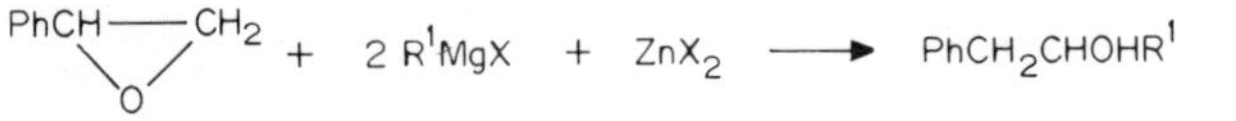

$$PhCH\underset{O}{\overset{}{\diagdown\!\!\!\diagup}}CH_2 + 2\ R^1MgX + ZnX_2 \longrightarrow PhCH_2CHOHR^1 \quad (36)$$

36)[102]. Since dialkylmagnesium affords exclusively the primary alcohol $PhCHR^1CH_2OH$, it was assumed that the reaction actually took place with phenylacetaldehyde formed by MgX_2-catalysed isomerization of styrene oxide rather than with styrene oxide itself. Even without metal halides, dialkylzinc compounds will react with epoxides to give ring opening in dimethyl sulphoxide[103].

2. Carbon—carbon bond formation

Organozincs prepared from α-allenic bromides react as dialkylzinc. These organometallics correspond to a mixture of allenic and dienic structures (reaction 37)[104,105]. They react readily with aldehydes to lead principally to β-allenic secondary alcohols (reaction 38)[105]. On the other hand, they show 40–50% low reactivity towards ketones and do not react with esters.

$$R^1CH=C=CR^2CH_2Br \xrightarrow{Zn} R^1CH=C=CR^2CH_2ZnBr + R^1CH=C(ZnBr)CR^2=CH_2 \quad (37)$$

$$R^1CH=C=CR^2CH_2ZnBr + R^3CHO \longrightarrow R^1CH=C=CR^2CH_2CHOHR^3 \quad (38)$$
$$40\text{–}95\%$$

b. Imines

Normal dialkylzinc compounds are relatively unreactive towards imines. However, the reagent formed *in situ* from $2R^1MgX + ZnX_2$ readily adds across the C=N bond to afford the expected amines[106–108].

$$R^1_2Zn + R^2CH=NR^3 \xrightarrow[\text{2. H}_2\text{O}]{\text{1. MgBr}_2} R^1R^2CHNHR^3 \quad (39)$$
$$55\text{–}63\%$$
$$R^2, R^3 = \text{aryl}$$

c. Terminal alkynes [61,109,110]

Di-*tert*-butylzinc is able to add to the triple bond of terminal alkynes to produce only the mono-addition compound and, as it is a bulky reagent, only the anti-Markovnikoff regioisomer is observed[109,110]. When the reaction is carried out in refluxing diethyl ether, an *anti*-addition is observed[109], giving the sterically crowded (Z)-alkene (reaction 40). The

$$Bu^t_2Zn + HC\equiv CPh \xrightarrow{48 \text{ h}, 35 \,^\circ C} (Z)\text{-}Bu^tCH=CHPh \quad (40)$$
$$70\%$$

same reaction, performed in a higher boiling solvent (tetrahydrofuran instead of diethyl ether) affords, however, a mixture of Z- and E-stereoisomers, probably by isomerization of the Z-isomer[110].

Di-*tert*-butylzinc reacts in refluxing tetrahydrofuran with propargylic or homopropargylic alcohols, ethers, and amines to produce always the anti-Markovnikoff regioisomer. Both Z- and E-stereoisomers are formed in variable ratios, depending on the heteroatom (reaction 41).

$$Bu^t_2Zn + HC\equiv C(CH_2)_nX \xrightarrow{24 \text{ h}, 65 \,^\circ C} (Z)\text{-}Bu^tCH=CH(CH_2)_nX + (E)\text{-}Bu^tCH=CH(CH_2)_nX$$
$$20\text{–}65\%$$
$$n = 1, 2 \,;\, X \quad OH, OC_4H_9, \text{tetrahydropyranyloxy}, NHC_2H_5, N(C_2H_5)_2 \quad (41)$$

d. Conjugated enynes [61,109,111]

Simple and functionalized conjugated terminal enynes easily react with di-*tert*-butylzinc in the same manner as with alkynes. In refluxing diethyl ether, only the anti-addition product is obtained (reaction 42)[109]. This reaction constitutes a regio- and stereoselective method for preparing conjugated dienes. In refluxing tetrahydrofuran[111], the reaction is regioselective but not stereoselective: the formation of four stereoisomers, ZE, EZ, ZZ and EE, (or of two or three isomers only) is observed. This result can be explained by an allylic rearrangement of the intermediate organozinc compound.

$$Bu^t_2Zn + HC{\equiv}CCH{=}CHCH_2Y \xrightarrow{24\,h,\,35\,°C} (Z)\text{-}Bu^tCH{=}CHCH{=}CHCH_2Y \quad (42)$$
$$40\text{-}55\%$$

Y = alkyl, OH, OC$_4$H$_9$, NHC$_2$H$_5$, N(C$_2$H$_5$)$_2$

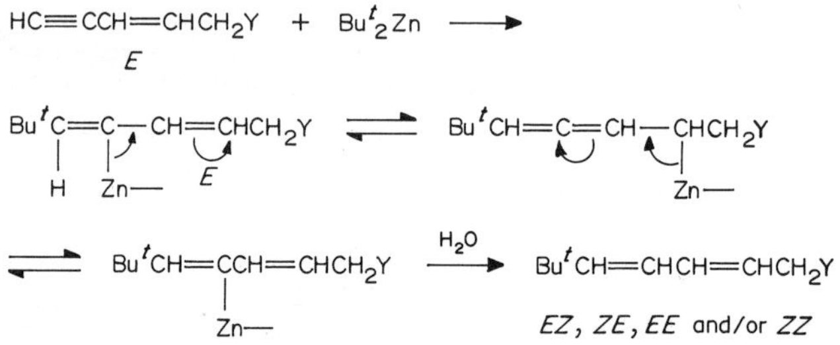

$$HC{\equiv}CCH{=}CHCH_2Y + Bu^t_2Zn \longrightarrow$$
$$E$$

$$\underset{H\quad Zn-}{Bu^tC{=}C{-}CH{=}CHCH_2Y} \rightleftharpoons \underset{Zn-}{Bu^tCH{=}C{=}CH{-}CHCH_2Y} \quad (43)$$

$$\rightleftharpoons \underset{Zn-}{Bu^tCH{=}CCH{=}CHCH_2Y} \xrightarrow{H_2O} Bu^tCH{=}CHCH{=}CHCH_2Y$$
$$EZ,\,ZE,\,EE\text{ and/or }ZZ$$

B. Substitution Reactions

1. Reactions of allylic organozinc compounds

Allylic organozinc compounds react with many derivatives[1-5].

a. Derivatives with a mobile halogen atom

Allylic organozinc compounds easily react with allylic halides (reactions 44 and 45)[49,112-114]. **6** is always obtained as the major product (60–90%) and **7** the minor product

$$R^1CH{=}CHCH_2ZnBr + BrCH_2CH{=}CH_2 \longrightarrow CH_2{=}CHCHR^1CH_2CH{=}CH_2 \quad (44)$$
$$57\text{-}82\%$$

$$R^1CH{=}C(R^2)CH_2ZnBr + ClCH_2CH{=}CHCH_2OH \longrightarrow CH_2{=}C(R^2)CH(R^1)CH_2CH{-}CHCH_2OH +$$
$$\mathbf{6}$$

R^1, R^2 = H, alkyl

$$CH_2{=}C(R^2)CH(R^1)CH(CH_2OH)CH{=}CH_2 +$$
$$\mathbf{7}$$

$$CH_2{=}C(R^2)CH(R^1)CHOHCH_2CH{=}CH_2$$
$$\mathbf{8}$$
$$70\text{-}83\% \quad (45)$$

(3–6%). Allylzinc compounds also react with α-allenic bromides in 26–35% yield[104]. Allylzinc bromide reacts with chloromethyl methyl ether in a similar manner (reaction

46)[55]. With acyl chlorides, the reaction leads to tertiary alcohols, as for the acid anhydrides[4]. Under certain conditions, the reaction with ethyl chloroformate gives only the β-ethylenic esters in good yields (reaction 47)[115], but allylzinc bromide leads only to triallylmethanol[115].

$$R^1CH=CHCH_2ZnBr + ClCH_2OCH_3 \longrightarrow CH_2=CHCHR^1CH_2OCH_3 \quad (46)$$
$$60-70\%$$

$$R^1CH=CHCH_2ZnBr + ClCOOC_2H_5 \xrightarrow[\text{inverse addition}]{-10\ °C} CH_2=CHCHR^1COOC_2H_5 \quad (47)$$
$$53-70\%$$

Recently, reactions of allylic organozinc derivatives with immonium salts have been studied (reaction 48)[116]. With R^1 = alkyl, **9** are the major products; with $R^1 = C_6H_5$ or vinyl, a mixture of **9** and **10** is obtained.

$$R^1CH=CHCH_2ZnBr + [R^2CH=\overset{+}{N}R^3_2, Cl^- \rightleftharpoons R^2CHClNR^3_2] \longrightarrow$$

$$\underset{\textbf{9}}{CH_2=CHCHR^1CHR^2NR^3_2} + \underset{\textbf{10}}{R^1CH=CHCH_2CHR^2NR^3_2} \quad (48)$$
$$40-82\%$$

b. Derivatives with a reactive alkoxy group

Allylic organozinc compounds easily react with *gem*-aminoethers and *gem*-aminothioethers to give β-unsaturated tertiary amines (reaction 49)[53,73,117,118]. These

$$R^1CH=CHCH_2ZnBr + R^2OCHR^3NR^4_2 \longrightarrow CH_2=CHCHR^1CHR^3NR^4_2 \quad (49)$$
$$65-75\%$$

organometallics also react with methoxyaminoesters to give unsaturated aminoesters (reaction 50)[119]. Finally allylzinc bromide reacts with α-ethylenic orthoesters to give a mixture of a ketal and a ketene acetal (reaction 51)[120].

$$CH_2=CHCH_2ZnBr + CH_3OCH[N(C_2H_5)_2]COOCH_3 \longrightarrow CH_2=CHCH_2CH[N(C_2H_5)_2]COOCH_3$$
$$38\% \quad (50)$$

$$CH_2=CHCH_2ZnBr + R^2CH=CHC(OR^3)_3 \longrightarrow R^2CH=CHC(CH_2CH=CH_2)(OR^3)_2 +$$
$$R^2CH(CH_2CH=CH_2)CH=C(OR^3)_2$$
$$64-71\% \quad (51)$$

Substitution of the cyano group of *gem*-aminonitriles has also been observed, with good yields[53].

2. Reactions of propargylic and allenic organozinc compounds

Propargylic and allenic organozinc compounds react with many derivatives[2,82].

a. Derivatives with a mobile halogen atom

Organozinc compounds prepared from propargyl halides react readily with immonium salts to give a mixture of acetylenic and allenic amines (reaction 52)[121].

$$R^1C{\equiv}CCHR^2X \xrightarrow[\text{2. } R^3CH={\overset{+}{N}}(R^4)_2, Cl^-]{\text{1. Zn/thf}} \underset{\mathbf{11}}{R^1C{\equiv}CCHR^2CHR^3N(R^4)_2} + \underset{\mathbf{12}}{R^2CH{=}C{=}CR^1CHR^3N(R^4)_2}$$

$R^1 = H$, $R^2 = CH_3$; yield = 40%, **11/12** = 50:50
$R^1 = CH_3$, $R^2 = H$; yield = 73%, **11/12** = 1:99 (52)

b. Derivatives with a reactive alkoxy group

Propargylic and allenic organozinc compounds react readily with *gem*-aminoethers (reaction 53)[73,118].

Similar reactions (54 and 55) are observed with methoxyaminoesters[122].

$$R^1C{\equiv}CCHXR^2 \xrightarrow[\text{2. } R^2OCHR^3NR^4_2]{\text{1. Zn/thf}} \underset{\mathbf{13}}{R^1C{\equiv}CCHR^2CHR^3NR^4_2} + \underset{\mathbf{14}}{R^2CH{=}C{=}CR^1CHR^3NR^4_2}$$

$R^1, R^2, R^3 = H$; yield = 60%, **13/14** = 100:0 (53)
$R^1 = CH_3$, $R^2 = H$, $R^3 = H$ or C_6H_5; yield = 38–49%, **13/14** = 10:90

$$CH_2{=}C{=}CHZnBr + CH_3OCH[N(C_2H_5)_2]COOCH_3 \longrightarrow HC{\equiv}CCH_2CH[N(C_2H_5)_2]COOCH_3$$
$$36\%$$ (54)

$$R^1C{\equiv}CCH_2Br \xrightarrow[\text{2. } CH_3OCH[N(C_2H_5)_2]COOCH_3]{\text{1. Zn/thf}} CH_2{=}C{=}C(R^1)CH[N(C_2H_5)_2]COOCH_3$$
$$40\text{–}43\%$$ (55)

3. Reactions of saturated and miscellaneous organozinc compounds

a. Derivatives with a mobile halogen atom

Dialkyl- and diaryl-zincs are reported to react with *tert*-alkyl chlorides and with alkyl iodides[1–3,123] to form hydrocarbons in 25–51% yield; this reaction gives much better yields in the presence of titanium(IV) catalysts[124]. Cyclopropane and cyclobutane derivatives can be prepared from 1,3-dihalopropanes and 1,4-dihalobutanes, using metallic zinc[3,125] or dialkylzincs[3,126]; these reactions presumably proceed via the intermediate formation of organozinc compounds. Saturated alkylzinc halides react with chloromethyl methyl ether[2] and similar compounds (reaction 56)[2].

$$ClCH(OC_2H_5)COOC_2H_5 + C_3H_7ZnI \longrightarrow C_3H_7CH(OC_2H_5)COOC_2H_5 \qquad (56)$$

$$R^2COCl + R^1ZnI \longrightarrow R^1COR^2 + ZnClI \qquad (57)$$

2. Carbon—carbon bond formation

Dialkylzinc reagents react with acyl chlorides to give ketones[1-3], but metallic halides play an important role in these reactions. Introduction of $AlCl_3$, $MgBr_2$, or $ZnBr_2$ increases the yield of ketones[127,128]. Use of the organozinc compounds R^1ZnI enables various ketones to be prepared easily from acyl chlorides (reaction 57) (Blaise's method)[1,2,129-131].

b. Palladium-catalysed reaction with aryl, alkenyl, and allenic halides

Recently, several interesting palladium-catalysed coupling reactions have been described. Thus, aryl iodides or bromides react with alkynyl-, benzyl-, and aryl-zinc halides to lead to arylalkynes[132], diarylmethanes[133], and biphenyl derivatives[133], respectively. Selective mono-alkylation and -arylation of aromatic dihalides containing the same halogen atoms has been achieved[134]. Similarly, the palladium-catalysed coupling reaction of alkenyl halides with alkynylzinc reagents[135] and homoallylic[136] and homopropargylic[137] zinc halides constitutes a new selective route to conjugated enynes, 1,5-dienes, and 1,5-enynes, respectively. Lastly, a new route to aryl-, vinyl-, and 1-alkynyl-allenes and diallenes via the [Pd(PPh$_3$)$_4$]-promoted reaction of propargylic or allenic halides with appropriate organozinc halides has been described (reaction 58)[138,139].

$$R^2R^3C=C=CHX \text{ or } R^2R^3C(X)C\equiv CH \xrightarrow[[Pd(PPh_3)_4]]{R^1ZnCl} R^2R^3C=C=CHR^1$$

R^1 = Ph, $CH_2=CH$, $(CH_3)_3SiC\equiv C$, $PhC\equiv C$, $H_2C=C(CH_3)C\equiv C$, $HC\equiv CC\equiv C$,

$Bu^tCH=C=CH$ \hfill (58)

c. Derivatives with a reactive alkoxy group

Saturated organozinc compounds react with alkyl alkoxydialkylaminoacetates to give aminoesters (reaction 59)[140]. Vinylic and alkynylic organozinc compounds also react with these derivatives[83,140].

$$R^1_2Zn + CH_3OCH[N(C_2H_5)_2]COOCH_3 \longrightarrow (C_2H_5)_2NCHR^1COOCH_3 \quad (59)$$
$$64-76\%$$

C. Carbene and Carbenoid Intermediates from Organozinc Compounds

1. Addition and insertion reactions

a. Carbon—Carbon double bonds

Halogenomethyl organozincs are reagents of outstanding utility for the preparation of cyclopropanes and halogenocyclopropanes from alkenes (Simmons–Smith reaction, 60)[2,3,141-146]. The active intermediate of the reaction must involve the Zn—CH_2I linkage

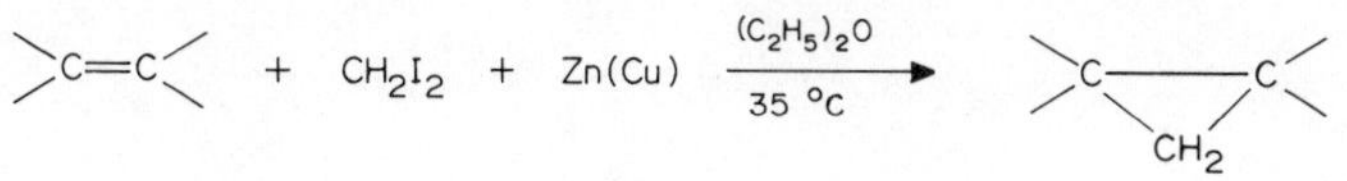

$$\text{>C=C<} + CH_2I_2 + Zn(Cu) \xrightarrow[35\ ^\circ C]{(C_2H_5)_2O} \text{>C—C<} \text{ (with } CH_2\text{ bridge)} \quad (60)$$

and could be either **15** or **16** or a mixture of both (reaction 61). Cyclopropane derivatives are formed in a stereospecific way[2,3,141-146]: reactions with (Z)- and (E)-olefins give the

$$2 \text{ ICH}_2\text{ZnI} \rightleftharpoons (\text{ICH}_2)_2\text{Zn} + \text{ZnI}_2 \qquad (61)$$
$$\quad\;\; \textbf{15} \qquad\qquad\quad \textbf{16}$$

corresponding (Z)- and (E)-cyclopropane derivatives, respectively. A Z-addition mechanism is generally accepted for the reaction and the transition state **17** has been suggested on the basis of experimental observations[141,143-145].

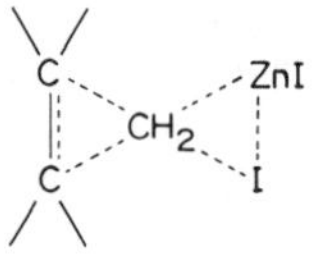

17

The presence of functional groups such as an ether, ester, hydroxy, ketone, ketal, amine or enamine in olefins frequently facilitates the Simmons–Smith reaction; many examples have been studied, often with determination of the stereochemistry of the products[1-3,141,146-166]. The Simmons–Smith reaction often starts very slowly and proceeds slowly at room temperature. The preparation of an active zinc–copper couple has been improved[167-169]. Zinc dust[170] and activated zinc[171] prepared by ultrasonic irradiation have recently been used. Zinc/copper(I) chloride[172,173] or a zinc–silver couple[174-176] may be used instead of the zinc–copper couple to give higher yields.

Several modified Simmons–Smith reactions have been proposed[2,3]. The Wittig–Schwarzenbach reagent ($CH_2N_2 + ZnI_2$) gives rise to the formation of the same carbenoid as the Simmons–Smith reaction and reacts with various olefins to produce the corresponding cyclopropane derivatives[177-184]. The Furukawa method[185-190] uses diethylzinc instead of the zinc–copper couple (reaction 62). This reaction is much more

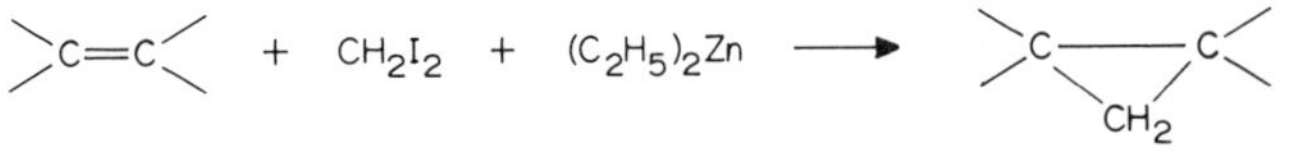

$$\text{>C=C<} + \text{CH}_2\text{I}_2 + (\text{C}_2\text{H}_5)_2\text{Zn} \longrightarrow \text{>C—C<}\;\;(\text{CH}_2) \qquad (62)$$

rapid than the Wittig–Schwarzenbach reaction and proceeds stereospecifically: (Z)- and (E)-olefins afford cyclopropanes whose configuration are Z and E with respect of the substituents of the starting alkenes. The active species of this reaction may include $ICH_2ZnC_2H_5$, ICH_2ZnI, and/or $(ICH_2)_2Zn$ or associated complexes containing these molecules. This reaction leads to much better results than the Simmons–Smith reaction when methyldiiodomethane, aryldiiodomethane, or various polyhalomethanes are used[187,191-200]. It can be used to produce norcaradiene derivatives from aromatic hydrocarbons[192]. The reaction of alkylbenzenes with diethylzinc and 1,1-diiodoethane gives 7-methylcyclohepta-1,3,5-triene derivatives in 31–44% yields, via a norcaradiene intermediate[201]. The Sawada modified reaction (63)[202] gives better yields in certain cases.

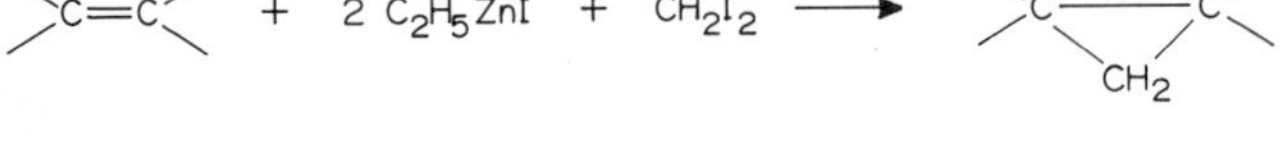

$$\text{>C=C<} + 2\,\text{C}_2\text{H}_5\text{ZnI} + \text{CH}_2\text{I}_2 \longrightarrow \text{>C—C<}\;\;(\text{CH}_2) \qquad (63)$$

b. Carbon—Carbon triple bonds

Terminal acetylenic compounds give methylacetylenes and allene derivatives with the Simmons–Smith reagent[203–205], probably via an insertion reaction (reaction 64). With disubstituted alkynes, the reaction forms cyclopropenes and their isomers (reaction

$$R^1C\equiv CH \xrightarrow[(C_2H_5)_2O,\ 8-15\ days]{CH_2I_2/Zn/Cu} R^1C\equiv CCH_3 + R^1CH=C=CH_2 + R^1CH_2C\equiv CH$$

R^1 = alkyl, aryl   (64)

$$R^1CH_2C\equiv CCH_2R^2 \xrightarrow{CH_2I_2 \atop Zn(Cu)} R^1CH_2\underset{CH_2}{\overset{}{C=CCH_2R^2}} + R^1CH_2\underset{CH_2}{\overset{}{CH-C=CHR^2}} +$$

$$R^1CH=\underset{CH_2}{\overset{}{C-CH-CH_2R^2}}$$ (65)

65)[2,3,206]. With cyclooctyne, the Simmons–Smith reagent undergoes a trans-annular reaction to lead to bicyclo[3.3.0]octane derivatives[207].

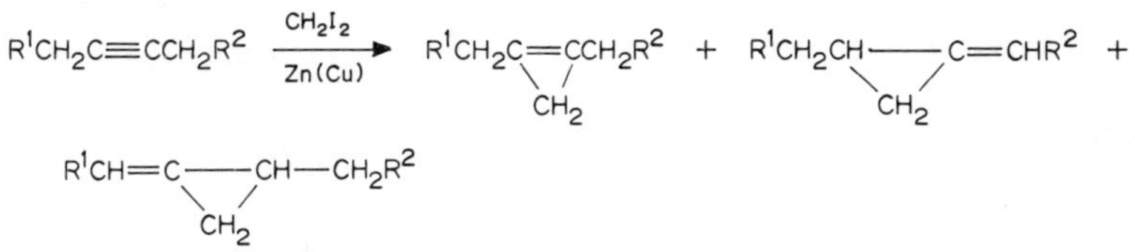

c. C=N derivatives

The reaction of Simmons–Smith reagent is not applicable to simple imines, but gives an aziridine with an iminoester[208].

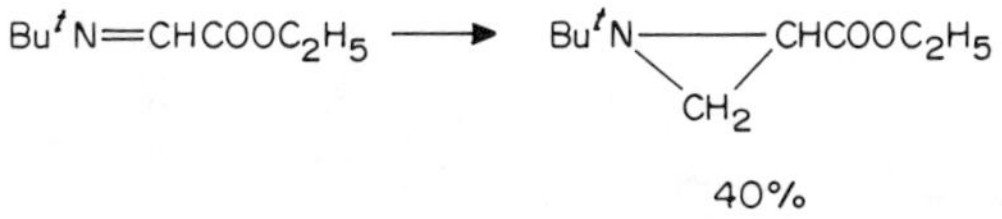

$$Bu^tN=CHCOOC_2H_5 \longrightarrow Bu^tN\underset{CH_2}{\overset{}{-CHCOOC_2H_5}}$$ (66)

40%

2. Wittig alkene synthesis

The reaction of aldehydes with the Simmons–Smith reagent, in the presence of excess of zinc dust, gives the corresponding olefins in 29–63% yields[2,3,209–211]. This reaction is not applicable to ketones[210], except in particular cases[212,213]. Two modified Simmons–Smith reagents[214], $CH_2I_2/Zn/(CH_3)_3Al$ and $CH_2Br_2/Zn/TiCl_4$, appear to be excellent methylenation reagents for aldehydes and ketones, respectively (reaction 67).

$$R^1COR^2 \longrightarrow R^1R^2C=CH_2$$

62–86% with $CH_2I_2/Zn/(CH_3)_3Al$

55–89% with $CH_2I_2/Zn/TiCl_4$ (67)

R^1 = alkyl, aryl ; R^2 = H, alkyl, aryl

D. Reformatsky and Related Reactions

The reaction between a carbonyl compound and an α-haloester, in the presence of zinc, is commonly known as the Reformatsky reaction (reaction 68). Four reviews[1,2,215,216]

$$XCR^1R^2COOR^3 + R^4COR^5 + Zn \longrightarrow R^4R^5C(OH)CR^1R^2COOR^3 \qquad (68)$$

covering the literature up to and including 1972 show that this reaction has been developed extensively. In this reaction, various halogen compounds can be used instead of α-haloesters: γ-halo-α,β-unsaturated esters, α-halopolyesters, α-polyhalo esters, α-polyhalo polyesters, α-halonitriles, α-haloamides, α-halothioesters, and α-halolactones. Also, the reaction can be applied not only to aldehydes and ketones, but also to esters, acid halides, nitriles, imines, nitrones, imides, ketenes, and epoxides; 1,2- and 1,4-additions to α,β-unsaturated compounds have also been reported.

During the period 1973–82, several new applications of the Reformatsky reaction in synthetic organic chemistry have been found, in addition to the continuation of research concerning new experimental conditions[170,217-219] or mechanistic characteristics (stereochemistry, reversibility of the reaction)[220-222].

1. Addition reactions

a. Carbonyl derivatives

The Reformatsky reaction continues to find numerous applications in synthetic chemistry, especially in the field of steroid synthesis[223-228]. Reactions with aldehydes and ketones of Reformatsky reagents such as derivatives of α-bromo salts[229] or of tetrahydropyranyl esters of α-bromo acids[230] or α-bromoaryl ketones[231] have been studied. Reformatsky reagents react with α- or β-aminoketones[232,233]. Asymmetric inductions in Reformatsky reactions with Mannich-type ketones[234-236] and with α-aminoketones[237] have been described. New C-4 substituted cephalosporins have been prepared via the Reformatsky reaction[238]. Reactions with the reagents prepared from alkyl halocrotonates have been developed in several cases (reaction 69)[218,222,239,240]. ^{14}C-labelled butyrolac-

$$R^2CH=C(COOC_2H_5)_2 + BrCH_2CH=CHCOOCH_3 \xrightarrow{Zn} \qquad (69)$$

$$R^2CH[CH(COOC_2H_5)_2]CH_2CH=CHCOOCH_3 \;+$$

$$R^2CH[CH(COOC_2H_5)_2]CH(COOCH_3)CH=CH_2$$

75–85 %

tones have been prepared by the Reformatsky reaction of ^{14}C-labelled bromoacrylates with acetone and cyclohexanone[241].

b. C=N derivatives and nitriles

The reactivity and stereochemistry of the reaction between aldimines and the reagent derived from α-bromamides have been studied[242]. The Reformatsky reaction of $BrZnCH_2COOR^1$ (R^1 = l-menthyl) with aldimines ($R^2CH=NR^3$) has been used as a primary step in the asymmetric synthesis of β-amino acids and aspartic acid[243]. With azirines, the Reformatsky reaction leads to β-aziridino esters exclusively when the reaction is carried out in a benzene–diethyl ether mixture[244,245]. Ketonitrones react with organozinc compounds derived from ethyl bromoacetate to give isoxazolidinones (reaction 70)[246]. New reactions have also been observed with nitriles[247,248], for example reaction 71.

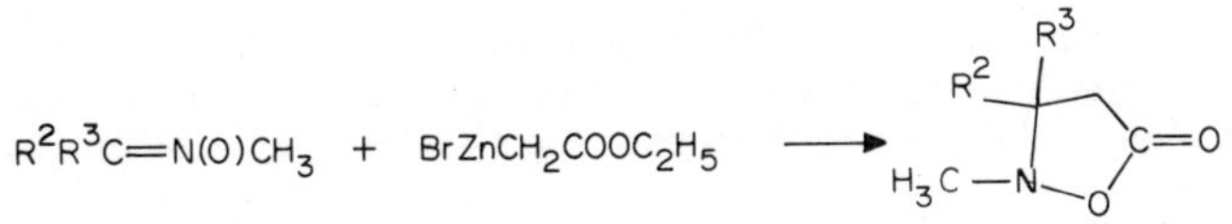

2. Carbon—carbon bond formation

$$R^2R^3C{=}N(O)CH_3 \;+\; BrZnCH_2COOC_2H_5 \;\longrightarrow\; \underset{H_3C-N\diagdown O}{\overset{R^2\diagup R^3}{\diagdown C{=}O}} \quad (70)$$

$$R^1R^2C(C{\equiv}N)ZnBr \;+\; R^3C{\equiv}N \;\longrightarrow\; R^3COCR^1R^2C{\equiv}N \quad (71)$$
$$47\text{-}82\%$$

R^1, R^2 = H, alkyl; R^3 = alkyl, aryl

c. Terminal alkynes

Malonic-type organozinc halides add to the triple bond of non-functionalized alk-1-ynes[249-251] to lead only to the Markovnikoff-type product (reactions 72 and 73). The

$$BrZnC(CH_3)(COOC_2H_5)_2 \;+\; HC{\equiv}CR^2 \;\xrightarrow{4\text{ h, }42\ °C}\; CH_2{=}CR^2C(CH_3)(COOC_2H_5)_2 \quad (72)$$
$$50\%$$

$$BrZnC(CH_3)(CN)COOC_2H_5 \;+\; HC{\equiv}CR^2 \;\xrightarrow{23\text{ h, }42\ °C}\; CH_2{=}CR^2C(CH_3)(CN)COOC_2H_5$$
$$20\text{-}60\% \quad (73)$$

reaction also occurs readily with functionalized alk-1-ynes, **18**[250-252]. In the **18** propargylic series, the orientation of these addition reactions is generally that observed in

$$HC{\equiv}C(CH_2)_n R^2$$
18
$$R^2 = OH, OR^3, NHR^3, N(R^3)_2$$

the addition to simple alk-1-ynes, although sometimes anti-Markovnikoff addition products are also obtained if the R^2 group is bulky (reaction 74). With homopropargylic

$$BrZnC(CH_3)(COOC_2H_5)_2 \;+\; HC{\equiv}CCH_2OBu^n \;\longrightarrow\; \quad (74)$$

$$CH_2{=}C(CH_2OBu^n)C(CH_3)(COOC_2H_5)_2 \;+$$
$$70 \hspace{4cm} 59\%$$

$$Bu^nOCH_2CH{=}CHC(CH_3)(COOC_2H_5)_2$$
$$30$$

$$BrZnC(CH_3)(COOC_2H_5)_2 \;+\; HC{\equiv}CCH_2CH_2OC_2H_5 \;\xrightarrow{23\text{ h, }42\ °C}\; \quad (75)$$

$$CH_2{=}C(CH_2CH_2OC_2H_5)C(CH_3)(COOC_2H_5)_2$$
$$56\%$$

substrates[252], the reaction always affords the Markovnikoff-type regioisomer (reaction 75). It is interesting that these reactions, when they are effected with alcohols and secondary amines, lead directly to the corresponding lactones and lactams (reactions 76 and 77)[250-252].

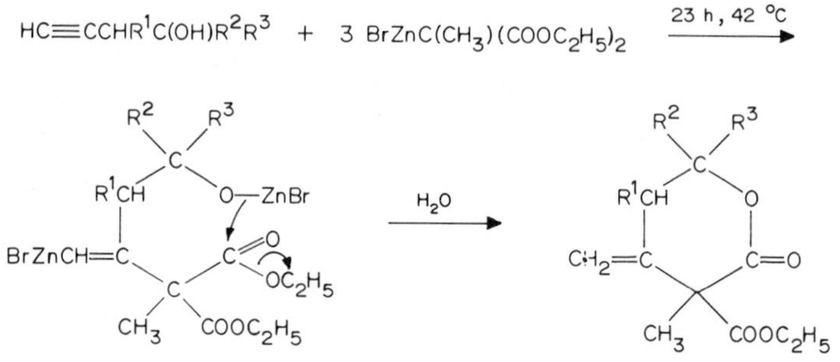

$$HC{\equiv}CCHR^1C(OH)R^2R^3 \; + \; 3 \; BrZnC(CH_3)(COOC_2H_5)_2 \xrightarrow{23 \; h, \; 42 \; °C}$$

[intermediate structure with ZnBr] $\xrightarrow{H_2O}$ [cyclic product] (76)

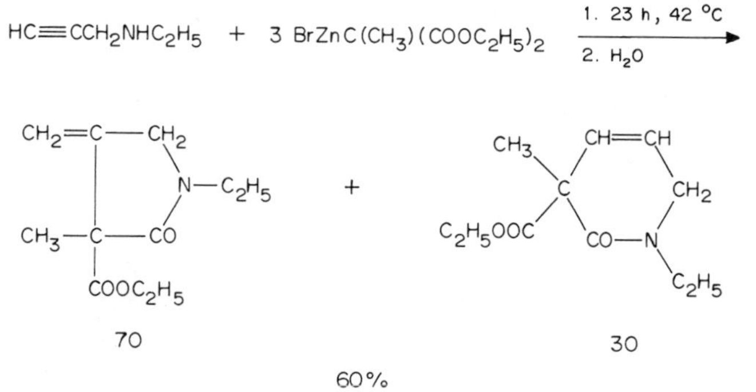

$$HC{\equiv}CCH_2NHC_2H_5 \; + \; 3 \; BrZnC(CH_3)(COOC_2H_5)_2 \xrightarrow[2. \; H_2O]{1. \; 23 \; h, \; 42 \; °C}$$

[two products shown] (77)

70   30

60%

2. Substitution reactions

a. Derivatives with a mobile halogen atom

Reformatsky reagents generally do not react with simple halides, but readily give substitution products with compounds having a mobile halogen atom (reaction 78)[2,216,253-255]. Chloromethyl ethers[253,256-261] and chloromethyl thioethers[262,263] also react.

$$CH_2{=}CHCH_2Br \; + \; ClZnC(Cl)(COOC_2H_5)_2 \longrightarrow CH_2{=}CHCH_2C(Cl)(COOC_2H_5)_2 \quad (78)$$

64%

The reagent derived from ethyl bromoacetate can be arylated by simple aromatic and vinylic halides in the presence of a transition metal catalyst and a dipolar aprotic solvent (reaction 79)[264,265]. Acid chlorides give functional ketones with Reformatsky reagents

$$ArX \; + \; BrZnCH_2COOC_2H_5 \xrightarrow{[Ni(PPh_3)_4]} ArCH_2COOC_2H_5 \; + \; ZnBrX \quad (79)$$

(reaction 80)[266]. A similar reaction is observed with alkyl chloroformate, leading to

2. Carbon—carbon bond formation

$$PhCOCR^1R^2ZnBr + R^3COCl \longrightarrow PhCOCR^1R^2COR^3 \quad (80)$$

functional esters[266]. Lastly, Reformatsky reagents may produce tertiary functional amines with immonium salts (reaction 81)[267].

$$BrZnCH_2COR^1 + R^2CH\overset{+}{=}NR^3_2, X^- \longrightarrow R^1COCH_2CHR^2NR^3_2 \quad (81)$$
$$60-68\%$$
$$R^1 = OC_2H_5, N(C_2H_5)_2$$

b. Derivatives with an alkoxy or acyloxy group

gem-Aminoethers lead to tertiary functional amines with Reformatsky reagents (reaction 82)[267,268]. Substitution of acyloxy groups has been observed under certain

$$BrZnCH_2COR^1 + R^2OCH_2NR^3_2 \longrightarrow R^1COCH_2NR^3_2 \quad (82)$$
$$68-91\%$$
$$R^1 = OC_2H_5, N(C_2H_5)_2$$

conditions (reaction 83)[269]. Finally, Reformatsky reagents enable ketoesters[270,271] to be prepared from acid anhydrides (reaction 84).

$$R^2COOCHPh_2 + BrCH_2COOC_2H_5 + Zn \longrightarrow R^2COOH + Ph_2CHCH_2COOC_2H_5 \quad (83)$$
$$(excess)$$

$$BrZnC(CH_3)_2COOC_2H_5 + R^2COOCOCF_3 \longrightarrow R^2COC(CH_3)_2COOC_2H_5 \quad (84)$$

III. REACTIONS OF CADMIUM ORGANOMETALLICS

Several general reviews[272-274] have covered the literature on organometallic compounds of cadmium up to and including 1976.

A. Addition Reactions

Organocadmium compounds are less reactive than the corresponding magnesium or zinc compounds towards keto groups[272-274]; however, allylic or benzylic organocadmium compounds are more reactive than saturated derivatives and, under certain conditions, saturated organocadmium compounds may show good reactivity towards carbonyl and other unsaturated products.

1. Reactions of allylic organocadmium compounds and analogues

Allylic organocadmium reagents react readily with aldehydes and ketones to give β-ethylenic alcohols (reaction 85)[13,99,272-276]. Substituted allylic organocadmium

$$(CH_2=CHCH_2)_2Cd + R^2COR^3 \longrightarrow R^2R^3C(OH)CH_2CH=CH_2 \quad (85)$$
$$50-90\%$$

compounds react with complete allylic rearrangement[272-276]. With esters, they readily form diallyl alkyl-methanols (reaction 86)[13].

$$CH_3COOC_2H_5 + (CH_2\!\!=\!\!CHCH_2)_2Cd \longrightarrow CH_3C(OH)(CH_2\!\!=\!\!CHCH_2)_2 \quad (86)$$
$$70\%$$

Allylic organocadmium compounds give only 1,2-addition with various α-enones (reaction 87)[277]. They also react with carbon dioxide to give β-ethylenic acids (reaction 88)[14,278].

$$R^2CH\!\!=\!\!CHCOR^3 + R^1CH\!\!=\!\!CHCH_2CdCl \longrightarrow R_1CH\!\!=\!\!CHC(OH)R^2CHR^1CH\!\!=\!\!CH_2 \quad (87)$$
$$80\text{-}90\%$$

$$(CH_2\!\!=\!\!CHCH_2)_2Cd + CO_2 \xrightarrow[H_2O/H^+]{-20\ °C \rightarrow 0\ °C} CH_2\!\!=\!\!CHCH_2COOH \quad (88)$$

Benzylic organocadmium compounds have been less studied than allylic derivatives; they normally react with carbonyl derivatives[279,280] and lead to 1,4-addition with dialkyl alkylidene malonates (reaction 89)[281].

$$R^1CH\!\!=\!\!C(COOC_2H_5)_2 + ArCH_2CdCl \longrightarrow R^1CH(CH_2Ar)CH(COOC_2H_5)_2 \quad (89)$$
$$60\text{-}77\%$$

Finally, the organocadmium compound prepared from propargyl bromide reacts with ketones to give a mixture of propargylic and allenylic alcohols, but in very low yields[85].

2. Reactions of saturated and phenylic organocadmium compounds

These organocadmium reagents react readily with aromatic aldehydes in the presence of metal halides such as MgI_2, $MgBr_2$, $MgCl_2$, LiBr, $ZnBr_2$, or $AlCl_3$ (reaction 90)[272-274,282-286]. Aliphatic aldehydes also react, but lead to several products (reaction 91)[283,286]. In similar manner, tertiary alcohols are formed with ketones, but generally in

$$(C_2H_5)_2Cd + MgBr_2 + PhCHO \xrightarrow[1\ h,\ 25\ °C]{(C_2H_5)_2O} PhCHOHC_2H_5 \quad (90)$$
$$86\%$$

$$R^1_2Cd + 2\ Pr^nCHO + 2\ MgBr_2 \longrightarrow \begin{cases} Pr^nCHOHR^1 & \text{(addition 20-40\%)} \\ Pr^nCOR^1 & \text{(Verley-Pondorf reduction)} \\ Pr^nCH_2OH & \text{(aldehyde reduction)} \end{cases} \quad (91)$$

low yields[99,287,288]. The stereochemistry of the addition of organocadmium reagents to 4-*tert*-butylcyclohexanone has been studied[13,99-101,274].

Organocadmium derivatives are more selective than other organometallic compounds and are used for mono-reactions from aldehydes or ketones which possess other functional groups such as COOR, NO_2, CH_2Cl, $CHCl_2$, or CCl_3 (reaction 92)[283-285,289-295].

2. Carbon—carbon bond formation

$$R^1_2Cd + C_2H_5OCOCHO \longrightarrow R^1CHOHCOOC_2H_5 \quad (92)$$
$$50\text{-}56\%$$

Saturated and phenylic organocadmium reagents generally give the product resulting from an 1,4-addition[274,296-299] with α-unsaturated ketones, α-unsaturated nitro compounds and dialkyl alkylidene malonates (reactions 93 and 94). They may react in the same

$$R^1_2Cd + R^2CH=CHNO_2 \longrightarrow R^1R^2CHCH_2NO_2 \quad (93)$$
$$55\text{-}76\%$$

$$(\alpha\text{-}C_{10}H_7)_2Cd + R^2CH=C(COOC_2H_5)_2 \longrightarrow \alpha\text{-}C_{10}H_7CHR^2CH(COOC_2H_5)_2 \quad (94)$$
$$60\text{-}77\%$$

manner with alkyl alkylidene cyanoacetates[300], but the main product formed in this reaction results of the reduction of the C=C bond.

Saturated and phenylic organocadmium compounds do not react with epoxides; however, in the presence of MgX_2, isomerization of the epoxide to a carbonyl derivative occurs and a reaction is then observed (reaction 95)[301]. These organocadmium

$$R^1_2Cd + PhCH\underset{O}{-\!\!-\!\!-}CH_2 + MgX_2 \longrightarrow [PhCH_2CHO] \longrightarrow PhCH_2CHOHR^1 \quad (95)$$
$$55\text{-}68\%$$

compounds do not react with simple esters[302], but give a hydroxyester with oxalic acid diesters (reaction 96)[273,283,303]. A reaction (97) has been also observed with pro-

$$R^1_2Cd + C_2H_5OCOCOOC_2H_5 + MgBr_2 \longrightarrow R^1_2C(OH)COOC_2H_5 \quad (96)$$

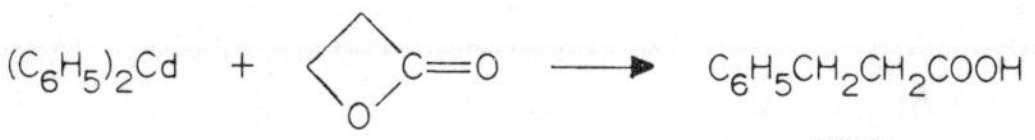

$$(C_6H_5)_2Cd + \underset{O}{\square}C=O \longrightarrow C_6H_5CH_2CH_2COOH \quad (97)$$
$$77\%$$

piolactone[304]. With acid anhydrides, saturated and phenylic organocadmium compounds form ketones (reaction 98)[272-274,283,295,305-307].

Saturated and phenylic organocadmium derivatives add to the imine bond (reaction 99)[106-108,274]. The effect of substituents ($p\text{-}OCH_3$, $p\text{-}CH_3$, $p\text{-}Cl$, $m\text{-}Cl$) in both rings on the

$$R^1_2Cd + 2R^2COOCOR^2 + MgBr_2 \longrightarrow R^1COR^2 + (R^2COO)_2Cd \quad (98)$$
$$50\text{-}60\%$$

$$PhCH=NPh + R^1_2Cd + MgBr_2 \longrightarrow PhCHR^1NHPh \quad (99)$$

rate of addition has been studied. Saturated and phenylic organocadmium derivatives also react with iminoglyoxylates to lead regiospecifically to secondary amines (reaction 100)[274]. On the other hand, saturated organocadmium compounds are unreactive

$$R^2N=CHCOOR^3 + R^1_2Cd \longrightarrow R^2NHCHR^1COOR^3 \quad (100)$$

towards nitriles[273,274]. An example of the inertness of the nitrile group is provided by 3-trichlorosilylpropanenitrile, which undergoes displacement at silicon in 90% yield (reaction 101).

$$Cl_3SiCH_2CH_2CN + (CH_3)_2Cd \longrightarrow CH_3SiCl_2CH_2CH_2CN \qquad (101)$$

3. Reactions of organocadmium reagents prepared from α-bromoesters

Organocadmium derivatives prepared from α-bromoesters react readily with aldehydes and ketones to produce β-hydroxyesters (reactions 102 and 103)[216,273,274,308].

$$Bu^tOCOCH_2CdBr + Pr^nCHO \xrightarrow[\text{or hmpa}]{\text{dmso}} Pr^nCHOHCH_2COOBu^t \qquad (102)$$
$$45\text{-}65\%$$

$$(CH_3CHCOOC_2H_5)_2Cd + n\text{-}C_6H_{13}COCH_3 \longrightarrow n\text{-}C_6H_{13}C(OH)(CH_3)CHCH_3COOC_2H_5$$
$$75\% \qquad (103)$$

B. Substitution Reactions

1. Alkylation

a. Alkyl, allylic, and benzylic halides

Organocadmium compounds do not generally react with primary and secondary alkyl halides[273,274,282,308] and give an 1,2-elimination with tertiary chlorides[273,274,308]. However dipropylcadmium in dmso[273] and diphenylcadmium[309] react readily with allyl bromide. Saturated and phenylic organocadmium compounds also give a reaction with benzylic halides[309-311].

The reaction of triphenylchloromethane with saturated organocadmium reagents yields alkylation and/or reduction products, depending on the nature of the R^1 group of the organometallic compound[312]. Organocadmium reagents (R^1 = alkyl, phenyl, benzyl) form principally (Z)-alk-2-enes by reaction with tetracarbonyl(methylallyl)iron cationic complexes[313]; alkylation of tricarbonylcyclohexadienyliron cationic complexes has been achieved[314].

b. Halides with other functional groups

Organocadmium compounds give a substitution reaction (104) with α-bromoketones[315] and β-chlorovinyl ketones[316,317]. With gem-chloroalkoxy[274,309,318-320] or

$$R^1_2Cd + 2 R^2COCH=CHCl \longrightarrow 2 R^2COCH=CHR^1 + CdCl_2 \qquad (104)$$
$$30\text{-}60\%$$
$$R^1 = \text{alkyl, phenyl, benzyl}; \; R^2 = \text{alkyl}$$

gem-chloroacyloxy[274,306,321-324] compounds, substitution of the only halogen atom generally occurs (reaction 105) although, in particular cases, both substitution of the halogen atom and an alkoxy or acyloxy group is observed[319,320,325].

$$R^1_2Cd + R^2OCHClR^3 \longrightarrow R^2OCHR^1R^3 \qquad (105)$$

2. Carbon—carbon bond formation

tert-Butyl-1-chloroethyl peroxide reacts with dimethylcadmium to afford the displacement product in 50% yield (reaction 106)[274]. With α-bromoesters, one of two reactions (107a or 107b) may be observed[308,309,315,326], the second often being predominant.

$$Bu^tOOCHClCH_3 + (CH_3)_2Cd \longrightarrow Bu^tOOCH(CH_3)_2 \qquad (106)$$

$$R^1_2Cd + 2\ R^2R^3CXCOOR^4 \begin{array}{c} \nearrow (R^4OCOCR^2R^3)_2Cd + R^1X \qquad (107a) \\ \searrow R^1R^2R^3CCOOR^4 + CdX_2 \qquad (107b) \end{array}$$

2. Acylation

Many simple or functional ketones may be prepared by reaction between acid halides and organocadmium derivatives[272-274,327-329]. The yields are moderate with dialkylcadmium compounds free of metal halides, but they increase by catalysis with magnesium halides or aluminium chloride[127,128,282,283,285,295,330-332]. Divinylcadmium compounds[273] and alkynylcadmium chlorides[333,334] also react with acid chlorides (reaction 108). With aliphatic acid chlorides, diallylcadmium compounds often lead to diallyl-alkyl-methanols (reaction 109)[273]. Dialkylcadmium compounds also react with

$$CH_2{=}CHC{\equiv}CCdCl + R^2COCl \longrightarrow H_2C{=}CHC{\equiv}CCOR^2 \qquad (108)$$
$$24-45\%$$

$$(CH_2{=}CHCH_2)_2Cd + R^2COCl \longrightarrow (CH_2{=}CHCH_2)_2C(OH)R^2 \qquad (109)$$

$ClCO(CH_2)_nCOCl$ to produce generally diketones[282,335,336], except in particular cases[337] (reaction 110). Similar diketones may be also prepared by reaction between acid halides and saturated organodicadmium halides (reaction 111)[338].

$$R^1_2Cd + ClCO(CH_2)_nCOCl \longrightarrow R^1CO(CH_2)_nCOR^1 \qquad (110)$$
$$n = 0,\ \text{yield} = 37\%\ ;\ n = 1,\ \text{yield} = 21-27\%$$

$$ClCd(CH_2)_nCdCl + R^1COCl \longrightarrow R^1CO(CH_2)_nCOR^1 \qquad (111)$$
$$n \geqslant 4, \qquad 10-85\%$$

Finally, α-unsaturated[297] or functional[274,339-342] acid chlorides react readily with organocadmium derivatives to produce α-unsaturated or functional ketones (reactions 112-114).

$$C_6H_5CH{=}CHCOCl + R^1_2Cd \longrightarrow C_6H_5CH{=}CHCOR^1 + CdCl_2 \qquad (112)$$
$$40-60\%$$

$$ClCH_2COCl + Bu^n_2Cd \longrightarrow ClCH_2COBu^n \qquad (113)$$
$$26\%$$

$$R^1_2Cd + o\text{-}ClSeC_6H_4COCl \longrightarrow o\text{-}R^1SeC_6H_4COR^1 \qquad (114)$$
$$75\%$$

C. Carbene and Carbenoid Intermediates from Organocadmium Compounds

Cadmium derivatives are much less often used than zinc compounds in the Simmons–Smith reaction and its analogues. However, several interesting results are obtained in modified Simmons–Smith reactions[273,274]. Diethylcadmium and diiodomethane react readily with simple or functional olefins to produce cyclopropane derivatives in good yields (reaction 115)[273,343]. When R^1CHI_2 (R^1 = alkyl, phenyl) is employed in this

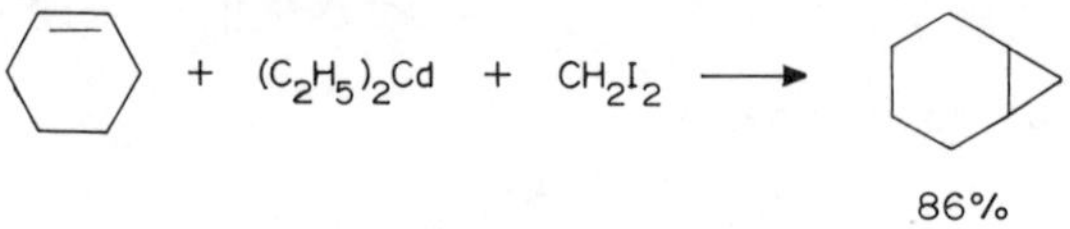

$$\text{cyclohexene} + (C_2H_5)_2Cd + CH_2I_2 \longrightarrow \text{norcarane} \quad (115)$$

86%

reaction, the yields are low (20%), whereas they are good (70%) in the reaction using $(C_2H_5)_2Zn$; with CCl_4, the reaction leads to 7,7-dichloronorcarane, among other compounds[344].

The dimethoxyethane complex of bis(trifluoromethyl)cadmium has been found to be an efficient low-temperature source of difluorocarbene (reaction 116)[345]. Diethylcadmium

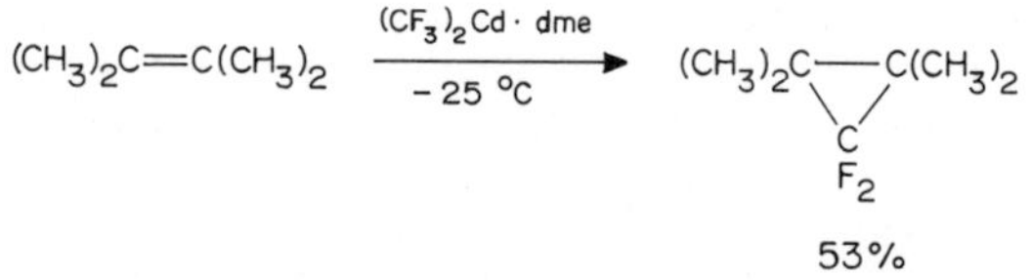

$$(CH_3)_2C=C(CH_3)_2 \xrightarrow[-25\,°C]{(CF_3)_2Cd \cdot dme} (CH_3)_2C\underset{CF_2}{\overset{}{-\!\!\!-\!\!\!-}}C(CH_3)_2 \quad (116)$$

53%

and 1,1-diiodoalkanes react with terminal alkynes to give substituted alkyne and isomeric allene compounds[274], probably via an insertion reaction in the C≡C—H bond (40–42% yield) (reaction 117).

$$PhC\equiv CH + (C_2H_5)_2Cd + R^1CHI_2 \longrightarrow PhC\equiv CCH_2R^1 + PhCH=C=CHR^1$$

R^1 = H, CH_3 (117)

Lastly, diethylcadmium and 1,1-diiodoethane react with alkylbenzenes to form 7-methylcyclohepta-1,3,5-triene derivatives (reaction 118)[201].

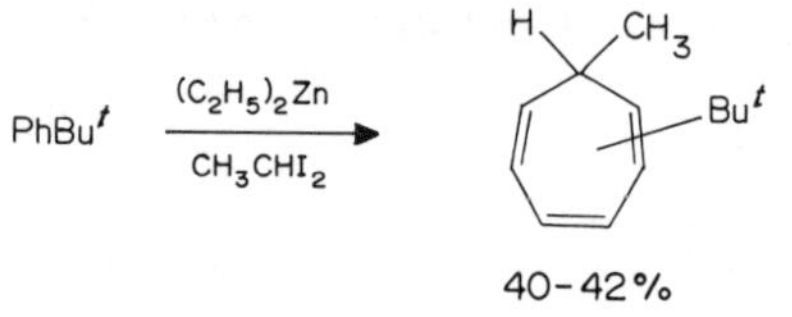

$$PhBu^t \xrightarrow[CH_3CHI_2]{(C_2H_5)_2Zn} \text{7-methyl-7-Bu}^t\text{-cyclohepta-1,3,5-triene} \quad (118)$$

40–42%

IV. REACTIONS OF MERCURY ORGANOMETALLICS

Several general reviews[346–349] covering the literature up to and including 1980 have been published recently on this subject; organomercurials undergo mild carbon—carbon bond formation and yet tolerate all important organic functional groups.

A. Alkene and Alkyne Additions and Heck Reaction

There are few examples of the direct addition of organomercurials to carbon—carbon double or triple bonds[346-349]. Primary, secondary, and benzylic dialkylmercurials will add to tetracyanoethylene (reaction 119)[350-352].

$$R_2Hg + (NC)_2C{=}C(CN)_2 \longrightarrow RC(CN)_2C(CN)_2HgR \tag{119}$$

Di-*tert*-butylmercury adds stereo- and regio-selectively to a variety of electron-deficient alkenes and alkynes (reaction 120)[353]. A wide variety of electron-deficient olefins have

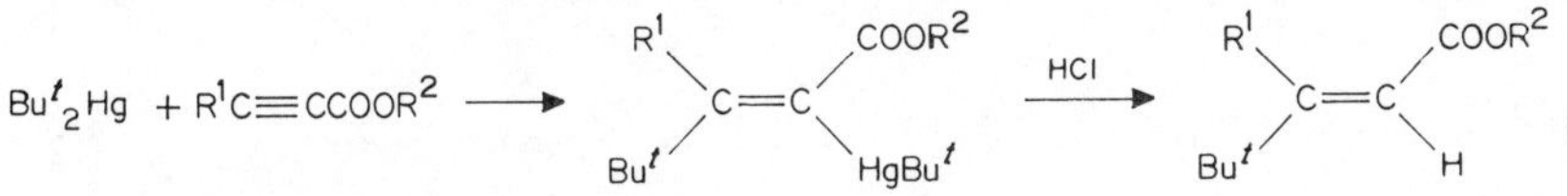

$$Bu^t_2Hg + R^1C{\equiv}CCOOR^2 \longrightarrow \underset{Bu^t}{\overset{R^1}{>}}C{=}C\underset{HgBu^t}{\overset{COOR^2}{<}} \xrightarrow{HCl} \underset{Bu^t}{\overset{R^1}{>}}C{=}C\underset{H}{\overset{COOR^2}{<}} \tag{120}$$

been alkylated by generating free radicals from the reaction of alkylmercurials and NaBH$_4$ or NaHB(OCH$_3$)$_3$ (reaction 121)[354-365]. This alkene addition reaction is complemented

$$RHgX + R^1CH{=}CR^2R^3 \xrightarrow[\text{NaHB(OCH}_3)_3]{\text{NaBH}_4 \text{ or}} RR^1CHCHR^2R^3 \tag{121}$$

by the Heck reaction[366-369], in which vinyl hydrogen substitution is effected by treating olefins with organomercurials and palladium(II) salts (reaction 122). In this reaction, the

$$RHgCl + H_2C{=}CHR^1 \xrightarrow{Li_2[PdCl_4]} \underset{30-88\%}{RCH{=}CHR^1} \tag{122}$$

intermediacy of organopalladium compounds has been established by isolating such compounds in some cases or by observing their dimerization or carbonylation products[347-349].

Many examples of the Heck reaction have been reported. It can be run at room temperature, in air, and in a wide variety of solvents. Methyl, benzyl, neopentyl, neophyl, carboalkoxy, aryl, polynuclear aromatic, heterocyclic, and organometallic organomercurials can be employed. Neither alkylmercurials bearing beta-hydrogen atoms nor vinylmercurials are accommodated. In the alkene, electron-withdrawing groups increase the rate of reaction and generally the yield. The less hindered the double bond is, the more reactive it will be.

These organopalladium reactions can also be used to prepare alkanes as in reaction 123[370]. When the arylation of arylolefins is carried out under aqueous conditions, hydroxyarylation is effected (reaction 124)[371]. While many functional groups are

$$RHgX \xrightarrow[Li_2[PdCl_4]]{H_2C{=}CHR^1} \xrightarrow{NaBH_4} RCH_2CH_2R^1 \tag{123}$$

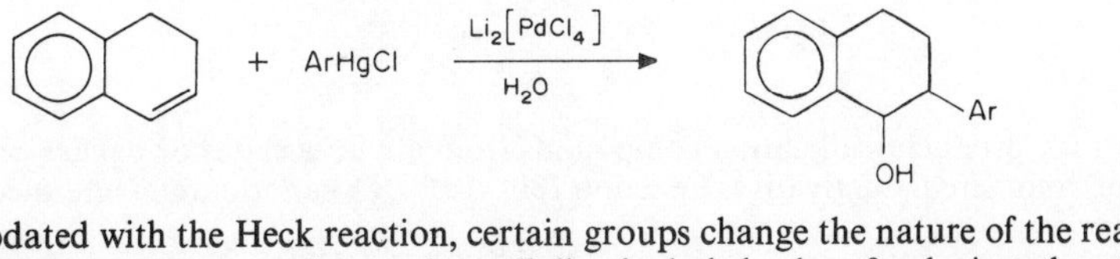

accommodated with the Heck reaction, certain groups change the nature of the reaction completely; thus, primary and secondary allylic alcohols lead to β-substituted carbonyl compounds (reaction 125)[372].

When the Heck reaction is carried out in the presence of high concentrations of CuCl$_2$,

$$H_2C=CHCHOHR^1 + ArHgCl \xrightarrow{Li_2[PdCl_4]} ArCH_2CH_2COR^1 \quad (125)$$

still another type of product is observed[373] (chloro-alkylation or -arylation) (reactions 126 and 127). Finally, Heck-type reactions also afford a valuable route to π-palladium

$$ArHgCl + H_2C=CHR^1 \xrightarrow[CuCl_2/LiCl]{Li_2[PdCl_4]} ArCH_2CHClR \quad (126)$$

$$PhHgCl + H_2C=CHCOCH_3 \xrightarrow[CuCl_2/LiCl]{Li_2[PdCl_4]} PhCH_2CHClCOCH_3 \quad (127)$$

compounds from aryl- or alkyl-mercury compounds and 1,3-dienes[374,375] or from vinylmercurials and alkenes[376,377]; these compounds have found considerable application in organic synthesis.

In summary, few organomercurials will directly add to alkenes or alkynes, but many useful reactions have been described involving the generation of free radicals or organopalladium compounds from organomercurials, and their subsequent addition to alkenes or substitution of vinyl hydrogens.

B. Substitution Reactions

1. Alkylation

The direct replacement of mercury in organomercurials by an organic group is a reaction of considerable synthetic importance. Unfortunately, the reaction of organomercurials with organic halides (reaction 128) is fairly restricted[346-349]. This reaction is very

$$R_2Hg + R^1X \longrightarrow R-R^1 + RHgX \quad (128)$$

difficult with alkyl, allyl, phenyl, vinyl, and alkynyl halides and almost all successful examples of this reaction have been achieved using triphenylmethyl or benzylic halides; these halides react with alkyl-[378-382], benzyl-[383,384], aryl-[381,382,384-388], ferrocenyl-[389], and vinyl-mercury compounds[384,390]. They also react with α-mercuriated carbonyl compounds[382,391-398] and, in this case, alkylation occurs on either carbon or oxygen depending on the electronic nature of the organic halide[391], the solvent[382,392,393], and the substituents of the organomercurial[392]. Chemical-induced dynamic nuclear polarization experiments suggest that many of these alkylation reactions proceed via radical intermediates[380,383,390].

Several catalytic or organometallic procedures have been reported recently for effecting carbon—carbon bond formation via organomercurials. Aluminium bromide promotes the alkylation of diphenyl and dibenzylmercury (reaction 129)[399]. $Li_2[PdCl_4]$ is a good

$$R_2Hg + R^1X \xrightarrow[CH_2Cl_2]{AlBr_3} RR^1 \quad (129)$$
$$20-60\%$$

catalyst for producing allylation compounds from allylic acetates or halides and aryl or vinyl organomercury derivatives (reaction 130)[400,401]. Organorhodium intermediates can

$$(E)\text{-}CHR=CHHgCl + ClCH_2CH=CH_2 \xrightarrow{Li_2[PdCl_4]} (E)\text{-}CHR=CH(CH_2CH=CH_2) \quad (130)$$
$$45-98\%$$

2. Carbon—carbon bond formation

be employed to effect the methylation of vinyl, alkynyl, and arylmercurials (reaction 131)[402], and [RhCl(PPh$_3$)$_3$] catalysts have been used for reactions between aryl

$$RHgCl + [CH_3RhI_2(PPh_3)_2] \xrightarrow[hmpa]{heat} RCH_3 \quad (131)$$
$$91-99\%$$

organomercurials and vinyl halides[402]. Copper[403,404] and organolithium[405] reagents can be also useful in the alkylation reactions of organomercurials (reactions 132 and 133).

$$(CF_3)_2Hg + ArI \xrightarrow[nmp]{Cu} ArCF_3 \quad (132)$$
$$38-88\%$$

$$RHgBr \xrightarrow{1.\ (Bu^n)_3PCuI}_{2.\ Bu^tLi\ ;\ 3.\ R^1I} RR^1 \quad (133)$$
$$21-70\%$$

Finally, recent experiences show that the use of simple organocuprate reagents is the most general method for the alkylation of a wide range of organomercury compounds (reactions 134 and 135)[348,406].

$$ArHgCl + LiCuR^1_2 \text{ or } Li_2CuR^1_3 \longrightarrow ArR^1 \quad (134)$$
$$35-92\%$$

$$(E)\text{-}CHR=CHHgCl \xrightarrow{LiCuR^1_2} (E)\text{-}CHR=CHR^1 \quad (135)$$
$$57-66\%$$

2. Acylation

The acylation of organomercurials provides a useful route to ketones[346–349]. Alkylmercurials react with acid halides only under forcing conditions to give ketones, but this reaction can be catalysed by addition of aluminium bromide[407] in CH$_2$Cl$_2$ or of [Pd(PPh$_3$)$_4$] in hmpa[408].

$$R_2Hg + R^1COCl \xrightarrow[20\ °C]{AlBr_3/CH_2Cl_2} RCOR^1 \quad (136)$$
$$70-100\%$$

The acylation of benzylmercury(II) chloride occurs, but two aryl ketones are formed[409] via initial acylation of the aromatic ring (reaction 137).

$$PhCH_2HgCl + CH_3COCl \longrightarrow p\text{-}CH_3C_6H_4COCH_3 + p\text{-}CH_3COC_6H_4CH_2COCH_3 \quad (137)$$

With allylmercury(II) iodide, the reaction does not lead to a β-ethylenic ketone, but to an enol ester (reaction 138)[410] which is apparently formed as shown in reaction 139 by

$$CH_2=CHCH_2HgI + 2\ R^1CH_2COCl \longrightarrow CH_2=CHCH_2C(OCOCH_2R^1)=CHR^1 \quad (138)$$

$$H_2C=CHCH_2HgI + R^1CH=C=O \longrightarrow IHgCH(R^1)COCH_2CH=CH_2 + \quad (139)$$
$$R^1CH_2COCl \longrightarrow R^1CH=C(OCOCH_2R^1)CH_2CH=CH_2$$

generation of a ketene from the acid chloride. Thus, the acylation of α-mercuriated carbonyl compounds provides a very useful route to enol esters (reactions 140–142)[349,411–417].

$$RCOCH_2HgCl + R^1COCl \longrightarrow R^1COOC(R)=CH_2 \quad (140)$$

$$RCOCH_2HgCl + ClCOOR^1 \longrightarrow R^1OCOOC(R)=CH_2 \quad (141)$$

$$RCOCH_2HgCl + COCl_2 \longrightarrow ClCOOC(R)=CH_2 + CH_2=C(R)OCOOC(R)=CH_2 \quad (142)$$

Vinylmercurials react readily with acid chlorides in the presence of AlCl$_3$ (reaction 143)[418,419]. Dialkynylmercurials undergo acylation by acid chlorides on simply refluxing

$$(Z)\text{-}CHR^1=CR^2HgCl + R^3COCl \xrightarrow{AlCl_3/CH_2Cl_2} (Z)\text{-}CHR^1=CR^2(COR^3) \quad (143)$$
$$64\text{-}100\%$$

in heptane (reaction 144), or by acid anhydrides in the presence of AlBr$_3$[420] (reaction 145). Finally, arylmercurials react with acid chlorides to produce ketones, generally in the presence of AlBr$_3$[349,379] or [Pd(PPh$_3$)$_4$] catalysts[349,380] (reaction 146).

$$(RC\equiv C)_2Hg \xrightarrow{R^1COCl} RC\equiv CCOR^1 \quad (144)$$
$$40\text{-}70\%$$

$$(RC\equiv C)_2Hg \xrightarrow[AlBr_3]{(R^1CO)_2O} RC\equiv CCOR^1 \quad (145)$$
$$51\text{-}70\%$$

$$Ar_2Hg + R^1COCl \longrightarrow ArCOR^1 \quad (146)$$

These results show that a wide variety of ketones can be prepared by the acylation of organomercurials. The acylation of α-mercuriated carbonyl compounds proceeds by reaction at oxygen atom and can lead to a wide variety of enol esters.

C. Dimerization

The dimerization of organomercurials (reactions 147 and 148) can be effected by thermolysis, by photolysis, or more commonly by using transition metal reagents[346–349].

$$R_2Hg \longrightarrow RR + Hg \quad (147)$$

$$2\,RHgX \longrightarrow RR + HgX_2 + Hg \quad (148)$$

The dimerization of simple alkylmercurials is difficult and the yields are generally low[349], whereas benzylmercurials are more easily dimerized by thermolysis[421], photolysis[422], or transition metal-promoted dimerization[423].

Vinylmercurials lead readily to 1,3-dienes using a variety of transition metal reagents including palladium(0)[424,425], palladium(II)[425,426], and rhodium(III)[427] complexes (reactions 149 and 150). Arylmercurials can also be readily dimerized to biaryls[347–349]; transition metal complexes such as [RhCl(PPh$_3$)$_3$] in hmpa[428], [RhCl(CO)$_2$]$_2$[427], Li$_2$[PdCl$_4$][425,429], Li$_2$[PdCl$_4$], and CuCl$_2$ or Cu[430,431], and PdCl$_2$/Cu/C$_5$H$_5$N[431] are

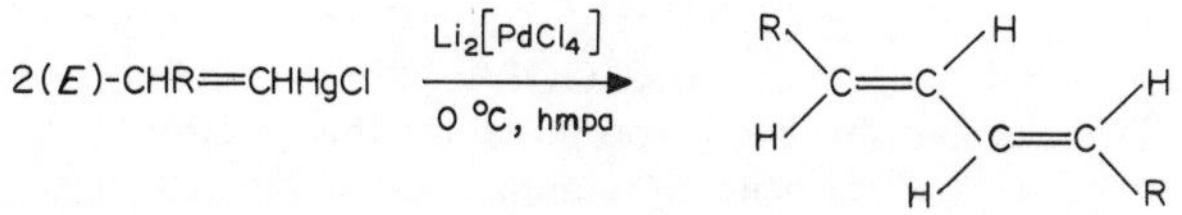

2. Carbon—carbon bond formation

$$2(E)\text{-CHR=CHHgCl} \xrightarrow[\text{0 °C, hmpa}]{\text{Li}_2[\text{PdCl}_4]} \begin{array}{c} R\diagdown/H \\ C=C \\ H/\diagdown C=C\diagup H \\ H/\diagdown R \end{array} \quad (149)$$

'symmetrical' dimer

$$2(E)\text{-CHR=CHHgCl} \xrightarrow[\text{C}_6\text{H}_6/\text{Et}_3\text{N}]{\text{PdCl}_2} \begin{array}{c} R\diagdown/H \\ C=C \\ H/\diagdown C=C\diagup H \\ R/\diagdown H \end{array} \quad (150)$$

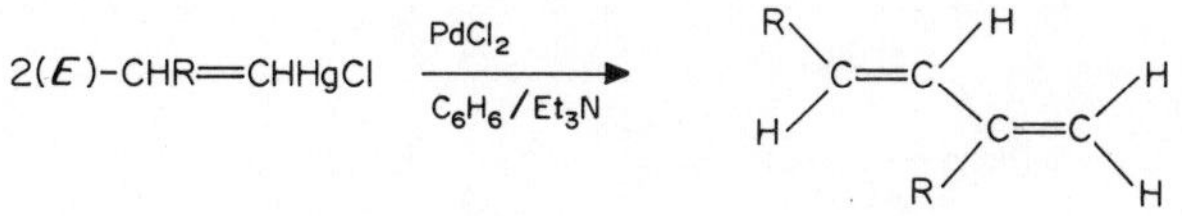

'head-to-tail' dimer

once again very effective in this reaction (reactions 151 and 152). Arylmercurials have also frequently been dimerized by the use of silver at elevated temperatures (240–360 °C)[432–434].

$$2\text{ ArHgCl} \xrightarrow[\text{4 LiCl/hmpt, 24 h, 80 °C}]{0.5\% \, [\text{RhCl(CO)}_2]_2} \underset{40-96\%}{\text{ArAr}} \quad (151)$$

$$2\text{ ArHgX} \xrightarrow[\text{C}_5\text{H}_5\text{N, 115 °C}]{\text{PdCl}_2/\text{Cu cat.}} \underset{47-95\%}{\text{ArAr}} \quad (152)$$

In summary, dimerization is an effective method of carbon—carbon bond formation from benzyl-, vinyl-, and aryl-organomercurials.

D. Carbene and Carbenoid Intermediates from Organomercury Compounds

Halogenomethyl organomercurials can either give rise to 'free' carbenes or can act as 'carbenoid' precursors[346–349,435–466]; they can be used for addition to the multiple bonds, for insertion reactions into a large variety of single bonds, and also for an alternative to the Wittig alkene synthesis.

1. Addition reactions

a. Carbon—carbon double bonds

Halogenomethyl organomercurials are reagents of outstanding utility for the preparation of cyclopropanes and halogenocyclopropanes from alkenes, according to the general scheme shown in reactions 153 and 154. Cyclopropanes have been obtained, generally in

$$\text{PhHgCX}_3 \xrightarrow{\text{heat}} \text{PhHgX} + [:\text{CX}_2] \quad (153)$$

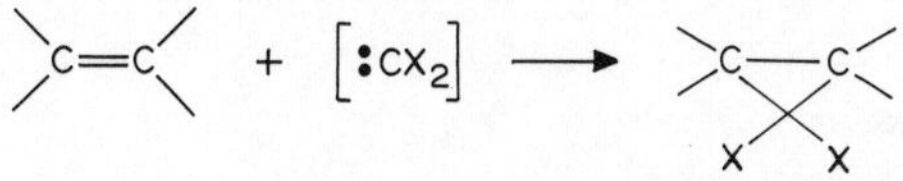

$$\diagup C=C\diagdown + [:\text{CX}_2] \longrightarrow \begin{array}{c} \diagup C\text{---}C\diagdown \\ XX \end{array} \quad (154)$$

good yields (60–90%), from a variety of carbenes, including :CH_2[435–438,466].

:CF_2[439-442], :$CFCl$[443-445], :CCl_2[446-449], :$CHCl$[450,451], :$CHBr$[451], $CClBr$[446,452], :$CFBr$[452], :CBr_2[446,453,467-469], :$CXCOOR$[453-456], :$CXPh$[456], and :$CClSi(CH_3)_3$[457,458]. Two mechanisms can be proposed for this reaction:

(*i*) Formation of a 'free' carbene by elimination of phenylhalogenomercury and addition to the alkene. Thus, phenyl (bromodichloromethyl) mercury leads to dichlorocarbene and phenylbromomercury; the extrusion of dichlorocarbene is thought to be concerted, proceeding via a cyclic transition state, **19** or **20**. This picture provides a

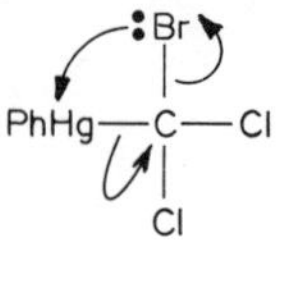

satisfactory explanation for the preferential elimination of phenylbromomercury, since intramolecular attack at mercury by bromine should be more favourable than attack by chlorine, and the carbon—bromine bond is weaker than the carbon—chlorine bond.

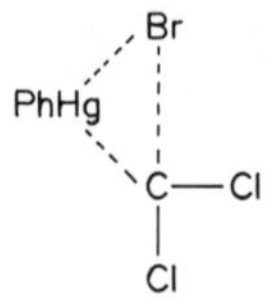

(*ii*) Direct reaction between the organomercury compound and the alkene. The addition of bis(bromomethyl)mercury to an alkene does not proceed via free carbene, :CH_2; experimental evidence favours a mechanism involving direct reaction.

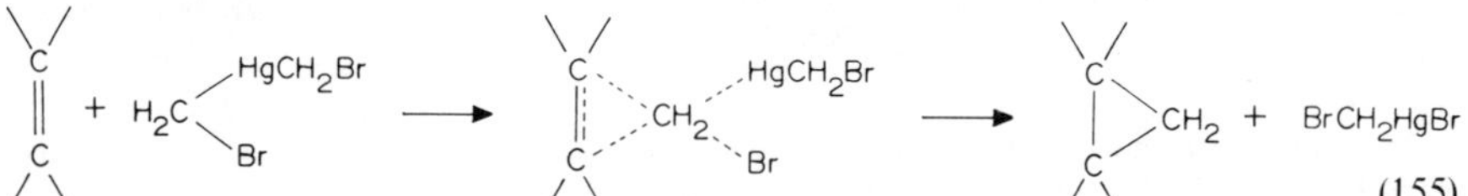

(155)

b. Carbon—carbon triple bonds

This addition can produce cyclopropenes[470,471] and derivatives, for example reaction 156.

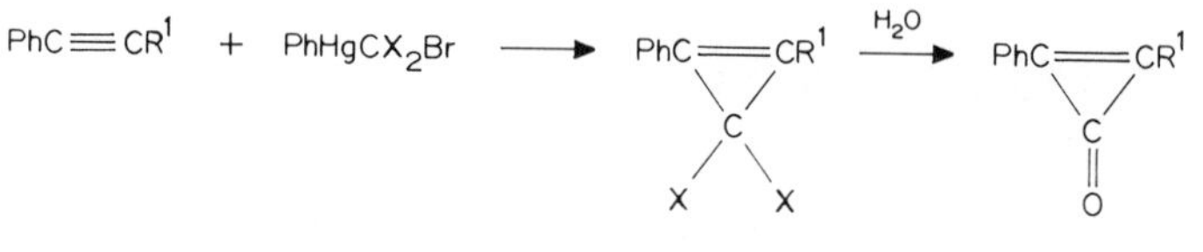

(156)

c. Other double bond systems

Carbenes formed from organomercurials react with a variety of other double bond systems including $C=N$[445,472], $C=S$[445,472-475], and $C=O$[474,476], leading to aziridines (in 29–43% yield), thiiranes (in 75% yield), and oxiranes (in 25–74% yield), respectively.

2. Insertion reactions

Halogenomethylmercury compounds can also be used in insertion reactions with single bonds such as C—H, B—C, Si—C and Ge—C bonds.

2. Carbon—carbon bond formation

a. C—H bonds

Many examples of insertion reactions (such as reaction 157) with benzylic compounds[476-480] and ethers [446,480,481] (40–80% yield) are known.

$$PhCH(CH_3)_2 + PhHgCCl_2Br \longrightarrow PhC(CH_3)_2CHCl_2 + PhHgBr \quad (157)$$

b. B—C bonds[347-349,482]

The (1,1-dichloroalkyl)borane formed in reaction 158 leads to an alkene by an intramolecular rearrangement (reaction 159).

$$(RCH_2CH_2)_3B + PhHgCCl_2Br \longrightarrow (RCH_2CH_2)_2BCCl_2CH_2CH_2R + PhHgBr \quad (158)$$

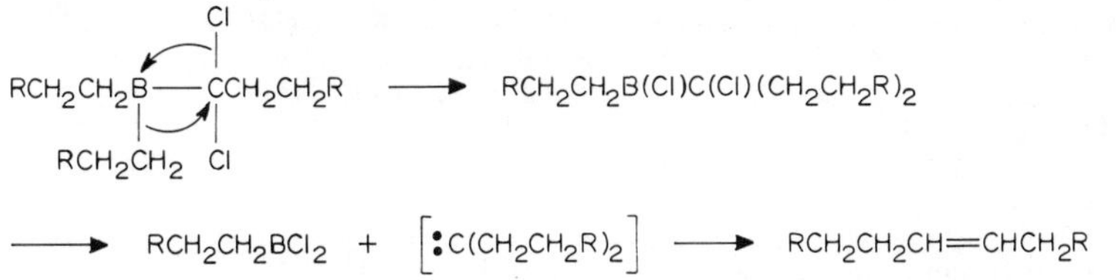

$$RCH_2CH_2B\text{—}CCH_2CH_2R \longrightarrow RCH_2CH_2B(Cl)C(Cl)(CH_2CH_2R)_2$$

$$\longrightarrow RCH_2CH_2BCl_2 + [:C(CH_2CH_2R)_2] \longrightarrow RCH_2CH_2CH=CHCH_2R \quad (159)$$

c. Si—C and Ge—C bonds

Dichlorocarbene insertion reactions have been observed with silacyclobutane and germacyclobutane derivatives[483-485].

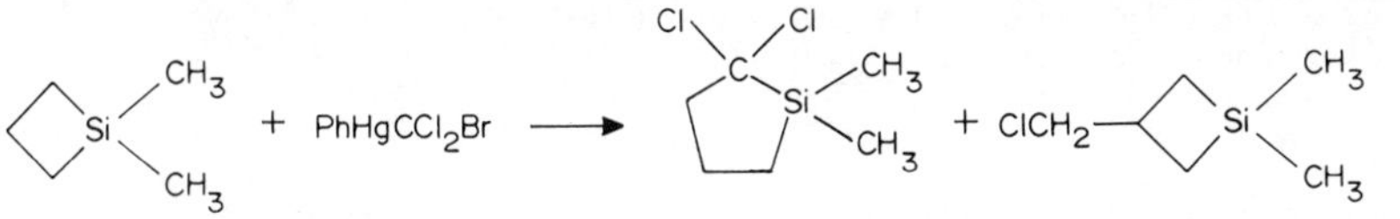

(160)

3. Wittig alkene synthesis

Halogenomethylmercury compounds can react with various derivatives[486-488] such as diazo compounds[486] or azides[487] to produce alkenes or imines, but the most interesting case is their reaction with triphenylphosphine and carbonyl compounds to yield alkenes (reaction 161). This reaction provides an alternative to the Wittig alkene synthesis[451,455,489].

$$PhHgCX_2Br + (Ph)_3P \longrightarrow [(Ph)_3\overset{+}{P}\text{—}\bar{C}X_2 \longleftrightarrow (Ph)_3P=CX_2]$$

$$\xrightarrow{R^1COR^2} (Ph)_3PO + R^1R^2C=CX_2 \quad (161)$$

$$40\text{–}82\%$$

V. REFERENCES

1. N. I. Sheverdina and K. A. Kocheshkov, *Methods of Elemento-Organic Chemistry*, Vol. 3, North-Holland, Amsterdam, 1967, pp. 1–163.
2. K. Nützel, 'Organo-Zink-Verbindungen', in Houben-Weyl, *Methoden der Organischen Chemie*, Vol. 13/2a, Verlag Chemie, Stuttgart, 1973, pp. 709–805.
3. J. Furukawa and N. Kawabata, *Adv. Organomet. Chem.*, **12**, 83 (1974).
4. M. Gaudemar, *Bull. Soc. Chim. Fr.*, 974 (1962); 1475 (1963).
5. G. Courtois and L. Miginiac, *J. Organomet. Chem.*, **69**, 1 (1974), and references cited therein.
6. L. Miginiac, Ph. Miginiac, and Ch. Prévost, *C. R. Acad. Sci., Ser. C*, **272**, 1682 (1971).
7. L. Miginiac, Ph. Miginiac, and Ch. Prévost, *Bull. Soc. Chim. Fr.*, 3560 (1965).
8. M. Gaudemar and S. Travers, *C. R. Acad. Sci., Ser. C*, **262**, 139 (1966).
9. G. Peiffer, *C. R. Acad. Sci., Ser. C*, **262**, 501 (1966).
10. B. Gross and Ch. Prevost, *Bull. Soc. Chim. Fr.*, 3610 (1967).
11. F. Gérard and Ph. Miginiac, *C. R. Acad. Sci., Ser. C*, **273**, 674 (1971).
12. F. Gérard and Ph. Miginiac, *C. R. Acad. Sci., Ser. C*, **275**, 1129 (1972).
13. D. Abenhaïm, E. Henry-Basch, and P. Fréon, *Bull. Soc. Chim. Fr.*, 4038 and 4043 (1969).
14. K. H. Thiele, J. Köhler, and P. Zdunneck, *Z. Chem.*, **7**, 307 (1967).
15. J. A. Katzenellenbogen and R. S. Lenox, *J. Org. Chem.*, **38**, 326 (1973).
16. E. Öhler, K. Reininger, and U. Schmidt, *Angew. Chem., Int. Ed. Engl.*, **9**, 457 (1970).
17. L. Miginiac and M. Lanoiselée, *Bull. Soc. Chim. Fr.*, 2716 (1971).
18. G. Daviaud and Ph. Miginiac, *Bull. Soc. Chim. Fr.*, 1617 (1970).
19. G. Daviaud, M. Massy-Barbot, and Ph. Miginiac, *C. R. Acad. Sci., Ser. C*, **272**, 969 (1971).
20. M. Bellassoued, Y. Frangin, and M. Gaudemar, *Synthesis*, 205 (1977).
21. F. Barbot and Ph. Miginiac, *C. R. Acad. Sci., Ser. C*, **272**, 1682 (1971).
22. Ph. Miginiac and C. Bouchoule, *Bull. Soc. Chim. Fr.*, 4675 (1968).
23. Ph. Miginiac, *Bull. Soc. Chim. Fr.*, 1070 (1970).
24. F. Gérard and Ph. Miginiac, *Bull. Soc. Chim. Fr.*, 1924 (1974).
25. F. Barbot and Ph. Miginiac, *Tetrahedron Lett.*, 3829 (1975).
26. F. Gérard and Ph. Miginiac, *Bull. Soc. Chim. Fr.*, 2527 (1974).
27. F. Barbot and Ph. Miginiac, *Bull. Soc. Chim. Fr.*, 113 (1977).
28. F. Barbot and Ph. Miginiac, *J. Organomet. Chem.*, **132**, 445 (1977).
29. F. Gérard and Ph. Miginiac, *J. Organomet. Chem.*, **155**, 271 (1978).
30. F. Barbot and Ph. Miginiac, unpublished results.
31. D. Abenhaïm, E. Henry-Basch, and P. Fréon, *C. R. Acad. Sci., Ser. C*, **267**, 655 (1968).
32. D. Abenhaïm, E. Henry-Basch, and P. Fréon, *Bull. Soc. Chim. Fr.*, 179 (1970).
33. D. Abenhaïm, G. Boireau, and J. L. Namy, *Bull. Soc. Chim. Fr.*, 985 (1972).
34. G. Boireau, J. L. Namy, D. Abenhaïm, E. Henry-Basch, and P. Fréon, *C. R. Acad. Sci., Ser. C*, **269**, 1565 (1969).
35. D. Abenhaïm, J. L. Namy, and G. Boireau, *Bull. Soc. Chim. Fr.*, 3254 (1971).
36. B. Mauzé and L. Miginiac, *Bull. Soc. Chim. Fr.*, 3832 (1968).
37. B. Mauzé and L. Miginiac, *Bull. Soc. Chim. Fr.*, 1082 (1973).
38. B. Mauzé and L. Miginiac, *Bull. Soc. Chim. Fr.*, 4673 (1968).
39. B. Mauzé and L. Miginiac, *Bull. Soc. Chim. Fr.*, 1832 (1973).
40. B. Mauzé and L. Miginiac, *Bull. Soc. Chim. Fr.*, 1838 (1973).
41. J. Pornet and L. Miginiac, *C. R. Acad. Sci., Ser. C*, **271**, 381 (1970).
42. N. Koga, G. Koga and J. P. Anselme, *Tetrahedron Lett.*, 3309 (1970).
43. J. Pornet and L. Miginiac, *Tetrahedron Lett.*, 967 (1971).
44. J. Pornet and L. Miginiac, *C. R. Acad. Sci., Ser. C*, **273**, 1763 (1971).
45. J. Pornet and L. Miginiac, *Bull. Soc. Chim. Fr.*, 989 (1974).
46. J. Pornet and L. Miginiac, *Bull. Soc. Chim. Fr.*, 994 (1974).
47. J. Pornet and L. Miginiac, *Bull. Soc. Chim. Fr.*, 841 (1975).
48. J. Pornet and L. Miginiac, *Bull. Soc. Chim. Fr.*, 1849 (1975).
49. C. Bouchoule and Ph. Miginiac, *C. R. Acad. Sci., Ser. C*, **266**, 1614 (1968).
50. B. Mauzé, G. Courtois, and L. Miginiac, *C. R. Acad. Sci., Ser. C*, **269**, 1225 (1969).
51. B. Mauzé and L. Miginiac, *Bull. Soc. Chim. Fr.*, 462 (1968).
52. M. Gaudemar, *C. R. Acad. Sci., Ser. C*, **273**, 1669 (1971).

53. B. Mauzé, C. Nivert, and L. Miginiac, *J. Organomet. Chem.*, **44**, 69 (1972).
54. Ph. Miginiac and G. Daviaud, *J. Organomet. Chem.*, **104**, 139 (1976).
55. G. Courtois and L. Miginiac, *Bull. Soc. Chim. Fr.*, 3330 (1969).
56. G. Courtois and L. Miginiac, *J. Organomet. Chem.*, **52**, 241 (1973).
57. H. Lehmkuhl and H. Nehl, *J. Organomet. Chem.*, **60**, 1 (1973).
58. H. Lehmkuhl, I. Döring, and H. Nehl, *J. Organomet. Chem.*, **221**, 123 (1981).
59. H. Lehmkuhl and H. Nehl, *J. Organomet. Chem.*, **221**, 131 (1981).
60. H. Lehmkuhl and R. McLane, *Justus Liebigs Ann. Chem.*, 736 (1980).
61. J. F. Normant and A. Alexakis, *Synthesis*, 841 (1981).
62. F. Bernadou, B. Mauzé, and L. Miginiac, *C. R. Acad. Sci., Ser. C*, **276**, 1645 (1973).
63. Y. Frangin and M. Gaudemar, *C. R. Acad. Sci., Ser. C*, **278**, 885 (1974).
64. F. Bernadou and L. Miginiac, *C. R. Acad. Sci., Ser. C*, **280**, 1473 (1975).
65. Y. Frangin and M. Gaudemar, *Bull. Soc. Chim. Fr.*, 1173 (1976).
66. F. Bernadou and L. Miginiac, *Tetrahedron Lett.*, 3083 (1976).
67. G. Courtois, A. Masson, and L. Miginiac, *C. R. Acad. Sci., Ser. C*, **286**, 265 (1978).
68. Y. Frangin and M. Gaudemar, *J. Organomet. Chem.*, **142**, 9 (1977).
69. Y. Frangin and M. Gaudemar, *C. R. Acad. Sci., Ser. C*, **280**, 1389 (1975).
70. F. Bernadou and L. Miginiac, *J. Organomet. Chem.*, **125**, 23 (1977).
71. M. Bellassoued and Y. Frangin, *Synthesis*, 838 (1978).
72. B. Mauzé, C. Nivert, and L. Miginiac, *C. R. Acad. Sci., Ser. C*, **271**, 698 (1970).
73. C. Nivert and L. Miginiac, *C. R. Acad. Sci., Ser. C*, **272**, 1996 (1971).
74. L. Miginiac, *J. Organomet. Chem.*, **238**, 235 (1982).
75. G. Courtois, B. Mauzé, and L. Miginiac, *J. Organomet. Chem.*, **72**, 309 (1974).
76. D. Mesnard and L. Miginiac, *J. Organomet. Chem.*, **117**, 99 (1976).
77. B. Mauzé, *J. Organomet. Chem.*, **131**, 321 (1977).
78. G. Courtois, B. Mauzé, and L. Miginiac, *C. R. Acad. Sci., Ser. C*, **274**, 658 (1972).
79. G. Courtois and L. Miginiac, *J. Organomet. Chem.*, **117**, 201 (1976).
80. B. Mauzé, *J. Organomet. Chem.*, **134**, 1 (1977).
81. D. Mesnard, J. P. Charpentier, and L. Miginiac, *J. Organomet. Chem.*, **214**, 135 (1981).
82. J. L. Moreau, in *The Chemistry of Ketenes, Allenes and Related Compounds* (Ed. S. Patai), Wiley, New York, 1980, pp. 372–386.
83. R. Golse, A. Liermain, and H. Bussière, *Bull. Soc. Pharm. Bordeaux*, **101**, 73 (1962); *Chem. Abstr.*, **58**, 6689c (1963).
84. J. L. Pansard and M. Gaudemar, *Bull. Soc. Chim. Fr.*, 3332 (1968).
85. J. L. Moreau and M. Gaudemar, *Bull. Soc. Chim. Fr.*, 2171 and 2175 (1970).
86. M. Gaudemar and J. L. Moreau, *Bull. Soc. Chim. Fr.*, 5037 (1968).
87. J. L. Moreau, *Bull. Soc. Chim. Fr.*, 1248 (1975).
88. J. L. Moreau, Y. Frangin, and M. Gaudemar, *Bull. Soc. Chim. Fr.*, 4511 (1970).
89. M. Bellassoued, Y. Frangin, and M. Gaudemar, *Synthesis*, 150 (1978).
90. G. Daviaud and Ph. Miginiac, *Tetrahedron Lett.*, 3251 (1971).
91. J. Huet, *Bull. Soc. Chim. Fr.*, 952 (1964).
92. J. L. Moreau and M. Gaudemar, *Bull. Soc. Chim. Fr.*, 3071 (1971).
93. J. L. Moreau and M. Gaudemar, *Bull. Soc. Chim. Fr.*, 2549 (1973).
94. J. L. Moreau and M. Gaudemar, *C. R. Acad. Sci., Ser. C*, **274**, 2015 (1972).
95. J. L. Moreau and M. Gaudemar, *Bull. Soc. Chim. Fr.*, 1211 (1975).
96. Y. Frangin, E. Favre, and M. Gaudemar, *C. R. Acad. Sci., Ser. C*, **282**, 277 (1976).
97. M. Bellassoued, Y. Frangin, and M. Gaudemar, *J. Organomet. Chem.*, **166**, 1 (1979).
98. B. Marx, E. Henry-Basch, and P. Fréon, *C. R. Acad. Sci., Ser. C*, **264**, 527 (1967); **266**, 1646 (1968).
99. D. Abenhaïm, *C. R. Acad. Sci., Ser. C*, **267**, 87 (1968); **267**, 1426 (1968).
100. P. R. Jones, E. J. Goller, and W. J. Kauffman, *J. Org. Chem.*, **34**, 3566 (1969).
101. P. R. Jones, W. J. Kauffman, and E. J. Goller, *J. Org. Chem.*, **36**, 186 (1971); **36**, 3311 (1971).
102. J. P. Deniau, E. Henry-Basch, and P. Fréon, *Bull. Soc. Chim. Fr.*, 4414 (1969).
103. J. Furukawa, N. Kawabata, and A. Kato, *J. Polym. Sci., Part A1*, **5**, 3139 (1967).
104. J. Pornet, B. Randrianoelina, and L. Miginiac, *J. Organomet. Chem.*, **174**, 1 (1979).
105. J. Pornet, B. Randrianoelina, and L. Miginiac, *J. Organomet. Chem.*, **174**, 15 (1979).
106. J. Thomas, E. Henry-Basch, and P. Fréon, *C. R. Acad. Sci., Ser. C*, **267**, 176 (1968).

107. J. Thomas and P. Fréon, *C. R. Acad. Sci., Ser. C*, **267**, 1850 (1968).
108. J. Thomas, E. Henry-Basch, and P. Fréon, *Bull. Soc. Chim. Fr.*, 109 (1969).
109. J. Auger, G. Courtois, and L. Miginiac, *J. Organomet. Chem.*, **133**, 285 (1977).
110. G. Courtois and L. Miginiac, *C. R. Acad. Sci., Ser. C*, **285**, 207 (1977).
111. G. Courtois and L. Miginiac, *J. Organomet. Chem.*, **195**, 13 (1980).
112. C. Prévost, Ph. Miginiac, and L. Miginiac, *Bull. Soc. Chim. Fr.*, 2485 (1964).
113. M. Andrac, F. Gaudemar, M. Gaudemar, B. Gross, L. Miginiac, Ph. Miginiac, and Ch. Prévost, *Bull. Soc. Chim. Fr.*, 1385 (1963).
114. C. Prévost and Ph. Miginiac, *Bull. Soc. Chim. Fr.*, 704 (1966).
115. Ph. Miginiac and G. Zamlouty, *J. Organomet. Chem.*, **96**, 163 (1975).
116. G. Courtois, M. Harama, and Ph. Miginiac, *J. Organomet. Chem.*, **218**, 275 (1981).
117. L. Miginiac and B. Mauzé, *Bull. Soc. Chim. Fr.*, 2544 (1968).
118. G. Courtois, M. Harama, and L. Miginiac, *J. Organomet. Chem.*, **198**, 1 (1980).
119. M. Bourhis, R. Golse, and J. J. Bosc, *C. R. Acad. Sci., Ser. C*, **289**, 379 (1979).
120. B. P. Keuk, J. Pornet, and L. Miginiac, *C. R. Acad. Sci., Ser. C*, **284**, 399 (1977).
121. G. Courtois, M. Harama, and Ph. Miginiac, *J. Organomet. Chem.*, **218**, 1 (1981).
122. M. Bourhis, R. Golse, and J. J. Bosc, *C. R. Acad. Sci., Ser. C*, **284**, 399 (1977).
123. C. R. Noller, *J. Am. Chem. Soc.*, **51**, 594 (1929).
124. M. T. Reetz, B. Wenderoth, R. Peter, R. Steinbach, and J. Westermann, *J. Chem. Soc., Chem. Commun.*, 1202 (1980).
125. R. C. Hahn, T. F. Corbin, and H. Shechter, *J. Am. Chem. Soc.*, **90**, 3404 (1968).
126. B. Teichmann, *Z. Chem.*, **4**, 387 (1964); *Chem. Abstr.*, **62**, 5199e (1965).
127. J. Michel, E. Henry-Basch, and P. Fréon, *Bull. Soc. Chim. Fr.*, 4898 (1968).
128. J. Michel, E. Henry-Basch, and P. Fréon, *Bull. Soc. Chim. Fr.*, 4902 (1968).
129. F. L. Breusch and F. Baykut, *Chem. Ber.*, **86**, 684 (1953).
130. G. Geiseler and P. Richter, *Chem. Ber.*, **93**, 2511 (1960).
131. H. Klein and H. Neff, *Angew. Chem.*, **68**, 681 (1956).
132. A. O. King, E. Negishi, F. J. Villani, Jr., and A. Silveira, Jr., *J. Org. Chem.*, **43**, 358 (1978).
133. E. Negishi, A. O. King, and N. Okukado, *J. Org. Chem.*, **42**, 1821 (1977).
134. A. Minato, K. Tamao, T. Hyashi, K. Suzuki, and M. Kumada, *Tetrahedron Lett.*, 845 (1980).
135. A. O. King, N. Okukado, and E. Negishi, *J. Chem. Soc., Chem. Commun.*, 683 (1977).
136. M. Kobayashi and E. Negishi, *J. Org. Chem.*, **45**, 5223 (1980).
137. E. Negishi, L. F. Valente, and M. Kobayashi, *J. Am. Chem. Soc.*, **102**, 3298 (1980).
138. K. Ruitenberg, H. Kleijn, C. J. Elsevier, J. Meijer, and P. Vermeer, *Tetrahedron Lett.*, 1451 (1981).
139. K. Ruitenberg, H. Kleijn, H. Westmijze, and P. Vermeer, *Recl. Trav. Chim. Pays-Bas*, **101**, 405 (1982).
140. R. Golse, M. Bourhis, and J. J. Bosc, *C. R. Acad. Sci., Ser. C*, **287**, 585 (1978).
141. D. Seyferth, *Pure Appl. Chem.*, **23**, 391 (1971).
142. H. E. Simmons and R. D. Smith, *J. Am. Chem. Soc.*, **80**, 5323 (1958); **81**, 4256 (1959).
143. E. W. Doering and P. Laflamme, *Tetrahedron*, **2**, 75 (1958).
144. H. E. Simmons, E. P. Blanchard, and R. D. Smith, *J. Am. Chem. Soc.*, **86**, 1347 (1964).
145. E. P. Blanchard and H. E. Simmons, *J. Am. Chem. Soc.*, **86**, 1337 (1964).
146. E. P. Blanchard, H. E. Simmons, and J. S. Taylor, *J. Org. Chem.*, **30**, 4321 (1965).
147. S. W. Staley and F. L. Wiseman, Jr., *J. Org. Chem.*, **35**, 3868 (1970).
148. E. C. Friedrich and R. L. Holmstead, *J. Org. Chem.*, **37**, 2550 (1972).
149. H. O. House and P. D. Weeks, *J. Am. Chem. Soc.*, **97**, 2778 (1975).
150. H. O. House and K. A. J. Snoble, *J. Org. Chem.*, **41**, 3077 (1976).
151. J. C. Limasset, P. Amice, and J. M. Conia, *Bull. Soc. Chim. Fr.*, 3981 (1969).
152. R. Perraud and P. Arnaud, *Bull. Soc. Chim. Fr.*, 1540 (1968).
153. G. E. Cartier and S. C. Bunce, *J. Am. Chem. Soc.*, **85**, 932 (1963).
154. J. J. Sims, *J. Am. Chem. Soc.*, **87**, 3511 (1965).
155. G. M. Rubottom and M. I. Lopez, *J. Org. Chem.*, **38**, 2097 (1973).
156. I. Ryu, S. Muraï, and N. Sonoda, *Tetrahedron Lett.*, 1995 (1977).
157. P. A. Tardella and L. Pellacani, *Gazz. Chim. Ital.*, **107**, 107 (1977).
158. I. M. Takakis and Y. E. Rhodes, *J. Org. Chem.*, **43**, 3496 (1978).
159. H. Dohlaine and G. Haegele, *Phosphorus Sulfur*, **4**, 123 (1978).

160. O. P. Vig and G. L. Kad,. *Indian J. Chem.*, **16B**, 452 (1978).
161. O. P. Vig and G. L. Kad, *Indian J. Chem.*, **16B**, 455 (1978).
162. J. Fajkos and J. Joska, *Collect. Czech. Chem. Commun.*, **44**, 251 (1979).
163. M. Ratier, M. Castaing, J. Y. Godet, and M. Pereyre, *J. Chem. Res.*, (S) 179 (1978), (M) 2309 (1978).
164. J. Joska and J. Fajkos, *Collect. Czech. Chem. Commun.*, **45**, 1850 (1980).
165. M. S. Ahmad, Jr., and S. M. Osman, *J. Am. Oil Chem. Soc.*, **57**, 363 (1980).
166. H P. Albrecht, G. von Philipsborn, M. Raschack, and H. U. Siebeneick, *Ger. Pat.*, 2841 044 (1980); *Chem. Abstr.*, **93**, 186684z (1980).
167. T. F. Corbin, R. C. Hahn, and H. Shechter, *Org. Synth.*, **44**, 30 (1964).
168. E. Legoff, *J. Org. Chem.*, **29**, 2048 (1964).
169. R. D. Smith and H. E. Simmons, *Org. Synth.*, **41**, 72 (1961).
170. R. D. Rieke, P. Tzu-Jung Li, T. P. Burns, and S. T. Uhm, *J. Org. Chem.*, **46**, 4324 (1981).
171. O. Repic and S. Vogt, *Tetrahedron Lett.*, 2729 (1982).
172. R. J. Rawson and I. T. Harrison, *J. Org. Chem.*, **35**, 2057 (1970).
173. E. C. Friedrich and G. Biresaw, *J. Org. Chem.*, **47**, 1615 and 2426 (1982).
174. J. M. Denis, C. Girard, and J. M. Conia, *Synthesis*, 549 (1972).
175. J. M. Denis and J. M. Conia, *Tetrahedron Lett.*, 4593 (1972).
176. J. E. Baldwin and B. M. Broline, *J. Org. Chem.*, **47**, 1385 (1982).
177. H. Hoberg, *Justus Liebigs Ann. Chem.*, **656**, 15 (1962).
178. U. Schöllkopf and A. Lerch, *Angew. Chem.*, **73**, 27 (1961).
179. G. Wittig and K. Schwarzenbach, *Angew. Chem.*, **71**, 652 (1959).
180. G. Wittig and K. Schwarzenbach, *Justus Liebigs Ann. Chem.*, **650**, 1 (1961).
181. G. Wittig and F. Wingler, *Justus Liebigs Ann. Chem.*, **656**, 18 (1962).
182. G. Wittig and F. Wingler, *Chem. Ber.*, **97**, 2139 and 2146 (1964).
183. G. Wittig and M. Jautelat, *Justus Liebigs Ann. Chem.*, **702**, 24 (1967).
184. U. Burger and R. Huisgen, *Tetrahedron Lett.*, 3057 (1970).
185. J. Furakawa, N. Kawabata, and J. Nishimura, *Tetrahedron Lett.*, 3353 (1966).
186. J. Furakawa, N. Kawabata, and J. Nishimura, *Tetrahedron*, **24**, 53 (1968).
187. J. Furakawa, N. Kawabata, and J. Nishimura, *Tetrahedron Lett.*, 3495 (1968).
188. J. Jacobus, Z. Majerski, K. Mislow, and P. Schleyer, *J. Am. Chem. Soc.*, **91**, 1998 (1969).
189. J. Nishimura, J. Furukawa, N. Kawabata, and M. Kitayama, *Tetrahedron*, **27**, 1799 (1971).
190. J. Vais, J. Burkhard, and S. Landa, *Z. Chem.*, **8**, 303 (1968); *Chem. Abstr.*, **69**, 76726m (1968).
191. J. Nishimura, N. Kawabata, and J. Furukawa, *Tetrahedron*, **25**, 2647 (1969).
192. J. Nishimura, N. Kawabata, and J. Furukawa, *Bull. Chem. Soc. Jpn.*, **43**, 2195 (1970).
193. J. Nishimura, N. Kawabata, and J. Furukawa, *Bull. Chem. Soc. Jpn.*, **44**, 1127 (1971).
194. J. Nishimura and J. Furakawa, *J. Chem. Soc., Chem. Commun.*, 1375 (1971).
195. S. Miyano and H. Hashimoto, *J. Chem. Soc., Chem. Commun.*, 1418 (1971).
196. S. Miyano and H. Hashimoto, *Bull. Chem. Soc. Jpn.*, **45**, 1946 (1972).
197. M. E. Kuehne and J. C. King, *J. Org. Chem.*, **38**, 304 (1973).
198. N. Kawabata, T. Nagagawa, T. Nakao, and S. Yamashita *J. Org. Chem.*, **42**, 3031 (1977).
199. I. Ryu, S. Muraï, and N. Sonoda, *Tetrahedron Lett.*, 4611 (1977).
200. S. Miyano, Y. Izumi, H. Fujii, and H. Hashimoto, *Synthesis*, 700 (1977).
201. J. Nishimura, J. Furukawa, N. Kawabata, and T. Fujita, *Tetrahedron*, **26**, 2229 (1970).
202. S. Sawada and Y. Inouye, *Bull. Chem. Soc. Jpn.*, **42**, 2669 (1969).
203. W. Kirmse, *Carbene Chemistry*, Academic Press, New York, 1964.
204. L. Vo-Quang and P. Cadiot, *Bull. Soc. Chim. Fr.*, 1525 (1965).
205. G. Emptoz, L. Vo-Quang, and Y. Vo-Quang, *Bull. Soc. Chim. Fr.*, 2653 (1965).
206. N. T. Castellucci and C. E. Griffin, *J. Am. Chem. Soc.*, **82**, 4107 (1960).
207. G. Wittig and J. J. Hutchinson, *Justus Liebigs Ann. Chem.*, **741**, 79 (1970).
208. P. Baret, H. Buffet, and H. P. Lankelma, *Bull. Soc. Chim. Fr.*, 825 (1972).
209. H. Hashimoto, M. Hida, and S. Miyano, *J. Chem. Soc. Jpn.*, **69**, 174 and 2134 (1966).
210. H. Hashimoto, M. Hida, and S. Miyano, *J. Organomet. Chem.*, **10**, 518 (1967).
211. S. Miyano, M. Hida, and H. Hashimoto, *J. Organomet. Chem.*, **12**, 263 (1968).
212. P. Turnbull, K. Syhora, and J. H. Fried, *J. Am. Chem. Soc.*, **88**, 4764 (1966).
213. I. T. Harrison, R. J. Rawson, R. J. Turnbull, and J. H. Fried, *J. Org. Chem.*, **36**, 3515 (1971).
214. K. Takaï, Y. Hotta, K. Oshima, and H. Nozaki, *Tetrahedron Lett.*, 2417 (1978).

215. R. L. Shriner, *Organic Reactions*, Vol. I, Wiley, New York, 1942, pp. 1–37.
216. M. Gaudemar, *Organomet. Chem. Rev., Sect. A*, **8**, 183 (1972).
217. R. D. Rieke and S. T. Uhm, *Synthesis*, 452 (1975).
218. K. Maruoka and H. Nozaki, *J. Am. Chem. Soc.*, **99**, 7705 (1977).
219. E. Santaniello and A. Manzocchi, *Synthesis*, 698 (1977).
220. A. Balsamo, P. L. Barili, P. Crotti, M. Ferretti, B. Macchia, and F. Macchia, *Tetrahedron Lett.*, 1005 (1974).
221. M. Mladenova, B. Blagoev, and B. Kurtev, *Bull. Soc. Chim. Fr.*, 1464 (1974); 77 (1979).
222. M. Gaudemar, *J. Organomet. Chem.*, **96**, 149 (1975).
223. A. H. Akhrem, A. V. Kamernitskii, R. P. Litvinovskaya, and I. G. Reshetova, *Izv. Akad. Nauk SSSR, Ser. Khim.*, 161 (1976).
224. L. Nedelec, V. Torelli, and G. Costerousse, *Bull. Soc. Chim. Fr.*, 2037 (1975).
225. R. Bucourt, L. Nedelec, J. C. Gasc, and M. Vigneau, *Bull. Soc. Chim. Fr.*, 2043 (1975).
226. I. G. Reshetova, A. V. Kamernitskii, V. A. Krivoruchko, and R. P. Litvinovskaya, *Izv. Akad. Nauk SSSR, Ser. Khim.*, 2073 (1975).
227. Z. H. Israeli and E. E. Smissman, *J. Org. Chem.*, **41**, 4070 (1976).
228. A. M. Seldes, C. R. Anding, and E. G. Gros, *Steroids*, **36**, 575 (1980).
229. M. Bellassoued and M. Gaudemar, *J. Organomet. Chem.*, **93**, 9 (1975); **102**, 1 (1975).
230. M. Bogavac, L. Arsenijevic, and V. Arsenijevic, *Bull. Soc. Chim. Fr.*, 145 (1980).
231. I. I. Lapkin, F. G. Saitkulova, G. G. Abashev, and V. V. Fotin, *Izv. Vyssh. Uchebn. Zaved., Khim. Khim. Tekhnol.*, **23**, 793 (1980).
232. J. P. Guetté and M. Lucas, *Bull. Soc. Chim. Fr.*, 2759 (1975).
233. J. P. Guetté and M. Lucas, *J. Chem. Res.*, (S) 214 (1978), (M)2510 (1978).
234. M. Lucas and J. P. Guetté, *Tetrahedron*, **34**, 1675 (1978).
235. M. Lucas and J. P. Guetté, *Tetrahedron*, **34**, 1681 (1978).
236. M. Lucas and J. P. Guetté, *Tetrahedron*, **34**, 1685 (1978).
237. M. Lucas and J. P. Guetté, *J. Chem. Res.*, (S) 53 (1980), (M) 721 (1980).
238. A. Balsamo, P. Crotti, B. Macchia, F. Macchia, A. Rosai, P. Domiano, and G. Nannini, *J. Org. Chem.*, **43**, 3036 (1978).
239. R. N. Gedye, P. Arora, and H. Khabil, *Can. J. Chem.*, **53**, 1943 (1975).
240. G. Daviaud and Ph. Miginiac, *Bull. Soc. Chim. Fr.*, 2325 (1971).
241. G. Schlewer, J. L. Stampf, and C. Benezra, *J. Labelled Compd. Radiopharm.*, **17**, 297 (1980).
242. F. Dardoize and M. Gaudemar, *Bull. Soc. Chim. Fr.*, 1561 (1976).
243. M. Furukawa, T. Okawara, Y. Noguchi, and Y. Terawaki, *Chem. Pharm. Bull.*, **26**, 260 (1978).
244. B. Kryczka, A. Laurent, and B. Marquet, *Tetrahedron*, **34**, 3291 (1978).
245. G. Alvernhe, S. Lacombe, A. Laurent, and B. Marquet, *J. Chem. Res.*, (S) 54 (1980), (M) 858 (1980).
246. H. Stamm and H. Steudle, *Arch. Pharm. (Weinheim, Ger.)*, **310**, 873 (1977).
247. N. Goasdoué and M. Gaudemar, *J. Organomet. Chem.*, **71**, 325 (1974).
248. S. Cetkovic, L. Arsenijevic, and V. Arsenijevic, *Arch. Farm.*, **28**, 189 (1978); *Chem. Abstr.*, **90**, 121 365q (1979).
249. K. E. Schülte, G. Rücker, and J. Feldkamp, *Chem. Ber.*, **105**, 24 (1972).
250. M. T. Bertrand, G. Courtois, and L. Miginiac, *Tetrahedron Lett.*, 1945 (1974).
251. M. T. Bertrand, G. Courtois, and L. Miginiac, *Tetrahedron Lett.*, 3147 (1975).
252. M. T. Bertrand, G. Courtois, and L. Miginiac, *C. R. Acad. Sci., Ser. C*, **280**, 999 (1975).
253. B. Castro and J. Villieras, *Bull. Soc. Chim. Fr.*, 3521 (1969).
254. J. Villieras, *J. Organomet. Chem.*, **34**, 209 (1972).
255. A. Jean and M. Lequan, *C. R. Acad. Sci., Ser. C*, **273**, 1662 (1971).
256. I. I. Lapkin and Z. D. Alekseeva, *Zh. Org. Khim.*, **2**, 393 (1966).
257. I. I. Lapkin and V. N. Musikhina, *Zh. Org. Khim.*, **3**, 998 (1967).
258. I. I. Lapkin and Z. D. Belykh, *Zh. Org. Khim.*, **4**, 1165 (1968).
259. I. I. Lapkin and F. G. Saitkulova, *Zh. Org. Khim.*, **4**, 1566 (1968); **6**, 450 (1970).
260. I. I. Lapkin and L. S. Kozlova, *Zh. Org. Khim.*, **6**, 453 (1970).
261. J. Zitsman and P. J. Johnson, *Tetrahedron Lett.*, 4201 (1971).
262. I. I. Lapkin and V. V. Fotin, *Zh. Org. Khim.*, **12**, 537 (1976).
263. F. G. Saitkulova, G. G. Abashev, and I. I. Lapkin, *Izv. Vyssh. Uchebn. Zaved. Khim. Khim. Tekhnol.*, **18**, 873 (1975).

264. J. F. Fauvarque and A. Jutand, *J. Organomet. Chem.*, **132**, C17 (1977).
265. J. F. Fauvarque and A. Jutand, *J. Organomet. Chem.*, **177**, 273 (1979).
266. F. G. Saitkulova, G. G. Abashev, T. P. Kadyrmatova, and I. I. Lapkin, *Izv. Vyssh. Uchebn. Zaved., Khim. Khim. Tekhnol.*, **20**, 669 and 1078 (1977).
267. G. Courtois and Ph. Miginiac, *Bull. Soc. Chim. Fr.*, II-395 (1982).
268. J. Canceill and J. Jacques, *Bull. Soc. Chim. Fr.*, 903 (1965).
269. F. Hénin-Vichard and B. Gastambide, *Bull. Soc. Chim. Fr.*, 1154 (1977).
270. I. I. Lapkin, F. G. Saitkulova, and V. V. Fotin, *Izv. Vyssh. Uchebn. Zaved. Khim. Khim. Tekhnol.*, **21**, 1072 (1978).
271. I. I. Lapkin, Z. D. Belykh, V. V. Fotin, and L. D. Orlova, *Sint. Metody Osn. Metallorg. Soedin.*, 46 (1977).
272. N. I. Sheverdina and K. A. Kocheshkov, *Methods of Elemento-Organic Chemistry*, Vol. 3, North-Holland, Amsterdam, 1967, pp. 181–239.
273. K. Nützel, 'Organo-Cadmium-Verbindungen', in Houben-Weyl, *Methoden der Organischen Chemie*, Vol. 13/2a, Verlag Chemie, Stuttgart, 1973, pp. 916–946.
274. P. R. Jones and P. J. Desio, *Chem. Rev.*, **78**, 491–516 (1978).
275. D. Abenhaïm, E. Henry-Basch, and P. Fréon, *C. R. Acad. Sci., Ser. C*, **264**, 213 and 1313 (1967).
276. E. J. Corey, *J. Am. Chem. Soc.*, **97**, 2287 (1975).
277. D. A. Evans, D. J. Baillargeon, and J. V. Nelson, *J. Am. Chem. Soc.*, **100**, 2242 (1978).
278. K. H. Thiele and J. Köhler, *J. Organomet. Chem.*, **7**, 365 (1967).
279. P. R. Jones, P. D. Sherman, and K. Schwarzenberg, *J. Organomet. Chem.*, **10**, 521 (1967).
280. C. Bernardon, *Tetrahedron Lett.*, 1581 (1979).
281. B. Riegel, S. Siegel, and W. M. Lilienfeld, *J. Am. Chem. Soc.*, **68**, 984 (1946).
282. J. Kollonitsch, *J. Chem. Soc. A*, 453 and 456 (1966); *Nature (London)*, **188**, 140 (1960).
283. F. Huet, E. Henry-Basch, and P. Fréon, *C. R. Acad. Sci., Ser. C*, **262**, 598, 954 and 1328 (1966).
284. L. Le Guilly and F. Tatibouet, *C. R. Acad. Sci., Ser. C*, **262**, 217 (1966).
285. F. Huet, E. Henry-Basch, and P. Fréon, *Bull. Soc. Chim. Fr.*, 1415 and 1426 (1970).
286. G. Soussan, *C. R. Acad. Sci., Ser. C*, **263**, 954 (1966); **268**, 267 (1969).
287. E. Henry-Basch, J. Michel, and P. Fréon, *C. R. Acad. Sci.*, **260**, 3695 and 5809 (1965).
288. L. Le Guilly, J. Chenault, and F. Tatibouet, *C. R. Acad. Sci.*, **260**, 6634 (1965).
289. P. Fréon and F. Tatibouet, *C. R. Acad. Sci.*, **249**, 1361 (1959); **250**, 145 (1960).
290. R. Golse and A. Liermain, *C. R. Acad. Sci.*, **252**, 3076 (1961).
291. J. Michel, E. Henry-Basch, and P. Fréon, *C. R. Acad. Sci.*, **258**, 6171 (1964).
292. G. W. Stacy and R. M. Curdy, *J. Am. Chem. Soc.*, **76**, 1914 (1954); **79**, 3587 (1957).
293. W. G. Dauben and J. W. Colette, *J. Am. Chem. Soc.*, **81**, 967 (1959).
294. M. Langlais, H. Buzas, and P. Fréon, *C. R. Acad. Sci.*, **253**, 2364 (1961); **254**, 1452 (1962).
295. F. Tatibouet and P. Fréon, *Bull. Soc. Chim. Fr.*, 1496 (1963).
296. G. Wittig, F. J. Meyer, and G. Lange, *Justus Liebigs Ann. Chem.*, **571**, 167 (1951).
297. M. Langlais and P. Fréon, *C. R. Acad. Sci.*, **261**, 2920 (1965).
298. M. Gocmen, G. Soussan, and P. Fréon, *Bull. Soc. Chim. Fr.*, 562 (1973); *J. Organomet. Chem.*, **61**, 19 (1973).
299. J. Michel and E. Henry-Basch, *C. R. Acad. Sci., Ser. C*, **262**, 1274 (1966).
300. F. S. Prout, *J. Am. Chem. Soc.*, **74**, 5915 (1952); **76**, 1911 (1954).
301. J. P. Deniau, E. Henry-Basch, and P. Fréon, *C. R. Acad. Sci., Ser. C*, **264**, 1560 (1967).
302. W. M. Hoehn and R. B. Moffett, *J. Am. Chem. Soc.*, **67**, 740 (1945).
303. C. Bernardon, E. Henry-Basch, and P. Fréon, *C. R. Acad. Sci., Ser. C*, **266**, 1502 (1968).
304. C. G. Stuckwisch and J. V. Bailey, *J. Org. Chem.*, **28**, 2362 (1963).
305. P. L. De Benneville, *J. Org. Chem.*, **6**, 462 (1940).
306. P. R. Jones and A. A. Lavigne, *J. Org. Chem.*, **25**, 2020 (1960).
307. D. S. Tarbell and J. R. Price, *J. Org. Chem.*, **21**, 144 (1956); **22**, 245 (1957).
308. J. Cason and R. J. Fessenden, *J. Org. Chem.*, **22**, 1326 (1957).
309. P. R. Jones and S. J. Costanzo, *J. Org. Chem.*, **38**, 3189 (1973).
310. E. Müller, H. Fettel and M. Sauerbier, *Synthesis*, 82 (1970).
311. G. Emptoz and F. Huet, *Bull. Soc. Chim. Fr.*, 1695 (1974).
312. R. F. Galliulina, V. N. Pankratova, L. P. Stepovik, and A. D. Chernova, *Zh. Obshch. Khim.*, **46**, 98 (1976).
313. A. J. Pearson, *Tetrahedron Lett.*, 3617 (1975).

314. A. J. Birch and A. J. Pearson, *Tetrahedron Lett.*, 2379 (1975).
315. P. R. Jones and J. R. Young, *J. Org. Chem.*, **33**, 1675 (1968).
316. G. Martin, *Ann. Chim. Fr.*, [13] **4**, 541 (1959).
317. J. P. Pradère and H. Quiniou, *C. R. Acad. Sci., Ser. C*, **273**, 1013 (1971).
318. H. Gross and J. Freiberg, *Chem. Ber.*, **99**, 3260 (1966).
319. R. K. Summerbell and L. N. Bauer, *J. Am. Chem. Soc.*, **58**, 759 (1936).
320. C. D. Hurd and R. P. Holysz, *J. Am. Chem. Soc.*, **72**, 2005 (1950).
321. F. N. Jones and C. R. Hauser, *J. Org. Chem.*, **27**, 3364 (1962).
322. C. F. Koelsch, *J. Org. Chem.*, **25**, 642 (1960).
323. M. Renson, *Bull. Soc. Chim. Belg.*, **69**, 236 (1960); **71**, 245 (1962).
324. R. A. Sharma, M. Bobek, and A. Bloch, *J. Med. Chem.*, **18**, 473 (1975).
325. P. R. Jones, C. J. Jarboe, and R. Nadeau, *J. Organomet. Chem.*, **8**, 361 (1967).
326. M. H. Palmer and J. A. Reid, *J. Chem. Soc.*, 931 (1960).
327. J. Cason, *Chem. Rev.*, **40**, 15 (1947).
328. D. A. Shirley, *Organic Reactions*, Vol. VIII, Wiley, New York, 1954, pp. 28–58.
329. M. Cais and A. Mandelbaum, in *The Chemistry of the Carbonyl Group* (Ed. S. Patai), Interscience, London, 1966, pp. 303–330.
330. G. L. B. Carlson, F. H. Quina, B. M. Zarnegar, and D. G. Whitten, *J. Am. Chem. Soc.*, **97**, 347 (1975).
331. J. M. Watson, J. L. Irvine, and R. M. Roberts, *J. Am. Chem. Soc.*, **95**, 3348 (1973).
332. T. Denzel and H. Höhm, *Arch. Pharm. (Weinheim, Ger.)*, **309**, 486 (1976).
333. O. G. Yashina, T. V. Zavara, and L. I. Vereschagin, *Zh. Org. Khim.*, **3**, 219 (1967); *Chem. Abstr.*, **66**, 94664g (1967).
334. O. G. Yashina, T. V. Zavara, and L. I. Vereschagin, *Zh. Org. Khim.*, **4**, 1904 (1968); *Chem. Abstr.*, **70**, 28341f (1969); *Zh. Org. Khim.*, **4**, 2104 (1968); *Chem. Abstr.*, **70**, 67812e (1969).
335. M. Renson and J. Bonhomme, *Bull. Soc. Chim. Belg.*, **68**, 437 (1959).
336. M. Renson and J. Beetz, *Bull. Soc. Chim. Belg.*, **70**, 77 and 537 (1961).
337. J. Cason and E. J. Reist, *J. Org. Chem.*, **23**, 1668 (1958).
338. G. Soussan and P. Fréon, *C. R. Acad. Sci., Ser. C*, **262**, 933 (1966).
339. M. Schmeisser and M. Weidenbruch, *Chem. Ber.*, **100**, 2306 (1967).
340. A. Ruwett and M. Renson, *Bull. Soc. Chim. Belg.*, **75**, 157 (1966).
341. A. Kirrmann and C. Wakselman, *C. R. Acad. Sci., Ser. C*, **262**, 1325 (1966).
342. K. G. Deshpande, K. S. Naragund, and S. M. Kulkarni, *J. Indian Chem. Soc.*, **55**, 813 (1978).
343. J. Furakawa, N. Kawabata, and T. Fujita, *Tetrahedron*, **26**, 243 (1970).
344. V. I. Shcherbakov, *Zh. Obshch. Khim.*, **41**, 2043 (1971); *Chem. Abstr.*, **76**, 34372g (1972).
345. L. J. Krause and J. A. Morisson, *J. Chem. Soc., Chem. Commun.*, 671 (1980).
346. L. G. Makarova and A. N. Nesmeyanov, *Methods of Elemento-Organic Chemistry*, Vol. 4, North-Holland, Amsterdam, 1967, pp. 337–455.
347. K. P. Zeller and H. Straub, 'Quecksilber-organische Verbindungen', in Houben-Weyl, *Methoden der Organischen Chemie*, Vol. 13/2b, Verlag Chemie, Stuttgart, 1974, pp. 314–379.
348. R. C. Larock, *Angew. Chem., Int. Ed. Engl.*, **17**, 27 (1978).
349. R. C. Larock, *Tetrahedron*, **38**, 1713 (1982).
350. V. A. Nikanorov, V. I. Rozenberg, G. V. Gavrilova, Yu. G. Bundel, and O. A. Reutov, *Izv. Akad. Nauk SSSR, Ser. Khim.*, 1675 (1975); *Bull. Acad. Sci. USSR, Div. Chem. Sci.*, 1568 (1975).
351. H. C. Gardner and J. K. Kochi, *J. Am. Chem. Soc.*, **98**, 2460 (1976).
352. O. A. Reutov, V. I. Rozenberg, G. V. Gavrilova, and V. A. Nikanorov *J. Organomet. Chem.*, **177**, 101 (1979).
353. U. Blaukat and W. P. Neumann, *J. Organomet. Chem.*, **49**, 323 (1973).
354. B. Giese and J. Meister, *Angew. Chem., Int. Ed. Engl.*, **16**, 178 (1977).
355. B. Giese and J. Meixner, *Angew. Chem., Int. Ed. Engl.*, **18**, 154 (1979).
356. B. Giese and J. Meister, *Chem. Ber.*, **110**, 2588 (1977).
357. B. Giese and J. Meixner, *Tetrahedron Lett.*, 2783 (1977).
358. H. F. Grützmacher and R. Schmuck, *Chem. Ber.*, **113**, 1192 (1980).
359. B. Giese and K. Heuck, *Tetrahedron Lett.*, 1829 (1980).
360. B. Giese and K. Heuck, *Chem. Ber.*, **112**, 3759 (1979).
361. B. Giese and W. Zwick, *Tetrahedron Lett.*, 3569 (1980).

362. B. Giese and W. Zwick, *Chem. Ber.*, **112**, 3766 (1979).
363. B. Giese, G. Kretzchmar, and J. Meixner, *Chem. Ber.*, **113**, 2787 (1980).
364. B. Giese and J. Meixner, *Tetrahedron Lett.*, 2779 (1977).
365. B. Giese and J. A. Gonzalez-Gomez, *Tetrahedron Lett.*, 2765 (1982).
366. R. F. Heck, *J. Am. Chem. Soc.*, **90**, 5518 (1968).
367. R. F. Heck, *J. Am. Chem. Soc.*, **91**, 6707 (1969).
368. R. F. Heck, *J. Am. Chem. Soc.*, **93**, 6896 (1971).
369. R. F. Heck, *J. Organomet. Chem.*, **37**, 389 (1972); *J. Am. Chem. Soc.*, **94**, 2712 (1972).
370. D. E. Bergstrom and M. K. Ogawa, *J. Am. Chem. Soc.*, **100**, 8106 (1978).
371. H. Horino, M. Arai, and N. Inoue, *Bull. Chem. Soc. Jpn.*, **47**, 1683 (1974).
372. R. F. Heck, *J. Am. Chem. Soc.*, **90**, 5526 (1968).
373. R. F. Heck, *J. Am. Chem. Soc.*, **90**, 5538 (1968).
374. R. F. Heck, *J. Am. Chem. Soc.*, **90**, 5542 (1968).
375. A. Kasahara and T. Izumi, *Bull. Chem. Soc. Jpn.*, **45**, 1256 (1972).
376. R. C. Larock and M. A. Mitchell, *J. Am. Chem. Soc.*, **100**, 180 (1978).
377. R. C. Larock, K Takagi, S. Hershberger, and M. Mitchell, *Tetrahedron Lett.*, 5231 (1981).
378. O. A. Reutov, E. V. Uglova, and V. D. Makhaev, *Dokl. Akad. Nauk SSSR*, **188**, 833 (1969); *Proc. Acad. Sci. USSR*, **188**, 808 (1969).
379. O. A. Reutov, E. V. Uglova, V. D. Makhaev, and V. S. Petrosyan, *Zh. Org. Khim.*, **6**, 2153 (1970); *J. Org. Chem. USSR*, **6**, 2164 (1970).
380. I. P. Beletskaya, S. V. Rykov, V. B. Vol'eva, A. L. Buchachenko, and A. V. Kessenikh, *Izv. Akad. Nauk SSSR, Ser. Khim.*, 684 (1972); *Bull. Acad. Sci. USSR, Div. Chem. Sci.*, 653 (1972).
381. F. C. Whitmore and E. N. Thurman, *J. Am. Chem. Soc.*, **51**, 1491 (1929).
382. I. P. Beletskaya, O. A. Maksimenko, and O. A. Reutov, *Dokl. Akad. Nauk SSSR*, **168**, 333 (1966); *Proc. Acad. Sci. USSR*, **168**, 473 (1966).
383. I. P. Beltskaya, V. B. Vol'eva, S. V. Rykov, A. L. Buchachenko, and A. V. Kessenikh, *Izv. Akad. Nauk SSSR, Ser. Khim.*, 454 (1971); *Bull. Acad. Sci. USSR, Div. Chem. Sci.*, 397 (1971).
384. I. P. Beletskaya, S. V. Rykov, and A. L. Buchachenko, *Org. Magn. Reson.*, **5**, 595 (1973).
385. I. P. Beletskaya, V. B. Vol'eva, and O. A. Reutov, *Dokl. Akad. Nauk SSSR*, **195**, 360 (1970); *Proc. Acad. Sci. USSR*, **195**, 808 (1970).
386. A. Kekulé and A. Franchimont, *Chem. Ber.*, **5**, 907 (1872).
387. O. A. Maksimenko, I. P. Beletskaya, and O. A. Reutov, *Zh. Org. Khim.*, **2**, 1137 (1966); *J. Org. Chem. USSR*, **2**, 1131 (1966).
388. W. D. Schroeder and R. Q. Brewster, *J. Am. Chem. Soc.*, **60**, 751 (1938).
389. A. N. Nesmeyanov, E. G. Perevalova, and O. A. Nesmeyanova, *Dokl. Akad. Nauk SSSR*, **119**, 228 (1958); *Proc. Acad. Sci. USSR*, **119**, 215 (1958).
390. I. P. Beletskaya, V. B. Vol'eva, V. B. Golubev, and O. A. Reutov, *Izv. Akad. Nauk SSSR, Ser. Khim.*, 1197 (1969); *Bull. Acad. Sci. USSR, Div. Chem. Sci.*, 1108 (1969).
391. A. N. Nesmeyanov and E. G. Perevalova, *Izv. Akad. Nauk SSSR, Otd. Khim. Nauk*, 1002 (1954); *Bull. Acad. Sci. USSR, Div. Chem. Sci.*, 873 (1954).
392. I. P. Beletskaya, O. A. Maksimenko, and O. A. Reutov, *Zh. Org. Khim.*, **2**, 1129 (1966); *J. Org. Chem. USSR*, **2**, 1124 (1966).
393. O. A. Maksimenko, I. P. Beletskaya, and O. A. Reutov, *Izv. Akad. Nauk SSSR, Ser. Khim.*, 662 (1966); *Bull. Acad. Sci. USSR, Div. Chem. Sci.*, 627 (1966).
394. U. Schöllkopf and N. Rieber, *Chem. Ber.*, **102**, 488 (1969).
395. A. N. Nesmeyanov, I. F. Lutsenko, and Z. M. Tumanova, *Izv. Akad. Nauk SSSR. Otd. Khim. Nauk*, 601 (1949); *Chem. Abstr.*, **44**, 7225c (1950).
396. D. Y. Curtin and M. J. Hurwitz, *J. Am. Chem. Soc.*, **74**, 5381 (1952).
397. G. A. Artamkina, I. P. Beletskaya, and O. A. Reutov, *J. Organomet. Chem.*, **42**, C17 (1972).
398. I. Rhee, I. Ryu, H. Omura, S. Murai, and N. Sonoda, *Chem. Lett.*, 1435 (1979).
399. I. P. Beletskaya, V. B. Vol'eva, and O. A. Reutov, *Dokl. Akad. Nauk SSSR*, **204**, 93 (1972); *Proc. Acad. Sci. USSR*, **204**, 383 (1972).
400. R. F. Heck, *J. Am. Chem. Soc.*, **90**, 5531 (1968).
401. R. C. Larock, J. C. Bernhardt, and R. J. Driggs, *J. Organomet. Chem.*, **156**, 45 (1978).
402. R. C. Larock and S. S. Hershberger, *Tetrahedron Lett.*, 2443 (1981); *J. Organomet. Chem.*, **225**, 31 (1982).
403. I. P. Beletskaya, Y. A. Artamkina, and O. A. Reutov, *J. Organomet. Chem.*, **99**, 343 (1975).

404. N. V. Kondratenko, E. P. Vechirko, and L. M. Yagupolski, *Synthesis*, 932 (1980).
405. D. E. Bergbreiter and G. M. Whitesides, *J. Am. Chem. Soc.*, **96**, 4937 (1974).
406. R. C. Larock and D. R. Leach, *Tetrahedron Lett.*, 3435 (1981); *Organometallics*, **1**, 74 (1982).
407. A. L. Kurts, I. P. Beletskaya, I. A. Savchenko, and O. A. Reutov, *J. Organomet. Chem.*, **17**, P21 (1969).
408. K. Takagi, T. Okamoto, Y. Sakakibara, A. Ohno, S. Oka, and N. Hayama, *Chem. Lett.*, 951 (1975).
409. Yu. G. Bundel, V. I. Rozenberg, A. L. Kurts, N. D. Antonova, and O. A. Reutov, *J. Organomet. Chem.*, **18**, 209 (1969).
410. I. Kuwajima, K. Narasaka, and T. Mukaiyama, *Tetrahedron Lett.*, 4281 (1967).
411. H. O. House, R. A. Auerbach, M. Gall, and N. P. Peet, *J. Org. Chem.*, **38**, 514 (1973).
412. A. N. Nesmeyanov and E. G. Perevalova, *Izv. Akad. Nauk SSSR, Otd. Khim. Nauk*, 1002 (1954); *Bull. Acad. Sci. USSR, Div. Chem. Sci.*, 873 (1954).
413. R. A. Olofson, B. A. Bauman, and D. J. Wancowicz, *J. Org. Chem.*, **43**, 752 (1978).
414. D. Rhum and G. L. Moore, *Brit. Pat.*, 1 129 229 (1968); *Chem. Abstr.*, **69**, 105926j (1968).
415. Y. Tamura, J. Haruta, S. Okuyama, and Y. Kita, *Tetrahedron Lett.*, 3737 (1978).
416. I. F. Lutsenko, V. L. Foss and A. N. Nesmeyanov, *Dokl. Akad. Nauk SSSR*, **169**, 117 (1966); *Chem. Abstr.*, **65**, 13533h (1966).
417. Y. Kita, J. Haruta, H. Tagawa, and Y. Tamura, *J. Org. Chem.*, **45**, 4519 (1980).
418. R. C. Larock and J. C. Bernhardt, *J. Org. Chem.*, **43**, 710 (1978).
419. R. C. Larock and J. C. Bernhardt, *Tetrahedron Lett.*, 3097 (1976).
420. B. P. Gusev, E. A. El'perina and V. F. Kucherov, *Izv. Akad. Nauk SSSR, Ser. Khim.*, 600 (1980); *Bull. Acad. Sci. USSR, Div. Chem. Sci.*, 418 (1980).
421. G. A. Razuvaev and M. M. Koton, *Chem. Ber.*, **65**, 613 (1932).
422. G. A. Razuvaev and Yu. A. Ol'dekop, *Zh. Obshch. Khim.*, **19**, 1487 (1949); *J. Gen. Chem. USSR*, **19**, 1489 (1949).
423. G. A. Razuvaev and M. M. Koton, *Chem. Ber.*, **66**, 854 (1933).
424. E. Vedejs and P. D. Weeks, *Tetrahedron Lett.*, 3207 (1974).
425. R. C. Larock, *J. Org. Chem.*, **41**, 2241 (1976).
426. R. C. Larock and B. Riefling, *J. Org. Chem.*, **43**, 1468 (1978).
427. R. C. Larock and J. C. Bernhardt, *J. Org. Chem.*, **42**, 1680 (1977).
428. K. Takagi, N. Hayama, T. Okamoto, Y. Sakakibara, and S. Oka; *Bull. Chem. Soc. Jpn.*, **50**, 2741 (1977).
429. T. Izumi and A. Kasahara, *Bull. Chem. Soc. Jpn.*, **48**, 1955 (1975).
430. R. F. Heck, *US Pat.*, 3 539 622 (1970); *Chem. Abstr.*, **74**, 12795d (1971).
431. R. A. Kretchmer and R. Glowinski, *J. Org. Chem.*, **41**, 2661 (1976).
432. G. Wittig and W. Herwig, *Chem. Ber.*, **87**, 1511 (1954); **91**, 883 (1958); **95**, 431 (1962).
433. M. D. Rausch, *Inorg. Chem.*, **1**, 414 (1962).
434. R. F. Kovar and M. D. Rausch, *J. Org. Chem.*, **38**, 1918 (1973).
435. D. Seyferth, M. A. Eisert, and L. J. Todd, *J. Am. Chem. Soc.*, **86**, 121 (1964); **91**, 5027 (1969).
436. D. Seyferth and S. B. Andrews, *J. Organomet. Chem.*, **18**, P21 (1969); **30**, 151 (1971).
437. D. Seyferth and C. K. Haas, *J. Organomet. Chem.*, **39**, C41 (1972).
438. R. Scheffold and U. Michel, *Angew. Chem., Int. Ed. Engl.*, **11**, 231 (1972).
439. D. Seyferth and S. P. Hopper, *J. Organomet. Chem.*, **26**, C62 (1971).
440. D. Seyferth, S. P. Hopper and G. J. Murphy, *J. Organomet. Chem.*, **46**, 201 (1972); *J. Org. Chem.*, **37**, 4070 (1972).
441. I. L. Knunyants, Y. F. Komissarov, B. L. Dyatkin, and L. T. Lantseva, *Izv. Akad. Nauk SSSR*, 943 (1973); *Chem. Abstr.*, **79**, 42635x (1973).
442. D. Seyferth, S. P. Hopper, and K. V. Darragh, *J. Am. Chem. Soc.*, **91**, 6536 (1969).
443. D. Seyferth and K. V. Darragh, *J. Organomet. Chem.*, **11**, P9 (1968).
444. D. Seyferth and K. V. Darragh, *J. Org. Chem.*, **35**, 1297 (1970).
445. D. Seyferth and G. J. Murphy, *J. Organomet. Chem.*, **49**, 117 (1973).
446. D. Seyferth, J. M. Burlitch, R. J. Minasz, J. Y. P. Mui, H. D. Simmons, and S. R. Dowd, *J. Am. Chem. Soc.*, **87**, 4259 (1965).
447. D. Seyferth and C. K. Haas, *J. Organomet. Chem.*, **30**, C38 (1971).
448. V. I. Shcherbakov, *Zh. Obshch. Khim.*, **41**, 1095 (1971); *J. Gen. Chem. USSR*, **41**, 1100 (1971).
449. D. Seyferth and C. K. Haas, *J. Organomet. Chem.*, **46**, C33 (1972).

450. D. Seyferth, H. D. Simmons, and H. M. Shih, *J. Organomet. Chem.*, **29**, 359 (1971).
451. D. Seyferth, H. D. Simmons, and G. Singh, *J. Organomet. Chem.*, **3**, 337 (1965).
452. D. Seyferth, C. K. Haas, and S. P. Hopper, *J. Organomet. Chem.*, **33**, C1 (1971).
453. D. Seyferth, D. C. Mueller, and R. L. Lambert, *J. Am. Chem. Soc.*, **91**, 1562 (1969).
454. D. Seyferth, R. A. Woodruff, D. C. Mueller, and R. L. Lambert, *J. Organomet. Chem.*, **43**, 55 (1972); *J. Org. Chem.*, **38**, 4031 (1973).
455. D. Seyferth and D. C. Mueller, *J. Am. Chem. Soc.*, **93**, 3714 (1971).
456. D. Seyferth and D. C. Mueller, *J. Organomet. Chem.*, **25**, 293 (1970).
457. D. Seyferth and E. M. Hanson, *J. Am. Chem. Soc.*, **90**, 2438 (1968).
458. D. Seyferth and E. M. Hanson, *J. Organomet. Chem.*, **27**, 19 (1971).
459. T. Sakakibara, Y. Odaira, and S. Tsutsumi, *Tetrahedron Lett.*, 503 (1968).
460. D. Seyferth, M. E. Gordon, J. Y. P. Mui, and M. Burlitch, *J. Am. Chem. Soc.*, **89**, 959 (1967).
461. D. Seyferth, J. Y. P. Mui, and J. M. Burlitch, *J. Am. Chem. Soc.*, **89**, 4953 (1967).
462. D. Seyferth, J. Y. P. Mui, and R. Damrauer, *J. Am. Chem. Soc.*, **90**, 6182 (1968).
463. D. Seyferth, R. Damrauer, J. Y. P. Mui, and T. F. Jula, *J. Am. Chem. Soc.*, **90**, 2944 (1968).
464. D. Seyferth, J. Y. P. Mui, M. E. Gordon, and J. M. Burlitch, *J. Am. Chem. Soc.*, **87**, 681 (1965).
465. D. Seyferth and J. M. Burlitch, *J. Am. Chem. Soc.*, **86**, 2730 (1964).
466. D. Seyferth, R. M. Turkel, M. A. Eisert, and L. J. Todd, *J. Am. Chem. Soc.*, **91**, 5027 (1969).
467. P. J. Van Vuuren, R. Fletterick, J. Meinwald, and R. E. Hughes, *J. Am. Chem. Soc.*, **93**, 4394 (1971).
468. E. Rosenberg and J. J. Zuckerman, *J. Organomet. Chem.*, **33**, 321 (1971).
469. M. C. Sacquet, B. Graffe, and P. Maitte, *Bull. Soc. Chim. Fr.*, 2557 (1971).
470. D. Seyferth and R. Damrauer, *J. Org. Chem.*, **31**, 1660 (1966).
471. E. V. Dehmlow, *J. Organomet. Chem.*, **6**, 296 (1966).
472. D. Seyferth and W. Tronich, *J. Organomet. Chem.*, **21**, P3 (1970).
473. D. Seyferth and W. Tronich, *J. Am. Chem. Soc.*, **91**, 2138 (1969).
474. D. Seyferth and W. Tronich, *J. Organomet. Chem.*, **18**, P8 (1969).
475. C. W. Martin and J. A. Landgrebe, *J. Chem. Soc., Chem. Commun.*, 15 (1971); 1438 (1971).
476. D. Seyferth and W. E. Smith, *J. Organomet. Chem.*, **26**, C55 (1971).
477. D. Seyferth and J. M. Burlitch, *J. Am. Chem. Soc.*, **85**, 2667 (1963).
478. D. Seyferth, S. P. Hopper, and T. F. Jula, *J. Organomet. Chem.*, **17**, 193 (1969).
479. D. Seyferth and Y. M. Cheng, *J. Am. Chem. Soc.*, **95**, 6763 (1973).
480. D. Seyferth, J. M. Burlitch, K. Yamamoto, S. S. Washburne, and C. J. Attridge, *J. Org. Chem.*, **35**, 1989 (1970).
481. D. Seyferth, V. A. Mai, and M. E. Gordon, *J. Org. Chem.*, **35**, 1993 (1970).
482. D. Seyferth and B. Prokai, *J. Am. Chem. Soc.*, **88**, 1834 (1966).
483. D. Seyferth, R. Damrauer, S. B. Andrews, and S. S. Washburne, *J. Am. Chem. Soc.*, **93**, 3709 (1971).
484. D. Seyferth, H. M. Shih, J. Dubac, P. Mazerolles, and B. Serres, *J. Organomet. Chem.*, **50**, 39 (1973).
485. D. Seyferth, S. S. Washburne, T. F. Jula, P. Mazerolles, and J. Dubac, *J. Organomet. Chem.*, **16**, 503 (1969).
486. D. Seyferth, J. D. H. Paetsch, and R. T. Marmor, *J. Organomet. Chem.*, **16**, 185 (1969).
487. H. H. Gibson, J. R. Cast, J. Henderson, C. W. Jones, B. F. Look, and J. B. Hunt, *Tetrahedron Lett.*, 1825 (1971).
488. E. O. Fisher and K. H. Dötz, *Chem. Ber.*, **103**, 1273 (1970).
489. D. Seyferth, J. K. Heeren, G. Singh, S. O. Grim, and W. B. Hughes, *J. Organomet. Chem.*, **5**, 267 (1966).

CHAPTER 3

Carbon–carbon bond formation using η³-allyl complexes

Part 1. η³-Allylnickel complexes

G. PAOLO CHIUSOLI
Istituto di Chimica Organica dell'Università, via M. D'Azeglio 85, 43100 Parma, Italy

GIUSEPPE SALERNO
Dipartimento di Chimica, Università della Calabria, Arcavacata di Rende, Cosenza, Italy

---

I. INTRODUCTION	143
II. GENERAL REACTIVITY PATTERNS	144
III. REACTIONS	151
A. C—C Coupling of Coordinated Allyl Groups	151
B. Insertion of Double or Triple Bonds	152
C. Addition to Activated Olefins, Carbonyl Compounds, Epoxides, and Quinones	154
D. Carbonylation and Carboxylation Reactions	156
E. Multiple Insertion Processes	157
F. Carbenoid Insertion Reactions	158
G. Addition to Dienes. HCN Addition	159
H. Reactions with C—Mg Bonds	159
IV. REFERENCES	159

I. INTRODUCTION

During the last three decades, allylnickel complexes as such or as intermediates formed *in situ* have been widely used in organic syntheses directed to the formation of C—C bonds.

The success of allylnickel complexes in synthetic work depends both on their high reactivity and on the ease of preparation. Nickel(0) can be directly attacked to form allylnickel complexes by allyl halides, alcohols, carboxylates, etc., or by other compounds able to form the allyl group *in situ*, such as diene epoxides, vinyllactones, vinylcyclopropanes, and above all conjugated dienes, the last group giving rise to bis(allyl)nickel complexes. Nickel(II) salts also can be easily transformed into allylnickel complexes. On the other hand, nickel(II) compounds obtained from stoichiometric reactions of these complexes can be reduced again to nickel(0). These processes occur spontaneously in catalytic reactions.

Allylnickel complexes are complementary to the widely used allylpalladium complexes. The nickel-bonded allyl group accepts more electron density than the corresponding palladium-bonded species, so its chemistry mainly derives from reactions with electrophiles (coupling and insertion reactions) or from radical-type coupling, whereas in palladium chemistry nucleophilic attack on coordinated allyl groups predominates.

The preparation and general features of the allyl—metal bond, including theoretical and structural aspects and spectroscopic properties, have been treated in Volumes 1 and 2 of this series. We deemed it useful to organize the material in this part as a general survey of reactivity patterns, followed by descriptions of single classes of reactions. In view of the large amount of material available, we preferred to concentrate on recent work and to summarize earlier results only briefly, referring the reader to reviews[1-10] that have appeared in the last few years.

II. GENERAL REACTIVITY PATTERNS

We consider here some elementary modes of reaction of nickel-coordinated allylic systems of the general type (**1**) or (**2**)[1]:

$$\text{\textlangle—NiX or } [(\eta^3\text{-CH}_2\text{CHCH}_2)\text{NiX}] \qquad \diagup\!\!\!\diagdown\!\!\!\diagup\text{NiX or } [(\eta^1\text{-CH}_2\!\!=\!\!\text{CHCH}_2)\text{NiX}]$$

$$\qquad\qquad (\mathbf{1}) \qquad\qquad\qquad\qquad\qquad\qquad (\mathbf{2})$$

where X can be another allyl group, a halide, a carboxylato, a cyclopentadienyl, or other atoms or groups. Coordinative unsaturation gives rise to X-bridged dimeric forms (**3**) when X is a halide or carboxylato group:

$$\text{\textlangle—Ni}\!\!\stackrel{X}{\diagdown\!\!\!\diagup}\!\!\stackrel{}{\diagup\!\!\!\diagdown}\!\!\text{Ni—\textrangle} \quad (\mathbf{3})$$
$$\qquad\qquad X$$

Coordinating molecules L, including solvents, can easily open the bridge to give the monomeric form (**4**).

$$\text{\textlangle—Ni}\!\!\stackrel{X}{\diagdown\!\!\!\diagup}\!\!\stackrel{}{\diagup\!\!\!\diagdown}\!\!\text{L}$$

$$(\mathbf{4})$$

Formulae of type (**1**) will be used in this chapter without indicating L. Equilibria of the type shown in equation 1 are present[11] and can be exploited for synthetic purposes.

$$2[(\eta^3\text{-CH}_2\text{CHCH}_2)\text{NiBr}] \rightleftharpoons [(\eta^3\text{-CH}_2\text{CHCH}_2)_2\text{Ni}] + \text{NiBr}_2 \qquad (1)$$

In principle the η^3-allyl group could undergo several dynamic processes, but the only one which is relevant to synthesis is the passage to η^1 form (equation 2). The position of

$$(\mathbf{1}) \rightleftharpoons (\mathbf{2}) \qquad (2)$$

3. Carbon—carbon bond formation using η^3-allyl complexes

this equilibrium depends on the solvent, ligand, substituent on the allyl group, and temperature and can be studied by n.m.r. techniques[1,12-14]. Exchange of coordinated allyl groups also occurs[15] (equation 3).

$$(1) + CH_2=\overset{*}{C}HCH_2X \rightleftharpoons (\overset{*}{1}) + CH_2=CHCH_2X \quad (3)$$

The nature of the allyl group, as far as its reactivity is concerned, also depends on the variables mentioned above.

Although formally the nickel-bonded allyl group should be regarded as nucleophilic in character, its properties can be largely modified so that it can also behave as an electrophile. An interesting case is offered by (η^3-allyl)nickel bromide, which can be attacked by acids to give propylene or by methanol to give methyl allyl ether[16].

Coupling of two nickel-coordinated allyl groups should be considered a radical-type reaction (equation 4). On the other hand, the same complex with HBr gives complex 1 and

$$[(\eta^3\text{-}CH_2CHCH_2)_2Ni] \longrightarrow CH_2=CHCH_2CH_2CH=CH_2 + Ni \quad (4)$$

$$\xrightarrow{\text{HBr}} (1, X = Br) + CH_2=CHMe$$

propylene[17]. Allylnickel complexes with triketone enolates can give rise either to hexa-1,5-diene or to bis(allyl enol ethers), depending on whether monometallic or bimetallic complexes are involved[18].

Carbanionic character can be conferred to bis(η^3-allyl)nickel complexes by reaction with organolithium compounds RLi and tetramethylethylendiamine, which leads to anionic η^3,η^1-bis(allyl)nickel complexes[19] (equation 5).

$$[(\eta^3\text{-}CH_2CHCH_2)_2Ni] + RLi + tmeda \longrightarrow [Li(tmeda)_2]^+ \left[\overset{R}{\underset{\diagdown}{\text{Ni}}} \right]^- \quad (5)$$

Reaction of allylnickel halides, containing electron-withdrawing substituents such as COOMe on the allyl group, with halide ions X^- in coordinating solvent such as hexamethylphosphoric acid triamide, gives rise to species probably of type 5, which react

$$[(\eta^3\text{-MeOOCCHCHCH}_2)NiX_2]^-$$

(5)

with electrophiles as if they were coordinated carbanions of the Reformatski type, $MeOO\overset{*}{C}CH=CHCH_2ZnX^{20}$. Thus allyl halides and ketones $RCOR^1$ mainly react at the internal position of the allylic system of 5 to give 6 and 7, respectively.

$$CH_2=CHCH(COOMe)CH_2CH=CH_2 \qquad CH_2=CHCH(COOMe)C(OH)RR^1$$

(6) (7)

On the other hand, the use of poorly coordinating anions X^-, such as trifluoroacetate, favours the formation of cationic complexes where coordinating substrates L can easily penetrate. Complexes of type 8 and 9 (X = halide, CF_3COO^-, PF_6^- etc., Y = alkyl, MX_nY_m = Lewis acid, $n+m$ = number of σ-bonds of M) are able to coordinate different L molecules, including olefins, butadiene, alkenes, and CO[21,22]. Thus the complex obtained from butadiene and $[HNi\{P(OEt)_3\}_4]^+PF_6^-$ easily reacts with ethylene to give the

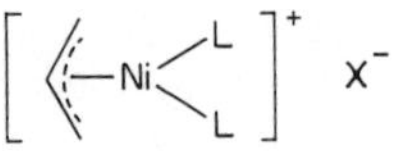

$$[\text{Ni}\overset{L}{\underset{L}{<}}]^+ \quad X^- \quad \text{or} \quad \text{Ni}\overset{L}{\underset{XM(X_nY_m)}{<}}$$

(8) 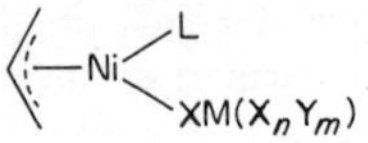　　　　　　　　　　(9)

linear isomer, (E)-hexa-1,4-diene. The intermediate **10** in the olefin insertion process, which becomes catalytic at 100 °C, has been isolated in the case of the insertion of hexa-1,5-diene into the allyl—nickel bond[22]. Analogous intermediates are present in

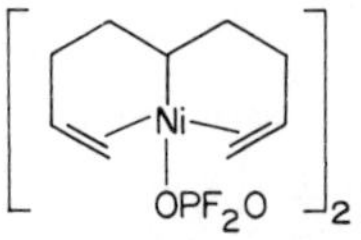

$$\left[\text{Ni} \atop \overline{OPF_2O}\right]_2$$

(10)

carbonylation reactions[8] where a CO ligand is inserted, for example in **8** to give **11**. The acetate anion promotes olefin insertion into the allyl—nickel bond[23].

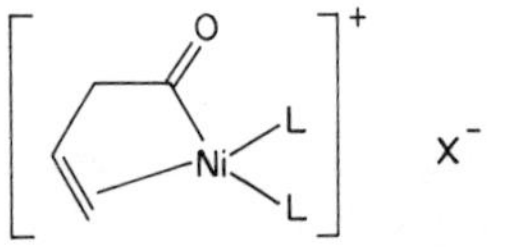

$$\left[\text{Ni}\overset{L}{\underset{L}{<}}\right]^+ \quad X^- \quad (11)$$

Oxidative elimination processes on carbalkoxyallylnickel complexes can give rise to formation of conjugated dienes as in the case of isoprene from the acetoxymethylbutenyl ligand[24] (equation 6).

$$[(\eta^3\text{-AcOCH}_2\text{CHC(Me)CH}_2)\text{NiX}] \longrightarrow CH_2{=}C(Me)CH{=}CH_2 + Ni(X)OAc \quad (6)$$

Electrophilic attack is likely to occur through the η^1 form. The latter also allows a rotation, which places the substituents on the Ni—C bond on the opposite site (syn⇌anti isomerization). In this way a syn complex (3-cyanoallylnickel halide) can react with an electrophile (allyl halide) giving the Z-product (hepta-2,5-dienoic nitrile) instead of the expected E-product[24,25] (equation 7). Ligands and solvents can influence the stereochem-

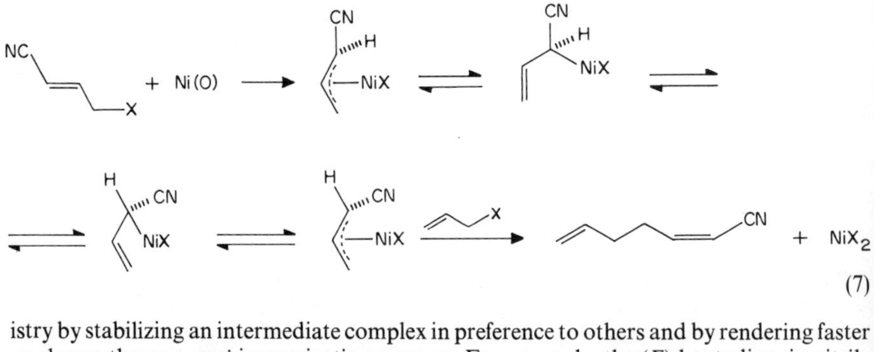

$$NC{\sim}{\sim}_X + Ni(0) \longrightarrow \cdots \rightleftharpoons \cdots \rightleftharpoons \cdots \rightleftharpoons \cdots \rightleftharpoons \cdots \longrightarrow {\sim}{\sim}{\sim}CN + NiX_2 \quad (7)$$

istry by stabilizing an intermediate complex in preference to others and by rendering faster or slower the syn–anti isomerization process. For example, the (E)-heptadienoic nitrile can be made to predominate in reaction 7 using acetonitrile instead of benzene and 2,6-lutidine as ligand[24].

3. Carbon—carbon bond formation using η^3-allyl complexes

TABLE 1. Regiochemistry of crotyl group attack on unsaturated systems

Reagents[a]	Product distribution (%)	
CH$_2$=CHCH$_2$COOCH$_2$CH=CHMe	MeCH=CHCH$_2$CH$_2$CH=CHCOOH	80
	CH$_2$=CHCH(Me)CH$_2$CH=CHCOOH	20
MeCOOCH$_2$CH=CHMe + PhC≡CH	MeCH=CHCH$_2$C=CPh	52
	CH$_2$=CHCH(Me)C=CPh	48
MeCOOCH$_2$CH=CHMe + PhNHN=CHPh	CH$_2$=CHCH(Me)CH(Ph)N=NPh	~100
MeCOOCH$_2$CH=CHMe + PhCHO	CH$_2$=CH(Me)CH(OH)Ph	98
	MeCH=CHCH$_2$CH(OH)Ph	2

[a]Similar results are obtained starting from the branched allylic isomer —CH(Me)CH=CH$_2$.

The site of attack on coordinated allyl systems also strongly depends on the type of electrophile used. In Table 1 a comparison is reported among different substrates in the same reaction type, based on oxidative addition of crotyl esters RCOOCH$_2$CH=CHMe to [Ni{P(OPr^i)$_3$}$_4$] at 80 °C to give crotylnickel carboxylates (12) able to insert C=C, C≡C, C=O or C=N bonds[26]. It can be seen that the trend towards formation of the

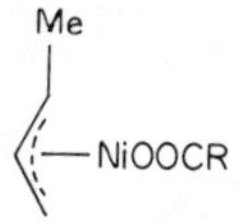

Me
|
⟨—NiOOCR

(12)

branched product follows the electrophilicity of the substrates (olefins < alkynes < aldehydes, ketones, phenylhydrazones). This behaviour suggests that the reaction is controlled by the difference in energy between the most nucleophilic centre of the nickel-bonded crotyl group, which is placed on the internal carbon atom of the allylic system, and the most electrophilic carbon of the substrate. When, however, the formation of the new bond is energetically less favourable, no longer is the reaction with the most nucleophilic centre preferred, steric hindrance being a more important factor than ability to release electrons.

Another problem in this field is whether the reaction centre on the allylic group reacts in an electrocyclic way (a) as proposed for allylpalladium complexes[27] or a non-electrocyclic way (b) (equation 9, Y = substituent, E = electrophile). There are some reasons for

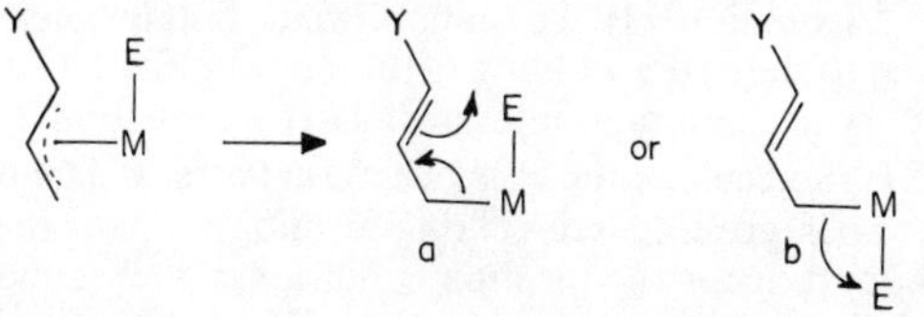

$$\text{(9)}$$

believing that the electrocyclic process takes place when alkyl-substituted allyl groups give rise to branched products. In particular, for the case of allylpalladium halide complexes it was observed[28] that alkyl-substituted η^3-allyl complexes pass to the η^1 form by opening the bond between palladium and the most substituted end of the allyl group preferentially. If one assumes that the same is true for the previously mentioned nickel–crotyl complexes the reaction with the electrophile should occur via an electrocyclic path when the branched product is formed. The same information from n.m.r. studies would also bring us to hypothesize that when the linear product is formed the crotyl group reacts in a non-electrocyclic way if the way of opening the C—Ni bond is still the same. If, however, there is an equilibration with the other allylic carbon the electrocyclic mode could also be at work.

As to the type of attack on allylic systems by nucleophiles, it appears that the situation is different. It has been shown[29,30] that when two different η^3 and η^1 sites are available in the same complex, electrophiles such as aldehydes or ketones prefer the η^1 site and nucleophiles such as amines or malonates attack the η^3 site preferentially. The complex obtained from three molecules of allene and a nickel atom should be considered on n.m.r. grounds as an η^3,η^1-complex. The main reaction product with aldehydes derives from attack on the η^1 site (equation 10).

$$(10)$$

The way in which the η^3-system is attacked is not yet clear. Most investigations have been carried out on allylpalladium systems, which offer better stability, and we refer to the results of these studies assuming that the situation would not change dramatically on passing to nickel.

Attack by stabilized carbon nucleophiles on allylpalladium complexes has been shown to occur with inversion of configuration at chiral centres, and this has been interpreted as external nucleophilic attack on the η^3-allyl complex[31]. Recent work on nickel complexes also points to the possibility of external attack. Amines have been shown to attack cationic allylnickel complexes from outside[32].

On the other hand, the intermediates in methoxylation of a bis(η^3-allyl)palladium complex obtained from butadiene and palladium have been isolated and the methoxy group shown to be bonded to palladium before forming the C—O bond[33]. A similar pattern can be postulated for the analogous nickel complexes and possibly for C-nucleophiles RH such as malonic esters. The latter add to butadiene in presence of nickel(0) catalysts[1,33,34] (equation 11). The intermediate bis(allyl)nickel complexes (**13a**) and (**13b**) can be induced to shift from addition of the 'nucleophile' RH (to give **15**) to linear triene (**16**) formation[35], via a common intermediate (**14**) (equation 11) which could be attacked by another RH molecule in the case of external attack. The two forms **13a** and **13b** can also give rise to an interesting series of ring-forming coupling reactions, depending on which form is stabilized under the reaction conditions[1,19,36] (equation 12). Another bis(allyl)nickel complex, formed by reaction of three molecules of butadiene with nickel(0), can also lead to a selective and stereoselective coupling[1,19] (equation 13).

Ligands, particularly tertiary phosphines and trialkyl or triaryl phosphites and hydrogen-active compounds, influence the formation of these products in different ways:

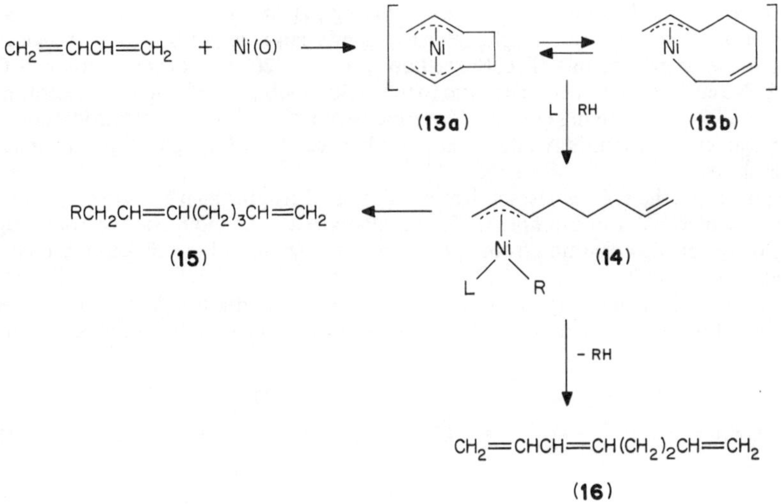

3. Carbon—carbon bond formation using η^3-allyl complexes

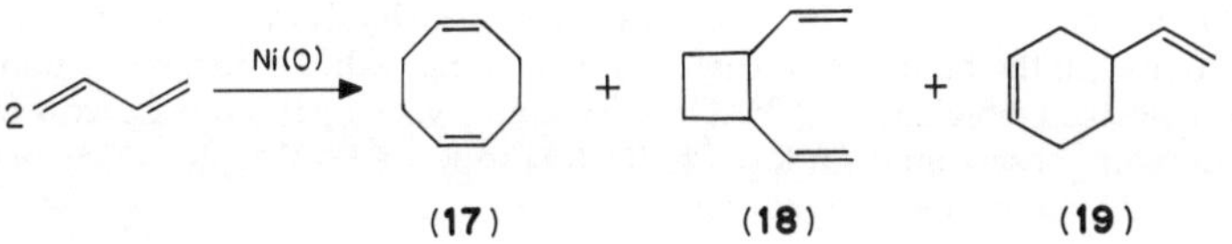

$$CH_2=CHCH=CH_2 + Ni(0) \longrightarrow [\text{(13a)} \rightleftharpoons \text{(13b)}] \quad (11)$$

$$\text{(13b)} \xrightarrow{L, RH} \text{(14)} \longrightarrow RCH_2CH=CH(CH_2)_3CH=CH_2 \text{ (15)}$$

$$\text{(14)} \xrightarrow{-RH} CH_2=CHCH=CH(CH_2)_2CH=CH_2 \text{ (16)}$$

$$2 \diagup\!\!\!\diagdown \xrightarrow{Ni(0)} \text{(17)} + \text{(18)} + \text{(19)} \quad (12)$$

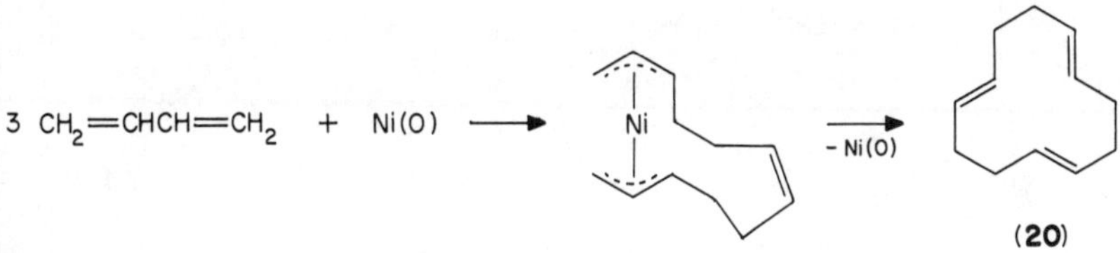

$$3\, CH_2=CHCH=CH_2 + Ni(0) \longrightarrow \text{(20)} \xrightarrow{-Ni(0)} \text{(13)}$$

(a) By favouring either β-hydride elimination to form open-chained compounds or C—C coupling. For example, when RH = secondary amine, phenol, or other compounds, containing active hydrogen, the reaction can be caused to give both the open-chain compound (16) and the R—C coupling product (15). With C-nucleophiles the latter is formed[1,33,37,38].

(b) By altering the coupling sites (internal or terminal) of the allyl group[1,39]. For example, 96% of (Z,Z)-cycloocta-1,5-diene (17) is obtained on complete conversion of butadiene with $P(OC_6H_4\text{-}o\text{-}C_6H_5)_3$ as ligand, but 40% of vinylcyclohexene (19) is formed with $P(\text{cyclohexyl})_3$. This has been ascribed to stabilization of the diene-derived bis(η^3-allyl)nickel (13a) or η^1,η^3-bis(allyl)nickel complexes (13b), respectively. There is no simple correlation with ligand electronegativity (Tolman's χ factors[40]), another effect connected with orbital control of ring-forming elimination being operative[39]. The former intermediate (13a) can give rise to 36% divinylcyclobutane (18) with the phosphite ligand at conversions up to 85%. This compound is in equilibrium with butadiene[36] and has also been shown to convert into cyclooctadiene[1,41]; the latter predominates on complete conversion of butadiene.

(c) By determining the number of molecules taking part in ring formation. Ligand-free nickel or nickel with certain phosphorus ligands can react with three molecules of butadiene to give mainly (E,E,E)-cyclododecatriene (20) and small amounts of the E,E,Z- and Z,Z,Z-isomers. The ratio to the cyclodimers (17–19) is mainly determined by steric effects (Tolman's cone angle or θ parameter[40]), whereas the electronic control (χ parameter[40]) essentially affects the ratio between the η^3,η^1- and η^3,η^3-allylnickel complexes[42].

(d) By governing the *syn–anti* isomerization of the allyl group which gives rise to E- or Z-stereoisomers[1,41]. For example, (E,E,E)-cyclododecatriene should derive from coupling of *syn* groups resulting from *anti* → *syn* isomerization of the nickel-coordinated precursor.

(e) With particular complexes as catalysts intramolecular double bond insertion has been observed to form, for example (21)[1,43] (equation 14). Olefin insertion processes into

$$2\ CH_2=CHCH=CH_2 \xrightarrow{\left[\begin{array}{c}-Ni\diagup^{PBu_3}_{\diagdown Br}\end{array}\right]}_{MeOH} \left[\begin{array}{c}\text{complex}\\ \text{Ni}\\ Bu_3P\end{array}\right] \longrightarrow \text{(product)} \quad (14)$$

(21)

these bis(allyl)nickel complexes have been shown to be orbital controlled, the terminal allylic carbon with the highest electron density reacting with the olefinic carbon with the highest electron deficiency[1,44,45]. Thus, by simply varying the substituents on the olefin undergoing insertion from R = Ph, R^1 = H to R = COOMe, R^1 = Me one can obtain 22 or 23, respectively (equation 15).

$$\left[\begin{array}{c}\text{Ni}\end{array}\right] + CH_2=CRR^1 \longrightarrow \left[\begin{array}{c}\text{Ni}\\ R\\ R^1\end{array}\right] \quad \left[\begin{array}{c}R\\ R^1\ \text{Ni}\end{array}\right]$$

(22)   (23)

(15)

This short survey of elementary reactivity patterns of nickel-coordinated allyl groups cannot ignore formation of nickel hydride species from allyl—nickel bonds. Hydrides can be generated from reactions of the allyl group that give rise to H-elimination (for example olefin insertion reactions). These hydrides are active in many reactions, particularly in olefin isomerization[46].

Isomerization reactions of dienes can go via allylnickel complexes in the presence of hydride sources[47]. In fact, this is also a way of preparing allylnickel complexes[48] (equation 16). Hydrides are involved in other reactions such as dimerization[1,49] and skeletal

$$CH_2=CHCH=CHCH_2R + NiH \longrightarrow \underset{\underset{Ni}{|}}{MeCH=CHCHCH_2R} \longrightarrow MeCH=CHCH=CHR + NiH$$

(16)

rearrangements of olefins[50] and vinylcyclopropanes[51]. We shall briefly mention here the norbornene–ethylene dimerization reaction which is catalysed by $[(\eta^3-CH_2=CHCH_2)Ni(L)Cl\ldots AlCl_2Me]$. Using an optically active ligand L [(−)-dimenthyl(methyl)phosphine], asymmetric induction has been obtained[52] (equation 17).

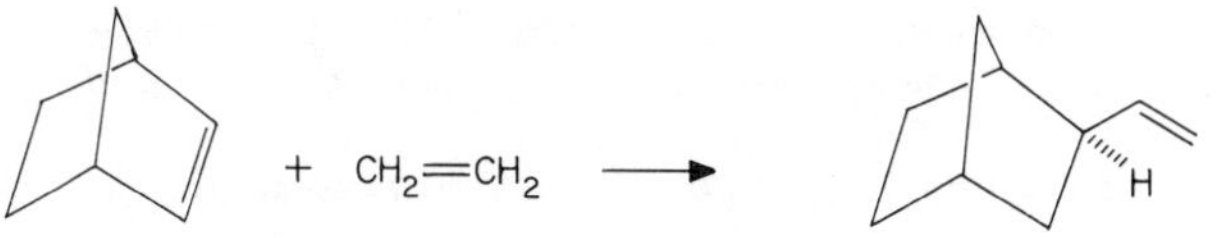

3. Carbon—carbon bond formation using η^3-allyl complexes

$$\text{norbornene} + CH_2=CH_2 \longrightarrow \text{norbornyl-CH=CH}_2 \quad (17)$$

III. REACTIONS

Reactions leading to C—C bond formation and involving allylnickel complexes are numerous and have been the subject of several reviews (see, for example refs. 1–10). In the limited space available we shall only consider some typical reactions not covered by existing reviews and the latest developments. C—C bond formation reactions represent the widest synthetic application of allylnickel complexes. We can distinguish the following classes, which include both stoichiometric reactions and catalytic reactions involving allylnickel complexes:

A. C—C coupling of coordinated allyl groups.
B. Insertion of double and triple bonds.
C. Addition to activated olefins, carbonyl compounds, epoxides, and quinones.
D. Carbonylation and carboxylation reactions.
E. Multiple insertion processes.
F. Carbenoid insertion reactions.
G. Addition to dienes. HCN addition.
H. Reactions with C—Mg Bonds.

A. C—C Coupling of Coordinated Allyl Groups

Although the first coupling reaction, later recognized as involving allylnickel complexes, was discovered in 1941[53] (equation 18), it was only after 1955 that this reaction

$$2CH_2=CHCH_2Cl + [Ni(CO)_4] \longrightarrow CH_2=CHCH_2CH_2CH=CH_2 + NiCl_2 + 4CO \quad (18)$$

began to be utilized in synthesis[54,55] and the relevant allylnickel complexes were discovered[56,56a]. It was soon realized that the true intermediate was an allylnickel halide and that the two allylic moieties involved in coupling were not behaving in the same way, only one of them giving rise to the η^3-coordinated group[57,58] (equation 19). The reaction was applied to cross-coupling of different moieties.

$$[(\eta^3\text{-}CH_2CHCH_2)NiCl] + CH_2=CHCH_2Cl \longrightarrow CH_2=CHCH_2CH_2CH=CH_2 + NiCl_2 \quad (19)$$

Since exchange reactions between different allyl groups are very easy, either a stable allyl group such as one having an electron-withdrawing substituent Y is used[58] or a second non-allylic moiety R is coupled to the nickel-coordinated allylic one[59] to obtain good selectivity (X = halide) (equation 20).

$$[(\eta^3\text{-}CH_2CHCH_2)NiX] + RX \longrightarrow CH_2=CHCH_2R + NiX_2 \quad (20)$$

The coupling reaction is general and has been applied to the synthesis of a variety of compounds[5,24,25,60–62b], including natural compounds and large ring compounds resulting from intramolecular coupling (equation 21). The ring size is influenced by ligands, such as tertiary phosphines, which can affect the access of the substrate to

$$XCH_2CH=CH(CH_2)_nCH=CHCH_2X + Ni(0) \longrightarrow \underline{CH_2CH=CH(CH_2)_nCH=CHCH_2} + NiX_2 \quad (21)$$

coordination sites. The ready availability of nickel(0), for example as $Ni(cod)_2$[63,64], has contributed to the development of very simple procedures[8]. From the synthetic point of view, the use of allyl groups which mask other functions is also interesting. Thus the 2-methoxyallyl group is a masked acetonyl function which has been used for coupling with several carbon compounds[65] (equation 22).

$$[(\eta^3\text{-}CH_2C(OMe)CH_2)NiBr] + PhBr \xrightarrow{H_3O^+} PhCH_2COMe + NiBr_2 \quad (22)$$

The use of diolefins, particularly butadiene, to prepare a variety of unsaturated rings or straight chains has been treated exhaustively[1] and was mentioned in the preceding section to illustrate the typical reactivity of nickel-coordinated allyl groups.

B. Insertion of Double or Triple Bonds

Allylnickel complexes obtained by oxidative addition of allylic halides or allylic esters of organic acids to zerovalent nickel easily insert alkynes[66] at room temperature (equation 23). The dimer of allylnickel bromide inserts two molecules of acetylene stereoselectively[67] (equation 24). Bis(η^3-allyl)nickel behaves analogously[68].

$$MeCOOCH_2CH=CH_2 + RC\equiv CH \xrightarrow{Ni(0)} CH_2=CHCH_2C\equiv CR + MeCOOH \quad (23)$$

$$[(\eta^3\text{-}CH_2CHCH_2)NiBr]_2 + 2CH\equiv CH \longrightarrow$$

$$(Z,Z)\text{-}CH_2=CHCH_2CH=CHCH=CHCH_2 + CH=CH_2 + NiBr_2 + Ni \quad (24)$$

The reaction of ethylene with butadiene-derived crotylnickel complexes has been mentioned in Section II. For recent extensions, see references[68a and 68b]. The reaction of higher olefins is less easy but occurs smoothly with chelating substrates. Thus allyl 3-butenoate gives rise to heptadienoic acids regioselectively (attack on the terminal carbon only). $[Ni\{P(OPr^i)_3\}_4]$ is an efficient catalyst[69,70] (equation 25).

$$CH_2=CHCH_2COOCH_2CH=CH_2 \xrightarrow{Ni(0)} CH_2=CHCH_2CH_2CH=CHCOOH +$$

$$CH_2=CHCH_2CH=CHCH_2COOH \quad (25)$$

Bis(allyl)nickel complexes derived from butadiene or dienes are particularly useful for obtaining cyclic insertion products of alkynes and alkenes[1,71,72] (equations 26 and 27). Acyclic compounds can also be obtained, as shown in Section II for butadiene dimerization.

The complex α,ω-dodecatrienediylnickel also inserts allene to give **24** or alkynes RC≡CR to afford **25** together with other isomers. Compound **24** in its turn can insert alkynes to give **26**[73] (equation 28). Sometimes insertion does not occur easily because of an

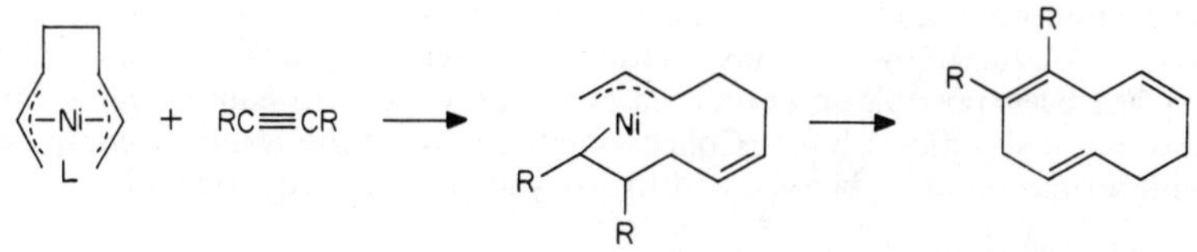

3. Carbon—carbon bond formation using η^3-allyl complexes

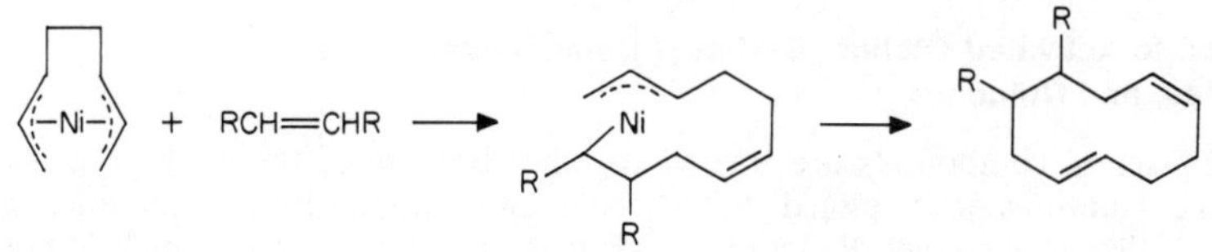

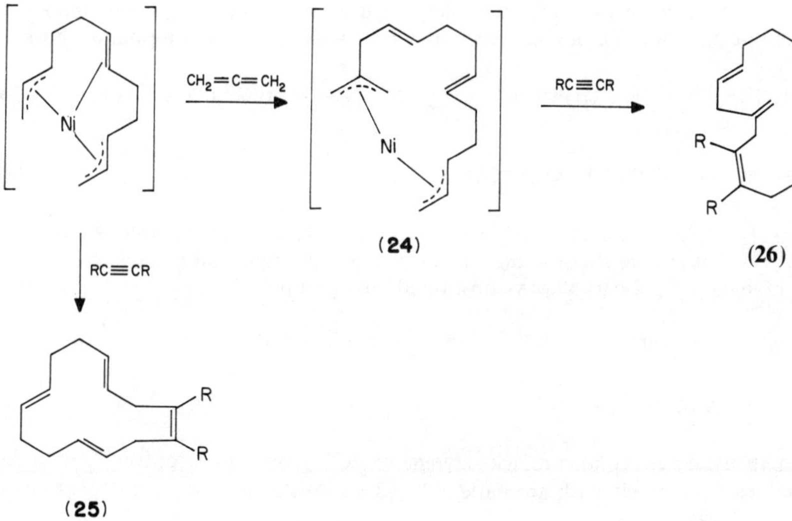

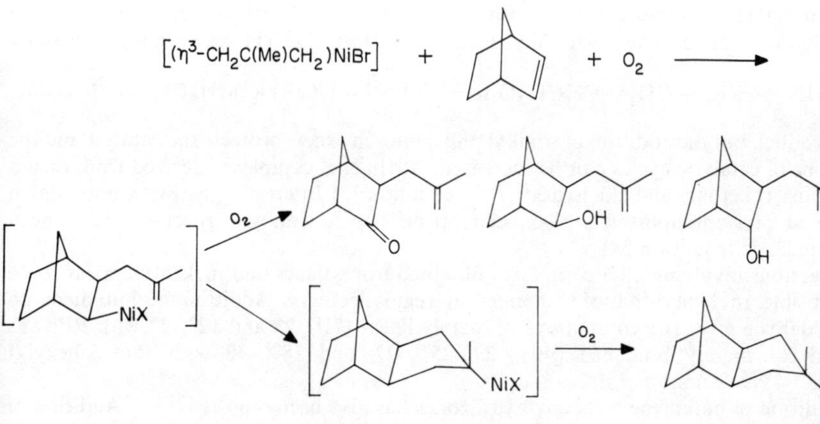

(26)

(27)

(24)

(26)

(25)

(28)

(29)

unfavourable equilibrium and lack of a final irreversible step. In these cases, however, insertion can be induced by oxidants. Although nickel complexes are sensitive to oxidation it has been possible in certain cases to use oxygen to cause double bond insertion into an allyl—nickel bond. Concomitantly, most of the resulting complex is oxidized with formation of carbonylic and hydroxylic groups[74] (equation 29).

C. Addition to Activated Olefins, Carbonyl Compounds, Epoxides, and Quinones

These classes of compounds are treated together because of the analogy between nucleophilic additions (exemplified by Michael and Grignard-type reactions) and reactions of allylnickel species. Reactions of allylic halides in presence of nickel(0) or of allylnickel complexes with acrylic compounds occur at room temperature with formation of intermediate complexes which can undergo hydride elimination or proton uptake[9,75], for example equation 30. Similar reactions occur with carbonyl compounds, both on

$$[(\eta^3\text{-}CH_2CHCH_2)NiBr]_2 + 2\,CH_2\text{=}CHCN \xrightarrow{-NiBr_2} [\{CH_2\text{=}CH(CH_2)_2CH(CN)\}_2\,Ni] \longrightarrow$$

$$CH_2\text{=}CHCH_2CH\text{=}CHCN + CH_2\text{=}CH(CH_2)_3CN \qquad (30)$$

allylnickel complexes and on nickel(0)[20,59,76] and allyl halides, for example (R = alkyl or aryl, R^1 = H, alkyl, or aryl (equation 31). The biologically important α-vinyllactones have been obtained using 2-carbalkoxy-substituted allyl groups[76a,b].

$$[(\eta^3\text{-}CH_2CHCH_2)NiBr] + RCOR^1 \longrightarrow [CH_2\text{=}CHCH_2C(R)(R^1)ONiBr] \xrightarrow{H_2O}$$

$$CH_2\text{=}CHCH_2C(R)(R^1)OH \qquad (31)$$

η^3-Methallylnickel bromide attacks styrene oxide[59] giving **26a** selectively. Allylnickel acetates react selectively with aromatic aldehydes in methanol even in the presence of

$$PhCH(CH_2OH)CH_2C(Me)\text{=}CH_2 \qquad \text{(26a)}$$

aliphatic aldehydes and ketones[77]. Phenylhydrazones and hydrazones in general have been found to be reactive[78] with allyl acetates (equation 32). The reaction is inhibited by

$$PhNHN\text{=}CRR^1 + CH_2\text{=}CHCH_2OAc \xrightarrow{Ni(0)} PhN\text{=}NC(R)(R^1)CH_2CH\text{=}CH_2 \qquad (32)$$

the product, but the addition of trialkyl phosphites in excess protects the catalyst and the number of catalytic cycles can be increased. Allylnickel complexes derived from dienes easily insert ketones and aldehydes[1,79–81] (equation 33). Interesting synthetic potential in the field of isoprenoids has also been opened up[79]. Catalytic reactions have been reported[81,82] (equation 34).

Reactions involving allyl complexes obtained from dienes and nickel(0) appear to be susceptible to ligand-induced changes in regioselectivity. Addition of butadiene to acetaldehyde gives rise to mixtures of mainly linear (71% **28** and 12% **27** with PPh$_3$ as ligand) or mainly branched [40% **29**, 25% **27**, and 18% **30** with P(cyclohexyl$_3$)] products[80,81] (equation 35):

Addition of butadiene to phenylhydrazones has also been reported[83,83a]. Addition of allylnickel halides to quinones[84] is an interesting reaction from both the synthetic and

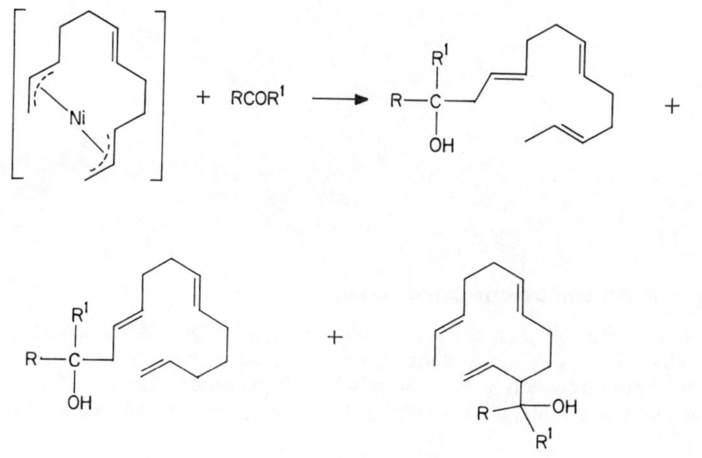

3. Carbon—carbon bond formation using η^3-allyl complexes

$$\text{[Ni complex]} + RCOR^1 \longrightarrow R-\underset{\underset{OH}{|}}{\overset{\overset{R^1}{|}}{C}}-\text{(allyl chain)} +$$

(33)

$$R-\underset{\underset{OH}{|}}{\overset{\overset{R^1}{|}}{C}}-\text{(chain)} \quad + \quad \text{(chain)}-\underset{\underset{R^1}{|}}{\overset{\overset{}{|}}{C}}(R)(OH)$$

$$2\ CH_2{=}CHCH{=}CH_2 + RCOR^1 \xrightarrow{Ni\ cat} RR^1C(OH)CH_2CH{=}CHCH_2CH{=}CHCH{=}CH_2 \quad (34)$$

$$CH_2{=}CHCH{=}CH_2 + MeCHO \xrightarrow{Ni(0)} CH_2{=}CHCH[CH(OH)Me]CH_2CH{=}CHCH{=}CH_2 +$$
(27)

$MeCH(OH)(CH_2CH{=}CH)_2CH{=}CH_2 \quad + \quad CH_2{=}CHCH[CH(OH)Me](CH_2)_3CH{=}CH_2 \ +$
(28) (29)

$CH_2{=}CHCH[CH(OH)Me](CH_2)_2CH{=}CHMe$
(30)

(35)

mechanistic points of view. At $-50\,°C$ in dimethylformamide, reaction 36 occurs. Under particular conditions (use of CO or triphenylphosphine in excess), reduction to hydroquinone can be suppressed. Reaction of 1,1-dimethylallylnickel bromide with 2,3-

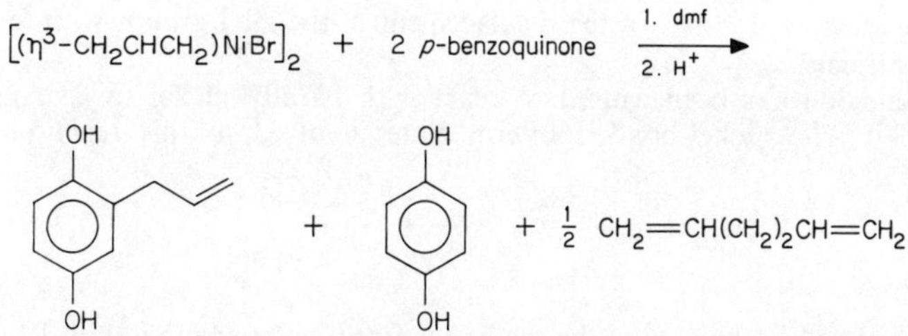

$$[(\eta^3\text{-}CH_2CHCH_2)NiBr]_2 + 2\ p\text{-benzoquinone} \xrightarrow[2.\ H^+]{1.\ dmf}$$

$$\text{2-allylhydroquinone} \quad + \quad \text{hydroquinone} \quad + \quad \tfrac{1}{2}\ CH_2{=}CH(CH_2)_2CH{=}CH_2$$

(36)

dimethoxy-5-methylbenzoquinone in formamide gives coenzyme Q1 (31) after oxidation with $FeCl_3$. From the mechanistic point of view the reaction appears to involve an electron transfer process from nickel to the quinone, followed by a highly regioselective radical coupling with the allyl group[84].

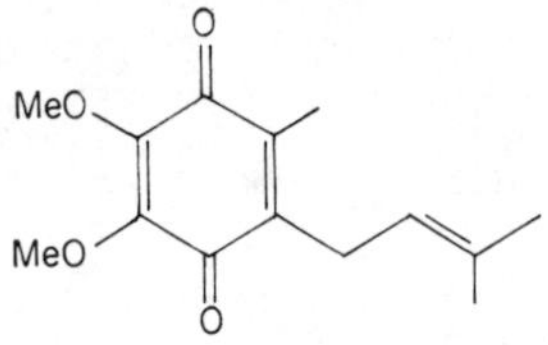

(31)

D. Carbonylation and Carboxylation Reactions

The reaction with CO of an allylnickel chloride complex formed *in situ* from $[Ni(CO)_4]$ and allyl chloride takes place at room temperature under conditions which differ from those of the coupling reaction by a slightly higher pressure (1–2 atm)[8,85] (equation 37, R = H or alkyl). Alternatively, active nickel or allyl(chloro) (thiourea)nickel complexes

$$CH_2=CHCH_2Cl + CO + ROH \xrightarrow{[Ni(CO)_4]} CH_2=CHCH_2COOR + HCl \quad (37)$$

can operate at atmospheric pressure and at low temperature[8]. Higher CO pressures inhibit the reaction and the temperature has to be correspondingly increased. This behaviour is explained by the need for dissociating CO, thus creating the coordination sites required to form the allyl—nickel bond. The following step probably involves a pentacoordinated intermediate, as shown by a model reaction on allyl(chloro)bis(triphenylphosphine)nickel[8,86]. Carbonylation affects the less hindered site of a substituted allyl group.

Bis(allyl)nickel complexes form ketones with CO at low temperature[1,87], for example reaction 38. CO attacks one terminal and one internal carbon of the two allyl groups. A

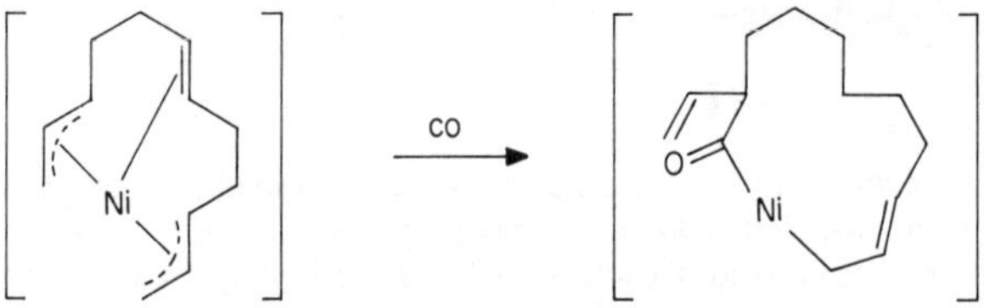

(38)

competitive reaction consists of the displacement of the allyl groups with formation of cyclododecatriene[87].

Carbon dioxide has been caused to react with bis(allyl)nickel to give γ-lactones[88] (equation 39). The nickel-bonded intermediate involved in this reaction had been

$$[(\eta^3\text{-}CH_2CHCH_2)_2Ni] + CO_2 \longrightarrow \text{(γ-lactone)} + \text{(γ-lactone)} \quad (39)$$

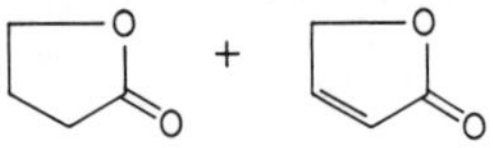

previously isolated in the case of the methallyl group by blocking it with PMe_3 to form **32**.[89]

$$[(\eta^3\text{-}CH_2C(Me)CH_2)Ni\{OOCCH_2C(Me)=CH_2\}(PMe_3)]$$

(**32**)

3. Carbon—carbon bond formation using η^3-allyl complexes

E. Multiple Insertion Processes

Allylnickel complexes can be used to start a sequence of insertions involving different molecules and groups until a termination step is reached. In the simplest case a catalytic reaction involving acetylene and CO insertion, followed by cleavage with ROH (R = H or alkyl, aryl) was observed at room temperature[85] (equation 40). The reaction is

$$CH_2=CHCH_2Cl + CH\equiv CH + CO + ROH \xrightarrow{Ni(0)} (Z)\text{-}CH_2=CHCH_2CH=CHCOOR + HCl$$

(40)

chemioselective, acetylene being inserted before CO. Competition between different molecules for insertion and termination can be effectively controlled by ligands, solvents, and pH. Allyl(chloro)(thiourea)nickel can be advantageously used as a catalyst[8,90,91]. Under phase transfer conditions no insertion of acetylene is observed, probably because more stable anionic butenylnickel complexes are involved[89a]. As shown above cis-addition of the allyl group and of carbon monoxide to acetylene occurs.

When suitable double bonds are present, cyclization is preferred to solvent attack and further CO attack can follow. Examples of different multiple insertion processes which can be carried out at room temperature and atmospheric pressure in acetone as solvent are given in equations 41–44[8,90–92]. The first process is stoichiometric and involves eight

$$CH_2=CHCH_2Cl + 2HC\equiv CH + 3CO + MeCOMe + H_2O + Ni(0) \longrightarrow$$

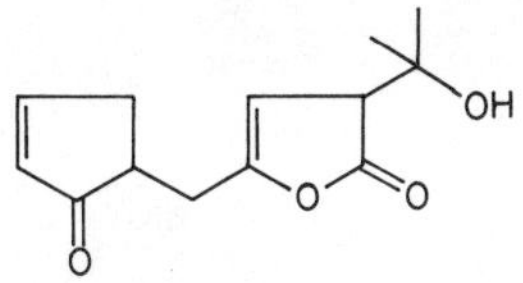

+ Ni(OH)Cl   (41)

$$CH_2=C(CH_3)CH_2Cl + CH_2=CH(CH_2CH_2CH=CH)_2CH_2CH_2CH=CH_2 + 4CO + H_2O \xrightarrow{Ni(0)}$$

$$CH_2=C(Me)CH_2CH_2R_3CH_2CO_2H + HCl$$

R =   (42)

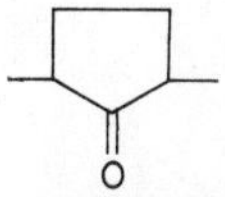

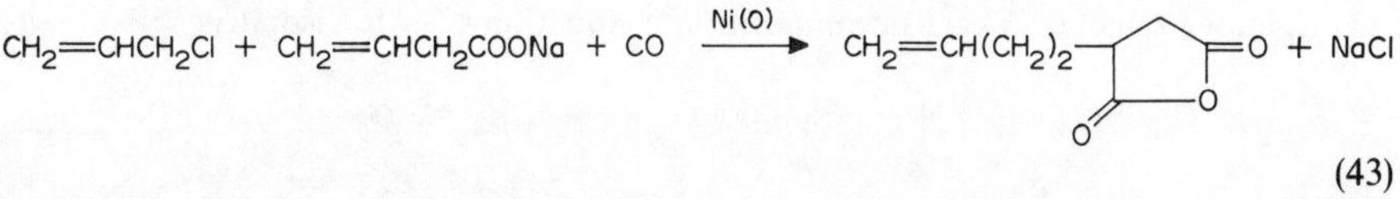

$$CH_2=CHCH_2Cl + CH_2=CHCH_2COONa + CO \xrightarrow{Ni(0)} CH_2=CH(CH_2)_2\text{—}\!\!\!<\!\!\!\text{image} \!\!\!> + NaCl$$

(43)

molecules, five being different from each other, the second is a type of cooligomerization which gives rise to a slightly helicoidal molecule stereoselectively, and the third is a double phase process terminated by internal attack on the acyl group by the carboxylate anion.

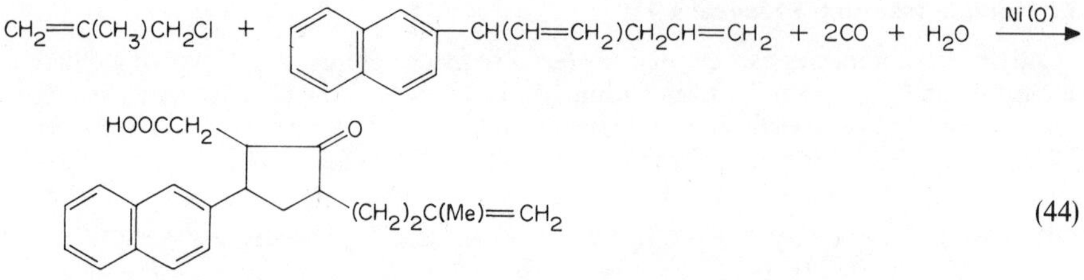

$$CH_2=C(CH_3)CH_2Cl + \text{[naphthyl]}-CH(CH=CH_2)CH_2CH=CH_2 + 2CO + H_2O \xrightarrow{Ni(0)}$$

[structure: HOOCCH$_2$-substituted cyclopentanone with naphthyl group and (CH$_2$)$_2$C(Me)=CH$_2$ side chain] (44)

The last is an example of the synthesis of more complex molecules related to the steroid class. The second and the last reactions are best carried out in the presence of poorly coordinating anions such as PF_6^-, which favour the formation of a cationic complex, thus allowing the substrate to penetrate the coordination sphere.

Multiple insertion process can be governed by rendering difficult the metal elimination process until a favourable arrangement for elimination is reached. In the following example, elimination is difficult after the first insertion of norbornene into the allylnickel bond because *cis, exo*-addition does not allow the required *syn*-elimination, so the insertion process continues until the correct arrangement for elimination is reached[93] (equation 45).

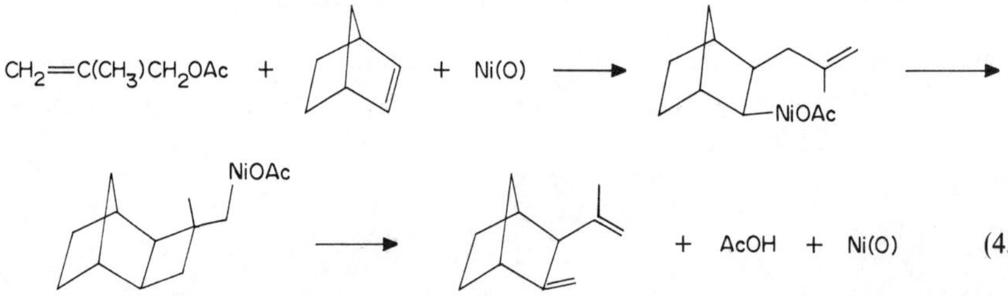

$$CH_2=C(CH_3)CH_2OAc + \text{[norbornene]} + Ni(0) \longrightarrow \text{[insertion product with NiOAc]} \longrightarrow$$

[structure with NiOAc] $\longrightarrow$ [bicyclic alkene product] + AcOH + Ni(0) (45)

Another sequential insertion process is based on generation of a hydroxy group by allyl attack on a formyl group, followed by cyclocarbonylation to lactone[93a] (equation 46).

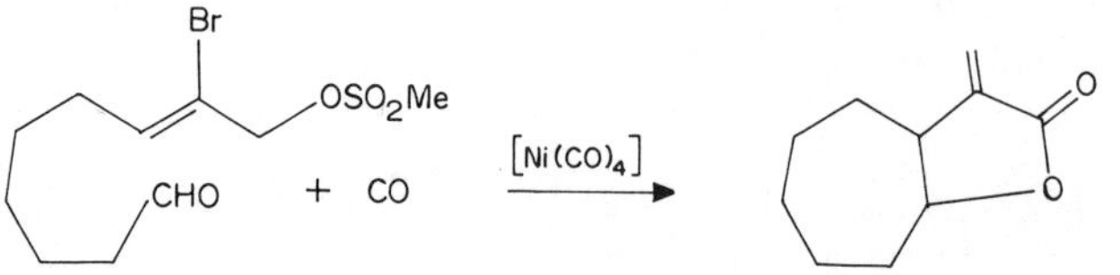

[structure with Br, OSO$_2$Me, CHO groups] + CO $\xrightarrow{[Ni(CO)_4]}$ [bicyclic lactone] (46)

F. Carbenoid Insertion Reactions

In addition to CO, allylnickel complexes are able to attack carbenoids. These species can originate from diazoalkane decomposition[94] (equation 47). Also, isonitriles RNC can

$$[(\eta^3\text{-}CH_2CHCH_2)NiBr] + CH(N_2)COOEt \longrightarrow CH_2=CHCH=CHCOOEt$$
$$(E:Z = 69:19) \quad (47)$$

be inserted into bis(η^3-allyl)nickel complexes to give macrocycles[95]. Muscone was obtained in this way after hydrolysis to a ketone and hydrogenation of the double bonds of the insertion product of allene and MeNC into (dodecatrienediyl)nickel[96] (equation 48).

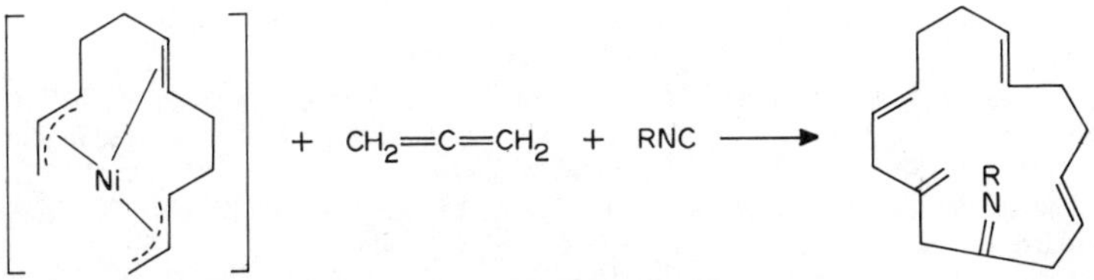

3. Carbon—carbon bond formation using η^3-allyl complexes

$$[\text{Ni(cyclooctadienyl)}] + CH_2{=}C{=}CH_2 + RNC \longrightarrow \text{product} \qquad (48)$$

G. Addition to Dienes. HCN Addition

We have shown in Section II that nucleophiles such as malonic and acetoacetic esters can attack coordinated butadiene, probably via a bis(allyl)nickel complex[1,33,34].

Hydrogen cyanide also can attack butadiene via a different mechanism involving a crotylnickel complex. Crotyl(cyano)(ligand)nickel complexes can be prepared directly from dienes and HCN in the presence of nickel(0)[97]. Trialkyl or triaryl phosphites are the best ligands. Reductive elimination gives a nitrile. With butadiene reaction 49 occurs. This

$$CH_2{=}CHCH{=}CH_2 + HCN + Ni(0) \longrightarrow [(\eta^3\text{-MeCHCHCH}_2)NiCN] \longrightarrow$$

$$NCCH_2CH{=}CHMe + Ni(0) \qquad (49)$$

reaction has formed the basis for one of the most important industrial processes for the production of adiponitrile. In fact, by adding a Lewis acid such as $ZnCl_2$ to the catalyst it is possible to shift the double bond to the terminal position and to add a second molecule of HCN to form $NC(CH_2)_4CN$[98,99]. The stereochemistry of the attack by HCN is probably cis, as suggested by recent results on HCN addition to olefins[100].

H. Reactions with C—Mg Bonds

Allylic compounds couple with C—Mg bonds in the presence of nickel complexes. The reaction of allylic alcohols with Grignard reagents in the presence of $[NiCl_2(PPh_3)_2]$[101] gives olefins, for example reaction 50. The reaction has been shown to involve allylnickel

$$MeCH{=}CHCH_2OH + PhMgX \xrightarrow{cat.} CH_2{=}CHCH(Me)Ph \qquad (50)$$

complexes as intermediates[101]. Cyclic alcohols are attacked by MeMgX stereospecifically on the side opposite to the hydroxyl group[102]. Allylic ethers can be attacked regioselectively at the internal position by Grignard reagents in presence of nickel complexes with 1,1'-bis(diphenylphosphino)ferrocene[103].

IV. REFERENCES

1. P. W. Jolly and G. Wilke, in *The Organic Chemistry of Nickel*, Vols. 1 and 2, Wiley, New York, 1974, 1975.
2. *Gmelin's Handbuch der Anorganischen Chemie*, Ergänzungswerk zur 8. Auflage, Vols. 16 and 17, Springer, Berlin 1974, 1975.
3. P. Heimbach, P. W. Jolly, and G. Wilke, *Adv. Organomet. Chem.*, **8**, 29 (1970).
4. W. Keim, in *Transition Metals Homogeneous Catalysis* (Ed. G. N. Schrauzer) Marcel Dekker, New York, 1971, pp. 59.
5. M. F. Semmelhack, *Org. React.*, **19**, 115 (1972).
6. R. Baker, *Chem. Rev.*, **73**, 487 (1973); *Chem. Ind. (London)*, 816 (1980).
7. L. S. Hegedus, *J. Organomet. Chem. Libr.*, **1**, 329 (1976).
8. G. P. Chiusoli and L. Cassar, in *Organic Syntheses via Metal Carbonyls* (Eds. I. Wender and P. Pino), Wiley, New York, 1977, pp. 297.

9. G. P. Chiusoli and G. Salerno, *Adv. Organomet. Chem.*, **17**, 195 (1979).
10. M. Ryang, *Organomet. Chem. Rev., Sect. A*, **5**, 67 (1970).
11. E. J. Corey, L. S. Hegedus, and M. F. Semmelhack, *J. Am. Chem. Soc.*, **90**, 2417 (1968).
12. B. Henc, P. W. Jolly, R. Salz, G. Wilke, R. Benn, E. G. Hoffman, R. Mynott, G. Schroth, K. Seevogel, J. C. Sekutowski, and C. Krüger, *J. Organomet. Chem.*, **191**, 425 (1980).
13. B. Henc, P. W. Jolly, R. Salz, S. Stobbe, G. Wilke, R. Benn, R. Mynott, K. Seevogel, R. Goddard, and C. Krüger, *J. Organomet. Chem.*, **191**, 449 (1980).
14. M. Julémont and Ph. Teyssié, *Aspects Homogeneous Catal.*, **4**, 100 (1982), and references cited therein.
15. E. J. Corey, M. F. Semmelhack, and L. S. Hegedus, *J. Am. Chem. Soc.*, **90**, 2416 (1968).
16. G. P. Chiusoli and S. Merzoni, *Z. Naturforsch., Teil B*, **17**, 850 (1962), and *unpublished results*.
17. G. Wilke, B. Bogdanovic, P. Hardt, P. Heimbach, W. Keim, M. Kröner, W. Oberkirch, K. Tanaka, E. Steinrücke, D. Walter, and H. Zimmermann, *Angew. Chem., Int. Ed. Engl*, **5**, 151 (1966).
18. B. Bogdanovic and M. Yus, *Angew. Chem., Int. Ed. Engl.*, **18**, 681, (1979).
19. S. Holle, P. W. Jolly, R. Mynott, and R. Salz, *Z. Naturforsch., Teil B*, **37**, 675 (1982).
20. L. Cassar and G. P. Chiusoli, *Tetrahedron Lett.*, 3295 (1965).
21. C. A. Tolman, *J. Am. Chem. Soc.*, **92**, 6777, 6785 (1970).
22. T. Taube, V. Schmidt, and H. Schwind, *Z. Anorg. Allg. Chem.*, **458**, 273 (1979).
23. M. C. Gallazzi, L. Porri, and G. Vitulli, *J. Organomet. Chem.*, **97**, 131 (1975).
24. G. P. Chiusoli, *Aspects Homogeneous Catal.*, **1**, 77 (1970); *Proceedings of 3rd International Symposium on Reactivity and Bonding in Transition Organometallic Compounds*, Inorganica Chimica Acta, Padova, 1970, Paper C5.
25. F. Guerrieri, G. P. Chiusoli, and S. Merzoni, *Gazz. Chim. Ital.*, **104**, 557 (1974).
26. M. Catellani, G. P. Chiusoli and G. Salerno, *Proc. XIX Conference on Coordination Chemistry, II, 25D*, Prague 1978.
27. R. P. Hughes and J. Powell, *J. Am. Chem. Soc.*, **94**, 7723 (1972).
28. K. Vrieze, A. P. Praat, and P. Cossee, *J. Organomet. Chem.*, **12**, 533 (1968).
29. R. Baker, A. H. Cook, and M. J. Crimmin, *J. Chem. Soc. Chem. Commun.*, 727 (1975).
30. R. Baker, A. Onions, R. J. Popplestone, and T. N. Smith, *J. Chem. Soc., Perkin Trans. 2*, 1133 (1975).
31. B. M. Trost and T. R. Verhoeven, *J. Org. Chem.*, **41**, 3215 (1976).
32. C. Moberg, *Tetrahedron Lett.*, **21**, 4539 (1980).
33. A. Döhring, P. W. Jolly, R. Mynott, K. P. Schick, and G. Wilke, *Z. Naturforsch., Teil B*, **36**, 1198 (1981).
34. R. Baker, A. H. Cook, and T. N. Smith, *J. Chem. Soc. Perkin Trans. 2*, 1517 (1974).
35. H. Müller, D. Wittenberg, H. Seibt, and E. Scharf, *Angew. Chem., Int. Ed. Engl.*, **4**, 327 (1965).
36. P. Heimbach and W. Brenner, *Angew. Chem.*, **6**, 800 (1967).
37. R. Baker, D. E. Halliday, and T. N. Smith, *J. Organomet. Chem.*, **35**, C61 (1972).
38. P. Heimbach, *Angew. Chem., Int. Ed. Engl.*, **7**, 882 (1968).
39. P. Heimbach, J. Kluth, H. Schenkluhn, and B. Weimann, *Angew. Chem., Int. Ed. Engl.*, **19**, 570 (1980).
40. C. A. Tolman, *Chem. Rev.*, **77**, 313 (1977).
41. P. Heimbach, *Angew. Chem. Int. Ed. Engl.*, **12**, 975 (1973).
42. P. Heimbach, J. Kluth, H. Schenkluhn, and B. Weimann, *Angew. Chem. Int. Ed. Engl.*, **19**, 569 (1980).
43. J. Furukawa, J. Kiji, K. Yamamoto, and T. Tojo, *Tetrahedron*, **29**, 3149 (1973).
44. P. Heimbach, A. Roloff, and H. Shenkluhn, *Angew. Chem., Int. Ed. Engl.*, **16**, 252 (1977).
45. P. Heimbach and H. Schenkluhn, *Top. Curr. Chem.*, **92**, 45 (1980).
46. M. J. D'Aniello, Jr., and E. K. Barefield, *J. Am. Chem. Soc.*, **100**, 1474 (1978) and references cited therein.
47. H. Lehmkuhl, H. Rufinska, R. Benn, G. Schroth, and R. Mynott, *Justus Liebigs Ann. Chem.*, 317 (1981).
48. H. Lehmkuhl, H. Rufinska, K. Mehler, R. Benn, G. Schroth, and R. Mynott, *Justus Liebigs Ann. Chem.*, 744 (1980).
49. B. Bogdanovic, *Angew. Chem., Int. Ed. Engl.*, **12**, 954 (1973).
50. R. G. Miller, P. A. Pinke, R. D. Stauffer, H. J. Golden, and D. J. Baker, *J. Am. Chem. Soc.*, **96**, 4211 (1974).
51. P. A. Pinke, R. D. Stauffer and R. G. Miller, *J. Am. Chem. Soc.*, **96**, 4229 (1974).

3. Carbon—carbon bond formation using η^3-allyl complexes

52. B. Bogdanovic, *Adv. Organomet. Chem.*, **17**, 105 (1979), and references cited therein.
53. I. G. Farbenindustrie, *Belg. Pat., 448884* (1943); *Chem. Abstr.*, **41**, 6576 (1947).
54. I. D. Webb and G. T. Borcherdt, *J. Am. Chem. Soc.*, **73**, 2654 (1951).
55. N. L. Bauld, *Tetrahedron Lett.*, 859 (1962).
56. E. O. Fisher and G. Bürger, *Z. Naturforsch., Teil B*, **16**, 77 (1961).
56a. T. Yamamoto, J. Ishizu, and A. Yamamoto, *J. Am. Chem. Soc.*, **103**, 6863 (1981).
57. M. Dubini, G. P. Chiusoli, and F. Montino, *Tetrahedron Lett.*, 1591 (1963).
58. G. P. Chiusoli and G. Cometti, *Chim. Ind. (Milan)*, **45**, 401 (1963).
59. E. J. Corey and M. F. Semmelhack, *J. Am. Chem. Soc.*, **89**, 2755 (1967).
60. L. S. Hegedus and S. Varaprath, *Organometallics*, **1**, 259 (1982).
61. K. Sato, S. Inoue, S. Ota, and J. Fujita, *J. Org. Chem.*, **37**, 462 (1972).
62. S. Inoue, K. Saito, K. Kato, S. Nozaki, and K. Sato, *J. Chem. Soc., Perkin Trans. 1*, 2097 (1974).
62a. P. A. Collins and D. Wege, *Aust. J. Chem.*, **32**, 1819 (1974).
62b. K. Sato, S. Inoue, and K. Watanabe, *J. Chem. Soc., Perkin Trans. 1*, 2411 (1981).
63. F. Guerrieri and G. Salerno, *J. Organomet. Chem.*, **114**, 339 (1974).
64. B. Bogdanovic, M. Kröner, and G. Wilke, *Justus Liebigs Ann. Chem.*, **699**, 1 (1966).
65. L. S. Hegedus and R.K. Stiverson, *J. Am. Chem. Soc.*, **96**, 3250 (1974).
66. M. Catellani, G. P. Chiusoli, G. Salerno, and F. Dallatomasina, *J. Organomet. Chem.*, **149**, C19 (1978).
67. F. Guerrieri and G. P. Chiusoli, *J. Organomet. Chem.*, **19**, 453 (1969).
68. J. J. Eisch and G. A. Damasewitz, *J. Organomet. Chem.*, **96**, C19 (1975).
68a. H. Fuellbier, W. Gaube, R. Lange, and M. Tarnow, *J. Prakt. Chem.*, **322**, 655 (1980).
68b. H. Fuellbier, W. Gaube, and R. Lange, *J. Prakt. Chem.*, **322**, 663 (1980).
69. G. P. Chiusoli, G. Salerno, and F. Dallatomasina, *J. Organomet. Chem.*, **146**, C19 (1978).
70. U. Bersellini, M. Catellani, G. P. Chiusoli, W. Giroldini, and G. Salerno, *Fund. Res. Homogeneous Catal.*, **3**, 893 (1979).
71. B. Bogdanovic, P. Heimbach, H. Hey, E. Müller, and G. Wilke, *Justus Liebigs Ann. Chem.*, **727**, 161 (1969).
72. W. Brenner, P. Heimbach, and G. Wilke, *Justus Liebigs Ann. Chem.*, **727**, 194 (1969).
73. R. Baker, P. C. Bevan, R. C. Cookson, A. H. Copeland, and A. D. Gribble, *J. Chem. Soc., Perkin Trans. 1*, 480 (1978).
74. G. P. Chiusoli, E. Dradi, G. Salerno, and G. Campari, *Bull. Soc. Chim. Belg.*, **89**, 869 (1980).
75. G. P. Chiusoli, *Chim. Ind. (Milan)*, **43**, 365 (1961).
76. G. Agnés, G. P. Chiusoli and A. Marraccini, *J. Organomet. Chem.*, **49**, 239 (1973).
76a. L. S. Hegedus, S. D. Wagner, E. L. Waterman, and K. Siirala-Hansen, *J. Org. Chem.*, **40**, 593 (1975).
76b. M. F. Semmelhack and E. S. C. Wu, *J. Am. Chem. Soc.*, **98**, 3384 (1976).
77. G. P. Chiusoli, G. Salerno, U. Bersellini, F. Dallatomasina, and G. Preseglio, *Trans. Met. Chem.*, **3**, 174 (1978).
78. U. Bersellini, G. P. Chiusoli and G. Salerno, *Angew. Chem., Int. Ed. Engl.*, **17**, 353 (1978).
79. G. Wilke, *Pure Appl. Chem.*, **17**, 179 (1968).
80. R. Baker, B. N. Blackett, R. C. Cookson, R. C. Cross, and P. D. Madden, *J. Chem. Soc., Chem. Commun.*, 343 (1974).
81. S. Akutagawa, *Bull. Chem. Soc. Jpn.*, **49**, 3646 (1976).
82. R. Baker and M. J. Crimmin, *J. Chem. Soc., Perkin Trans. 1*, 1264 (1979).
83. R. Baker, S. Nobbs, and P. M. Winton, *J. Organomet. Chem.*, **137**, C43 (1976).
83a. H. U. Blaser and D. Reinher, *Helv. Chim. Acta*, **60**, 208 (1977).
84. L. S. Hegedus, B. R. Evans, D. E. Korte, E. L. Waterman, and K. Siöberg, *J. Am. Chem. Soc.*, **98**, 3901 (1976), and references cited therein.
85. G. P. Chiusoli, *Gazz. Chim. Ital.*, **89**, 1332 (1959).
86. F. Guerrieri and G. P. Chiusoli, *J. Organomet. Chem.*, **15**, 209 (1968).
87. G. Wilke, F. W. Müller, M. Kröner and B. Bogdanovic, *Angew. Chem., Int. Ed. Engl.*, **2**, 105 (1963).
88. T. Tsuda, Y. Chujo and T. Saegusa, *Synth. Commun.*, **9**, 427 (1979).
89. P. W. Jolly, S. Stobbe, G. Wilke R. Goddard, C. Krüger, C. Sekutowski, and Y. Tsay, *Angew. Chem., Int. Ed. Engl.*, **17**, 124 (1978).
89a. M. Foà and L. Cassar, *Gazz. Chim. Ital.*, **109**, 619 (1979).
90. L. Cassar, G. P. Chiusoli, and F. Guerrieri, *Synthesis*, 509 (1973).

91. G. P. Chiusoli, *Proc. Int. Congr. Pure Appl. Chem., 23rd*, **6**, 196 (1971).
92. G. P. Chiusoli, G. Salerno, and E. Bergamaschi, *J. Organomet. Chem.*, **177**, 245 (1979).
93. M. Catellani, G. P. Chiusoli, E. Dradi, and G. Salerno, *J. Organomet. Chem.*, **177**, C29 (1979).
93a. M. F. Semmelhack and S. J. Brickman, *J. Am. Chem. Soc.*, **103**, 3945 (1981).
94. I. Moritani, Y. Yamamoto, and H. Konishi, *J. Chem. Soc., Chem. Commun.*, 1457 (1969).
95. H. Breil and G. Wilke, *Angew. Chem. Int. Ed. Engl.*, **9**, 376 (1970).
96. R. Baker, R. C. Cookson, and J. R. Vinson, *J. Chem. Soc., Chem. Commun.*, 515 (1974).
97. J. D. Druliner, A. D. English, J. P. Jesson, P. Meakin, and C. A. Tolman, *J. Am. Chem. Soc.*, **98**, 2156 (1976).
98. E. S. Brown, in *Organic Syntheses via Metal Carbonyls* Vol. 2 (Eds. I. Wender and P. Pino), Wiley, New York, 1977, p. 655.
99. G. W. Parshall, *J. Mol. Catal.*, **4**, 256 (1978).
100. J. E. Bäckvall and O. S. Andell, *J. Chem. Soc., Chem. Commun.*, 1098 (1981).
101. H. Felkin, M. Joly-Goudget and S. Davies, *Tetrahedron Lett.*, **22**, 1157 (1981).
102. G. Consiglio, F. Morandini, and O. Piccolo, *J. Am. Chem. Soc.*, **103**, 1846 (1981).
103. T. Hayashi, M. Konishi, K. Yokota, and M. Kumada, *J. Chem. Soc., Chem. Commun.*, 313 (1981).

After Completion of the Chapter the Following References Appeared

Allyl coupling
R. Baker, N. Ekanayake and S. A. Johnson, *J. Chem. Res.* (S), **74** (1983).

Allyl alkylations
T. Cuvigny and M. Julia, *J. Organomet. Chem.*, **250**, C21 (1983).
M. Catellani, G. P. Chiusoli and A. Mari, *J. Organomet. Chem.*, **275**, 129 (1984).

Octadienoic acids from butadiene and 3-butenoic acids
G. P. Chiusoli, L. Pallini and G. Salerno, *J. Organomet. Chem.*, **238**, C85 (1982).

Dienoic amides from dienes, CO and amines
H. J. Riegel and H. Hoberg, *J. Organomet. Chem.*, **260**, 121 (1984).
F. J. Fañanás and H. Hoberg, *J. Organomet. Chem.*, **275**, 249 (1984).

Unsaturated acids from dienes and CO_2
H. Hoberg and D. Schaefer, *J. Organomet. Chem.*, **255**, C15 (1983).
H. Hoberg, D. Schaefer and B. W. Oster, *J. Organomet. Chem.*, **266**, 313 (1984).
H. Hoberg and B. W. Oster, *J. Organomet. Chem.*, **266**, 321 (1984).

Hydrocyanation of butadiene
C. A. Tolman, W. C. Seidel, J. D. Druliner and P. J. Domaille, *Organometallics*, **3**, 33 (1984).
W. Keim, A. Behr, H. Luehr and J. Weisser, *J. Catal.*, **78**, 209 (1982).

Part 2. η^3-Allylpalladium complexes

JIRO TSUJI

Department of Chemical Engineering, Tokyo Institute of Technology, Meguro, Tokyo 152, Japan

I. INTRODUCTION	163
II. PREPARATION OF η^3-ALLYLPALLADIUM COMPLEXES.	164
A. From Various Allylic Compounds	164
B. From Various Olefinic Compounds	165
C. From Diolefins and $PdCl_2$	166
D. Miscellaneous Methods.	167
III. STOICHIOMETRIC CARBON—CARBON BOND FORMATION.	168
A. Reaction of Carbon Monoxide and Isocyanides	168
B. Reactions of Carbonucleophiles.	170
C. Miscellaneous Reactions	174
IV. CATALYTIC REACTIONS	175
A. Reactions of Conjugated Dienes.	175
1. Introduction	175
2. Oligomerization	176
3. Dimerization with carbonucleophiles	176
4. Carbonylation reactions	180
5. Cocyclization reactions.	181
6. Reaction with carbon dioxide	183
B. Reactions of Allenes.	183
C. Reactions of Various Allylic Compounds	184
1. Reactions of allylic ethers, esters, alcohols, amines, and nitroalkanes with carbonucleophiles.	184
2. Reactions of allylic halides.	193
V. REFERENCES	194

I. INTRODUCTION

η^3-Allylpalladium complexes play an important role in carbon—carbon bond formation by reacting with various carbonucleophiles[1-4]. The reaction can be carried out as both stoichiometric and catalytic reactions. In the former reaction, η^3-allylpalladium complexes, isolated as stable compounds, react with various carbonucleophiles to give allylated products. This reaction(1) was first reported by Tsuji et al.[5] in 1965. In catalytic reactions, various allylic compounds and conjugated dienes react with palladium(0)

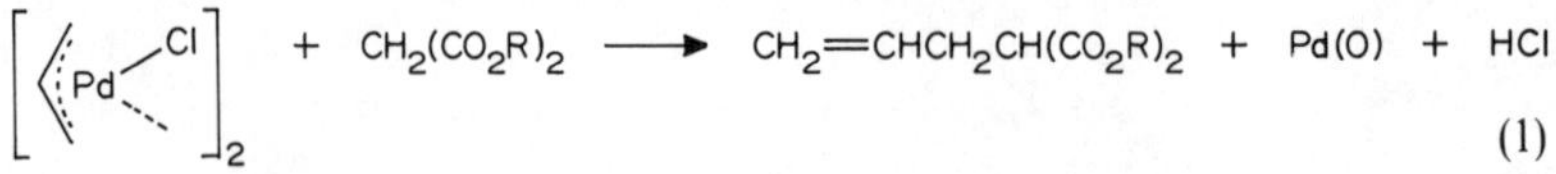

$$\left[\begin{array}{c}\diagup Pd\diagdown Cl\end{array}\right]_2 + CH_2(CO_2R)_2 \longrightarrow CH_2{=}CHCH_2CH(CO_2R)_2 + Pd(0) + HCl \quad (1)$$

complexes to form η^3-allylpalladium complexes as intermediates. These η^3-allylpalladium complexes, formed *in situ* and without isolation, react with carbonucleophiles to form new carbon—carbon bonds. These two reactions are treated in this part. Only fundamental aspects and synthetic applications are described, with typical examples. A complete literature survey is not the purpose of this review.

II. PREPARATION OF η^3-ALLYLPALLADIUM COMPLEXES

η^3-Allylic complexes of palladium are prepared by a number of methods from various allylic and olefinic compounds. Except for bis(η^3-allyl)palladium, which is airsensitive and rarely used in organic synthesis, π-allylic complexes are stable, soluble in organic solvents, and can be handled easily.

A. From Various Allylic Compounds

The first synthesis of η^3-allylpalladium chloride was carried out by the reaction of $PdCl_2$ with allyl alcohol[6] and allyl chloride[7], but the yields were not satisfactory. Mechanistically η^3-allyl complexes are formed by oxidative addition of allyl compounds to palladium(0) generated from $PdCl_2$ with appropriate reducing agents. Satisfactory results were obtained by using carbon monoxide (84%)[8-10], ethylene[11], $SnCl_2$ (90%)[12], $TiCl_3$, Fe, Cu, and Zn[13] as the reducing agents of $PdCl_2$. Photocatalysed allylic chlorination of terminal

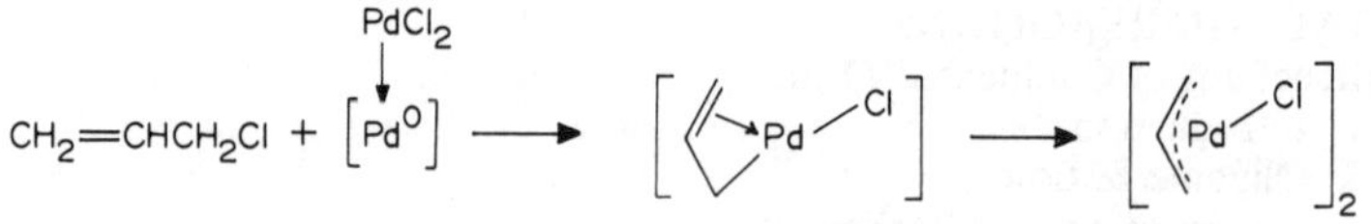

$$CH_2{=}CHCH_2Cl + \begin{bmatrix}PdCl_2 \\ \downarrow \\ Pd^0\end{bmatrix} \longrightarrow \left[\diagup Pd\diagdown Cl\right] \longrightarrow \left[\diagup Pd\diagdown Cl\right]_2 \quad (2)$$

double bonds, followed by reaction with $PdCl_2$ and CO, offers a convenient synthetic method for η^3-allylpalladium complexes[14] (reaction 3). Reaction of allylsilanes with $PdCl_2$

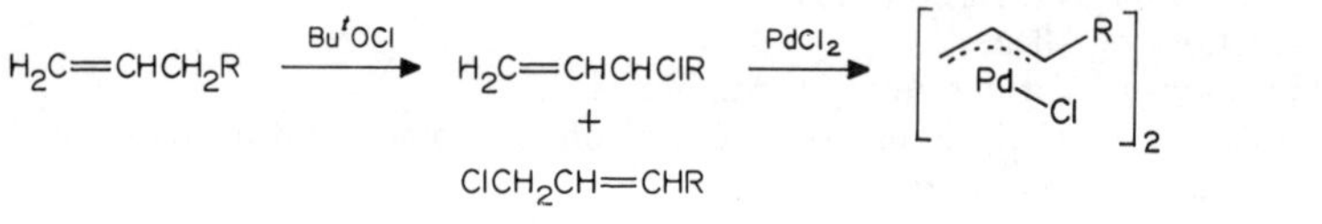

$$H_2C{=}CHCH_2R \xrightarrow{Bu^tOCl} \begin{array}{c}H_2C{=}CHCHClR \\ + \\ ClCH_2CH{=}CHR\end{array} \xrightarrow{PdCl_2} \left[\diagup Pd\diagdown Cl \diagup R\right]_2 \quad (3)$$

in alcohol affords the complex[15,16] (reaction 4). Diallyl ether is another source of the complex[17] (reaction 5).

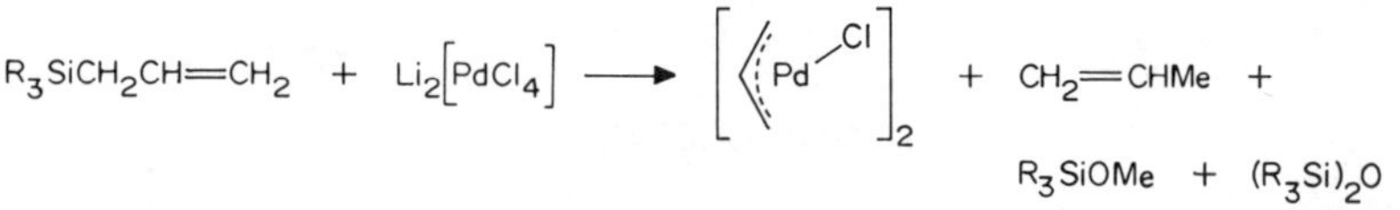

$$R_3SiCH_2CH{=}CH_2 + Li_2[PdCl_4] \longrightarrow \left[\diagup Pd\diagdown Cl\right]_2 + CH_2{=}CHMe + R_3SiOMe + (R_3Si)_2O \quad (4)$$

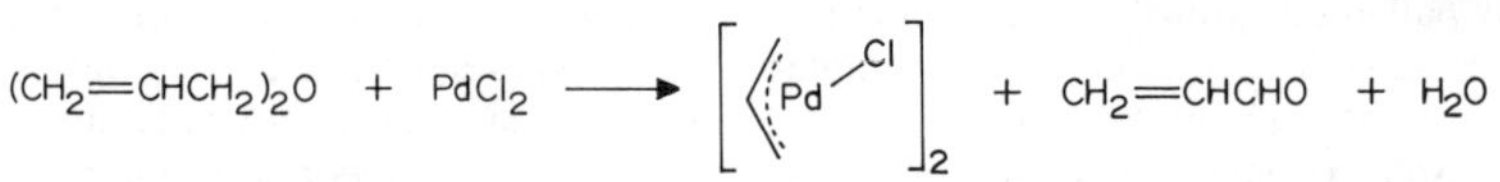

$$(CH_2{=}CHCH_2)_2O + PdCl_2 \longrightarrow \left[\diagup Pd\diagdown Cl\right]_2 + CH_2{=}CHCHO + H_2O \quad (5)$$

3. Carbon—carbon bond formation using η^3-allyl complexes

The complex can be prepared directly from palladium(0) complexes. Facile oxidative addition of allylic chlorides to zerovalent tris(dibenzylideneacetone)dipalladium is a good method of synthesizing η^3-allyl complexes[18,19]. η^3-Allylpalladium bromide was obtained in 90% yield by the reaction of highly active palladium metal powder, produced by the potassium reduction of $PdCl_2$ with allyl bromide. Allyl chloride afforded the corresponding complex in 6% yield[20-22]. η-Allylpalladium acetate was prepared by the reaction of allyl acetate with the zerovalent palladium phosphine complex $[Pd(PCy_3)_2]$[23].

$$CH_2\!=\!CHCH_2Br \;+\; Pd(0) \;\longrightarrow\; \left[\begin{array}{c} \diagup\!\!\!\!\!\!\!\!\to Pd\!\!-\!\!Br \end{array} \right]_2 \qquad (6)$$

$$Pd(PCy_3)_2 \;+\; CH_2\!=\!CHCH_2OAc \;\longrightarrow\; \left[\begin{array}{c} \diagup\!\!\!\!\!\!\!\!\to Pd\!\!\begin{array}{c}PCy_3\\OAc\end{array} \end{array} \right]_2 \qquad (7)$$

B. From Various Olefinic Compounds

Reaction of substituted olefins with $PdCl_2$ produces η^3-allyl complexes by the elimination of HCl[24-27]. The reaction (8) proceeds smoothly in acetic acid[28-30], chloroform, dichloromethane[31], methanol[32], and dmf in the presence of bases[33-35].

$$R^1CH\!=\!CHCHR^2 \;+\; PdCl_2 \;\longrightarrow\; \left[\begin{array}{c}R^2\\ \diagup\!\!\!\!\!\!\!\!\to Pd\!\!-\!\!Cl \\ R^1\end{array} \right]_2 \;+\; HCl \qquad (8)$$

Various modifications of this reaction have been reported. Addition of a weak oxidizing agent such as $CuCl_2$ not only improves the generality of the approach, but also affects the regiochemistry[36,37]. The main products of the reaction are Markownikoff-like products; that is, the hydrogen abstracted is normally allylic to the more substituted end of the olefins. Reaction 9 was observed with 2-methylbut-1-ene[38]. Palladium trifluoroacetate

$$CH_2\!=\!C(Me)Et \;\xrightarrow{PdCl_2}\; \left[Cl\!-\!Pd\!\leftarrow\!\!\!\diagup\!\!\!\!\!\!\!\right]_2 \;+\; \left[\diagup\!\!\!\!\!\!\!\!\to Pd\!\!-\!\!Cl\right]_2 \qquad (9)$$

No additive: 29%   71%
$CuCl_2$:   74%   26%

was used for the preparation of η^3-allylpalladium complexes from cyclic and monosubstituted olefins[39]. Complex formation from unsaturated steroids was carried out[40-42].

The substitution reaction of mesityl oxide[43,44], or more generally α,β- and β,γ-unsaturated carbonyl compounds, with $PdCl_2$ in the presence of bases to afford η^3-allyl complexes proceeds more easily owing to the high acidity of the hydrogen which is eliminated (reaction 10)[45,46]. Complexes were prepared from cyclic unsaturated ketones

$$RC(O)CH_2CH\!=\!CHR^1 + PdCl_2 \xrightarrow{base} \left[\begin{array}{c}COR\\C\!-\!H\\ \diagup\!\!\!\!\!\!\!\!\to Pd\!\!\begin{array}{c}Cl\\Cl\end{array}\\R^1\end{array}\right] \longrightarrow \left[\begin{array}{c}R\\C\!=\!O\\ \diagup\!\!\!\!\!\!\!\!\to Pd\!\!-\!\!Cl\\R^1\end{array}\right]_2 + HCl \qquad (10)$$

such as cyclohexenone and isophorone[47] and steroidal, unsaturated ketones[48-51] (reaction 11). Alkylisopropenylmalonic acids form the complexes via decarboxylation

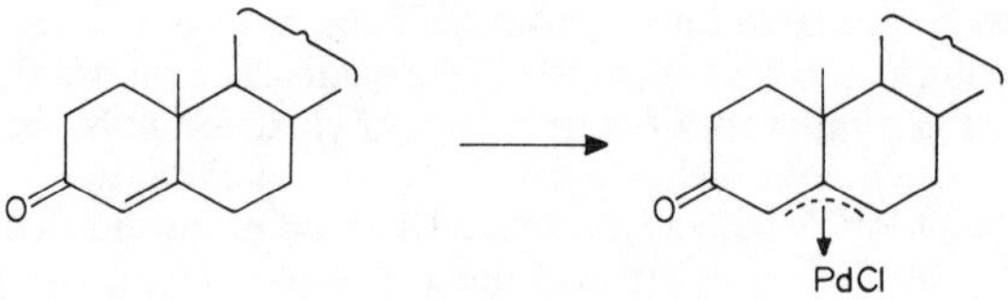
(11)

(reaction 12)[52]. The formation of diacyl-η^3-allylpalladium complexes from pyrilium salts via ring opening is the same type of reaction[53].

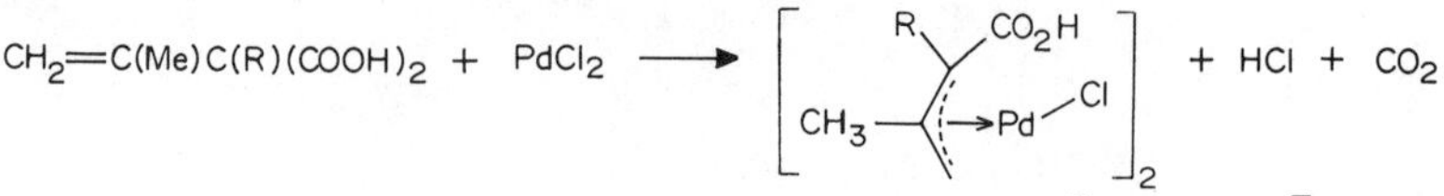
(12)

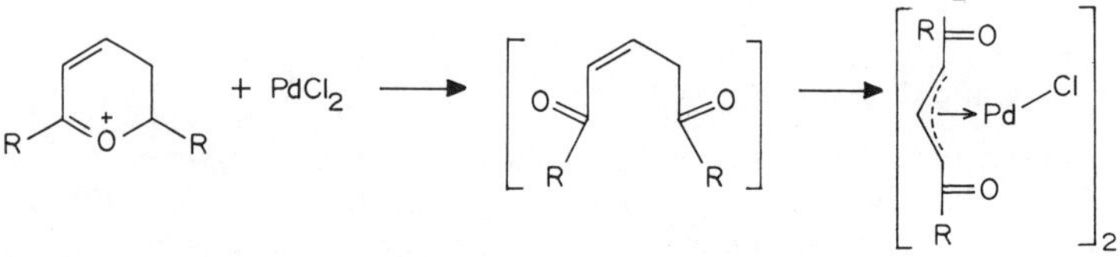
(13)

C. From Diolefins and $PdCl_2$

Mechanistically, palladation of one double bond of conjugated dienes produces η^3-allyl complexes substituted with nucleophiles. Typically, butadiene reacts with $PdCl_2$ in the presence of nucleophiles to give substituted η^3-allyl complexes (reaction 14)[54–57].

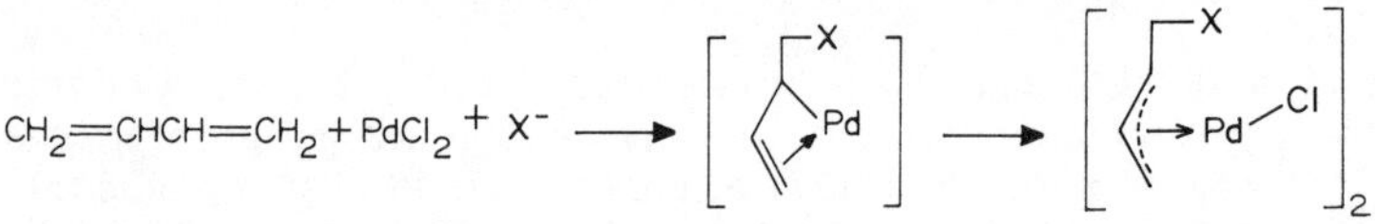
X = nucleophiles, Cl, OR, OAc (14)

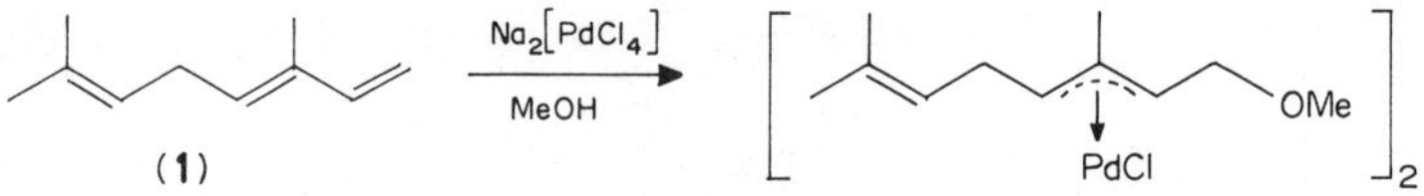
(15)

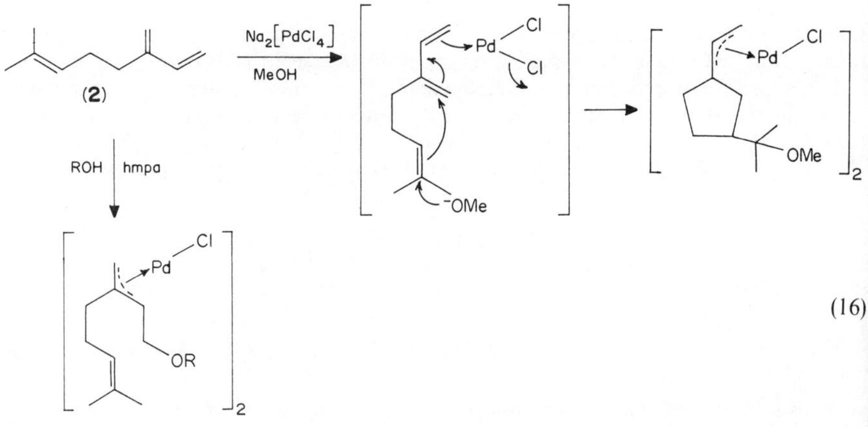
(16)

R = H, Et, Ac

3. Carbon—carbon bond formation using η^3-allyl complexes

Ocimene (1) reacted to give the expected η^3-allyl complex in methanol, whereas myrcene (2) was converted into a η^3-allyl complex after cyclization[58]. However, mainly the terminal position of myrcene was attacked by O-nucleophiles in hmpa[59].

Nucleophiles attack the central carbon of allene to form the two 2-substituted η^3-allyl complexes (reaction 17)[60-63]. Divinylmethanols are a source of η^3-allyl complexes (reaction 18)[64].

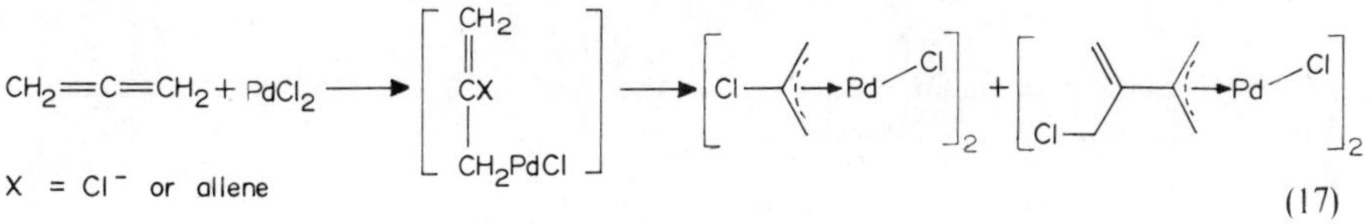

$$CH_2=C=CH_2 + PdCl_2 \longrightarrow \left[\begin{array}{c} CH_2 \\ \| \\ CX \\ | \\ CH_2PdCl \end{array} \right] \longrightarrow \left[Cl-\!\!\!\overset{\diagup}{\underset{\diagdown}{\cdot}}\!\!\!\!\rightarrow Pd\overset{Cl}{\diagdown} \right]_2 + \left[\overset{Cl-}{\underset{}{}}\!\!\!\overset{\diagup}{\underset{\diagdown}{\cdot}}\!\!\!\!\rightarrow Pd\overset{Cl}{\diagdown} \right]_2$$

X = Cl$^-$ or allene
(17)

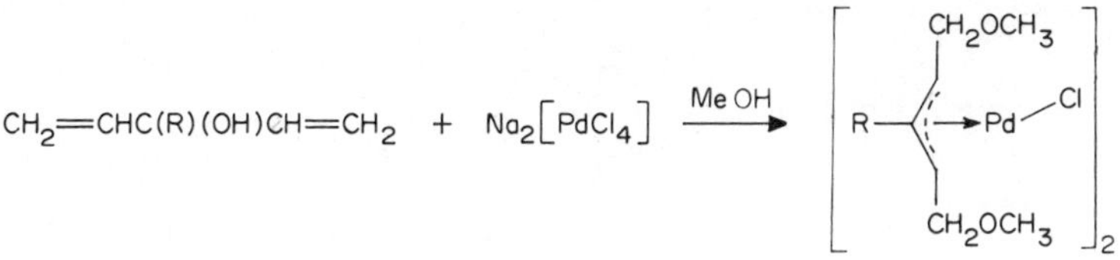

$$CH_2=CHC(R)(OH)CH=CH_2 + Na_2[PdCl_4] \xrightarrow{MeOH} \left[R-\!\!\!\overset{CH_2OCH_3}{\underset{CH_2OCH_3}{\cdot}}\!\!\!\!\rightarrow Pd\overset{Cl}{\diagdown} \right]_2 \quad (18)$$

D. Miscellaneous Methods

Cyclopropanes are a source of η^3-allyl complexes. The ring-opening reactions of vinylcyclopropanes[65-67], methylenecyclopropanes[68], bicyclopropyl, and dicyclopropylmethanes[69] with PdCl$_2$ afford η^3-allyl complexes (reaction 19). 3-Phenylcyclopropene was

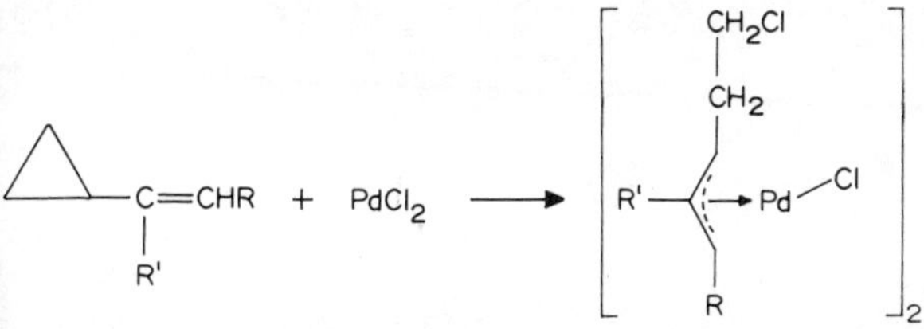

$$\triangle\!\!-\!\!\underset{R'}{\overset{}{C}}\!\!=\!\!CHR + PdCl_2 \longrightarrow \left[R'-\!\!\!\overset{CH_2Cl}{\underset{R}{\overset{CH_2}{\cdot}}}\!\!\!\!\rightarrow Pd\overset{Cl}{\diagdown} \right]_2 \quad (19)$$

converted quantitatively into 1-chloro-3-phenyl-η^3-allylpalladium chloride (reaction 20)[70,71] and tetraphenylcyclopropene was converted into the η^3-allyl complex of indene

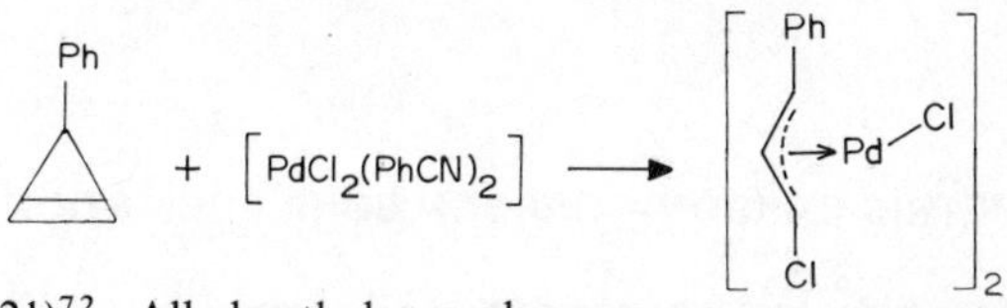

$$\overset{Ph}{\triangle} + [PdCl_2(PhCN)_2] \longrightarrow \left[\overset{Ph}{\underset{Cl}{\cdot}}\!\!\!\!\rightarrow Pd\overset{Cl}{\diagdown} \right]_2 \quad (20)$$

(reaction 21)[72]. Alkylmethylenecyclopropanes are converted into 2-chloro-η^3-allyl-

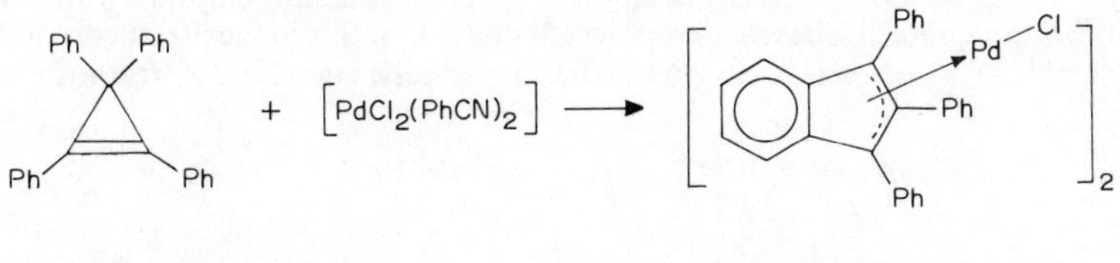

$$\underset{Ph}{\overset{Ph\quad Ph}{\triangle}}\!\!\!\!Ph + [PdCl_2(PhCN)_2] \longrightarrow \left[\begin{array}{c} Ph \\ \\ Ph \end{array}\!\!\!\!\!\!\!\overset{Ph}{\underset{Ph}{\cdot}}\!\!\!\!\overset{Cl}{\underset{}{Pd}}\!\!\!\!\!\!\!\!\!\!\!\!\!\!\!\!\!\!\!\right]_2 \quad (21)$$

palladium complexes (reaction 22)[73]. η^3-Allyl complexes are prepared by the reaction of alkenylmercury(II) chloride with olefins in the presence of $PdCl_2$ in thf. The reaction

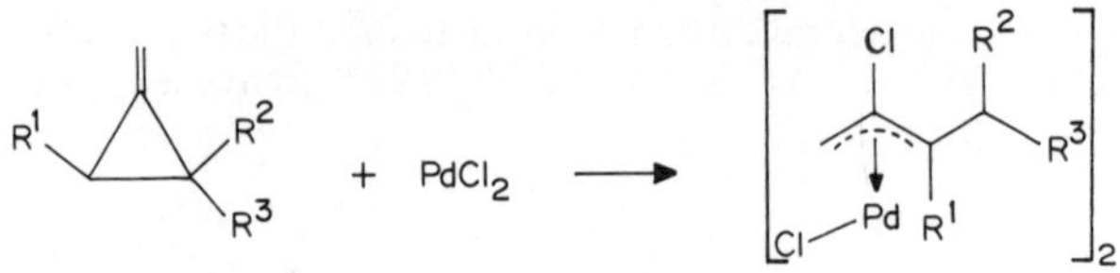

$$\underset{R^3}{\overset{R^1}{\triangle}}\!\!\!\!\!\!\!\!\!\!\!\!\!\!\!\overset{R^2}{} + PdCl_2 \longrightarrow \left[\begin{array}{c} Cl \quad R^2 \\ \diagup\!\!\!\diagup\!\!\!\diagdown\!\!\!R^3 \\ Cl\!\!-\!\!Pd\;R^1 \end{array}\right]_2 \quad (22)$$

involves insertion of olefin to the palladium—carbon bond, followed by palladium hydride rearrangement[74,75] (reaction 23).

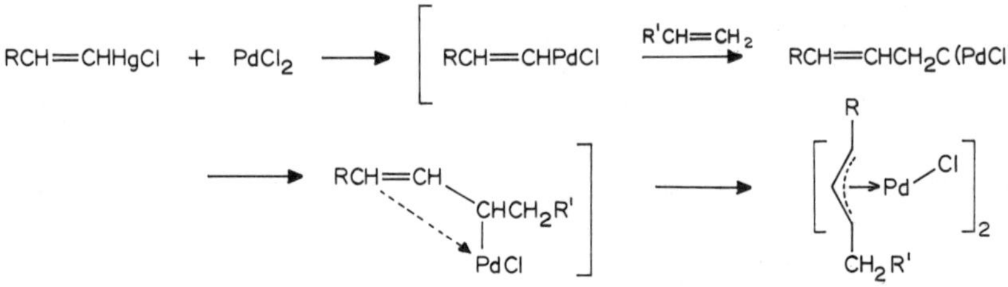

$$RCH\!=\!CHHgCl + PdCl_2 \longrightarrow \left[RCH\!=\!CHPdCl\right] \xrightarrow{R'CH=CH_2} RCH\!=\!CHCH_2C(PdCl)CHR'$$

$$\longrightarrow \left[\begin{array}{c} RCH\!=\!CH \\ \diagdown CHCH_2R' \\ | \\ PdCl \end{array}\right] \longrightarrow \left[\begin{array}{c} R \\ \diagup\!\!\!\diagup\!\!\!\diagdown\!\!\!\overset{Cl}{Pd} \\ CH_2R' \end{array}\right]_2 \quad (23)$$

Addition reaction of acetylenes to olefins promoted by $PdCl_2$ in chloroform at room temperature leads to the formation of η^3-allyl complexes (reaction 24)[76]. Reaction of diketene with $Na_2[PdCl_4]$ in ethanol produces a η^3-allyl complex of acetoacetate[77–79].

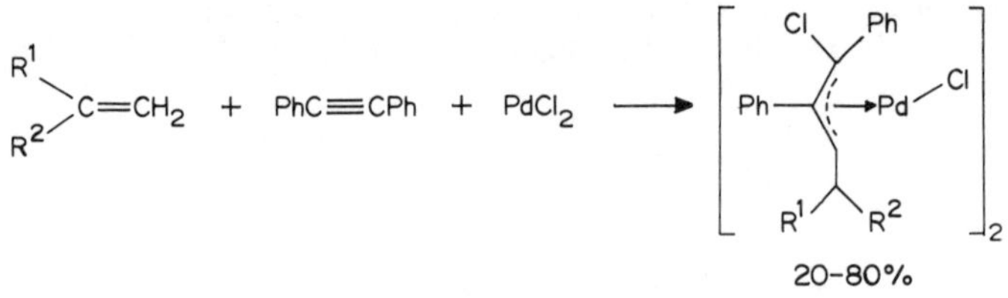

$$\underset{R^2}{\overset{R^1}{\diagdown}}\!\!\!C\!=\!CH_2 + PhC\!\equiv\!CPh + PdCl_2 \longrightarrow \left[\begin{array}{c} Cl \quad Ph \\ Ph\!-\!\!\!\diagup\!\!\!\diagup\!\!\!\diagdown\!\!\!\overset{Cl}{Pd} \\ R^1 \quad R^2 \end{array}\right]_2 \quad (24)$$

$$20\text{–}80\%$$

$$\underset{O\diagdownO}{CH_2\!=\!C\!\!-\!\!CH_2} + Na_2 PdCl_4 \xrightarrow{EtOH} \left[\begin{array}{c} CO_2Et \\ HO\!-\!\!\!\diagup\!\!\!\diagup\!\!\!\diagdown\!\!\!\overset{}{Pd}\!\!-\!\!Cl \end{array}\right]_2 \quad (25)$$

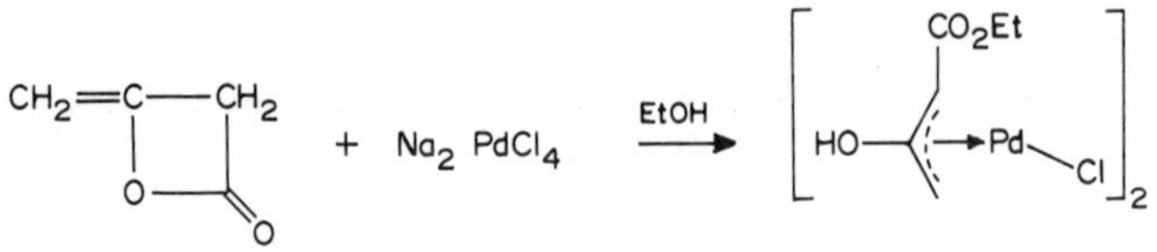

III. STOICHIOMETRIC CARBON—CARBON BOND FORMATION

A. Reaction of Carbon Monoxide and Isocyanides

A but-3-enoate is formed by the reaction of η^3-allylpalladium chloride with carbon monoxide in an alcohol[80]. The reaction proceeds more smoothly in the presence of sodium carboxylate[81]. This ester was converted to another η^3-allyl complex **3**. Carbonylation of

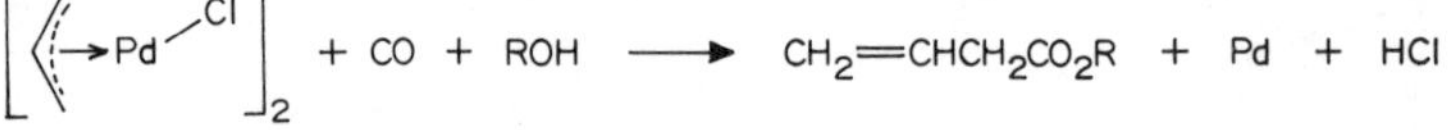

$$\left[\diagup\!\!\!\diagup\!\!\!\diagdown\!\!\!\overset{Cl}{Pd}\right]_2 + CO + ROH \longrightarrow CH_2\!=\!CHCH_2CO_2R + Pd + HCl \quad (26)$$

3. Carbon—carbon bond formation using η^3-allyl complexes

the complex **3** gave the diester **4**, which was then converted further to the η^3-allyl complex **5**[46].

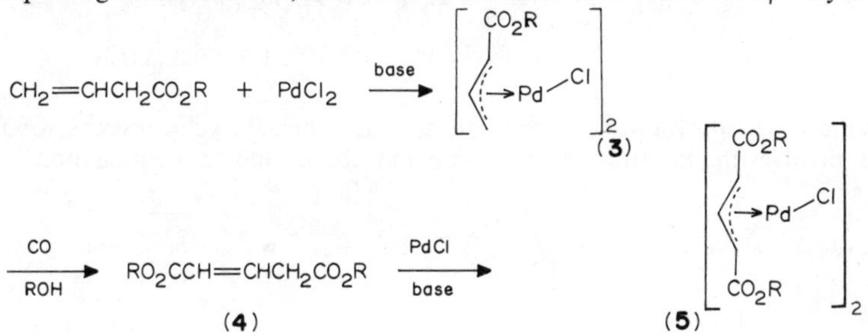

$$CH_2=CHCH_2CO_2R + PdCl_2 \xrightarrow{base} \left[\begin{array}{c} CO_2R \\ \langle\cdots\rightarrow Pd \diagdown Cl \end{array}\right]_2 \quad (27)$$
$$(3)$$

$$\xrightarrow[ROH]{CO} RO_2CCH=CHCH_2CO_2R \xrightarrow[base]{PdCl} \left[\begin{array}{c} CO_2R \\ \langle\cdots\rightarrow Pd \diagdown Cl \\ CO_2R \end{array}\right]_2$$
$$(4) \qquad\qquad (5)$$

Carbonylation of the butadiene complex produced 5-chloropent-3-enoate (reaction 28)[82].

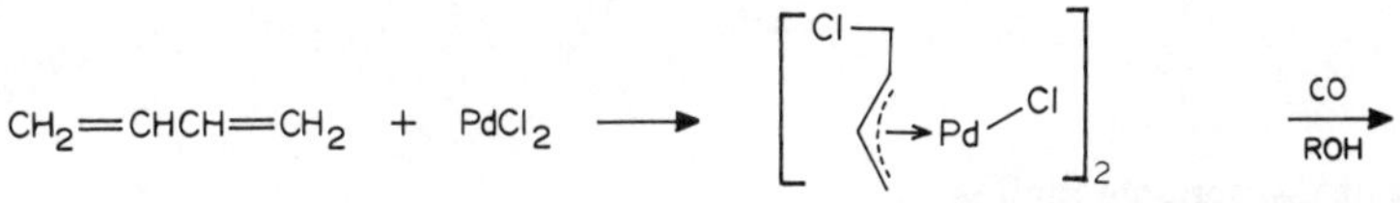

$$CH_2=CHCH=CH_2 + PdCl_2 \longrightarrow \left[\begin{array}{c} Cl \\ \langle\cdots\rightarrow Pd \diagdown Cl \end{array}\right]_2 \xrightarrow[ROH]{CO}$$
$$(E)\text{-}ClCH_2CH=CHCH_2COOR \quad (28)$$

The oxidative carbonylation of butadiene to give unsaturated mono- and die-esters has been reported (reaction 29)[83,84]. Detailed studies on the stoichiometric carbonylation of

$$CH_2=CHCH=CH_2 + CO + MeOH \xrightarrow{PdCl_2/CuCl_2/O_2} (E)\text{-}MeOCH_2CH=CHCH_2CO_2Me +$$
$$(E)\text{-}MeO_2CCH_2CH=CHCH_2CO_2Me \quad (29)$$

the isoprene complex to give a mixture of mono- and di-esters and lactone have been carried out[85]. Carbonylation of η^3-allyl complexes derived from allene afforded the esters shown in reactions 30 and 31[86]. Direct reaction of allene, carbon monoxide, and $PdCl_2$ in an alcohol produced itaconate in low yield[86].

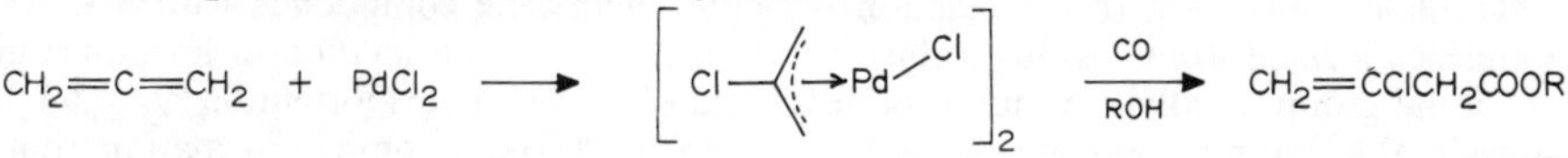

$$CH_2=C=CH_2 + PdCl_2 \longrightarrow \left[Cl-\langle\cdots\rightarrow Pd\diagdown Cl\right]_2 \xrightarrow[ROH]{CO} CH_2=CClCH_2COOR$$
$$(30)$$

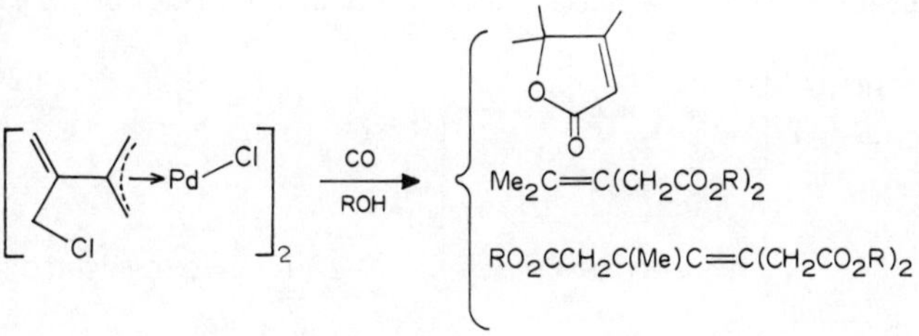

$$\left[\begin{array}{c} \langle\cdots\rightarrow Pd\diagdown Cl \\ Cl \end{array}\right]_2 \xrightarrow[ROH]{CO} \begin{cases} \text{(lactone)} \\ Me_2C=C(CH_2CO_2R)_2 \\ RO_2CCH_2C(Me)C=C(CH_2CO_2R)_2 \end{cases} \quad (31)$$

$$CH_2=C=CH_2 + CO + ROH + PdCl_2 \longrightarrow CH_2=C(CO_2R)CH_2CO_2R + Pd$$
$$(32)$$

Isocyanides, which are isoelectronic with carbon monoxide, also reacted with η^3-allyl complex in alcohol to give an imido ester, which was converted into a but-3-enoate by mild hydrolysis (reaction 33)[87,88]. η^3-Allyl complexes formed from olefins were converted to

$$[Pd(\eta^3\text{-}CH_2CHCH_2)Cl]_2 + RN\equiv C \longrightarrow CH_2=CHCH_2(OR)=NR \xrightarrow{H_2O}$$

$$CH_2=CHCH_2CO_2R \qquad (33)$$

ketenimine by treatment with isocyanide and then diazabicycloundecene (dbu). The reaction involves the insertion of isocyanide and dbu promoted β-elimination[89].

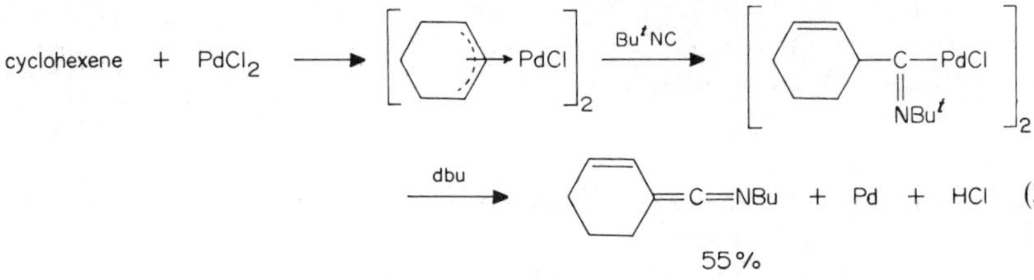

cyclohexene + PdCl$_2$ $\longrightarrow$ [cyclohexene–PdCl]$_2$ $\xrightarrow{Bu^tNC}$ [cyclohexene–C(=NBu^t)–PdCl]$_2$

$\xrightarrow{dbu}$ cyclohexene=C=NBu + Pd + HCl   (34)

55%

B. Reactions of Carbonucleophiles

η^3-Allyl complexes react smoothly with soft nucleophiles derived from malonates, acetoacetate, acetylacetone, β-keto sulphoxides, β-keto sulphones, and β-keto sulphides. Enamines also react easily. Malonate and acetoacetate were allylated with η^3-allylpalladium chloride in dmso in the presence of bases[5,90]. Similarly, acetylacetone was allylated (reaction 35)[91]. Reaction of carbonucleophiles with substituted η^3-allyl com-

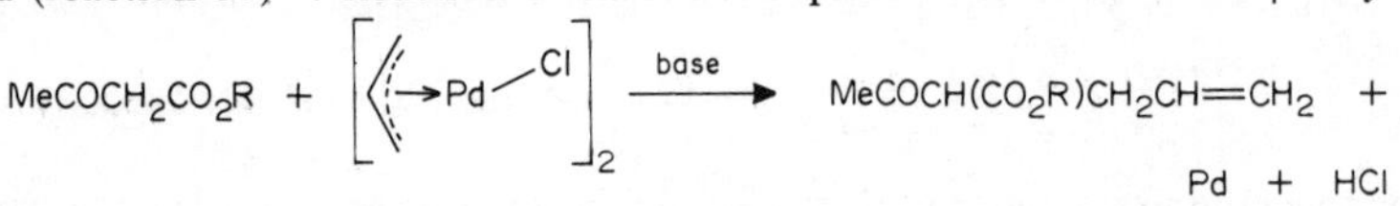

$$MeCOCH_2CO_2R + [\text{allyl-Pd-Cl}]_2 \xrightarrow{base} MeCOCH(CO_2R)CH_2CH=CH_2 +$$

$$Pd + HCl \qquad (35)$$

plexes proceeds in thf, or dmf at room temperature in the presence of an excess of PPh$_3$, or PBu$_3$[92,93].

Synthetic application of the reaction of active methylene compounds with η^3-allyl complexes formed from various olefins has been explored as a method of introducing functional groups at allylic positions of olefins via η^3-allyl complex formation[31,36,92-97]. Considerable, but not complete, regioselectivity was observed in complex formation from substituted olefins[38]. Attack by nucleophiles occurs at both sides of the allyl system.

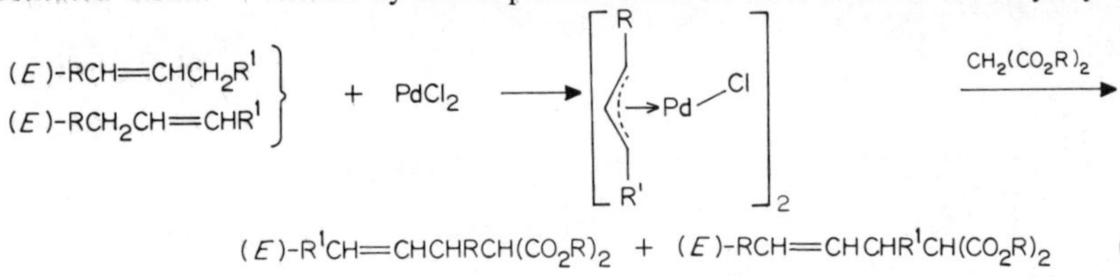

(E)-RCH=CHCH$_2$R^1
(E)-RCH$_2$CH=CHR^1 } + PdCl$_2$ $\longrightarrow$ [R-allyl-Pd-Cl, R^1]$_2$ $\xrightarrow{CH_2(CO_2R)_2}$

(E)-R^{1}CH=CHCHRCH(CO$_2$R)$_2$ + (E)-RCH=CHCHR^{1}CH(CO$_2$R)$_2$ (36)

Regioselectivity depends on the nature of attacking nucleophiles, the structure of the η^3-allyl complexes, and the ligands[92,97]. Attack at the less substituted side of the η^3-allyl moiety seems to be general, but some exceptions were observed. For example, different results were observed in the reaction of nucleophiles such as methyl methylsulphonylacetate and methyl malonate with the η^3-allyl complex of methylenecyclohexane. The attack at the primary carbon occurred in hmpa. On the other hand, the attack took place selectively at the secondary carbon in the presence of stericallyl hindered tri-*o*-

3. Carbon—carbon bond formation using η^3-allyl complexes

tolyphosphine (tot) as a ligand. Also in the latter case the introduced nucleophile took an axial direction.

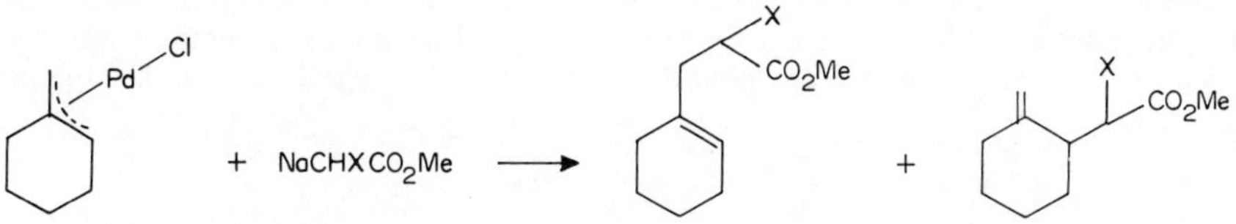

X = CO$_2$Me, hmpa (70% yield)   79%   21%
X = CO$_2$Me, tot (57% yield)   26%   74%   (37)
X = SO$_2$Me hmpa (58–90% yield)   100%   0%

Alkylation of methyl groups of geranylacetone (**6**) without the protection of the carbonyl group has been achieved by converting it in 70–85% yield into a mixture of the η^3-allyl complexes **7** and **8**, which were treated with methyl methylsulphonylacetate in the

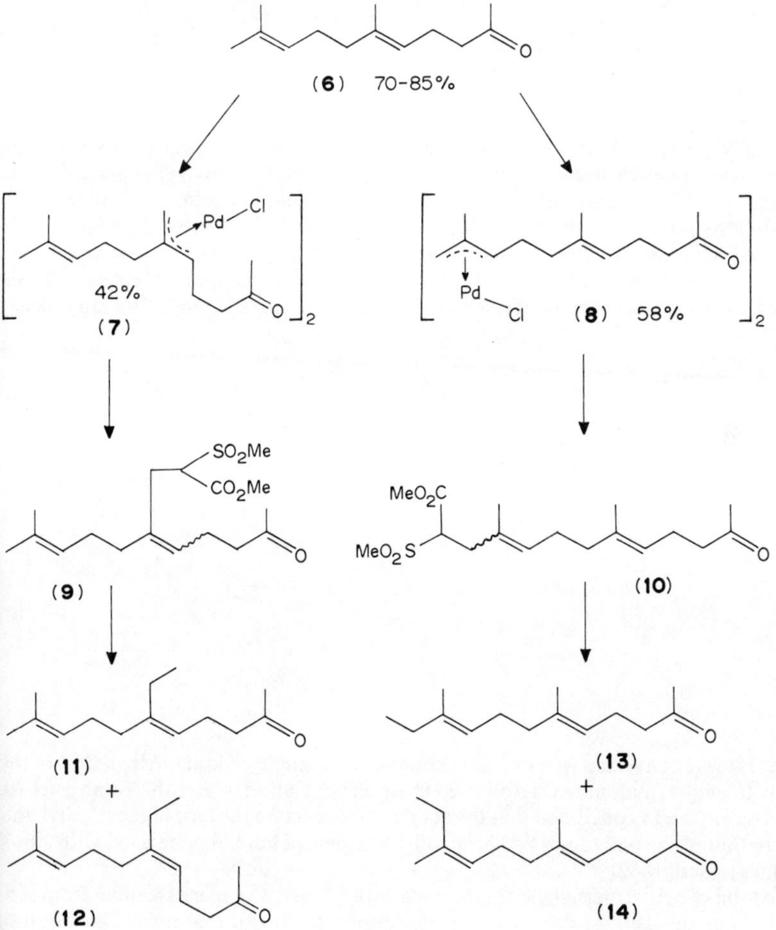

(38)

presence of PPh_3 to give the esters **9** and **10** in 24–85% yields. Finally, the functional groups were removed to give the methylated products **11–14**[93]. Another example is the formation of geranylgeraniol from farnesoate[98]. By the same methodology, farnesoate and the pheromone of Monarch butterfly were synthesized from methyl geraniate[99].

Vitamin A (**15**) and related compounds were synthesized by the reaction of sulphones and the η^3-allyl complex **16** derived from prenyl acetate. Reaction of the η^3-allyl complex with 3-methyl-1-(phenylsulphonyl)-5-(2,6,6-trimethylcyclohex-1-en-1-yl)penta-2,4-diene (**17**) in dmf in the presence of PPh_3 gave in 52% yield 1-acetoxy-3,7-dimethyl-5-(phenylsulphonyl)-9-(2,6,6-trimethylcyclohex-1-en-1-yl)nona-2,6,8-triene, which was converted to vitamin A (**15**)[100]. Reaction of phenyl sulphinylacetate with η^3-allyl complexes, followed by oxidative elimination of the phenylsulphinyl group, affords conjugated dienecarboxylates (reaction 40)[94]. A mixture of (E)- and (Z)-2-ethylidene-nopinane (**18**) was converted to its η^3-allyl complex as a single product. Alkylation of the complex with methyl malonate was followed by decarboxylation to give the mono ester **19**. From this result, it was concluded that the alkylation occurs on the face of the η^3-allyl unit opposite to that of the palladium[95]. The η^3-allyl complex of butadiene reacted with 2 mol of malonate (reaction 42)[101].

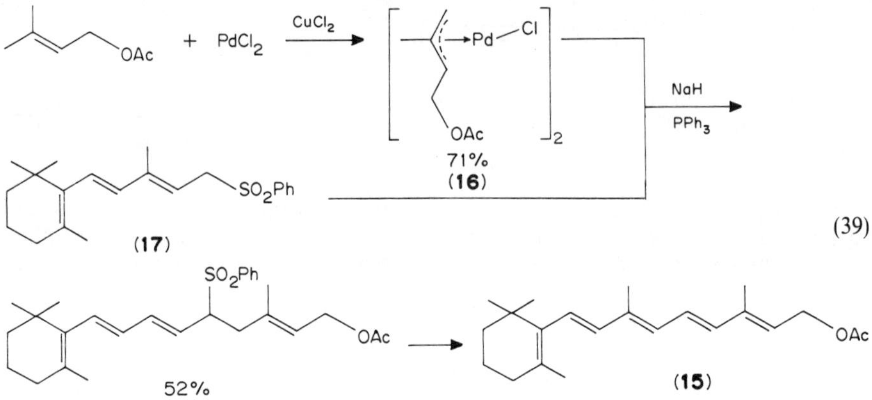

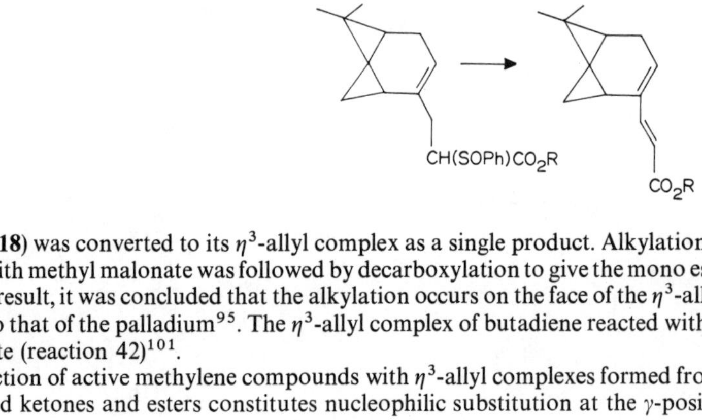

The reaction of active methylene compounds with η^3-allyl complexes formed from α,β-unsaturated ketones and esters constitutes nucleophilic substitution at the γ-position of

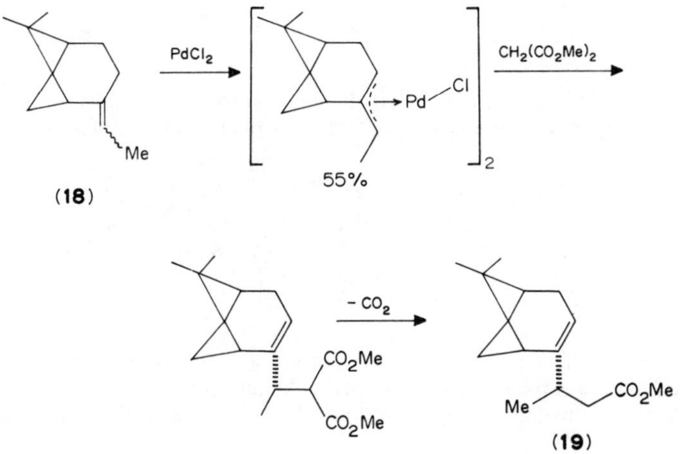

3. Carbon—carbon bond formation using η^3-allyl complexes

(18) → [PdCl complex] 55% → with $CH_2(CO_2Me)_2$ →

→ (19) $-CO_2$ → (41)

$[Pd(\eta^3-CH_2CHCHCH_2Cl)Cl]_2 + RCH(CO_2Et)_2 \longrightarrow (E)-RC(CO_2Et)_2CH_2CH=CHCH_2CR(CO_2Et)_2$ (42)

the unsaturated ketones (reaction 43)[102-104]. The complexes formed from cholest-4-en-3-one, testosterone, and progesterone reacted with methyl malonate in dmso to give 6β-dicarboxymethyl derivatives in high yields with complete stereospecificity[105].

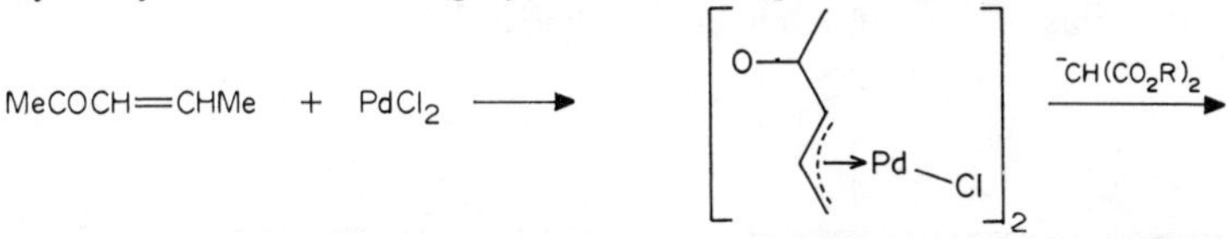

$MeCOCH=CHMe + PdCl_2 \longrightarrow$ [Pd complex] $\xrightarrow{\bar{C}H(CO_2R)_2}$ $(E)-MeCOCH=CHCH_2CH(CO_2R)_2$ (43)

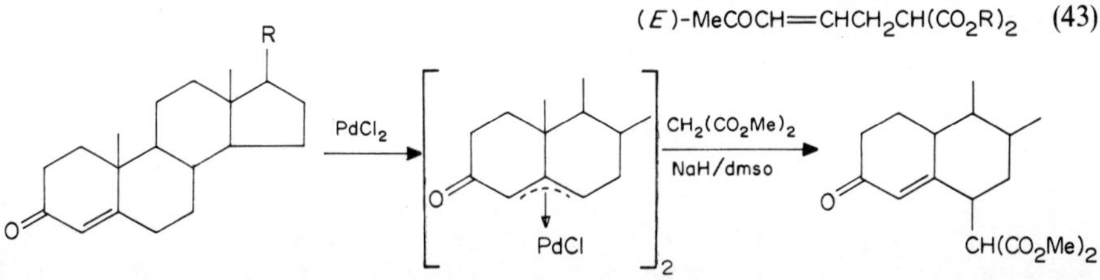

(44)

Enamines as a nucleophile react with η^3-allylpalladium chloride to give α-allyl ketones after hydrolysis (reaction 45)[5,90].

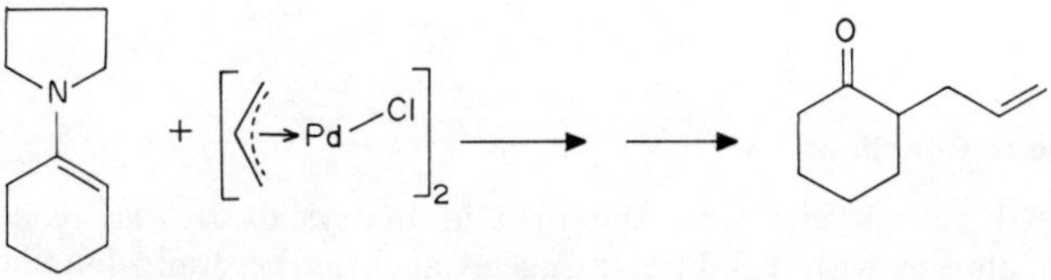

(45)

The stereochemistry and scope of the addition reaction of ketone enolates with η^3-allylpalladium chloride (reaction 46) have been studied[106]. Unusual cyclopropanation

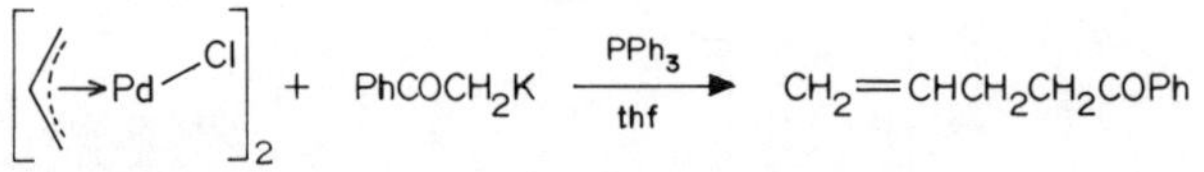

$$\left[\underset{}{\overset{}{\langle\!\!\!-\!\!\!\text{Pd}}}\!\!-\!\!\text{Cl}\right]_2 + \text{PhCOCH}_2\text{K} \xrightarrow[\text{thf}]{\text{PPh}_3} \text{CH}_2\!\!=\!\!\text{CHCH}_2\text{CH}_2\text{COPh} \qquad (46)$$

took place by the reaction of η^3-allylpalladium chloride with methyl cyclohexanecarboxylate in the presence of hmpa and triethylamine (reaction 47)[107]. Palladium induced the

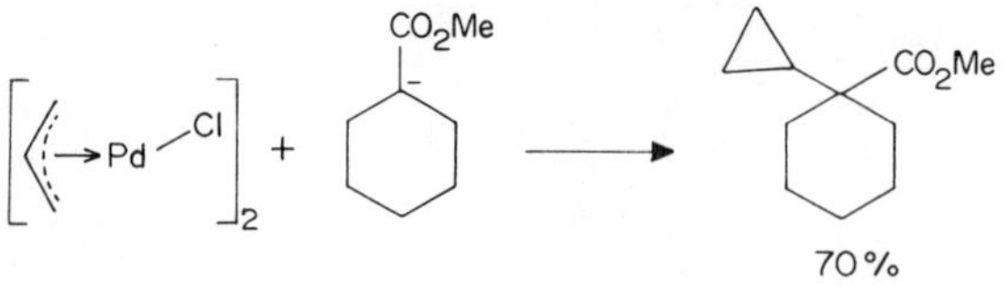

$$\left[\underset{}{\overset{}{\langle\!\!\!-\!\!\!\text{Pd}}}\!\!-\!\!\text{Cl}\right]_2 + \underset{\text{}}{\text{[cyclohexane-CO}_2\text{Me]}} \longrightarrow \underset{70\%}{\text{[cyclohexane-C(cyclopropyl)(CO}_2\text{Me)]}} \qquad (47)$$

conjugated diene formation from allylic alcohols and aldehydes in the presence of PPh_3. One explanation is the reaction of η^3-allylpalladium complex formed from allyl alcohol with the carbanion of the aldehyde (reaction 48)[108]. The regiocontrolled coupling of an η^3-

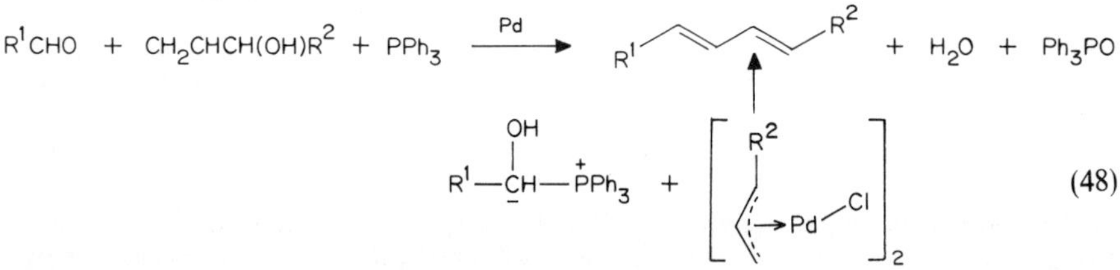

$$\text{R}^1\text{CHO} + \text{CH}_2\text{CHCH(OH)R}^2 + \text{PPh}_3 \xrightarrow{\text{Pd}} \text{R}^1\!\!-\!\!\text{CH}\!\!=\!\!\text{CH}\!\!-\!\!\text{CH}\!\!=\!\!\text{CH}\!\!-\!\!\text{R}^2 + \text{H}_2\text{O} + \text{Ph}_3\text{PO}$$

$$\text{R}^1\!\!-\!\!\underset{\text{OH}}{\text{CH}}\!\!-\!\!\overset{+}{\text{PPh}_3} + \left[\underset{}{\overset{\text{R}^2}{\langle\!\!\!-\!\!\!\text{Pd}}}\!\!-\!\!\text{Cl}\right]_2 \qquad (48)$$

allylpalladium complex with organozirconium species was used for the introduction of steroid side-chains[109-111] and humulene synthesis[112].

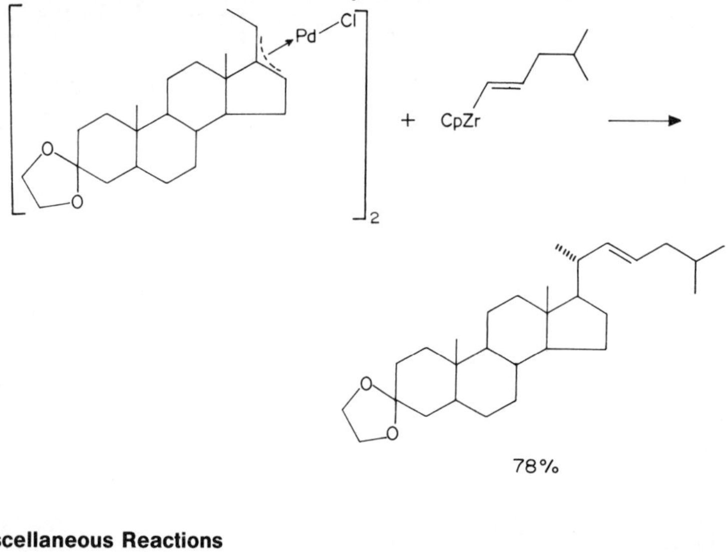

$$\left[\text{steroid-Pd-Cl}\right]_2 + \text{CpZr-CH}_2\text{CH}\!\!=\!\!\text{CHCH}(\text{CH}_3)_2 \longrightarrow$$

$$\text{steroid side-chain product} \qquad (49)$$

78%

C. Miscellaneous Reactions

Five-membered heterocycles were obtained in low yields by the reaction of η^3-allylpalladium chloride with 1,3-dipolar species such as benzonitriloanilide or benzonitrile N-oxide[113]. Bis(η^3-allyl)palladium complexes react with carbon dioxide to give β,γ-unsaturated acids. For example, carbonylation of bis(η^3-butenyl)palladium afforded

3. Carbon—carbon bond formation using η^3-allyl complexes

but-1-ene-3-carboxylic acid (reaction 50)[114,115]. 1,5-Dienes are formed by the photolytic coupling of η^3-allylpalladium complexes (reaction 51)[116]. η^3-Allylpalladium chlorides do

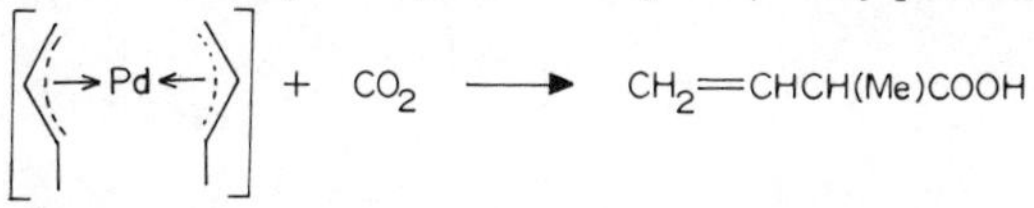

$$[(\text{allyl})\text{Pd}(\text{allyl})] + CO_2 \longrightarrow CH_2=CHCH(Me)COOH \qquad (50)$$

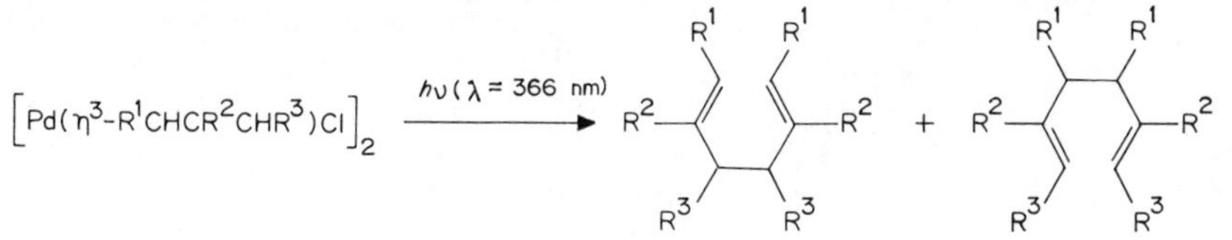

$$[Pd(\eta^3\text{-}R^1CHCR^2CHR^3)Cl]_2 \xrightarrow{h\nu (\lambda = 366 \text{ nm})} \text{products} \qquad (51)$$

not add to olefins directly, but the corresponding η^3-allylpalladium hexafluoroacetylacetonates add readily to bicyclic olefins[117].

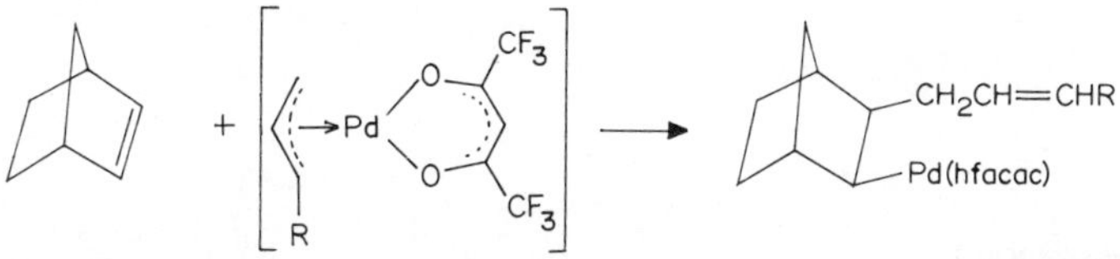

$$\text{(norbornene)} + [\text{Pd(hfacac allyl)}] \longrightarrow \text{product-CH}_2\text{CH=CHR} \qquad (52)$$

IV. CATALYTIC REACTIONS

Allylic halides, esters, ethers, alcohols, amines, sulphones, and nitro compounds undergo catalytic substitution reactions with carbonucleophiles. Conjugated dienes also react with carbonucleophiles in the presence of palladium catalysts. All of these reactions are believed to proceed via the formation of η^3-allylpalladium complexes as reactive intermediates.

A. Reactions of Conjugated Dienes

1. Introduction

Conjugated dienes undergo dimerization reactions in the presence of palladium–phosphine complexes as catalysts[118,119]. It is apparent from mechanistic considerations that an active species in the palladium-catalysed dimerization of butadiene is the palladium(0) complex, which forms the bis-η^3-allyl complex **20**. [Pd(PPh$_3$)$_4$] or [Pd(PPh$_3$)$_2$] coordinated by dienophiles was used[120,121]. Instead of palladium(0) complexes, which are tedious to prepare and unstable in air, readily available and stable palladium(II) compounds are reduced *in situ* to palladium(0), and stabilized by coordination of PPh$_3$ to act as the true catalyst.

There are two carbon—carbon bond-forming reactions of conjugated dienes catalysed by palladium complexes. The first type is linear dimerization. Octa-1,3,7-triene (**21**) is formed by a simple dimerization. The most characteristic and useful reaction is the dimerization with incorporation of certain nucleophiles. Some carbonucleophiles react with butadiene to form dimeric telomers in which nucleophiles are introduced mainly at the terminal position to form 1-substituted octa-2,7-diene (**22**). As a minor product, 3-substituted octa-1,7-diene (**23**) is formed. The second reaction characteristic of palladium catalysts is cocyclization of butadiene with the C=O bond of aldehydes and C=N bonds

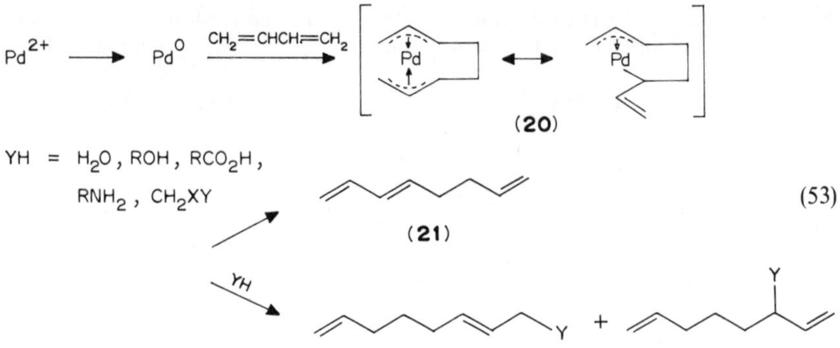

$$Pd^{2+} \longrightarrow Pd^0 \xrightarrow{CH_2=CHCH=CH_2} \left[\begin{array}{c} Pd \end{array} \longleftrightarrow \begin{array}{c} Pd \end{array} \right] \qquad (53)$$

(20)

YH = H_2O, ROH, RCO_2H, RNH_2, CH_2XY

(21)

(22)   +   (23)

of isocyanates and Schiff bases to form six-membered heterocyclic compounds (24) with two vinyl groups, as expressed by the general scheme 54.

$$\text{A=B} \quad + \quad \longrightarrow \quad \text{(24)} \qquad (54)$$

2. Oligomerization

Butadiene is converted into octa-1,3,7-triene (21) in aprotic solvents such as benzene, thf, and acetone[120]. Some protic solvents react with butadiene and cannot be used. The dimerization to give octa-1,3,7-triene proceeds more smoothly in propan-2-ol[121].

The main path of the palladium-catalysed reaction of butadiene is the dimerization. However, trimerization to form n-dodeca-1,3,6,10-tetraene takes place with certain palladium complexes in the absence of a phosphine ligand. The reaction of butadiene in benzene solution at 50 °C using η^3-allylpalladium acetate as a catalyst yielded n-dodeca-1,3,6,10-tetraene in 22 h with a selectivity of 79% at a conversion of 30% based on butadiene[122,123]. The reaction carried out at 70 °C using η^3-allyl complexes stabilized by chelating ligands in dmf or dmso produced dodeca-1,3,6,10-tetraene in 60% yield at 30% conversion[124].

Octa-1,3,7-triene was converted into n-hexadeca-1,5,7,10,15-pentaene (25) with 70% selectivity by the catalytic action of bis(η^3-allyl)palladium. On the other hand, 4-(but-3-enyl)dodeca-1,6,8,11-tetraene (26) was formed as the main product when PPh_3 was added in a 1:1 ratio[125].

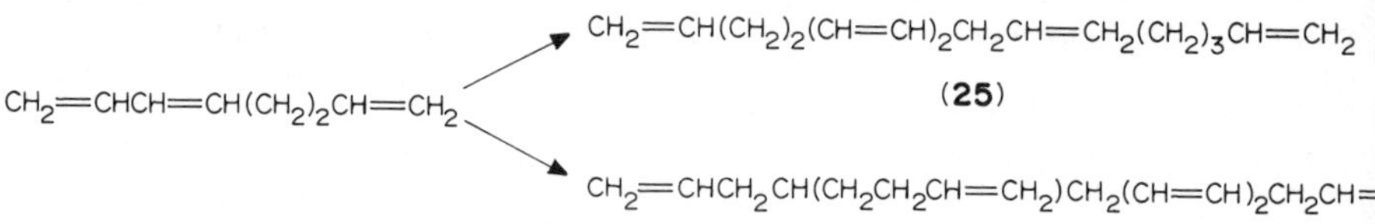

$$CH_2=CHCH=CH(CH_2)_2CH=CH_2 \begin{array}{c} \nearrow \\ \searrow \end{array} \begin{array}{l} CH_2=CH(CH_2)_2(CH=CH)_2CH_2CH=CH_2(CH_2)_3CH=CH_2 \\ (25) \\ \\ CH_2=CHCH_2CH(CH_2CH_2CH=CH_2)CH_2(CH=CH)_2CH_2CH= \\ (26) \end{array}$$

(55)

3. Dimerization with carbonucleophiles

Butadiene reacts with various carbonucleophiles via η^3-allylpalladium complex formation. Enamines react with butadiene and the octadienyl group is introduced[126]. The

3. Carbon—carbon bond formation using η^3-allyl complexes

pyrrolidine enamine of cyclohexanone was allowed to react with butadiene using $Pd(OAc)_2$ and PPh_3 at 80 °C for 3 h and the product was hydrolysed with dilute acid. As the main product, 2-(octa-2,7-dienyl)cyclohexanone was obtained in a high yield, accompanied by a small amount of 2,6-di(octa-2,7-dienyl)cyclohexanone.

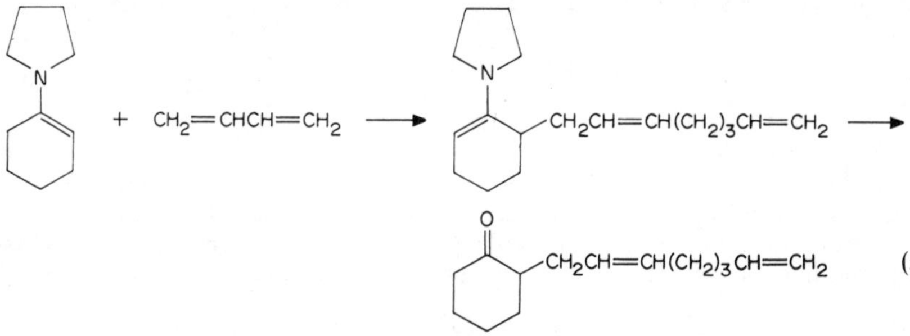

$$\text{pyrrolidine-cyclohexene} + CH_2=CHCH=CH_2 \longrightarrow \text{enamine-}CH_2CH=CH(CH_2)_3CH=CH_2 \longrightarrow$$

$$\text{2-substituted cyclohexanone-}CH_2CH=CH(CH_2)_3CH=CH_2 \quad (56)$$

Compounds with methylene and methyne groups to which two electronegative groups, such as carbonyl, alkoxycarbonyl, formyl, cyano, nitro, and sulphonyl, are attached, react readily with butadiene and their acidic hydrogens are replaced with the octa-2,7-dienyl group to give mono- and di-substituted compounds[127,128]. Branched products are also formed as byproducts. The reaction of active methylene compounds gives two kinds of products, namely the 1:2 adduct (**27**) and the 1:4 adduct (**28**). The reaction was carried out

$$YXCH_2 + CH_2=CHCH=CH_2 \longrightarrow YXCHCH_2CH=CH(CH_2)_3CH=CH_2 \xrightarrow{CH_2=CHCH=CH_2}$$
$$\quad\quad\quad\quad\quad\quad\quad\quad\quad\quad\quad\quad\quad\quad\quad\quad\quad\quad\quad\quad\quad (\mathbf{27})$$

$$YXC\left[CH_2CH=CH(CH_2)_3CH=CH_2\right]_2 \quad (57)$$
$$\quad\quad\quad\quad\quad\quad\quad (\mathbf{28})$$

with β-keto esters, β-diketones, malonate, α-formyl ketones, α-cyano and α-nitro esters cyanoacetamide, and phenylsulphonylacetate. $Pd(OAc)_2$ combined with PPh_3 is a good catalyst. When a bidentate ligand is used, a 1:1 rather than a 1:2 addition reaction takes place[129]. For example, dppe produced a mixture of addition products (reaction 58).

$$YXCH_2 + CH_2=CHCH=CH_2 \xrightarrow{[Pd], dppe} YXCHCH_2CH=CHMe + CH_2=CHCH(Me)CHXY$$
$$\quad\quad\quad\quad\quad\quad\quad\quad\quad\quad\quad\quad\quad\quad\quad\quad\quad\quad\quad\quad\quad\quad\quad\quad\quad (58)$$

Synthesis of amino acids was carried out by the reaction of butadiene with acetaminomalonate[130,131].

$$MeCONHCH(CO_2R)_2 + CH_2=CHCH=CH_2 \longrightarrow$$

$$CH_2=CH(CH_2)_3CH=CHCH_2(NHCOCH_3)(CO_2R)_2 \longrightarrow Me(CH_2)_7CH(NH_2)CO_2H \quad (58a)$$

One mole of isoprene reacted with one mole of acetoacetate by using dppe. Reaction of 2,3-dimethylbutadiene with acetoacetate was carried out by using $PdCl_2$ in the presence of sodium phenoxide. When PPh_3 was used, a 1:2 adduct was obtained. On the other hand, use of 1-phenyl-1-phospha-3-methylcyclopent-3-ene (**29**) at 100 °C caused 1:1 addition to give 3-carbomethoxy-5,6-dimethylhept-5-en-2-one (**30**), from which 5,6-dimethylhept-5-

en-2-one (**31**) was formed[133]. Reactions of active methylene compounds with myrcene (**2**) catalysed by $PdCl_2$ and dppe or PBu_3 in the presence of sodium phenoxide afforded 1:1

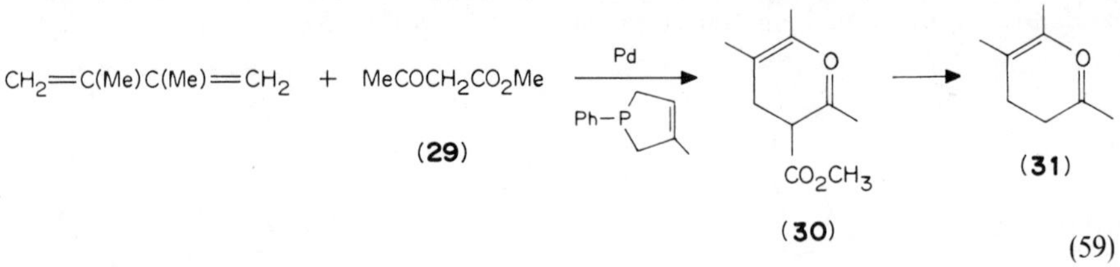

$$CH_2=C(Me)C(Me)=CH_2 + MeCOCH_2CO_2Me \xrightarrow[Ph-P]{Pd} \text{(30)} \rightarrow \text{(31)}$$

(**29**)

(59)

adducts. When PPh_3 was used in the reaction of acetylacetone, a 1:1 mixture of the two monoketones and three diketones shown in equation 60 (conversion 53%) was obtained in the proportions shown. The former is the product of the reverse Claisen reaction[134].

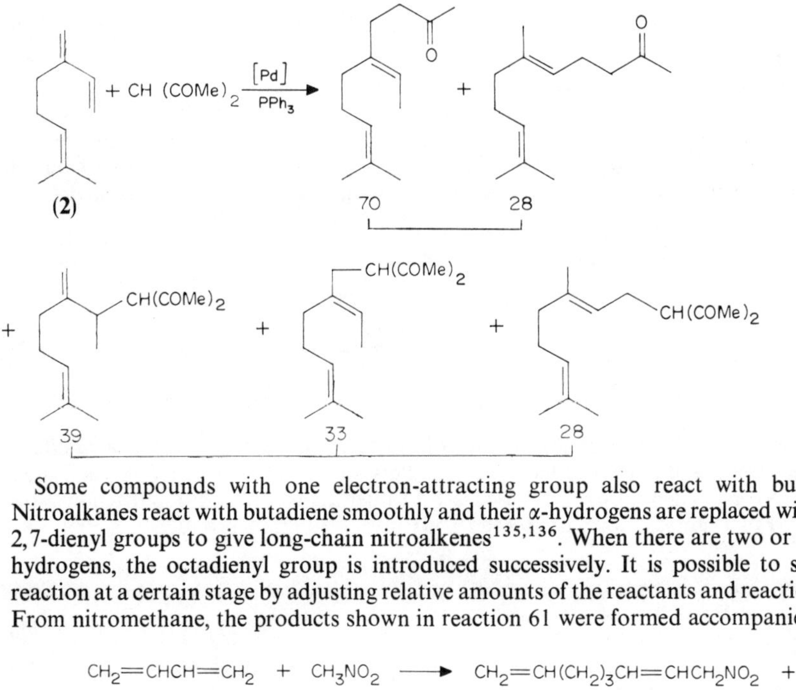

$$(\mathbf{2}) + CH(COMe)_2 \xrightarrow[PPh_3]{[Pd]}$$

70   28

39   33   28   (60)

Some compounds with one electron-attracting group also react with butadiene. Nitroalkanes react with butadiene smoothly and their α-hydrogens are replaced with octa-2,7-dienyl groups to give long-chain nitroalkenes[135,136]. When there are two or three α-hydrogens, the octadienyl group is introduced successively. It is possible to stop the reaction at a certain stage by adjusting relative amounts of the reactants and reaction time. From nitromethane, the products shown in reaction 61 were formed accompanied by 3-

$$CH_2=CHCH=CH_2 + CH_3NO_2 \longrightarrow CH_2=CH(CH_2)_3CH=CHCH_2NO_2 +$$

$$\left[CH_2=CH(CH_2)_3CH=CHCH_2\right]_2 CHNO_2 + \left[CH_2=CH(CH_2)_3CH=CHCH_2\right]_3 CNO_2 \quad (61)$$

substituted octa-1,7-diene as minor products. Acylamino ketones reacted with butadiene using sulphinate as a cocatalyst (reaction 62)[131]. Acyloins also reacted with butadiene in

$$RCOCHRNHCOR + CH_2=CHCH=CH_2 \xrightarrow[tosNa]{PdCl_2} CH_2=CH(CH_2)_3CH=CHCH_2CR(NHCOR)COR$$

(62)

3. Carbon—carbon bond formation using η^3-allyl complexes

the presence of sulphinate (reaction 63)[137] and 3-sulpholene reacted with 4 mol of butadiene (reaction 64)[138]. Oxazolidines reacted with 4 mol of butadiene. After hydrolysis,

$$RCOCH(OH)R + CH_2=CHCH=CH_2 \longrightarrow CH_2=CH(CH_2)_3CH=CHCH_2C(OH)RCOR \quad (63)$$

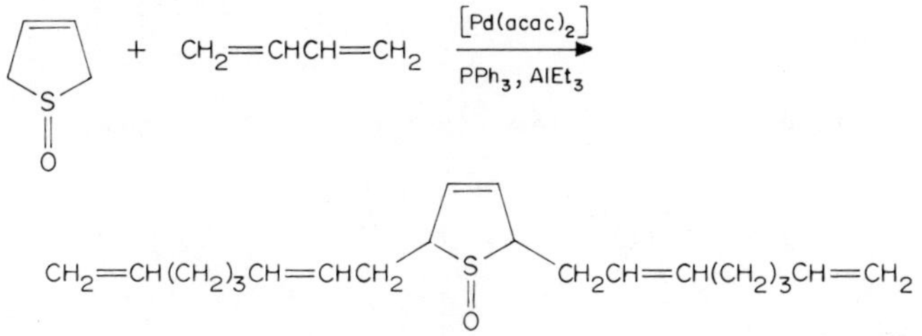

<chemical reaction 64 showing sulpholene + butadiene with [Pd(acac)$_2$], PPh$_3$, AlEt$_3$ giving CH$_2$=CH(CH$_2$)$_3$CH=CHCH$_2$—S(O)—CH$_2$CH=CH(CH$_2$)$_3$CH=CH$_2$>

(64)

di(octa-2,7-dienyl) ketone was obtained (reaction 65)[139]. Reaction of phenylhydrazones with butadiene gave three products (reaction 66)[140,141]. Two azo products (**32** and **33**)

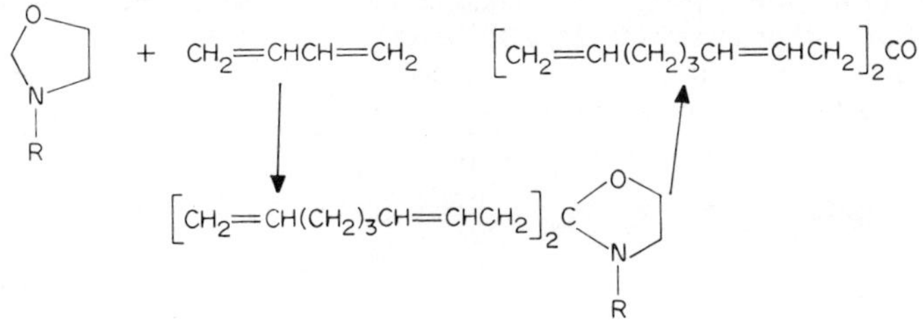

<reaction 65: oxazolidine + CH$_2$=CHCH=CH$_2$ → [CH$_2$=CH(CH$_2$)$_3$CH=CHCH$_2$]$_2$CO, via intermediate [CH$_2$=CH(CH$_2$)$_3$CH=CHCH$_2$]$_2$C(oxazolidine)>

(65)

$$CH_2=CHCH=CH_2 + R^1R^2C=NNHPh \xrightarrow{86\%} CH_2=CH(CH_2)_3CH=CHCH_2CR^1R^2N=NPh +$$

(**32**)

$$CH_2=CH(CH_2)_3CH(CR^1R^2N=NPh)CH=CH_2 +$$

(**33**)

$$CH_2=CH(CH_2)_3CH=CHCH_2N(Ph)N=CR^1R^2$$

(66)

were formed by the reaction of the η^3-allyl group at the electrophilic carbon atom followed by hydrogen shift.

tert-Allylic amines as a source of η^3-allyl complex reacted with 2 mol of butadiene (reaction 67)[142]. The palladium-catalysed reactions of conjugated dienes with aryl or

$$CH_2=CHCH_2NMe_2 + 2CH_2=CHCH=CH_2 \xrightarrow[AlEt_3]{[PdCl_2(PPh_3)_2]}$$
$$CH_2=CHCH_2CCH(CH=CH_2)(CH_2)_2CH=CHCH_2NMe_3 \quad (67)$$

alkenyl bromides in the presence of secondary amines afford a series of products (reactions 68 and 69). The tertiary amine was formed by the reaction of the secondary amines with η^3-allylpalladium intermediate formed by the reaction of the dienes with bromides[143–145].

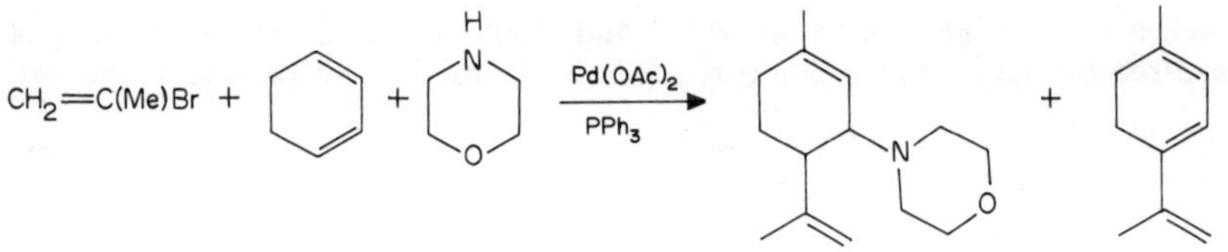

$CH_2=C(Me)Br$ + [cyclohexadiene] + [morpholine] $\xrightarrow{\text{Pd(OAc)}_2, \text{PPh}_3}$ [product with morpholinyl and isopropenyl] + [methyl-isopropenyl-cyclohexadiene] (68)

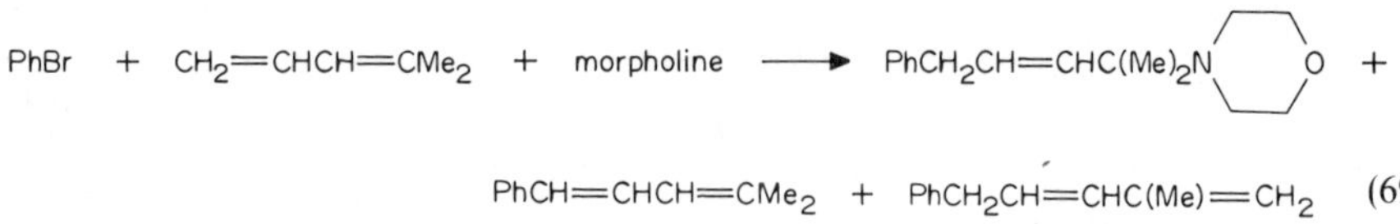

PhBr + $CH_2=CHCH=CMe_2$ + morpholine $\longrightarrow$ $PhCH_2CH=CHC(Me)_2N$[morpholine] +

$PhCH=CHCH=CMe_2$ + $PhCH_2CH=CHC(Me)=CH_2$ (69)

4. Carbonylation reactions

Butadiene is carbonylated in the presence of a palladium catalyst. In the catalytic carbonylation using $PdCl_2$ as a catalyst in alcohol, methyl-η^3-allylpalladium chloride (**34**) is formed by the insertion of one double bond of the diene into the Pd—H bond of the catalytic species. Carbon monoxide attack then affords the pent-3-enonate[146,147]. On the other hand, dimerization–carbonylation takes place with $Pd(OAc)_2$ and PPh_3 via the bis-η^3-allyl complex (**35**) to give the nona-3,8-dienoate (reactions 70 and 71)[148,149]. The

$MeCH=CHCH_2CO_2R$ $\xleftarrow{\text{PdCl}_2}$ $CH_2=CHCH=CH_2$ + CO + ROH $\xrightarrow{\text{Pd(OAc)}_2, \text{PPh}_3}$

$CH_2=CH(CH_2)_3CH=CHCH_2CO_2R$ (70)

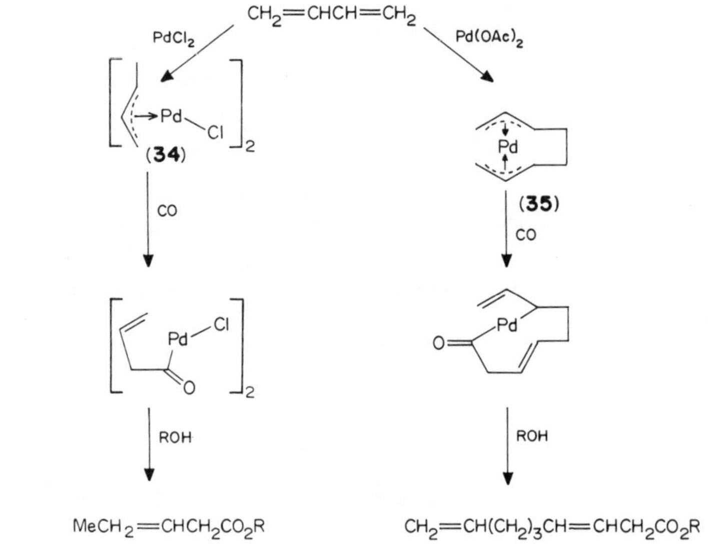

$CH_2=CHCH=CH_2$

PdCl$_2$ ↙   ↘ Pd(OAc)$_2$

[Pd-Cl complex (**34**)]$_2$ [Pd complex (**35**)]

↓ CO   ↓ CO

[Pd-Cl acyl complex]$_2$ [Pd acyl complex]

↓ ROH   ↓ ROH

$MeCH_2=CHCH_2CO_2R$ $CH_2=CH(CH_2)_3CH=CHCH_2CO_2R$ (71)

presence of chloride ion in the coordination sphere of palladium seems to inhibit the formation of bis-η^3-allyl system. Carbonylation of isoprene catalysed by $PdCl_2$ gave the 4-

methylpent-3-enoate as the sole product. The carbonylation using Pd(OAc)$_2$ and PPh$_3$ also afforded the 4-methylpent-3-enoate and no dimerization–carbonylation of isoprene took place[150] (reaction 72).

$$CH_2{=}C(Me)CH{=}CH_2 + CO + ROH \longrightarrow Me_2C{=}CHCH_2CO_2R \qquad (72)$$

5. Cocyclization reactions

A characteristic reaction catalyzed by palladium is the cocyclization of 2 mol of butadiene with one hetero double bond such as a C=N or C=O bond to give a six-membered ring with two vinyl groups (36). A typical reaction is the formation of 2-

$$\text{(diagram, reaction 73)} \qquad (73)$$

(36)

substituted 3,6-divinyltetrahydropyrans (37) by the reaction of butadiene with aldehydes[151–154]. In this reaction, unsaturated non-cyclized alcohols (38) are also formed. The selectivity towards the pyrans and alcohols can be controlled by a ratio of palladium to PPh$_3$ in the catalyst system.

$$CH_2{=}CHCH{=}CH_2 + RCHO \longrightarrow \text{(37)} + \text{RCHOH (38)} \qquad (74)$$

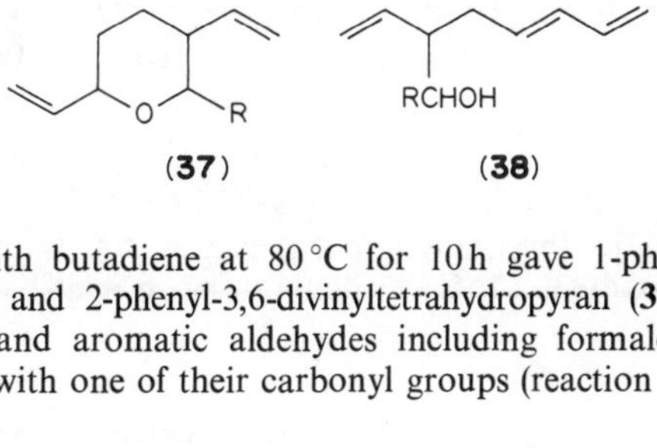

The reaction of benzaldehyde with butadiene at 80 °C for 10 h gave 1-phenyl-2-vinylhepta-4,6-dien-1-ol (38, R = Ph) and 2-phenyl-3,6-divinyltetrahydropyran (37, R = Ph) in 90% yield. Both aliphatic and aromatic aldehydes including formaldehyde behave similarly. α-Diketones react with one of their carbonyl groups (reaction 75)[152].

$$CH_2{=}CHCH{=}CH_2 + RCOCOR \longrightarrow \text{(pyran product)} \qquad (75)$$

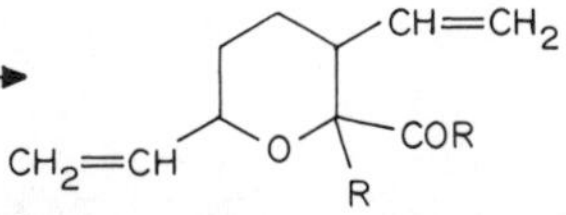

Perfluoroacetone also (39) reacted with butadiene to give 2,2-bis(trifluoromethyl)-3,6-divinyltetrahydropyran (40) and 1,1,1-trifluoro-2-trifluoromethylhexa-3,5-dien-2-ol (41).

$$CH_2{=}CHCH{=}CH_2 + CF_3COCF_3 \longrightarrow \text{(40)} + \text{(41)}$$

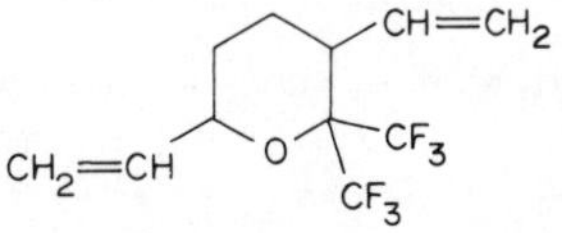

(39)

$$CH_2{=}CHCH{=}CHC(OH)(CF_3)_2 \qquad (76)$$

(41)

Reaction of acetone gave the products in reaction 77 as minor products when PEt_3 was

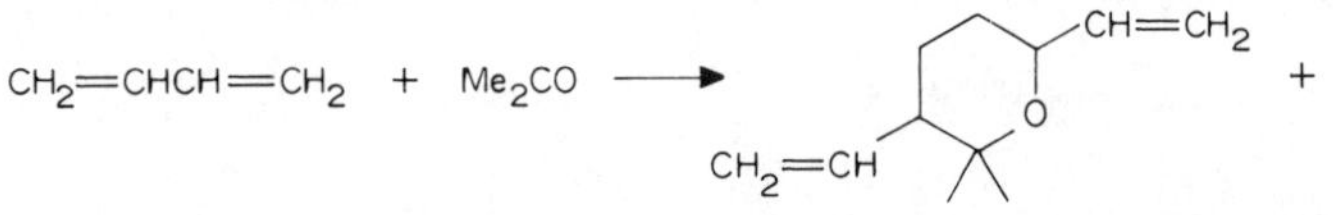

$$CH_2=CHCH=CH_2 + Me_2CO \longrightarrow$$

$$MeCO(CH_2)_2CH=CH(CH_2)_3CH=CH_2 + MeCOCH_2CH(CH=CH_2)(CH_2)_3CH=CH_2 \quad (77)$$

used as the ligand, whereas when tricyclohexylphosphine was the ligand the products shown in reaction 78 were obtained in low yield[155,156].

$$CH_2=CHCH=CH_2 + Me_2CO \longrightarrow CH_2=CHCH=CHCH_2CH(CMe_2OH)CH=CH_2 \quad (78)$$

Another group of molecules which take part in the cocylization are the aryl isocyanates. The C=N double bond, rather than C=C double bond, reacts with butadiene to give 3-ethylidene-1-phenyl-6-vinyl-2-piperidone (**42**) in 75% yield. In this reaction, double bond migration to the conjugated position took place[157]. With isoprene, selective head-to-head

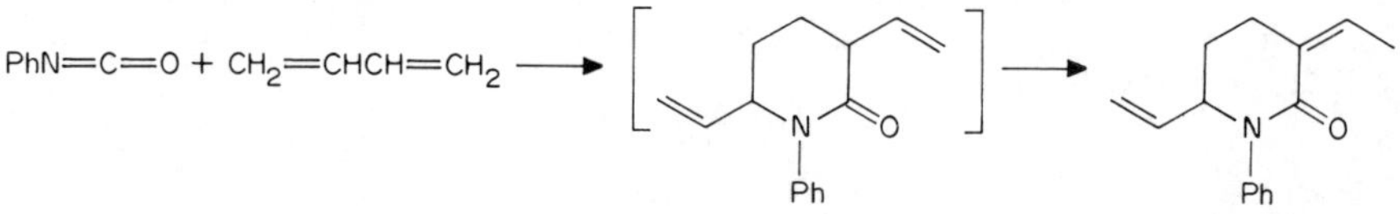

$$PhN=C=O + CH_2=CHCH=CH_2 \longrightarrow$$

(**42**)

(79)

dimerization–cyclization took place at 100 °C to give 3,6-diisopropenyl-1-phenyl-2-piperidone (**43**). The C=N double bond in Schiff bases reacted with butadiene to give 3,6-

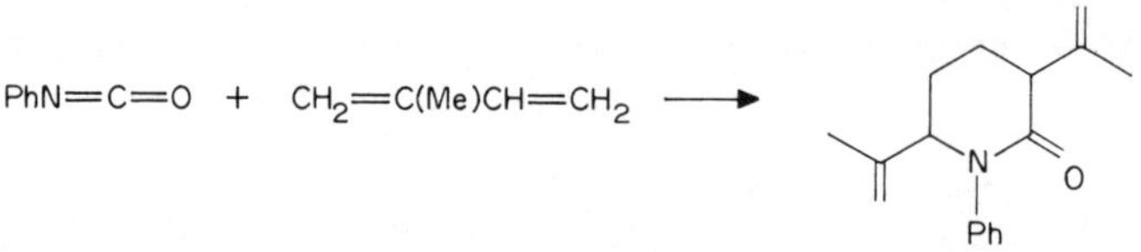

$$PhN=C=O + CH_2=C(Me)CH=CH_2 \longrightarrow$$

(80)

(**43**)

divinylpiperidines (**44**)[158]. Only the Schiff bases of aromatic aldehydes took part in the reaction.

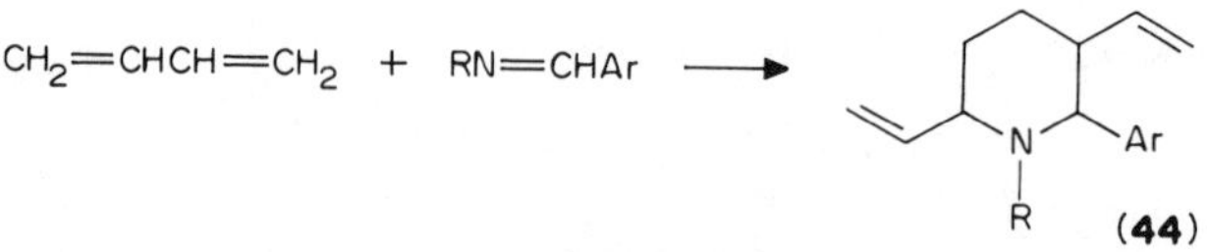

$$CH_2=CHCH=CH_2 + RN=CHAr \longrightarrow$$

(81)

(**44**)

Benzoquinone, when coordinated to palladium(0), does not undergo the usual Diels–Alder reaction with butadiene, but the unusual cyclization shown in reaction 82 took place with 2 mol of butadiene at 60 °C to give the tricyclic compound in 75% yield[159].

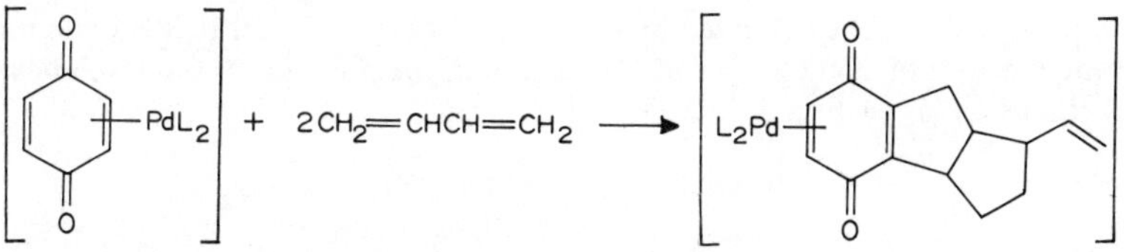

3. Carbon—carbon bond formation using η^3-allyl complexes

$$\left[\begin{array}{c}\text{quinone-PdL}_2\end{array}\right] + 2\,CH_2{=}CHCH{=}CH_2 \longrightarrow \left[\begin{array}{c}L_2Pd\text{-complex}\end{array}\right] \quad (82)$$

L = PPh$_3$

6. Reaction with carbon dioxide

The reaction of butadiene and carbon dioxide carried out at 120 °C in dmf using a palladium catalyst coordinated by dppe produced 2-ethylidenehepta-4,6-dienoic acid (**45**) in 4–12% yield. When isolated, the acid lactonized easily to form the lactone **46**[160–163]. A similar lactone was obtained from isoprene in a low yeild[164].

$$CH_2{=}CHCH{=}CH_2 + CO_2 \longrightarrow MeCH{=}C(CO_2H)CH_2CH{=}CHCH{=}CH_2 \longrightarrow$$
$$(\mathbf{45})$$

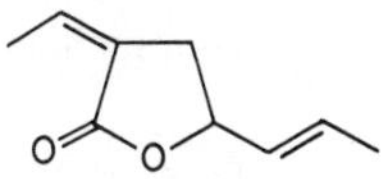

(**46**)   (83)

B. Reactions of Allenes

Like butadiene, allene is dimerized by palladium catalysts with incorporation of nucleophiles to form 2,3-dialkylbutadiene derivatives. Reaction of allene with amines produced 3-methyl-2-methylenebut-3-enylamine (**47**) (reaction 84)[165]. With acetic acid, the acetate of 3-methyl-2-methylenebut-3-enyl alcohol was obtained[166]. Two moles of

$$CH_2{=}C{=}CH_2 + RNH_2 \xrightarrow{Pd(PPh_3)_2} CH_2{=}C(Me)C({=}CH_2)CH_2NHR \xrightarrow{CH_2{=}C{=}CH_2}$$
$$(\mathbf{47})$$
$$\left[CH_2{=}C(Me)C({=}CH_2)CH_2\right]_2 NR \quad (84)$$

allene reacted at 120 °C with butadiene using [Pd(PPh$_3$)$_2$(maleic anhydride)] as the catalyst to give *cis*- and *trans*-2-methyl-3-methyleneocta-1,5,7-triene in 39% yield (reaction 85)[167]. The six-membered ring dimer **48** was obtained as the main product by the

$$2\,CH_2{=}C{=}CH_2 + CH_2{=}CHCH{=}CH_2 \longrightarrow$$
$$CH_2{=}CHCH{=}CHCH_2C({=}CH_2)C(Me){=}CH_2 \quad (85)$$

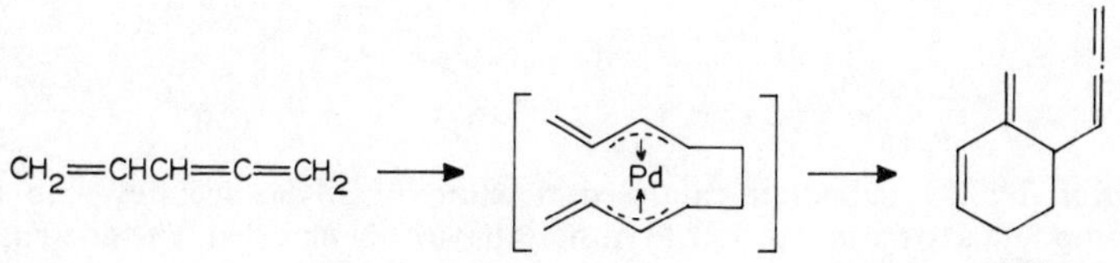

$$CH_2{=}CHCH{=}C{=}CH_2 \longrightarrow \left[\text{Pd complex}\right] \longrightarrow \quad (86)$$
$$(\mathbf{48})$$

reaction of penta-1,2,4-triene catalysed by $[Pd(acac)_2]/Et_2Al(OEt)/PPr^i_3$[168]. Palladium catalysed the reaction of allene with carbon dioxide in the presence of bisdicyclohexylphosphinoethane as in reaction 87[169].

$$CH_2=C=CH_2 + CO_2 \longrightarrow CH_2=C(Me)C(=CH_2)CH_2OCOC(Me)=CH_2 +$$

$$CH_2=C(Me)C(=CH_2)CH_2OCOCH=CHMe +$$

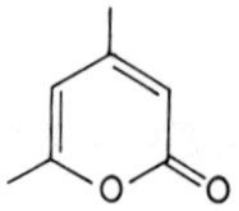

(87)

C. Reactions of Various Allylic Compounds

1. Reactions of allylic ethers, esters, alcohols, amines, and nitroalkanes with carbonucleophiles

Following allylic exchange, reaction of various allylic compounds with carbonucleophiles takes place smoothly to offer a catalytic method for carbon—carbon bond formation in the presence of palladium complexes via the formation of η^3-allylpalladium complexes as intermediates (reaction 88)[170–173]. For example, allyldiethylamine reacted

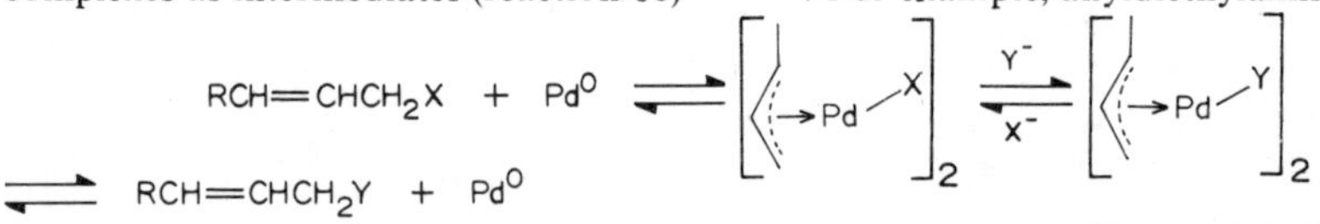

$$RCH=CHCH_2X + Pd^0 \rightleftharpoons [\text{Pd-X}]_2 \rightleftharpoons [\text{Pd-Y}]_2$$

$$\rightleftharpoons RCH=CHCH_2Y + Pd^0$$

(88)

X = OR, OCOR, OH OPh, NR$_2$; YH = nucleophiles

with acetylacetone at 85 °C to give 3-allylacetylacetone (**49**) in 70% yield and 3,3-diallylacetylacetone (**50**) in 20% yield (reaction 89)[173]. Reactions of acetylacetone with

$$CH_2=CHCH_2NEt_2 + CH_2COCH_2COMe \longrightarrow MeCOCH(CH_2CH=CH_2)COMe +$$
(**49**)

$$MeCOC(CH_2CH=CH_2)_2COMe \quad (89)$$
(**50**)

various allylic alcohols and acetates have been carried out[174]. Bicyclo[4.3.1]decenone (**51**) and vinylbicyclo[3.2.1]octanone (**52**) were obtained by the reaction of cyclohexanone enamine with diacetate of butenediol[175]. A number of applications of carbon—carbon

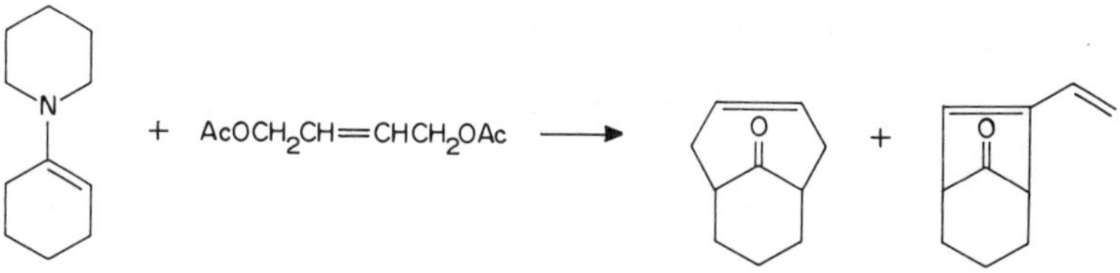

(90)

(**51**)   (**52**)

bond formation by the palladium-catalysed reaction of allylic acetates with active methylene compounds to natural product synthesis have been reported. The side-chains of ecdysone and steroids were prepared by the stereocontrolled displacement of phenylsul-

3. Carbon—carbon bond formation using η^3-allyl complexes

phonylacetate at the allylic acetate moiety catalysed by [Pd(PPh$_3$)$_4$] as a key step (reaction 91)[176]. The method was applied to the stereocontrolled synthesis of vitamin D metabolite[177].

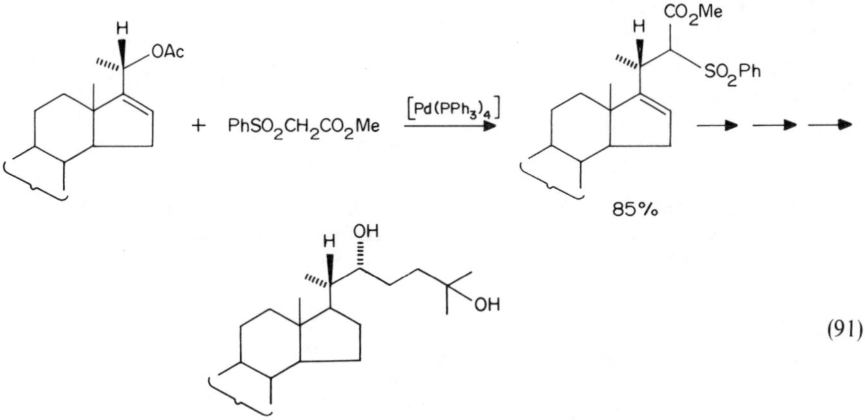

(91)

It was found that the replacement of the acetate group by carbonucleophiles proceeds with retention of configuration. Thus the conversion of 4,5-α-dihydrotestosterone (53) to 5-α-cholestan-3-one (54) was carried out by the reaction of the allylic acetate with methyl phenylsulphonylacetate catalysed by a palladium(0) complex, followed by alkylation and removal of the phenylsulphonyl and carbomethoxy groups[178,179]. The stereochemical course of allylic alkylation of six-membered rings has been studied. η^3-Allyl complex formation takes place with inversion and the nucleophilic attack also proceeds with inversion, thus resulting in retention of the stereochemistry (reaction 92)[180-183]. Chirality

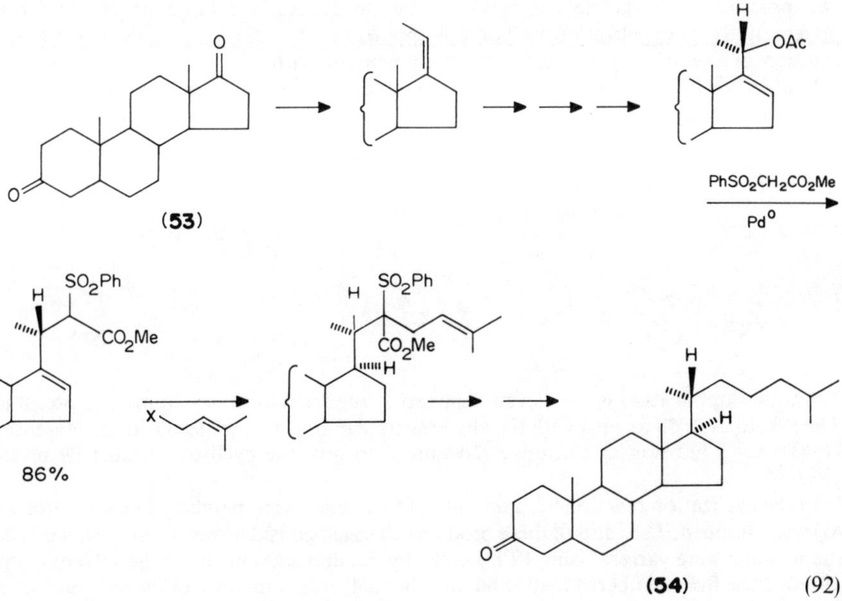

(92)

transfer is possible by the reaction of optically active allylic compounds with nucleophiles, and a vitamin E side-chain was prepared from glucose[184,185].

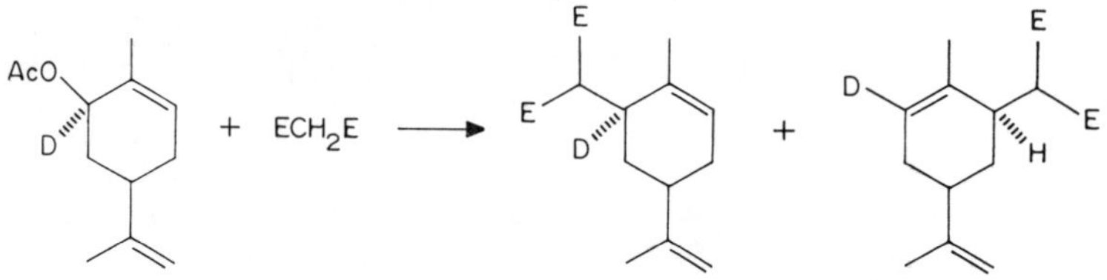

$$\text{AcO-[cyclohexene-D]} + \text{ECH}_2\text{E} \longrightarrow \text{[cyclohexene-D,E]} + \text{[cyclohexene-D,E,H]} \quad (93)$$

Intramolecular reaction offers a cyclization method. Macrolide skeletons (**55**) were constructed by the reaction of carbanions generated from a phenylsulphonylacetate with the allylic acetate moiety in **56** by using [Pd(PPh$_3$)$_4$] (reaction 94)[186,187]. The formation of

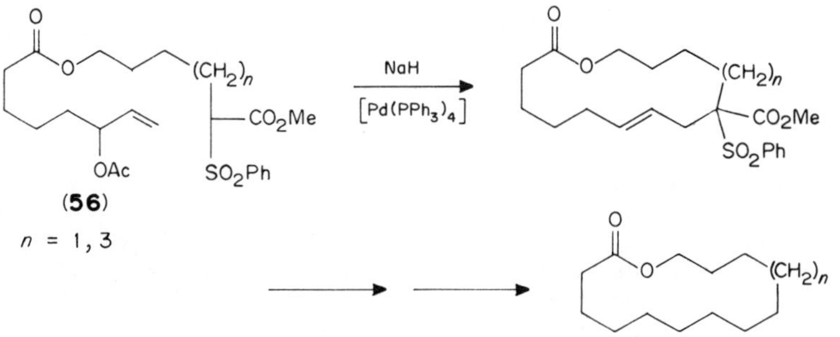

$$\text{(56)} \xrightarrow[\text{[Pd(PPh}_3)_4]}{\text{NaH}} \longrightarrow \longrightarrow \text{(55)} \quad (94)$$

$n = 1, 3$

two possible allylic isomers is expected, but in general the larger rings are formed preferentially. The cyclization method was applied to the synthesis of recifeiolide (**57**). A thf solution of the anion generated from the precursor with NaH was added slowly to a solution of [Pd(PPh$_3$)$_4$] (9 mol-%) at reflux temperature. The lactone was obtained in 78% yield stereoselectively as the E isomer and regioselectively without ten-membered lactone formation[188].

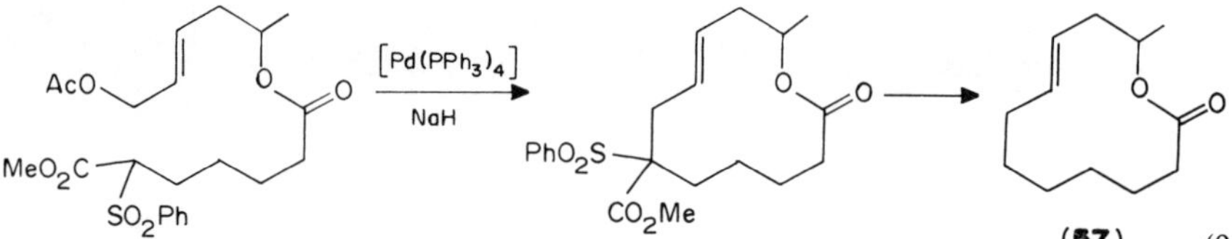

$$\xrightarrow[\text{NaH}]{[\text{Pd(PPh}_3)_4]} \longrightarrow \text{(57)} \quad (95)$$

Another application is the eleven-membered ring formation of a humulene precursor. The β-keto ester **58** reacted with the allylic acetate in the same molecule in the presence of [Pd(PPh$_3$)$_4$] (20 mol-%) and dppe (20 mol-%) to give the cyclized product **59** in 45% yield[189].

In the cyclization reaction 97, a mixture of five- and seven-membered products (**60** and **61**) was obtained. The ratio of these products changed considerably when the ligands and the solvents were varied. Using PPh$_3$ as the ligand and acetonitrile as the solvent mainly afforded the five-membered ketone **60**. In other solvents, a mixture of the five- and seven-

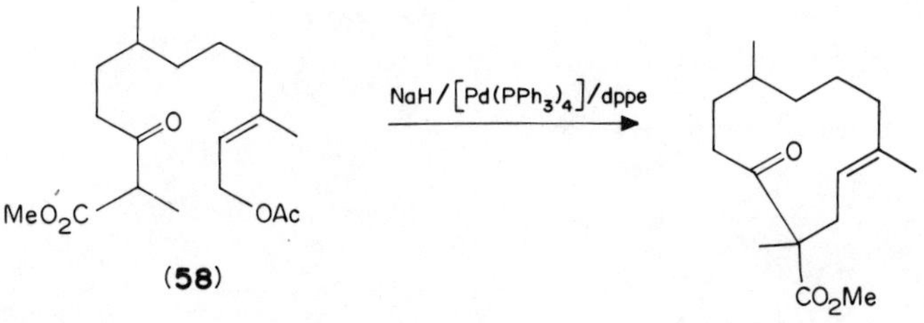

3. Carbon—carbon bond formation using η^3-allyl complexes

(96)

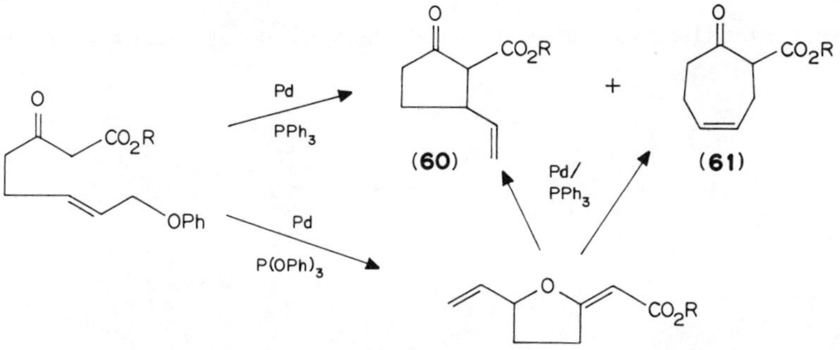

(97)

membered ketones **60** and **61** was obtained. Triphenyl phosphite afforded the five-membered O-alkylation product **62** selectively, which was converted into **60** and **61** with Pd–PPh$_3$ catalyst[190]. Only the six-membered ketone was obtained in reaction 98, with no eight-membered ketone[190].

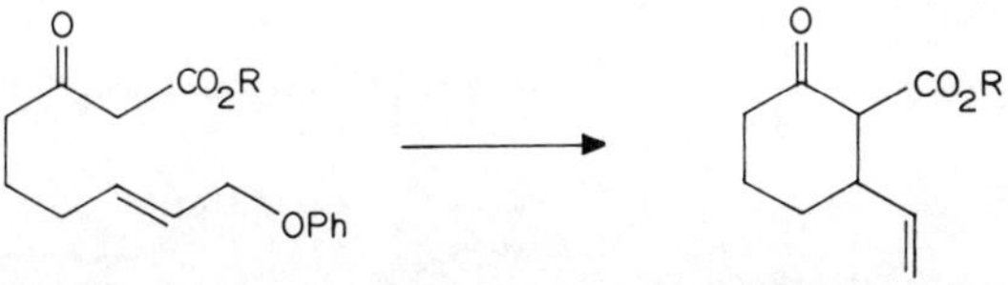

(98)

As a related reaction, the palladium-catalysed 1,3-oxygen to carbon alkyl shift reaction has been studied as a synthetic method for cyclopentanone and cycloheptanone derivatives, together with its stereo- and regiochemistry (reactions 99 and 100)[191,192]. An intramolecular reaction of allylic acetates with carbonuclophiles offers a convenient

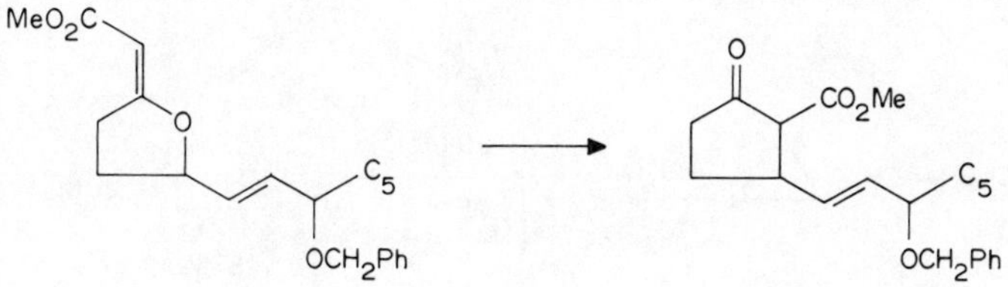

(99)

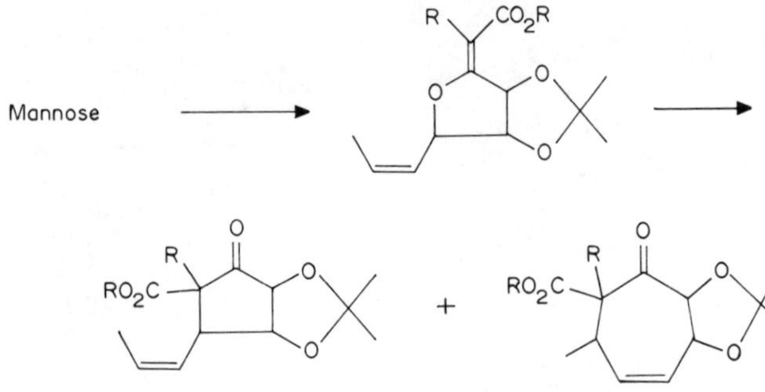

Mannose → [intermediate] → [cyclopentanone product] + [cycloheptenone product] (100)

synthetic method for spirocycles without forming bicyclo compounds (reactions 101 and 102)[193,194].

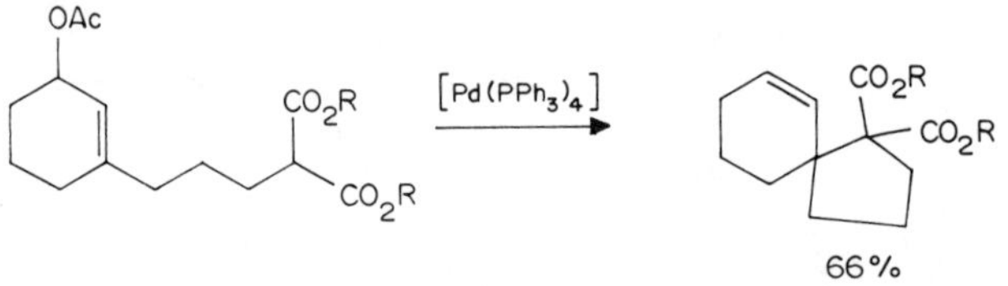

[reaction scheme] (101) 66%

[reaction scheme] (102)

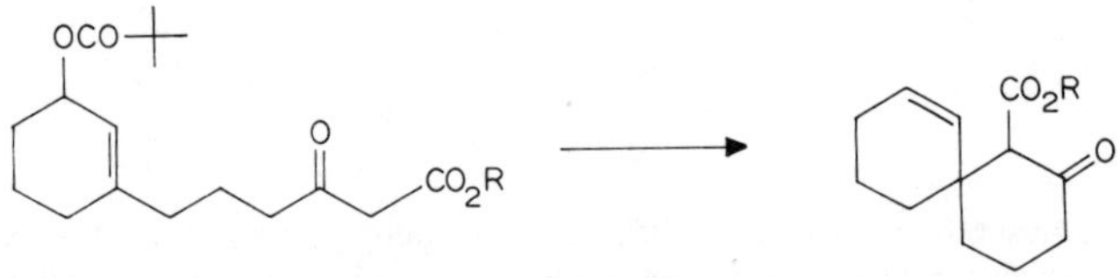

Cyclopropanation can be effected by the palladium-catalysed reaction of the biallylic compound with malonate (reaction 103)[195–197]. On the other hand, an interesting

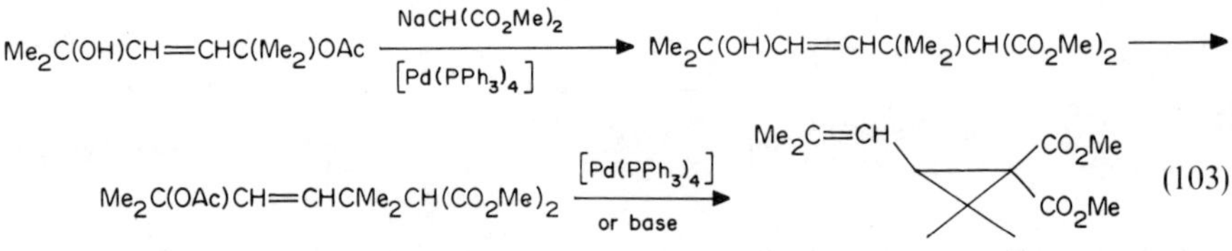

$Me_2C(OH)CH=CHC(Me_2)OAc \xrightarrow[{[Pd(PPh_3)_4]}]{NaCH(CO_2Me)_2} Me_2C(OH)CH=CHC(Me_2)CH(CO_2Me)_2 \longrightarrow$

$Me_2C(OAc)CH=CHCMe_2CH(CO_2Me)_2 \xrightarrow[\text{or base}]{[Pd(PPh_3)_4]}$ [cyclopropane product] (103)

rearrangement or ring expansion reaction of 2-(buta-1,3-dienyl)cyclopropane-1,1-dicarboxylate esters to 2-alkenylcyclopent-3-ene-1,1-dicarboxylate esters took place with a palladium catalyst. The reaction is explained by η^3-allyl complex formation by ring opening, followed by intramolecular reaction of malonate (reaction 104)[198].

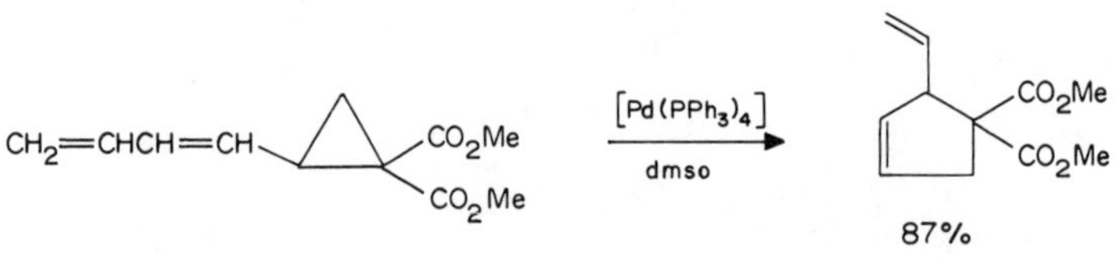

[reaction scheme] (104) 87%

3. Carbon—carbon bond formation using η^3-allyl complexes

Some functional groups in allylic compounds not only tolerate the reaction, but also offer synthetic methods for some interesting compounds. Reaction of 2-ethoxyallylic acetate with a sulphonylacetate, followed by hydrolysis, affords functionalized ketones. Thus the overall reaction offers methods for fumaroylation and succinoylation. The method was applied to the synthesis of pyrenophorin (reaction 105)[199]. As another

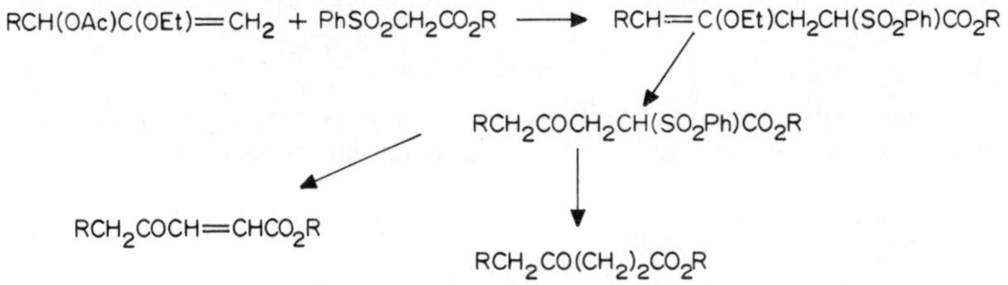

$$RCH(OAc)C(OEt)=CH_2 + PhSO_2CH_2CO_2R \longrightarrow RCH=C(OEt)CH_2CH(SO_2Ph)CO_2R$$

$$\downarrow$$

$$RCH_2COCH_2CH(SO_2Ph)CO_2R$$

$$\downarrow \quad (105)$$

$$RCH_2COCH=CHCO_2R \quad\longleftarrow\quad RCH_2CO(CH_2)_2CO_2R$$

application, 2-ethoxy-3-acetoxypropene was used for cyclopentenone annulation (reaction 106)[200]. The palladium-catalysed displacement of acetoxydihydropyran with malonate proceeded regio- and stereo-selectively (reaction 107)[201].

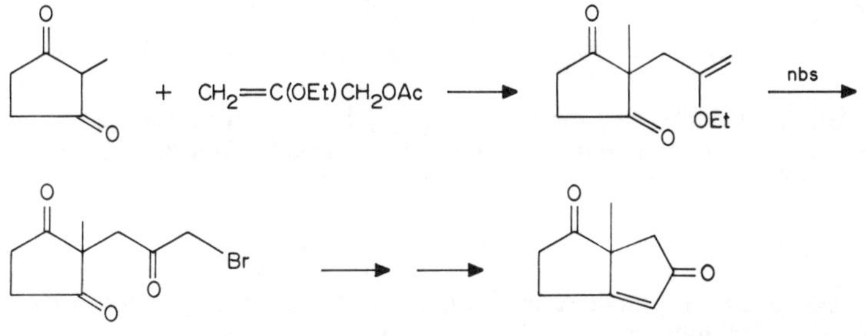

(106)

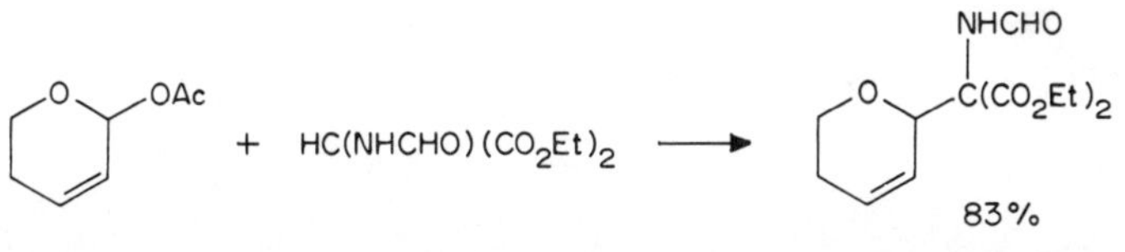

(107) 83%

2-Acetoxymethyl-3-allyltrimethylsilane reacts with palladium(0) to form the trimethylenemethane–palladium complex **63**, which undergoes methylenecyclopentane annulation with α,β-unsaturated ketones[202-204]. The reaction was extended to the

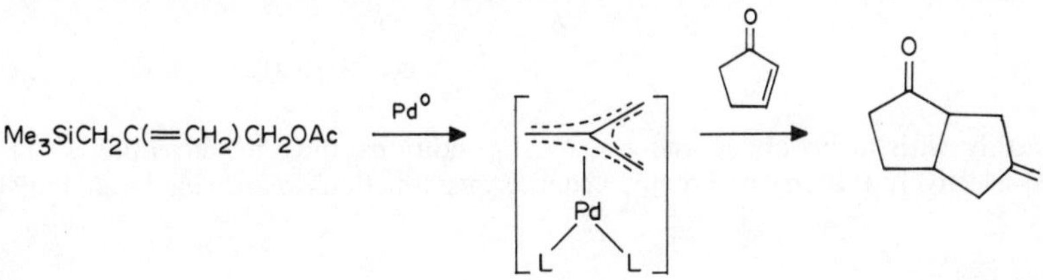

$$Me_3SiCH_2C(=CH_2)CH_2OAc \xrightarrow{Pd^0} \quad (108)$$

(63)

intramolecular carbocyclic [3 + 2] cycloaddition reaction to form bicyclic systems in one step (reaction 109)[205]. Vinylsilanes are formed by the palladium-catalysed regioselective

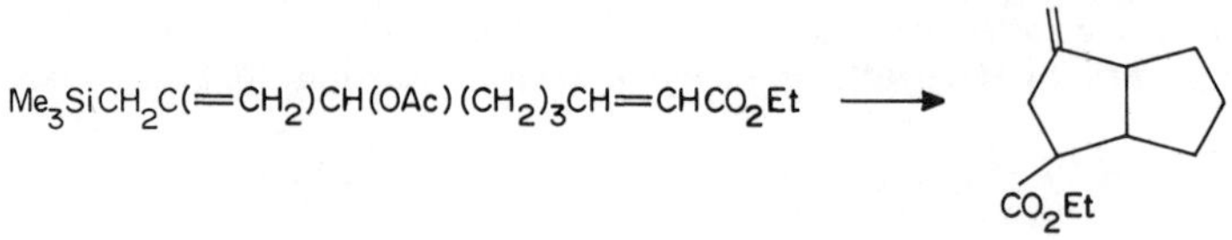

$$Me_3SiCH_2C(=CH_2)CH(OAc)(CH_2)_3CH=CHCO_2Et \longrightarrow \quad (109)$$

reaction of 1-trimethylsilylallyl acetates with carbonucleophiles (reaction 110)[206]. When electron-attracting groups are present in the allylic acetates, nucleophiles are introduced

$$\left.\begin{array}{l} CH_2=CHCH(OAc)SiMe_3 \\ \\ Me_3SiCH=CHCH_2OAc \end{array}\right\} + RCH(CO_2R)_2 \xrightarrow{[Pd(PPh_3)_4]} Me_3SiCH=CHCH_2CH(CO_2R)_2 \quad (110)$$

regioselectively at the carbon which is remote from the electron-attracting groups (reaction 111)[207].

$$RCH=CHCH(CN)OAc + CH_2(CO_2R)_2 \longrightarrow RCH\{CH(CO_2R)_2\}CH=CHCN \quad (111)$$

C-Allylation takes place by the reaction of allyl acetate with phenylsulphonylnitromethane, α-nitro esters, and primary and secondary nitroalkanes using alkaline methoxide (reaction 112)[208-211].

$$LiCH(NO_2)SO_2Ph + Me_2C=CHCH_2OAc \longrightarrow Me_2C=CHCH_2CH(NO_2)SO_2Ph \quad (112)$$

Allylic carbonates are more reactive than acetates, and the reaction with carbonucleophiles proceeds under neutral conditions at room temperature. The η^3-allyl complex formation is followed by decarboxylation to generate alkoxide ion, which abstracts a proton from nucleophiles (reaction 113)[212]. For example, only the allylic carbonate moiety

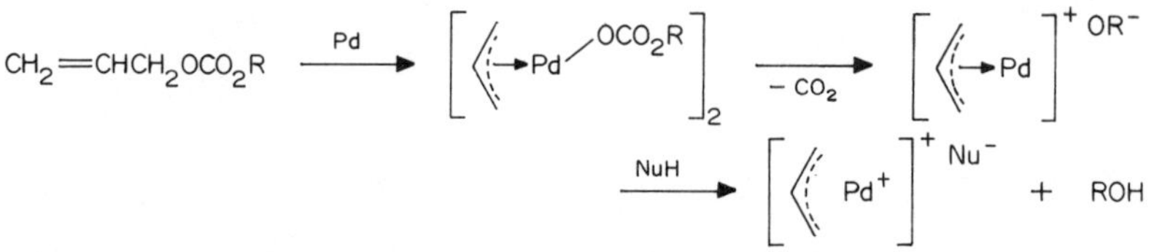

$$CH_2=CHCH_2OCO_2R \xrightarrow{Pd} \left[\begin{array}{c} \\ \end{array} \rightarrow Pd \diagup OCO_2R \right] \xrightarrow{-CO_2} \left[\begin{array}{c} \\ \end{array} \rightarrow Pd \right]^+ OR^-$$

$$\xrightarrow{NuH} \left[\begin{array}{c} \\ \end{array} Pd^+ \right]^+ Nu^- + ROH \quad (113)$$

of the compound shown in reaction 114 reacted with the nucleophile at room temperature in the absence of a base, leaving the allylic acetate moiety intact. α,β-Unsaturated epoxides

$$AcOCH_2CH=CHCH_2OCO_2Me + CH_3CH_2COCH(CH_3)CO_2Me \longrightarrow AcOCH_2CH=CHCH_2CMe$$

$$(CO_2Me)COCH_2CH_3 + CO_2 \quad (114)$$

react smoothly with nucleophiles under neutral conditions and the nucleophiles are introduced selectively at the carbon remote from oxygen functions to form allylic alcohols (reaction 115)[213,214].

$$RCH_2CH\overset{O}{\frown}CHCH=CHR' + CH_2(CO_2R)_2 \longrightarrow RCH_2CH(OH)CH=CHCHR'CH(CO_2R)_2 \quad (115)$$

3. Carbon—carbon bond formation using η^3-allyl complexes

Allylic esters of acetoacetic acid and β-keto acids undergo smooth decarboxylation to give allylated ketones. This is the palladium-catalysed Carroll rearrangement (reaction 116)[215,216]. Allyl enol carbonates undergo smooth palladium-catalysed arrangement and

$$\text{MeCOCH}_2\text{COOCH}_2\text{CH}=\text{CH}_2 \longrightarrow \text{MeCO(CH}_2)_2\text{CH}=\text{CH}_2 + \text{CO}_2 \quad (116)$$

decarboxylation to give 2-allylated ketones (reaction 117)[217]. Certain simple ketones react with allylic acetate only as their lithium enolates (reaction 118)[218].

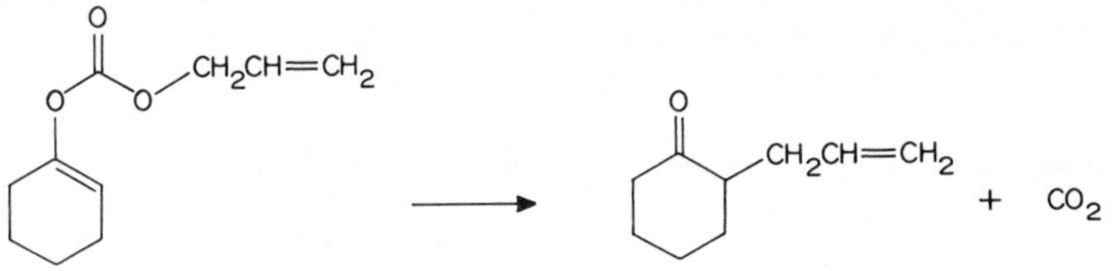

(117)

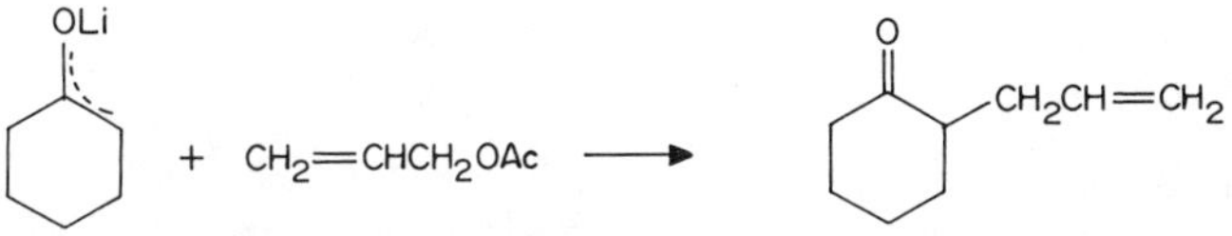

(118)

η^3-Allylpalladium reacts with various organometallic compounds. Palladium-catalysed allylation of ketones is possible by the reaction of enol stannanes (reaction 119)[219]. Allyl-

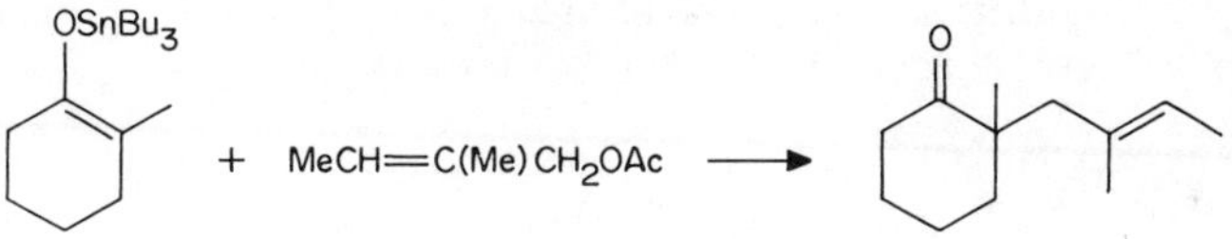

(119)

and phenyl-stannanes react with allylic acetates (reaction 120)[220,221]. Allylic esters and

$$\text{PhCH}=\text{CHCH}_2\text{OAc} + \text{CH}_2=\text{CHCH}_2\text{SnR}_3 \longrightarrow \text{PhCH}=\text{CH(CH}_2)_2\text{CH}=\text{CH}_2 \quad (120)$$

ethers (also allylic halides) react with alkenyl and aryl compounds of Mg, Al, Zn, and Zr in the presence of palladium phosphine complexes to give coupled products (reactions 121 and 122)[222,223]. It was confirmed that the reaction of allylic acetate with alkenylalu-

$$\text{RCH}=\text{CHCH}_2\text{OPh} + \text{PhMgAr} \xrightarrow{[\text{Pd(PPh}_3)_4]} \text{RCH}=\text{CHCH}_2\text{Ph} \quad (121)$$

$$\text{BuC(Me)}=\text{CHAlMe}_2 + \text{MeC}\equiv\text{CH(CH}_2)_2\text{C(Me)}=\text{CHCH}_2\text{OAc} \longrightarrow$$

$$\text{MeC}\equiv\text{CH(CH}_2)_2\text{C(Me)}=\text{CHCH}_2\text{C}=\text{C(Me)Bu} \quad (122)$$

minium in the presence of a η^3-allylpalladium complex proceeds with inversion of the stereochemistry (reaction 123)[224]. 3,4-Epoxybut-1-ene as a source of an η^3-allyl complex reacted with 1-alkenylboranes to give a mixture of allylic and homoallylic alcohols

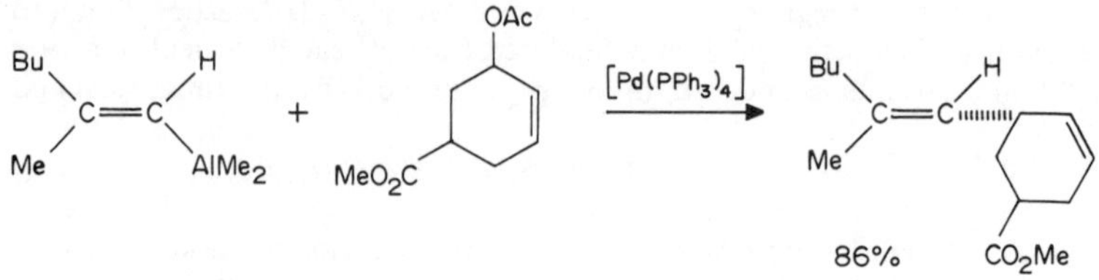

$$\underset{Me}{\overset{Bu}{>}}C=C\underset{AlMe_2}{\overset{H}{<}} + \text{[cyclohexenyl OAc with MeO}_2\text{C]} \xrightarrow{[Pd(PPh_3)_4]} \underset{Me}{\overset{Bu}{>}}C=C\underset{\text{cyclohexenyl}}{\overset{H}{<}} \quad (123)$$

86%   CO₂Me

(reaction 124)[225]. The anion of cyclopentadiene reacts with allylic acetate in the presence of palladium catalyst (reaction 125)[226].

$$RCH{=}CRBX_2 + CH_2{=}CHCH\overset{O}{-}CH_2 \xrightarrow{[Pd(dba)_2]} RCH{=}CRCH(CH{=}CH_2)CH_2OH +$$

74

$$RCH{=}CRCH_2CH{=}CHCH_2OH \quad (124)$$

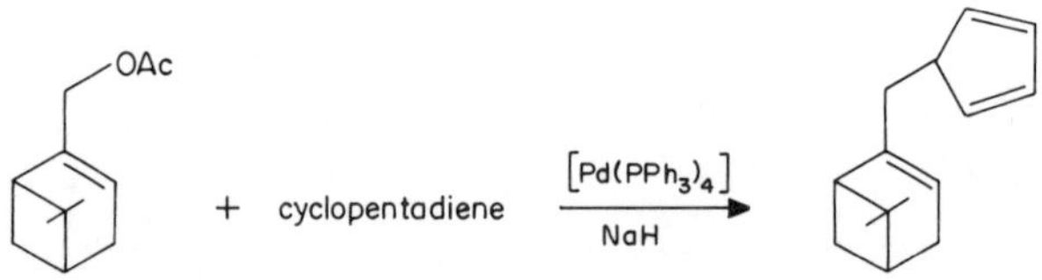

26

$$\text{[cubyl-CH}_2\text{OAc]} + \text{cyclopentadiene} \xrightarrow[\text{NaH}]{[Pd(PPh_3)_4]} \text{[cubyl-CH}_2\text{-cyclopentadienyl]} \quad (125)$$

η^3-Allylpalladium complex formation takes place from allylsulphones. Thus the palladium-catalysed displacement of sulphone with carbonucleophiles is possible (reaction 126)[227]. Allylnitroalkanes can also be used for η^3-allyl complex formation and the nitro group is displaced with nucleophiles[228,229].

$$CH_2{=}C(R)CH_2SO_2Ph + CH_2(CO_2R)_2 \longrightarrow CH_2{=}C(R)CH_2CH(CO_2R)_2 \quad (126)$$

$$CH_2{=}CHCH_2NO_2 + NuH \xrightarrow{[Pd(PPh_3)_4]} CH_2{=}CHCH_2Nu \quad (127)$$

Allylic esters, ethers, and alcohols were carbonylated under high pressure by using PdCl₂ as a catalyst[230,231]. The carbonylation of diallyl ether proceeded in two steps to form but-3-enoic anhydride as the final product (reaction 128). The carbonylation of but-

$$(CH_2CH{=}CH_2)_2O \xrightarrow{CO} CH_2{=}CHCH_2COOCH_2CH{=}CH_2 \xrightarrow{CO} (COCH_2CH{=}CH_2)_2O \quad (128)$$

1-en-3-ol (**64**) and but-2-en-1-ol (**65**) in ethanol afforded only ethyl pent-3-enoate (**66**)[230].

$$\left.\begin{array}{c} MeCH{=}CHOH \\ (\textbf{64}) \\ \\ CH_2{=}CHCH(OH)Me \\ (\textbf{65}) \end{array}\right\} + CO + EtOH \longrightarrow MeCH{=}CHCH_2CO_2Et \quad (129)$$

(**66**)

3. Carbon—carbon bond formation using η^3-allyl complexes

The best method for the carbonylation of allylic compounds to give β,γ-unsaturated esters under mild conditions is the use of allylic carbonates. The decarboxylation–carbonylation proceeded even under carbon monoxide at atmospheric pressure at 50 °C using a palladium–phosphine complex as a catalyst[232].

$$RCH=CHCH_2OCO_2R' + CO \longrightarrow RCH=CHCH_2CO_2R' + CO_2 \quad (130)$$

2. Reactions of allylic halides

Allylic halides react with various palladium(0) compounds to form η^3-allyl complexes, which then undergo insertion reactions. In the final stage of the reaction, the palladium(0) species is regenerated, making the whole process catalytic.

Allylic halides are carbonylated to give β,γ-unsaturated esters in an alcohol via η^3-allyl complexes[230,231,233,234]. From allyl chloride, a but-3-enoate is formed (reaction 131). But-

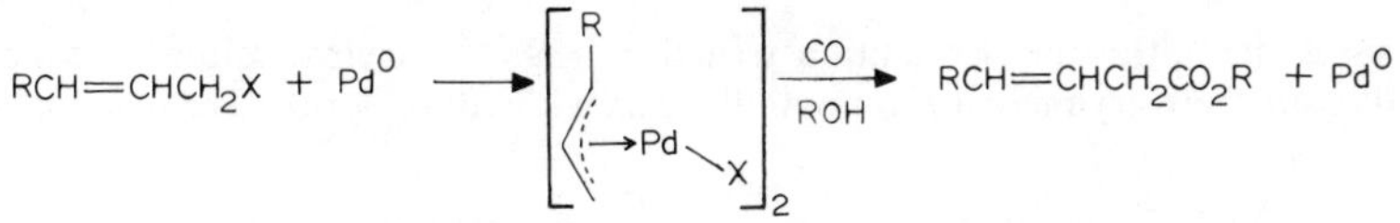

$$RCH=CHCH_2X + Pd^0 \longrightarrow \left[\begin{array}{c} R \\ \diagdown \\ Pd \diagdown X \end{array}\right]_2 \xrightarrow[ROH]{CO} RCH=CHCH_2CO_2R + Pd^0 \quad (131)$$

3-enoyl chloride was formed at 90 °C and 85 atm by the carbonylation of allyl chloride catalysed by η^3-allylpalladium chloride or $PdCl_2$ in aprotic solvents such as dimethoxyethane and benzene (reaction 132). Carbonylation of a mixture of allyl chloride and

$$CH_2=CHCH_2Cl + CO \longrightarrow CH_2=CHCH_2COCl \quad (132)$$

butadiene with $PdCl_2$ produced an octa-3,7-dienoate as one product among others (reaction 133)[234].

$$CH_2=CHCH=CH_2 + CH_2=CHCH_2Cl + CO + ROH \longrightarrow$$
$$CH_2=CH(CH_2)_2CH=CHCH_2CO_2R \quad (133)$$

Allyl halides and (Z)-alkenylpentafluorosilicates reacted in the presence of $Pd(OAc)_2$ (10 mol-%) at room temperature to give 1,4-dienes in good yields. Alkenyl-

$$RC\equiv CH + HiSiCl_3 \longrightarrow (E)\text{-}RCH=CHSiCl_3 \xrightarrow[H_2O]{KF} K_2[(E)\text{-}RCH=CHSiF_5]$$

$$\xrightarrow[Pd(OAc)_2]{CH_2=CHCH_2Cl} (E)\text{-}RCH=CHCH_2CH=CH_2 \quad (134)$$

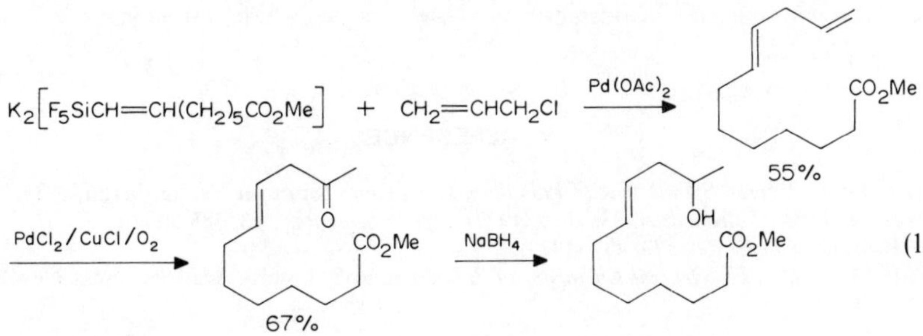

$$K_2[F_5SiCH=CH(CH_2)_5CO_2Me] + CH_2=CHCH_2Cl \xrightarrow{Pd(OAc)_2}$$

(135)

pentafluorosilicates are prepared from terminal alkynes by platinum-catalysed hydrosilylation, followed by the treatment with KF (reaction 134)[235,236]. The method was applied to the synthesis of (E)-11-hydroxydodecenoate (reaction 135).

Allylic halides react with organoboranes[237], stannanes[238], and zirconium[239] compounds (reactions 136–138). Perfluorozinc iodides react with allyl halides with palladium

$$CH_2{=}CHCH_2Br + (E)\text{-}RCH{=}CHBX_2 \xrightarrow[\text{base}]{[Pd(PPh_3)_4]} (E)\text{-}RCH{=}CHCH_2CH{=}CH_2 \quad (136)$$

$$RCH_2C(Me){=}CHCH_2Br + CH_2{=}CHCH_2SnR_3 \longrightarrow RCH_2C(Me){=}CH(CH_2)_2CH{=}CH_2 \quad (137)$$

$$MeCH{=}CHCH_2Cl + Cp_2ZrClR \longrightarrow MeCH{=}CHCH_2R + MeCH(R)CH{=}CH_2 \quad (138)$$

catalysis under ultrasonic irradiation (reaction 139)[240]. Arylpalladium intermediates formed *in situ* from arylmercury compounds react with allyl chloride. The reaction is only

$$R_fI \xrightarrow{Zn} R_fZnI \xrightarrow[\text{[Pd(PPh}_3)_4]]{RCH_2{=}CHCH_2X} CH_2{=}CHCHRR_f \quad (139)$$

partially catalytic (reaction 140)[241–243]. Reaction of allyl chloride with acetylene catalysed by PdCl$_2$ and LiCl affords hexa-1,5-dienyl chloride. However, a mechanism involving the

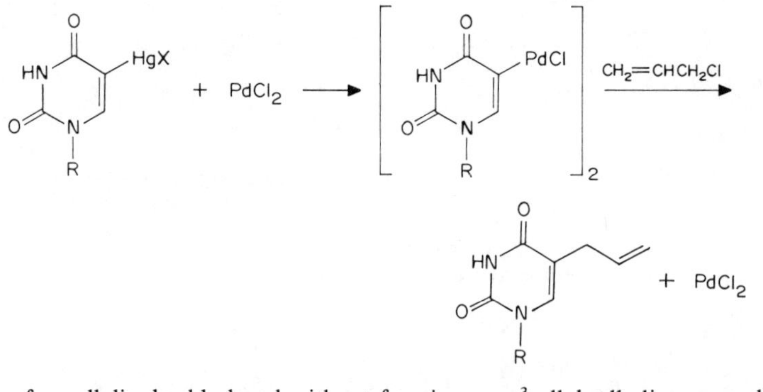

(140)

insertion of an allylic double bond without forming an η^3-allylpalladium complex was proposed[244,245].

$$CH_2{=}CHCH_2Cl + HC{\equiv}CH \xrightarrow[\text{LiCl}]{PdCl_2} CH_2{=}CH(CH_2)_2CH{=}CHCl \quad (141)$$

V. REFERENCES

1. J. Tsuji, *Organic Synthesis with Palladium Compounds*, Springer Verlag, Berlin, 1981.
2. B. M. Trost, *Tetrahedron*, **33**, 2615 (1977); *Acc. Chem. Res.*, **13**, 385 (1980).
3. R. Baker *Chem. Rev.*, **73**, 487 (1973).
4. P. M. Maitlis, *The Organic Chemistry of Palladium*, Vols. 1 and 2, Academic Press, New York, 1971.

3. Carbon—carbon bond formation using η^3-allyl complexes

5. J. Tsuji, H. Takahashi, and M. Morikawa, *Tetrahedron Lett.*, 4387 (1965).
6. J. Smidt and W. Hafner, *Angew. Chem.*, **71**, 284 (1959).
7. R. Hüttel, J. Kratzer, and M. Bechter, *Angew. Chem.*, **71**, 456 (1959).
8. W. T. Dent, R. Long, and J. Wilkinson, *J. Chem. Soc.*, 1585 (1964).
9. J. K. Nicholson, J. Powell, and B. L. Shaw, *Chem. Commun.*, 174 (1966).
10. Y. Tatsuno, T. Yoshida, and S. Otsuka, *Inorg. Synth.*, **19**, 220 (1979).
11. F. R. Hartley, and S. R. Jones, *J. Organomet. Chem.*, **66**, 465 (1974).
12. M. Sakakibara, Y. Takahashi, S. Sakai, and Y. Ishii, *Chem. Commun.*, 396 (1969).
13. J. H. Lukes and J. E. Blom, *J. Organomet. Chem.*, **26**, C25 (1971).
14. R. C. Larock and J. P. Burkhart, *Synth. Commun.*, **9**, 659 (1979).
15. J. M. Kliegman, *J. Organomet. Chem.*, **29**, 73 (1971).
16. K. Itoh, M. Fukui, and Y. Kurachi, *J. Chem. Soc., Chem. Commun.*, 501 (1977).
17. R. Pietropaolo, F. Faraone, and S. Sergi, *J. Organomet. Chem.*, **42**, 177 (1972).
18. K. Ito, S. Hasegawa, Y. Takahashi, and Y. Ishii, *J. Organomet. Chem.*, **73**, 401 (1974).
19. Y. Ishii, S. Hasegawa, S. Kimura, and K. Ito, *J. Organomet. Chem.*, **73**, 411 (1974).
20. E. O. Fischer and G. Bürger, *Z. Naturforsch., Teil B*, **16**, 702 (1961).
21. R. D. Riecke and A. V. Kavaliunas, *J. Org. Chem.*, **44**, 3069 (1979).
22. R. D. Riecke, A. V. Kavaliunas, and D. J. J. Fraser, *J. Am. Chem. Soc.*, **101**, 246 (1979).
23. T. Yamamoto, O. Saito, and A. Yamamoto, *J. Am. Chem. Soc.*, **103**, 5600 (1981).
24. R. Hüttel and H. Christ, *Chem. Ber.*, **97**, 1439 (1964).
25. R. Hüttel, J. Kratzer, and M. Bechter, *Chem. Ber.*, **94**, 766 (1961).
26. R. Hüttel and H. Christ, *Chem. Ber.*, **96**, 3101 (1963).
27. R. Hüttel and H. Dietl, *Chem. Ber.*, **98**, 1753 (1965).
28. R. Hüttel, H. Dietl, and H. Christ, *Chem. Ber.*, **97**, 2037 (1964).
29. H. C. Volger, *Ind. Eng. Chem., Prod. Res. Dev.*, **9**, 311 (1970).
30. H. S. Volger, *Rec. Trav. Chim. Pays-Bas*, **88**, 225 (1969).
31. B. M. Trost and T. J. Fullerton, *J. Am. Chem. Soc.*, **95**, 292 (1973).
32. A. D. Ketley and J. Braatz, *Chem. Commun.*, 169 (1968).
33. D. Morelli, R. Ugo, F. Conti, and M. Donati, *Chem. Commun.*, 801 (1967).
34. F. Conti, M. Donati, and G. F. Pregaglia, *J. Organomet. Chem.*, **30**, 421 (1971).
35. M. Donati and F. Conti, *Inorg. Nucl. Chem. Lett.*, **2**, 343 (1966).
36. B. M. Trost and P. E. Strege, *Tetrahedron Lett.*, 2603 (1974).
37. R. Hüttel and M. McNiff, *Chem. Ber.*, **106**, 1789 (1973).
38. B. M. Trost, P. E. Strege, L. Weber, T. J. Fullerton, and T. J. Dietsche, *J. Am. Chem. Soc.*, **100**, 3407 (1978).
39. B. M. Trost and P. J. Metzner, *J. Am. Chem. Soc.*, **102**, 3572 (1980).
40. D. N. Jones and S. D. Knox, *J. Chem. Soc., Chem. Commun.*, 165 (1975).
41. I. J. Harvie and F. J. McQuillin, *J. Chem. Soc., Chem. Commun.*, 747 (1978).
42. J. Y. Satoh and C. A. Horiuchi, *Bull. Chem. Soc. Jpn.*, **52**, 2653 (1979).
43. G. W. Parshall and G. Wilkinson, *Inorg. Chem.*, **1**, 896 (1962).
44. G. W. Parshall and G. Wilkinson, *Chem. Ind. (London)*, 261 (1962).
45. J. Tsuji and S. Imamura, *Bull. Chem. Soc. Jpn.*, **40**, 197 (1967).
46. J. Tsuji, S. Imamura, and J. Kiji, *J. Am. Chem. Soc.*, **86**, 4491 (1964).
47. A. Kasahara, K. Tanaka, and K. Asamiya, *Bull. Chem. Soc. Jpn.*, **40**, 351 (1967).
48. R. W. Howsam and F. J. McQuillin, *Tetrahedron Lett.*, 3667 (1968).
49. K. H. Henderson and F. J. McQuillin, *Chem. Commun.*, 15 (1978).
50. J. Y. Satoch and C. A. Horiuchi, *Bull. Chem. Soc. Jpn.*, **52**, 2653 (1979).
51. D. J. Collins, B. M. K. Gathhouse, W. R. Jackson, A. G. Kakos, and R. N. Timms, *Chem. Commun.*, 138 (1980).
52. R. Hüttel and H. Schmid, *Chem. Ber.*, **101**, 252 (1968).
53. L. Yu. Ukhim, V. I. Ilin, Zh. I. Orlova, N. G. Bokii, and Yu. T. Struchkov, *J. Organomet. Chem.*, **113**, 167 (1976).
54. B. L. Shaw, *Chem. Ind. (London)*, 1190 (1962).
55. M. Donati and F. Conti, *Tetrahedron Lett.*, 1219 (1966).
56. S. D. Robinson and B. L. Shaw, *J. Chem. Soc.*, 4806 (1963).
57. J. M. Rowe and D. A. White, *J. Chem. Soc., A*, 1451 (1967).
58. K. Dunne and F. J. McQuillin, *J. Chem. Soc., C*, 2196, 2200 (1970).

59. M. Takahashi, H. Suzuki, Y. Morooka, and T. Ikawa, *Chem. Lett.*, 53 (1979).
60. R. G. Schultz, *Tetrahedron Lett.*, 301 (1964).
61. M. S. Lupin, J. Powell, and B. L. Shaw, *J. Chem. Soc., A,* 1687 (1966).
62. R. G. Schultz, *Tetrahedron*, **20**, 2809 (1964).
63. M. S. Lupin and B. L. Shaw, *Tetrahedron Lett.*, 883 (1964).
64. K. Tsukiyama, Y. Takahashi, S. Sakai, and Y. Ishii, *J. Chem. Soc., A,* 3112 (1971).
65. A. D. Ketley and J. A. Braatz, *J. Organomet. Chem.*, **9**, P5 (1967).
66. T. Shono, T. Yoshimura, Y. Matsumura, and R. Oda, *J. Org. Chem.*, **33**, 876 (1968).
67. A. D. Ketley and J. Braatz, *Chem. Commun.*, 959 (1968).
68. R. Noyori and H. Takaya, *Chem. Commun.*, 525 (1969).
69. A. D. Ketley, J. A. Braatz, and J. Craig, *Chem. Commun.*, 1117 (1970).
70. P. Mushak and M. A. Battiste, *J. Organomet. Chem.*, **17**, P46 (1969).
71. M. A. Battiste, L. E. Friedrich, and R. A. Fiato, *Tetrahedron Lett.*, 45 (1979).
72. R A. Fiato, P. Mushak, and M. A. Battiste, *J. Chem. Soc., Chem. Commun.*, 869 (1975).
73. R. P. Hughes, D. E. Hunton, and K. Schumann, *J. Organomet. Chem.*, **169**, C37 (1979).
74. R. C. Larock and M. A. Mitchell, *J. Am. Chem. Soc.*, **98**, 6718 (1976).
75. R. C. Larock and M. A. Mitchell, *J. Am. Chem. Soc.*, **100**, 180 (1978).
76. S. Staicu, I. G. Dinulescu, F. Chiraleu, and M. Avram, *J. Organomet. Chem.*, **113**, C69 (1976).
77. S. Baba, T. Sobata, T. Ogura, and S. Kawaguchi, *Bull. Chem. Soc. Jpn.*, **47**, 2792 (1974).
78. Y. Tezuka, T. Ogura, and S. Kawaguchi, *Bull. Chem. Soc. Jpn.*, **42**, 443 (1969).
79. S. Baba, T. Sobata, T. Ogura, and S. Kawaguchi, *Inorg. Nucl. Chem. Lett.*, **8**, 605 (1972).
80. J. Tsuji, J. Kiji, and M. Morikawa, *Tetrahedron Lett.*, 1811 (1963).
81. D. Milstein, *J. Organomet. Chem.*, **1**, 888 (1982).
82. J. Tsuji, J. Kiji, and S. Hosaka, *Tetrahedron Lett.*, 605 (1964).
83. J. K. Stille and R. Divakarumi, *J. Org. Chem.*, **44**, 3474 (1979).
84. W. Funakoshi, T. Urasaki, and H. Fujimoto, *Jpn. Kokai*, 130, 714, 1975; *Chem. Abstr.*, **84**, 43356 (1976).
85. J. Tsuji and S. Hosaka, *J. Am. Chem. Soc.*, **87**, 4075 (1965).
86. J. Tsuji and T. Susuki, *Tetrahedron Lett.*, 3027 (1965).
87. T. Kajimoto, H. Takahashi, and J. Tsuji, *J. Organomet. Chem.*, **23**, 275 (1970).
88. T. Boschi and B. Crociani, *Inorg. Chim. Acta* **5**, 477 (1971).
89. Y. Itoh, T. Hirao, N. Ohta, and T. Saegusa, *Tetrahedron Lett.*, 1009 (1977).
90. J. Tsuji, H. Takahashi, and M. Morikawa, *Kogyo Kagaku Zasshi*, **69**, 920 (1966).
91. W. R. Jackson and J. U. Strauss, *Aust. J. Chem.*, **31**, 1073 (1978).
92. B. M. Trost, L. Weber, P. E. Strege, T. J. Fullerton, and T. J. Dietsche, *J. Am. Chem. Soc.*, **100**, 3416 (1978).
93. B. M. Trost, T. J. Dietsche, and T. J. Fullerton, *J. Org. Chem.*, **39**, 737 (1974).
94. B. M. Trost, W. P. Conway, P. E. Strege, and T. J. Dietsche, *J. Am. Chem. Soc.*, **96**, 7165 (1974).
95. B. M. Trost and L. Weber, *J. Am. Chem. Soc.*, **97**, 1611 (1975).
96. B. M. Trost and T. J. Fullerton, *J. Am. Chem. Soc.*, **95**, 292 (1973).
97. B. M. Trost and P. E. Strege, *J. Am. Chem. Soc.*, **97**, 2534 (1975).
98. B. M. Trost and L. Weber, *J. Org. Chem.*, **40**, 3617 (1975).
99. B. M. Trost, L. Weber, P. E. Strege, T. J. Fullerton, and T. J. Dietsche, *J. Am. Chem. Soc.*, **100**, 3426 (1978).
100. P. S. Manchand, H. S. Wong, and J. F. Blount, *J. Org. Chem.*, **43**, 4769 (1978).
101. B. Akermark, A. Ljungqvist, and M. Panunzio, *Tetrahedron Lett.*, **22**, 1055 (1981).
102. W. R. Jackson and J. U. G. Strauss, *Tetrahedron Lett.*, 2591 (1975).
103. D. J. Collins, W. R. Jackson, and R. N. Timms, *Tetrahedron Lett.*, 495 (1976).
104. W. R. Jackson and J. U. Strauss, *Aust. J. Chem.*, **30**, 553 (1977).
105. D. J. Collins, W. R. Jackson, and R. N. Timms, *Aust. J. Chem.*, **30**, 2167 (1977).
106. B. Akermark and A. Justand, *J. Organomet. Chem.*, **217**, C41 (1981).
107. L. S. Hegedus, W. H. Darlington, and C. E. Russell, *J. Org. Chem.*, **45**, 5193 (1980).
108. M. Moreo-Monas and A. Trius, *Tetrahedron Lett.*, **22**, 3109 (1981).
109. J. S. Temple and J. Schwartz, *J. Am. Chem. Soc.*, **102**, 7381 (1980).
110. M. Riediker and J. Schwartz, *Tetrahedron Lett.*, **22**, 4655 (1981).
111. J. S. Temple, M. Riediker, and J. Schwartz, *J. Am. Chem. Soc.*, **104**, 1310 (1982).

3. Carbon—carbon bond formation using η^3-allyl complexes

112. J. E. McMurry and J. R. Matz, *Tetrahedron Lett.*, **23**, 2723 (1982).
113. M. C. Aswraf, T. G. Burrowes, and W. R. Jackson, *Aust. J. Chem.*, **29**, 2643 (1976).
114. T. Ito, Y. Kindaichi, and Y. Takami, *Chem. Ind. (London)*, 83 (1980).
115. T. Hung, P. W. Jolly, and G. Wilke, *J. Organometl. Chem.*, **190**, C5 (1980).
116. J. Muzart, P. Pale, and J. P. Pete, *Chem. Commun.*, 668 (1981).
117. R. C. Larock, J. P. Burkhart, and K. Oertle, *Tetrahedron Lett.*, **23**, 1071 (1982).
118. J. Tsuji, *Acc. Chem. Res.*, **6**, 8 (1973).
119. J. Tsuji, *Adv. Organomet. Chem.*, **17**, 141 (1979).
120. S. Takahashi, T. Shibano, and N. Hagihara, *Tetrahedron Lett.*, 2451 (1967).
121. S. Takahashi, T. Shibano, and N. Hagihara, *Bull. Chem. Soc. Jpn.*, **41**, 454 (1968).
122. D. Medema and R. van Helden, *Rec. Trav. Chim. Pays-Bas*, **90**, 324 (1971).
123. D. Medema and R. van Helden, *Rec. Trav. Chim. Pays-Bas*, **90**, 304 (1971).
124. A. S. Astakhova, A. S. Berenblyum, L. G. Korableva, I. P. Lavrent'ev, B. G. Rogachev, and M. L. Khidekel, *Izv. Akad. Nauk SSSR, Ser. Khim.*, 1362 (1972); *Chem. Abstr.*, **75**, 88750b (1971).
125. W. Keim and H. Chung, *J. Org. Chem.*, **37**, 947 (1972).
126. J. Tsuji, *Bull. Chem. Soc. Jpn.*, **46**, 1896 (1973).
127. K. Takahashi, A. Miyake, and G. Hata, *Chem. Ind. (London)*, 488 (1971).
128. G. Hata, K. Takahashi, and A. Miyake, *J. Org. Chem.*, **36**, 2116 (1971).
129. K. Takahashi, A. Miyake, and G. Hata, *Bull. Chem. Soc. Jpn.*, **45**, 1183 (1972).
130. J. P. Haudegond, Y. Chauvin, and D. Commereuc, *J. Org. Chem.*, **44**, 3063 (1979).
131. Y. Tamaru, R. Suzuki, M. Kagotani, and Z. Yoshida, *Tetrahedron Lett.*, **21**, 3791 (1980).
131. K. Takahashi, A. Miyake, and G. Hata, *Bull. Chem. Soc. Jpn.*, **45**, 1183 (1972).
133. S. Watanabe, K. Suga, and T. Fujita, *Can. J. Chem.*, **51**, 848 (1973).
134. R. Baker and R. J. Popplestone, *Tetrahedron Lett.*, 3575 (1978).
135. T. Mitsuyasu, M. Hara, and J. Tsuji, *Chem. Commun.*, 345 (1971).
136. T. Mitsuyasu and J. Tsuji, *Tetrahedron*, **30**, 831 (1974).
137. Y. Tamaru, R. Suzuki, M. Kagotani, and Z. Yoshida, *Tetrahedron Lett.*, **21**, 3787 (1980).
138. P. V. Kunakova, F. V. Sharipova, G. A. Tolstikova, L. M. Zelenova, A. A. Panasenko, L. V. Spirikhin, and U. M. Dzhemilev, *Izv. Akad. Nauk SSSR*, 1833 (1980); *Chem. Abstr.*, **94**, 65408 (1981).
139. R. V. Kunakova, G. A. Tolstikov, N. Z. Baibulatova, S. S. Zlotskii, D. L. Rakhmankulov, V. V. Zorin, and U. M. Dzhemilev, *Zh. Org. Khim.*, **16**, 1775 (1980); *Chem. Abstr.*, **94**, 65060e (1981).
140. R. Baker, M. S. Nobbs, and D. T. Robinson, *J. Chem. Soc., Chem. Commun.*, 723 (1975).
141. R. Baker, M. S. Nobbs, and P. M. Winton, *J. Organomet. Chem.*, **137**, C43 (1977).
142. C. Moberg, *Tetrahedron Lett.*, **22**, 4827 (1981).
143. B. A. Patel, L. C. Kao, N. A. Cortese, J. V. Minkiewicz, and R. F. Heck, *J. Org. Chem.*, **44**, 918 (1979).
144. F. G. Stakem and R. F. Heck, *J. Org. Chem.*, **44**, 3584 (1980).
145. D. D. Bender, F. G. Stakem, and R. F. Heck, *J. Org. Chem.*, **47**, 1278 (1982).
146. S. Brewis and P. R. Hughes, *Chem. Commun.*, 157 (1965).
147. S. Hosaka and J. Tsuji, *Tetrahedron*, **27**, 3821 (1971).
148. J. Tsuji, Y. Mori, and M. Hara, *Tetrahedron*, **28**, 3721 (1972).
149. W. E. Billups, W. E. Walker, and T. C. Shield, *Chem. Commun.*, 1067 (1971).
150. J. Tsuji and H. Yasuda, *Bull. Chem. Soc. Jpn.*, **50**, 553 (1977).
151. K. Ohno, T. Mitsuyasu, and J. Tsuji, *Tetrahedron Lett.*, 67 (1971).
152. K. Ohno, T. Mitsuyasu, and J. Tsuji, *Tetrahedron*, **28**, 3075 (1972).
153. R. M. Manyik, W. E. Walker, K. E. Atkins, and E. S. Hammack, *Tetrahedron Lett.*, 3813 (1970).
154. P. Haynes, *Tetrahedron Lett.*, 3687 (1970).
155. A. Musco, *Inorg. Chim. Acta*, **11**, L11 (1974).
156. R. Bortolin, G. Gatti, and A. Musco, *J. Mol. Catal.*, **14**, 95 (1982).
157. K. Ohno and J. Tsuji, *Chem. Commun.*, 247 (1971).
158. J. Kiji, K. Yamamoto, H. Tomita, and J. Furukawa, *J. Chem. Soc., Chem. Commun.*, 506 (1974).
159. H. Minematsu, S. Takahashi, and N. Hagihara, *J. Chem. Soc., Chem. Commun.*, 466 (1975).

160. Y. Sasaki, Y. Inoue, and H. Hashimoto, *J. Chem. Soc., Chem. Commun.*, 605 (1976).
161. Y. Sasaki, Y. Inoue, and H. Hashimoto, *Bull. Chem. Soc. Jpn.*, **51**, 2375 (1978).
162. A. Musco, C. Perego, and V. Tartiari, *Inorg. Chim. Acta*, **28**, L147 (1978).
163. A. Musco, *J. Chem. Soc., Perkin Trans. 1*, 693 (1980).
164. Y. Inoue, S. Sekiya, Y. Sasaki, and H. Hashimoto, *Yukigosei Kyokaishi*, **36**, 328 (1978).
165. D. R. Coulson, *J. Org. Chem.*, **38**, 1483 (1973).
166. G. D. Shier, *J. Organomet. Chem.*, **10**, P15 (1967).
167. D. R. Coulson, *J. Org. Chem.*, **37**, 1253 (1972).
168. H. Siegel, H. Hopf, A. Germer, and P. Binger, *Chem. Ber.*, **111**, 3112 (1978).
169. A. Dohring and P. W. Jolly, *Tetrahedron Lett.*, **21**, 3021 (1980).
170. K. Takahashi, G. Hata, and A. Miyake, *Bull. Chem. Soc. Jpn.*, **46**, 1012 (1973).
171. K. Takahashi, A. Miyake, and G. Hata, *Bull. Chem. Soc. Jpn.*, **45**, 230 (1972).
172. G. Hata, K. Takahashi, and A. Miyake, *Chem. Commun.*, 1392 (1970).
173. K. E. Atkins, W. E. Walker, and R. M. Manyik, *Tetrahedron Lett.*, 3821 (1970).
174. M. M. Manas and A. Trius, *Tetrahedron*, **37**, 3009 (1981).
175. H. Onoue, I. Moritani, and S. Murahashi, *Tetrahedron Lett.*, 121 (1973).
176. B. M. Trost and T. Matsumura, *J. Org. Chem.*, **42**, 2036 (1977).
177. B. M. Trost, P. B. Bernstein, and P. C. Funfschilling, *J. Am. Chem. Soc.*, **101**, 4378 (1979).
178. B. M. Trost and T. R. Verhoeven, *J. Am. Chem. Soc.*, **98**, 630 (1976).
179. B. M. Trost and T. R. Verhoeven, *J. Am. Chem. Soc.*, **100**, 3435 (1978).
180. B. M. Trost and N. R. Schmuff, *Tetrahedron Lett.*, **22**, 2999 (1981).
181. J. Backvall, R. E. Nordberg, E. E. Bjorkman, and C. Moberg, *J. Chem. Soc., Chem. Commun.*, 943 (1980).
182. J. C. Fiaud and J. L. Malleron, *Tetrahedron Lett.*, **22**, 1399 (1981).
183. B. M. Trost, T. R. Verhoeven, and J. M. Fortunak, *Tetrahedron Lett.*, 2301 (1979).
184. B. M. Trost and T. P. Klun, *J. Am. Chem. Soc.*, **103**, 1864 (1981).
185. B. M. Trost and T. P. Klun, *J. Am. Chem. Soc.*, **101**, 6756 (1979).
186. B. M. Trost and T. R. Verhoeven, *J. Am. Chem. Soc.*, **99**, 3867 (1977).
187. B. M. Trost and T. R. Verhoeven, *J. Am. Chem. Soc.*, **101**, 1595 (1979).
188. B. M. Trost and T. R. Verhoeven, *Tetrahedron Lett.*, 2275 (1978).
189. Y. Kitagawa, A. Itoh, S. Hashimoto, H. Yamamoto, and H. Nozaki, *J. Am. Chem. Soc.*, **99**, 3865 (1977).
190. J. Tsuji, Y. Kobayashi, H. Kataoka, and T. Takahashi, *Tetrahedron Lett.*, **21**, 1475 (1980).
191. B. M. Trost and T. A. Runge, *J. Am. Chem. Soc.*, **103**, 2485, 7550, 7559 (1981).
192. B. M. Trost, T. A. Runge, and L. N. Jungheim, *J. Am. Chem. Soc.*, **102**, 2840 (1980).
193. S. A. Godleski, J. D. Meinhart, D. J. Miller, and S. V. Wallendael, *Tetrahedron Lett.*, **22**, 2247 (1981).
194. S. A. Godleski and R. S. Valpey, *J. Org. Chem.*, **47**, 381 (1982).
195. J. P. Genet, F. Piau, and J. Ficini, *Tetrahedron Lett.*, **21**, 3183 (1980).
196. J. P. Genet and F. Piau, *J. Org. Chem.*, **46**, 2414 (1981).
197. J. P. Genet, M. Balabane, and Y. Legras, *Tetrahedron Lett.*, **23**, 331 (1982).
198. Y. Morizawa, K. Oshima, and H. Nozaki, *Tetrahedron Lett.*, **23**, 2871 (1982).
199. B. M. Trost and F. W. Gowland, *J. Org. Chem.*, **44**, 3448 (1979).
200. B. M. Trost and D. P. Curran, *J. Am. Chem. Soc.*, **102**, 5699 (1980).
201. L. V. Dunkerton and A. J. Serino, *J. Org. Chem.*, **47**, 2812 (1982).
202. B. M. Trost and D. M. T. Chan, *J. Am. Chem. Soc.*, **101**, 6429, 6432 (1979).
203. B. M. Trost and D. M. T. Chan, *J. Am. Chem. Soc.*, **102**, 6359 (1980).
204. B. M. Trost and D. M. T. Chan, *J. Am. Chem. Soc.*, **103**, 5972 (1981).
205. B. M. Trost and D. M. T. Chan, *J. Am. Chem. Soc.*, **104**, 3733 (1982).
206. T. Hirao, J. Enda, Y. Ohshiro, and T. Agawa, *Tetrahedron Lett.*, **22**, 3079 (1981).
207. J. Tsuji, H. Ueno, Y. Kobayashi, and H. Okumoto, *Tetrahedron Lett.*, **22**, 2573 (1981).
208. P. A. Wade, S. D. Morrow, S. A. Hardinger, M. S. Saft, and H. R. Hinney, *J. Chem. Soc., Chem. Commun.*, 287 (1980).
209. P. A. Wade, H. R. Hinney, N. V. Amin, P. D. Vail, S. D. Morrow, S. A. Hardinger, and M. S. Saft, *J. Org. Chem.*, **46**, 765 (1981).
210. P. W. Wade, S. D. Morrow, and S. A. Hardinger, *J. Org. Chem.*, **47**, 365 (1982).
211. P. Aleksandorozicz, H. Piotrowska, and W. Sas, *Tetrahedron*, **38**, 1321 (1982).

3. Carbon—carbon bond formation using η^3-allyl complexes

212. J. Tsuji, I. Shimizu, I. Minami, and K. Ohashi, *Tetrahedron Lett.*, **23**, 4809 (1982).
213. J. Tsuji, H. Kataoka, and Y. Kobayashi, *Tetrahedron Lett.*, **22**, 2575 (1981).
214. B. M. Trost and G. A. Molander, *J. Am. Chem. Soc.*, **103**, 5969 (1981).
215. I. Shimizu, T. Yamada, and J. Tsuji, *Tetrahedron Lett.*, **21**, 3199 (1980).
216. T. Tsuda, Y. Chujo, S. Nishi, K. Tawara, and T. Saegusa, *J. Am. Chem. Soc.*, **102**, 6381 (1980).
217. J. Tsuji, I. Minami, I. Shimizu, *Tetrahedron Lett.*, **24**, 1973 (1983).
218. J. C. Fiaud and J. L. Malleron, *J. Chem. Soc., Chem. Commun.*, 1159 (1981).
219. B. M. Trost and E. Keinan, *Tetrahedron Lett.*, **21**, 2591 (1980).
220. B. M. Trost and E. Keinan, *Tetrahedron Lett.*, **21**, 2595 (1980).
221. I. P. Beletskaya, *Izv. Akd. Nauk SSSR*, 2414 (1981).
222. T. Hayashi, M. Konishi, K. Yokota, and M. Kumada, *J. Chem. Soc., Chem. Commun.*, 313 (1981).
223. E. Negishi, H. Chatterjee, and H. Matsushita, *Tetrahedron Lett.*, **22**, 3737 (1981).
224. H. Matsushita and E. Negishi, *J. Chem. Soc., Chem. Commun.*, 160 (1982).
225. N. Miyaura, Y. Tanable, H. Suginome, and A. Suzuki, *J. Organomet. Chem.*, **233**, C13 (1982).
226. J. C. Fiaud and J. L. Malleron, *Tetrahedron Lett.*, **21**, 4437 (1980).
227. B. M. Trost, N. R. Schmuff, and M. J. Miller, *J. Am. Chem. Soc.*, **102**, 5979 (1980).
228. R. Tamura and L. S. Hegedus, *J. Am. Chem. Soc.*, **104**, 3727 (1982).
229. N. Ono, I. Hamamoto, and A. Kaji, *J. Chem. Soc. Chem. Commun.*, 820 (1982).
230. J. Tsuji, J. Kiji, S. Imamura, and M. Morikawa, *J. Am. Chem. Soc.*, **86**, 4350 (1964).
231. J. F. Knifton, *J. Organomet. Chem.*, **188**, 223 (1980).
232. J. Tsuji, K. Sato, and H. Okumoto, *Tetrahedron Lett.*, **23**, 5189 (1982); *J. Org. Chem.*, **49**, 1341 (1984).
233. W. T. Dent, R. Long, and G. H. Whitfield, *J. Chem. Soc.*, 1588 (1964).
234. D. Medema, R. van Helden, and C. F. Kohll, *Inorg. Chim. Acta*, **3**, 255 (1969).
235. J. Yoshida, K. Tamao, M. Takahashi, and M. Kumada, *Tetrahedron Lett.*, 2161 (1978).
236. J. Yoshida, K. Tamao, H. Yamamoto, T. Kakui, T. Uchida, and M. Kumada, *Organometallics*, **1**, 542 (1982).
237. N. Miyaura, T. Yano, and A. Suzuki, *Tetrahedron Lett.*, **21**, 2865 (1980).
238. J. Godschalx and J. K. Stille, *Tetrahedron Lett.*, **21**, 2599 (1980).
239. Y. Hayashi, M. Riediker, J. S. Temple, and J. Schwartz, *Tetrahedron Lett.*, **22**, 2629 (1981).
240. T. Kitazume and N. Ishikawa, *Chem. Lett.*, 137 (1982).
241. R. F. Heck, *J. Am. Chem. Soc.*, **90**, 5531 (1968).
242. R. F. Heck, *J. Organomet. Chem.*, **33**, 399 (1977).
243. D. E. Bergstrom, J. L. Ruth, and P. Warwick, *J. Org. Chem.*, **46**, 1432 (1981).
244. K. Kaneda, T. Uchiyama, Y. Fujiwara, T. Imanaka, and S. Teranishi, *J. Org. Chem.*, **44**, 55 (1979).
245. T. Imanaka, T. Kimura, K. Kaneda, and S. Teranishi, *J. Mol. Catal.*, **9**, 103 (1980).

Part 3. Other η^3-allyl transition metal complexes

FUMIE SATO

Department of Chemical Engineering, Tokyo Institute of Technology, Meguro, Tokyo 152, Japan

I. OTHER η^3-ALLYL TRANSITION METAL COMPLEXES 201
II. REFERENCES 204

I. OTHER η^3-ALLYL TRANSITION METAL COMPLEXES

η^3-Allyltitanium(III) complexes, [Cp$_2$Ti(η^3-allyl)], are readily prepared by reaction of [Cp$_2$TiCl$_2$] with allyl Grignard reagents[1] or with an alkyl Grignard reagent and diolefins (reactions 1 and 2)[2]. The titanium—allyl bond in these complexes is very reactive and a variety of reaction types are observed.

$$Cp_2TiCl_2 + 2\ RCH=CHCH_2MgBr \longrightarrow \left[Ti(\eta^3\text{-}CH_2CHCHR)Cp_2\right] \quad (1)$$

$$Cp_2TiCl_2 + CH_2=C(R)CH=CH_2 + 2\ Pr^iMgBr \longrightarrow \left[Ti(\eta^3\text{-}CH_2CRCHMe)Cp_2\right] \quad (2)$$

The η^3-allyltitanium complexes undergo a highly regio- and stereo-selective addition reaction with aldehydes or ketones under mild conditions to produce, after hydrolysis, homoallyl alcohols in excellent yields (reaction 3)[3,4]. The more substituted carbon atom of the allyl group is attached by the carbonyl carbon, and in the reaction with aldehydes the product possessing the *threo*-configuration is obtained with high preference. The complexes also show high chemoselectivity towards the carbonyl group of aldehydes or ketones, other functionalities being unaffected and α,β-unsaturated systems undergoing exclusive 1,2-addition. [Cp$_2$TiCl$_2$] is recovered in high yields after the reaction.

$$\left[Ti(\eta^3\text{-}CH_2CHCHMe)Cp_2\right] + PhCHO \longrightarrow \left[Cp_2TiOCH(Ph)CH(Me)CH=CH_2\right]$$

$$\xrightarrow[93\%]{HCl/O_2} CH_2=CHCH(Me)CH(OH)Ph + \left[Cp_2TiCl_2\right] \quad (3)$$

(*threo / erythro* = 95 : 5)

Silyltitanation of 1,3-dienes by [$Cp_2TiSiMe_2Ph$] followed by treatment of the resulting silylated η^3-allyltitanium complexes with aldehydes or ketones also affords the corresponding homoallyl alcohols (reaction 4)[5].

$$[Cp_2Ti(SiMe_2Ph)] + CH_2=C(Me)CH=CH_2 \longrightarrow [Ti(\eta^3\text{-}CH_2C(Me)CHCH_2SiMe_2Ph)Cp_2]$$

$$\xrightarrow[\text{ii. } H_3O^+]{\text{i. } CH_3CHO} CH_2=C(Me)CH(CH_2SiMe_2Ph)CH(OH)Me \quad (4)$$

η^3-Trimethylsilylallyltitanium complex, formed *in situ* by reaction of [Cp_2TiCl] and trimethylsilylallyllithium, reacts with various aldehydes such as primary, secondary, or tertiary aldehydes or aryl aldehydes to yield ($\pm$) – (R, S)-3-trimethylsilyl-4-hydroxyalk-1-enes in excellent yields. The resulting products can be deoxysilylated stereospecifically by the methods of Hudrlik and Peterson[6] to yield either E- or Z-terminal dienes (reaction 5)[6a]. As the Wittig reaction to produce 1,3-dienes is very complicated, this reaction seems to be useful as a simple alternative.

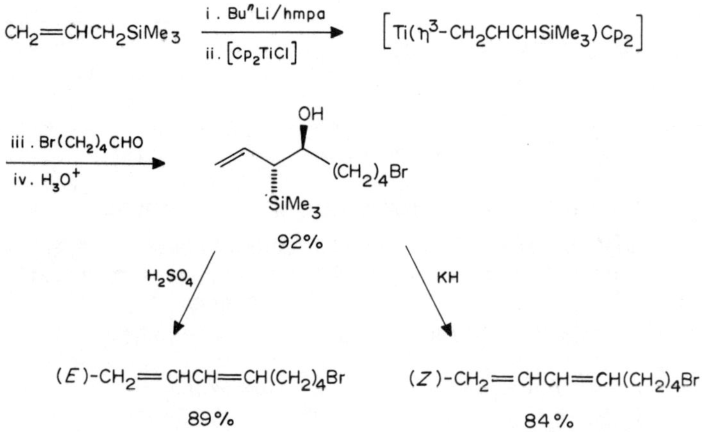

$$CH_2=CHCH_2SiMe_3 \xrightarrow[\text{ii. } [Cp_2TiCl]]{\text{i. } Bu^nLi/hmpa} [Ti(\eta^3\text{-}CH_2CHCHSiMe_3)Cp_2]$$

$$\xrightarrow[\text{iv. } H_3O^+]{\text{iii. } Br(CH_2)_4CHO} \underset{92\%}{\overset{OH}{\text{CH}_2=CH-CH(SiMe_3)-CH(OH)-(CH_2)_4Br}} \quad (5)$$

$H_2SO_4 \swarrow \qquad \searrow KH$

(E)-$CH_2=CHCH=CH(CH_2)_4Br$ (Z)-$CH_2=CHCH=CH(CH_2)_4Br$

89%   84%

The η^3-allyltitanium complexes react with carbon dioxide under ordinary pressure at room temperature to form β,γ-unsaturated carboxylic acids[4,7]. Again, insertion takes place in the most substituted part of the allyl moiety. Use of a η^3-allyltitanium complex containing a chiral cyclopentadienyl ligand to effect this carbon dioxide fixation lead to asymmetric induction. Thus, the chiral titanium complex, prepared *in situ* from dichlorobis(neomenthylcyclopentadienyl)titanium[8], butadiene, and isopropylmagnesium bromide, reacted with carbon dioxide to give (S) – (+)-2-methylbut-3-enoic acid in 18.9% optical yield[7].

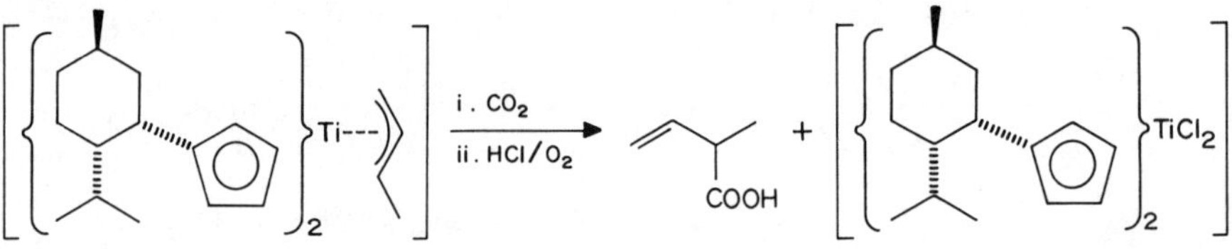

$$[\{(neomenthylCp)_2\}Ti\text{---}\|] \xrightarrow[\text{ii. } HCl/O_2]{\text{i. } CO_2} CH_2=CH\text{-}CH(Me)\text{-}COOH + [\{(neomenthylCp)_2\}TiCl_2] \quad (6)$$

3. Carbon—carbon bond formation using η^3-allyl complexes

Insertion into a titanium—allyl bond of the η^3-allyltitanium complexes with PhNCO, PhN=CHPh, and CH$_3$CN proceeds analogously to afford, after hydrolysis, the corresponding amides, amines, and ketones, respectively (reactions 7–9)[4].

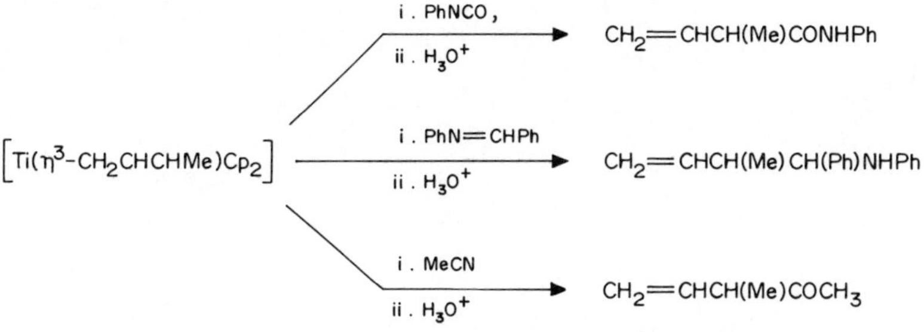

$$[\text{Ti}(\eta^3\text{-CH}_2\text{CHCHMe})\text{Cp}_2] \xrightarrow{\begin{array}{c}\text{i. PhNCO,}\\ \text{ii. H}_3\text{O}^+\end{array}} \text{CH}_2\text{=CHCH(Me)CONHPh} \quad (7)$$

$$\xrightarrow{\begin{array}{c}\text{i. PhN=CHPh}\\ \text{ii. H}_3\text{O}^+\end{array}} \text{CH}_2\text{=CHCH(Me)CH(Ph)NHPh} \quad (8)$$

$$\xrightarrow{\begin{array}{c}\text{i. MeCN}\\ \text{ii. H}_3\text{O}^+\end{array}} \text{CH}_2\text{=CHCH(Me)COCH}_3 \quad (9)$$

The reaction of the η^3-allyltitanium complexes with allylic halides, in contrast to η^3-allylnickel halide complexes, results in an oxidative addition reaction with formation of η^1-allyltitanium complexes and the dimers of the coordinated allyl ligand. Since the reaction proceeds essentially quantitatively, it offers a convenient synthesis of the [Cp$_2$TiX(η^1-allyl)] η^1-allyltitanium complexes. The resulting η^1-allyltitanium complexes add to aldehydes with high *threo*-selectivity wherein the halogen ligand affects the diastereoselectivity (reaction 10)[9].

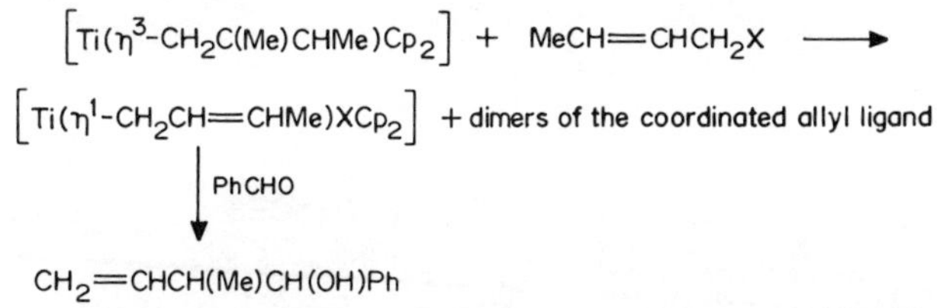

$$[\text{Ti}(\eta^3\text{-CH}_2\text{C(Me)CHMe})\text{Cp}_2] + \text{MeCH=CHCH}_2\text{X} \longrightarrow$$

$$[\text{Ti}(\eta^1\text{-CH}_2\text{CH=CHMe})\text{XCp}_2] + \text{dimers of the coordinated allyl ligand}$$

$$\downarrow \text{PhCHO}$$

CH$_2$=CHCH(Me)CH(OH)Ph
threo/erythro ratio:
X = Cl, Br, I   (10)
   60:40 100:0 94:6

[Cp$_2$TiCl$_2$] is an effective catalyst for regioselective isoprene insertion into an allyl—magnesium bond. The actual reagent is probably an η^3-allyltitanium complex. Various natural terpenes have been synthesized using this reaction (reaction 11)[10].

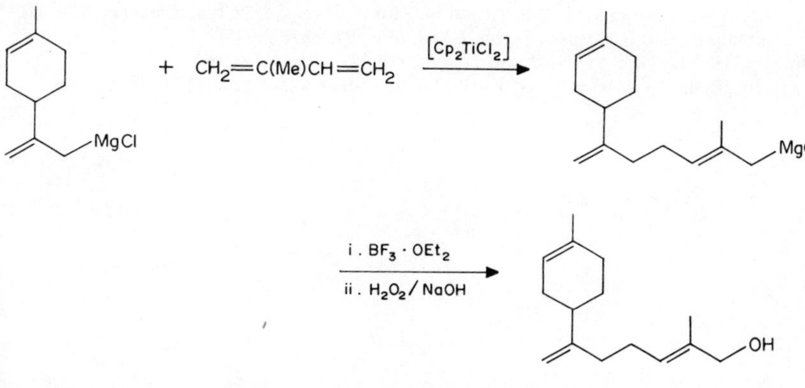

(11)

1,3-Dienes react with propylmagnesium bromide in the presence of a catalytic amount of $[Cp_2TiCl_2]$ to form allyl-Grignard reagents (reaction 12)[11]. It was suggested that the reaction involves the intermediacy of an η^3-allyltitanium complex.

$$CH_2=C(Me)CH=CH_2 + Pr^nMgCl \xrightarrow{[Cp_2TiCl_2]} CH_3CH=C(Me)CH_2MgCl + MeCH=CH_2$$

$$\downarrow Me_2CO$$

$$CH_2=C(Me)CH(Me)C(OH)Me_2$$
$$95\% \qquad\qquad (12)$$

Alkylation of allylic substrates (chlorides, acetates, formate) with diethyl malonate anion is catalysed by the η^3-allyliron complex $[(\eta^3\text{-crotyl})Fe(CO)_2NO]$, pre-formed or formed *in situ* by reaction of $Na^+Fe(CO)_3NO]^-$ and allylic substrates (reaction 13). $[(\eta^3\text{-Crotyl})Co(CO)_3]$ is also an active catalyst[12].

$$CH_2=CHCH(Me)Cl + {}^-CH(COOEt)_2 \xrightarrow[84\%]{[(\eta^3\text{-crotyl})Fe(CO)_2NO]} CH_2=CHCH(Me)CH(CO_2Et)_2 +$$
$$74$$

$$MeCH=CHCH_2CH(CO_2Et)_2$$
$$26 \qquad\qquad (13)$$

II. REFERENCES

1. H. A. Martin and F. Jellinek, *J. Organomet. Chem.*, **8**, 115 (1967).
2. H. A. Martin and F. Jellinek, *J. Organomet. Chem.*, **12**, 149 (1968).
3. F. Sato, S. Iijima, and M. Sato, *Tetrahedron Lett.*, **22**, 243 (1981).
4. B. Klei, J. H. Teuben, and H. J. L. Meijer, *J. Chem. Soc., Chem. Commun.*, 342 (1981).
5. H. Tamao, M. Akita, R. Kanatani, N. Ishida, and M. Kumada, *J. Organomet. Chem.*, **226**, C9 (1982).
6. P. F. Hudrlik and D. Peterson, *J. Am. Chem. Soc.*, **97**, 1464 (1975).
6a. F. Sato, Y. Suzuki, and M. Sato, *Tetrahedron Lett.*, 4589 (1982).
7. F. Sato, S. Iijima, and M. Sato, *J. Chem. Soc., Chem. Commun.*, 180 (1981).
8. E. Cesarotti and H. B. Kagan, *J. Organomet. Chem.*, **162**, 297 (1978).
9. F. Sato, K. Iida, S. Iijima, H. Moriya, and M. Sato, *J. Chem. Soc., Chem. Commun.*, 1140 (1981).
10. S. Akutagawa and S. Otsuka, *J. Am. Chem. Soc.*, **97**, 6870 (1975).
11. F. Sato, H. Ishikawa, and M. Sato, *Tetrahedron Lett.*, **21**, 365 (1980).
12. J. L. Roustan, J. Y. Merour, and F. Houlihan, *Tetrahedron Lett.*, 3721 (1979).

CHAPTER 4

Olefin oligomerization

OLAV-T. ONSAGER

Laboratory of Industrial Chemistry, Norwegian Institute of Technology, University of Trondheim, N-7034 Trondheim-NTH, Norway

JON EIGILL JOHANSEN

Division of Applied Chemistry, SINTEF, N-7034 Trondheim-NTH, Norway

I. INTRODUCTION.	206
II. SCOPE.	206
III. THE OLEFIN BOND	206
IV. PRIMARY PRODUCTS OF CATALYTIC DIMERIZATION—GENERAL ASPECTS	208
V. BASE CATALYSIS	211
A. Dimerization of Propene.	211
B. Ethylation of Olefins	213
VI. ACID CATALYSIS	214
A. Oligomerization of Propene.	214
B. Dimerization of Isobutene	215
VII. ORGANOMETALLIC SYNTHESIS	215
VIII. COORDINATIVE TRANSITION METAL COMPLEX CATALYSIS.	219
A. Introduction.	219
B. General Mechanistic Aspects	220
1. Olefin activation via complex formation	220
2. The migratory insertion of olefins	225
3. β-H elimination.	227
4. The rate of reaction	228
C. The Mechanism of Coordinative Transition Metal Catalysis.	231
1. Nichel complex catalysis.	231
a. The effect of Lewis acids	233
b. The effect of Lewis bases	236
c. Kinetics	249
2. Rhodium complex catalysis.	251
3. Titanium complex catalysis .	252
4. Tantalum complex catalysis	254
IX. REFERENCES.	255

I. INTRODUCTION

Oligomerization of olefins is a general term applied to the chemical process of converting lower olefins into a higher molecular weight olefinic product. From a chemical and

$$n(C=C) \to C_2 \cdot \text{olefin} \tag{1}$$

mechanistic point of view, oligomerization is closely related to polymerization, the only difference being the number of monomer units, n, incorporated into the final product. The borderline between the two types of reactions is not sharply defined. When n is large we move into the area of polymerization. For all the systems which will be discussed in this chapter n is less than 20. By far the most important type of oligomerization is dimerization, where $n = 2$. The dimerization of lower olefins is a thermodynamically feasible process accompanied by significant overall decrease in free energy (ΔG^0). From a practical point of view the conversion of lower olefins into higher molecular weight products at temperatures below 200 °C may be regarded as irreversible. Despite the fact that oligomerization is indeed a feasible process, pure ethene, propene, butene, and mixtures thereof are almost indefinitely stable at reasonable temperatures. In the absence of a suitable catalyst oligomerization is a very slow process. However, a large number of active anionic, cationic, and coordination catalysts are known, some of which will catalyse this type of carbon—carbon (C—C) bond formation at high rates even at temperatures well below 0 °C.

Of special interest is the stereoselective coordination catalysis observed with homogeneous transition metal complexes, where the mode of linking of monomer units can be widely controlled by careful selection of the complexing ligands. In our opinion detailed studies of the mechanism in this type of system will contribute significantly to the understanding of the true nature of catalytic C—C bond formation.

II. SCOPE

The literature concerning the dimerization of alkenes was reviewed by Fel'dblyum and Obeschalova in 1968[1] and by Lefebvre and Chauvin in 1970[2].

The scope of this review is limited to simple olefins having one olefinic bond and the conversion of such substrates into open-chain olefinic oligomers. The dimerization of olefins by thermal treatment through the 'ene' reaction is outside the scope of this review, as is the photochemically initiated cyclodimerization of olefins. Anionic and cationic catalyst systems as well as the organometallic synthesis reaction as developed by Ziegler will briefly be discussed.

The main emphasis of this chapter will be on the attractive method of achieving olefin oligomerization via homogeneous transition metal coordination complex catalysis. This type of catalytic oligomerization was intensively investigated during the 1960s and a large number of patents and publications have appeared on the subject. The initial technical developments are covered in the reviews in refs. 1 and 2. This analysis will therefore not deal exhaustively with all the existing literature but will concentrate mainly on newer developments and information related to the reaction mechanism and the chemical nature of catalytically active species.

Both Lewis acids and Lewis bases markedly influence the activity and the selectivity of transition metal complex catalysts. The origin of these effects will be analysed.

III. THE OLEFIN BOND

The bonding orbitals of ethene are made up of a π-bond and a 'σ-framework' consisting of five bonds composed of the 1s hydrogen orbitals and the $2s$, $2p_x$, and $2p_y$ orbitals of the two carbon atoms. The π-bond, which is higher in energy than the σ-bonding orbitals, is made up by the overlap of the two p_z orbitals of the carbon atoms (Figure 1). The π-overlap

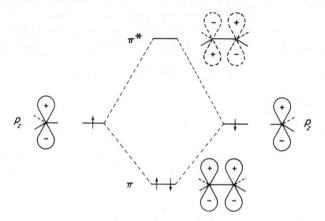

FIGURE 1. π-Bond MO diagram

strengthens the C—C bond in ethene compared with that of ethane, but is much less effective in lowering the energy than a σ-overlap.

The π-electron density distribution of ethene is symmetrical. In propene and higher α-olefins the olefin bond is polarized. For instance, owing to the electron repulsive effect of the methyl group, the relative π-electron density distribution in propene is:

$$\underset{\text{CH}_3\text{—CH}=\text{CH}_2}{\overset{0.972 \quad\quad 1.043}{}}$$

(1)

The bonding and antibonding π-MOs of ethene and propene are illustrated in Figure 2.

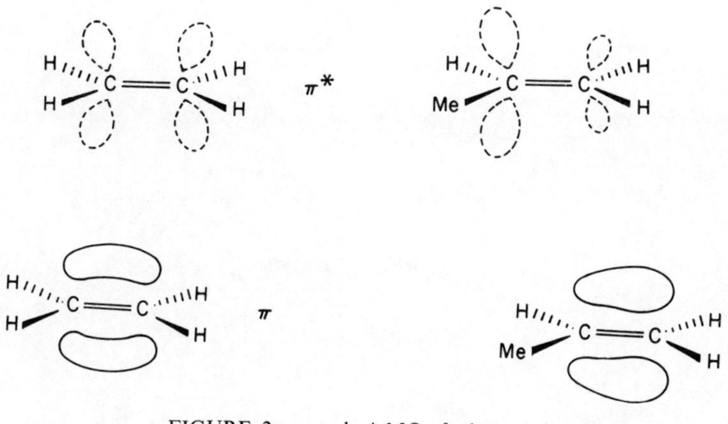

FIGURE 2. π and π* MO of ethene and propene

IV. PRIMARY PRODUCTS OF CATALYTIC DIMERIZATION— GENERAL ASPECTS

The catalytic dimerization of olefins may yield a number of different products, depending on the catalyst used and secondary reactions that might take place. Possible open-chain C_4, C_5, and C_6 olefins and their abbreviations are given in Table 1.

The primary products of anionic dimerization are mainly formed via allylic anion intermediates which, in accordance with Markownikoff's rule, add to the second

TABLE. 1. Possible C_4, C_5, and C_6 olefin products

Carbon structure of product	Abbreviation	B.p. (°C) (760 mmHg)
C=CCC	B1	−6.3
CC=CC	B2(Z/E)	3.7/0.88
C=CCCC	P1	29.96
CC=CCC	P2(Z/E)	36.9/33.35
C=CCC C	2MB1	31.16
CC=CC C	2MB2	38.57
C=CCC C	3MB1	20
C=CCCC	H1	63.35
CC=CCCC	H2(Z/E)	68.84/68 (750 mmHg)
CCC=CCC	H3(Z/E)	66.44/67.08
C=CCCC C	2MP1	60.7
CC=CCC C	2MP2	67.29
C=CCCC C	3MP1	51.14
CC=CCC C	3MP2	70.45
CC=CCC C	4MP2(Z/E)	56.3/58.55
C=CCCC C	4MP1	53.88
CC || C=CCC	2,3DMB1	55.67
CC=CC C C	2,3DMB2	73.2
CC | C=CCC	2EB1	64.7

monomer unit. In propene dimerization, where $R = H$, the primary product of reaction is 4MP1, and the reaction can be highly selective.

$$RCH_2CH{=}CH_2 \xrightarrow{-H^+} R\bar{C}HCH{=}CH_2 \xrightarrow{RCH_2CH{=}CH_2} \underset{R}{\underset{|}{RCH_2\overset{\overset{\displaystyle ^-CH_2}{|}}{C}HCHCH{=}CH_2}}$$

$$\downarrow +H^+ \qquad (2)$$

$$\underset{R}{\underset{|}{RCH_2\overset{\overset{\displaystyle Me}{|}}{C}HCHCH{=}CH_2}}$$

The acid-catalysed dimerization of olefins will, however, produce a number of isomeric products. For instance, the dimerization of propene via cationic mechanisms will, in addition to 4MP and 2MP, yield 3MP as a result of intermediate carbonium ion rearrangements. Further, acid catalysts tend to give products with a fairly broad molecular weight distributions, while anionic and coordinative systems often selectively yield dimer products.

During the coordinative dimerization of ethene, B1 is the single primary product (Scheme 1). When higher olefins are used, the reaction scheme becomes more complex owing to the stereochemistry of the insertion reaction. Metal complex hydride and alkyl

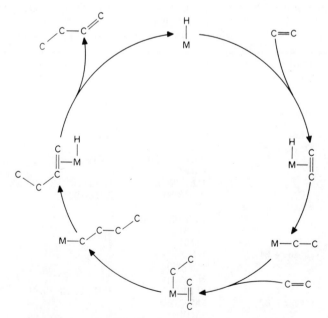

SCHEME 1. Dimerization of ethene via coordinative catalysis

groups may add to the C_α as well as the C_β of the complexed olefin (equation 3). Based on the formalism that H and R are added to the olefin in the form of negative (nucleophilic)

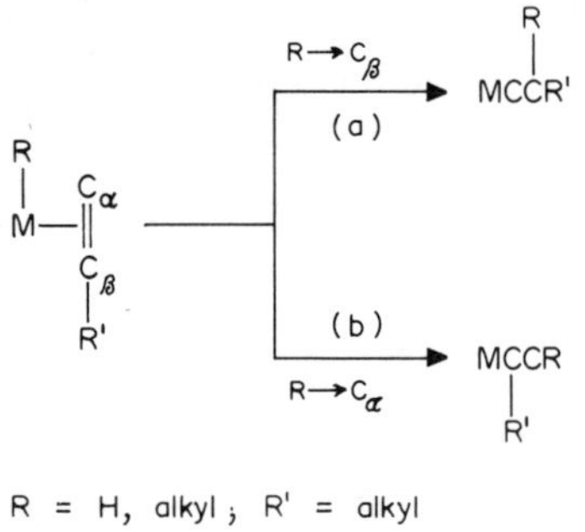

$$\begin{array}{c} R \\ | \\ M—C_\alpha \\ \| \\ C_\beta \\ | \\ R' \end{array} \begin{array}{c} \xrightarrow{R \to C_\beta} \text{(a)} \\ \\ \xrightarrow{R \to C_\alpha} \text{(b)} \end{array} \begin{array}{c} R \\ | \\ MCCR' \\ \\ \\ MCCR \\ | \\ R' \end{array} \qquad (3)$$

R = H, alkyl; R' = alkyl

groups, the reactions 3a and 3b are often referred to in the literature as either Markownikoff (reaction 3a) or anti-Markownikoff (reaction 3b) addition. In this chapter, however, we prefer to use the terms $R \to C_\beta$ or $R \to C_\alpha$ addition since, from a mechanistic point of view, the migratory insertion reactions taking place within the electron orbital sphere of a transition metal complex are completely different in nature from the free ionic addition of H^+X^- to olefinic bonds.

The different modes of insertion as indicated in reactions 3 combined with the mode of reaction as outlined for ethene in Scheme 1 give rise to primary dimer products with different structures. The key to controlling the selectivity with respect to the carbon chain structure of the product lies in controlling the mode of the insertion reaction. Whereas ethene yields a single dimer product, B1, the dimerization of propene gives H, 2MP, 4MP, and 2,3DMB isomers (Table 2).

An inherent feature of the coordination catalysts is their ability to isomerize the product with respect to Z/E configuration as well as the position of the double bond. From a mechanistic point of view (Z)-olefins are the most likely primary product, but owing to the

TABLE. 2. Primary and isomerized products obtained via coordinative mechanisms

Monomer	Primary products	Isomerized products
Ethene	B1	B2(Z/E)
Propene	2MP1	
	4MP2(Z/E)	2MP2
	4MP1	
	H2(Z/E)	H3(Z/E)
	2,3DMB1	2,3DMB2
Ethene + propene	P1	
	P2(Z/E)	
	2MB1	2MB2
	3MB1	
Ethene + B1	H1	
	H2(Z/E)	
	H3(Z/E)	
	2EB1	3MP2(Z/E) 3MP1
Ethene + B2	3MP2(Z/E)	
	3MP1	

high rate of Z/E conversion, equilibrated mixtures are normally obtained. The double bond isomerization is, however, a slower process and can in many systems be controlled to such a degree that conclusive information regarding the chemical structure of the primary products is obtained.

V. BASE CATALYSIS

The first anionic oligomerization of lower olefins was reported in 1956 using sodium/anthracene to initiate the reaction at elevated temperatures, 280–300 °C[3]. As indicated by the high reaction temperature, this type of catalysis is a fairly slow process. Even with the most active systems temperatures in excess of 150 °C are necessary to obtain reasonable rates of reaction. The subject was reviewed by Pines in 1974[4].

A. Dimerization of Propene

Of particular interest is the dimerization of propene via proton transfer. A systematic study of alkali metal-initiated reactions revealed important differences with regard to the dimer selectivities when different alkali metals were employed[5,6]. Potassium, rubidium, and caesium yielded 4MP1 as the predominant dimer, whereas sodium was found to give significant amounts of 4MP2 (isomerized product) and H.

During the initiation process, two different types of protons may in principle be removed from the monomer, leading to allylic or vinylic intermediates (reaction 4).

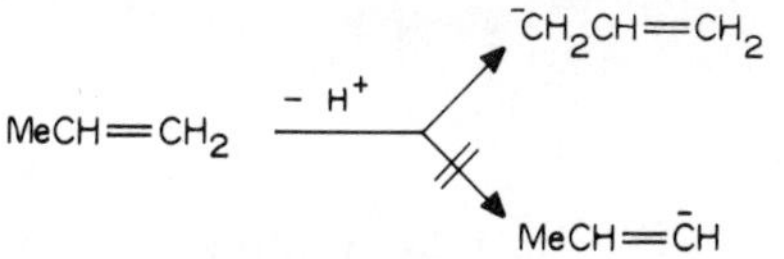

$$MeCH=CH_2 \xrightarrow{-H^+} \begin{array}{c} ^-CH_2CH=CH_2 \\ MeCH=\bar{C}H \end{array} \qquad (4)$$

Owing to the acidic character of the allylic hydrogen caused by the resonance stability of allylic systems, allylic anions are preferentially formed. The new C—C bond of the dimer product is then obtained via a nucleophilic addition of the allylic anion to a propene molecule:

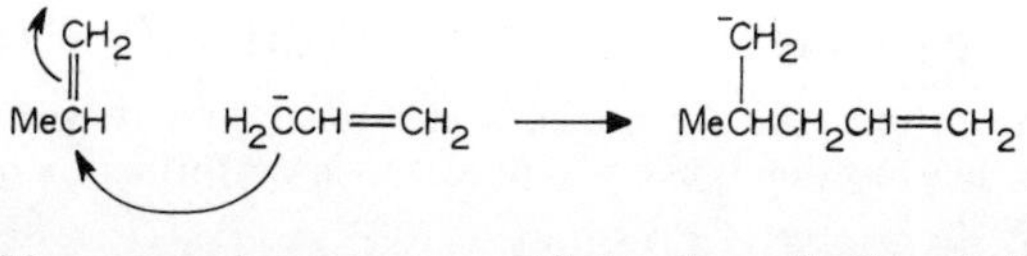

$$\begin{array}{c} CH_2 \\ \| \\ MeCH \end{array} \quad H_2\bar{C}CH=CH_2 \longrightarrow Me\overset{^-CH_2}{\underset{|}{C}}HCH_2CH=CH_2 \qquad (5)$$

The preferred addition occurs in agreement with Markownikoff's rule. The primary anion product can add to a second molecule of propene, so forming a trimer, or abstract a proton from the monomer, yielding the dimer and a new allylic anion. Again, owing to the high acidity of the allylic hydrogens the second mode of reaction is strongly preferred (reaction 6).

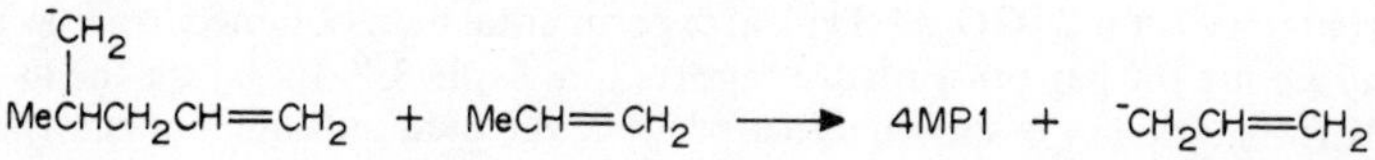

$$Me\overset{^-CH_2}{\underset{|}{C}}HCH_2CH=CH_2 + MeCH=CH_2 \longrightarrow 4MP1 + {}^-CH_2CH=CH_2 \qquad (6)$$

As a natural result of the mechanism, propene is dimerized with high selectivity to 4MP1 via allylic anion intermediates even at the elevated temperatures employed. The primary product, 4MP1, is fairly readily isomerized to 4MP2 (Z/E). Further isomerization to

2MP2 and 2MP1 is kinetically a slow process. The main secondary products found in alkali metal-catalysed processes are therefore the 4MP2 isomers.

The active allylic catalyst may be formed via two different types of chemical reactions, as outlined in reaction 7 and Scheme 2. Strong bases (M^+R^-) such as alkali metal

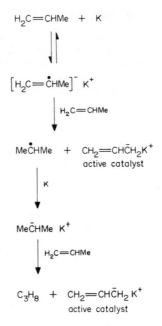

SCHEME 2. Formation of allylpotassium catalyst[5].

hydrides[7-9], alkyls[10], amides[11,12], and carbides[13] react with propene via proton transfer forming the allylic propenyl anion and RH.

$$M^+R^- + MeCH=CH_2 \rightarrow M^{+-}CH_2CH=CH_2 + RH \qquad (7)$$

At elevated temperatures, the alkali metals also react directly with propene, forming allylic compounds. The reaction is likely to proceed via the formation of anion radicals as outlined in Scheme 2.

The formation of significant amounts of organoalkali metal compounds during the dimerization processes has been demonstrated. Using potassium as the initiator, 60% of the metal was recovered after reaction in the form of allylic compounds[14].

The dimer selectivity and the relative composition of C_6 product were found to be significantly influenced by the presence of catalyst supports, e.g. K_2O[15], graphite[16], $MgO-K_2CO_3$[17], and Al_2O_3[18]. The most effective catalyst for the production of 4MP1 was reported to be K on K_2CO_3[19]. Typical experimental data obtained in a flow system at 150 °C and about 100 bar pressure are reported in Table 3[20]. Based on the assumption that 4MP1 is the primary branched dimer product, the data indicate that catalyst systems containing a graphite support exhibit a much higher isomerization activity than K_2CO_3-supported systems. It is further interesting that Na metal on K_2CO_3 support gives a selectivity to 4MP1 which is similar to the potassium systems, while unsupported Na catalyst gives higher yields of isomerized products and *n*-hexenes.

4. Olefin oligomerization

TABLE 3. Dimerization of propene using supported alkali metal catalysts[20]

Catalyst (wt.-% metal)	Rate of reaction [g(g-atom metal)$^{-1}$h^{-1}]	Selectivity to dimer	Product distribution (%)				
			4MP1	4MP2	2MP2	2MP1	H
K/graphite (28.8)	63	97	79	12	1	4	4
K/graphite (11.9)	120	98	62	24	6	4	4
K/graphite (5.1)	97	98	30	50	12	4	4
K/K$_2$CO$_3$ (4.4)	150	98	75	16	1	0	8
Na/graphite (3.0)	155	89	21	56	15	5	3
Na/K$_2$CO$_3$ (3.7)	91	98	74	18	1	0	7

The technical synthesis of 4MP1 is of industrial importance[21]. The stereoregular polymers which can be prepared from 4MP1 monomer are superior to other thermoplastics in their overall properties[22].

B. Ethylation of Olefins

Owing to the instability of vinylic anions and the inert character of vinylic CH groups towards bases, ethene is not oligomerized to any significant extent by alkali metal catalysts below 200 °C. On the other hand, ethene is an excellent anion acceptor and is therefore efficiently co-reacted with other olefins. The receptivity of anions by alkenes is reported[23] to follow the order $C_2H_4 > C_3H_6 > n\text{-}C_4H_8 > i\text{-}C_4H_8$. By passing an equimolar mixture of ethene and propene through an alkali metal-supported catalyst bed at 80–120 °C, a product mixture containing 92% of pentenes (codimer) and only 5.2% hexenes (propene dimer) was obtained. The reaction is initiated by the formation of allylic anions, which in the C—C bond formation are added to the ethene:

Initiation

$$\text{MeCH}=\text{CH}_2 \xrightarrow{-\text{H}^+} {}^-\text{CH}_2\text{CH}=\text{CH}_2 \qquad (8)$$

C—C bond formation:

$$\text{CH}_2=\text{CH}_2 + {}^-\text{CH}_2\text{CH}=\text{CH}_2 \rightarrow {}^-\text{CH}_2\text{CH}_2\text{CH}_2\text{CH}=\text{CH}_2 \qquad (9)$$

Product formation:

$$^-\text{CH}_2\text{CH}_2\text{CH}_2\text{CH}=\text{CH}_2 \xrightarrow{-\text{H}^+} \text{P1} \qquad (10)$$

A large number of codimerized products using a 1:1 (molar ratio) Li–K-supported catalyst are reported in ref. 23. Of particular interest is the selective low-temperature codimerization of ethene and n-butenes. At 80 °C, B1 and B2 are reacted with ethene to give hexenes composed of about 90% of 3MP1, a product which easily might be isomerized to 3MP2 and cracked to isoprene with high yields. The formation of 3MP1 as the primary product is well understood, based on the mode of reaction outlined above for the ethene–propene system. In summary, the characteristic feature of base-catalysed olefin oligomerization is the formation of intermediate allylic anions.

VI. ACID CATALYSIS

Despite the fact that cationic processes are less selective than anionic and coordinative reactions, acid catalysis is extensively used for the industrial production of olefin oligomers, e.g. isobutene dimers, 2,2,4-trimethylpentenes (2,2,4TMP), propene trimers and tetramers, and other products used as motor fuel constituents. Typical catalysts are mineral acids such as sulphuric and phosphoric acids. These may be used in the free form or chemically fixed to the surface of solid supports, e.g. pumice, silica, alumina, and similar systems. In addition, Lewis acids such as BF_3, $AlCl_3$, $AlBr_3$, $ZnCl_2$, and $ZnBr_2$ are reported to catalyse the oligomerization. In the presence of a proton source these catalysts are, however, likely to be converted into super acids, which probably represent the true catalytic species in most of the so-called Lewis acid systems. A common feature for the catalytically active systems is the presence of free protons. The catalysts may therefore correctly be denoted by the general form H^+A^-. During the initiation process, a proton is added to the olefin. This addition, being ionic in character, occurs in accordance with Markownikoff's rule and will lead to the formation of the most stable carbonium ion. The order of carbonium ion stability is well known to be $RCH_2^+ < RCHMe < R_2CMe$. In the C—C bond formation step, the carbonium ions are added to the olefinic bond. The tendency of olefins to react with ions of this type follows the general order $CH_2=CH_2 < RCH=CH_2 < CR_2=CH_2$. It is therefore easily understood that internal olefins most readily undergo oligomerization reactions via acid catalysis, whereas ethene remains unconverted under the given conditions.

A. Oligomerization of Propene

The most commonly used catalyst for the tri- and tetra-merization of propene is solid phosphoric acid. The reaction is carried out at 175–225 °C and 30–70 bar. Under these conditions only minor amounts of dimer (2–5%) is formed[24]. The trimer and tetramer fractions were found[25] to contain mainly isomers of structure **2**.

$$Me_2CH[CH_2CH(Me)]_nCH=CHMe$$
(2)
Propene trimer, $n = 1$; tetramer, $n = 2$

Under milder conditions (< 170 °C), increasing amounts of dimer product are formed. The main dimer product was determined to be 4MP2. The experimental data are

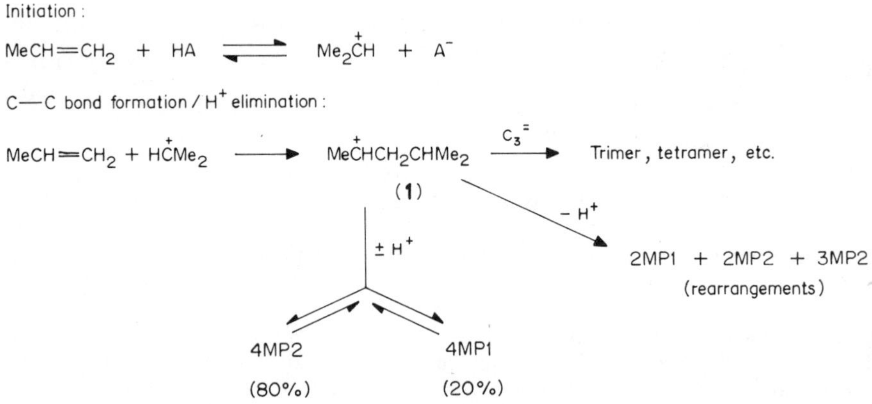

Initiation:

$$MeCH=CH_2 + HA \rightleftharpoons Me_2\overset{+}{C}H + A^-$$

C—C bond formation / H^+ elimination:

$$MeCH=CH_2 + H\overset{+}{C}Me_2 \longrightarrow Me\overset{+}{C}HCH_2CHMe_2 \xrightarrow{C_3^=} \text{Trimer, tetramer, etc.}$$
(1)

$\pm H^+$ → 4MP2 (80%) + 4MP1 (20%)

$-H^+$ → 2MP1 + 2MP2 + 3MP2 (rearrangements)

SCHEME 3. Acid-catalysed oligomerization of propene

4. Olefin oligomerization

rationalized in Scheme 3. The addition of protons and carbonium ions to the olefin bonds occurs in accordance with Markownikoff's rule, while the elimination of protons follows Saytseff's rule, giving mainly the 4MP2 isomer in which the double bond carries the largest possible number of alkyl groups, in addition to minor amounts of 4MP1 and 2MP and 3MP isomers, the latter products being formed via isomerization and rearrangement of the intermediate dimer carbonium ion **3**.

B. Dimerization of Isobutene

The acid-catalysed dimerization of isobutene yields 2,2,4TMP1 and 2,2,4TMP2 as primary products. The intermediate carbonium ion **4** is formed via the same type of initiation and C—C bond formation reactions as outlined for the propene system (Scheme 3). Owing to the highly branched structure of the carbonium ion **4**, the H^+

$$Me_3CCH_2\overset{+}{C}(Me)_2 \;\;\begin{array}{c}\xrightarrow{-H^+} 2,2,4TMP1 \quad 80\%\\ \\ \xrightarrow{-H^+} 2,2,4TMP2 \quad 20\%\end{array}$$
(**4**)

(11)

(12)

elimination will occur in accordance with Hofmann's rule, giving preferentially the product 2,2,4TMP1, in which the double bond carries the least number of alkyl groups. By comparing the mode of H^+ elimination in the propene and *i*-B dimerization systems, it becomes clear that this type of elementary reaction is a function of the environment of the carbonium ion.

VII. ORGANOMETALLIC SYNTHESIS

By reacting Al hydride and Al alkyl compounds with ethene, Ziegler et al.[26] synthesized organoaluminium compounds with growing alkyl groups attached to the metal (reaction 13). This 'aufbau' reaction proceeds with high selectivity as long as the reaction

$$\text{>AlH} + C_2H_4 \longrightarrow \text{>AlC}_2H_5 \xrightarrow{n(C_2H_4)} \text{>Al(CH}_2CH_2)_n C_2H_5 \quad (13)$$

temperature is limited to about 200 °C. Chemically, it is closely related to the so-called living polymer systems in which each metal—carbon bond starts a growing chain and continues to grow as long as monomer is present in the reaction mixture. A well known system of this type is the organolithium-initiated synthesis of (z)-1,4-polyisoprene[27]. Based on kinetic evidence[28] and the fact that highly stereoselective polymer products are formed in non-polar hydrocarbon solvents, the organolithium-initiated polymer reactions are concluded to consist of two distinct elementary steps: (a) activation of the isoprene monomer through coordination and (b) migratory insertion of the coordinated monomer into the 'hair-like' growing carbon chain. Associated organolithium compounds (dimer, trimer, and hexamer) are reported to take part in the C—C bond formation process[29].

Based on their extensive work, Ziegler et al. concluded that the insertion of monomer into Al—C bonds takes place with monomeric aluminium species. The exact mechanism is, however, not known and other research groups view the participation of dimer species as likely[29]. Both types of reaction will therefore be discussed.

The organoaluminium compounds are unique in Group III in forming stable dimers. Organo-B, -Ga, -In, and -Tl compounds are monomeric in the vapour phase and in solution. The bridging is accomplished by means of Al—C—Al three-centre bonds where each aluminium atom supplies an sp^3 hybrid orbital and so also does the carbon atom. By linear combination of atomic orbitals in accordance with the LCAO-MO theory, one bonding (ψ_b), one non-bonding (ψ_n), and one antibonding (ψ_a) orbital may readily be constructed. Since only two electrons are available for the system, only the bonding orbital is occupied by electrons. The situation is depicted in Figure 3. The two electrons in the

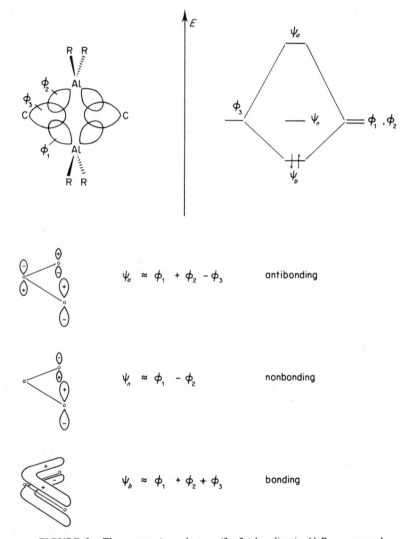

FIGURE 3. Three-centre/two-electron (3c–2e) bonding in Al_2R_6 compounds

bonding orbital ψ_b which serve to unite three atoms are distributed uniformly throughout the Al—C—Al system. There are also cases of three-centre bonding known in which a total of four electrons participate. The second pair of electrons is then located in the ψ_n orbital; these electrons are entirely concentrated on the end aluminium atoms, as may be seen from Figure 3. Assuming that the ethene insertion reaction takes place on the dimer species, the following route of reaction appears to be open:

(a) ethene coordinates to aluminium via the empty ψ_n orbital and thereby donates two electrons to the three-centre bonding system;
(b) owing to the nature of the non-bonding ψ_n orbital, the electron density on aluminium is significantly increased. In turn this electron density is passed on and distributed to the R groups, resulting in increased nucleophilicity and reactivity towards the polarized coordinated olefin.

A possible reaction mechanism is outlined in Scheme 4. The C—C bond formation may take place via a migratory insertion type of elementary reaction or as a nucleophilic

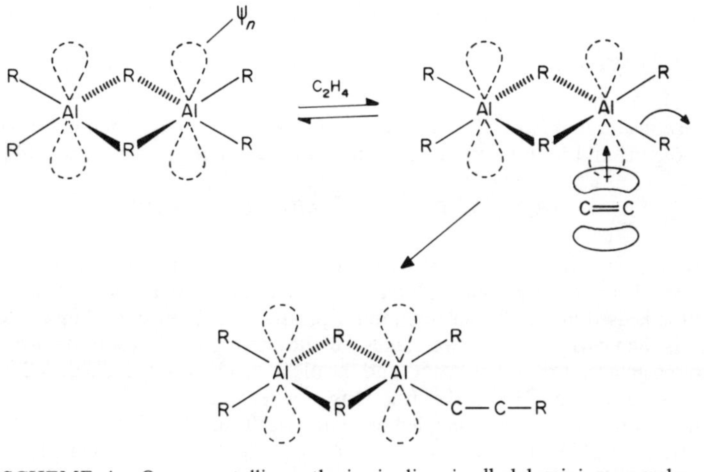

SCHEME 4. Organometallic synthesis via dimeric alkylaluminium complexes

addition of dissociated R⁻ to the coordinated olefin. A true anionic mechanism where the C—C bond formation takes place via addition of R⁻ to free ethene molecules is unlikely because the presence of Lewis bases reduces the rate of reaction. An important feature of coordinatively unsaturated alkyl compounds such as LiR, MgR_2, BeR_2, and AlR_3 is the moderately rapid exchange of alkyl groups via dimeric intermediates[30]. The rates of exchange reactions are usually slowed by the presence of donor ligands and, if the donor is sufficiently strong to block the coordination sites, the exchange is stopped. Since the organometallic syntheses are fairly slow reactions, the net result is parallel growth of all alkyl groups. It is further noteworthy that a dimer mechanism would easily explain the fact that diborane readily reacts with olefins (even internal double bonds) forming alkyl boron compounds, while the monomeric boron alkyls do not react with olefins under organometallic synthesis conditions.

In a monomeric mechanism the coordinatively unsaturated AlR_3 could activate ethene via interaction of the olefin π-electron system with an sp^3 Al hybrid orbital as outlined in

Scheme 5. Based on the information available, both mechanisms appear to be possible. More detailed kinetic analyses are necessary in order to distinguish between the two modes of organometallic synthesis.

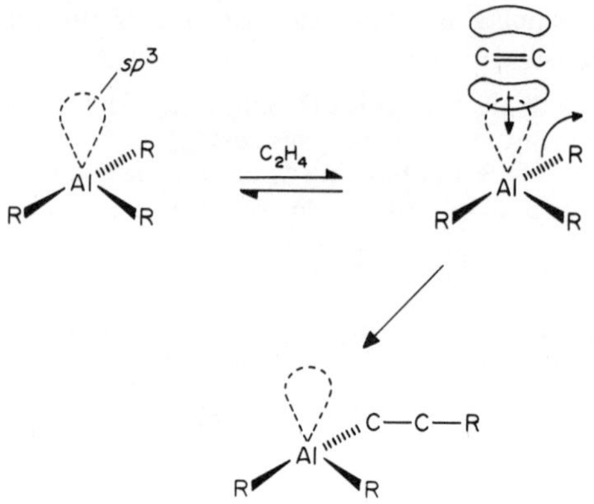

SCHEME 5. Organometallic synthesis via monomeric AlR_3

By subjecting higher alkylaluminium compounds to increased temperatures (250–300 °C), α-olefins and aluminium hydride compounds are formed (reaction 14).

$$AlCH_2CH_2R \xrightarrow{250-300\,°C} AlH + CH_2{=}CHR \qquad (14)$$

Empty orbitals on aluminium, i.e. sp^3 hybrid orbitals present in monomeric systems and the ψ_n orbital of dimer complexes, are likely to take part in the Al hydride-forming process. This reaction is used in the Ziegler method for production of linear α-olefins. The overall procedure is then carried out in two stages, i.e. the chain growth reaction with ethene is effected in the neighbourhood of 110 °C and the olefin product is then displaced by raising the temperature to about 300 °C for fractions of a second.

The dimerization of propene is carried out catalytically at 180–200 °C and 170–200 bar. The selectivity for 2MP1 is reported to exceed 99%. This reaction (Scheme 6) represents

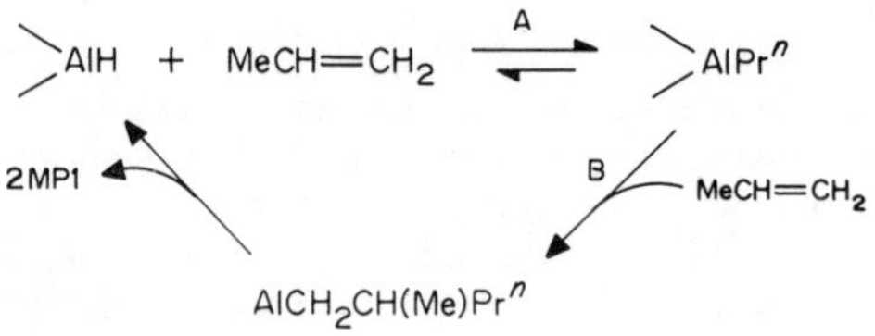

SCHEME 6. Catalytic dimerization of propene

the first stage of the Goodyear Scientific Design isoprene process[31], in which the primary product 2MP1 is isomerized over acidic catalysts to 2MP2, and then demethanized to isoprene in the presence of small amounts of hydrobromic acid at 650–800 °C. Owing to the high rate of β-H elimination from the substituted β-C atom of the dimer adduct, only very little polyaddition forming higher products occurs. The C—C bond formation reaction is extremely selective and occurs formally in accordance with Markownikoff's rule, only minor amounts of hydrogen being formed. This observation may indicate that the addition reaction involves the participation of polar intermediates. The presence of free ions in the system is unlikely, however, based on the fact that isomerization does not occur and that Al—H as well as Al—C bonds are known to have a high degree of covalent character. The rate of dimerization is reported to be first order with respect to the concentration of propene[32]. This is in agreement with a reaction scheme in which the C—C bond formation step (reaction B, Scheme 6) is rate limiting and reaction A is a pre-established equilibrium lying essentially on the product side. Under technical process conditions the hourly production is about 20 g of dimer per gram of Al present in the system.

Higher α-olefins can be dimerized with aluminium catalysts. In all cases β-C-branched α-olefins (5) are obtained. Surprisingly little appears to be known about the true mechanism of organometallic synthesis reaction. Through this discussion of the topic, we hope that some questions have been raised which will stimulate further mechanistic investigations of this most fundamental reaction within the field of organometallic catalysis.

$$CH_2 = C(R)CH_2CH_2R$$
(5)

VIII. COORDINATIVE TRANSITION METAL COMPLEX CATALYSIS

A. Introduction

On 28 July 1924 a most remarkable experiment was reported. Job and Reich had treated ethene with a complex mixture of phenylmagnesium bromide and $NiCl_2$ in diethyl ether at 6 °C and 1 bar and concluded that a catalytic conversion of ethene had taken place on an organometallic nickel species[33]. After hydrolysis, ethane, ethylbenzene, styrene, and biphenyl in addition to 'significant amounts of carbon product with a boiling point exceeding 270 °C' were identified. Based on current knowledge, it seems reasonable to conclude that this was the first experiment to demonstrate important catalytic elementary reactions such as migratory insertion, β-hydrogen elimination, reductive coupling, and polymerization of ethene on bimetallic coordinative transition metal complexes.

About 30 years later, the famous 'nickel effect' on the organometallic synthesis reaction was discovered by Ziegler, which led to the development of the Ziegler–Natta systems for the polymerization of olefins. In 1961 a method of ethene oligomerization with soluble catalytic systems was proposed in the USA, based on alkylaluminium halides and nickel compounds[34]. In subsequent years a large number of patents were filed on the subject[2].

Technical reports on the dimerization and oligomerization of lower olefins using soluble bimetallic nickel complexes of different kinds were published almost simultaneously in West Germany[35,36], France[37], and the USSR[38]. Ewers[35] and Chauvin et al.[37] used highly active systems composed of nickel acetylacetonate and dialkylaluminium chloride or monoalkylaluminium dichloride, while Fel'dblyum et al.[38] investigated the activity of organic nickel salts in similar types of systems. A paper by Wilke et al.[36] described the dimerization of lower olefins under the influence of π-allylnickel chloride in combination with Lewis acids and tertiary phosphines. Later, Onsager et al.[39a–e] reported on the kinetics and mechanism of tetramethylcyclobutadienenickel chloride–tertiary phosphine–alkylaluminium chloride initiated reactions. More recently, Keim and Kowaldt[40] presented a novel phosphorus ylid–nickel complex which converted ethene with high selectivity and activity into higher linear α-olefins or high polymers in the absence of organometallic cocatalysts such as are required in the Ziegler–Natta-type systems.

Although it can be asserted that nickel is *the* element of olefin dimerization and oligomerization owing to its high specific activity and selectivity characteristics, other transition metal systems containing e.g. titanium[41], cobalt[42], ruthenium[43], rhodium[44], iridium[45], palladium[46], platinum[47], and tantalum[48] have also been reported. Most catalytic systems contain transition elements from group VIII, but also early transition elements, such as titanium, zirconium, and tantalum, can give low molecular weight products under specific conditions. The early transition elements are, however, mostly used in polymerization reactions, the classical system being $TiCl_3-Et_3Al$.

B. General Mechanistic Aspects

Although each individual catalytic system has its own characteristics of activity and selectivity, the types of elementary reaction steps which are involved in the dimerization cycle are of the same chemical nature for a large number of systems. Except for tantalum, which is reported to react via metallocyclic intermediates, the following elementary steps are involved:

1. Olefin activation via complex formation

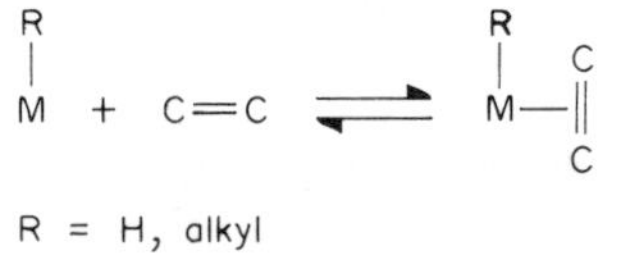

$$M-R + C=C \rightleftharpoons M(R)-\|_C^C$$

R = H, alkyl (15)

2. Migratory insertion

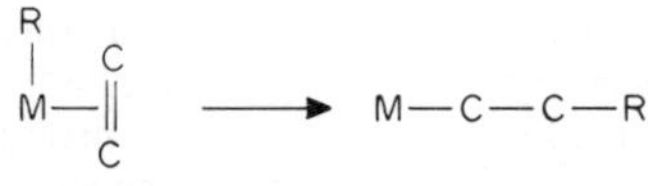

$$M(R)-\|_C^C \longrightarrow M-C-C-R$$ (16)

3. β-H elimination

$$M-CH_2-C_\beta(H)(R)(H) \rightleftharpoons M-H + CH_2=CHR$$ (17)

Transition metal hydrides (MH) and alkyl complexes (MR) are distinct intermediates in the catalytic cycle.

1. Olefin activation via complex formation

The first known olefin transition metal complex $K[Pt(C_2H_4)Cl_3]$, was prepared by the Danish pharmacist Zeise in 1827, and is commonly known as Zeise's salt[49]. The development of olefin coordination chemistry proceeded slowly until more than 100 years later. In about 1950 the theoretical understanding and preparative chemistry of olefin complexes started to blossom. Dewar[50] and Chatt and Duncanson[51] provided a constructive bonding model for the interaction between an olefin and a transition metal. In principle, two different types of chemical bonding are involved:

(a) *Sigma* (σ)-*type bond.* The cloud of electron density in one of the π-orbitals of the olefin bond overlaps with a vacant σ-orbital of the metal, forming a chemical bond of σ-symmetry.

(b) *Pi* (π)-*type bond.* A bond of π-symmetry is formed by overlap of a filled d or dp hybrid orbital of the metal with the empty antibonding π^* orbital of the olefin.

The two types of bonding modes, which are complementary and synergic in the transfer of electron density from the olefin to the metal and back again, are depicted in Figure 4.

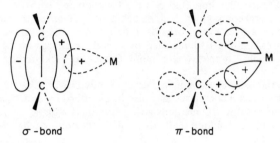

FIGURE 4. Molecular orbital views of the olefin—metal bond

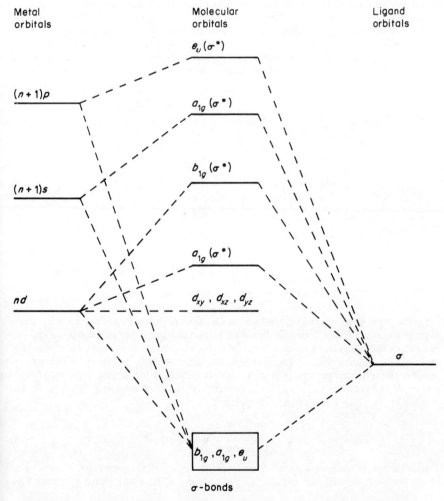

FIGURE 5. Qualitative MO diagram for σ-bonding in a square-planar [ML$_4$] complex

The overall effect is a bond of moderate stability. The lengthening of the C—C bond observed in a number of complexes is explained by the population of the π^*-antibonding olefin orbital. Of particular importance is the backbonding with low oxidation state organotransition metal compounds. Such π-backbonding stabilizes the metal complex by distributing electrons over ligand orbitals, thus reducing the negative charge on the central metal. Backbonding is diminished by a positive charge at the central metal. Consequently, anionic complexes are more prone to backbonding than cationic complexes.

In addition to the basic bonding model developed by Dewar, Chatt, and Duncanson, the transition metal olefin complexes may also be described by molecular orbital (MO) theories. These range from MO methods giving essentially a quantitative description of the bond via *ab initio* calculation methods to qualitative approaches based mainly on symmetry properties.

A number of the most active dimerization catalyst systems contain d^8 complexes of e.g. Ni(II), Pd(II), Pt(II), Rh(I), and Ir(I). In such complexes, square-planar geometry is very commonly encountered. The qualitative MO energy level diagram for σ-bonding in square-planar d^8 complexes is shown in Figure 5. In this type of complex, the orbitals p_z, d_{xz}, and d_{yz} have the right symmetry for forming π-bonding perpendicular to the xy-plane, and d_{xy} and p_x, p_y for forming the π-backbonding to empty π^* orbitals to olefins oriented in the xy-plane[52]. The net result of such π-bonding is a stabilization of the complex as indicated in Figure 6, where the d_{xz} orbital is shown to be responsible for the π-bond.

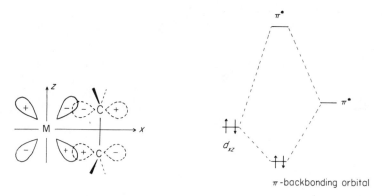

FIGURE 6. Orbital overlap for π-backbonding

Owing to the number of orbitals which may be used for π-bonding either in the xy-plane or perpendicular to it, two coordination sites, *cis* and *trans*, may be occupied by any pair of equivalent ligands. Thus configurationally stable *cis* and *trans* isomers are commonly observed for diamagnetic d^8 complexes in the square-planar geometry. When two π-acceptor ligands are present, however, the *cis* configuration appears to be strongly preferred. The instability of *trans* 'π,π' complexes is evidenced by the rapid isomerization of *trans*-$[(\eta^2\text{-}C_2H_4)_2PtCl_2]$ to the *cis* isomer[53].

$$\textit{trans-}[Pt(\eta^2\text{-}C_2H_4)_2Cl_2] \xrightarrow[-10\,°C]{\text{diethyl ether}} \textit{cis-}[Pt(\eta^2\text{-}C_2H_4)_2Cl_2] \qquad (18)$$

The distortion of square-planar d^8 complexes into tetrahedral geometry is accompanied by conversion from a D_{4h} diamagnetic singlet state to a T_d paramagnetic triplet state. This

4. Olefin oligomerization

is observed for 'weak ligand-field' complexes, especially in the case of first-row 3d metals, e.g. [NiL$_2$X$_2$] complexes where L represents a mixed alkylarylphosphine[54]. Palladium(II) and platinum(II) strongly prefer to form square-planar complexes.

In square-planar complexes, the olefinic C—C axis is usually oriented nearly perpendicular to the ligand plane when three other ligands are present, but almost in the plane of the square planar complex with two other ligands:

(6)   (7)

Representative examples of such structures are e.g. Zeise's salt (8) compared with nickel(0) ethene complexes of the type [(R$_3$P)$_2$Ni(C$_2$H$_4$)] (9)[55–57].

(8)   (9)

Thorn and Hoffmann[58] explained the upright olefin orientation in d^8-type complexes by filled orbital–filled orbital repulsions in the planar structure. For the complex [PtH(PH$_3$)$_2$(C$_2$H$_4$)], an energetic preference of 0.3 eV in favour of the upright structure was calculated. In solution the olefin can rotate about the Pt—olefin axis, as demonstrated for a number of noble metal olefin complexes. The barrier to rotation of ethene in [RhCp(C$_2$H$_4$)$_2$] is reported to have a activation energy of 15 ± 2 kcal mol^{-1} [59].

The true length of the C—C bond in transition metal olefin complexes was discussed by Åkermark et al.[60]. For Zeise's salt C—C bond distances ranging from 1.35 to 1.48 Å were reported, but a recent neutron diffraction study gave 1.354 Å, which the authors suggest is the most reliable result. Thus, it appears that the C—C bond length in the platinum(II) complex is only slightly longer than that (1.33 Å) in ethene itself. Other recent studies give C—C olefin ligand bond lengths for palladium(II) and platinum(II) compounds in the range 1.30–1.40 Å, with a majority around 1.35 Å, which is in good agreement with value of 1.354 Å for Zeise's salt. Based on these values it seems reasonable to conclude that the π-backbonding is of only minor importance in tetracoordinated d^8 transition metal olefin complexes. Stable neutral nickel(II) complexes of the type [NiX$_2$(C$_2$H$_4$)] have not yet been isolated. It appears that the acceptor properties of nickel are too low for the formation of stable compounds. As is evident from their second ionization potentials. Ni 18.15 eV, Pt 18.56 eV, and Pd 19.9 eV, palladium(II) and platinum(II) are better electron acceptors than nickel(II) and thus give more stable complexes. The metal—olefin bond energies for palladium(II) and platinum(II) complexes have been calculated to range from 6 to 10 kcal mol^{-1} [61].

In contrast, several complexes of zerovalent nickel of the general form 7 have been isolated and their structures determined. The olefin in such complexes is usually oriented in the ligand plane. The metal—ethene bond energy for PR$_3$ = P(o-Tol)$_3$ and PR$_3$ = PH$_3$ were calculated to be 33 and 36 kcal mol^{-1}, respectively[60,62]. For the metal(0) complexes

TABLE 4. Equilibrium constants for the replacement of ethene in acetylacetonatobisethenerhodium(I) by another olefin (from ref. 64)

Olefin	K(at 25 °C)
$CH_2=CH_2$	1.0
$CH_2=CHMe$	0.08
$CH_2=CHEt$	0.09
(Z)-MeCH=CHMe	0.004
(E)-MeCH=CHMe	0.002
$CH_2=CMe_2$	0.00035

significantly longer C—C bond lengths have been determined, typically in the range 1.40–1.48 Å, with predominance around 1.43 Å[56], indicating significant π-backbonding in such systems.

Recent GVB-CI calculations[63] on 'naked' nickel(0) ethene complexes indicate that delocalization of the ligand π orbital into the nickel and delocalization of the nickel d_π orbital are both small (C—C bond length increase by ca. 0.03 Å), mainly owing to the repulsive interaction with the 4s orbital in neutral nickel ($4s^1 3d^9$) systems. Bringing the bond of ethene up to Ni^+ ($3d^9$) is reported to lead to a significant bond energy (ca 60 kcal mol^{-1}) owing to the net attraction of the nickel centre for the unshielded π pair.

X-ray diffraction studies of $[Rh(PPh_3)Cp(C_2H_4)]$ indicate a strong π component in the Rh—ethene bond[59]. Compared with higher olefins, ethene is a strongly preferred ligand. An instructive series of equilibrium constants reported by Cramer[64] for the rhodium(I) system in reaction 19 are given in Table 4.

$$[Rhacac(C_2H_4)_2] + \text{olefin} \overset{K}{\rightleftharpoons} [Rhacac(C_2H_4)(\text{olefin})] + C_2H_4 \quad (19)$$

A similar trend of stability was observed for palladium(II) olefin complexes[65], based on equilibrium measurements in the following system:

$$[PdCl_4]^{2-} + \text{olefin} \overset{K'}{\rightleftharpoons} [Pd(\text{olefin})Cl_3]^- + Cl^- \quad (20)$$

At 25 °C $K\cdot$ was determined to be ethene 17.4 propene 14.5, B1 11.5, (Z)-B2 807, and (E)-B2 4.5.

From a chemical reactivity point of view, election donation from the olefin to the transition metal will leave the olefin with a positive charge and thus increase its reactivity towards nucleophiles. π-Back donation of negative charge on to the substrate will have the opposite effect. For nickel, palladium, and platinum the reactivity of the complexed olefin in external nucleophilic addition reactions is found to correlate well with the acceptor properties of the metal in the divalent state[60]. That σ-bonding is the predominant factor in

$$M^+ \overset{C}{\underset{C}{-\|}} + Nu^- \longrightarrow M \overset{C}{\underset{C}{\diagup \diagdown}} Nu \quad (22)$$

such systems is strongly suggested by calculations performed on the relative contributions of σ- and π-bonding in Zeise's salt[66]. These calculations show that even in the anionic platinum(II) complex which should favour π-backdonation, the σ donation bond energy is about three times as large as that of π-back donation, leaving the olefin with a considerable positive charge. Consequently, in neutral or even more so in positively charged complexes σ donation is expected to predominate even more.

Based on the assumption that H and R groups attached to transition metals react as nucleophiles, the reactivity of complexed olefins in migratory insertion reactions is expected to follow the same general pattern as outlined for the external nucleophilic addition. From a mechanistic point of view, the insertion process is far more complex. By taking place within the orbital sphere of the complex, between two groups both attached to the same metal and preferably located in *cis* positions, the understanding requires knowledge about the factors influencing the reactivity of both the reacting groups simultaneously, conditions favouring the Z configuration, as well as the energetically favourable reaction path. Such topics are discussed in the next section.

2. The migratory insertion of olefins

The general concept of olefin insertion into transition metal—H and —C bonds originated with the Ziegler–Natta-type polymerization systems. Elementary processes of the insertion (complex) type were believed to be responsible for the initiation and chain growth reaction. In its general form, the transition state of reaction is as shown in equation 23. Calculations by Thorn and Hoffmann[58] demonstrated clearly that for this

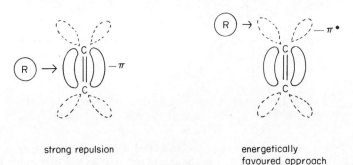

R = H or alkyl

reaction to occur, the M—R and the olefin π-bond must be coplanar with R approaching one end of the olefin bond. Only in this way is the antibonding π* orbital of the olefin able to mix into the process, thus diminishing the antibonding repulsion between R and C_β. As R approaches the centre of the olefin bond, the π* orbital cannot mix in and the repulsive R–π combination rises much more steeply. Therefore, only the 'end-on' coplanar approach is believed to be able to permit insertion (Figure 7). This stereochemical

strong repulsion   energetically
                          favoured approach

FIGURE 7. Evaluation of the migratory insertion process

requirement naturally leads to the conclusion that the metal and the ligand add *cis* to the olefin. In contrast, an external nucleophilic addition will normally be *trans*.

During the insertion process, the following change in bonding structure is taking place: the metal—R σ-bond, the metal—olefin σ- and π-bonds and the C=C double bond are broken, while a σ-bond between the migrating ligand and the C atom closest to it and a σ-bond between the metal and the other C atom of the olefin are formed.

Based on the MO perturbation theory, Fukui and Inagaki[67] showed that the insertion process might be accomplished by delocalization of electrons from the HOMO of R to the LUMO of the inserting olefins as indicated in Figure 8. A similar view was proposed by

FIGURE 8. Electron transfer from HOMO of R to the olefin via π*-antibonding orbital

Furukawa[68], in which electrons from R are first partly delocalized to an unoccupied orbital of the metal and then subsequently to the π* orbital of the olefin.

The basic idea of empty d metal orbital participation in the process was also favoured by Cossee[69] for the titanium chloride-catalysed polymerization reaction. The most important feature of this mechanism is depicted in Figure 9. By mixing of Ψ_1 and Ψ_3, made

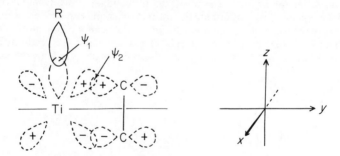

FIGURE 9. Empty or half-filled d-orbital participation in the insertion process

possible by the presence of an empty or half-filled d_{yz}, the energy barrier of R–olefin approach is reduced. The d_{yz} orbital will be able to perform this task optimally when it has a good interaction with both σ_R and π^*. Its energy level should be comparable to those of σ_R and π^*.

An extensive MO study has been reported for the insertion of ethene into Pt—H bonds[58]. Letting the insertion process take place in the xy-plane, the main feature of the proposed mechanism is as shown in Figure 10. During the insertion process the $d_{x^2-y^2}$ orbital develops smoothly into a Pt—C_α bonding orbital. H—C_β repulsion is observed, but the π^* mixing helps both to relieve the repulsion and to form the Pt—C_α bond. The

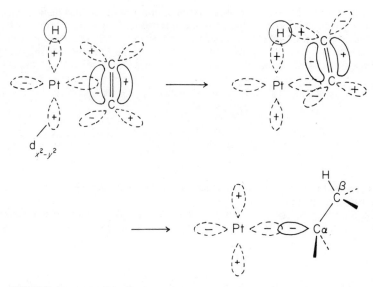

FIGURE 10. Insertion of ethene into a P—H bond. The d_{xy} Pt—olefin π-backbond has been omitted for clarity

$d^*_{x^2-y^2}$ orbitals 'resents' the conversion strongly and climbs to high energy in the transition state region. It is therefore concluded that this d-orbital must be empty for insertion to occur. The other d-block orbitals show little change in energy as the insertion progresses, except for a slight destablization of d_{xy} as some back bonding into the ethene π^* is lost. Replacing H by an alkyl group will not change the path of insertion, but is likely to reduce the rate of conversion.

The intramolecular migratory insertion of an olefin into M—H bonds is facile and consequently, stable olefin complexes which contain hydride in a cis position are rare. One such complex is $[MoCp_2H(C_2H_4)](PF_6)$[70]. In the presence of added PPh_3, however, the insertion readily takes place. The trans complex trans-$[Pt(PEt_3)_2H(C_2H_4)]^+$ can be isolated[71], which indicates that insertion from a trans orientation of ligands is a significantly slower process in platinum(II) complexes than cis insertion.

For nickel(II) complexes, which have different high-spin, low-spin characteristics, the trans insertion is regarded as a possible mechanism, although cis insertion via the four-coordinate mechanism is considered the most favourable of all[58]. The insertion of ethene into the Pt—H bond of neutral $[Pt(PR_3)_2(H)Cl]$ is a slow process[72]. Abstracting the Cl^- ion by silver cations[73] or adding $SnCl_2$[74] forming $[Pt(PR_3)_2H]^+(SnCl_3)^-$ speeds up the rate of insertion significantly. In view of later discussions concerning the chemical nature of catalytically active species, it is noteworthy that cationic complexes appear to undergo more rapid insertion than neutral compounds. The activation of a number of catalytic systems by Lewis acids is related to this phenomena.

3. β-H Elimination

The β-H elimination reaction (Volume 2, Chapter 8) is the reverse reaction of inserting olefins into M—H bonds, and accordingly proceeds via the same transition state as indicated in reaction 23. Since the coordination number of the central metal is increased by one unit during the process, the presence of a free coordination site on the metal is considered a precondition for the reaction to occur. It is generally known that Lewis bases added to the system will reduce the rate of reaction. For the decomposition of $[Pt(PPh_3)_2(Bu^n)_2]$ a dissociative mechanism is proposed (reactions 24 and 25)[75]. Added

$$[Pt(PPh_3)_2(Bu^n)_2] \rightleftharpoons [Pt(PPh_3)(Bu^n)_2] + PPh_3 \qquad (24)$$

$$[Pt(PPh_3)(Bu^n)_2] \rightleftharpoons \begin{bmatrix} Ph_3P_{\cdots\cdots}Pt^{\cdots\cdots H} \\ Bu^n \qquad C_4H_8 \end{bmatrix} \xrightarrow{red.\ elimination} \qquad (25)$$

PPh_3 is reported to retard the rate of reaction without changing the energy of activation for the dissociation step.

The equilibrium reaction 26 between M—alkyl and M—H olefin complexes was proted by Werner and Feser[76]. By using n.m.r., the state of equilibrium was established.

$$\begin{bmatrix} Cp \\ | \\ Rh^+ \\ Me_3P \quad H \quad -C- \\ -C \\ | \end{bmatrix} \rightleftharpoons \begin{bmatrix} Cp \\ | \\ Rh^+ \\ Me_3P \quad \square \quad C-C \\ \quad \quad \quad H \end{bmatrix} \qquad (26)$$

The free coordination site on Rh^+ is likely to be partially occupied by solvent (CH_3NO_2) molecules.

4. The rate of reaction

In order to understand both the activity and the selectivity of catalytic dimerization/oligomerization systems, studies related to the rate of reaction are of the utmost importance. The identification of rate-determining steps is a precondition for influencing the rate and tailoring of the catalyst. The reader is also reminded that the selectivity is a result of *relative* rates in the system. Thus, all the factors which influence the relative rates, will change the selectivity.

Based on the information given in Section VIII. B. 1–3 and the fact that the formation of transition metal olefin complexes of the type in question is a very rapid process, it appears reasonable to analyse the general kinetics of the system based on the assumption that the migratory insertion of olefins to M—C bonds (reaction 4 in Figure 11) represents the rate-limiting step and that all other steps are established equilibria. In accordance with transition state theory and the third law of thermodynamics, the rate constant for the rate-limiting step is given by equation 27.

$$k_4 = \frac{kT}{h}\exp(-\Delta G^{\ddagger\circ}/RT) = \frac{kT}{h}\exp(-\Delta H^{\ddagger\circ}/RT)\exp(\Delta S^{\ddagger\circ}/R) \qquad (27)$$

where k = Boltzman's constant, h = Planck's constant, $\Delta G^{\ddagger\circ}$ = free energy of activation, $\Delta H^{\ddagger\circ}$ = enthalpy of activation, and $\Delta S^{\ddagger\circ}$ entropy of activation. Accordingly, there exist two ways of increasing the rate constant: (a) reduction of $\Delta H^{\ddagger\circ}$ and (b) increasing $\Delta S^{\ddagger\circ}$. In practice such effects are often coupled, so that ways designed to effect an increase in $\Delta S^{\ddagger\circ}$ simultaneously result in a reduction of $\Delta H^{\ddagger\circ}$. The kinetic explanation as to why *cis* insertion is a facile reaction and much more so than the *trans* process, is to be found in the

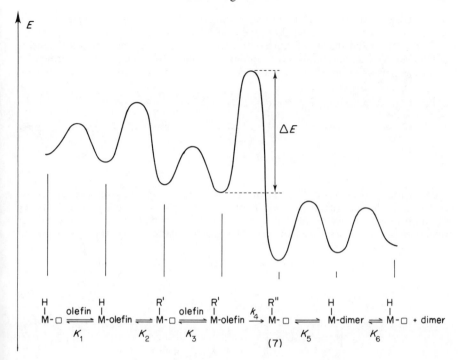

FIGURE 11. General scheme of dimerization via Z insertion mechanism and the potential energy diagram for the process. K = equilibrium constant; k_4 = rate constant for the rate limiting step; $\square$ = free coordination site

fact that *cis* orientation gives a very favourable combination of $\Delta H^{\ddagger \circ}$ and $\Delta S^{\ddagger \circ}$. In particular, it is easy to see that two ligands in *cis* positions are more likely to react than two *trans* ligands, the kinetic consequence being $\Delta S^{\ddagger \circ}$ (*cis*) > $\Delta S^{\ddagger \circ}$ (*trans*), resulting in improved rates for the *cis* insertion process.

From the mechanism discussed in Section VIII. B.2, it appears reasonable to conclude that a significant part of $\Delta G^{\ddagger \circ}$ is to be found in the repulsion between the migrating alkyl group and the olefin. By mixing in the participation of empty π^* orbitals of the olefin in the process, the repulsion is reduced. Consequently, the highest rate constants are to be expected for complexes in which the π-backbonding between the transition metal and the reacting olefin is low. With reference to the information given in Section VIII. B.1, this observation is in good agreement with reports stating cationic titanium[41] and cationic nickel[39e,77] complexes to be highly active catalysts.

Using steady-state kinetics on the reaction scheme shown in Figure 11, the rate of dimerization is given by equation 28, where $\square$ = free coordination site. Accordingly, the

$$r_2 = k_4[cis - M(R)\text{olefin}] = k_4 K_1 K_2 K_3 [M(\square)H][\text{olefin}]^2 \qquad (28)$$

reaction is first order with respect to the concentration of metal hydride complex having a free coordination site in the *cis* position, second order in reacting monomer olefin and directly proportional to the equilibrium constants K_1, K_2, and K_3. Anything that will reduce the concentration of active hydride complex will retard the reaction. Consequently, the dimerization process is self-inhibiting in that dimer olefins formed during the reaction will complex with M($\square$)H and thus reduce its concentration.

On the other hand, the rate of trimerization will increase as the concentration of dimer is building up. Trimers and codimers of monomer and dimer are formed via the intermediate species (Scheme 7) of the dimerization cycle. Again, assuming steady-state kinetics with

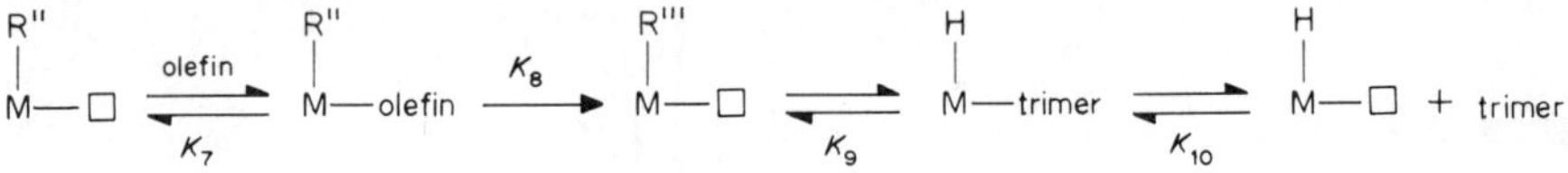

SCHEME 7. Trimerization

the insertion step 8 being rate limiting, the rate of trimerization, r_3, is given by equation 29. Accordingly, the relative rate of dimerization and trimerization in a given system will obey equation 30, and the overall selectivity to the trimer, S_3, for an experiment performed over time t can be calculated from equation 31. Based on the assumption that the rate constant

$$r_3 = k_8[cis - M(R'')\text{olefin}] = k_8 K_5 K_6 K_7 [M(\square)H][\text{olefin}][\text{dimer}] \qquad (29)$$

$$\frac{r_2}{r_3} = \frac{k_4 K_1 K_2 K_3 [\text{olefin}]}{k_8 K_5 K_6 K_7 [\text{dimer}]} \qquad (30)$$

$$S_3 = \int_0^t \frac{r_3}{r_2 + r_3} dt = \int_0^t \frac{1}{\frac{k_4 K_1 K_2 K_3 [\text{olefin}]}{k_8 K_5 K_6 K_7 [\text{dimer}]} + 1} dt \qquad (31)$$

of insertion is independent of the chain length of the migrating alkyl group ($k_4 \approx k_8$), the selectivity becomes primarily a question of relative values for the equilibrium constants, K_1, K_2, K_3, and K_5, K_6, K_7, and the relative concentration of monomer olefin and dimer.

The general observation that bulky ligands tend to favour dimerization over

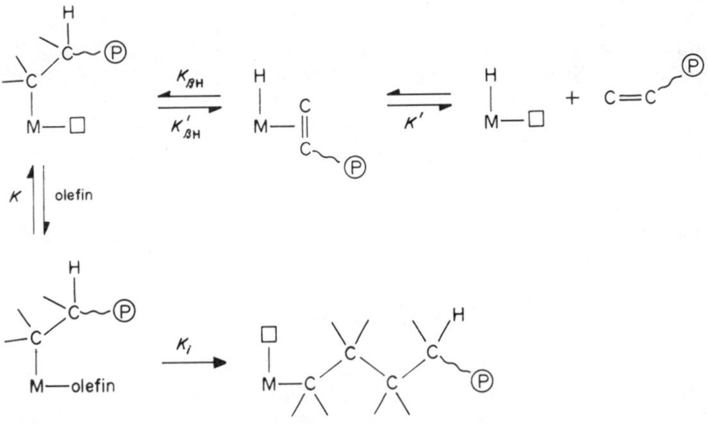

SCHEME 8. Kinetic scheme of polymerization

trimerization, finds its rational explanation in the selective effect of bulky ligands to reduce the value of K_6 (equation 32). Analogous arguments do of course also explain the fact that bulky ligands favour dimerization over oligomerization in general.

$$K_6 = \frac{[M(H)\,dimer]}{[M(\square)H][dimer]} \tag{32}$$

In systems in which high polymers, C=C~P, are formed, the rate of chain growth is much larger that the overall rate of β-H elimination. In accordance with the kinetic analysis of Scheme 8 this will be the case if

$$kK[M(\square)(C-C \sim P)][\text{olefin}] \gg K_{\beta H}[M(\square)(C-C \sim P)]$$
$$-k'_{\beta H}K'[M(\square)H][C=C \sim P] \tag{33}$$

Accordingly, the molecular weight of the product is influenced not only by the relative rates of olefin insertion into M—C bonds and the forward β-H elimination alone, but also the rate of olefin product back insertion into transition metal hydride bonds. For systems in which

$$k_{\beta H}[M(\square)(C-C\sim P)] > k'_{\beta H}K'[M(\square)H][C=C \sim P] \tag{34}$$

the number-average degree of polymerization, P_n, is given by equation 35.

$$P_n = \frac{k_i K[\text{olefin}]}{k_{\beta H}} \tag{35}$$

C. The Mechanism of Coordinative Transition Metal Catalysis

As indicated, nickel is *the* element of olefin dimerization and oligomerization owing to its high specific activity and selectivity characteristics. The catalytic performance of nickel complexes will therefore be analysed in some detail, followed by more general mechanistic aspects of rhodium, titanium, and tantalum complex-initiated reactions.

1. Nickel complex catalysis

A large number of extremely active nickel catalysts have been reported[2]. Basically, they can be classified into three categories:
I. nickel compound + Lewis acid;
II. nickel compound + Lewis acid + Lewis base;
III. nickel hydride complexes.

By far the most active systems contain $AlRCl_2$ and/or AlR_2Cl as the Lewis acid. Lewis bases, especially of the tertiary phosphine type, are used for tailoring of the catalytic properties.

In combination with $Al_2R_3Cl_3$ almost any soluble nickel compound will show catalytic activity. Nickel compounds, neutral, cationic, and anionic, with oxidation states ranging from +2 to zero have been used. Representative examples are nickel acetylacetonates[78,79], nickel acetates[80], nickel oleate[38], nickel naphthenate[81], nickel dipropylsalicylate[82], nickel alkylbenzenesulphonates[83], tetramethylcyclobutadienenickel dichloride[39], π-alkylnickel halides[84], $[Ni(PR_3)_2X_2]$ complexes in which X may be halides, haloacetates, sulphate, nitrite, nitrate, thiocyanate, thiophenolate, etc.[80], $[Ni\{(NR_2)_3PO\}_2X_2]^{85}$, $[Ni(PR_3)(NO)Br]^{80}$, $[Ni(PR_3)_3X]^{86}$, $[Ni(PCl_3)_4]^{87}$, $[Ni(PR_3)_4]^{87}$, $[Ni(CO)_2(PR_3)_2]^{87}$, $[Ni(\text{acrylonitrile})_2]^{87}$, $[Ni(dmso)_6]^{2+}[NiCl_4]^{2-}$, $[NiCp(PPh_3)_2]^+[SnCl_3]^{-88}$, and $[R_4P]^+[NiX_3(PR_3)]^{-89}$.

In the absence of Lewis acids unique catalytic properties are reported for square-planar

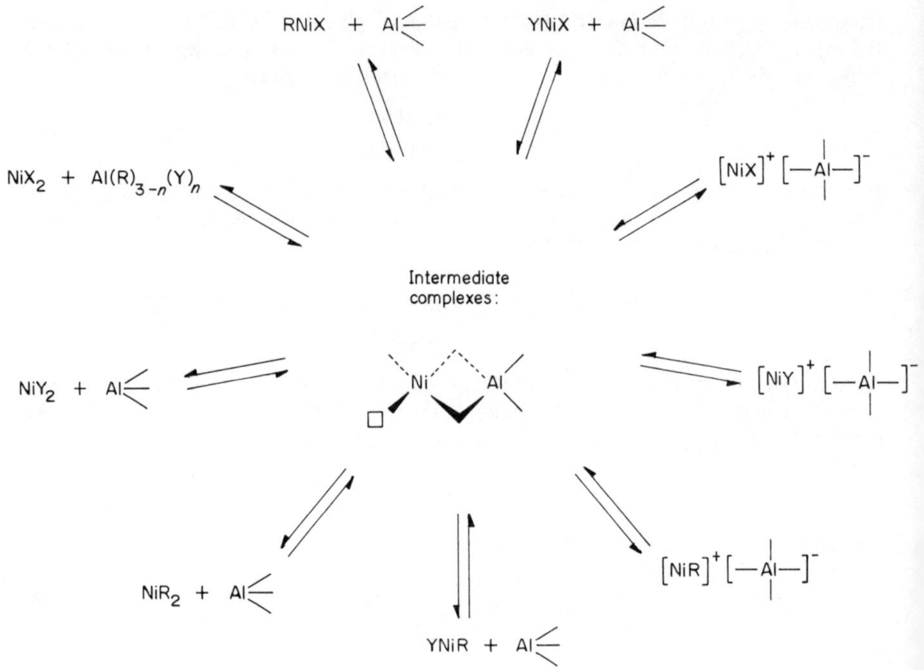

SCHEME 9. Exchange of ligands via bimetallic complexes during the reaction between nickel(II) compounds and Lewis acids

R = alkyl; X = monovalent anion; Y = halide; n = 1 or 2

nickel hydride complexes formed by the reaction of a nickel salt with $NaBH_4$ in the presence of PPh_3 and a chelating ligand such as diphenylphosphinoacetic acid[90] (reaction 36). Catalysts of this type are extremely selective for the preparation of linear α-olefins

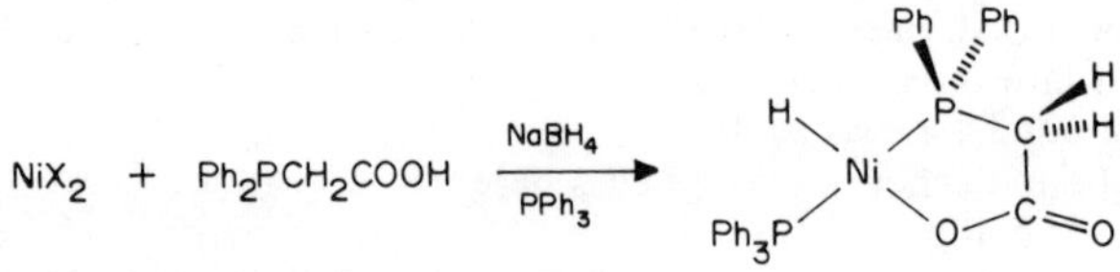

$$NiX_2 + Ph_2PCH_2COOH \xrightarrow[PPh_3]{NaBH_4} \text{(complex)} \quad (36)$$

from ethene[91]. In practice, ethene is allowed to react with the catalyst in a solvent such as ethene glycol or butane-1,4-diol at about 100 °C and 40 bar. In typical experiments[91c] the following distribution of linear α-olefins is obtained: C_4–C_8 41%, C_{10}–C_{18} 40.5%, and C_{20+} 18.5%. The linearity is reported to exceed 99% and 98% of the product consists of α-olefins. Similar results are also reported[92] with the catalyst **10**.

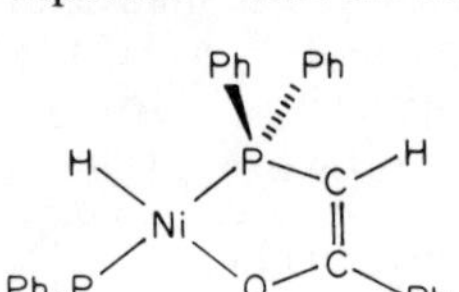

(10)

A common feature of all the active systems is the presence of Ni—H or Ni—alkyl species, which are believed to be responsible for the initiation of the catalytic reaction. Depending on the particular recipe used, such groups may be present initially or formed during the induction period[39c].

a. The effect of Lewis acids

When nickel(II) compounds are reacted with Lewis acids, in principle two different types of elementary reactions may occur: (1) complex formation and (2) exchange of ligands. Both reactions are, however, closely related since the exchange of ligands is likely to proceed via the intermediate formation of bimetallic complexes as indicated in Scheme 9 for a nickel salt–alkylaluminium halide system. In addition to a free site of coordination, the catalyst must possess an active R ligand bound to the nickel. Consequently, the following species represent potential catalysts for the reaction: RNiX, NiR_2, and $[NiR]^+ [Al]^-$.

Based on the fundamental information that the migratory insertion of olefins into M—H and M—C bonds is a more facile process with cationic complexes than with neutral compounds[58], the cationic species $[NiR]^+$ is expected to possess the highest catalytic activity. This theory is in agreement with the mechanistic views presented by a number of authors[39e,77,93], and further supported by the activity data reported in Table 5. On comparing the activity of nickel acetate and nickel haloacetate systems, the highest activity is obtained with nickel trifluoroacetate. It appears reasonable to conclude that this is caused by increased dissociation and the formation of cationic nickel species. This conclusion is also in agreement with the fact that improved rates are observed in the nickel acetate system when $AlEtCl_2$ is used as the Lewis acid component compared with $AlEt_2Cl$ (Table 6). It is well known that $AlEtCl_2$ is a stronger Lewis acid than $AlEt_2Cl$ and that the

TABLE 5. Dimerization of propene (30 min) using the catalyst $[Ni(RCO_2)_2]$ (5.0×10^{-4} mol l^{-1})–$AlEtCl_2$ (2.0×10^{-4} mol l^{-1}) at 20 °C and 1 bar in chlorobenzene (25 ml)[88]

R	Product (ml)	Dimer (%)	Dimer distribution		
			2MP	2,3DMB	H
Me	29	82.6	75.8	5.7	18.5
CH_2Cl	35	83.0	74.5	4.8	20.7
$CHCl_2$	38	82.5	75.9	5.1	19.0
CCl_3	39	80.3	74.5	4.7	20.8
CF_3	44	79.1	76.5	4.0	19.5

TABLE 6. Dimerization of propene (30 min) using the catalyst $[Ni(OAc)_2]$ (5.0×10^{-4} mol l^{-1})–AlR_nCl_{3-n} (2.0×10^{-2} mol l^{-1}) at 20 °C and 1 bar in chlorobenzene (25 ml)[88]

n	Product (ml)	Dimer (%)	Dimer distribution		
			2MP	2,3DMB	H
1	29	82.6	75.8	5.7	18.5
1.5	19	89.3	79.0	4.4	16.6
2	2.5	94.8	80.0	1.5	18.5

former compound for this reason should favour the formation of ionic species. With other nickel(II)-based systems it is commonly observed that a mixture of mono- and dialkylaluminium halides gives the highest catalytic activity[35,39e]. In addition to $AlCl_2$, which is necessary for ionization to occur, the presence of AlR_2Cl ensures that the nickel component is alkylated to an optimal degree. Thus the catalytic action of organometallic Lewis acids in combination with simple nickel(II) salts is well understood based on the theory of its dual functionality: (1) alkylation of the nickel(II) compound and (2) ionization of the catalyst with formation of the highly active $[NiR]^+$ species.

In addition to the active components discussed above, it should also be kept in mind that non-ionic bimetallic complexes might contribute to the total activity of the system. The general criteria for catalytic activity would again be the presence of a free coordination site and an alkyl group attached to nickel, as shown for example in the following formulae:

(11) (12) (13)

Further, $[NiR^+]$ is a highly electron-deficient species and is therefore likely to form complexes with excess free Lewis acid present in the system (reaction 37).

$$[NiR]^+ [-Al-]^- + Al\lessgtr \rightleftharpoons \left[\underset{R}{Ni} Al \right]^+ [-Al-]^- \quad (37)$$

The catalytic properties of such complexes are discussed in ref. 39e. In the absence of olefins, nickel salts, e.g. nickel oleate, are slowly reduced to nickel metal by $Al_2Et_3Cl_3$ at room temperature, owing to the low thermal stability of $EtNiCl$, $EtNi(RCO_2)$, and especially Et_2Ni^{77}.

A large number of active catalysts containing Lewis base (LB) complexes of nickel(II) compounds, $[Ni(LB)_2X_2]$ have been reported. When such complexes are reacted with alkylaluminium halides, $AlR_{3-n}(Y)_n$, in principle the same type of reactions as outlined above for simple nickel salts will occur. Consequently, the presence of organonickel complexes such as $[Ni(LB)_2R(X)]$, $[Ni(LB)_2R(Y)]$, $[Ni(LB)_2R_2]$, and $[Ni(LB)_2R]^+$ $[-Al-]^-$ are expected. Being tetracoordinated, the neutral complexes are not likely to complex with olefins and thus initiate insertion reactions. In a secondary reaction with Lewis acids, however, three coordinate complexes with catalytic activity may be formed:

$$(LB)_2Ni\lessgtr + Al\lessgtr \rightleftharpoons (LB)Ni\lessgtr + LB \cdot Al\lessgtr \quad (38)$$

Also, the cationic complex will react in an analogous fashion forming highly active $[Ni(LB)R]^+$ species:

$$[Ni(LB)_2R]^+ + Al\lessgtr \rightleftharpoons [Ni(LB)R]^+ + LB \cdot Al\lessgtr \quad (39)$$

The catalytic activation $[Ni(LB)_2X_2]$ systems by organometallic Lewis acids thus results from a three-fold action of the Lewis acid: (1) alkylation of nickel, (2) formation of cationic nickel complexes, and (3) complex formation with Lewis bases forming coordinatively unsaturated organonickel species. The alkylation process occurs preferentially with dialkylaluminium halide while complex formation and the formation of Lewis base complexes takes place preferentially with the stronger Lewis acid monoalkylaluminium

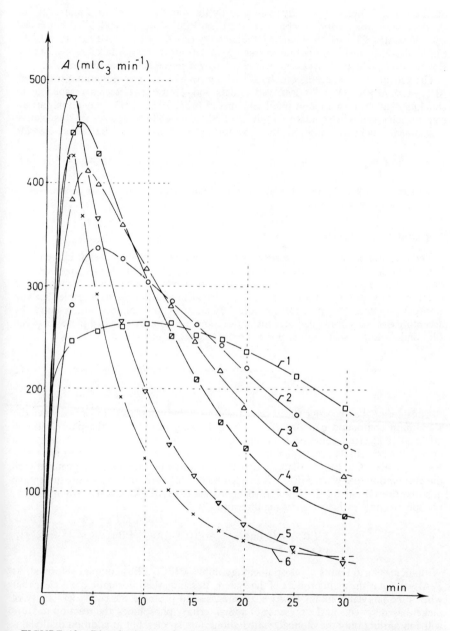

FIGURE 12. Dimerization of propene. The catalytic activity (A) of [Ni(PBu$_3^n$)$_2$(chloroacetate)$_2$] (7.5×10^{-5} mol l^{-1})–AlEt$_n$Cl$_{n-3}$ (2.0×10^{-2} mol l^{-1}) at 20 °C and 1 bar in chlorobenzene (25 ml). Curves: (1) $n = 1$; (2) $n = 1.1$; (3) $n = 1.25$; (4) $n = 1.5$; (5) $n = 1.75$; and (6) $n = 2$

dihalide. These effects are clearly evidenced by the rate data shown in Figure 12. The shortest induction period is obtained when [Ni(PBu$_3^n$)$_2$(chloroacetate)$_2$] is combined with pure AlEt$_2$Cl while the highest initial activity is observed with AlEtCl$_2$ to AlEt$_2$Cl ratios of 1:3. It thus appears that in this system the alkylation of nickel represents the slowest step of the three types of reactions involved in the activation process.

The activation of nickel(0) complexes with Lewis acids such as aluminium trihalides, AlY$_3$, and Al(R)$_{3-n}$(Y)$_n$[87] is less well understood. However, it seems reasonable to conclude that the formation of small amounts of Brönsted acid, HY, play an important role in such systems. By oxidative addition of HY to nickel(0) complexes via dissociative mechanisms nickel(II) catalysts may be formed (reaction 40). The nickel hydride

$$[\text{Ni(LB)}_4] + \text{HY} + 2\text{Al}{<} \longrightarrow [\text{Ni(LB)}_2(\text{H})\text{Y}] + 2\text{LB}\cdot\text{Al}{<} \qquad (40)$$

complexes so formed are, of course, equivalent catalysts to the organonickel(II) compounds discussed above.

b. *The effect of Lewis bases*

The presence of Lewis bases, especially of the PR$_3$ type, has a significant influence on both the activity and the selectivity of reaction. Tertiary phosphines form complexes with both transition metal acceptors and non-transition metal Lewis acids. Owing to the electron configuration and orbital structure of trivalent phosphorus, having filled sp^3 and empty $3d$ orbitals, PR$_3$ may form σ-bonds by electron pair donation and π-bonds by accepting electrons from filled orbitals of the complexing partner. In complexes between PR$_3$ and Al Lewis acids only σ-bonds are formed (equation 41).

$$\qquad (41)$$

Considering the electronic effects only, the most stable complexes are formed when the aluminium compound contains electron-withdrawing ligands and the phosphorus is attached to electron-donating R groups. Consequently, for a given phosphine system in which steric effects are negligible the complex stability will follow the order AlCl$_3$· PR$_3$ > (Al(alkyl)Cl$_2$·PR$_3$ > Al(alkyl)$_2$Cl·PR$_3$. This order of stability is in agreement with the observation that when AlEtCl$_2$ is reacted with PBu$_3^n$ at room temperature in heptane at molar ratios P:Al $\leqslant$ 1:2, AlEtCl$_2$ disproportionates to AlEt$_2$Cl and the insoluble AlCl$_3$· phosphine complex is formed[39a] (equilibrium 42).

$$\text{AlEtCl}_2 + \tfrac{1}{2}\text{PBu}_3^n \rightleftharpoons \tfrac{1}{2}\text{AlEt}_2\text{Cl} + \text{AlCl}_3\cdot\text{PBu}_3^n \qquad (42)$$
$$\text{P:Al} \leqslant 1:2$$

With increasing amounts of phosphine, the soluble AlEtCl$_2$·PBu$_3^n$ complex is formed. At P:Al = 1:1 it is the main product of the system. Based on this information, it is clear that when monoalkylaluminium Lewis acids are used to promote the catalytic activity of nickel-based systems in the presence of added tertiary phosphines, the reaction mixture will contain not only the monoalkylated aluminium species, but in addition dialkylaluminium compounds. Since mixtures of the two types of Lewis acids possess completely different catalytic properties to pure systems (see Figure 12), this effect has to be taken into account when the effect of tertiary phosphines on the catalytic activity of a system is considered. The activity of [Ni(Bu$_3^n$)$_2$(chloroacetate)$_2$] with pre-reacted mixtures of PBu$_3^n$ and AlEtCl$_2$ is shown in Figure 13. Comparing these data with the results reported in

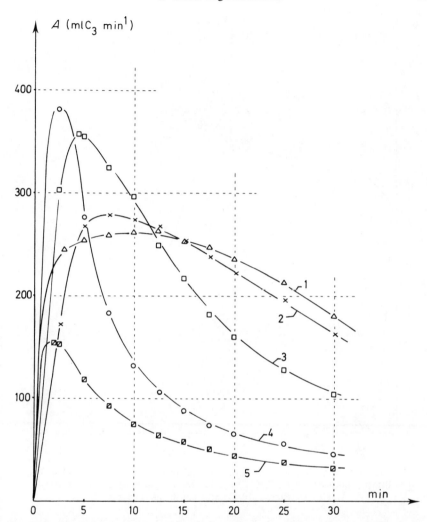

FIGURE 13. Dimerization of propene. The catalytic activity (A) of [Ni(PBu$_3^n$)$_2$(chloroacetate)$_2$] (7.5×10^{-5} mol l^{-1})–AlEtEtCl$^2 \cdot n$PBu$_3^n$ (2.0×10^{-2} mol l^{-2}) at 20 °C and 1 bar in chlorobenzene (25 ml)[88]. Curves: (1) $n = 0$; (2) $n = 1.25$; (3) $n = 0.25$; (4) $n = 0.5$; and (5) $n = 0.75$

Figure 12, it appears reasonable to conclude that a major part of the observed activity change of the catalyst is caused by the disproportionation of AlEtCl$_2$ with PBu$_3^n$. The addition of tertiary phosphines to Al(alkyl)$_2$Cl has a different effect. In this system, significant amounts of free PR$_3$ are in equilibrium with the complex, Al(alkyl)$_2$Cl·PR$_3$. At a molar ratio P:Al > 1:8 the catalytic activity of [Ni(PBu$_3^n$)$_2$(chloroacetate)$_2$] is reduced to approximately zero by the blocking of free coordination sites on nickel by the PR$_3$ (Figure 14). For a given Lewis acid, the complex stability will increase with increasing basicity of the Lewis base. Consequently, in such systems the following order of stability is expected based on electronic effects: P(t-alkyl)$_3$ > P(s-alkyl)$_3$ > P(n-alkyl)$_3$ > P(aryl)$_3$.

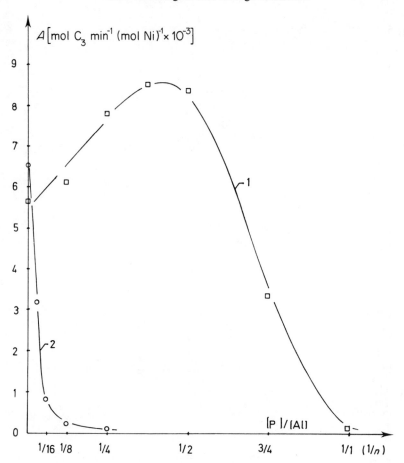

FIGURE 14. Dimerization of propene. The catalytic activity (A) of the catalysts (1) [Ni(PBu$_3^n$)$_2$(chloroacetate)$_2$]–AlEtCl$_2$·1/n PBu$_3^n$ and (2) [Ni(PBu$_3^n$)$_2$(chloroacetate)$_2$]–AlEt$_2$Cl·1/n PBu$_3^n$ at 20 °C and 1 bar in chlorobenzene[88]

When tertiary phosphines are complexed with nickel, π-bonds may be formed in addition to the σ-bonds:

(43)

Based on comparative spectroscopic data, π-ligands can be arranged in the following order with regard to their π-bonding properties: P(aryl)$_3$ > P(benzyl)$_3$ > P(n-alkyl)$_3$ > P(s-alkyl)$_3$ > P(t-alkyl)$_3$.

Since π-complexes are preferentially formed with nickel in low oxidation states, the most significant π-bonding contribution is found in nickel(0) P(aryl)$_3$ complexes. That some π-bonding also takes place when nickel(II) is complexed with PR_3 is in harmony with the observation that phosphines influence the catalytic behaviour of nickel/aluminium systems differently from what is observed with pure σ-donor amines. Tertiary amines will complex the Lewis acid component with high preference and are thus unable to influence the course of the catalytic reaction taking place on nickel species. The main contribution to the stability of nickel(II) phosphine complexes is, however, found in the σ-donor bonds. From a stereochemical point of view, it is further important to note that the structure of tetracoordinated nickel(II) complexes strongly depends on the identity of the ligands. In the case of $[Ni(PR_3)_2X_2]$ tetrahedral structures are found when PR_3 is PPh_3 or PPh_2Bu, whereas the complexes with $P(alkyl)_3$ or $P(c\text{-}Hex)_3$ are square-planar. In solution, mixed alkylarylphosphine complexes exist in an equilibrium mixture of tetrahedral and square-planar forms. Careful studies have shown that the influence of varying R groups in the phosphines is almost entirely electronic rather than steric[94]. By comparing the structures of the complexes of PPh_2Bu and $P(c\text{-}Hex)_3$ which do not differ greatly in size in CH_2Cl_2 solution the molar fractions of the tetrahedral form are close to 1.0 and 0.0, respectively. At 25 °C the rate constants for interconversion of tetrahedral into square-planar isomers are in the range 10^5–$10^6 \, s^{-1}$ with enthalpies of activation of around 45 kJ mol^{-1}. While steric effects appear to play only a minor role in nickel(II) complexes of this type, it is well known that the stability of more crowded nickel(0) complexes of the $[NiL_4]$ type strongly depend on the size and steric requirements of the ligands[95].

The fact that tertiary phosphines form nickel complexes, which are comparable in stability to PR_3 Lewis acid complexes makes phosphines suitable ligands for influencing not only the activity of the catalysts, but also the selectivity of reaction, the selectivity being mainly a function of the structure of the active nickel species.

During the dimerization of propene, 4MP1/4MP2, 2,3DM1, H1/H2, and 2MP1 are formed as primary products, in accordance with Scheme 10. Depending on the isomerization activity of the catalyst, the secondary products 2MP2, 2,3DMB2, and H3 are also produced. Since the second insertion reaction, the formation of C—C bonds, is the rate-limiting step and all other steps are established equilibria, the key to understanding the selectivity lies in the relative rates of reactions a, b, c, and d.

Studies by Wilke, Bogdanović and coworkers reveal that the reaction of nickel(II) hydride and propene gives 70–80%-isopropylnickel at 30–40 °C, independent of the type of ligands present[96], except for the strong Lewis base and stereochemically demanding ligand $PBu^t_2Pr^i$, which is reported[79] to give a Pr^i to Pr^n ratio of 19:81. It thus appears reasonable to assume that the ratio between Ni–Pr^n and Ni–Pr^i for other phosphine systems is constant. The unique effect of PBu^t_2Pr is discussed below.

The selectivity situation outlined in Scheme 10 may be summarized as follows:

Reaction	Primary product	Secondary product
(a) Ni–$Pr^n \xrightarrow{k_a} C_\beta$	2MP1	⇌ 2MP2/4MP2/4MP1
(b) Ni–$Pr^n \xrightarrow{k_b} C_\alpha$	H2/H1	⇌ H3
(c) Ni–$Pr^i \xrightarrow{k_c} C_\beta$	2,3DMB1	⇌ 2,3DMB2
(d) Ni–$Pr^i \xrightarrow{k_d} C_\alpha$	4MP2/4MP1	⇌ 2MP2/2MP1

In the absence of Lewis bases, the individual dimer selectivity in pure nickel(II) Lewis acid systems is independent of the type of nickel(II) compound and Lewis acid as well as the

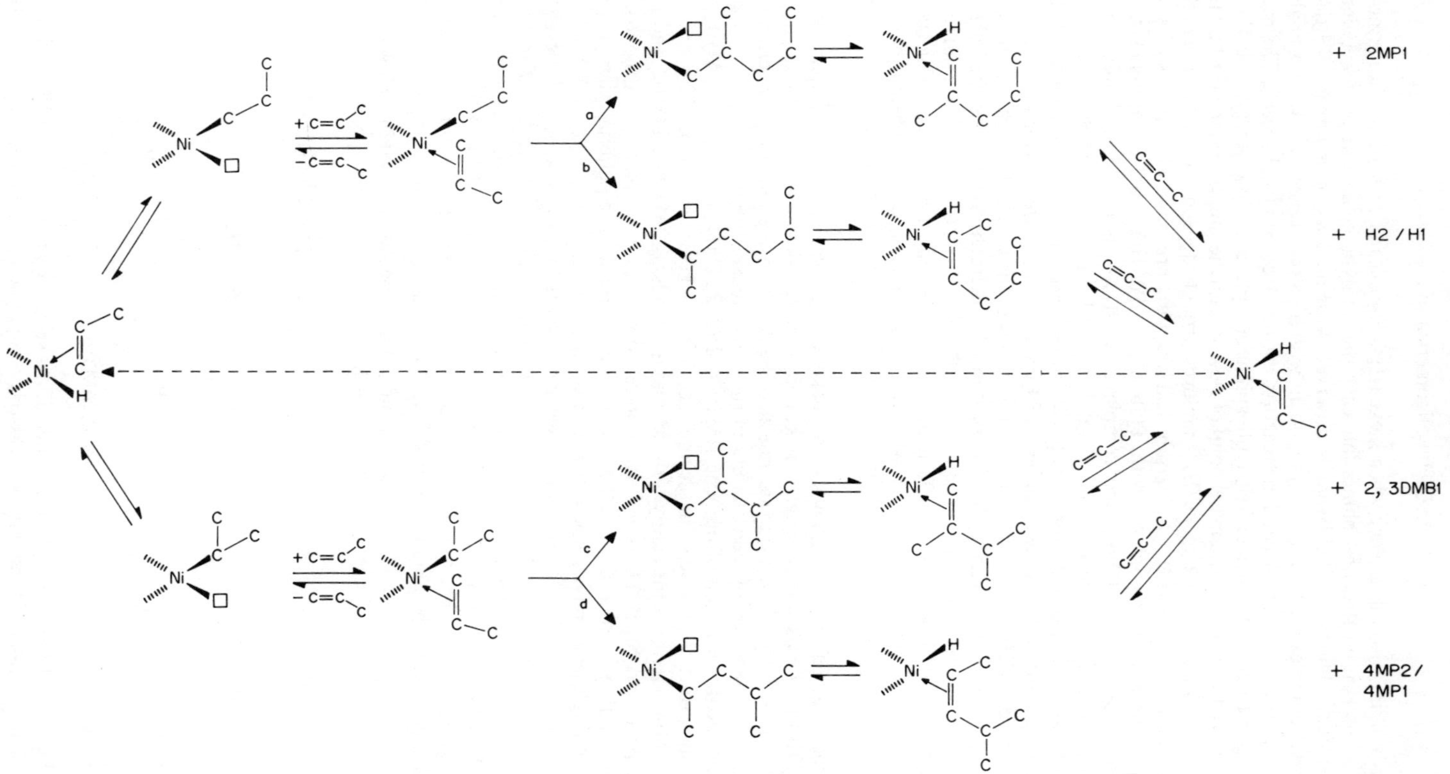

SCHEME 10. Reaction scheme for the dimerization of propene at a nickel(II) centre

conditions of reaction, including the concentrations of both catalyst components and the nature of solvent, and is defined by the product composition: ΣMP (74–80%), ΣH (18–22%), and Σ2,3DMB (2–6%). Consequently, if more than one type of nickel catalyst is responsible for the catalytic action, the different types of active centres do give approximately the same product distribution.

In agreement with fundamental theories concerning the migratory insertion reaction and the basic stereochemistry of nickel(II) compounds, the most probable structures for the active nickel centres are of the cationic square-planar (14) or tetrahedral (15) type. In

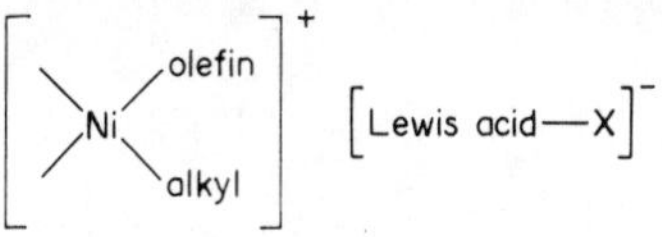

(14) 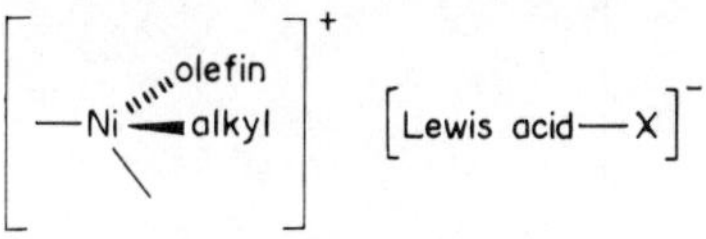   (15)

the absence of Lewis base, 4MP2 and H2 are the main primary products. This fact leads to the conclusion that in such systems the transition-state complexes 16 and 17 (shown for planar complexes) are favoured. In both reactions, the C—C bond formation takes place

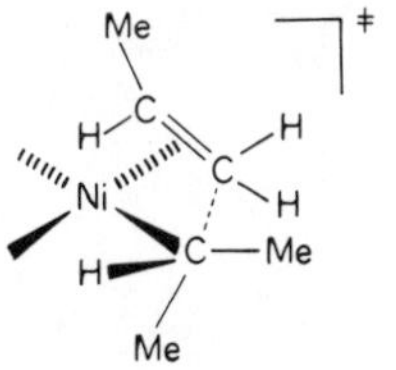

(16) 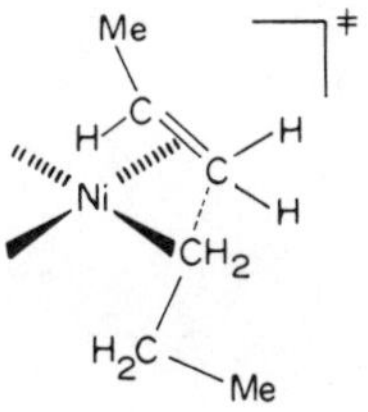   (17)

between non-polar Ni—alkyl groups and the C_α atom of propene (Scheme 10, reaction paths b and d), in agreement with fundamental theories of migratory insertion reactions. The complexed olefin is coplanar with the Ni—C bond, and the molecular overlap takes place at the olefin C_α atom with minimum repulsion. The olefin is attached to nickel via σ-bonds with only a minor contribution of π-backbonding. Hence the energy of activation for bringing the olefin from an upright position to the coplanar structure is expected to be low.

When tertiary phosphines are brought into the system they will complex with the free Lewis acid as well as compete with the [Lewis acid·X]$^-$ anion for the coordination sites on nickel. Depending on the amount of PR$_3$ added, one or two sites may be occupied (complexes 18–22). In the presence of PR$_3$, the selectivity is shifted from 4MP2 and H2

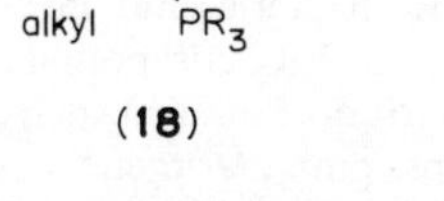

(18) 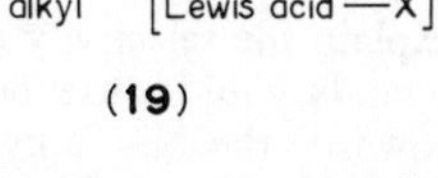   (19) 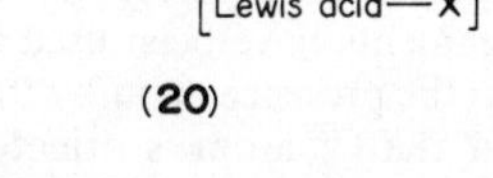   (20)

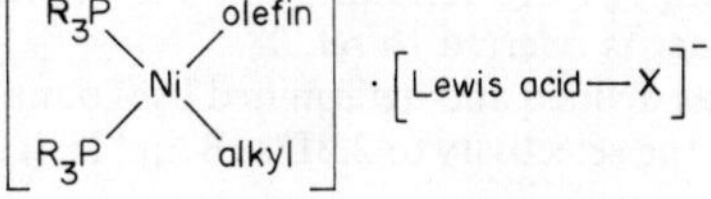

(21) 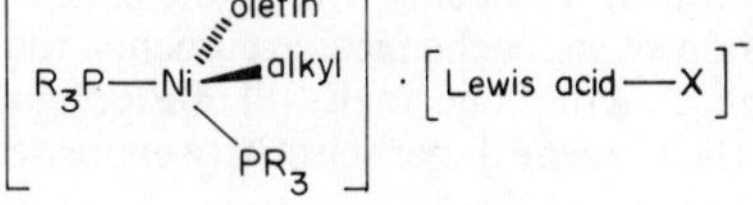   (22)

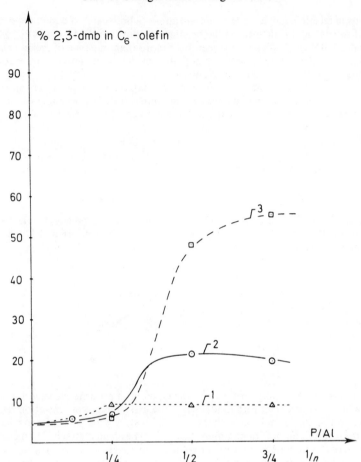

FIGURE 15. Dimerization of propene (30 min). The selectivity to 2,3DMB for the catalysts Ni(acetate)$_2$ (5×10^{-4} mol l^{-1})–AlEtCl$_2$·1/n PR$_3$ at 20 °C and 1 bar in chlorobenzene.[88] Curves: (1) R = Ph; (2) R = Bu^n; and (3) R = c-Hex

towards the formation of the primary products, 2,3DMB1 and 2MP1[39b], as the result of C—C bond formation via reactions a and c (Ni—C → C$_\beta$). The observed shift in selectivity is a function of the chemical nature of PR$_3$ as well as its concentration. Figure 15. Representative selectivity data are given in Table 7[98]. Both steric and electronic arguments have been used to explain the selectivity influence of PR$_3$ compounds.

If the presence of bulky PR$_3$ ligands would cause the olefin to coordinate in such a way that the C$_\beta$ atom is oriented towards the Ni—alkyl groups, pure *steric* effects could explain the observed shift in the mode of C—C bond formation. For the square-planar and tetrahedral systems, transition states **23** and **24** are relevant. For an extensive discussion of stereochemical arguments, the reader is referred to ref. 98.

By plotting the cone angles (θ) of PR$_3$ ligands as defined and determined by Tolman[99] versus the propene dimer selectivity expressed by the selectivity to 2,3DMB and H, it can

4. Olefin oligomerization

TABLE 7. Influence of phosphines on the dimerization of propene using the catalyst $[Ni(\pi-C_3H_5)X]_2$–$Et_3Al_2Cl_3$–PR_3 at $-20\,°C$ and 1 bar[98]

PR_3	H(%)	2MP(%)	2,3DMB (%)
—	19.8	76.0	4.2
PPh_3	21.6	73.9	4.5
$PPh_2benzyl$	19.2	75.4	5.1
PMe_3	9.9	80.3	9.8
PPh_2Pr^i	14.4	73.0	12.6
PEt_3	9.2	69.7	21.1
PBu^n_3	7.1	69.6	23.3
$P(benzyl)_3$	6.7	63.6	29.2
$P(c-Hex)_3$	3.3	37.9	58.8
PPr^i_3	1.8	30.3	67.9
PBu^t_2Me	1.2	24.5	74.0
PBu^t_2Et	0.6	22.3	77.0
$PBu^tPr^i_2$	0.1	19.0	80.9
$PBu^t_2Pr^i$	0.6	70.1	29.1

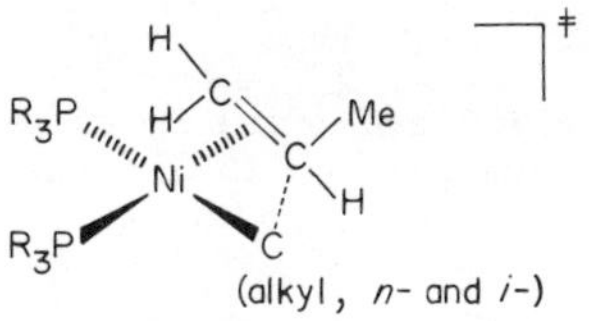

(23) 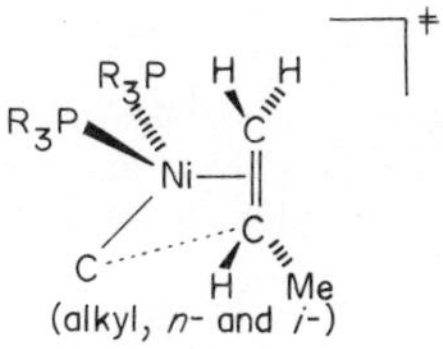   (24)

easily be seen that the increase in 2,3DMB selectivity does not simply follow the increase in PR_3 cone angles. For example, the less bulky ligand, PBu^n_3 ($\theta = 130\,°$) causes more shift in selectivity than the space-demanding ligand PPh_3 ($\theta = 145\,°$), and PPr^i_3 ($\theta = 160\,°$) gives a much higher 2,3DMB selectivity than $P(benzyl)_3$.

In order to explain the experimental observations based on steric effects, it is necessary to consider the stereochemistry of nickel(II) complexes as discussed above. With soft Lewis bases such as PPh_3 and $P(benzyl)_3$, tetrahedral structures are preferred, while $P(alkyl)_3$ give square-planar complexes. Since the steric influence of ligands is expected to be different in the two cases, the soft Lewis bases and the $P(alkyl)_3$ type ligands have to be examined separately. Based on this information, the selectivity data are well explained as shown in Figure 16.

The *electronic* effects expected when tertiary phosphines of high Lewis base strength complex with nickel are (1) increased π-backbonding between nickel and the olefin (especially in square-planar complexes) and (2) polarization of the Ni—C bond. Effects of both these types could cause drastic changes in the C—C bond formation mechanism. The first effect would cause the olefin to be oriented in an upright position and the second effect would favour the alkyl group formally leaving nickel as an anion. Hence these effects may be summarized in the transition state structures **25** and **26** (shown for planar complexes). Nucleophilic attack of i- and n-Pr anions on propene would preferentially occur with the C_β atom (Scheme 10 reactions a and c) giving 2MP1 and 2,3DMB1 as primary products, in harmony with the experimental observations. Since secondary

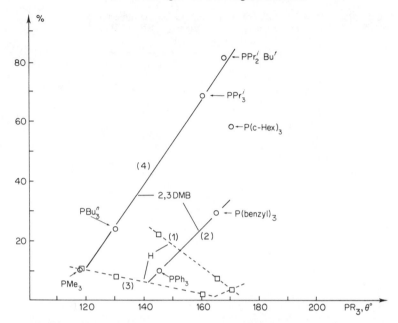

FIGURE 16. Relationship between selectivity and steric parameter for the PR_3 ligand[99] in nickel-catalysed dimerization of propene at $-20\,°C$. Curves: (1) and (2) tetrahedral centres; (3) and (4) square-planar centres. Data from Table 7 and ref. 88

anions are more reactive than primary, the formation of 2,3DMB should be favoured at low temperatures. This was actually observed with the catalyst, $[Ni(\pi\text{-allyl})Cl]_2-AlEtCl_2-PPr^i_2Bu^t$, which at $-60\,°C$ is reported[98] to give 96.3% of 2,3DMB.

As part of this discussion, the propene dimerization selectivity expressed by the selectivity to 2,3DMB and hexenes has been related[100] to the sum of Taft constants[101] for the phosphine substituents R, using the catalyst $[Ni(PR_3)_2X_2]$-$AlEt_2Cl$ at $0\,°C$ and compared with early selectivity data reported by Wilke et al.[102]. A most interesting relationship is observed, as shown in Figure 17. The selectivity to 2,3DMB passes through a maximum when $\Sigma\sigma^*$ is -0.5 to -0.6. The same general trend was observed with the catalyst system $[Ni(PR_3)_2Br_2]$-$AlEtCl_2$ at $20\,°C$ (Table 8). In addition, it is important to note that the selectivity to H with this system goes through a minimum for $\Sigma\sigma^*$ between -0.45 and -0.74 and back up again to the same level as obtained in phosphine-free systems (experiments 6–8, Table 8). Based on the fact that the 2MP composition is also typical for phosphine-free systems, the following explanation seems reasonable. The

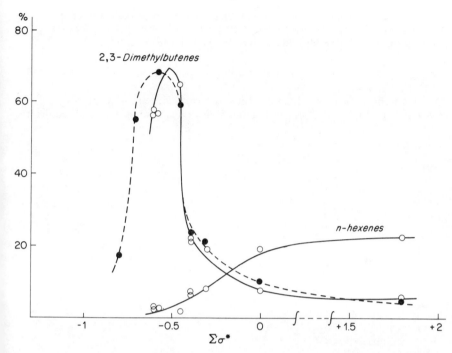

FIGURE 17. Selectivity to 2,3DMB and H during the dimerization of propene as the function of the Taft constants ($\Sigma\sigma^*$) of PR_3. Reproduced from ref. 100 with the permission of the publisher, Archives de l'Académie des Sciences de Paris

TABLE 8. Dimerization of propene (30 min) using the catalyst $[Ni(PR_3)_2Br_2]$ (5.0 × 10^{-4} mole l^{-1})–$AlEtCl_2$ (2.0 × 10^{-2} mol l^{-1}) at 20 °C and 1 bar in chlorobenzene (25 ml)[88]

Expt. No.	PR_3	Product (ml)	Dimers (%)	Dimer distribution			Taft constant
				2MP	2,3DMB	H	
1	PPh_3	22	86.9	70.6	9.6	19.9	+1.80
2	$PH(c\text{-}Hex)_2$	17	93.6	73.2	15.8	11.0	+0.19
3	PBu_3^n	31	91.6	71.0	19.8	9.2	−0.39
4	$P(n\text{-}oct)_3$	36	91.9	73.0	20.8	6.2	−0.42
5	$P(c\text{-}Hex)_3$	25	88.0	48.4	47.9	3.7	−0.45
6	$PEtBu_2^t$	23	84.4	76.8	5.7	17.5	−0.74
7	$P(c\text{-}Hex)Bu_2^t$	27	84.6	77.2	5.9	16.9	−0.75
8	PBu_3^t	23	84.4	76.9	5.5	17.6	−0.96

phosphines $PEtBu_2^t$, $P(c\text{-}Hex)Bu_2^t$, and PBu_3^t in these systems are dissociated from nickel to such a degree that their selectivity influence is lost. Since $NiBr_2$ as such exhibits only low catalytic activity, one phosphine unit remains on nickel while the other is removed and complexed with the Lewis acid, $AlEtCl_2$, present in the system. Consequently, two phosphine units have to be complexed with the active nickel centre in order for the catalyst to give high selectivities to the 2,3DMB propene dimer.

Strohmeier and Müller[103] showed that PR_3 ligands can be ranked in an electronic series based on the A_1 carbonyl stretching frequencies of $[NiPR_3(CO)_3]$. The electronic parameter v, for $PR^1R^2R^3$ ligands is given by the following equation:

$$v = 2056.1 + \sum_{i=1}^{3} R^i \tag{44}$$

The relationship between selectivity and the so determined electronic parameter v for PR_3 compounds in Table 7 is shown in Figure 18. Data for square-planar and tetrahedral

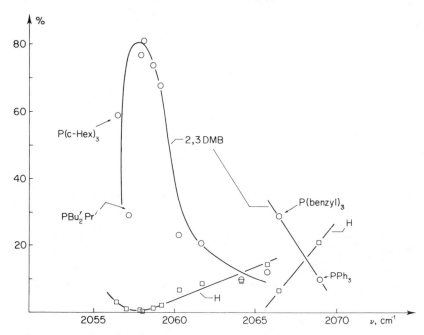

FIGURE 18. Relationship between selectivity and electronic parameter for the PR_3 ligands[103] in nickel-catalysed dimerization of propene at $-20\,°C$ Data from Table 7 and ref. 88

systems are treated here separately. The observation that electronic effects are more pronounced in tetrahedral than in square-planar structures is not easily explained. With $P(benzyl)_3$ a steric effect is very likely to be present. The reduced 2,3DMB selectivity observed with $P(c-Hex)_3$ could be caused by the effect that the ligand to a certain degree is dissociated from nickel.

The only phosphine which does not behave as expected relative to the other PR_3 compounds, is $PPr^iBu^t_2$. From the low H value, it can be concluded that it remains attached to the nickel centre. Here, the observation made by Bönnemann et al.[97] is relevant: nickel hydride complexed with the phosphine in question gives an 'abnormal' ratio of Ni—Pr^i and Ni—Pr^n of 19:81 at $-78\,°C$ compared with other systems including the strong Lewis base $P(c-Hex)_3$, which gives 80:20. Consequently, with the phosphine $PPr^iBu^t_2$ reaction a (the addition of Ni—$Pr^n \rightarrow C_\beta$ of propene giving 2MP1), competes with reaction c (the addition of Ni—$Pr^i \rightarrow C_\beta$ of propene giving 2,3DMB1), owing to an exceptionally high concentration of Ni–Pr^n species in this particular system.

TABLE 9. Codimerization (30 min) of ethene and propene (molar ratio $C_2:C_3 = 1:1$) at 20 °C and 1 bar in chlorobenzene (25 ml) using the catalyst (tetramethylcyclobutadiene)$NiCl_2$ (1.5 × 10^{-4} mol l^{-1})–$AlEtCl_2 \cdot 0.5 PR_3$ (2.0 × 10^{-2} mol l^{-1})[39d]

PR$_3$	Product (ml)	C_4+C_5 $+C_6$ (%)	C_4 (%)	C_5 (%)	C_6 (%)	Codimer distribution				
						3MB1	P1	2MB1	(E)-P2	(Z)-P2 2MB2
PPh$_3$	10	98	22.3	57.0	20.7	—	0.9	24.0	35.5	9.4 30.2
PBu$_3^n$	6	99	24.7	55.3	20.0	0.2	0.8	31.0	23.1	5.9 39.0
P(c-Hex)$_3$	5	99	37.3	44.8	17.9	4.4	1.3	58.7	11.2	6.6 17.8

In conclusion, the influence of phosphine ligands on the selectivity in the nickel-catalysed dimerization of propene is most readily explained on the basis of steric effects. Additional information concerning the influence of soft PR$_3$ compounds is necessary, however, before electronic effects can be neglected.

The codimerization of ethene and propene gives predominantly C_4, C_5, and C_6 olefin products. The primary C_5 products were found[39d] to be 2MB1, P1/P2, and 3MB1 formed in accordance with reactions 45–48.

$$Ni-Et \rightarrow C_\beta(\text{propene}): 2MB1 \quad (45)$$

$$Ni-Et \rightarrow C_\alpha(\text{propene}): P1/P2(Z) \quad (46)$$

$$Ni-Pr^n \rightarrow (\text{ethene}) \quad :P1 \quad (47)$$

$$Ni-Pr^i \rightarrow (\text{ethene}) \quad :3MB1 \quad (48)$$

The selectivity influence of phosphines is in harmony with the selectivity effects discussed for propene dimerization. In the presence of P(c-Hex)$_3$, 2BM1 is the main primary codimer and (Z)-P2 is present in the product mixture in an amount exceeding the thermodynamic value relative to (E)-P2, indicating Z products to be primarily formed. Representative selectivity data are given in Table 9. Representative data concerning the oligomerization of ethene are given in Table 10. The primary products of ethene–butene codimerization are explained by the fact that both B1 and B2 take part in the C_6 product formation[39c]. Further, these data clearly demonstrate the reduced isomerization activity of catalyst mixtures containing P(c-Hex)$_3$ compared with PPh$_3$ and PBu$_3^n$. Also, phosphine-free systems exhibit a lower isomerization activity than systems containing PPh$_3$ or PBu$_3^n$, as clearly demonstrated by the high B1 content in the C_4 product fraction reported in Table 10.

The influence of phosphines on the steric course of propene–butene codimerization is given in Table 11.

With Lewis bases of the P(NR$_2$)$_3$ type a similar selectivity influence is observed as with PR$_3$[80]. Other Lewis bases, e.g. phosphites, amines, pyridines, arsines, stibines, ethers, and thioethers, do not seem to influence the steric course of C—C bond formation. If steric effects alone are responsible for the selectivity influence observed with PR$_3$, these type of ligands must be removed from the active nickel centre to such a degree that their steric effects are lost. Under the assumption that two ligands remain attached to the catalytic species, this observation represents strong evidence for the theory that unique electronic effects play an important role in the selectivity controlling process with PR$_3$ systems.

The high selectivity to linear α-olefins obtained with the SHOP (Shell Higher Olefins Process) type catalysts during the oligomerization of ethene is readily explained by steric effects of the bidentate ligands. In such systems ethene is complexed to the active nickel

TABLE 10. Oligomerization of ethene (30 min) using the catalyst (tetramethylcyclobutadiene)NiCl$_2$ (1.5×10^{-4} mol.l^{-1})–AlEtCl$_2$·0.5 PR$_3$ at 20 °C and 1 bar in chlorobenzene (25 ml)[39c]

PR$_3$	Product (ml)	C$_4$ (%)	C$_6$ (%)	C$_8$ (%)	C$_4$ olefin distribution			C$_6$ olefin distribution							
					B1	(E)-B2	(Z)-B2	H1	H3	3MP1	2EB1	(E)-H$_2$	(E)-3MP2	(Z)-H2	(Z)-3MP2
PPh$_3$	4	76.1	16.7	7.2	7.3	69.5	23.2	1.5	27.7	3.4	—	51.4	—	14.4	1.6
PBu$^n_3$	13	72.0	24.9	2.1	2.3	71.5	26.2	0.3	3.4	0.8	15.1	14.6	19.0	4.9	41.9
PBu$^n_3$	9	89.0	10.5	0.5	2.3	71.7	26.0	—	2.4	2.1	19.8	10.0	21.0	3.3	41.4
P(c-Hex)$_3$	11	72.9	25.1	3.0	8.7	54.4	36.9	1.0	1.2	8.8	45.5	13.4	6.8	4.5	18.8

TABLE 11. Influence of PR_3 on propene–butene codimerization using the catalyst $[Ni(PR_3)_2Cl_2]$–$AlEt_2Cl$ without solvent at $0\,°C^2$

PR_3	n-Heptenes (%)	2-Methylhexenes (%)	3-Methylhexenes (%)	2,3-Dimethylpentenes (%)
$-^a$	16	15	43.5	25.3
PPh_3	20	31	36	13
PBu_3^n	13	27	39	21
$P(c\text{-Hex})_3$	11	13	43	33

$^a[Ni(acac)_2]$.

centre with high preference compared with higher olefins owing to steric factors. With regard to the formation of C_6 olefins the following situation is relevant:
(a) Ethene trimerization

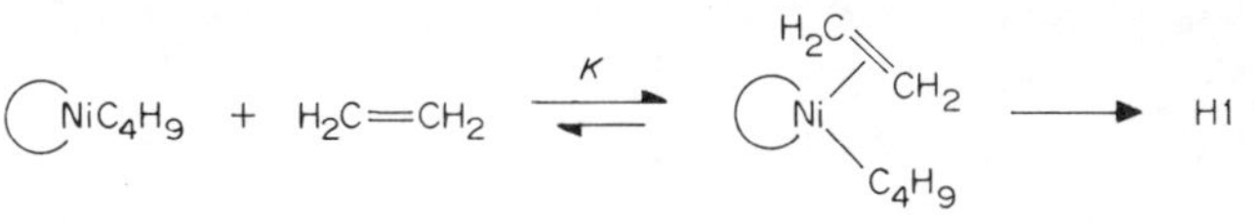

(b) Ethene–B1 codimerization

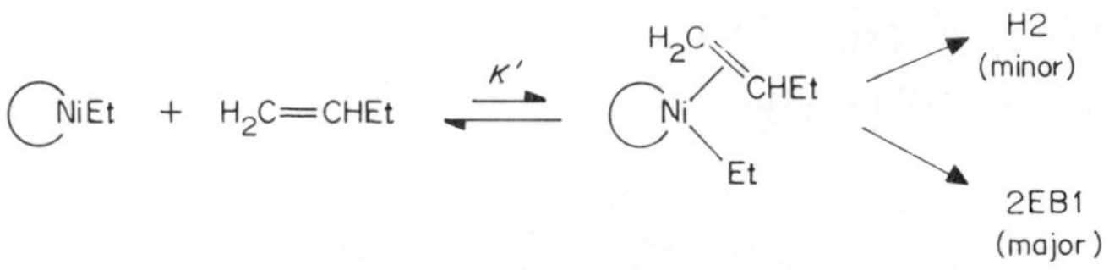

Owing to the kinetic consequence of $K \gg K'$ the rate of codimerization is much smaller than the rate of linear trimerization of ethene. Consequently, the selectivity of the C—C bond formation is kinetically controlled, giving linear α-olefin products.

When α-olefins are back-inserted into the catalytic reaction, insertion occurs during the reaction with Ni—H. This step is stereochemically controlled by the ligands giving Ni n-alkyls with high preference.

c. Kinetics

Under the conditions outlined in Section VIII. B.4, the rate of dimerization is expected to obey equation 49. Consequently, if the concentration of active nickel hydride having a free coordination site is proportional to the total concentration of nickel compound, the dimerization should be first order in nickel and second order with respect to the concentration of monomer:

$$r \sim k_4 K_1 K_2 K_3 [Ni(\Box)(H)][M]^2 \qquad (49)$$

$$r = f k_4 K_1 K_2 K_3 [Ni]_{total}[M]^2 \qquad (50)$$

where

$$f = [Ni(\Box)(H)]/[Ni]_{total} \qquad (51)$$

Since Lewis acids when used as promoters, are normally present in large excess relative to nickel, the rate of dimerization in such systems is expected to be independent of their concentration.

Only a few reports concerning kinetic data have been published[39]. The experimental kinetic analysis is complicated by the following factors: (1) most systems are characterized by the presence of an induction period (ca. 1–5 min); and (2) the rate of reaction depends strongly on the concentration of dimer product owing to chemical inhibition, reduction in monomer partial pressure, and solvent effects. Hence the true initial rates are not readily obtained, and rate comparisons have to be made at the same conversion level. Using the maximum rate at approximately the same conversion level, the following kinetic relationships have been established:

Catalyst I: (tetramethylcyclobutadiene)$NiCl_2$–$AlEtCl_2 \cdot 0.5PBu_3^n$ (excess) in chlorobenzene at 20 °C [39e] —

Reaction	Exp. rate ($mol\,l^{-1}$, min)	$E_{A(exp)}$ ($kJ\,mol^{-1}$)	$-\Delta S^{\ddagger}$ (e.u.)
$C_2 + C_2 \rightarrow C_4$	$r_{22} = 18.8 \times 10^4 [Ni][C_2]^2$	30	20.3
$C_2 + C_3 \rightarrow C_5$	$r_{23} = 7.6 \times 10^4 [Ni][C_2][C_3]$	33	19.0
$C_3 + C_3 \rightarrow C_6$	$r_{33} = 5.1 \times 10^3 [Ni][C_3]^2$	40	18.9
$C_2 + C_4 \rightarrow C_6$	$r_{24} = 6.5 \times 10^2 [Ni][C_2][C_4]$	44	19.6

Catalyst II: [$NiCp(PPh_3)Cl$]–$AlEt_2Cl$ (excess) in chlorobenzene at 20 °C [88] —

Reaction	Exp. rate ($mol\,l^{-1}$, min)	$E_{A(exp)}$ ($kJ\,mol^{-1}$)	$-\Delta S^{\ddagger}$ (e.u.)
$C_2 + C_2 \rightarrow C_4$	$r_{22} = 3.7 \times 10^4 [Ni][C_2]^2$	35	19.4

Catalyst III: [$Ni(PBu_3^n)_2(chloroacetate)_2$]–$AlEtCl_2$ (excess) in chlorobenzene at 20 °C [88] —

Reaction	Exp. rate ($mol\,l^{-1}$, min)	$E_{A(exp)}$ ($kJ\,mol^{-1}$)	$-\Delta S^{\ddagger}$ (e.u.)
$C_3 + C_3 \rightarrow C_6$	$r_{33} = 1.6 \times 10^5 [Ni]^{1.2} [C_3]^2$	77	19.4

Within given concentration limits, the catalyst systems I and II give rates which are related to the concentration of nickel compound and monomer with the expected orders. With catalyst III an order of reaction with respect to the concentration of nickel complex greater than unity was observed, and the experimental Arrhenius activation energy obtained in the temperature range 0–20 °C was exceptionally high. A reasonable exploration for these kinetic observations is that part of the PBu_3^n ligand is dissociated from nickel and reacted with $AlEtCl_2$ giving $AlEt_2Cl$ and $AlCl_3 \cdot PBu_3^n$. As discussed in Section VIII. B.4, this would give an additional contribution to the rate of dimerization.

The kinetic studies have further clearly shown that the rate of reaction is strongly dependent on the dielectric constant of the solvent[39b]. For instance, with catalyst I and with catalyst II, the dimerization of ethene is nearly three times faster in *o*-dichlorobenzene than in chlorobenzene[88].

$$\frac{fk_4 K_1 K_2 K_3 (\text{chlorobenzene})}{f' k'_4 k_1 K_2 K_3 (\text{benzene})} \approx 5 \qquad (52)$$

Consequently, it appears reasonable to conclude that ionic or highly polarized catalytic complexes are involved in the dimerization reactions. The observed kinetic orders with respect to the concentrations of monomer and nickel complex are in harmony with a 'double-insertion' mechanism in which the migratory insertion of monomer into N—C bonds represents the rate-limiting step[39e].

2. Rhodium complex catalysis

The mechanism of Rh-catalysed dimerization of olefins is closely related to the nickel system in that metal hydride and metal alkyls are formed as intermediates in the catalytic cycle. In accordance with the dimerization in Scheme 11 proposed by Cramer[104], based

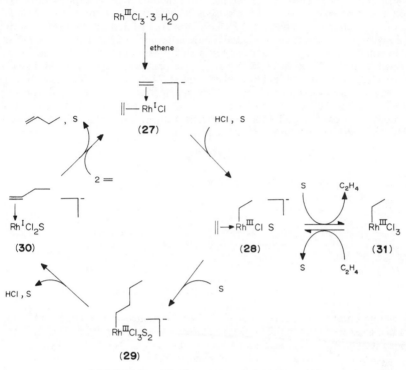

SCHEME 11. Rhodium-catalysed dimerization[104].

on extensive spectroscopic and kinetic evidence, both anionic rhodium(I) and rhodium(III) complexes take part in the catalytic reaction. The dimerization is usually performed in alcoholic HCl media at around room temperature and atmospheric pressure. The primary product of ethene dimerization is B1 but, owing to the high isomerization activity of the catalyst, (Z/E)-B1 is formed[105]. Dimer yields greater than 99% were reported[104]. Propene yields a mixture of n-hexenes and iso-hexenes. Its rate of dimerization is, however, much slower than the rates obtained with ethene.

The activity of the catalyst is halide dependent. Other types of acids such as H_2SO_4, HNO_3, CH_3COOH, or CF_3COOH result in low catalytic activity. When $RhCl_3 \cdot 3H_2O$ is used to initiate the dimerization of ethene near atmospheric pressure, there is an induction period of about 30 min. During this period rhodium(III) is reduced to rhodium(I) by ethene. Primarily the sparingly soluble $[Rh(C_2H_4)_2Cl]_2$ complex is formed, which subsequently in converted to the anionic complex $[Rh(C_2H_4)_2Cl_2]^-$ by reaction with HCl.

With reference to Scheme 11, complex **27** is rapidly converted by reaction with HCl into an ethylrhodium(III) compound, **28**. From an elementary reaction point of view, it appears likely that this process proceeds via oxidative addition of HCl to the hydride intermediate, which in a facile reaction is converted into the ethyl complex by ethene insertion into the Rh—H bond. Owing to the high rate of insertion, rhodium hydride could not be detected by n.m.r. measurements. **28** rearranges by a slow, rate-determining insertion reaction, giving the *n*-butylrhodium(III) complex **29**. **29** decomposes rapidly through loss of HCl to give the B1 complex **30** of rhodium(I). Again, this is likely to be a reaction composed of more than one elementary step, for example β-H elimination followed by reductive elimination of HCl. Coordinated B1 and solvent in **30** are rapidly displaced by ethene, reforming the initial rhodium(I) complex **27**. At high ethene pressures and in ethanol solution above about 0.1 M in HCl, most of the rhodium in the system is reported to be in the form of **28**, but at ethene pressures near 1 bar, **28** dissociates extensively, rapidly, and reversibly to ethene and **31**. The mechanism outlined above is in harmony with the experimental kinetics of the system, the rate of dimerization being first order in rhodium, H^+, Cl^-, and ethene concentrations, equation 53. Rate constants and thermodynamic parameters for the dimerization of ethene–ethanol solution 1.00 mol. l^{-1} in HCl and 0.67 mol. l^{-1} in rhodium are given in Table 12.

$$r = k[\text{Rh}][\text{H}^+][\text{Cl}^-][\text{C}_2\text{H}_4] \tag{53}$$

TABLE 12. Kinetic data for the dimerization of ethene catalysed by rhodium (I/III)[104]

Temperature (°C)	$10^4 k$ (s^{-1})	$E_{A(exp)}$ (kJ l mol^{-1})	$-\Delta S^{\ddagger}$ (cal l mol^{-1} K^{-1})
10	0.350, 0.371		
30	2.67, 2.79	72	20.1
50	16.8, 16.9, 17.9		

It is interesting to compare the values of $\Delta S^{\ddagger}$ obtained in the nickel and rhodium systems, since both refer to the same type of elementary process, the formation of C—C bonds through migratory insertion of olefin into metal—C bonds: $\Delta S^{\ddagger}$ (Rh at 30 °C = $-$ 20.1 e.u. and $\Delta S^{\ddagger}$ (Ni at 20 °C) = $-$ 19.4 and $-$ 20.3 e.u. These values are also close to the $\Delta S^{\ddagger} = -21.1$ e.u. determined for the insertion of CO into the Mn—C bond of methylmanganese pentacarbonyl[106].

Rhodium(I) complexes are activated by Lewis acids[87]. Using [RhCl(PPh$_3$)$_3$]–AlEtCl$_2$ in chlorobenzene solution at 20 °C and 5 bar, propene, which had proved to be one of the less reactive olefins in pure rhodium systems, was converted with good rates into a product mixture composed of 87.6% of C_6 olefins and 12.4% of C_9 and higher products. The C_6 fraction contained 68.3% of MP, 10.7% of DMB, and 21% of H, which is practically the same selectivity as obtained in the corresponding nickel complex system.

3. Titanium complex catalysis

The classical Ziegler systems composed of TiCl$_3$ or TiCl$_4$ and Et$_3$Al convert simple olefins such as ethene and propene into high molecular weight polymers. These catalysts are heterogeneous and the polymer chain grwoth reaction takes place on the surface of the insoluble catalyst particles. Even the TiCl$_4$–Et$_3$Al system, which is composed of two soluble components, becomes heterogeneous on mixing the two compounds, owing to reduction of titanium(IV) to insoluble titanium(III), and partly titanium(II). It is generally agreed that the C—C formation process occurs on the surface of the titanium(III) particles.

Mixtures of titanium esters and trialkylaluminium produce dimer (B1) or high polymer products from ethene depending on the catalyst component ratio. At R_3Al:$[Ti(OR)_4] > 20$ high polymers are formed, whereas systems having R_3Al:$[Ti(OR)_4] < 10$ produce predominantly B1 with selectivities of the order of 90%[107,108].

The use of halide-substituted organoaluminium compounds instead of R_3Al generally favours the formation of dimers and lower oligomers. For instance, when $TiCl_4$ is used in combination with $AlEt_2Cl$ or $AlEtCl_2$ in benzene solution at 5 °C and 1 bar, oligomer yields of 30% and 92% respectively, are obtained, the rest being solid polymer[109]. The selectivity in such systems is strongly dependent on the polarity of the solvent and the reaction temperature. In polar media such as halo-substituted hydrocarbons (e.g. methylene chloride, chloroform, and chlorobenzenes) and low temperatures, the dimerization predominates. All these observations support the theory that cationic titanium(IV) complexes are mainly responsible for the dimerization.

Bestian and coworkers[41,110] have investigated in detail the oligomerization of ethene with the $MeTiCl_3$–$MeAlCl_2$ catalyst at low temperatures (down to -100 °C) in methylene chloride solution and concluded that the active catalyst consists of the cationic titanium(IV) complex **32**. The function of the Lewis acid in this system is to promote the formation of cationic titanium(IV) centres. When $TiCl_4$ is used instead of $MeTiCl_3$, the Lewis acid has a dual function as outlined for the nickel system: (a) alkylation of Ti and (b) formation of ionic complexes. That cationic complexes favour dimerization over polymerization is well explained by the effect that such complexes selectively increase the rate of β-H elimination.

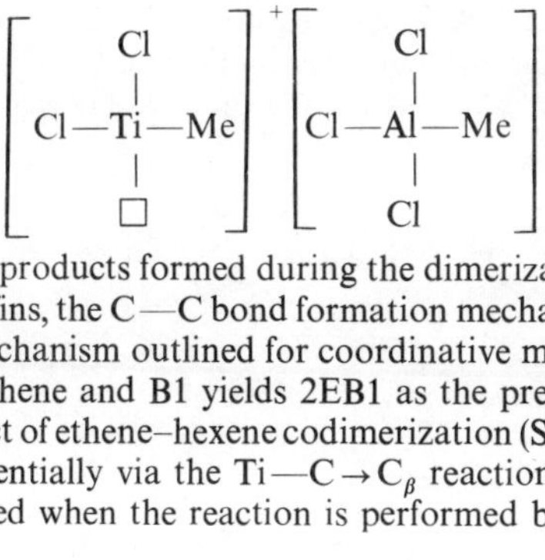

Based on the primary products formed during the dimerization and codimerization of ethene with higher α-olefins, the C—C bond formation mechanism with titanium is likely to follow the general mechanism outlined for coordinative metal catalysis. For instance, the codimerization of ethene and B1 yields 2EB1 as the predominant C_6 product, and 2EH1 is the main product of ethene–hexene codimerization (Scheme 12). The C—C bond formation occurs preferentially via the Ti—$C_\alpha \rightarrow C_\beta$ reaction, since less than 30% of n-olefin product is obtained when the reaction is performed below -70 °C.

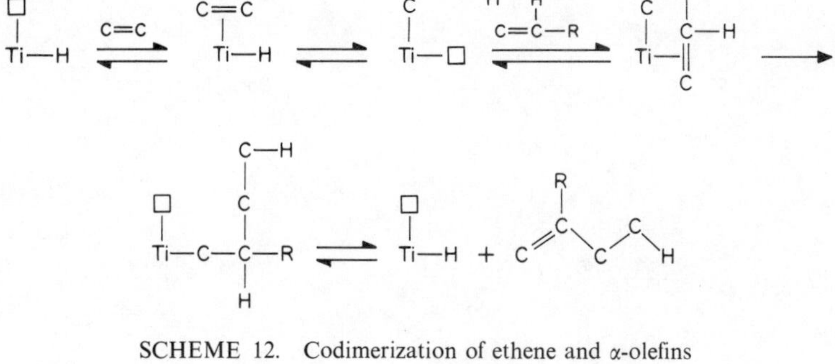

SCHEME 12. Codimerization of ethene and α-olefins

A kinetic study of the ethene oligomerization was carried out using the soluble catalyst system [(EtO)$_3$TiCl]–AlEtCl$_2$ in toluene solution[109]. For ethene pressures in the range 3–12 bar and with Al/Ti ⩾ 7 the rate of reaction was found to be first order both in Ti and in the concentration of monomer. This is in agreement with a kinetic scheme in which the insertion of ethene into the growing chain is rate determining and the insertion of ethene into Ti—H is a very facile reaction. In this respect the system is kinetically analogous to the rhodium system.

Compared with TiCl$_4$ under given conditions, [(EtO)$_3$TiCl] gives an oligomerized product with a much higher content of linear α-olefins[111]. Under certain restricted reaction conditions (low temperature and high concentration of ethene) the [(EtO)$_3$TiCl] catalyst is reported to give predominately (> 90%) linear α-olefins[109]. In this case, only a minor amount of intermediate product, RC=C, is inserted directly into the growing chain, owing to the effect that Ti(☐)R coordinates ethene with high preference. It appears likely that intermediate products are back-inserted into the catalytic cycle via the reaction with Ti—H species:

$$TiCH_2CH_2R \rightleftharpoons Ti(\square)H + CH_2\!=\!CHR \tag{54}$$

The observed selectivity is in agreement with such a mechanism.

4. Tantalum complex catalysis

A dimerization mechanism entirely different in nature from those discussed above was proposed by McLain and Schrock[112]. On reacting [TaCpCl$_2$(=CHCMe$_3$)] with propene in pentane at 0 °C, 2,4,4-trimethylpent-1-ene and the tantallocycle complex **33** (identified by n.m.r.) was obtained. At 35 °C in decane the metallocycle **33** rapidly decomposes,

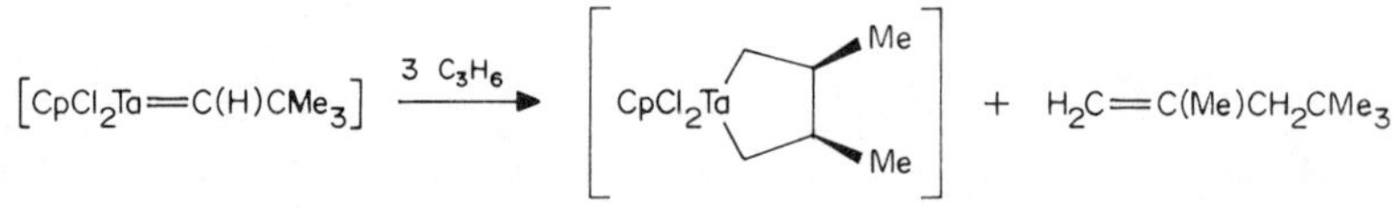

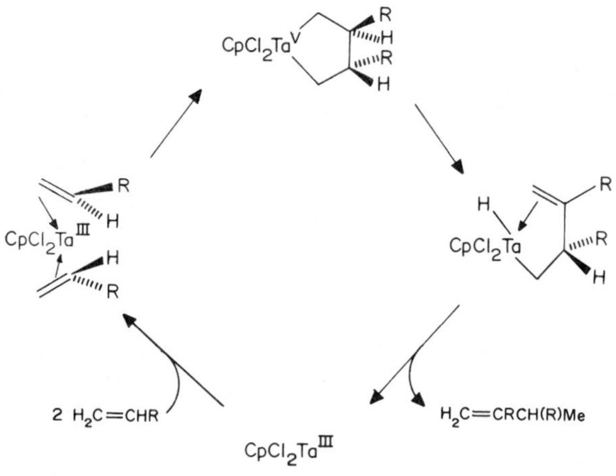

SCHEME 13. Tantalum-catalysed olefin dimerization[112]

yielding 68% of 2,3DMB1 and an unidentified brown powder. In the presence of propene, a catalytic dimerization of propene to 2,3DMB1 with a selectivity exceeding 90% occurs. At 45 °C and 40 psi of propene an initial rate corresponding to two turnovers per hour is observed. The catalyst activity steadily decreases as an orange, unidentified paramagnetic complex precipitates, with total turnovers of 20 under these conditions. Thus it is clearly demonstrated that the system can operate catalytically (Scheme 13). The rate-determining reaction is believed to be step 1, the transfer of β-H from the substrate to tantalum(V). The C—C bond formation occurs through the oxidative coupling of two olefin units, a type of elementary process which has been observed with halo-substituted olefins with other transition metal complexes, e.g. $[Fe(C_2F_4)_2(CO)_3]$[113] and $[Ni(C_2F_3H)_2(PPh_3)_2]$[114].

IX. REFERENCES

1. V. Sh. Fel'dblyum and N. V. Obeschalova, *Russ. Chem. Rev.*, **37**, 789 (1968).
2. G. Lefebvre and Y. Chauvin, in *Aspects of Homogenous Catalysis* (Ed. R. Ugo), Vol. 1, Carlo Manfredi, Milan, 1970, p. 107
3. V. Mark and H. Pines, *J. Am. Chem. Soc.*, **78**, 5946 (1956).
4. H. Pines, *Synthesis*, 309 (1974).
5. A. W. Shaw, C. W. Bittner, W. V. Bush, and G. Holzman, *J. Org. Chem.*, **30**, 3286 (1965).
6. California Research Corp., *U.S. Pat.*, 2 986 588, 1961.
7. Goodyear Tire & Rubber Co., *Belg. Pat.*, 629 861, 1963; *Chem. Abstr.*, **60**, 14384 (1964).
8. Toyo Rayon Co., Ltd., *Jap. Pat.*, 26 826, 1964; *Ref. Zh., Khim.*, 14N16 (1967).
9. Ethyl Corp., *Fr. Pat.*, 1 356 267, 1964; *Chem. Abstr.*, **61**, 4211 (1964).
10. Ethyl Corp., *Br. Pat.*, 933 700, 1963; *Chem. Abstr.*, **60**, 2749 (1964).
11. Universal Oil Products Co., *Br. Pat.*, 917 358, 1963; *Chem. Abstr.*, **59**, 2645 (1963).
12. Universal Oil Products Co., *U.S. Pat.*, 3 148 157, 1964; *Chem. Abstr.*, 15973 (1964).
13. British Petroleum Co., Ltd., *U.S. Pat.*, 3 084 206, 1961; *Off. Gaz.*, No. 789, 243 (1963).
14. J. B. Wilkes, *J. Org. Chem.*, **32**, 3231 (1967).
15. Ethyl Corp., *U.S. Pat.*, 3 185 745, 1965; *Chem. Abstr.*, **63**, 4160 (1965).
16. Goodyear Tire & Rubber Co., *Belg. Pat.*, 613 324, 1962; *Chem. Abstr.*, **57**, 16395f (1962).
17. Goodyear Tire and Rubber Co., *Neth. Pat.*, 6 402 598, 1964; *Chem. Abstr.*, **62**, 6393 (1965).
18. California Research Corp., *U.S. Pat.*, 3 175 020, 1965; *Chem. Abstr.*, **63**, 1696 (1965).
19. I. I. Pis'man, M. A. Dalin, V. R. Ansheles, and G. V. Vasil'kovskaya, *Dokl. Akad. Nauk* SSSR, **179**, 608 (1968).
20. J. K. Hambling, *Chem. Br.*, **5**, 354 (1969).
21. J. K. Hambling and R. P. Northcott, *Rubber Plast. Age*, **49**, 224 (1968).
22. J. Wilkes, paper presented at Seventh International Petroleum Congress, Mexico, April 1967.
23. W. V. Bush, G. Holzman, and A. W. Shaw, *J. Org. Chem.*, **30**, 3290 (1965).
24. M. Sitting, *Chemicals from Propene*, Chemical Process Monograph, No. 9, Noyes Development Corp., New York, 1965, p. 103.
25. E. Terres, *Brennst.-Chem.*, **34**, 335 (1953).
26. (a) K. Ziegler, H.-G. Gellert, H. Kühlhorn, H. Martin, K. Meyer, K. Nagel, H. Sauer and K. Zosel, *Angew. Chem.*, **64**, 323 (1952); (b) K. Ziegler, H.-G. Gellert, K. Zosel, E. Holzkamp, J. Schneider, M. Söll, and W.-R. Kroll, *Justus Liebigs Ann. Chem.*, **629**, 121 (1960).
27. F. E. Mathews and E. H. Strange, *Br. Pat.*, 24 790, 1910.
28. O. T. Onsager, *Thesis*, Technische Hochschule München, 1962.
29. H. Sinn, C. Lundberg, and O. T. Onsager, *Makromol. Chem.*, **70**, 22 (1964).
30. J. P. Oliver, *Adv. Organomet. Chem.*, **8**, 167 (1970).
31. *Chem. Week*, May 6, 73 (1961).
32. B. Bogdanović and G. Wilke, *Brennst.-Chem.*, **49**, 323 (1968).
33. A. Job and R. Reich, *C.R. Acad. Sci.*, **177**, 330 (1924).
34. Phillips Petroleum Co., *U.S. Pat.*, 2 969 408, 1961; *Chem. Abstr.*, **55**, 16009 (1961).
35. J. Ewers, *Angew. Chem.*, **78**, 593 (1966).
36. G. Wilke, B. Bogdanović, P. Hardt, P. Heimbach, W. Keim, M. Kröner, W. Oberkirch, K. Tanaka, E. Steinrücke, D. Walter, and H. Zimmermann, *Angew. Chem.*, **78**, 157 (1966).

37. J. Chauvin, N. H. Phung, N. Genchard, and G. Lefebvre, *Bull. Soc. Chim. Fr.*, 3223 (1966).
38. V. Sh. Fel'dblyum, N. V. Obeschalova, and A. I. Leshcheva, *Dokl. Akad. Nauk SSSR*, **172**, 111 (1967).
39. (a) O. T. Onsager, H. Wang, and U. Blindheim, *Helv. Chim. Acta*, **52**, 187 (1969); (b) **52**, 196 (1969). (c) **52**, 215 (1969). (d) **52**, 224 (1969). (e) **52**, 230 (1969).
40. W. Keim and F. H. Kowaldt, *Erdöl und Kohle, Erdgas, Petrochem.*, Compendium 78/79, 453 (1978).
41. H. Bestian and K. Clauss, *Angew. Chem., Int. Ed. Engl.*, **2**, 704 (1963).
42. G. Hata, *Chem. Ind. (London)*, 223 (1965).
43. T. Alderson, E. L. Jenner, and R. V. Lindsey, *J. Am. Chem. Soc.*, **87**, 5638 (1965).
44. R. Cramer, *Acc. Chem. Res.*, **1**, 186 (1968).
45. E. I. DuPont de Nemours and Co., *U.S. Pat.*, 3 013 066, 1961.
46. J. T. van Gemert and P. R. Wilkinson, *J. Phys. Chem.*, **68**, 645 (1964).
47. N. H. Phung and G. Lefebvre, *C.R. Acad. Sci., Ser. C*, **265**, 519 (1967).
48. S. J. McLain and R. R. Schrock, *J. Am. Chem. Soc.*, **100**, 1315 (1978).
49. W. C. Zeise, *Poggendorfs Ann. Phys. Chem.*, **9**, 632 (1827).
50. M. J. S. Dewar, *Bull. Soc. Chim. Fr.*, **18**, C79 (1951); M. J. S. Dewar and G. P. Ford, *J. Am. Chem. Soc.*, **101**, 783 (1979).
51. J. Chatt and L. A. Duncanson, *J. Chem. Soc.*, 2939 (1953).
52. J. P. Collman and L. S. Hegedus, *Principles and Applications of Organotransition Metal Chemistry*, University Science Books, Mill Valley, CA, 1980, p. 33.
53. G. E. Coates, M. L. H. Green, and K. Wade, *Organometallic Compounds: The Transition Elements*, Vol. II, Methuen, London, 1968, p. 32.
54. L. Fr. Que and L. H. Pignolet, *Inorg. Chem.*, **12**, 156 (1973).
55. J. A. J. Jarvis, B. T. Kilbourn, and P. G. Owston, *Acta Crystallogr., Sect. B*, **27**, 366 (1971).
56. M. Black, R. H. B. Mais, and P. G. Owston, *Acta Crystallogr., Sect. B.*, **25**, 1753 (1969).
57. R. A. Love, T. K. Koetzle, G. J. B. Williams, L. C. Andrews, and R. Ban, *Inorg. Chem.*, **14**, 2653 (1975).
58. D. L. Thorn and R. Hoffmann, *J. Am. Chem. Soc.*, **100**, 2079 (1978).
59. R. Cramer, J. B. Kline, and J. D. Roberts, *J. Am. Chem. Soc.*, **91**, 2519 (1969).
60. B. Åkermark, M. Almemark, J. Almlöf, J.-E. Bäckval, B. Roos and Å. Støgard, *J. Am. Chem. Soc.*, **99**, 4617 (1977).
61. (a) F. R. Hartley, *Chem. Rev.*, **73**, 163 (1973); (b) I. I. Moiseev and M. N. Vargaftik, *Dokl. Acad. Nauk. SSSR*, **152**, 147 (1963); (c) S. V. Pestrikov, I. I. Moiseev and T. N. Romanova, *Russ. J. Inorg. Chem.*, **10**, 1199 (1965).
62. C. A. Tolman, *J. Am. Chem. Soc.*, **96**, 2780, (1974).
63. T. H. Upton and W. A. Goddard, *J. Am. Chem. Soc.*, **100**, 321 (1978).
64. R. Cramer, *J. Am. Chem. Soc.*, **89**, 4621 (1967).
65. P. M. Henry, *J. Am. Chem. Soc.*, **88**, 1595 (1966).
66. N. Rösch, R. P. Messmer, and K. H. Johanson, *J. Am. Chem. Soc.*, **96**, 3855 (1974).
67. K. Fukui and S. Inagaki, *J. Am. Chem. Soc.*, **97**, 4445 (1975).
68. J. Furukawa, *Shokubai*, **13**, 107 (1971).
69. P. Cossee, *Recl. Trav. Chim. Pays-Bas*, **85**, 1151 (1966).
70. F. W. S. Benfield and M. L. H. Green, *J. Chem. Soc. D*, 1324 (1974).
71. A. J. Deeming, B. F. G. Johnson, and J. Lewis, *J. Chem. Soc. D*, 1848 (1973).
72. J. Chatt and B. L. Shaw, *J. Chem. Soc.*, 5075 (1962).
73. H. C. Clark and H. Kurosawa, *Inorg. Chem.*, **11**, 1275 (1972).
74. H. C. Clark, C. Jablonski, J. Halpern, A. Mantovani, and T. A. Weil, *Inorg. Chem.*, **13**, 1541 (1974).
75. G. M. Whitesides, J. F. Gaasch, and E. R. Stedronsky, *J. Am. Chem. Soc.*, **94**, 5258 (1972).
76. H. Werner and R. Feser, *Angew. Chem., Int. Ed. Engl.*, **18**, 157 (1979).
77. N. V. Obeshchalova, V. Sh. Fel'dblyum, and N. M. Pashchenko, *J. Org. Chem. USSR*, **4**, 982 (1968).
78. J. Ewers, *Angew. Chem., Int. Ed. Engl.*, **5**, 584 (1966).
79. Shell Oil Co., *U.S. Pat.*, 3 424 815, 1967.
80. Sentralinstituttet for Industriell Forskning, *Norw. Pat.*, 113 388, 1966.
81. ICI, *Neth. Pat. Appl.*, 68 12220, 1967.

82. Shell International Research Maatschappij, *Fr. Pat.*, 1 385 503, 1962.
83. Shell Oil Co., *U.S. Pat.*, 3 327 015, 1963.
84. Studiengesellschaft Kohle, *Br. Pat.* 1 058 680 1963.
85. Farbwerke Hoechst, *Neth. Pat. Appl.*, 67 06533, 1966.
86. G. Hata and A. Miyake, *Chem. Ind. (London)*, 921 (1967).
87. Sentralinstituttet for Industriell Forskning, *Norw. Pat.*, 116 245, 1966.
88. O. T. Onsager, U. Blindheim, and H. Wang, *Report to the Royal Norwegian Council for Scientific and Industrial Research*, No. 5, Project OH 660108, 1968.
89. Sun Oil Co., *U.S. Pat.*, 3 459 825, 1967, and 3 472 911, 1968.
90. Shell Development, *U.S. Pat.*, 3 644 563, 1972, and 3 737 475, 1973.
91. (a) *Chem. Week*, **70**, Oct. 23 (1974); (b) *Chem. Mark. Rep.*, April 18 (1977); (c) F. H. Kowaldt, *Thesis*, Rheinisch-Westfälische Technische Hochschule, Aachen, 1977.
92. W. Keim, *Chimia*, **35**, 344 (1981).
93. U. Birkenstock, H. Bönnemann, B. Bogdanović, D. Walter, and G. Wilke *Adv. Chem. Ser.*, No. 70, 250 (1968).
94. L. Que, Jr., and H. Pigmolet, *Inorg. Chem.*, **12**, 156 (1973).
95. C. A. Tolman, *Science*, **181**, 501 (1973).
96. B. Bogdanović, B. Henc, H. G. Karmann, H. G. Nüssel, D. Walter, and G. Wilke, *Ind. Eng. Chem.*, **62**, 34 (1970).
97. H. Bönnemann, C. Grard, W. Kopp and G. Wilke, *23rd Int. Congr. Pure Appl. Chem.*, **6**, 265 (1971).
98. B. Bogdanović, *Adv. Organomet. Chem.*, **17**, 195 (1979).
99. C. A. Tolman, *Chem. Rev.*, **77**, 313 (1977).
100. M. Uchino, Y. Chauvin, and G. Lefebvre, *C.R. Acad. Sci., Ser. C*, **265**, 103 (1967).
101. R. W. Taft, Jr., *J. Am. Soc.*, **75**, 4231 (1953).
102. G. Wilke, B. Bogdanović, P. Hardt, P. Heimbach, W. Keim, M. Kröner, W. Oberkirch, K. Tanaka, E. Steinrücke, D. Walter and H. Zimmermann, *Angew. Chem., Int. Ed. Engl.*, **5**, 151 (1966).
103. W. Strohmeier and F. J. Müller, *Chem. Ber.*, **100**, 2812 (1967).
104. R. Cramer, *J. Am. Chem. Soc.*, **87**, 4717 (1965).
105. R. Cramer, *J. Am. Chem. Soc.*, **88**, 2272 (1966).
106. F. Calderazzo and F. A. Cotton, *Inorg. Chem.*, **1**, 30 (1962).
107. G. Natta, *J. Polym. Sci.*, **34**, 151 (1959).
108. M. Farina and M. Ragazzini, *Chim. Ind. (Milan)*, **40**, 816 (1958).
109. G. Henrici-Olivé and S. Olivé, *Adv. Polym. Sci.*, **15**, 1 (1974).
110. H. Bestian, K. Clauss, H. Jensen, and E. Prinz, *Angew. Chem., Int. Ed. Engl.*, **2**, 32 (1963).
111. G. Henrici-Olivé and S. Olivé, *Chem.-Ing.-Tech.*, **43**, 906 (1971).
112. S. J. McLain and R. R. Schrock, *J. Am. Chem. Soc.*, **100**, 1315 (1978).
113. C. A. Tolman, *Chem. Soc. Rev.*, **1**, 337 (1972).
114. J. Ashley-Smith, M. Green, and F. G. A. Stone, *J. Chem. Soc. A*, 3019 (1969).

The Chemistry of the Metal—Carbon Bond, Vol. 3
Edited by F. R. Hartley and S. Patai
© 1985 John Wiley & Sons Ltd.

CHAPTER 5

Alkyne oligomerization

MARK J. WINTER

Department of Chemistry, The University, Sheffield S3 7HF, UK

I. INTRODUCTION	259
II. DIMERIZATION OF ALKYNES	261
A. Linear Dimerization of Alkynes	261
B. Cyclodimerization of Alkynes	262
III. DIMETAL COMPLEXES OF ALKYNES AND ALKYNE DIMERS AND TRIMERS	268
A. Complexes of a Single Alkyne	268
B. Complexes Containing Two Linked Alkynes	268
C. Dimetal Complexes of Alkyne Trimers	270
IV. METALLACYCLOHEPTATRIENE COMPLEXES	272
V. CATALYTIC CYCLOTRIMERIZATION OF ALKYNES	272
VI. REACTIONS OF DIYNES	277
VII. CYCLOPENTADIENYLCOBALT-INDUCED COCYCLIZATIONS	280
VIII. METAL COMPLEXES OF ALKYNE TETRAMERS	283
A. Molybdenum	283
B. Rhenium	286
C. Palladium	287
D. Manganese	288
IX. CYCLOTETRAMERIZATION OF ALKYNES; CYCLOOCTATRAENE AND OTHER TETRAMER FORMATIONS	288
X. REFERENCES	292

I. INTRODUCTION

One compelling reason for studying reactions of organotransition metal complexes with alkynes is the array of complexes and structural types available, many of which are reactive in their own right towards further alkyne. Interest in this area of chemistry was perhaps aroused as a result of the fascinating work by Reppe and his coworkers involving the catalytic cyclotetramerization of ethyne to cyclooctatetraene in the presence of nickel catalysts[1]. This was followed by a phase in which extensive investigations were undertaken into the wide array of complexes available from interaction of metal carbonyls with alkynes. Many isolated complexes contain ligands derived from oligomerized alkyne functions while others contain ligands incorporating other groups, particularly CO, as well as alkynes.

Owing to the vast quantity of published material in this area, it is not possible to present other than a fragmentary coverage and the exclusion of many interesting topics is unavoidable. The literature is covered up to late 1982.

There are several ways in which a monoalkyne may function as a ligand. The bonding in simple alkyne complexes has been discussed[2,3]. There are several different structural and bonding types for complexes involving a monoalkyne such as **1–6**. It is most difficult, in

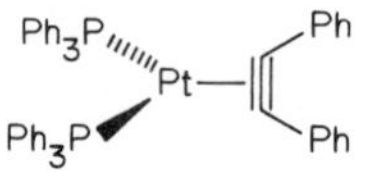

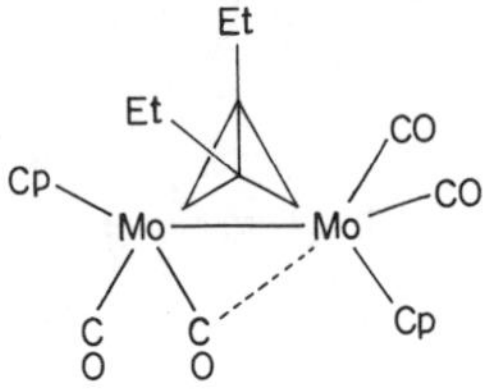

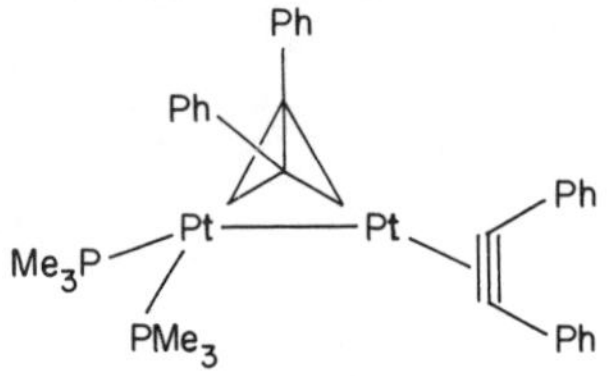

general, to predict a bonding mode for any particular system. Frequently complexes of a single alkyne (which may not necessarily be detected or isolated) will react further with excess of alkyne to generate complexes containing two or more alkyne functions. These are readily classified into three groups: (a) those in which the alkyne functions are not linked; (b) those in which the alkynes interact so as to form a more complex ligand; and (c) those in which non-alkyne groups (e.g. CO, alkene) are incorporated with the alkyne(s) to form a more complex ligand.

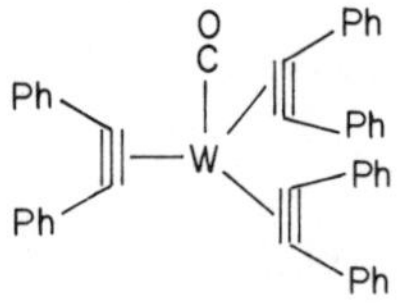

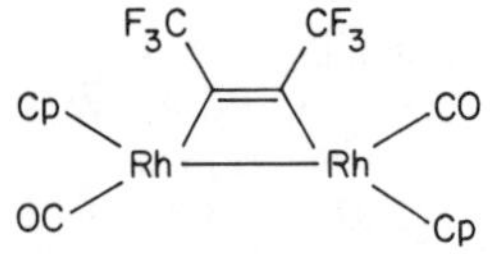

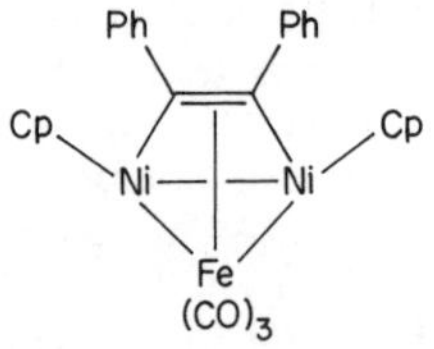

In addition, many reactions lead to quantities of organic materials, either stoichiometrically or catalytically, resulting from oligomerization of alkynes. Such products may contain other groups such as CO, nitriles, or alkenes. It is likely that many reported reactions do not contain reference to organic species formed as they were not recognized or isolated.

In general, alkynes react with transition metal complexes to form cyclic rather than linear products. This is because the mechanisms leading to linear products are probably different, involving insertion of the metal species into the terminal C—H bond of an alkyne, and not particularly favourable compared with cyclization pathways[10]. Common cyclic products formed are cyclobutadienes, arenes, cyclooctatetraenes, metallacycles, cyclopentadienones, or quinones.

Reactions of alkynes are characterized by their apparent non-selectivity, i.e. the multiplicity of products. Such groups of products may be produced from series of reactions within the same reaction vessel or through competing parallel reaction pathways, all of which suggests that there are many reactions requiring reinvestigation using modern analytical (spectroscopic and crystallographic) and separation techniques (such as high-performance liquid chromatography). For these reasons, our basic understanding of alkyne chemistry must be said to be fragmentary, although there is undoubtedly a vast quantity of published data.

II. DIMERIZATION OF ALKYNES

A. Linear Dimerization of Alkynes

This process is not observed at all frequently, requires a terminal alkyne, and involves a hydrogen shift which is probably initiated by an oxidative insertion into the terminal C—H bond of the alkyne (reaction 1). Typical products are 1,4-disubstituted butadienes or in

$$L_nM + RC{\equiv}CH \rightarrow L_n\underset{H}{M}{-}C{\equiv}CR \quad (1)$$

some cases the 2,4-disubstituted butadienes, but there are some reports of the formation of linear alkyne dimers, that is, conjugated enynes. Thus terminal alkynes react in the presence of $ZnBu_2^i$, PPh_3, and a nickel catalyst selectively to form the head-to-tail dimer 1,3-dialkylbutenyne[11]. The molar ratios $[RC{\equiv}CH]/[ZnBu_2^i] = 2$ and $[RC{\equiv}CH]/[Ni] = 120$ are most satisfactory; if too little $ZnBu_2^i$ is present, higher oligomers are also

$$2RC{\equiv}CH + ZnBu_2^i \xrightarrow[PPh_3]{[Ni]} RC{\equiv}CC(R){=}CH_2 \quad (2)$$

$R = alkyl$

formed. The terminal alkyne oct-1-yne dimerizes in the presence of catalytic quantities of $[RHCl(PPh_3)_3]$ to a mixture of (predominantly, 87%) the head-to-tail enyne **7** and the (E)-tail-to-tail isomer **8**[12]. A catalyst made from $[Cr(OBu^t)_4]$ and $ZnEt_2$ also leads to the head-

$$C_6H_{13}C{\equiv}CH \xrightarrow{[RhCl(PPh_3)_3]} \underset{(7)}{(E)\text{-}C_6H_{13}CH{=}CHC{\equiv}CC_6H_{13}} + \underset{(8)}{C_6H_{13}C{\equiv}CC(C_6H_{13}){=}CH_2}$$

$$(3)$$

to-tail isomers in good yield[13]. In the presence of a nickel catalyst and $AlBu_3^i$ the chiral alkyne (S)-3-methylpent-1-yne dimerizes to the (E)-head-to-tail diene and both possible isomeric arenes; in each case retention of configuration is observed.

$$RC{\equiv}CH \xrightarrow[AlBu_3^i]{[Ni]} (E)\text{-}RCH{=}CHC(R){=}CH_2 + 1,2,4\text{-}R_3C_6H_3 + 1,3,5\text{-}R_3C_6H_3$$

$$R = \begin{array}{c}Me\quad Et\\ \diagdown\!\diagup\\ H\end{array} \quad (4)$$

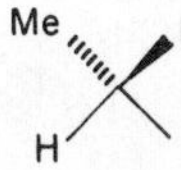

Other reactions lead to more unusual products. In the presence of $[RuH_2(CO)(PPh_3)_3]$, $Bu^tC{\equiv}CH$ is converted into a complex mixture of three enynes and two butatrienes, **9** (major) and **10** (minor)[14]. Since linear dimerizations are not observed for disubstituted

(Z)-Bu^{t}CH=C=C=CHBu^t (E)-Bu^{t}CH=C=C=CHBu^t (Me$_3$Si)$_2$C=C=C=C(SiMe$_3$)$_2$

(9)   (10)   (11)

alkynes, an insertion step into the terminal C—H bond may be important and it is certainly possible to write down mechanisms (Scheme 1) where precedents exist for the individual steps and which may be expanded readily to accommodate formation of other isomers. Butatriene (11) is formed in the reaction of [CoCp(CO)$_2$] with Me$_3$SiC=CSiMe$_3$ (Section II.B).

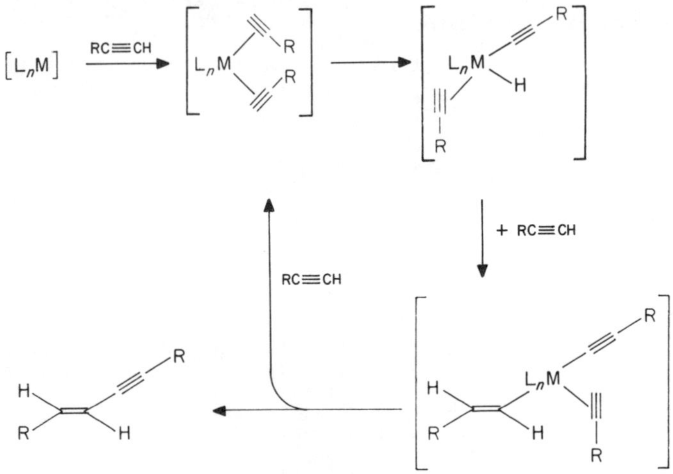

SCHEME 1

B. Cyclodimerization of Alkynes

By far the most important dimerization products are cyclobutadiene complexes. Interest in cyclobutadiene complexes was initially aroused because of the possibility of studying reactions of the four-membered carbocyclic ring which is not isolatable under normal conditions. They are also important because of a possible involvement in cyclotrimerization and cyclotetramerization processes. There are many synthetic routes now available to complexes of this type other than from alkyne precursors[15], but alkyne precursors are frequently convenient for their syntheses. The combined photo and thermal reactions of [M(CO)$_4$Cp](12)(M = V or Nb) (Scheme 2) lead to cyclobutadiene (R$_4$C$_4$) complexes as shown. Initial irradiation of the tetracarbonyl 12 leads to 13 or 14 in the presence of excess of alkyne[16-20]. Under these conditions the vanadium complex decomposes to tetraphenylcyclopentadienone[16]. Heating 13 with PhC=CPh leads to the cyclobutadiene 15, or a mixture of 15 and 16 when conducted under CO pressure[17]. This is likely to proceed via a vanadacyclopentadiene species. Reaction of 14, M = Nb, with PhC=CPh at 80 °C leads to the cyclobutadiene complex 17[19,21], which displays an intriguing reaction in that it decomposes to hexaphenylbenzene. This is possibly consistent with a cyclobutadiene pathway for arene formation, but the mechanism in this specific case is actually unknown.

The reaction of [Mo(CO)$_6$] with PhC=CPh is complicated[22,23]; several cyclobutadiene (R$_4$C$_4$) complexes are formed in benzene together with hexaphenylbenzene. Cyclobutadiene complexes are available, albeit in low yield, from ethyne itself; the pressure reaction of HC=CH with [Fe(CO)$_5$] at 110 °C affords the parent cyclobu-

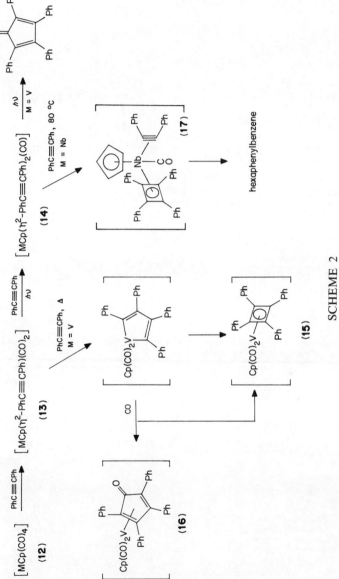

SCHEME 2

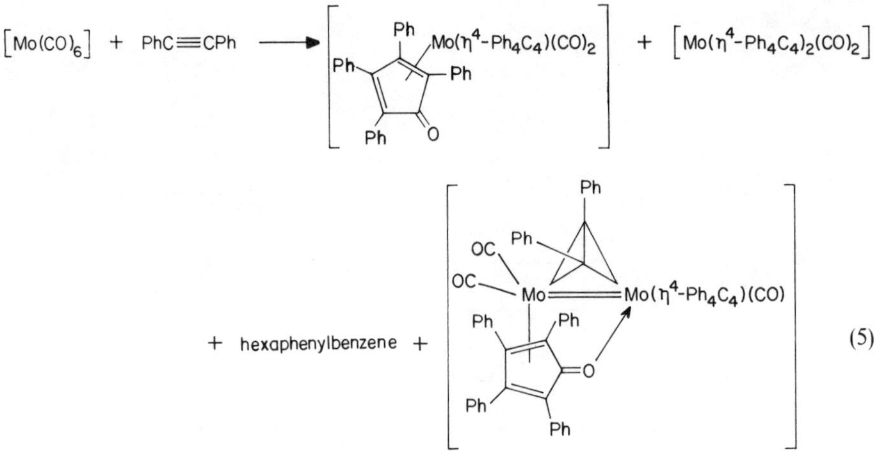

$$[Mo(CO)_6] + PhC{\equiv}CPh \longrightarrow \left[\begin{array}{c} Ph_4C_4(CO)(=O)\ complex \end{array} \right] + [Mo(\eta^4\text{-}Ph_4C_4)_2(CO)_2]$$

$$+ \text{ hexaphenylbenzene } + \left[\begin{array}{c} \text{Mo-Mo dimer complex} \end{array} \right] \quad (5)$$

tadiene complex **18** together with the ferrole **19**, which has been shown in separate experiments not to be a precursor of **18**[24].

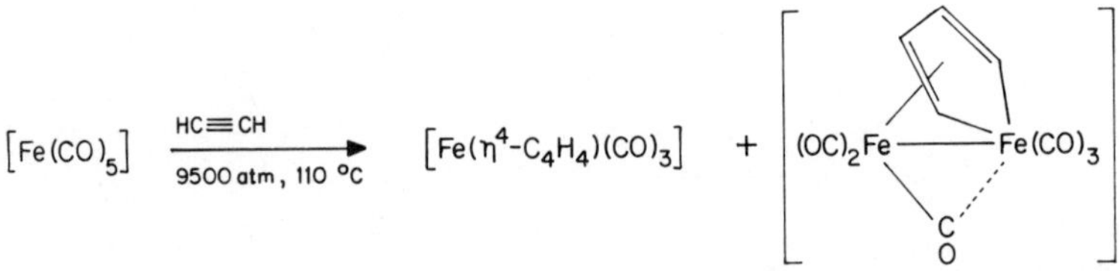

$$[Fe(CO)_5] \xrightarrow[9500\ atm,\ 110\ °C]{HC{\equiv}CH} [Fe(\eta^4\text{-}C_4H_4)(CO)_3] + \left[(OC)_2Fe\text{---}Fe(CO)_3 \text{ ferrole} \right] \quad (6)$$

(**18**) (**19**)

Cobalt cyclobutadiene complexes are available from several routes such as the interaction of $[CoCp(\eta^4\text{-}C_8H_{12})]$ with $PhC{\equiv}CPh$ to form **20**[25] or $PhC{\equiv}CSiMe_3$,

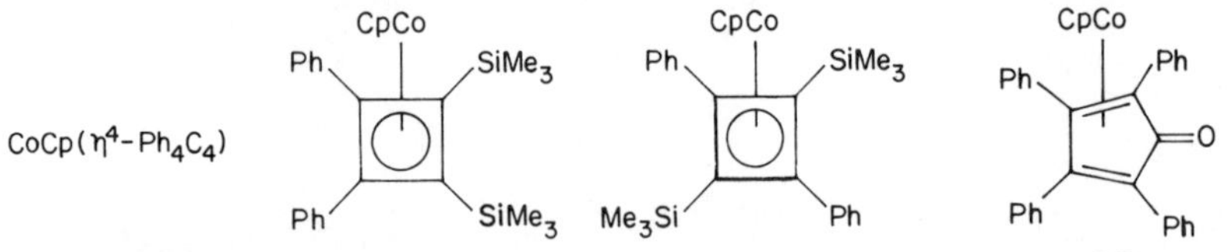

$CoCp(\eta^4\text{-}Ph_4C_4)$

(**20**)   (**21**)   (**22**)   (**23**)

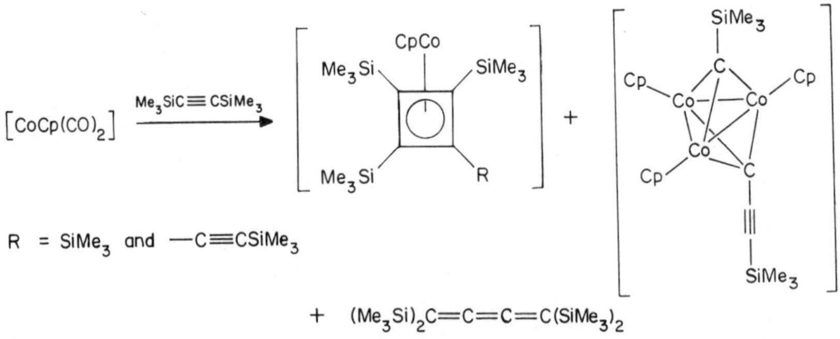

$$[CoCp(CO)_2] \xrightarrow{Me_3SiC{\equiv}CSiMe_3} \left[\text{CpCo cyclobutadiene complex} \right] + \left[Co_3Cp_3 \text{ cluster with } SiMe_3 \right] \quad (7)$$

R = $SiMe_3$ and $-C{\equiv}CSiMe_3$

$+ (Me_3Si)_2C{=}C{=}C{=}C(SiMe_3)_2$

which results in the isomeric pair **21** and **22**[26]. Choice of starting material is important; thus [CoCp(CO)$_2$] reacts with PhC≡CPh to produce the cyclopentadienone **23** as well as the cyclobutadiene **20**[27]. A more complex array of compounds results from reaction of the sterically demanding Me$_3$SiC≡CSiMe$_3$ with [CoCp(CO)$_2$][28].

The interactions of palladium and platinum complexes with alkynes are complex and much still remains to be learnt. Cyclobutadiene complexes may be formed in some instances, the steric requirements of the alkyne substituents being important. Thus [PdCl$_2$(NCPh)$_2$] reacts with Bu^{t}C≡CBu^t to give the palladium (II) complex **24** (Scheme 3) only, whereas Bu^{t}C≡CPh and PhC≡CPh lead to cyclobutadiene complexes[29]. More

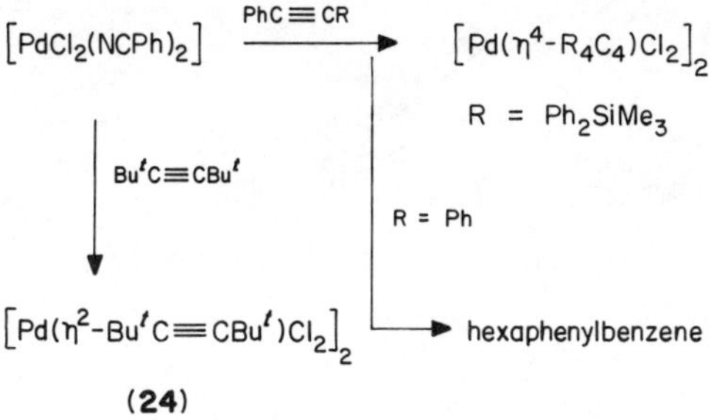

$$[PdCl_2(NCPh)_2] \xrightarrow{PhC \equiv CR} [Pd(\eta^4-R_4C_4)Cl_2]_2$$

R = Ph$_2$SiMe$_3$

Bu^{t}C≡CBu^t ↓

R = Ph

$$[Pd(\eta^2-Bu^tC \equiv CBu^t)Cl_2]_2 \longrightarrow \text{hexaphenylbenzene}$$

(24)

SCHEME 3

recent work shows these systems to be complex and has been concerned with mechanistic pathways for cyclobutadiene formation[30].

Other fascinating reactions appear to indicate that it is possible for coordinated alkynes to combine directly to a cyclobutadiene complex through a thermal rearrangement or by action of CO[31] (reaction 8).

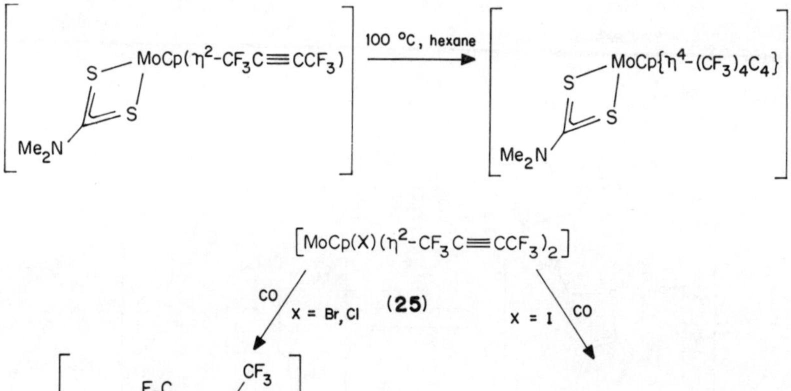

$$\left[\begin{array}{c} S \\ \diagdown \\ S \end{array} \!\!\!\! MoCp(\eta^2\text{-}CF_3C \equiv CCF_3) \right] \xrightarrow{100\ ^\circ C,\ hexane} \left[\begin{array}{c} S \\ \diagdown \\ S \end{array} \!\!\!\! MoCp\{\eta^4\text{-}(CF_3)_4C_4\} \right] \quad (8)$$

(with Me$_2$N groups)

$$[MoCp(X)(\eta^2\text{-}CF_3C \equiv CCF_3)_2]$$

X = Br, Cl / CO   **(25)**   X = I \ CO

(9)

[Cp(CO)XMo—cyclopentadienone with F$_3$C, CF$_3$, F$_3$C, CF$_3$, =O]

(26)

$$[MoCp(CO)I\{\eta^4\text{-}(CF_3)_4C_4\}]$$

(27)

Interestingly, complexes **25** (X = Br or Cl) lead to the cyclopentadienone **26** whereas the iodide **25** (X = I) leads to **27**, reflecting the subtle balance of factors influencing such pathways. It is not apparent whether these reactions are unimolecular[32]; this is critical since the formation of cyclobutadiene (C_4H_4) complexes from bis-alkyne species is a theoretically interesting problem. The question is whether a $(2+2)\pi$ cyclization of a monometallic bis-alkyne takes place directly (A → C, reaction 10) or via a metallacyclopentadiene intermediate (A → B → C). The former is said not to occur on theoretical grounds, although the participation of a second metal as shown would lift this restriction[33]. Others

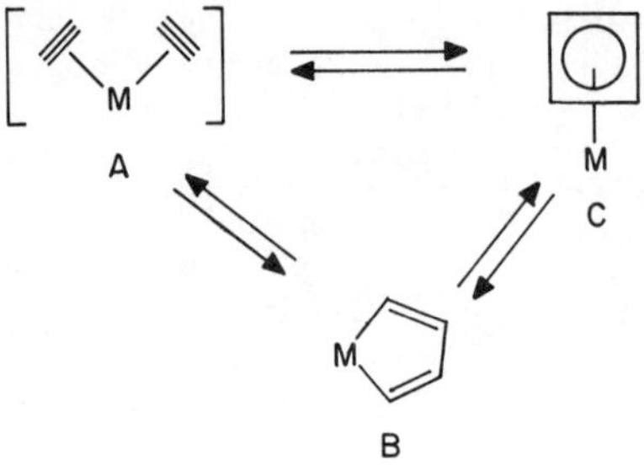

$$[\text{A}] \rightleftharpoons [\text{C}] \quad \text{via} \quad [\text{B}] \qquad (10)$$

$$[\text{bis-alkyne bridged}] \longrightarrow [\text{cyclobutadiene}] + M \qquad (11)$$

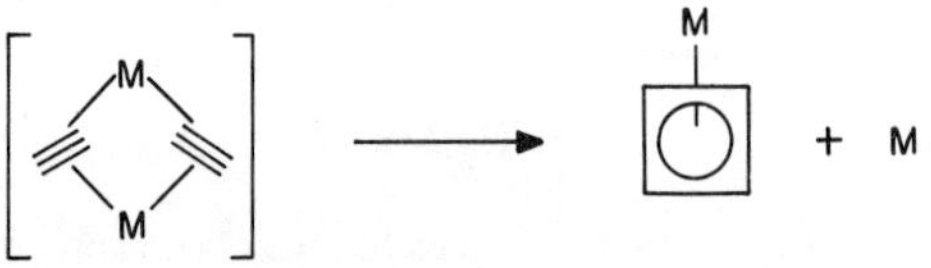

argue that the $(2+2)\pi$ mechanism is allowed even in the presence of only one metal[34]. Comparatively few examples of the metallacyclic route have been reported, perhaps because of the scarcity of suitable metallacyclic starting materials. The cobalt complexes **28**[35-37] and **29**[38] do form cyclobutadiene complexes on heating, probably after

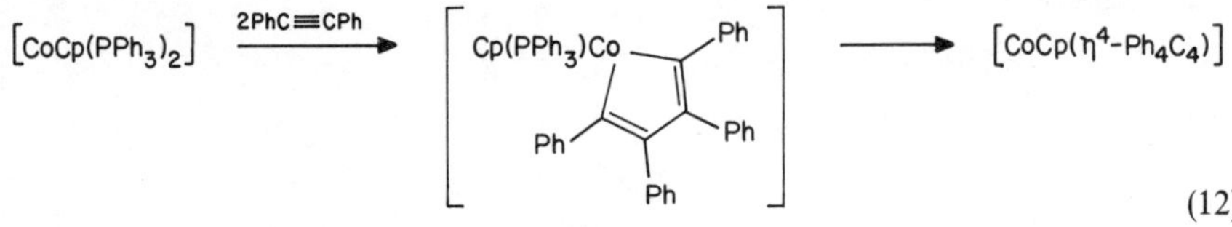

$$[\text{CoCp(PPh}_3)_2] \xrightarrow{2\text{PhC}\equiv\text{CPh}} \begin{bmatrix} \text{Cp(PPh}_3)\text{Co} & \text{Ph} \\ \text{Ph} & \text{Ph} \\ & \text{Ph} \end{bmatrix} \longrightarrow [\text{CoCp}(\eta^4\text{-Ph}_4\text{C}_4)] \qquad (12)$$

(28)

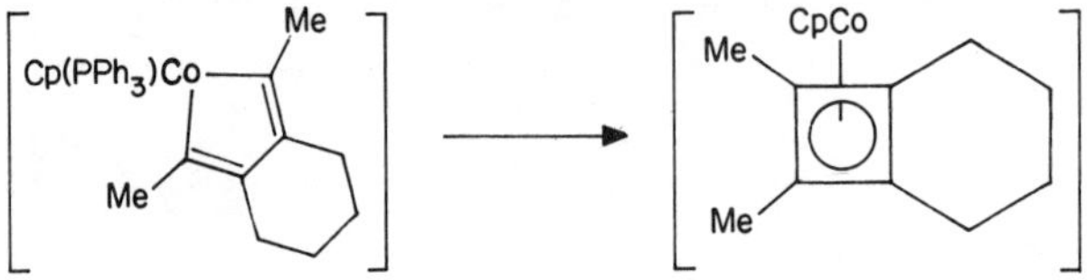

$$\begin{bmatrix} \text{Cp(PPh}_3)\text{Co} & \text{Me} \\ \text{Me} & \end{bmatrix} \longrightarrow \begin{bmatrix} \text{CpCo} \\ \text{Me} & \\ \text{Me} & \end{bmatrix} \qquad (13)$$

(29)

decomplexation of the phosphine donor ligand. However, it is not clear whether cyclobutadiene formation in this instance proceeds through initial retrocycliztion to a bis-alkyne complex (B → A → C) or directly (B → C).

Recent work has addressed the question of cyclobutadiene–metal, metallacyclopentadiene and bis-alkyne–metal complex interconversions[39] and shows that the diastereomeric pairs **30** and **31** interconvert through the intermediacy of the bis-alkyne **32** by a single alkyne rotation. Formation of a metallacycle at some stage in this process cannot be ruled out as being kinetically inaccessible, but if such a metallacycle is formed it cannot be planar and it must be configurationally stable.

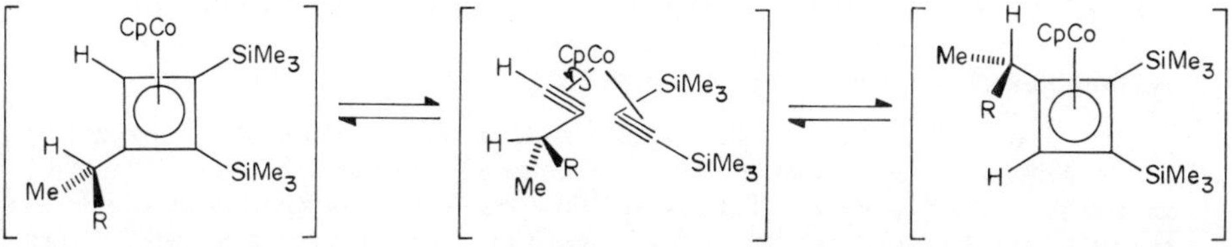

The mechanism of cobalt cyclobutadiene formation is still not absolutely defined but clearly differs from that in palladium systems. Under cyclotrimerization reaction conditions they are catalytically inert (Section V). However, the cyclobutadiene ring may be expanded in some systems; thus the niobium complex **17** does decompose to hexaphenylbenzene in a remarkable reaction while $Fe(CO)_3$ fragments may insert into an iron–cyclobutadiene system to form the isomeric ferroles **77** and **34**[40].

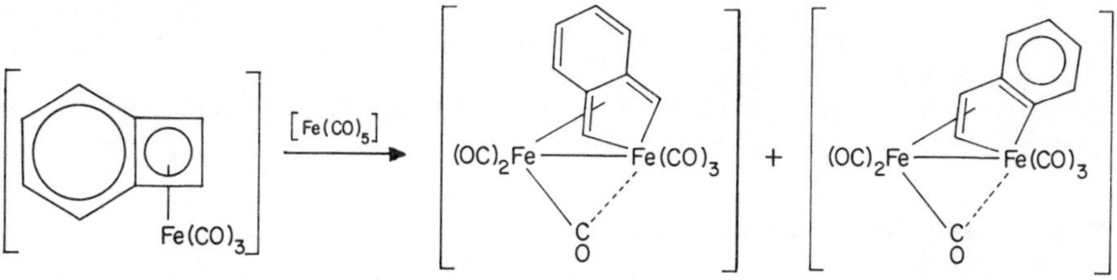

Although cyclobutadiene complexes appear inert under cyclotrimerization conditions, metallacyclopentadiene complexes are probably important intermediates in such reactions (Section V).

More exotic cyclodimerizations have been observed in cases where the alkyne substituents are not innocent. Thus an excess of PhC≡CPh reacts with $[RuH(NO)(PPh_3)_3]$ in toluene at reflux to give dimer **35** as the only isolated organic product[41]. The reaction of EtC≡CEt in the presence of $[NiH\{Ph_2P(CH_2)_2PPh_2\}]\cdot[CF_3CO_2]$ leads to the cyclic dimer **36** as well as some cyclotrimers[42,43].

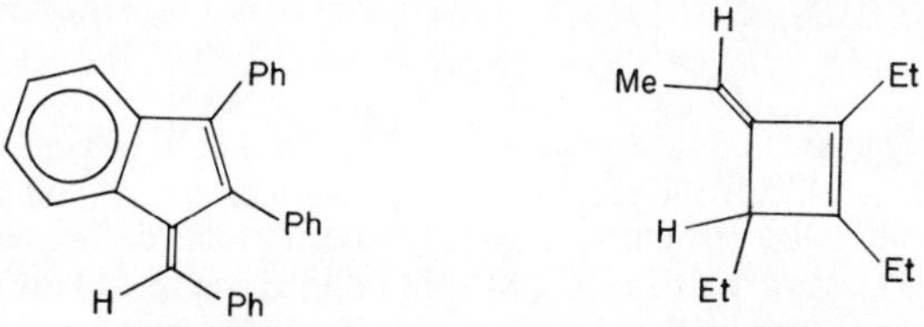

III. DIMETAL COMPLEXES OF ALKYNES AND ALKYNE DIMERS AND TRIMERS

A. Complexes of a Single Alkyne

Many transition metals have been shown to form dimetal complexes of alkynes or alkyne oligomers. The simplest are those containing a single alkyne bridging two metals, generally in a transverse sense. Thus $[Co_2(CO)_8]$ reacts with alkyne to form complexes $[Co_2(CO)_6(\mu\text{-RCCR})]$, 37, several of which have been structurally characterized[44]. Other examples include the molybdenum complex 2 and the unusual diiron species 38[45] containing bridging $Bu^tC{\equiv}CBu^t$ functions which do not appear to interact further, probably for steric reasons. The $F_3CC{\equiv}CCF_3$ complex 5 contains the ligand bridged in such a fashion that the Rh_2C_2 fragment is approximately planar, thus constituting a dimetallacyclobutene ring system.

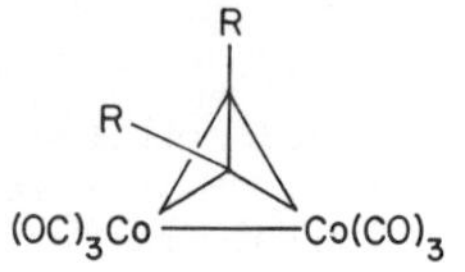

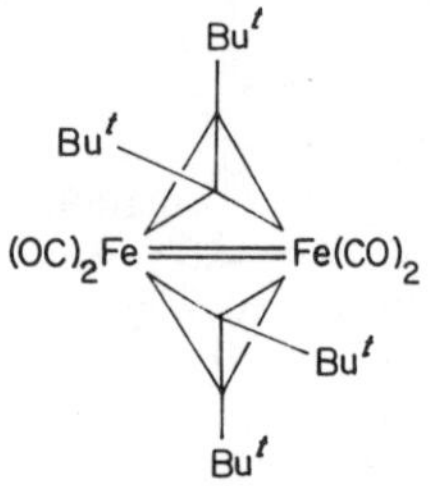

It is common for dimetal species of this type to react further, possibly under more forcing conditions to form complexes containing complexes of alkyne oligomers.

B. Complexes Containing Two Linked Alkynes

In many cases reactions producing complexes of an alkyne dimer proceed without isolation (or detection) of complexes of the monomer. The most common structures are of the ferrol type 39, which some regard as $Fe(CO)_3$-stabilized ferracyclopentadienes. Many structures of this type are known, the structures of several of these compounds have been established [46,47] and indicate the presence of a semi-bridging carbonyl on one of the iron atoms as shown. Related structures have also been isolated for macrocyclic diynes (Section VI). Some alkynes react with the $M{\equiv}M$ bonded species $[\{MoCp(CO)_2\}_2]$ (40) to give bis-alkyne species such as 41–43 in good yield, reactions which proceed via transverse bridged monoalkyne species of type 2. Some of these complexes are reactive in their own right towards alkynes (Section VIII). Another dimolybdenum complex recently noted is from a very different system in which ethyne reacts with $[Mo_2(OCH_2Bu^t)_6(NC_5H_5)(\mu\text{-}HC_2H)]$ to give 44[53].

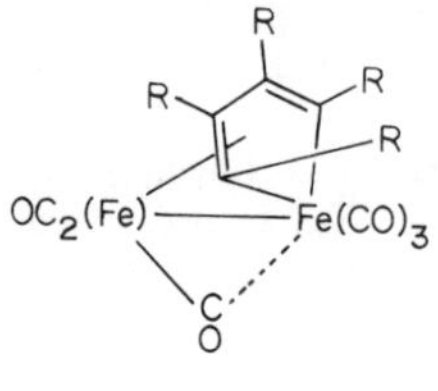

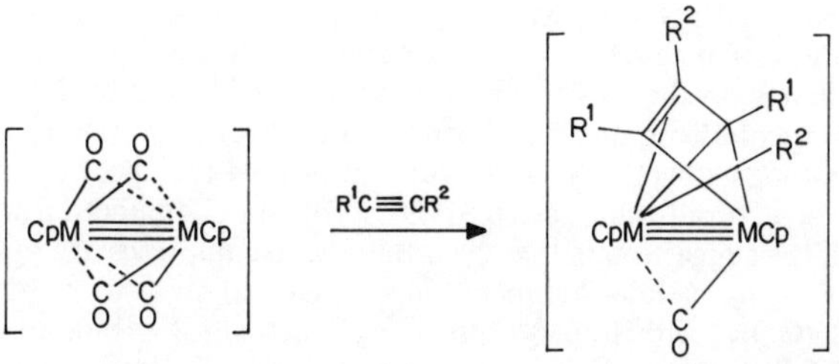

5. Alkyne oligomerization

(16)

(40)

(41) M = Cr, R^1 = R^2 = H, Ph[48,49]
(42) M = Cr, R^1 = Ph, R^2 = H[49,50]
(43) M = Mo, R^1 = R^2 = Ph, Et[49,51,52]

(17)

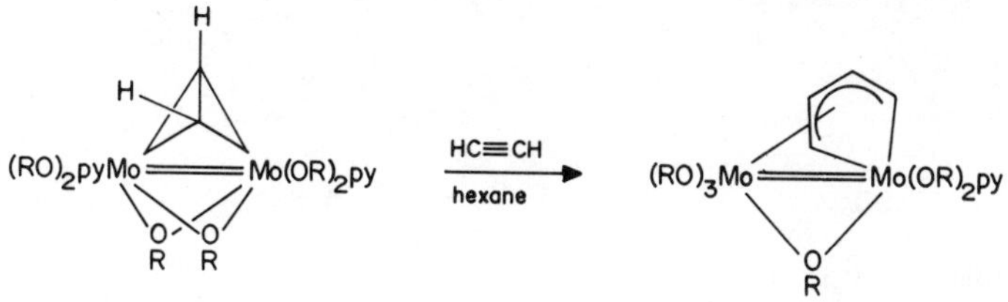

(44)

R = CH$_2$Bu^t

An unusual complex is produced in the reaction of cyclooctyne with [Co$_2$(CO)$_8$], the dimer **45**[54], which is formed together with the simple bridged species **46** which itself does

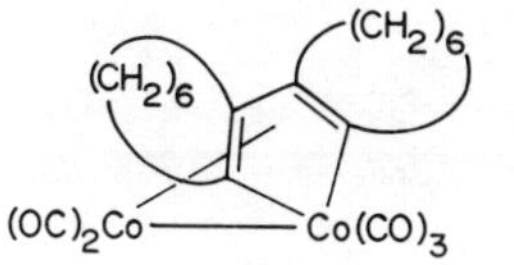

(45) 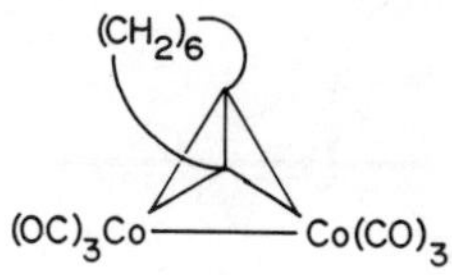   (46)

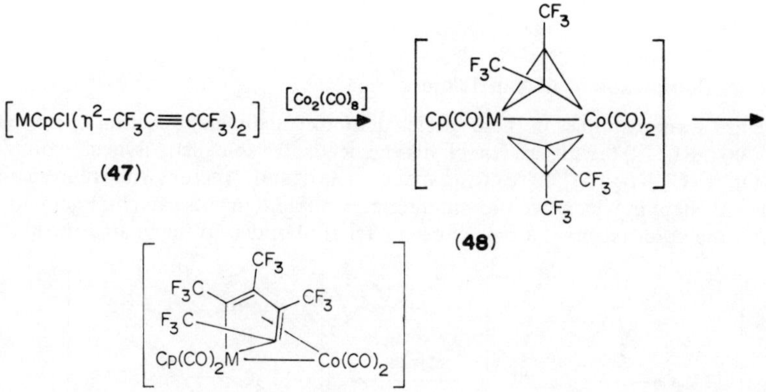

(47) → (48) → (49)

(18)

M = Mo, W

not appear to be the precursor of **45**. Cyclooctyne appears to be the only alkyne reported to form a complex of type **45** with cobalt carbonyls. Species of this type are suggested to be intermediates on the way to 'flyover' complexes[54] (Section III.C).

Recently a metallacyclopentadiene unit has been reported from the reaction of the formally electron deficient **47** with [$Co_2(CO)_8$], which produces the spectroscopically characterized **48** and the structurally characterized **49**[32]. Taken in conjunction with the formation of complex **27**, the suggestion is that cyclobutadiene complexes are produced via stepwise reactions involving metallacyclopentadiene intermediates.

The reaction of [$RhCp(CO)_2$] with $PhC\equiv CPh$ or $F_5C_6C\equiv CC_6F_5$ leads to several products, among them **50** (R = Ph or C_6F_5)[55], while $F_3CC\equiv CCF_3$ leads to **50** (R = CF_3)[56]. The unusual compound **51** is obtained from reaction of [$Ir(CO)_2(\eta^5\text{-}C_5Me_5)$]

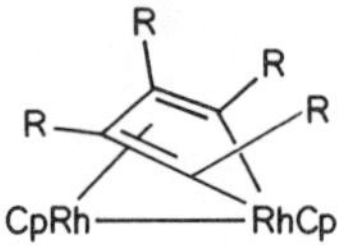

(50)   R = Ph, C_6F_5, CF_3

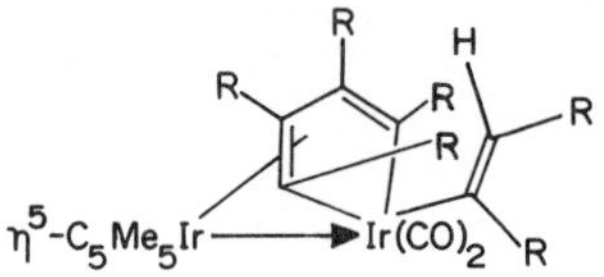

(51)   R = CF_3

with $F_3CC\equiv CCF_3$, an unusual molecule containing an iridium atom which has lost a C_5Me_5 group[57].

It is not unusual for carbonyls to become incorporated into the metallacyclic structures. Thus a complex of the unusual 'cross' ligand is obtained from [$Fe_2(CO)_9$] and $MeC\equiv CMe$[58], while complex **53** is produced in the reaction of [$\{FeCp(CO)_2\}_2$] with $F_3CC\equiv CCF_3$[59].

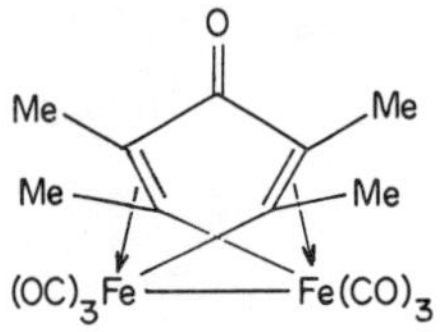

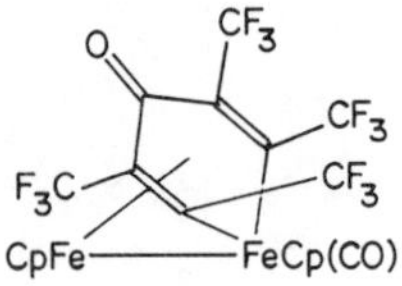

(52)

(53)

C. Dimetal Complexes of Alkyne Trimers

There are fewer examples of these species than alkyne dimers. Reaction of complexes [$Co_2(CO)_6(\mu\text{-}RCCR)$] with excess of alkyne leads to so-called 'flyover' complexes [$Co_2(CO)_4\{(RCCR)_3\}$] **54**[44], a reference made to the ligand structure when represented as a six-carbon chain in which two non-interacting π-allyl fragments each interact with one metal and the outer terminal atoms of each allyl are bonded to the second metal atom.

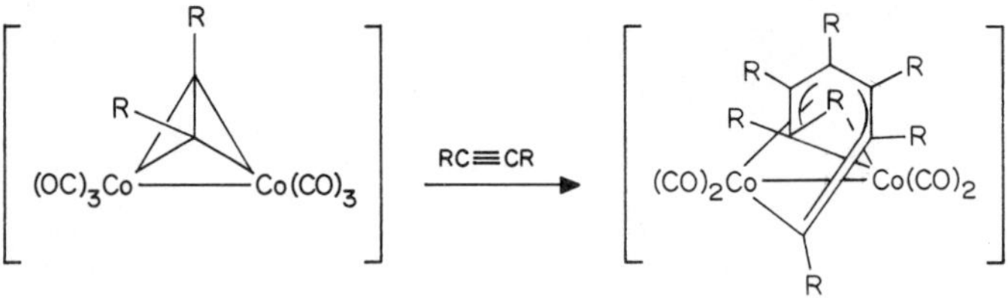

$$\left[\begin{array}{c}\text{(OC)}_3\text{Co}\text{—}\text{Co(CO)}_3\end{array}\right] \xrightarrow{RC\equiv CR} \left[\begin{array}{c}\text{(CO)}_2\text{Co}\text{—}\text{Co(CO)}_2\end{array}\right] \quad (19)$$

(54)

5. Alkyne oligomerization

This representation disguises the fact that the Co_2C_2 fragment is extremely symmetrical and planar with the Co–C distances equal in, for example, the complex $[Co_2(CO)_4\{(CF_3CCH)_3\}]^{60}$. In this complex and the analogous $[Co_2(CO)_4\{(Bu^tCCH)_3\}]$ the ligand substituents are situated on carbon atoms 1, 3 and 6 of the chain. On heating or treatment with Br_2 these two complexes eliminate 1,2,4-trisubstituted benzenes; this is consistent with the preservation of the substitution pattern of the complex and ring formation by the two ends of the ligand coming together[44].

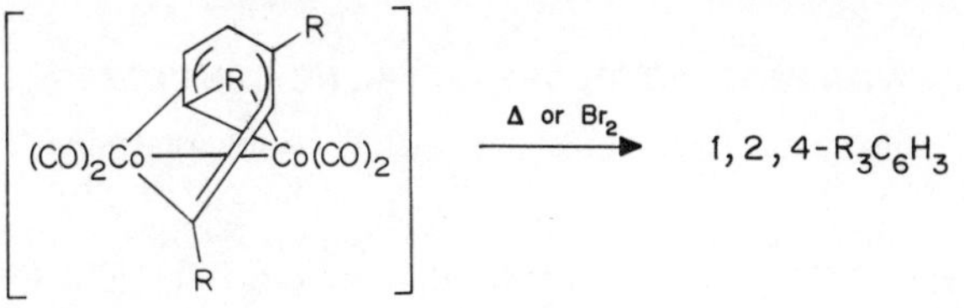

$$\left[\begin{array}{c} \text{(structure with } (CO)_2Co\text{—}Co(CO)_2 \text{ and R groups)} \end{array}\right] \xrightarrow{\Delta \text{ or } Br_2} 1,2,4\text{-}R_3C_6H_3 \quad (20)$$

$$R = CF_3, Bu^t$$

Related complexes are formed in some dimolybdenum systems. Complex **55** reacts with $PhC\equiv CPh^{49}$ to give the spectroscopically characterized **56**, probably formed by the stepwise insertion of two alkynes into one of the two terminal CMo_2 fragments of the molecule. It is interesting that on steric grounds the second $PhC\equiv CPh$ molecule might have been expected to attack at the chain atom not bearing a Ph group, leading to a

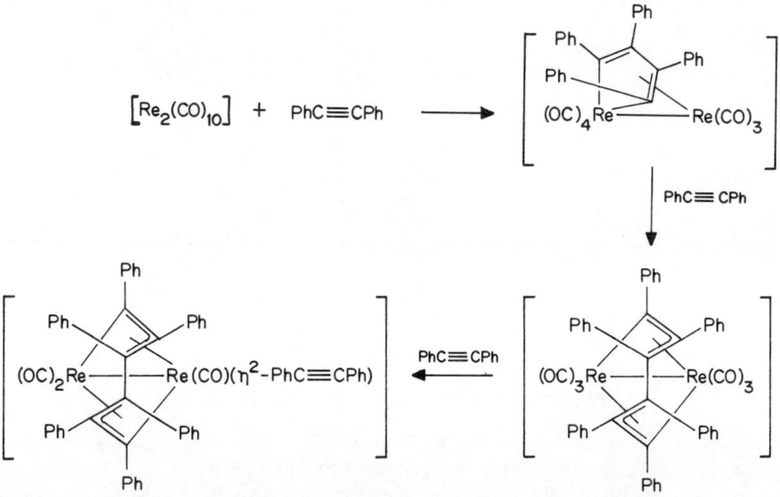

$$[Re_2(CO)_{10}] + PhC\equiv CPh \longrightarrow \left[(OC)_4Re\text{—}Re(CO)_3\right]$$

$$\downarrow PhC\equiv CPh$$

(57) $\left[(OC)_2Re\text{—}Re(CO)(\eta^2\text{-}PhC\equiv CPh)\right] \xleftarrow{PhC\equiv CPh} \left[(OC)_3Re\text{—}Re(CO)_3\right]$

SCHEME 4

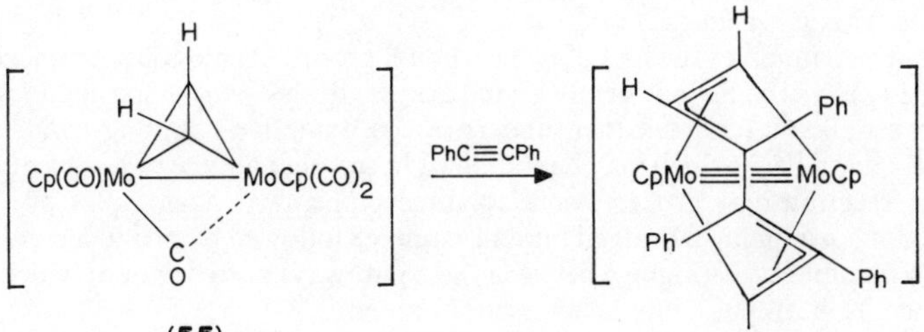

$$\left[Cp(CO)Mo\text{—}MoCp(CO)_2\right] \xrightarrow{PhC\equiv CPh} \left[CpMo\equiv\equiv MoCp\right] \quad (21)$$

(55) **(56)**

symmetrical structure. In contrast, a symmetrical substitution pattern is observed for reaction of the analogous complex of the sterically demanding $Me_3SiC\equiv CSiMe_3$.[61]

Few rhenium alkyne complexes are known but recently the reactions of $[Re_2(CO)_{10}]$ with $PhC\equiv CPh$ have been reported[62] and complexes of trimerized alkyne observed. Three complexes have been isolated from this reaction (Scheme 4) and shown to be formed sequentially. A derivative of the final product **57** has been crystallographically characterized in which two isonitrile ligands replace the lone monoyne.

IV. METALLACYCLOHEPTATRIENE COMPLEXES

There is little information available for metallacycloheptatriene complexes, but the nickel complex **58** does react with $F_3CC\equiv CCF_3$[63] to give the structurally characterized metallacyclohepta-(Z,E,Z)-triene complex **59**[64]; a different isomer, the metallacyclohepta-(Z,Z,Z)-triene, is obtained from $[Pt(PEt_3)_2(F_3CC\equiv CCF_3)]$ and $F_3CC\equiv CCF_3$.[63]

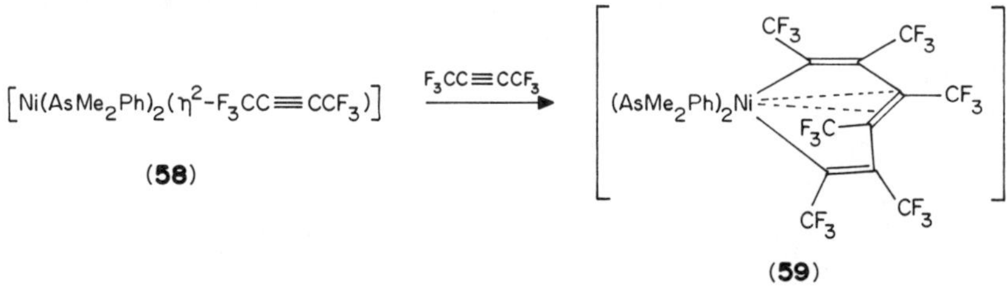

$$[Ni(AsMe_2Ph)_2(\eta^2\text{-}F_3CC\equiv CCF_3)] \xrightarrow{F_3CC\equiv CCF_3} [(AsMe_2Ph)_2Ni(\cdots)] \quad (22)$$

(**58**)   (**59**)

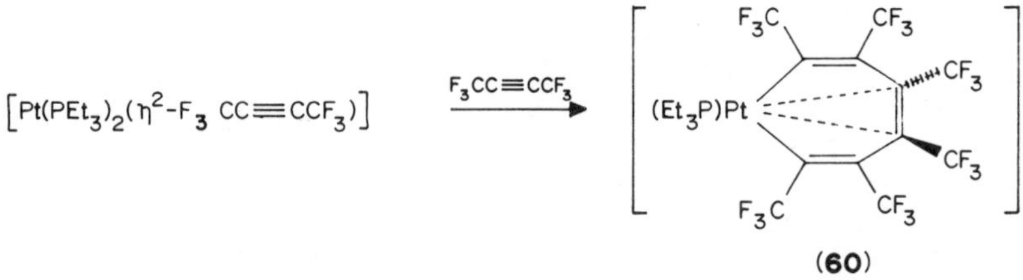

$$[Pt(PEt_3)_2(\eta^2\text{-}F_3CC\equiv CCF_3)] \xrightarrow{F_3CC\equiv CCF_3} [(Et_3P)Pt(\cdots)] \quad (23)$$

(**60**)

V. CATALYTIC CYCLOTRIMERIZATION OF ALKYNES

Cyclotrimerization reactions producing arenes from three alkyne molecules are very common. There are reviews listing catalysts and products available[3,65,66]. It is unlikely that there is a unifying mechanism for cyclotrimerization and information on most systems is sketchy. One of the more extensively studied systems is that involving $[CoCp(CO)_2]$[67]. Even here the mechanism is still not precisely understood, but a good working model is given in Scheme 5.

The initial step involves the loss of two carbonyl groups. Analogous complexes of, say, PMe_3 or PEt_3 have the ligands firmly bound and catalysis by these species does not take place. The bis-alkyne complex then undergoes an oxidative coupling to form a 16e$^-$ cobaltacyclopentadiene which may be stabilized by another alkyne or a solvent molecule. Subsequent reaction may proceed via a cobaltacycloheptatriene complex or a cycloaddition reaction, producing a bridged bicyclic complex followed by product elimination. It is not really possible to distinguish between these pathways or to decide at which points of the sequence fresh alkyne units become coordinated.

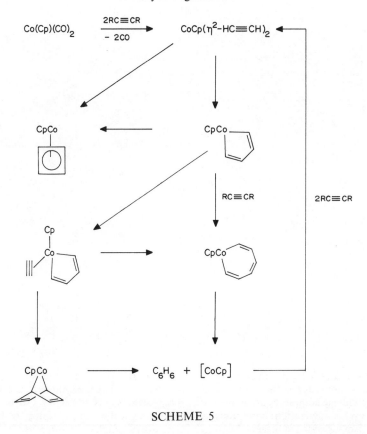

SCHEME 5

Earlier suggestions that cyclobutadiene complexes are involved in cyclotrimerization reactions are discounted, at least for most metals, by the observation that in general $H_3CC\equiv CCD_3$ is cyclotrimerized by metal catalysts to benzenes in which there are no products containing three adjacent CD_3 groups (reaction 24)[68]. In fact, such a product

(24)

was obtained for some palladium (II) systems but it is not clear whether a different process is operating. Bearing in mind these results, the observation that the niobium complex **17** decomposes smoothly to hexaphenylbenzene becomes even more interesting, and a repetition of this reaction using an appropriately labelled alkyne would be useful.

There are still other mysteries associated with these reactions. Complex **61** reacts with $Me_3SiC\equiv CSiMe_3$ to give cyclopentadienones as shown[69]; the minor isomer **62** and the arene **63** must be formed with cleavage of the carbon—carbon bond in $Me_3SiC\equiv CSiMe_3$ and the authors suggest that an intermediate with cyclobutadiene symmetry is involved. Cyclobutadiene complexes are byproducts from reactions of $[CoCp(CO)_2]$ with alkynes under cyclotrimerization conditions and are catalytically inert[67]. Support for the mechanism suggested in Scheme 5 comes from isolated examples of complexes of the types

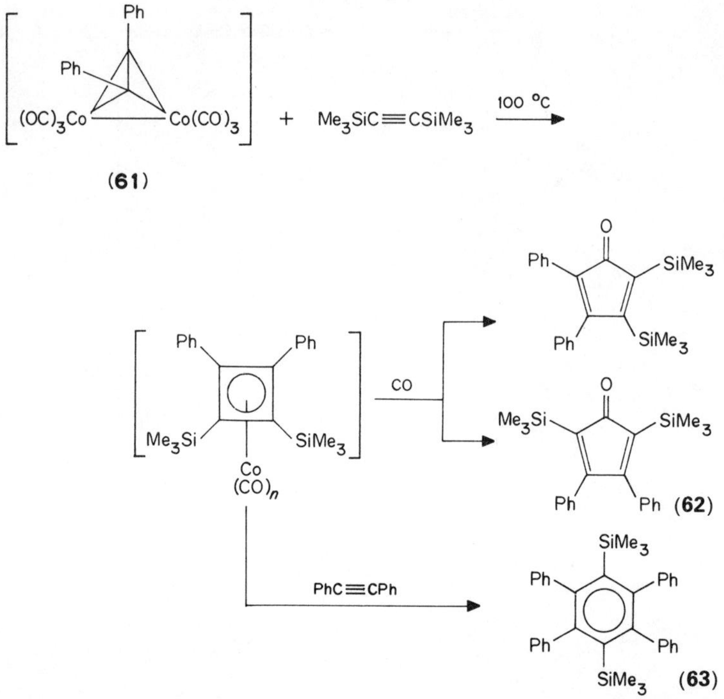

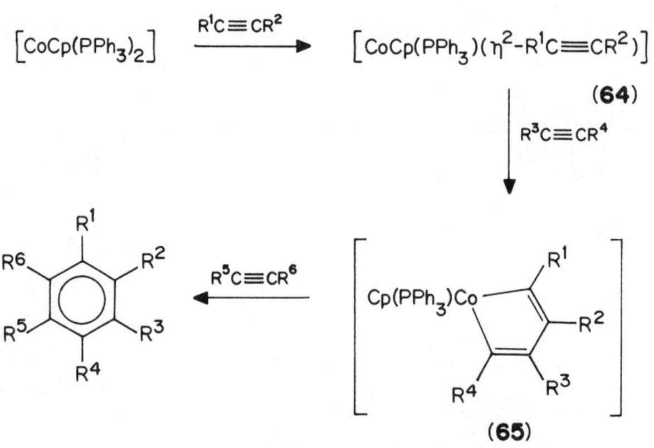

SCHEME 6

illustrated. Thus $[CoCp(PPh_3)_2]$ reacts with alkynes to form isolatable complexes **64** (Scheme 6)[70], which react further with alkynes to form cobaltacyclopentadiene complexes

5. Alkyne oligomerization

65 with considerable control over functionality. In turn these metallacycles react further to form substituted arenes. It is not possible as yet to say whether the final stage proceeds via a cobaltacycloheptatriene or a bicyclic intermediate. It is likely in this sequence that initial reaction of the metallacycle with alkyne involves prior loss of PPh_3 followed by coordination of the alkyne. However, this appears not to be the case in a related system Replacement of the phosphine in **66** by PEt_3 or PMe_3 leads to the metallacycle **67**[71] (Scheme 7), which does not itself react with but-2-yne to form hexamethylbenzene owing

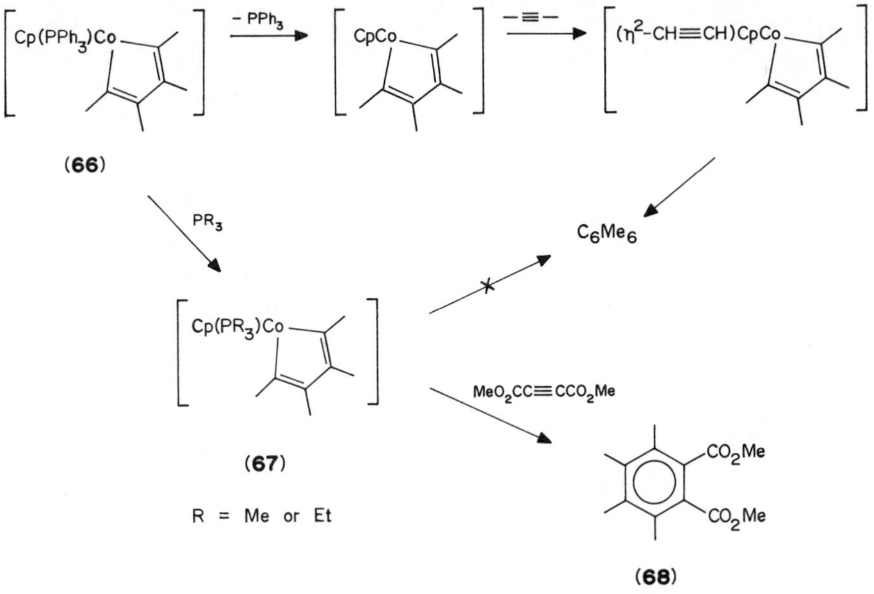

SCHEME 7

to the affinity of the $[CoCp(C_4Me_4)]$ fragment for PR_3 (i.e. the PR_3 does not dissociate). However, **67** does react with $MeO_2CC \equiv CCO_2Me$ to give the arene **68**. In contrast to the reaction of **66** with but-2-yne, which proceeds in a *dissociative* manner, complex **67** reacts *directly* with the metallacycle *without* prior coordination.

Many palladium complexes are reactive towards alkynes[72]. These reactions are complex and seem to proceed in different fashions to those indicated above, although similar arenes may be the end result. A proposed mechanism for palladium-induced alkyne cyclotrimerization is given in Scheme 8[73]. This series of reactions is based on several isolated examples along the reaction sequence. The extent of reaction appears to depend very much on the size of the alkyne substituent. The first step involves formation of a dimeric η^2-alkyne species **69**; the reaction stops at this point for $R_1 = R_2 = Bu^t$[74] whereas for the less sterically demanding alkyne $Me_3SiC \equiv CPh$ the cyclobutadiene complex **70** results[75,76]. Complexes of this type are inert to further reaction with alkynes to form arenes.

Smaller alkynes insert into the σ,η^2-butadienyl complex **71** eventually to form complex **72** which eliminates arene.

In support of the latter steps treatment of $PhC \equiv CMe$ with $[PdCl_2(NCPh)_2]$ in CH_2Cl_2 produces three isomeric arenes **73–75** catalytically[77]. This may proceed via complex **76**, which is isolatable when the reaction is carried out in benzene and which

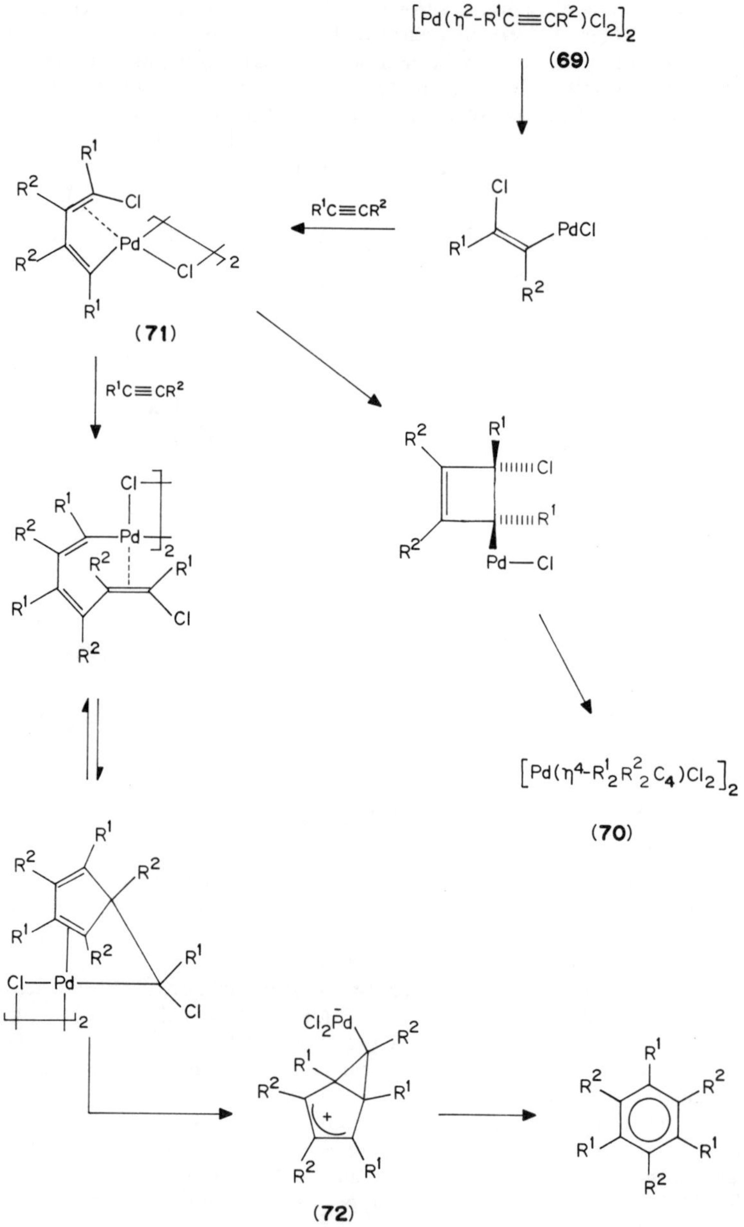

SCHEME 8.. Palladium-induced cyclotrimerization of alkynes.

decomposes to the three arenes in dichloromethane. The third arene (75) is most significant since it contains three adjacent Me groups and can only have been formed by C≡C cleavage of an alkyne.

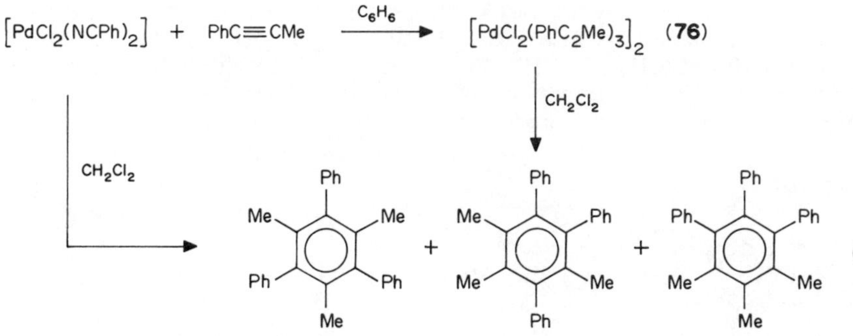

5. Alkyne oligomerization

$$[PdCl_2(NCPh)_2] + PhC{\equiv}CMe \xrightarrow{C_6H_6} [PdCl_2(PhC_2Me)_3]_2 \quad (76)$$

$$\downarrow CH_2Cl_2$$

(73) + (74) + (75) (26)

Indications have come from studies involving MeC≡CMe and the same metal complex that an intermediate of type **77** is involved (as in Scheme 9) and further that it is fluxional[78]. A dynamic process involving related complexes in the PhC≡CMe reaction would account for the formation of **75** as shown. This result is clearly relevant to the observations of Whitesides and Ehmann concerning palladium(II) cyclizations of $H_3CC{\equiv}CCD_3$ to arenes[68].

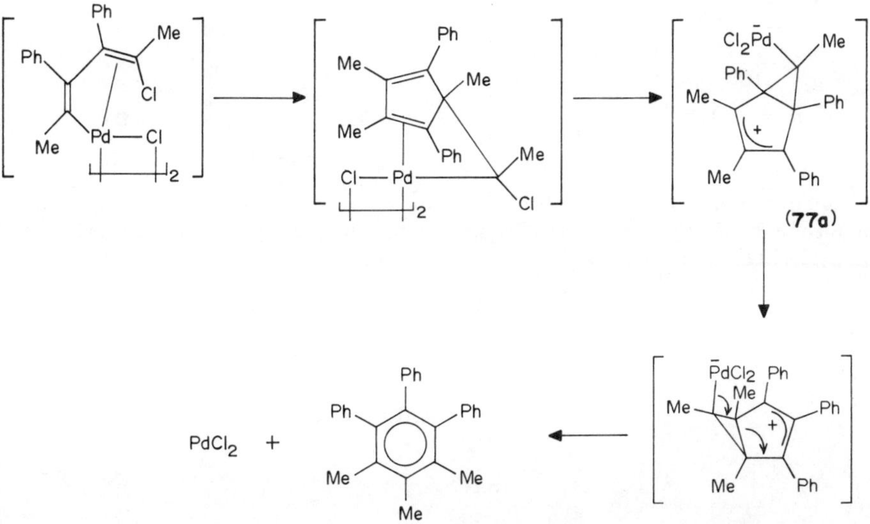

SCHEME 9

VI. REACTIONS OF DIYNES

Compared with the wealth of information available on reactions of monoalkynes with metal complexes, little concerning diynes is known. Under catalytic conditions the predominant products appear to be insoluble polymeric materials. Two types of compound are produced from $HC{\equiv}C(CH_2)_nC{\equiv}CH$ using Ziegler–Natta catalysts, depending on n, although for both cases only low yields of trimer **78** ($n = 2,3,4,5$) or **79** ($n = 5,6,7$) are formed[79]. The cobalt catalyst [CoCp(CO)$_2$] leads to higher yields of **78** ($n = 2,3,4$) on reaction with HC≡C(CH$_2$)$_n$C≡CH[80]. The mercury catalyst [Hg{Co(CO)$_4$}$_2$] results in low yields of the cyclized dimer product **80**[81,82]. The diyne diphenylbutadiyne is cyclized to **81** (R = Ph) together with a lesser quantity of the corresponding 1,2,4-substituted

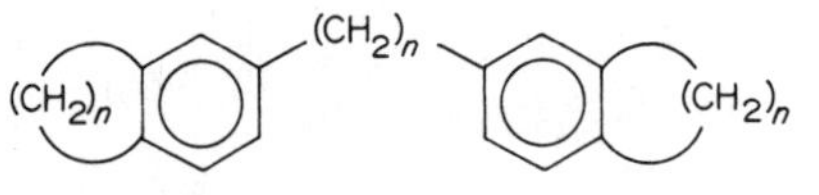

(**78**)

$n = 2, 3, 4, 5$

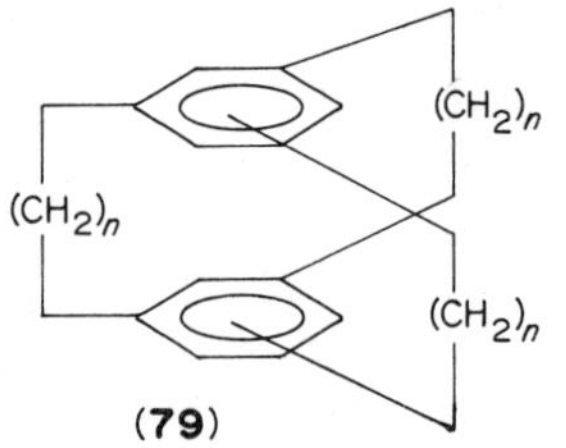

(**79**)

$n = 5, 6, 7$

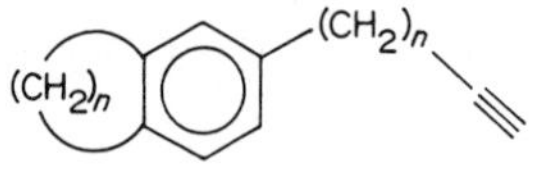

(**80**)

$n = 3, 4, 5$

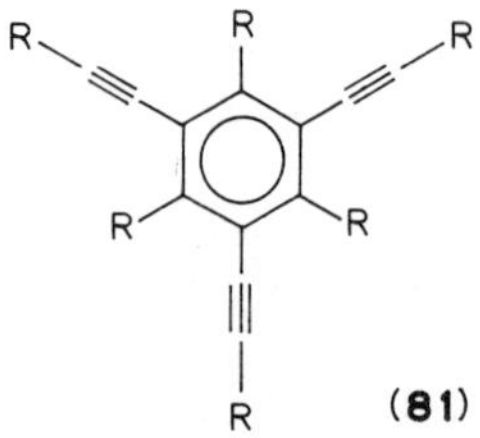

(**81**)

R = Ph, Me

compound; a similar product distribution is observed for hexa-2,4-diyne (R = Me)[83]. The non-conjugated diyne MeC≡C(CH$_2$)$_5$C≡CMe is dimerized to bicyclic **82** using [Hg{Co(CO)$_4$}$_2$] as catalyst[72].

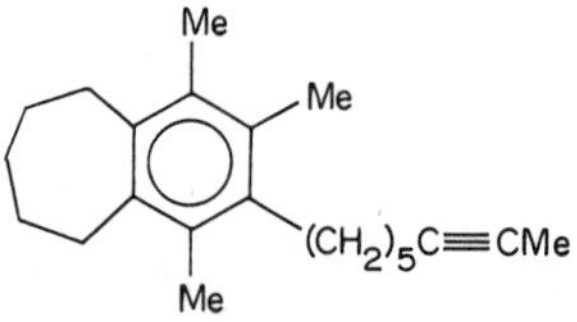

(**82**)

Diynes show a diverse chemistry in their stoichiometric reactions with organotransition metal complexes, many of the structures obtained being related to those obtained from monoynes, although there are a few surprises. Thus, while cyclotetradeca-1,8-diyne and [CoCp(CO)$_2$] lead to the cyclobutadiene complex **83**[83,84], the same diyne with [Fe(CO)$_5$] produces the unusual iron dimer **84**[84], in which formation of the cyclopentadienyl rings involves carbon—carbon bond formation and hydrogen migration reactions.

Other reactions of [Fe(CO)$_5$] with macrocyclic diynes illustrate the difficulty of predicting reaction products in this type of system and also show that the ultimate products are dependent on ring size and the relative positions of the triple bonds. Thus [Fe(CO)$_5$] with cyclotetradeca-1,7-diyne gives the cyclobutadiene complex **85**, with no sign of any dimeric species analogous to **34**, but cyclododeca-1,7-diyne leads only to a

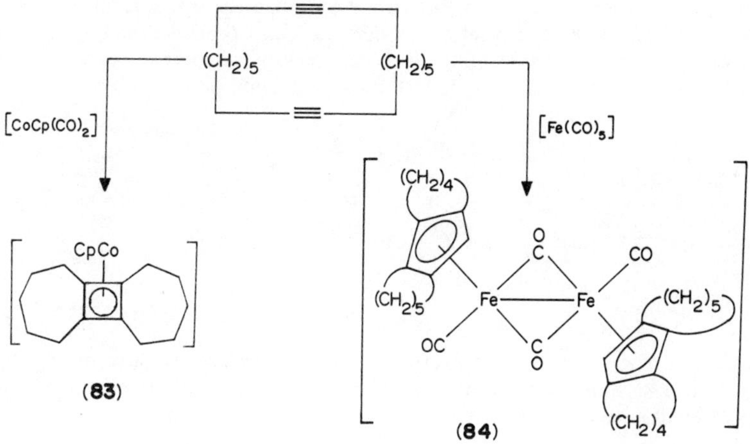

5. Alkyne oligomerization

(27)

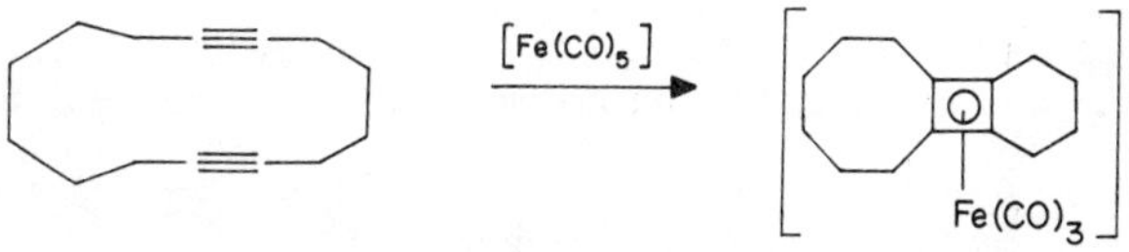

(28)

complex initially proposed as having the ferrole structure **86**[85]; the original structure proposed was shown to be incorrect through the results of an X-ray crystallographic study[86]. The corrected structure **87** is significant since it is the result of a gross skeletal

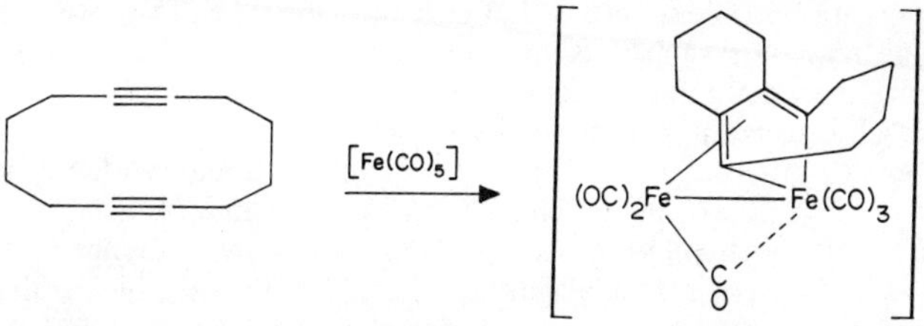

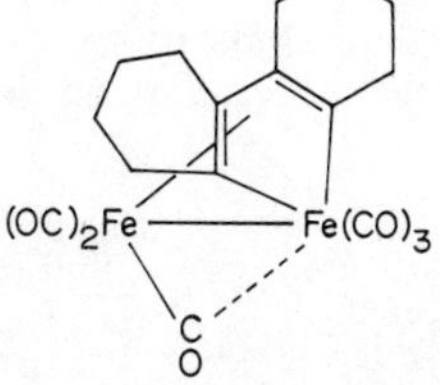

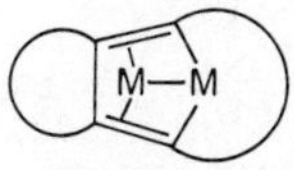

(29)

rearrangement of the ring and cleavage, in a clearly complicated reaction, of a carbon—carbon triple bond. It is speculated that initially a metallacycle of type **88** is formed in the reaction but that it rearranges, possibly via a cyclobutadiene iron intermediate **89** to the observed metallacycle **90**. The isolation of trace amounts of the cyclobutadiene complex **91** at least demonstrates the accessibility of such systems, but there is no evidence to suggest their conversion to metallacycles under the reaction conditions.

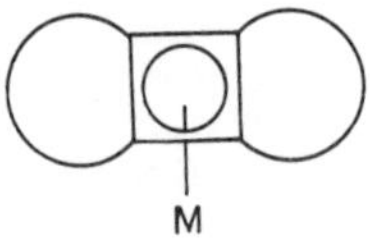

(89) 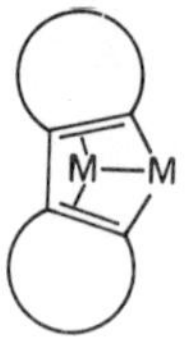   (90) 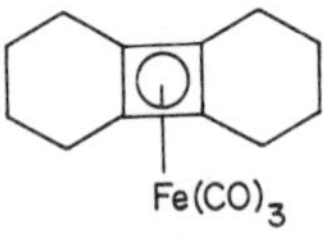   (91)

Irrespective of any intermediacy of cyclobutadiene complexes, these results do suggest a metallacycle intermediate in alkyne disproportionation reaction sequences and that saturated analogues may play a part in alkene disproportionation.

VII. CYCLOPENTADIENYLCOBALT-INDUCED COCYCLIZATIONS

The fact that [CoCp(CO)$_2$] is a good catalyst for alkyne cyclization reactions has now been put to good use in several cocyclization reactions. In particular, α,ω-diynes (**92**) may,

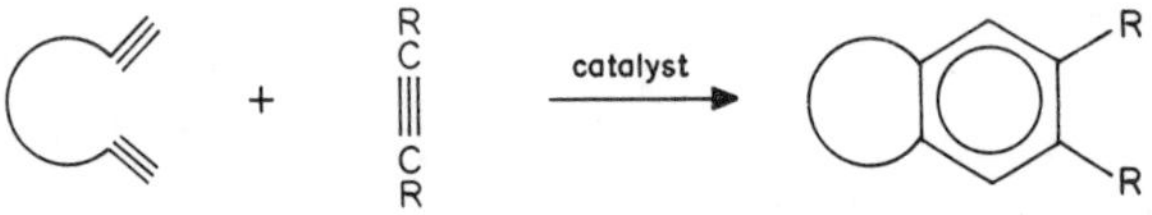

(92)   (93)

in principle, be cyclized with alkynes, leading to a general class of polycyclic compounds (**93**)[67]. The potential problem here is that care must be taken to prevent indiscriminate unwanted side reactions such as simple cyclotrimerization of RC≡CR or reactions involving more than one molecule of diyne.

One approach is to employ a monoalkyne incapable of trimerization itself but which will cocyclize with a diyne and still provide useful functionality after reaction. Such a monoalkyne is Me$_3$SiC≡CSiMe$_3$. This alkyne cannot trimerize itself for steric reasons but may react with other alkynes. Reactions of two or more diynes are prevented by extremely slow addition of mixtures of diyne, Me$_3$SiC≡CSiMe$_3$, and catalyst to refluxing Me$_3$SiC≡CSiMe$_3$. Instantaneous concentrations of diyne are therefore extremely low, preventing diyne combination, while from a practical point of view solvent Me$_3$SiC≡CSiMe$_3$ is recoverable after reaction.

This technique has been put to use in several applications. Thus, hex-1,5-diynes and monoalkynes may be cyclized to form benzocyclobutenes (**94**)[87]. This particular

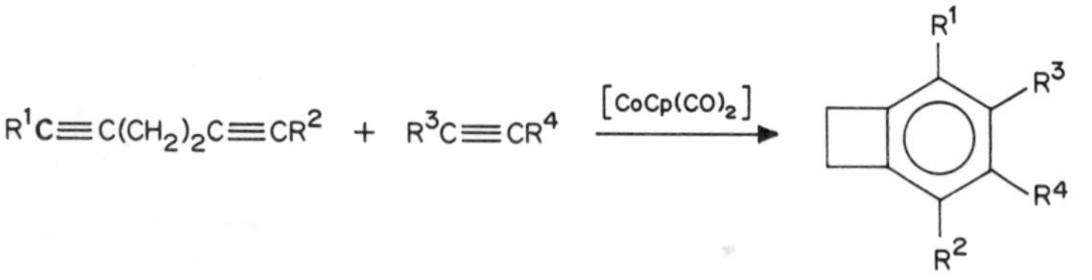

$$R^1C\equiv C(CH_2)_2C\equiv CR^2 + R^3C\equiv CR^4 \xrightarrow{[CoCp(CO)_2]}$$ (31)

(94)

5. Alkyne oligomerization

cyclization is extremely tolerant to functionality and a variety of R_1–R_4 other than $R = H$ or $SiMe_3$ is possible, allowing good control of substitution of the aromatic ring. These benzocyclobutenes are useful in their own right as a result of their reactivity and further reaction with a dienophile ($Me_3SiC\equiv CSiMe_3$) leads to ring formation via trapping of an intermediate o-xylylene[88].

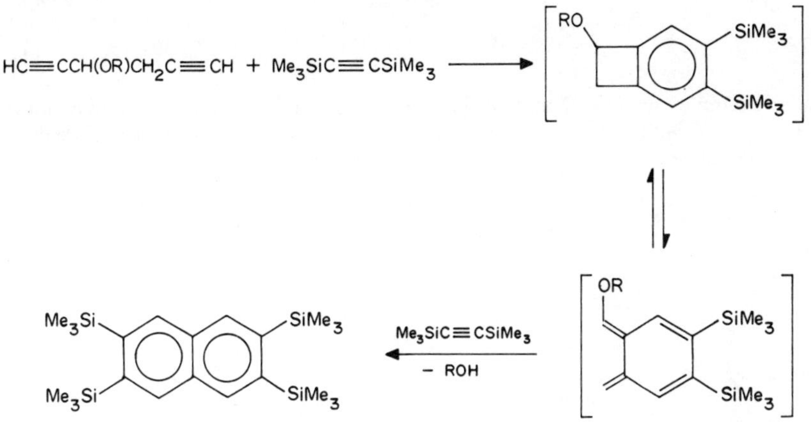

(32)

More interesting results are obtained if an intramolecular source of dienophile as in **95** is used. The reactions are then one step and lead to polycycles **96** in general and in

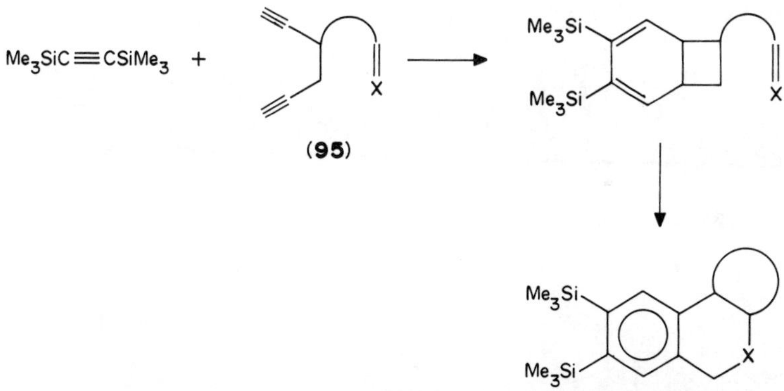

(33)

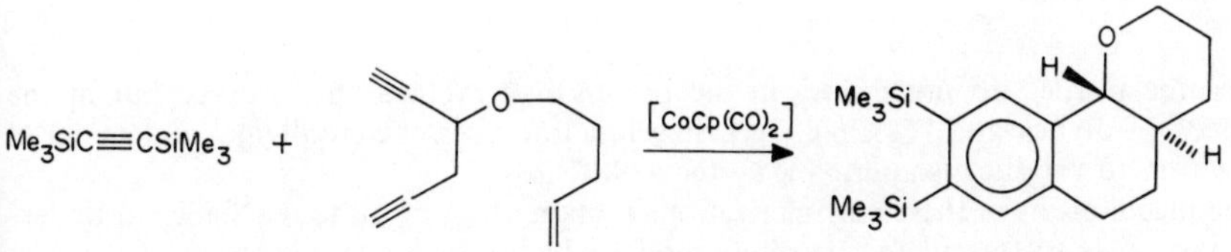

(34)

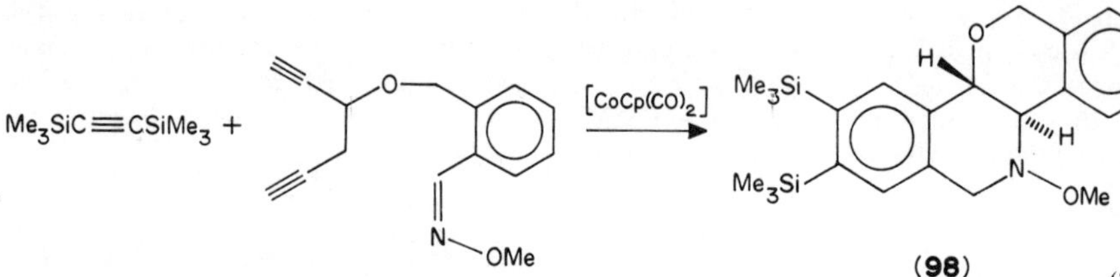

particular examples such as **97** and **98**. The stereochemistry at the new ring junction is generally *E*. Particularly interesting in this class of reactions is a synthesis of racemic oestrone[90] (Scheme 10).

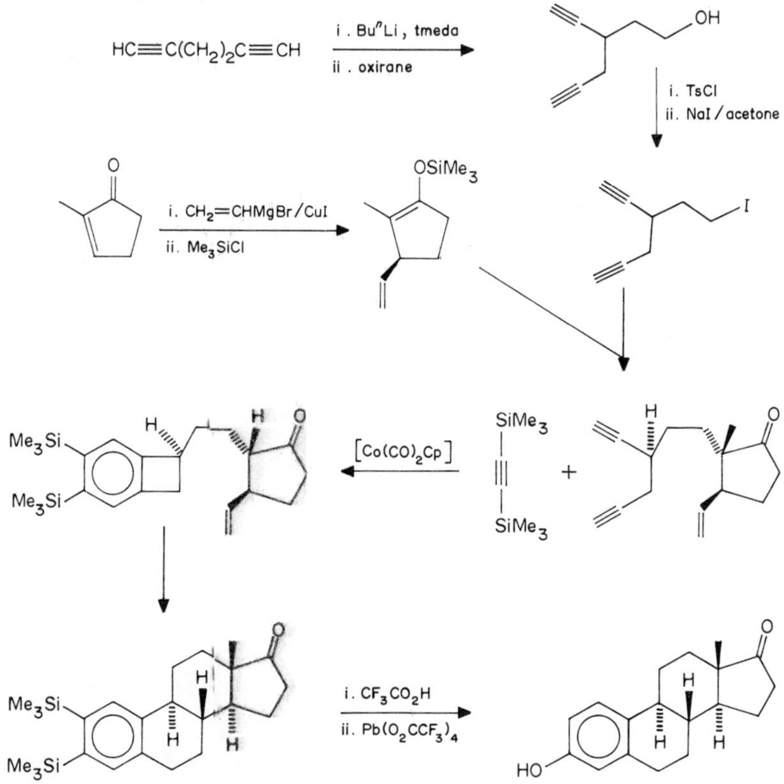

SCHEME 10. Cobalt-catalysed cotrimerization of alkynes to produce racemic oestrone.

Organic nitriles do not cyclize in the presence of cyclization catalysts, but in the presence of diynes and [CoCp(CO)$_2$] they function as a 'heteroalkyne' and undergo cyclization to produce isoquinoline systems (**98**)[91].

The mechanisms of these cotrimerizations remain obscure, although sufficient understanding exists to formulate a working model which allows the prediction of reaction products with reasonable confidence. Clearly there is a strong relationship between

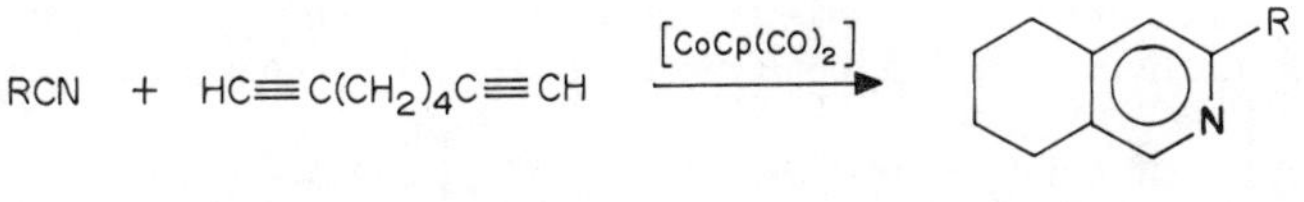

5. Alkyne oligomerization

$$\text{RCN} + \text{HC}\equiv\text{C(CH}_2)_4\text{C}\equiv\text{CH} \xrightarrow{[\text{CoCp(CO)}_2]} \text{(98)} \quad (36)$$

cotrimerizations and simpler cyclotrimerizations. However, it is not clear whether the initial bis-alkyne species (Scheme 11) is a complex of both alkyne units of the diyne or one of the monoalkyne together with a single alkyne unit for the diyne. Either pathway shown would generate the same products.

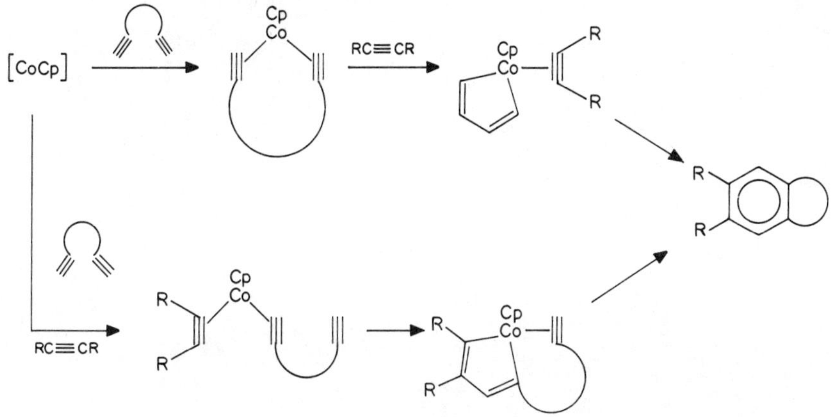

SCHEME 11. Possible cotrimerization mechanisms for cobalt.

VIII. METAL COMPLEXES OF ALKYNE TETRAMERS

A. Molybdenum

Recently, a series of reactions has been reported in which alkynes link sequentially at a dimolybdenum centre to form complexes containing two (Section III.B), three (Section III.C), and four alkyne units bridging the two molybdenum atoms. In general, under mild conditions complex **99** reacts with alkynes to give complexes **100** in which the alkyne

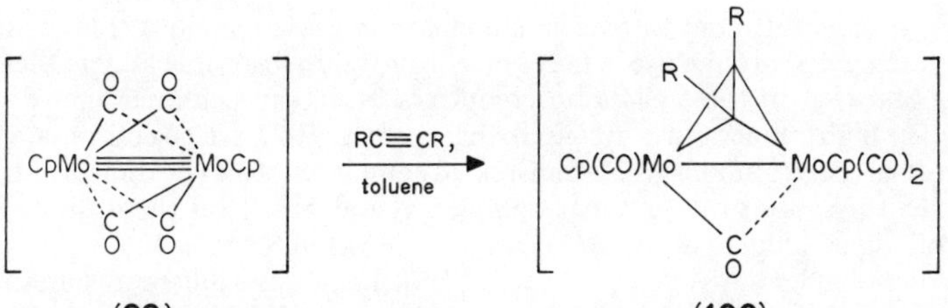

$$(99) \xrightarrow[\text{toluene}]{\text{RC}\equiv\text{CR}} (100) \quad (37)$$

R = H, Ph, Me, Et, etc.

transversely bridges the Mo—Mo bond, forming a tetrahedrane Mo_2C_2 core[92]. More forcing conditions (refluxing octane) are required to induce **100** to react further with alkynes, in which case complexes of alkyne oligomers result. Although the extent of chain extension is determined by the alkyne substituents, it is not easy to rationalize the observed substitution patterns in the products; clearly more than just steric influences are important.

The reaction of complex **101** with $MeO_2CC{\equiv}CCO_2Me$ affords three isolated products, **102–104**[48], two of which are isomeric and contain a ligand consisting of four linked alkynes. Related products are formed from complex **100** (R = H) and $MeO_2CC{\equiv}CCO_2Me$[48]. The structures of one example of each type of alkyne tetramer complex have been structurally characterized[93]; the two isomeric forms appear not to be interconvertible by heat. It is notable that these structures contain $Mo{\equiv}Mo$ bonds and that the terminal carbon atoms are carbenoid in nature. It appears critical to these reactions that dimolybdenum has the ability to undergo changes in formal metal—metal bond order.

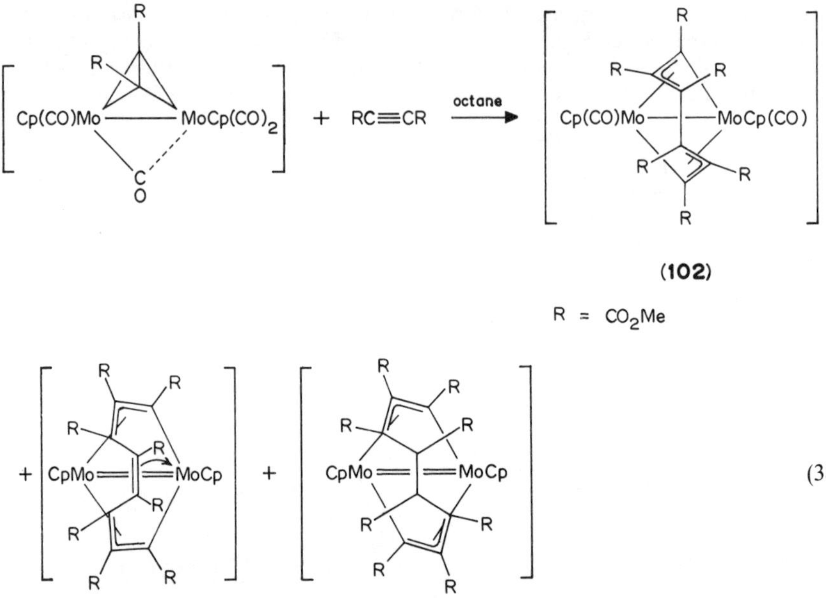

There is considerable control over functionality in these reactions. Maximum chain extension is achieved through use of the electronegatively substituted alkyne $MeO_2CC{\equiv}CCO_2Me$, but even in these cases no complexes of alkyne pentamers have yet been identified. The high temperatures are proabably necessary to drive off coordinated carbon monoxide. It has been found possible for isolated complexes of alkyne dimers or trimers to react further to give alkyne tetramer complexes, and based on these observations a sequence of alkyne linkage at dimetal centres is shown in Scheme 12[48].

Related dimolybdenum systems are available through quite different routes. Thermal loss of carbon monoxide from **105**[94] and replacement by MeCN gives cation **106**. Electrochemical studies show this species irreversibly accepts a single electron; a suitable chemical reagent to supply one electron is $[Na][FeCp(CO)_2]$, which accordingly reacts with **106** to give $[FeCp(CO)_2]_2$ together with **107**, which contains a similar eight-carbon chain to that of **104**. It is suggested that the first step in this oligomerization involves dimerization of a paramagnetic species to give an intermediate, **108**, which may react

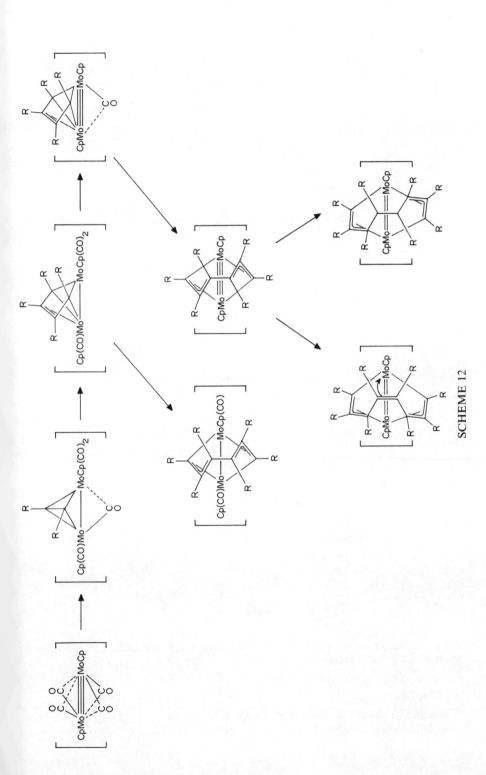

SCHEME 12

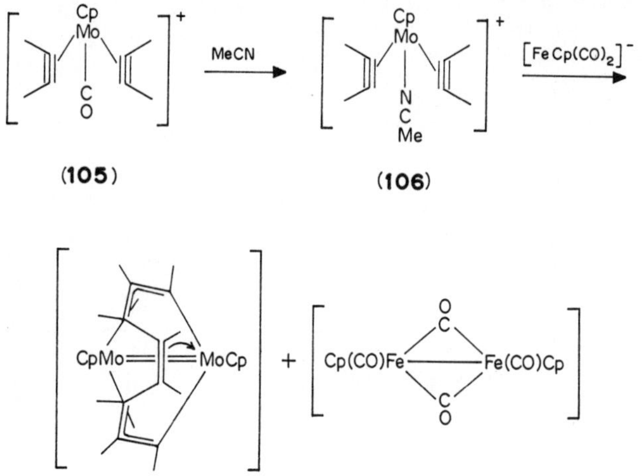

[SCHEME 13 with structures (105), (106), (107), (108) and equation (39)]

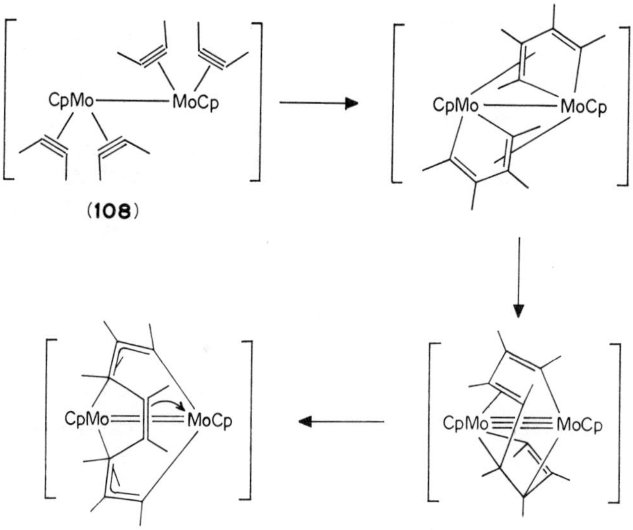

SCHEME 13

further as shown in Scheme 13. Clearly the reaction sequence leading to **107** is very different from that forming **103** or **104**.

B. Rhenium

Reaction of $[Re_2(CO)_{10}]$ with MeC≡CMe generates a compound whose structure is postulated as **109**[62] and clearly related to **103**, **104**, and **107**. Based on related work using

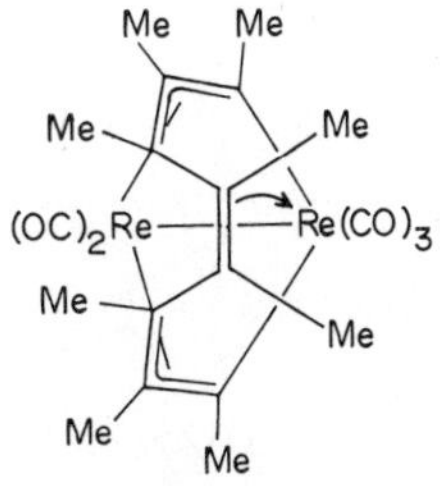

5. Alkyne oligomerization

(109)

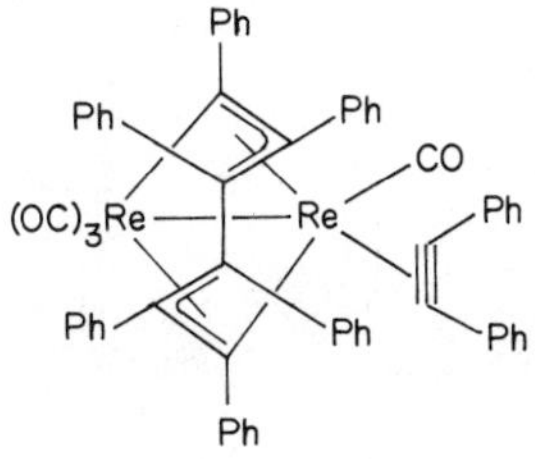

(110)

PhC≡CPh, it is likely that formation of this complex again involves sequential linking of alkynes and in which the final intermediate is of the structural type **110**.

Recently a most unusual ring-opening reaction of cyclooctatetraene has been observed in the reaction of chromocene with $[Na]_2[C_8H_8]$[95]. The orientation of the eight-carbon ligand in **111** is very similar to that of the minor isomer **104**. The chromium complex **111**

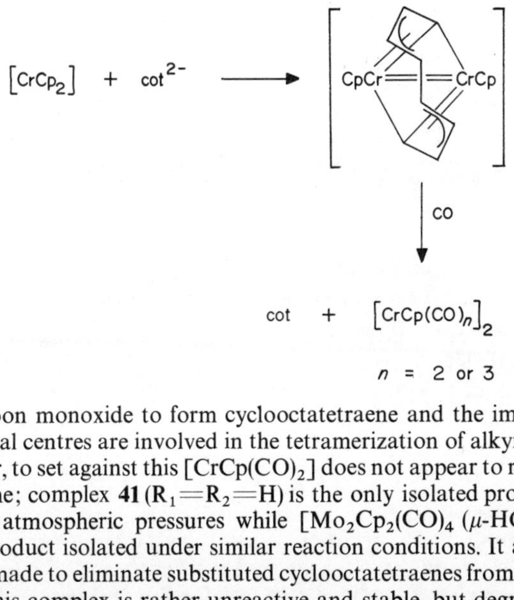

$$[CrCp_2] + cot^{2-} \longrightarrow \left[CpCr=\!\!=\!\!CrCp \right] \quad (111)$$

$$\downarrow CO$$

$$cot + [CrCp(CO)_n]_2 \qquad (40)$$

$$n = 2 \text{ or } 3$$

reacts with carbon monoxide to form cyclooctatetraene and the implication is that in some way dimetal centres are involved in the tetramerization of alkynes to cyclooctatetraenes. However, to set against this $[CrCp(CO)_2]$ does not appear to react thermally with cyclooctatetraene; complex **41** ($R_1=R_2=H$) is the only isolated product from reaction with ethyne at atmospheric pressures while $[Mo_2Cp_2(CO)_4\,(\mu\text{-HCCH})]$ is the only molybdenum product isolated under similar reaction conditions. It appears that efforts have only been made to eliminate substituted cyclooctatetraenes from the dimolybdenum complex **103**. This complex is rather unreactive and stable, but degradation is possible using cerium(IV) ions; unfortunately, no organic products were isolated from this reaction. It would clearly be interesting to know the results of similar degradations on complex **104** in which the carbon chain possesses the same orientation as that in the dichromium species. It is still not clear whether dimetal centres are implicated in the catalytic formation of cyclooctatetraenes from alkynes.

C. Palladium

The reaction products from treatment of $[PdCl_2(NCPh)_2]$ with the activated alkyne $MeO_2CC\equiv CCO_2Me$ are dependent on the solvent. Whereas reaction in benzene leads to trimer complexes (Section V), that in methanol leads to **112** (52%) containing a tetramer

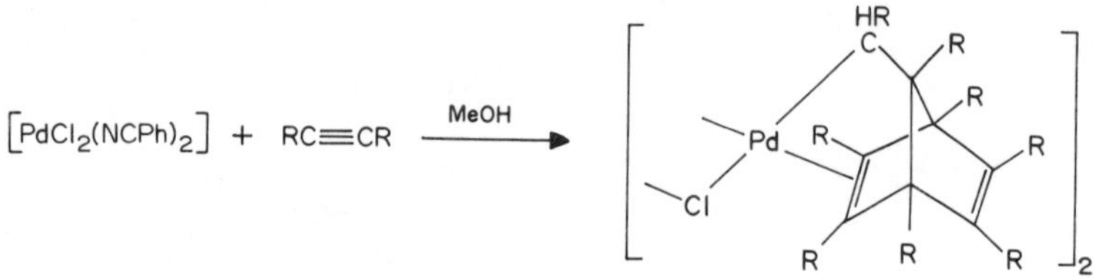

$$[PdCl_2(NCPh)_2] + RC{\equiv}CR \xrightarrow{MeOH} \left[\begin{array}{c} \text{(112)} \end{array} \right]_2 \quad (41)$$

$$R = CO_2Me$$

of the alkyne together with several organic products including the cyclopentadiene $C_5(CO_2Me)_5H$, the benzene $C_6(CO_2Me)_6$ and dimethyl oxalate[96]. The dimer is cleaved by action of donor ligands and the dipyridine adduct has been structurally characterized.

D. Manganese

The high-temperature reaction of ethyne with $[Mn_2(CO)_{10}]$ leads to the cymantrene derivative **113**[97] in 40% yield. The formation of this compound is mechanistically obscure.

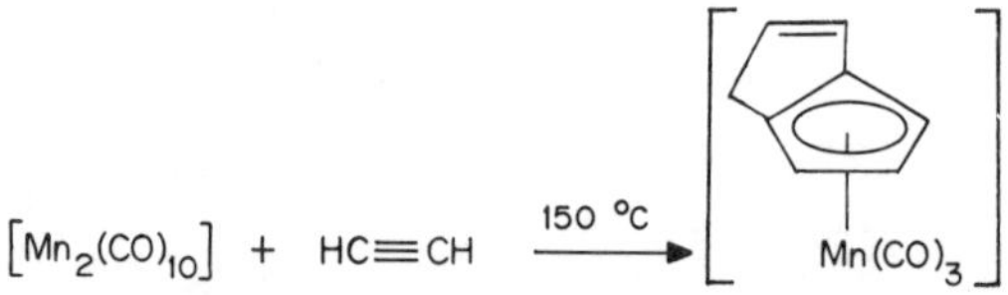

$$[Mn_2(CO)_{10}] + HC{\equiv}CH \xrightarrow{150\ °C} \left[\begin{array}{c} \text{Mn(CO)}_3 \end{array} \right] \quad (42)$$

(113)

IX. CYCLOTETRAMERIZATION OF ALKYNES; CYCLOOCTATRAENE AND OTHER TETRAMER FORMATIONS

One of the reasons for continued interest in reactions of alkynes with organotransition metal complexes is the observation by Reppe and coworkers that some nickel salts are able to catalyse the conversion of ethyne into the theoretically interesting tetramer cyclooctatetraene, together with benzene, styrene, and small amounts of other organic

$$HC{\equiv}CH \xrightarrow{[NiX_2]} cot + C_6H_6 + PhCH{=}CH_2 \quad (43)$$

oligomers such as phenylbutadiene, azulene, naphthalene, and vinylcyclooctatetraene[1]. With suitable catalysts yields of around 70% of cyclooctatetraene may be realized. Cyclooctatetraene was originally available only through a long and tedious organic synthesis[98]. The most efficient cyclooctatetraene formation catalysts are nickel(II) salts such as $[Ni(acac)_2]$ and $[Ni(CN)_2]$, but the nickel(0) complex $[Ni(PCl_3)_4]$ is very efficient for the cyclotetramerization of $HC{\equiv}CCO_2Me$, an 83% yield of the 1,2,4,6-tetrasubstituted cyclooctatetraene being isolated together with 1,2,4-tricarbomethoxybenzene as byproduct[99,100]. For ethylpropiolate a mixture of benzenes is the major product together with two isomers of tetrasubstituted cyclooctatetraenes.

$$HC{\equiv}CR \xrightarrow{[Ni(PCl_3)_4]} 1,2,4,6\text{-}R_4\text{-}cot + 1,3,5,7\text{-}R_4\text{-}cot + C_6H_3R_3 \quad (44)$$

$R = CO_2Me$	83%	0%	1,2,4-	17%
$R = CO_2Et$	28%	1%	1,2,4- + 1,3,5-	71%

Disubstituted alkynes do not form cyclooctatetraenes under these conditions, but they may be incorporated into a cyclooctatetraene ring by co-oligomerization reactions. The yields are generally moderate[101,102].

$$HC\equiv CH + R^1C\equiv CR^2 \xrightarrow{\text{Ni catalyst}} 1\text{-}R^1\text{-}2\text{-}R^2\text{-cot} + \text{cot} + 1\text{-}R^1\text{-}2\text{-}R^2\text{-}C_6H_4 + C_6H_6 \quad (45)$$

$R^1 = R^2 = Ph^{97}, Me^{98}$
$R^1 = H, R^2 = Me^{98}, Ph^{98}, Pr^{n98}, Bu^{n98}, vinyl^{103,104}$

The reactions of monosubstituted alkyne co-oligomerizations with ethyne have also been studied[102] and lead to substituted cyclooctatetraenes together with other aromatic products in relative yields depending on the nature of R.

The mechanism of cyclooctatetraene formation has been the subject of much speculation and argument, and the question has not been resolved. Reppe et al.[1] originally suggested a sequence (Scheme 14) involving a nickelanonatetraene (114), which is still a

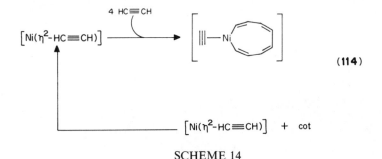

SCHEME 14

very reasonable suggestion; presumably the metallacycle is built up by successive insertion reactions of alkyne into a nickel—carbon bond of smaller metallacycles. Another mechanism suggests a concerted process involving the cyclization taking place at an octahedral nickel intermediate after adopting a suitable conformation[65,66,105,106]. This mechanism would find it difficult to account for the observed substitution patterns for

substituted alkyne tetramerizations and the fact that nickel tends to undergo insertion reactions fairly readily. It does have the merit of accounting for the preference for benzene formation in the presence of a donor ligand—presumably by competition with alkyne for a vacant coordination site[105].

Recent work implicates the importance of dinuclear metallacycles (Section VIII); indeed, the dinickel complex [$Ni_2(C_8H_8)_2$] (**104**)[107,108] (Scheme 15), whose structure has been determined crystallographically[109], has been shown to be a catalyst for cyclooctatetraene formation[109].

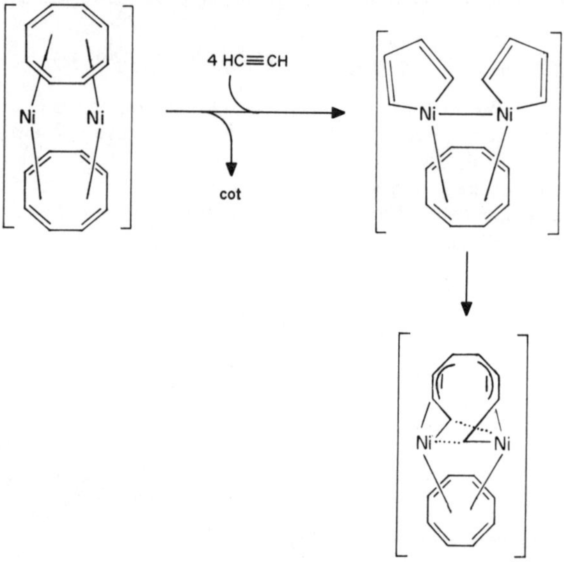

SCHEME 15

In order to cast ligh on the nickel-catalysed cyclooctatetraene formation, the labelled alkyne $HC{\equiv}^{13}CH$ has been cyclized in the presence of a catalyst[110] and a statistical analysis of the labelling carried out. Four possibilities were considered (Scheme 16), a

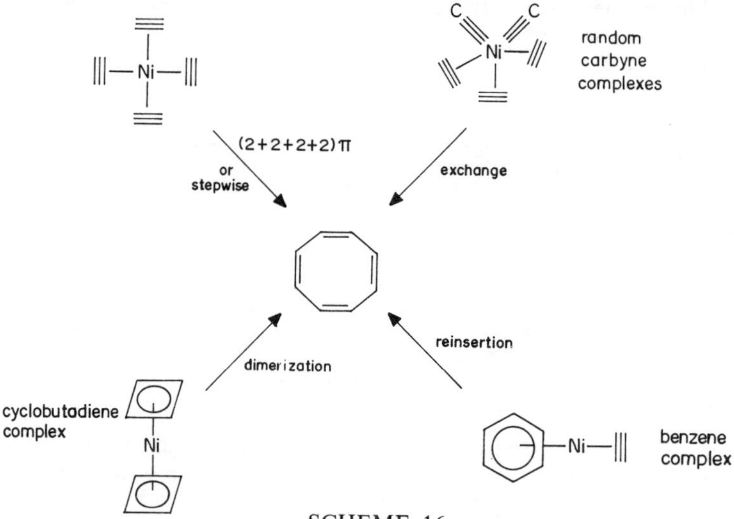

SCHEME 16

5. Alkyne oligomerization

concerted or stepwise mechanism in which the integrity of the initial carbon pairs is maintained and three other processes, all of which involve at some stage rupture of the alkyne carbon pairs either directly (a carbyne mechanism) or by insertion into complexed C_4 and C_6 ring systems (cyclobutadiene and benzene, respectively). Analysis of the reaction products demonstrates that the integrity of the initial C_2 units is maintained, a result only consistent with predominant stepwise or concerted mechanisms. Unfortunately, the experiment is incapable of distinguishing between the two. At present the evidence is very strong for stepwise insertion processes, but it is not clear whether one or two metals are involved and further experiments are still necessary.

Although it seems unlikely that intermediates of cyclobutadiene symmetry are involved in the catalytic formation of cyclooctatetraene, there are some examples of cyclobutadiene complexes which undergo decomposition to substituted cyclooctatetraenes. Thus the palladium cyclobutadiene complex **115**[111,112] reacts with PPh_3 and liberates the cyclooctatetraene **116** (Scheme 17). The nickel complexes **117**[113] and **118**[114] do so also,

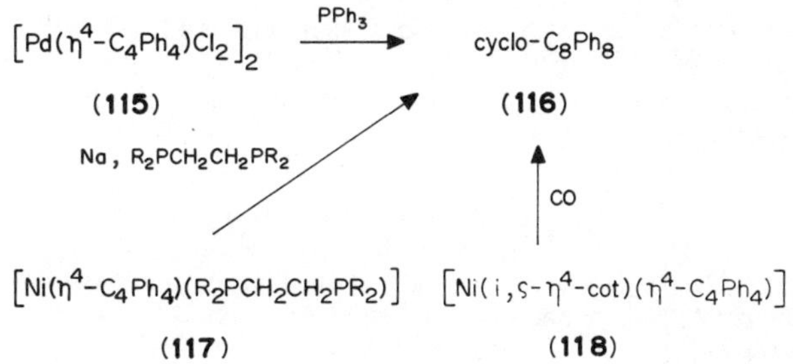

$$[Pd(\eta^4\text{-}C_4Ph_4)Cl_2]_2 \xrightarrow{PPh_3} \text{cyclo-}C_8Ph_8$$
(115) (116)

Na, $R_2PCH_2CH_2PR_2$ ↗ ↑ CO

$[Ni(\eta^4\text{-}C_4Ph_4)(R_2PCH_2CH_2PR_2)]$ $[Ni(i,s\text{-}\eta^4\text{-}cot)(\eta^4\text{-}C_4Ph_4)]$
(117) (118)

SCHEME 17

although in the latter two cases the cyclobutadiene does not originate directly from alkynes[14].

One recent example of a cyclotetramerization of a disubstituted alkyne is known but not involving a conventional catalyst: vapourized nickel and $MeC\equiv CEt$ lead to the cyclooctatetraene **119**[115].

$$4 \text{ MeC}\equiv\text{CEt} \xrightarrow{[Ni]} \text{cyclo-}C_8Me_4Et_4 \qquad (48)$$
(119)

The activated alkyne $MeO_2CC\equiv CCO_2Me$ is converted into the unusual tetramer **120** in the presence of uranium powder[116].

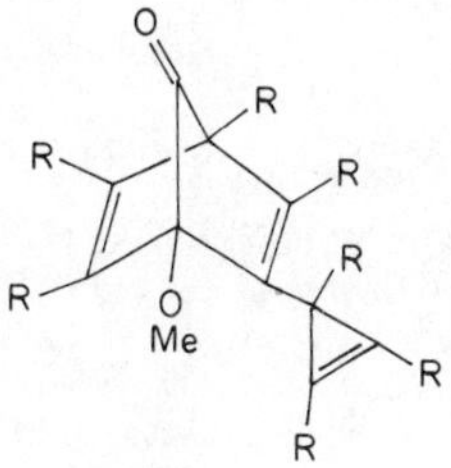

(120)

Compound **121** is a derivative of an intermediate in the reaction of [PdCl$_2$(NCPh)$_2$] with Bu^{t}C≡CH (Section V). Treatment with either of the terminal alkynes phenylacetylene or

$$\text{(121)} \xrightarrow[\text{benzene}]{2\ Bu^tC\equiv CH} \text{(122)} \tag{49}$$

(**121**)

R = Ph (**122**)

R = C$_6$H$_4$, Cl-p (**122**)

p-chlorophenylacetylene leads to dihydropentalene **119** or **120**. These are related to other dihydropentalenes obtained from the PdCl$_2$-induced oligomerization of phenylacetylene and their formations are explained in terms of mechanisms involving intermediates similar to those in Scheme 8.

X. REFERENCES

1. W. Reppe, O. Schlichting, K. Klager, and T. Toepel, *Justus Liebigs Ann. Chem.*, **560**, 1 (1948).
2. S. Otsuka and A. Nakamura, *Adv. Organomet. Chem.*, **14**, 245 (1976).
3. K. M. Nicholas, M. O. Nestle, and D. Seyferth, in *Transition Metal Organometallics in Organic Synthesis*, (ed. H. Alper) Vol. II, Academic Press, New York, 1978, p. 1.
4. J. O. Glanville, J. M. Stewart, and S. O. Grim, *J. Organomet. Chem.*, **7**, P9 (1967).
5. W. I. Bailey, Jr., D. M. Collins, and F. A. Cotton, *J. Organomet. Chem.*, **135**, C53 (1977).
6. M. Green, D. M. Grove, J. A. K. Howard, J. L. Spencer, and F. G. A. Stone, *J. Chem. Soc., Chem. Commun.*, 759 (1976).
7. R. M. Laine, R. E. Moriarty, and R. Bau, *J. Am. Chem. Soc.*, **94**, 1402 (1972).
8. R. S. Dickson, H. P. Kirsch, and D. J. Lloyd, *J. Organomet. Chem.*, **101**, C48 (1975).
9. J. F. Tilney-Bassett, *J. Chem. Soc.*, 4784 (1963).
10. R. F. Heck, *Organotransition Metal Chemistry, A Mechanistic Approach*, Academic Press, New York, 1974.
11. G. Giacomelli, F. Maracci, A. M. Caporusso, and L. Lardicci, *Tetrahedron Lett.*, 3217 (1979).
12. L. Carlton and G. Read, *J. Chem. Soc., Perkin Trans. 1*, 1631 (1978).
13. N. Hagihara, M. Tamura, H. Yamazaki, and M. Fujinara, *Bull. Chem. Soc. Jap.*, **34**, 892 (1961).
14. H. Yamazaki, *J. Chem. Soc., Chem. Commun.*, 841 (1976).
15. A. Efraty, *Chem. Rev.*, **77**, 691 (1977).
16. A. N. Nesmeyanov, K. N. Anisimov, N. E. Kolobova, and A. A. Pasynskii, *Dokl. Akad. Nauk SSSR, Ser. Khim.*, **182**, 112 (1968).
17. A. A. Pasynskii, K. N. Anisimov, N. E. Kolobova, and A. N. Nesmeyanov, *Dokl. Akad. Nauk SSSR, Ser. Khim.*, **185**, 610 (1969).
18. A. N. Nesmeyanov, K. N. Anisimov, N. E. Kolobova, and A. A. Pasynskii, *Izv. Akad. Nauk SSSR, Ser. Khim.*, 774 (1966).
19. A. N. Nesmeyanov, K. N. Anisimov, N. E. Kolobova, and A. A. Pasynskii, *Izv. Akad. Nauk SSSR, Ser. Khim.*, 100 (1969).
20. A. N. Nesmeyanov, A. I. Gusev, A. A. Pasynskii, K. N. Anisimov, N. E. Kolobova, and Yu. T. Struchov, *J. Chem. Soc., Chem. Commun.*, 277 (1969).
21. A. N. Nesmeyanov, A. I. Gusev, A. A. Pasynskii, K. N. Anisimov, N. E. Kolobova, and Yu. T. Struchov, *J. Chem. Soc., Chem. Commun.*, 739 (1969).
22. W. Hübel and R. Merényi, *J. Organomet. Chem.*, **2**, 213 (1964).
23. J. A. Potenza, R. J. Johnson, R. Chirico, and A. Efraty, *Inorg. Chem.*, **16**, 2354 (1977).

24. R. Bühler, R. Geist, R. Mündnich, and H. Plieninger, *Tetrahedron Lett.*, 1919 (1973).
25. A. Nakamura and N. Hagihara, *Bull. Soc. Chem. Jap.*, **34**, 452 (1961).
26. J. F. Helling, S. C. Rennison, and A. Merijan, *J. Am. Chem. Soc.*, **89**, 7140 (1967).
27. M. D. Rausch and R. A. Genetti, *J. Org. Chem.*, **35**, 3888 (1970).
28. J. R. Fritch, K. P. C. Vollhardt, M. R. Thompson, and V. W. Day, *J. Am. Chem. Soc.*, **101**, 2768 (1979).
29. P. M. Maitlis, *Acc. Chem. Res.*, **9**, 93 (1976).
30. E. A. Kelley and P. M. Maitlis, *J. Chem. Soc., Dalton Trans.*, 167 (1979).
31. J. L. Davidson, *J. Chem. Soc., Chem. Commun.*, 113 (1980).
32. J. L. Davidson, L. Manojlovic-Muir, K. W. Muir, and A. N. Keith, *J. Chem. Soc., Chem. Commun.*, 749 (1980).
33. F. D. Mango, *Coord. Chem. Rev.*, **15**, 109 (1975).
34. R. G. Pearson, *Symmetry Rules for Chemical Reactions*, Wiley, Chichester, New York, 1976, p. 431.
35. H. Yamazaki and Y. Wakatsuki, *J. Organomet. Chem.*, **139**, 157 (1977).
36. H. Yamazaki and N. Hagihara, *J. Organomet. Chem.*, **21**, 431 (1970).
37. R. G. Gastinger, M. D. Rausch, D. A. Sullivan, and G. J. Palenik, *J. Am. Chem. Soc.*, **98**, 719 (1976).
38. L. P. McDonnell-Bushnell, E. R. Evitt, and R. G. Bergman, *J. Organomet. Chem.*, **157**, 445 (1978).
39. G. Ville, K. P. C. Vollhardt, and M. J. Winter, *J. Am. Chem. Soc.*, **103**, 5267 (1981).
40. R. E. Davies, B. L. Basnett, R. G. Amiet, W. Mark, J. S. McKennis, and R. Pettit, *J. Am. Chem. Soc.*, **96**, 7108 (1974).
41. R. A. Sanchez-Delgado and G. Wilkinson, *J. Chem. Soc., Dalton Trans.*, 804 (1977).
42. Y. Inoue, Y. Itoh, and H. Hashimoto, *Chem. Lett.*, 633 (1978).
43. Y. Inoue, Y. Itoh, and H. Hashimoto, *Chem. Lett.*, 911 (1978).
44. R. S. Dickson and P. J. Fraser, *Adv. Organomet. Chem.*, **12**, 313 (1974).
45. K. Nicholas, L. S. Bray, R. E. Davis, and R. Pettit, *J. Chem. Soc., Chem. Commun.*, 608 (1971).
46. G. Dettlaf and E. Weiss, *J. Organomet. Chem.*, **108**, 213 (1976).
47. S. R. Prince, *Cryst. Struct. Commun.*, **5**, 451 (1976).
48. S. A. R. Knox, R. F. D. Stansfield, F. G. A. Stone, M. J. Winter, and P. Woodward, *J. Chem. Soc., Commun.*, 221 (1978).
49. S. A. R. Knox, R. F. D. Stansfield, F. G. A. Stone, M. J. Winter, and P. Woodward, *J. Chem. Soc., Dalton Trans.*, 173 (1982).
50. J. S. Bradley, *J. Organomet. Chem.*, **150**, C1 (1978).
51. S. Slater and E. Muetterties, *Inorg. Chem.*, **19**, 3337 (1980).
52. S. Slater and E. Muetterties, *Inorg. Chem.*, **20**, 946 (1981).
53. M. H. Chisholm, J. C. Huffman, and I. P. Rothwell, *J. Am. Chem. Soc.*, **103**, 4245 (1981).
54. Bennett and P. B. Donaldson, *Inorg. Chem.*, **17**, 1995 (1978).
55. S. A. Gardner, P. S. Andrews, and M. D. Rausch, *Inorg. Chem.*, **12**, 2396 (1973).
56. R. S. Dickson and H. P. Kirsch, *Aust. J. Chem.*, **25**, 2535 (1972).
57. P. A. Corrigan, R. S. Dickson, G. D. Fallon, L. J. Michel, and C. Mok, *Aust. J. Chem.*, **31**, 1937 (1978).
58. J. Piron, P. Piret, J. Meunier-Piret, and Y. Degrève, *Bull. Soc. Chem. Belg.* **78**, 121 (1969).
59. J. L. Davidson, M. Green, F. G. A. Stone, and A. J. Welch, *J. Chem. Soc., Chem. Commun.*, 286 (1975).
60. R. S. Dickson, P. J. Fraser, and B. M. Gatehouse, *J. Chem. Soc., Dalton Trans.*, 2278 (1972).
61. J. A. Beck, S. A. R. Knox, R. F. D. Stansfield, F. G. A. Stone, M. J. Winter, and P. Woodward, *J. Chem. Soc., Dalton Trans.*, 195 (1982).
62. M. J. Mays, D. W. Prest, and P. R. Raithby, *J. Chem. Soc., Dalton Trans.*, 771 (1981).
63. J. Browning, M. Green, J. L. Spencer, and F. G. A. Stone, *J. Chem. Soc., Dalton Trans.*, 97 (1974).
64. J. Browning, M. Green, B. R. Penfold, J. L. Spencer, and F. G. A. Stone, *J. Chem. Soc., Chem. Commun.*, 31 (1973).
65. C. W. Bird, *Transition Metal Intermediates in Organic Synthesis*, Logos Press, London, 1967.
66. C. Hoogzand and W. Hübel, in *Organic Synthesis via Metal Carbonyls* (Ed. I. Wender and P. Pino), Vol. I, Wiley, Chichester, New York, 1968, p. 343.

67. K. P. C. Vollhardt, *Acc. Chem. Res.*, **10**, 1 (1977).
68. G. M. Whitesides and W. J. Ehmann, *J. Am. Chem. Soc.*, **91**, 3800 (1969).
69. U. Krüerke, C. Hoogzand, and W. Hübel, *Chem. Ber.*, **94**, 2817 (1961).
70. Y. Wakatsuki, T. Kuramitsu, and H. Yamazaki, *Tetrahedron Lett.*, 4549 (1974).
71. D. R. McAlister, J. E. Bercaw, and R. G. Bergman, *J. Am. Chem. Soc.*, **99**, 1666 (1977).
72. P. M. Maitlis, *Acc. Chem. Res.*, **9**, 93 (1976).
73. P. M. Maitlis, *J. Organomet. Chem.*, **200**, 161 (1980).
74. T. Hosokawa, I. Montani, and S. Nishioka, *Tetrahedron Lett.*, 3833 (1969).
75. T. Hosokawa and I. Montani, *Tetrahedron Lett.*, 3021 (1969).
76. M. Avram, I. G. Dinulescu, G. D. Mateescu, E. Avram' and C. D. Nenitzescu, *Rev. Roum. Chim.*, **14**, 1181 (1969).
77. H. Dietl, H. Reinheimer, J. Moffatt, and P. M. Maitlis, *J. Am. Chem. Soc.*, **92**, 2276 (1970).
78. H. Reinheimer, J. Moffatt, and P. M. Maitlis, *J. Am. Chem. Soc.*, **92**, 2285 (1970).
79. A. J. Hubert and J. Dale, *J. Chem. Soc.*, 3160 (1965).
80. K. P. C. Vollhardt and R. G. Bergman, *J. Am. Chem. Soc.*, **96**, 4996 (1974).
81. W. Hübel and R. Merényi, *Chem. Ber.*, **96**, 930 (1963).
82. R. B. King and A. Efraty, *J. Am. Chem. Soc.*, **92**, 6071 (1970).
83. R. B. King and A. Efraty, *J. Am. Chem. Soc.*, **94**, 3021 (1972).
84. R. B. King and C. W. Eavenson, *J. Organomet. Chem.*, **16**, P75 (1969).
85. R. B. King, I. Haiduc, and C. W. Eavenson, *J. Am. Chem. Soc.*, **95**, 2508 (1973).
86. H. B. Chin and R. Bau, *J. Am. Chem. Soc.*, **95**, 5070 (1973).
87. R. L. Hillard, III, and K. P. C. Vollhardt, *J. Am. Chem. Soc.*, **99**, 4058 (1977).
88. R. L. Funk and K. P. C. Vollhardt, *J. Chem. Soc., Chem. Commun.*, 833 (1976).
89. R. L. Funk and K. P. C. Vollhardt, *J. Am. Chem. Soc.*, **98**, 6755 (1976).
90. R. L. Funk and K. P. C. Vollhardt, *J. Am. Chem. Soc.*, **101**, 215 (1979).
91. A. Naiman and K. P. C. Vollhardt, *Angew. Chem. Int. Ed. Engl.*, **16**, 708 (1977).
92. W. I. Bailey, M. H. Chisholm, F. A. Cotton, and L. A. Rankel, *J. Am. Chem. Soc.*, **100**, 5764 (1978).
93. A. M. Boileau, A. G. Orpen, R. F. D. Stansfield, and P. Woodward, *J. Chem. Soc., Dalton Trans.*, 187 (1982).
94. M. Green, N. C. Norman, and A. G. Orpen, *J. Am. Chem. Soc.*, **103**, 1269 (1981).
95. W. Geibel, G. Wilke, R. Goddard, C. Krüger, and R. Mynott, *J. Organomet. Chem.*, **102**, 2576 (1980).
96. A. Konietzny, P. M. Bailey, and P. M. Maitlis, *J. Chem. Soc., Chem. Commun.*, 78 (1975).
97. T. H. Coffield, K. G. Ihrman, and W. Burns, *J. Am. Chem. Soc.*, **82**, 4209 (1960).
98. L. A. Paquette, *Tetrahedron*, **31**, 2855 (1975).
99. J. R. Leto and M. F. Leto, *J. Am. Chem. Soc.*, **83**, 2944 (1961).
100. J. R. Leto and M. L. Fiene, *U.S. Pat.*, 3 076 016, 1963; *Chem. Abstr.*, **59**, 6276c (1963).
101. A. C. Cope and D. S. Smith, *J. Am. Chem. Soc.*, **74**, 5136 (1952).
102. A. C. Cope and H. C. Campbell, *J. Am. Chem. Soc.*, **74**, 179 (1952).
103. L. E. Craig and C. E. Larrabee, *J. Am. Chem. Soc.*, **73**, 1191 (1951).
104. A. C. Cope and S. W. Fenton, *J. Am. Chem. Soc.*, **73**, 1195 (1951).
105. G. N. Schrauzer and S. Eichler, *Chem. Ber.*, **95**, 550 (1962).
106. G. N. Schrauzer, P. Glockner, and S. Eichler, *Angew. Chem., Int. Ed. Engl.*, **3**, 185 (1964).
107. G. Wilke, *Angew. Chem.*, **72**, 581 (1960).
108. B. Bogdanović, M. Kröner, and G. Wilke, *Justus Liebigs Ann. Chem.*, **699**, 1 (1966).
109. G. Wilke, *Pure Appl. Chem.*, **50**, 677 (1978).
110. R. E. Colburn and K. P. C. Vollhardt, *J. Am. Chem. Soc.*, **103**, 6259 (1981).
111. P. M. Maitlis and F. G. A. Stone, *Proc. Chem. Soc.*, 330 (1962).
112. G. S. Pauley, W. N. Lipscomb, and H. H. Freedman, *J. Am. Chem. Soc.*, **86**, 4725 (1964).
113. H. Hoberg and W. Richter, *J. Organomet. Chem.*, **195**, 355 (1980).
114. H. Hoberg and C. Fröhlich, *Angew. Chem., Int. Ed. Engl.*, **19**, 45 (1980).
115. L. H. Simons and J. J. Lagowski, *Fund. Res. Homog. Catal.*, **2**, 73 (1978).
116. T. C. Wen, C. C. Chang, Y. D. Chuang, J. P. Chiu, and C. T. Chang, *J. Am. Chem. Soc.*, **103**, 4576 (1981).

The Chemistry of the Metal—Carbon Bond, Vol. 3
Edited by F. R. Hartley and S. Patai
© 1985 John Wiley & Sons Ltd.

CHAPTER **6**

Transition metal carbonyls in organic synthesis

JULIAN A. DAVIES and RANDY J. SHAVER

Department of Chemistry, College of Arts and Sciences, University of Toledo, Toledo, Ohio 43606, USA

I.	INTRODUCTION	297
II.	ALKANES AS SUBSTRATES	297
	A. Simple Alkanes	297
	1. Alkane Activation	297
	B. Strained Cycloalkanes	298
	1. Valence isomerization and carbonylation	298
III.	ALKENES AS SUBSTRATES	301
	A. Protection and Storage of Reactive Alkenes	301
	B. Rearrangement/Isomerization	302
	C. Carbonylation	303
	D. Cyclooligomerization	303
	1. Diene cyclooligomerization	303
	2. Allene cyclooligomerization	304
	E. Nucleophilic Attack on Coordinated Olefins	304
	F. Electrophilic Attack on Coordinated Olefins	305
	G. Diels–Alder Reactions of Coordinated Alkenes	306
	H. Cyclopropane Formation	306
IV.	ALKYNES AS SUBSTRATES	307
	A. Protection and Storage of Reactive Alkynes	307
	B. Carbonylation	308
	C. Cyclooligomerization	309
V.	ORGANIC HALIDES AS SUBSTRATES	311
	A. Alkyl and Aryl Halides	311
	1. Carbonylation and coupling reactions	311
	B. Allylic Halides	312
	1. Carbonylation and coupling reactions	312
	C. Acyl Halides	314
	1. Decarbonylation	314
VI.	ALCOHOLS AS SUBSTRATES	314
	A. Carbonylation	314

		B. Hydrosilylation (Silane Alcoholysis)	314
VII.	KETONES AND KETENES AS SUBSTRATES		315
	A.	Decarbonylation	315
	B.	Reductive Amination	316
	C.	Reductive Alkylation and Arylation	316
VIII.	ALDEHYDES AS SUBSTRATES		317
	A.	Carbonylation	317
	B.	Decarbonylation	317
	C.	Reductive Amination	317
	D.	N-Acylamino Acid Synthesis	318
	E.	Reductive Alkylation and Arylation	318
IX.	EPOXIDES AND ETHERS AS SUBSTRATES		318
	A.	Rearrangement	318
		1. Epoxides	318
	B.	Reduction	319
		1. Epoxides	319
	C.	Carbonylation	319
		1. Simple ethers	319
		2. Epoxides	320
		3. Cyclic ethers	320
	D.	Polymerization under Hydrosilylation Conditions	321
X.	CARBOXYLIC ACIDS AS SUBSTRATES		321
	A.	Decarbonylation	321
	B.	Hydrosilylation	321
XI.	ACID ANHYDRIDES AS SUBSTRATES		321
	A.	Decarbonylation	321
XII.	ESTERS AS SUBSTRATES		322
	A.	Carbonylation of Esters and Lactones	322
	B.	Decarboxylation of Pyrones	323
XIII.	DIALKYLACETALS AS SUBSTRATES		323
	A.	Carbonylation	323
XIV.	*ORTHO*-ESTERS AS SUBSTRATES		323
	A.	Carbonylation	323
XV.	SULPHOXIDES AS SUBSTRATES		324
	A.	Deoxygenation	324
XVI.	$>$C=S AND —N=S COMPOUNDS AS SUBSTRATES		324
	A.	Dehydrosulphuration and Desulphuration	324
XVII.	$>$C=N—,—N=N— AND —C≡N COMPOUNDS AS SUBSTRATES		326
	A.	Carbonylation	326
XVIII.	MAIN-GROUP ORGANOMETALLICS AS SUBSTRATES		327
	A.	Alkyllithium Reagents	327
	B.	Grignard Reagents and Organomercury(II) Halides	328
XIX.	TRANSITION METAL CARBENE COMPLEXES AS SUBSTRATES		329
	A.	Thermolysis Reactions	329
	B.	Carbene Cleavage	329
XX.	Concluding Remarks		329
XXI.	REFERENCES		329

I. INTRODUCTION

This chapter describes the utilization of metal carbonyls in stoichiometric and catalytic organic syntheses. Work is described which employs either homoleptic or substituted transition metal carbonyls, with emphasis on reagents in which the carbonyl moiety fulfills a specific function and is not merely present as an auxiliary ligand. Even with this limitation, the scope of the topic is vast, demonstrating the importance which organometallic reagents have gained in synthetic organic chemistry[1-5]. Within the limitations of the space available, we have chosen to divide the topic into a series of organic substrates, divided according to functionality, and to describe a limited number of transformations which can be performed for each class of substrate. Thus, the discussion begins with work on hydrocarbon substrates (alkanes, alkenes, and alkynes), followed by halogenated substrates and then substrates with oxygen-, sulphur-, and nitrogen-containing functional groups. Finally, organometallic substrates are considered. For each class of substrate, categories of reaction are described which were chosen to illustrate both typical and atypical reaction types. Reaction classes which have been discussed in detail elsewhere, or those in which there is only minimal synthetic interest, are described only briefly or omitted.

Several important topics are not covered here. Homogeneous hydrogenation of unsaturates[6,7] and the closely related hydrosilylation[8] and hydrocyanation[9] reactions have been discussed in detail previously and elsewhere in this series (Volume 4). Similarly, polymerization reactions[10,11] have been reviewed many times in the past and covered in this text in Chapters 4 and 5. Here, these reactions are mentioned only when they are relevant to other chemical reactions (e.g. the polymerization of tetrahydrofuran under hydrosilylation conditions). Several topics are of sufficient interest and scope to warrant detailed description in separate chapters. Thus the hydroformylation of alkenes is dealt with in Chapter 8 and the carbonylation of alkenes and alcohols in Chapter 7. Throughout this chapter we have employed the term carbonylation in its most general sense and thus the more specific designations, hydrocarboxylation, hydrocarbonylation, homologation, etc., are all considered under the same general heading.

Several major texts have contributed enormously to the chemistry of metal carbonyls. In particular, texts by Falbe[12,13] and those edited by Wender and Pino[14] are invaluable in this field.

II. ALKANES AS SUBSTRATES

A. Simple Alkanes

1. Alkane Activation

The lack of reactivity of normal alkanes has precluded their use as substrates for the rational synthesis of functionalized molecules. Considerable effort is being expended[15-18] in the field of alkane activation chemistry in order to develop the necessary background knowledge for the evolution of metal complexes capable of stoichiometric or, particularly, catalytic alkane functionalization. There is some indication[19] that low-valent carbonyls may have a role to play in this area in the future. One of the major classes of metal complex capable of interaction with an alkane is a group of highly electron-rich, nucleophilic species whose interaction with an alkane is presumably initiated by donation of electron density from the nucleophilic metal into the σ^* orbitals of the alkane C—H bonds. Such a nucleophile is believed to be formed upon irradiation of $[Ir(\eta^5\text{-}C_5Me_5)(CO)_2]$ by evolution of one equivalent of carbon monoxide[19]. The presumed intermediate undergoes oxidative addition reactions with the C—H bonds of alkanes (equation 1).

$$[\text{Ir}(\eta^5\text{-C}_5\text{Me}_5)(\text{CO})_2] \xrightarrow[-\text{CO}]{h\nu} [\text{Ir}(\eta^5\text{-C}_5\text{Me}_5)(\text{CO})] \xrightarrow{\text{RH}} [\text{Ir}(\eta^5\text{-C}_5\text{Me}_5)(\text{CO})(\text{R})\text{H}] \quad (1)$$

Reaction 1 has been observed for neopentane and cyclohexane, and the hydridoiridium(III) complex has been found to undergo exchange with carbon tetrachloride to yield the corresponding chloroiridium(III) species. These results represent an important step forward in the development of alkane activation processes but, as yet, the importance of the carbonyl moiety in promoting such chemistry is not known. The analogous nucleophile $[\text{Ir}(\eta^5\text{-C}_5\text{Me}_5)(\text{PMe}_3)]$, formed by irradiation of $[\text{IrH}_2(\eta^5\text{-C}_5\text{Me}_5)(\text{PMe}_3)]$ via reductive elimination of dihydrogen, undergoes nearly identical chemistry[20]. Subsequent functionalization studies may indicate differences in selectivity between the carbonyl and phosphine substituted complexes, however.

B. Strained Cycloalkanes

1. Valence isomerization and carbonylation

Strained cyclic alkanes exhibit a far greater reactivity than simple alkanes towards transition metal ions. The interactions of such cyclic alkanes with metal complexes have been studied in depth as examples of 'symmetry restricted processes'[21-23]. For example, the concerted thermal reaction of two molecules of ethylene to yield cyclobutane, and the corresponding cycloreversion, are forbidden processes by the Woodward–Hoffmann rules[24]. Accordingly, the mechanisms involved in rearrangements, or valence isomerization reactions, of molecules containing cyclobutane moieties have attracted much interest. From this work, several examples of synthetically useful routes have evolved which utilize rhodium(I) carbonyls as stoichiometric reagents. Catalysis of skeletal rearrangement processes, by species such as silver(I) and palladium(II), has been reviewed previously[21-23,25] and is outside the scope of this chapter.

The most notable syntheses based on the chemistry of strained cycloalkanes involve the carbonylation of cubane and its substituted analogues. Initially[26], it was observed that $[\text{RhCl(diene)}]_2$ catalysed the cycloreversion reactions[27] shown in equations 2 and 3.

For the same reactions, $[\text{RhCl(PPh}_3)_3]$ and $[\text{RhCl(CO)(PPh}_3)]$ were virtually inactive as catalysts, but the use of $[\text{RhCl(CO)}_2]_2$ as a stoichiometric reagent gave rise to the formation of a new product (equation 4). The complex **1**, presumably formed by oxidative addition of a C—C bond of cubane to rhodium(I), followed by carbonyl insertion, could be isolated in 90% yield and liberated the carbonylation product upon treatment with

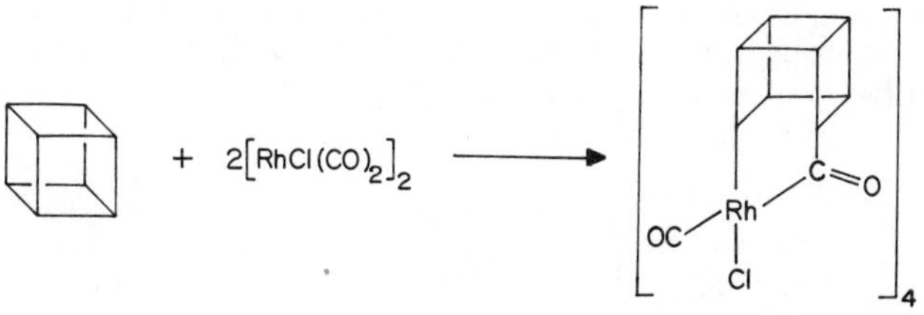

6. Transition metal carbonyls in organic synthesis

$$\text{cubane} + 2[\text{RhCl(CO)}_2]_2 \longrightarrow \left[\begin{array}{c} \text{cubyl-Rh(CO)(Cl)C=O} \end{array} \right]_4 \quad (4)$$

(1)

$$\mathbf{1} + 8\,\text{PPh}_3 \longrightarrow 4[\text{RhCl(CO)(PPh}_3)_2] + 4\,(\text{homocubanone}) \quad (+\ 5\text{-}10\%\ \text{cyclooctatetraene}) \quad (5)$$

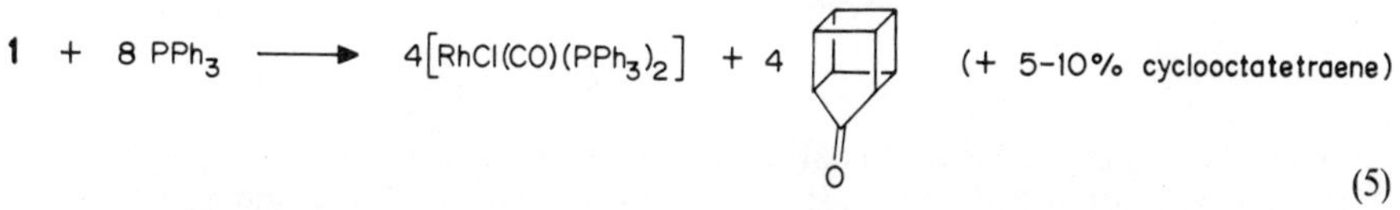

triphenylphosphine[26] (equation 5). Analogous procedures have been employed[26] in the carbonylation of substituted cubanes (equation 6 and 7).

$$\text{1,4-(COOMe)}_2\text{-cubane} \longrightarrow \text{carbonylated product} \quad (6)$$

$$\text{COOMe-cubane} \longrightarrow \text{product A (34\%)} + \text{product B (66\%)} \quad (7)$$

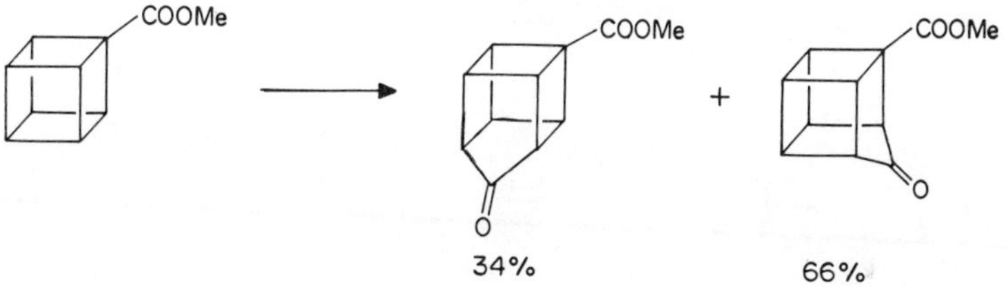

A much studied example of the cycloreversion reaction of a cyclobutane derivative is the rhodium(I)-catalysed valence isomerization of quadricyclane to norbornadiene[28] (equation 8). The stoichiometric reaction of quadricyclane with $[\text{RhCl(CO)}_2]_2$, however,

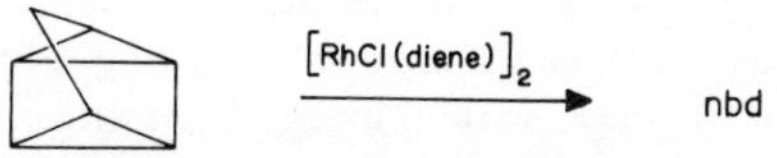

$$\text{quadricyclane} \xrightarrow{[\text{RhCl(diene)}]_2} \text{nbd} \quad (8)$$

(2)

proceeds via a different route[28], analogous to the cubane reaction, involving formation of a rhodium(III) acyl (equation 9). Treatment of **3** with triphenylphosphine causes further

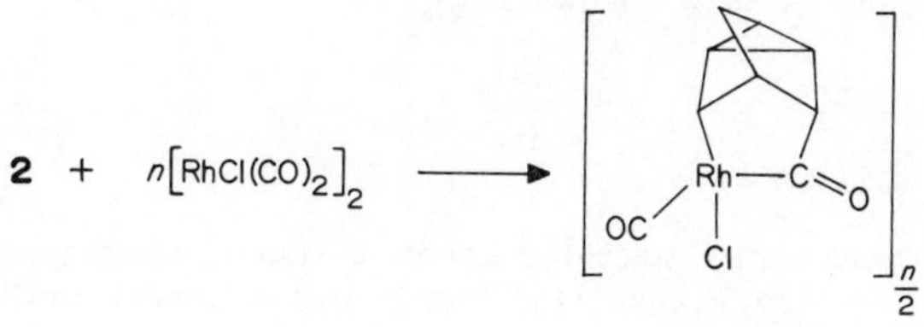

$$\mathbf{2} + n[\text{RhCl(CO)}_2]_2 \longrightarrow \left[\begin{array}{c} \text{Rh(CO)(Cl)C=O complex} \end{array} \right]_{n/2} \quad (9)$$

(3)

carbonyl insertion, producing **4**. The unsubstituted analogue of **4** may be prepared by direct treatment of quadricyclane with carbon monoxide in the presence of $[RhCl(CO)_2]_2$. The product has been assigned the structure **5**.

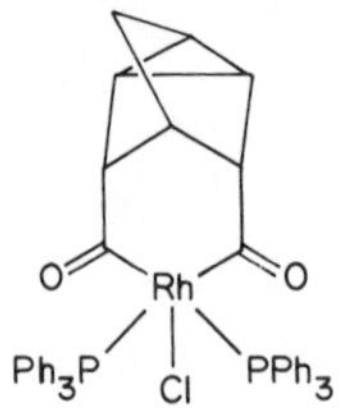

(4) 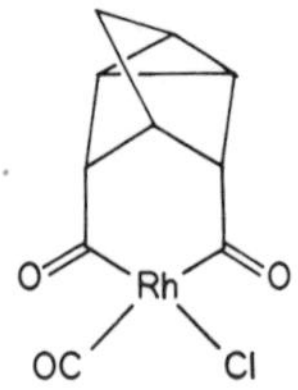   (5)

Other rhodium(I) carbonyls do not seem to show the same ability as $[RhCl(CO)_2]_2$ to participate in oxidative addition/carbonyl insertion sequences. For example, the valence isomerization of the 1,1'-bishomocubane derivative **6** occurs to produce the dicyclopropane isomer **7** when catalysed by silver(I) ion (equation 10) and the dienes **8** and **9** when catalysed by $[RhX(CO)(PPh_3)_2]$ (X = Cl, Br, I)[29] (equation 11). The isomer ratio of **8** and

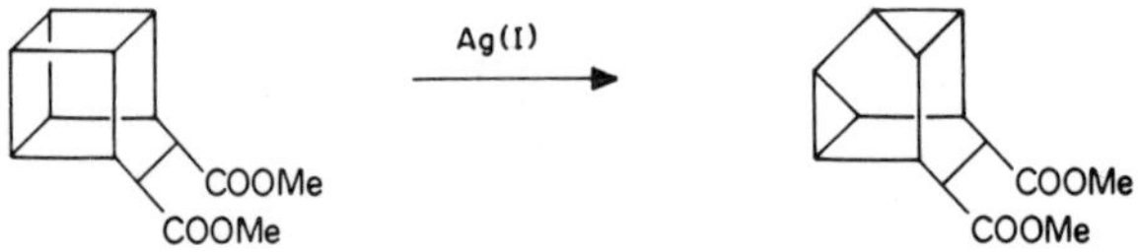

(6) → Ag(I) → (7) (10)

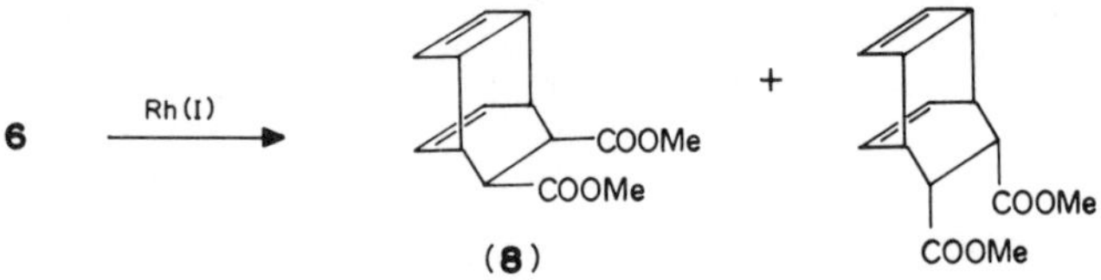

6 → Rh(I) → (8) + (9) (11)

9 (23% of **8** and 76% of **9**) does not vary with the different halorhodium complexes employed, nor is there any evidence for products resulting from carbonyl insertion.

Slightly different chemistry occurs with non-cage fused ring systems, as exemplified by comparison of the cubane reaction (equation 4) with that shown in equation 12[22]. The

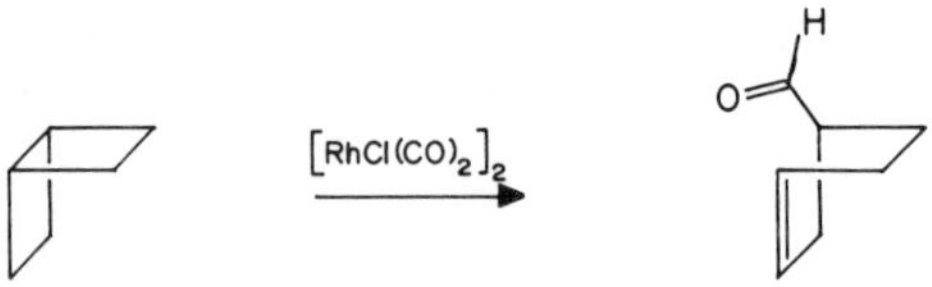

(10) → $[RhCl(CO)_2]_2$ →   (12)

mechanistic differences arise since the non-cage system has β-hydride transfer and reductive elimination steps available to it, which are not possible in the cubane system. Equation 13 outlines the steps involved. Simple non-fused cycloalkanes can also be

6. Transition metal carbonyls in organic synthesis

$$10 \xrightarrow{Rh(I)} \cdots \longrightarrow \cdots \longrightarrow \cdots \quad (13)$$

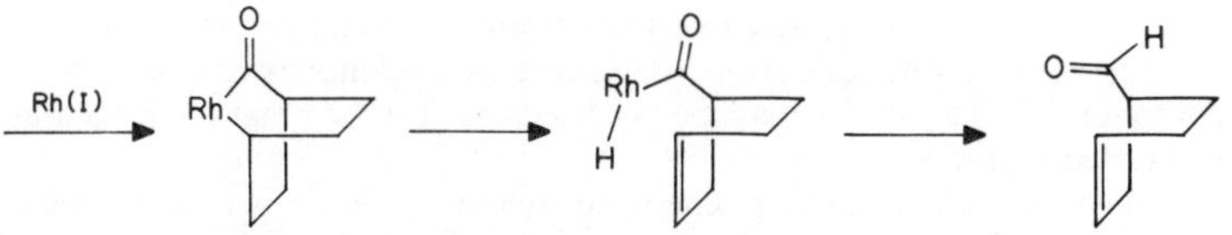

carbonylated with [RhCl(CO)$_2$]$_2$ to produce rhodium(III) acyls; equation 14 shows an example involving cyclopropane[30].

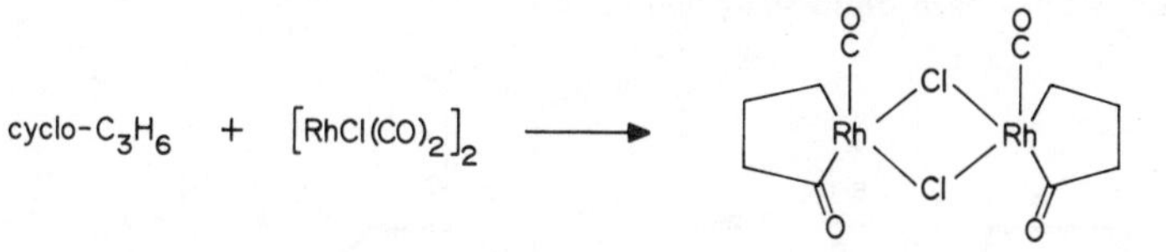

$$\text{cyclo-}C_3H_6 + [RhCl(CO)_2]_2 \longrightarrow \cdots \quad (14)$$

A typical carbonyl-promoted ring opening is observed in the reaction of the activated vinylcyclopropane **11** with [Fe(CO)$_5$] in dibutyl ether at 140 °C, which occurs to yield the complex of the conjugated diene formed by ring-opening[31] (equation 15). Conceivably, oxidation of **12** may yield the free, highly reactive, diene (see section III.A).

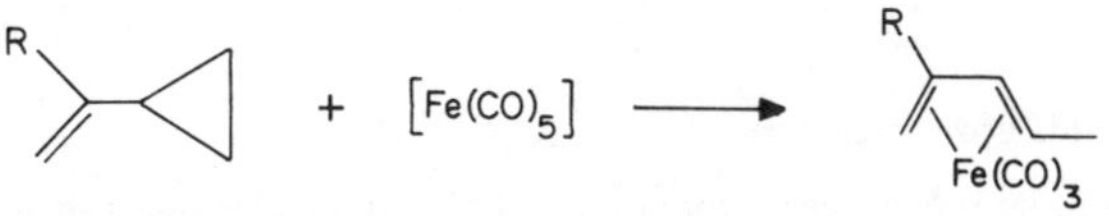

$$\cdots + [Fe(CO)_5] \longrightarrow \cdots \quad (15)$$

R = *p*-C$_6$H$_4$Cl

(**11**)   (**12**)

A most unusual reaction involving ring opening of a substituted cyclopropenium ion is shown in equation 16[32]. The release of ring strain is again likely to be a major factor in this process.

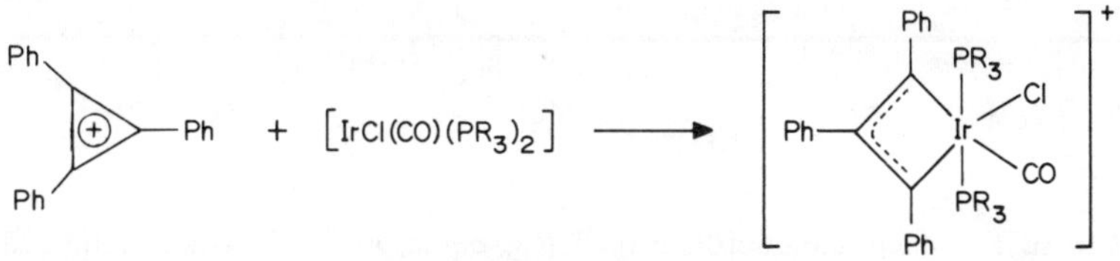

$$\cdots + [IrCl(CO)(PR_3)_2] \longrightarrow \cdots \quad (16)$$

Further details of metal-catalysed valence isomerizations and related 'symmetry forbidden' processes may be found in reviews[21-23]. From an organic synthetic viewpoint, these syntheses, with the possible exception of the cubane carbonylation reactions, largely remain synthetic curiosities. The brief mechanistic descriptions in terms of simple oxidative addition, insertion, and related fundamental processes, given above, may well be revised in the future. Results, such as the necessity that oxygen be present for the rhodium-catalysed opening of bicyclo[3.1.0]hex-2-enes[33], have yet to be fully understood[5].

III. ALKENES AS SUBSTRATES

A. Protection and Storage of Reactive Alkenes

Many highly reactive olefins form stable complexes with transition metals with no rearrangement. This allows trapping, storage, and protection of such reactive molecules to be achieved. Complexes of the reactive unsaturates tetramethylallene and hexafluoro (Dewar) benzene, for example, coordinate to the 'Fe(CO)$_4$' moiety in η^2-fashion, yielding stable compounds[5]. The olefin may be released for further reaction by a number of displacement techniques[5].

Protection of an olefin during chemical reactions has also been achieved by complexation[5]. The N-cyanation reaction of the [Fe(diene)(CO)$_3$] complex **13**, for example[34], may be followed by decomplexation with iron(III) chloride to yield N-cyanothebaine (**14**) (equation 17). Applications of this strategy in the synthesis of vitamin D precursors has been discussed previously[5].

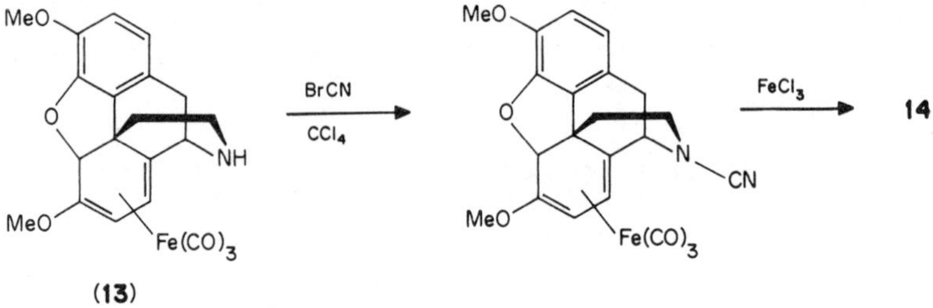

B. Rearrangement/Isomerization

The reactions of metal carbonyls, most notably [Fe(CO)$_5$][35], with dienes have led to a number of useful synthetic routes based on olefin isomerization. In general terms, isomerization of a given diene occurs to produce the diene which can bond most favourably to the Fe(CO)$_3$ unit in η^4-fashion. Thus, non-conjugated dienes tend to rearrange to conjugated dienes[36] (e.g. equation 18)[37]. s-(E)-Dienes may change confor-

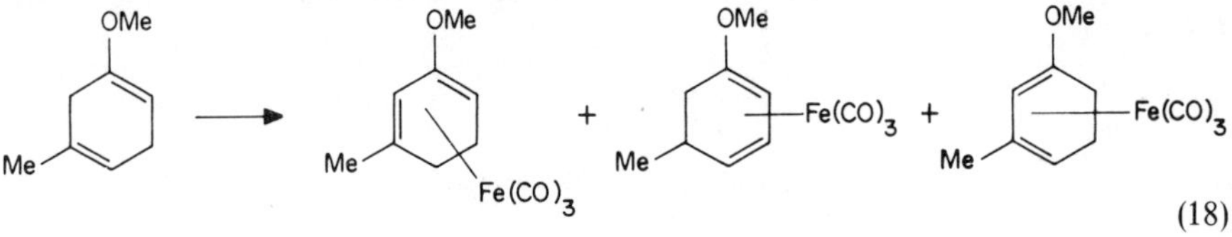

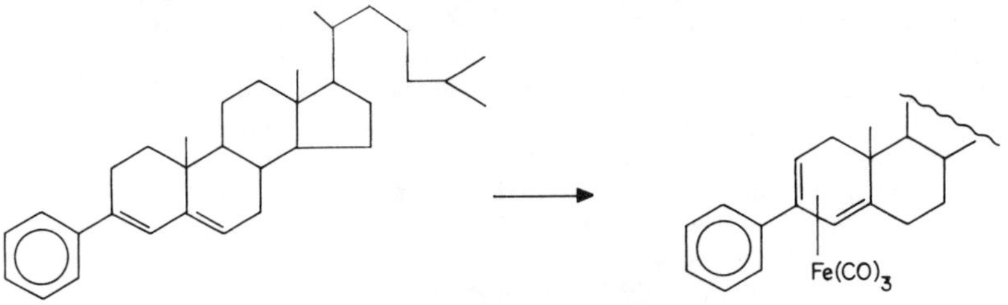

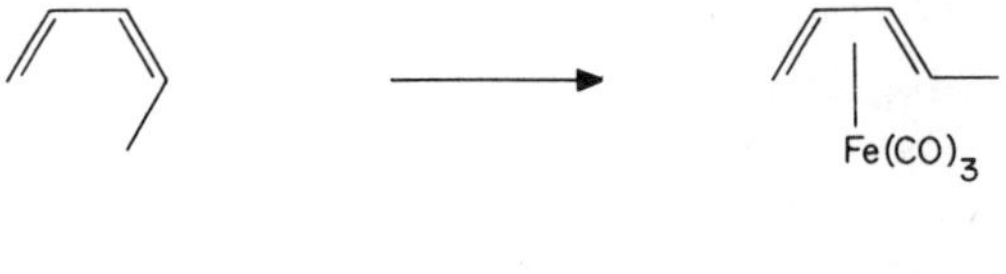

mation to s(Z)- to facilitate coordination[38] (e.g. equation 19). Sterically hindered (Z)-dienes may similarly isomerize to the E-geometry[39] (e.g. equation 20). In many cases, the free dienes may be liberated from the [Fe(diene)(CO)$_3$] complexes by iron(III) chloride or cerium(IV) ion[35,40,41], although further oxidation or isomerization of the diene may occur in specific cases[35].

C. Carbonylation

Many useful organic syntheses are based on the carbonylation of carbon—carbon double bonds. The major reaction types include syntheses with CO/H$_2$ (hydroformylation, equation 21) and with CO/HX (X = OH, OR, Cl, OOCR, SR, NR$_2$, etc.; the Reppe reaction, also known as hydrocarboxylation, where HX = H$_2$O, equation 22). Related

$$RCH{=}CH_2 + CO + H_2 \xrightarrow{\text{catalyst}} RCH_2CH_2CHO + RCH(CH_3)CHO \quad (21)$$

$$RCH{=}CH_2 + CO + HX \xrightarrow{\text{catalyst}} RCH_2CH_2COX \quad (22)$$

reactions with allylic compounds, dienes, allenes, and other cumulenes, etc., provide synthetic routes to a variety of carbonylated products. This chemistry is discussed in detail in Chapters 3 and 7.

D. Cyclooligomerization

Cyclooligomerization of unsaturates is an important class of ring synthesis. In this area, nickel(0) complexes have played a major role in promoting reactions that superficially appear to be symmetry-forbidden in many cases. The background chemistry for this class of reaction has been extensively documented[42-46] and so here we shall simply outline the types of synthesis possible with various olefinic substrates employing carbonyl complexes as reagents. Further discussion of this important topic may be found in Chapters 4 and 5. The actual role of the carbonyl ligands in these processes is minimal and they are frequently eliminated early in the reaction. The chemistry centres around the formation of π-allyl complexes with well known insertion sequences resulting in oligomerization. Full mechanistic discussion is available elsewhere[46].

1. Diene cyclooligomerization

The dimerization of buta-1,3-diene serves as a model reaction for diene cyclooligomerizations. It has been known for some years[47] that phosphine- and phosphite-modified nickel carbonyls promote this dimerization, yielding (Z,Z)-cycloocta-1,5-diene in good yield. Typically, [Ni(CO)$_2$L$_2$] [L = PPh$_3$, P(OR)$_3$] complexes are activated by treatment with acetylene prior to use in the cyclooligomerization. Some trimers, tetramers, and products resulting from the cooligomerization of butadiene and acetylene are also formed[47]. The polymer-supported analogue of this system has also been described[48].

[Ni(CO)$_4$] itself can be used as a catalyst precursor for diene oligomerization[49]. Buta-1,3-diene yields cycloocta-1,5-diene and (E,E,E)-cyclododeca-1,5,9-triene, with the formation of the cyclododecatriene being favoured at low temperatures and high butadiene concentrations. The reaction is extremely efficient[45], with 800 kg of butadiene being cyclized per mole of [Ni(CO)$_4$] in a continuous synthesis. The most efficient class of catalyst for these types of reaction contain no carbonyl ligands[50]; indeed, in the above-mentioned systems the carbonyl groups are displaced early in the reaction pathway[45]. Carbonyl-free systems are discussed in Chapter 4.

Linear oligomerization[46] of conjugated dienes is also known to be catalysed by [Ni(CO)$_4$]–PR$_3$ and [Co$_2$(CO)$_8$]–R$_3$Al systems, but is less interesting from a laboratory synthesis viewpoint. In terms of industrial synthesis, linear cooligomerizations of dienes and monoenes (e.g. hexadiene from butadiene and ethylene) are of great interest[46].

2. Allene cyclooligomerization

The thermal oligomerization of allene yields a complex mixture of dimers, trimers, and tetramers[51]. The cyclopolymerization can, however, be effected with slight selectivity by a number of nickel carbonyl and phosphine- or phosphite-modified complexes[52]. For example, at 105–110 °C, $[Ni(CO)_2(PPh_3)_2]$ promotes the formation of the two trimers **15** and **16** and the tetramer **17**. The yields were 35% (trimers) and 6% (tetramer). Other

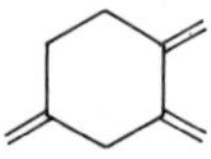

(15) 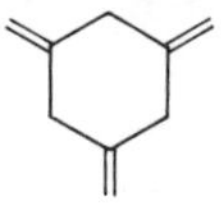   (16) 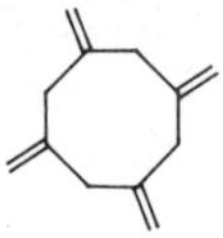   (17)

nickel(0) carbonyl complexes, such as $[Ni(CO)(P\{OPh\}_3)_3]$ and $[Ni(CO)_3(P\{OPh\}_3)]$, may also be employed in this synthesis[52].

A more unusual system[53] for allene cyclooligomerization is based on the gas-phase reaction of allene with a 'heterogeneous' nickel carbonyl complex, prepared from 1,4-$C_6H_4(PPh_2)_2$ and $[Ni(CO)_4]$, of empirical formula $[Ni(CO)_2(Ph_2P\{1,4-C_6H_4\}PPh_2)]$. At 200 °C, two dimers, **18** and **19**, are produced in 60% and 13% yields, respectively. Some trimeric products, largely **15**, are also formed (27%). In contrast, liquid-phase systems

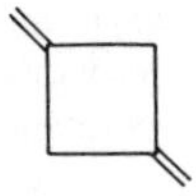

(18) 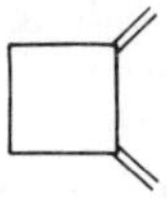   (19)

based on the same nickel(0) complex, yield largely tetramers, pentamers, and higher oligomers. The gas-phase reaction does have a number of difficulties associated with it, largely owing to the exothermicity of the oligomerization, requiring inert diluents for both the solid-phase complex and the gas-phase allene.

E. Nucleophilic Attack on Coordinated Olefins

Electrophilic olefin complexes, such as $[Fe(C_5H_5)(CO)_2(olefin)]^+$, are subject to nucleophilic attack at the uncoordinated face of the olefin[51,54,55]. The resulting alkyl may be decomplexed with a number of reagents, including bromine (e.g. equation 23). With an

$$[Fe(Cp)(CO)_2(\eta^2\text{-}C_2H_4)]^+ \xrightarrow{Nu^-} [Fe(Cp)(CO)_2(CH_2CH_2Nu)] \xrightarrow{Br_2} BrCH_2CH_2Nu \quad (23)$$
$$(20)$$

unsymmetrical olefin, such as but-1-ene, two possible alkyls, corresponding to **20**, may be formed (**21** and **22**). In certain cases, regio- and stereo-selective nucleophilic attack can occur, and these reactions have been described previously[5,52].

$[Fe(Cp)(CO)_2\{CH_2CH(Nu)Et\}]$ $[Fe(Cp)(CO)_2\{CH(Et)CH_2Nu\}]$
(21)   (22)

Similar chemistry may be employed to synthesize substituted monoenes from dienes via the intermediate formation of allyl complexes which are susceptible to nucleophilic attack[56,57]. An example is shown in equation 24, where the diene complex is protonated to yield the intermediate allyl complex, which can undergo nucleophilic attack at either of the positive centres of the allyl moiety. A wide variety of nucleophiles may be employed in such syntheses[5].

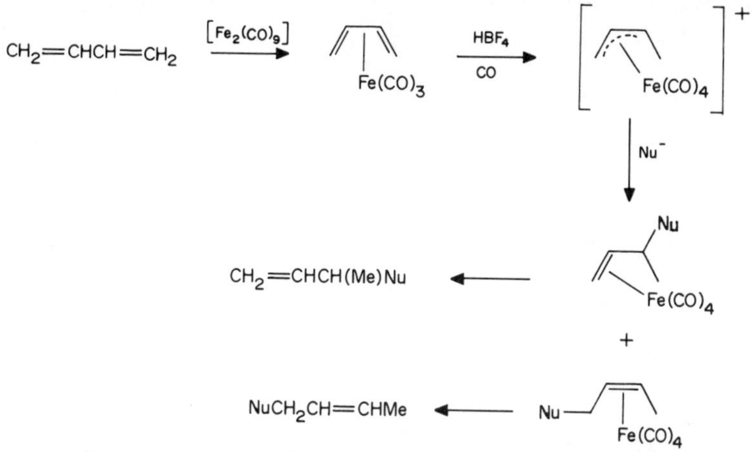

$$CH_2=CHCH=CH_2 \xrightarrow{[Fe_2(CO)_9]} \underset{Fe(CO)_3}{\text{(diene complex)}} \xrightarrow{HBF_4, CO} \left[\underset{Fe(CO)_4}{\text{(allyl)}} \right]^+ \xrightarrow{Nu^-} \tag{24}$$

$$CH_2=CHCH(Me)Nu \longleftarrow \underset{Fe(CO)_4}{\text{(Nu-substituted)}}$$

$$+$$

$$NuCH_2CH=CHMe \longleftarrow Nu-\underset{Fe(CO)_4}{\text{(allyl)}}$$

The sequence described in equation 24 is not restricted to simple allyl intermediates. Thus, dienyl complexes[5,58], available via protonation of coordinated trienes[59] or via hydride abstraction from diene complexes[37], behave in an analogous fashion (equation 25). Synthetic procedures based on this chemistry have been described[60].

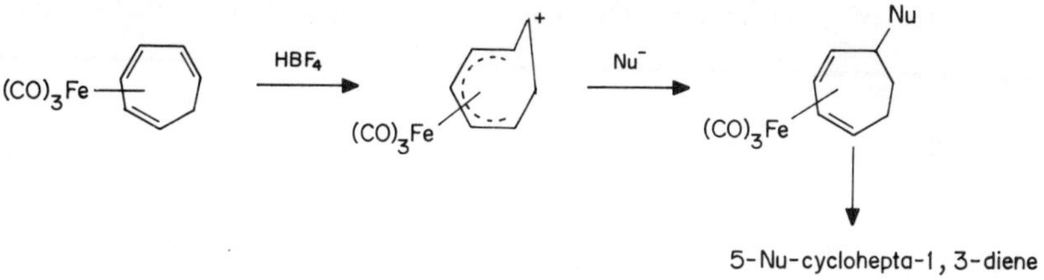

$$(CO)_3Fe-\text{(triene)} \xrightarrow{HBF_4} \underset{(CO)_3Fe}{\text{(dienyl)}}^+ \xrightarrow{Nu^-} \underset{(CO)_3Fe}{\text{(Nu-dienyl)}} \tag{25}$$

$$\downarrow$$

5-Nu-cyclohepta-1,3-diene

F. Electrophilic Attack on Coordinated Olefins

Conjugated dienes, cyclic trienes, and tetraenes tend to undergo polymerization under Friedel–Crafts conditions and thus electrophilic substitution reactions of the derived complexes are of interest synthetically. A typical example is the acylation of the butadiene complex **23**[61] via electrophilic attack of the acylium ion (equation 26). Alkylation

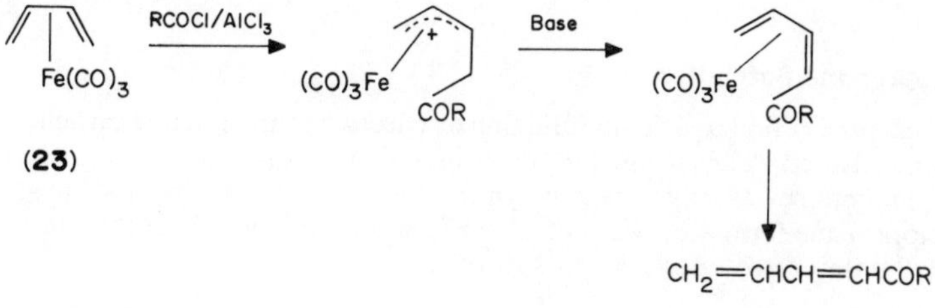

$$\underset{Fe(CO)_3}{\text{(butadiene)}} \xrightarrow{RCOCl/AlCl_3} \underset{(CO)_3Fe}{\overset{+}{\text{(acyl)}}} \xrightarrow{Base} \underset{(CO)_3Fe}{\text{(acylated diene)}} \tag{26}$$

(23)

$$\downarrow$$

$$CH_2=CHCH=CHCOR$$

reactions are similarly well known, with intramolecular alkylation providing a convenient route to ring synthesis (e.g. equation 27)[62]. Entirely analogous systems utilizing triene and

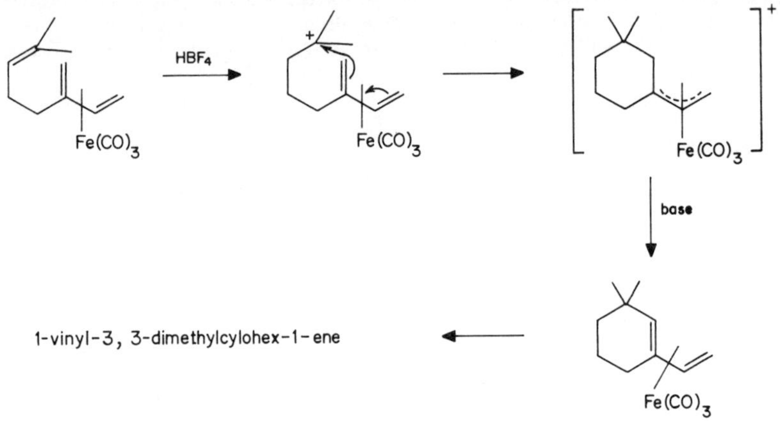

(27)

1-vinyl-3, 3-dimethylcylohex-1-ene

tetraene complexes as substrates have been described in detail[5] and have considerable synthetic utility.

G. Diels–Alder Reactions of Coordinated Alkenes

The iron carbonyl complex of cyclobutadiene has been extensively utilized as a source of the cyclobutadiene moiety in Diels–Alder reactions. The report that the optically active complex **24** undergoes a Diels–Alder reaction with tetracyanoethylene yielding a racemized product is evidence that uncomplexed cyclobutadiene is in fact involved[63] (equation 28). Use of the $[Fe(diene)(CO)_3]$ complex as a source of cyclobutadiene has

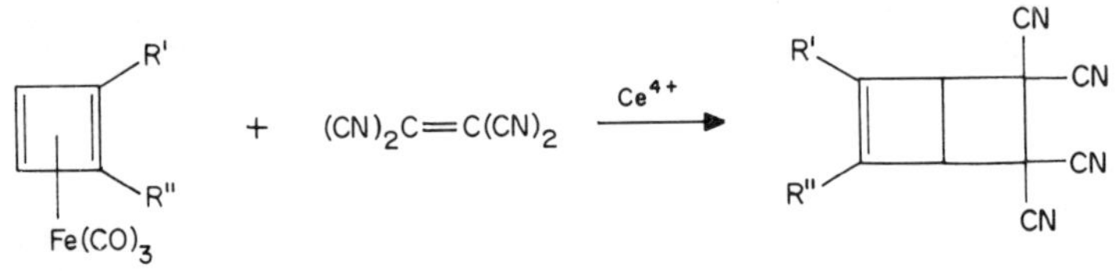

(28)

(24)

permitted a number of elegant syntheses via Diels–Adler sequences[5]. For example, equation 29[64] shows the synthesis of a cubane derivative via an initial Diels–Alder reaction of the cyclobutadiene moiety followed by a conventional intramolecular Diels–Alder reaction.

H. Cyclopropane Formation

Metal carbene complexes do not function as effective sources of free carbenes, but will react with activated olefins to yield the corresponding cyclopropanes[65] (e.g. equation 30). The carbene complex **25** may be readily synthesized by a number of routes[66] (e.g. equation 31). Cyclopropane formation via metal carbenes is generally only successful with vinyl ethers, α,β-unsaturated esters, etc. as substrates.

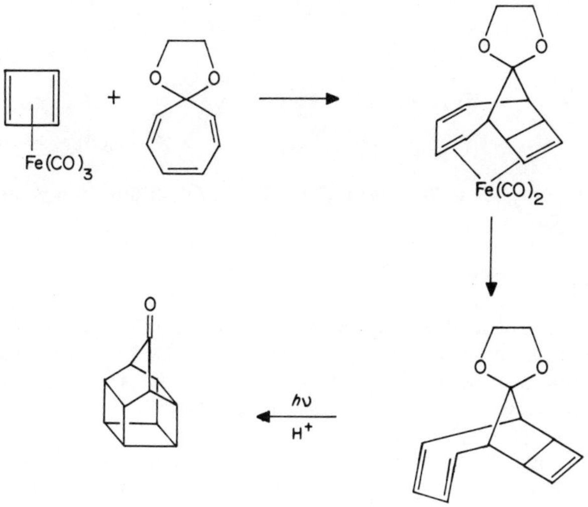

[structure with Fe(CO)₃ + dioxolane-cycloheptatriene → Fe(CO)₂ complex] (29)

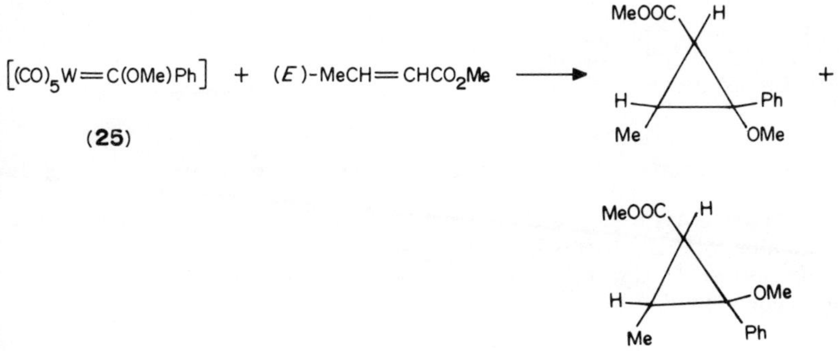

[(CO)₅W=C(OMe)Ph] + (E)-MeCH=CHCO₂Me → cyclopropane products (30)

(25)

$$W(CO)_6 \xrightarrow{PhLi} [(CO)_5WC(O)Ph]^- Li^+$$

$$+ R_3OBF_4 \downarrow$$

$$[(CO)_5W=C(OR)Ph]$$

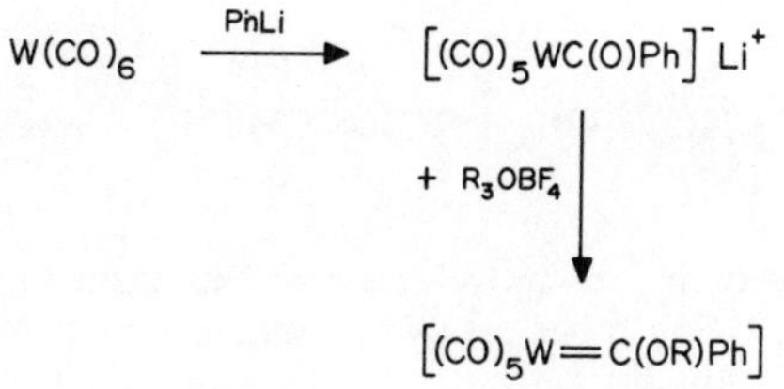

(31)

IV. ALKYNES AS SUBSTRATES

A. Protection and Storage of Reactive Alkynes

The concept of trapping reactive unsaturates as stable transition metal complexes applies to both alkynes and alkenes (see Section IIIA). An example is the capture of

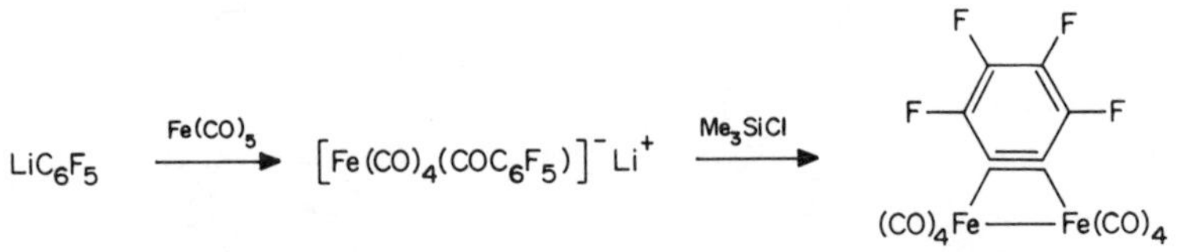

$$LiC_6F_5 \xrightarrow{Fe(CO)_5} [Fe(CO)_4(COC_6F_5)]^- Li^+ \xrightarrow{Me_3SiCl} (CO)_4Fe\text{—}\!\!\!\!<\!\!\!\!\begin{array}{c}C_6F_4\end{array}\!\!\!\!>\!\!\!\!\text{—}Fe(CO)_4 \quad (32)$$

tetrafluorobenzyne by [Fe(CO)$_5$], equation 32[67,68]. Other similar reactions have been described previously[5].

B. Carbonylation

The carbonylation of acetylenic substrates may give rise to a wide variety of products, depending on the conditions employed. Such syntheses have been exhaustively documented[69] in the past and so here only the major synthetic routes will be described in brief. Many transition metal catalysts have been employed for acetylene carbonylation, with [Ni(CO)$_4$] and [Co$_2$(CO)$_8$] being two of the most common.

In aprotic solvents, carbonylation of acetylenes generally occurs stoichiometrically, yielding cyclopentadienones or p-benzoquinones, depending on the substrate structure and the carbonyl reagent employed (e.g. equations 33–35)[70–75]. Catalytic carbonylation

$$CH\!\equiv\!CH \xrightarrow{[Ni(CO)_4]} \text{indan} \quad (33)$$

$$PhC\!\equiv\!CPh \xrightarrow{[Ni(CO)_4]} \text{tetraphenylcyclopentadienone} \quad (34)$$

$$MeC\!\equiv\!CMe \xrightarrow{[RhCl(CO)_2]_2} \text{tetramethyl-}p\text{-benzoquinone} \quad (35)$$

of acetylenes at low C$_2$H$_2$ to CO ratios proceeds with the formation of lactones in poor yield (e.g. equation 36)[43,76], whereas at higher C$_2$H$_2$ to CO ratios oligomers and polymers are generally formed[69]. Under hydroformylation conditions, aldehydes are formed according to equation 37[69].

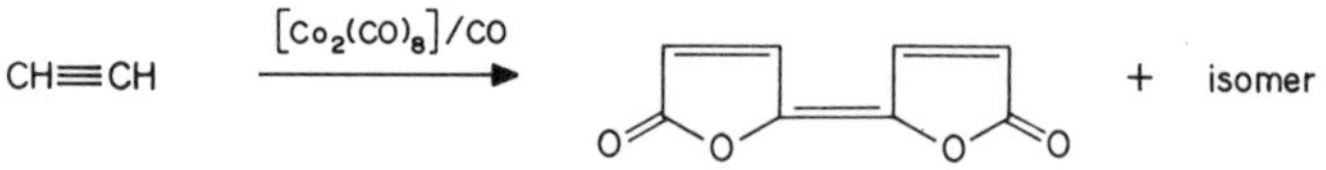

$$CH\!\equiv\!CH \xrightarrow{[Co_2(CO)_8]/CO} \text{(bifuranone)} + \text{isomer} \quad (36)$$

$$RC\!\equiv\!CH \xrightarrow[{[Co_2(CO)_8]}]{CO+H_2} RC(CHO)\!=\!CH_2 + RCH\!=\!CHCHO \xrightarrow{H_2} RCH(CHO)CH_3$$
$$+ RCH_2CH_2CHO \quad (37)$$

An alternative pathway, involving reduction to the olefin followed by hydroformylation, is also possible. Using other catalysts, cyclic products may be formed, such as hydroquinone (e.g. equation 38)[77].

$$CH\!\equiv\!CH \xrightarrow[{[Ru_3(CO)_{12}]}]{CO/H_2} p\text{-}C_6H_4(OH)_2 \quad (38)$$

In the presence of hydroxylic solvents, hydroformylation by [Co$_2$(CO)$_8$] gives rise to formation of β-formylcarboxylic acid esters as major products (e.g. equation 39)[78,79].

$$CH\!\equiv\!CH \xrightarrow{CO/ROH} [H_2C\!=\!CHCOOR] \xrightarrow{CO/H_2} OHCCH_2CH_2CO_2R \quad (39)$$

6. Transition metal carbonyls in organic synthesis

Carbonylation of acetylenic substrates in the presence of hydrogen donors generally leads to formation of acids (from water), esters (from alcohols), amides (from amines), etc. By-product formation, particularly ketone production, is often extensive, however. An example of the complexity of the possible reaction pathways is shown in equation 40[69].

$$HC\equiv CH \begin{cases} +ROH+CO \rightarrow \begin{cases} CH_2=CHCOOR \\ \frac{1}{2}ROOCCH=CHCH=CHCOOR \end{cases} \\ +2ROH+2CO \rightarrow \begin{cases} ROOCCH_2CH_2COOR \\ ROOCCH=CHCOOR \\ ROOCCH(Me)COOR \end{cases} \\ +3ROH+3CO \rightarrow ROOCCH(COOR)CH_2COOR \end{cases} \quad (40)$$

Under specific reaction conditions a high degree of selectivity for such processes is possible; for example, Reppe[80] obtained a 77% yield of the thioester derived from acetylene, benzyl mercaptan, and CO at 40–50 °C using $[Ni(CO)_4]$ as a stoichiometric reagent in toluene solution.

C. Cyclooligomerization

Transition metal-catalysed cyclooligomerization reactions of alkynes have been of interest since Reppe et al.'s report[81] that acetylene may be trimerized to benzene in the presence of certain nickel catalysts. Functionalized acetylenes behave similarly, allowing a wide variety of substituted benzenes to be synthesized by this route. In laboratory synthesis, the commercially available substituted carbonyl $[Co(Cp)(CO)_2]$ is a widely used catalyst for acetylene oligomerizations. The cooligomerization of an α,ω-diyne with an acetylene, as described by Vollhardt[82] and others, has found utility in the synthesis of new ring systems (equation 41). Of particular interest are the reactions of hexa-1,5-diynes with functionalized acetylenes which yield benzocyclobutenes, (equation 42).

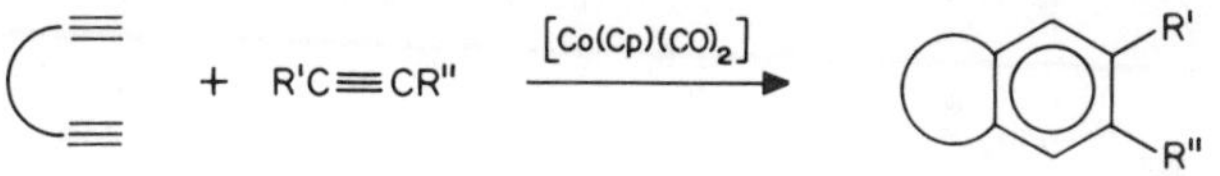

$$\text{diyne} + R'C\equiv CR'' \xrightarrow{[Co(Cp)(CO)_2]} \text{substituted benzene with } R', R'' \quad (41)$$

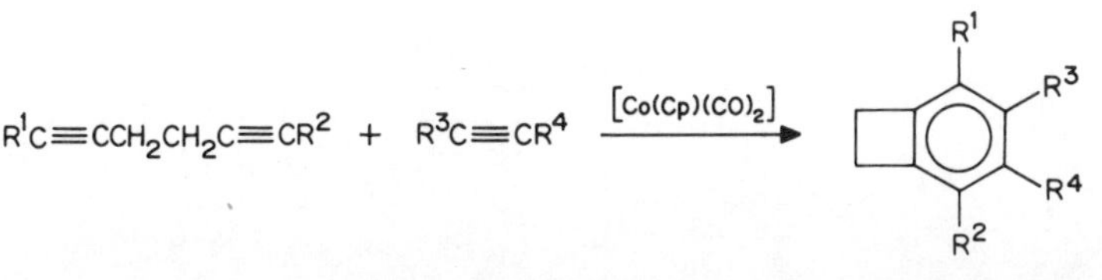

$$R^1C\equiv CCH_2CH_2C\equiv CR^2 + R^3C\equiv CR^4 \xrightarrow{[Co(Cp)(CO)_2]} \text{benzocyclobutene with } R^1, R^2, R^3, R^4 \quad (42)$$

The benzocyclobutenes may be utilized as a source of o-xylylenes, which can be trapped by further reaction with dienophiles (e.g. equation 43)[83]. A large number of

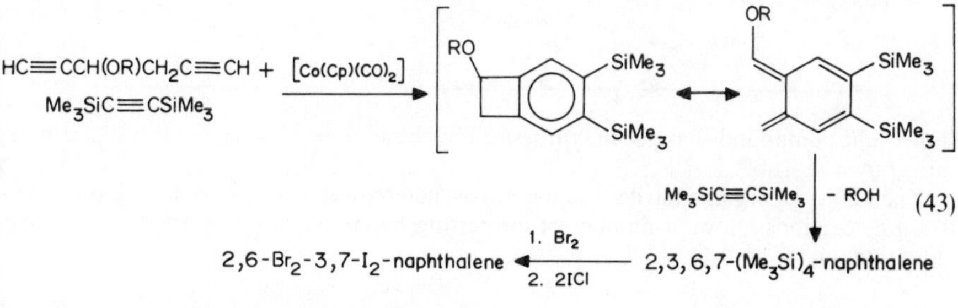

$$HC\equiv CCH(OR)CH_2C\equiv CH + [Co(Cp)(CO)_2] \longrightarrow \left[\text{benzocyclobutene} \leftrightarrow \text{o-xylylene} \right]$$

$$\xrightarrow{Me_3SiC\equiv CSiMe_3, -ROH} \quad (43)$$

$$2,6\text{-}Br_2\text{-}3,7\text{-}I_2\text{-naphthalene} \xleftarrow[\text{2. 2ICl}]{\text{1. }Br_2} 2,3,6,7\text{-}(Me_3Si)_4\text{-naphthalene}$$

Table 1. Polycycles by acetylene cyclization. Reprinted with permission from ref. 82. Copyright 1977 American Chemical Society.

Starting material	Product	Yield(%)
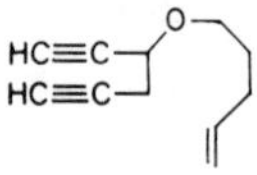 HC≡C–O–(CH₂)₃–CH=CH₂ type diyne-ene	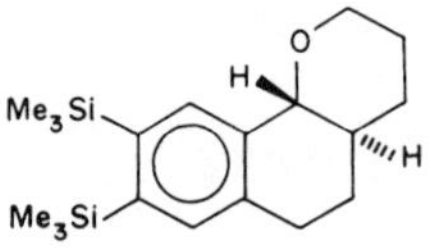 Tricyclic Me₃Si-substituted product with pyran ring	60
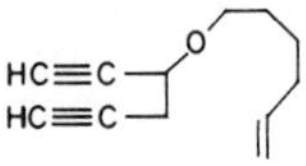 HC≡C–O–(CH₂)₄–CH=CH₂ type diyne-ene	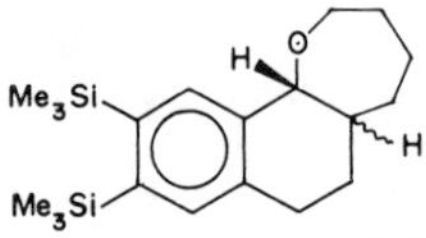 Tricyclic Me₃Si-substituted product with oxepane ring + regioisomers (unseparated mixture)	65
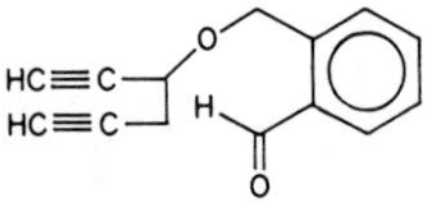 HC≡C–O–CH₂–(o-C₆H₄)–CHO diyne	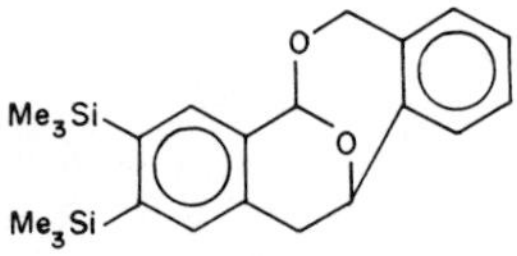 Tricyclic Me₃Si-substituted product with fused dioxine	50
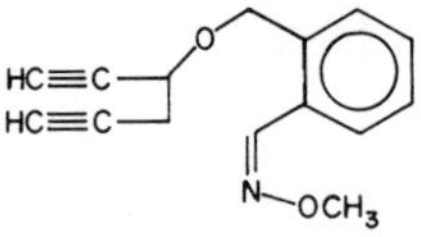 HC≡C–O–CH₂–(o-C₆H₄)–CH=N–OCH₃ diyne	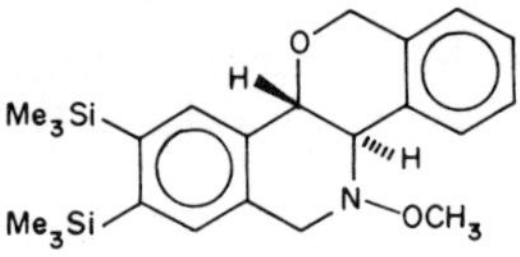 Tricyclic Me₃Si-substituted product with N–OCH₃	45
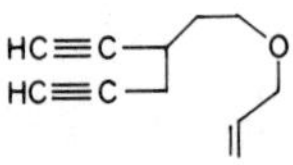 HC≡C–(CH₂)₂–O–CH₂–CH=CH₂ diyne-ene	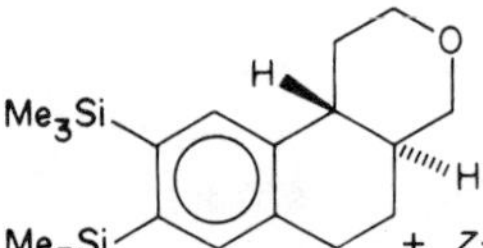 Tricyclic Me₃Si-substituted product + Z-isomer (ca. 15%)	90
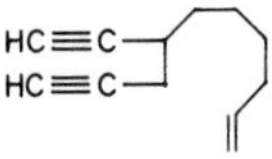 HC≡C–(CH₂)₃–CH=CH₂ diyne-ene	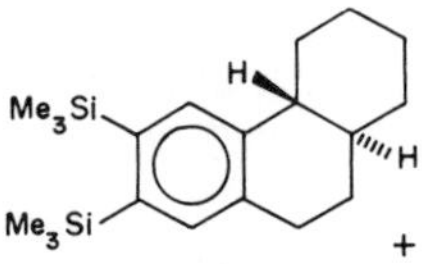 Tricyclic Me₃Si-substituted all-carbon product + Z-isomer (<5%)	80

polycyclic compounds have been synthesized by these routes[82] and a number of examples are shown in Table 1.

The ability of organic nitriles to function as 'heteroatom acetylenes' in cyclooligomerization reactions allows a number of interesting heterocycles to be prepared[84,85] (e.g. equation 44)[82].

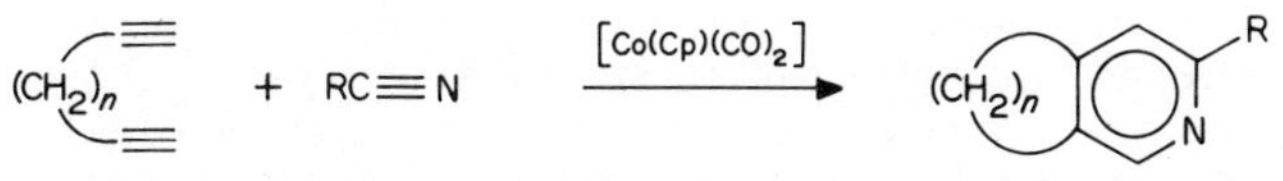

6. Transition metal carbonyls in organic synthesis

$$(CH_2)_n \overset{=\!=\!=}{\underset{=\!=\!=}{}} + RC\equiv N \xrightarrow{[Co(Cp)(CO)_2]} (CH_2)_n\text{-pyridine-}R \qquad (44)$$

V. ORGANIC HALIDES AS SUBSTRATES

A. Alkyl and Aryl Halides

1. Carbonylation and coupling reactions

Simple alkyl halides are carbonylated by cobalt carbonyls. Since the reaction is envisaged as passing through an alkylcobalt intermediate, β-hydrogen transfer allows formation of a cobalt–olefin complex and hence isomerization of the alkyl chain (e.g. equation 45)[86]. Isomerization in such reactions appears to be retarded by maintaining a low reaction temperature[86].

$$RCH_2CH_2X \xrightarrow[R'OH]{[Co_2(CO)_8]-CO} RCH_2CH_2COOR' + RCH(Me)COOR' \qquad (45)$$

The carbonylation of methyl iodide, as occurs in the iodide-promoted carbonylation of methanol (Monsanto acetic acid process), is catalysed by $[RhI_2(CO)_2]^-$. This industrially important synthesis is discussed fully in Chapter 7.

Unsaturated hydrocarbon chains may lead to cyclization in the carbonylation of organic halides by carbonyl metallate anions. For example, 5-bromopent-1-ene reacts with $[Fe(CO)_4]^{2-}$ to yield the alkyliron anion, which undergoes carbonyl insertion and intramolecular olefin insertion to yield cyclohexanone (equation 46)[87]. The cyclohexanone moiety is displaced by acidification.

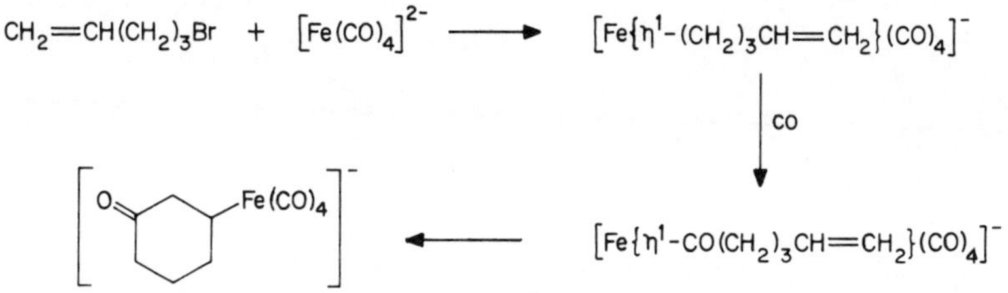

$$CH_2=CH(CH_2)_3Br + [Fe(CO)_4]^{2-} \longrightarrow [Fe\{\eta^1\text{-}(CH_2)_3CH=CH_2\}(CO)_4]^-$$

$$\downarrow CO \qquad (46)$$

$$\left[\overset{O}{\underset{}{\bigcirc}}\!\!-Fe(CO)_4 \right]^- \longleftarrow [Fe\{\eta^1\text{-}CO(CH_2)_3CH=CH_2\}(CO)_4]^-$$

Ethyl ketones may be prepared similarly by carbonylation of alkyl halides in the presence of ethylene (equation 47)[88]. Similar chemistry has been reported for oganic halides containing acetylenic hydrocarbon chains, where the reaction with $[Ni(CO)_4]$ in the presence of base leads to generation of unsaturated ring compounds[89].

$$RX + [Fe(CO)_4]^{2-} \rightarrow [FeR(CO)_4]^- \xrightarrow{CO} [RC(O)Fe(CO)_4]^-$$

$$\downarrow C_2H_4$$

$$RCOEt \xleftarrow{H^+} [RCOCH_2CH_2Fe(CO)_4]^- \qquad (47)$$

Benzylic halides are carbonylated by carbonyl metallate anions[90]. For example, benzyl chloride yields dibenzyl ketone on treatment with sodium tetracarbonylferrate (equation 48). $[Fe_3(CO)_{12}]$ has also been employed for this synthesis[91].

$$2PhCH_2Cl \xrightarrow{[Fe(CO)_4]^{2-}} (PhCH_2)_2CO \qquad (48)$$

Aromatic halides are carbonylated stoichiometrically by $[Ni(CO)_4]$ in the presence of primary or secondary amines to yield the corresponding amides. A more unusual synthesis involves the carbonylation of aromatic halides catalytically by $[Ni(CO)_4]$ in the presence of calcium hydroxide, yielding the calcium salt of the carboxylic acid[92,93]. Symmetrical ketones may be prepared by treatment of aryl iodides with $[Fe_3(CO)_{12}]$ (equation 49)[91].

$$PhI \xrightarrow{[Fe_3(CO)_{12}]} Ph_2CO \qquad (49)$$

Alcoholic solutions of $[Ni(CO)_4]$ react with aryl iodides to yield aryl esters, by alcoholysis of the nickel aroyl intermediate and benzil, via coupling of two aroyl groups by reductive elimination from a diaroylnickel intermediate (equation 50)[94].

$$PhI \xrightarrow[ROH]{[Ni(CO)_4]} PhCOOR + PhCOCOPh \qquad (50)$$

B. Allylic Halides

1. Carbonylation and coupling reactions

Coupling reactions of organic halides are frequently performed with organometallic reagents other than metal carbonyls[5], and yet $[Ni(CO)_4]$ has been successfully utilized in coupling allylic halides with alkyl and aryl iodides (equation 51)[95]. The coupling reaction

$$RCH=CHCH_2Br \xrightarrow[R'X]{[Ni(CO)_4]} RCH=CHCH_2R' \qquad (51)$$

of allyl chloride with itself to yield biallyl is promoted by $[Ni(CO)_4]$[96] and has been extended to cover a number of synthetically useful transformations[46,97,98]. Most notable is the tendency for coupling to occur at the least substituted position of the allyl group (e.g. equation 52). An example of the application of such allylic coupling to organic synthesis is

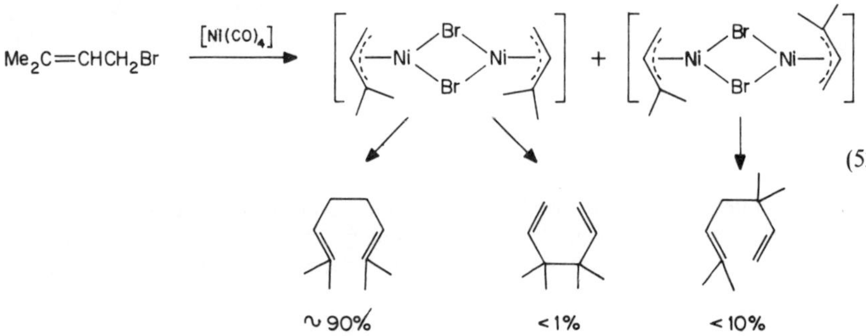

$$Me_2C=CHCH_2Br \xrightarrow{[Ni(CO)_4]} \qquad (52)$$

~90%   <1%   <10%

Corey's total synthesis of the fundamental monocyclic triisoprenoid, humulene (**28**). A key step involves the cyclization of a 1,11-dibromoundeca-2,5,9-triene derivative, **26**, by $[Ni(CO)_4]$ (equation 53). The 4,5-Z isomer of humulene (**27**) is produced, which is

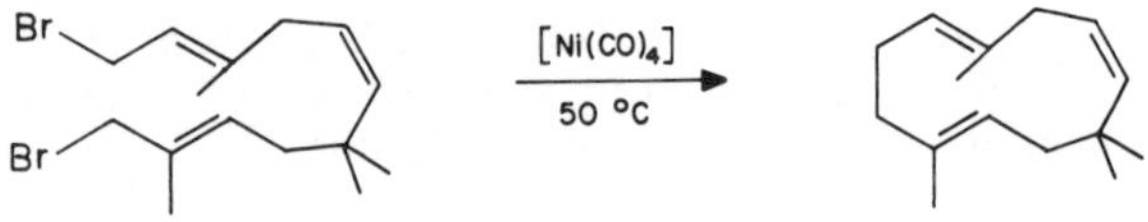

$$\text{(26)} \xrightarrow[50\,°C]{[Ni(CO)_4]} \text{(27)} \qquad (53)$$

conveniently isomerized by irradiation in the presence of diphenyl disulphide to yield humulene (equation 54).

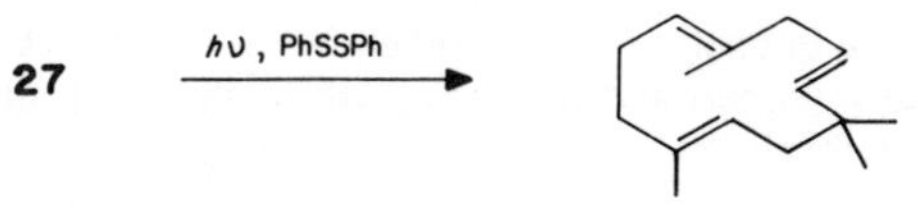

$$\mathbf{27} \xrightarrow{h\nu, \text{PhSSPh}} \mathbf{(28)} \quad (54)$$

The susceptibility of allylic halides to $[\text{Ni}(\text{CO})_4]$-promoted coupling reactions allows syntheses based on a variety of allylic substrates to be performed. For example, **29** may be synthesized by any of the three routes outlined in equation 55[99].

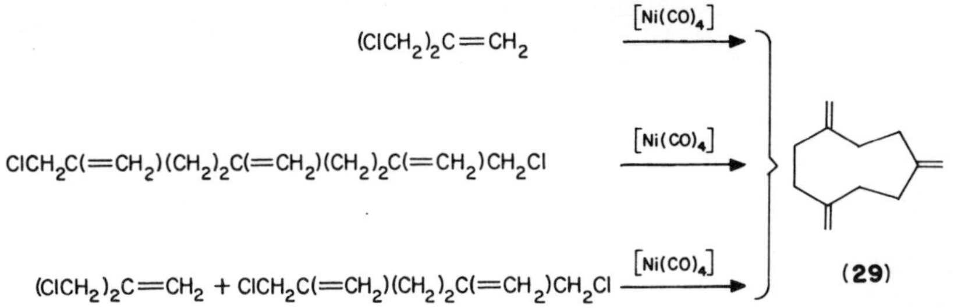

$$(\text{ClCH}_2)_2\text{C}=\text{CH}_2 \xrightarrow{[\text{Ni}(\text{CO})_4]}$$

$$\text{ClCH}_2\text{C}(=\text{CH}_2)(\text{CH}_2)_2\text{C}(=\text{CH}_2)(\text{CH}_2)_2\text{C}(=\text{CH}_2)\text{CH}_2\text{Cl} \xrightarrow{[\text{Ni}(\text{CO})_4]} \Biggr\} \mathbf{(29)} \quad (55)$$

$$(\text{ClCH}_2)_2\text{C}=\text{CH}_2 + \text{ClCH}_2\text{C}(=\text{CH}_2)(\text{CH}_2)_2\text{C}(=\text{CH}_2)\text{CH}_2\text{Cl} \xrightarrow{[\text{Ni}(\text{CO})_4]}$$

The carbonylation of allylic halides is a much studied process, from which many synthetically useful routes have been derived. The simple carbonylation of allyl bromide, for example, yields unsaturated acyl bromides when performed in aprotic solvents or unsaturated esters in the presence of alcohols (equation 56)[100-102]. In the presence of

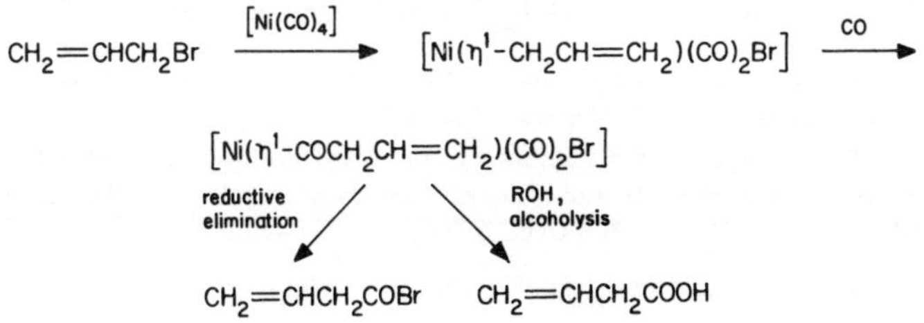

$$\text{CH}_2=\text{CHCH}_2\text{Br} \xrightarrow{[\text{Ni}(\text{CO})_4]} [\text{Ni}(\eta^1\text{-CH}_2\text{CH}=\text{CH}_2)(\text{CO})_2\text{Br}] \xrightarrow{\text{CO}}$$

$$[\text{Ni}(\eta^1\text{-COCH}_2\text{CH}=\text{CH}_2)(\text{CO})_2\text{Br}]$$

reductive elimination / ROH, alcoholysis

$$\text{CH}_2=\text{CHCH}_2\text{COBr} \quad \text{CH}_2=\text{CHCH}_2\text{COOH} \quad (56)$$

added acetylenes, further insertion reactions of the aroylnickel complex are observed (equation 57)[103]. With suitably oriented substituents, termination tends to occur after the

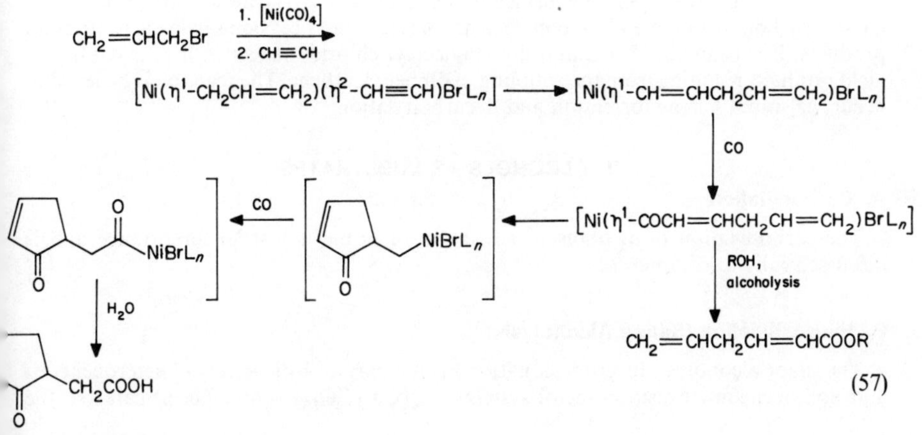

$$\text{CH}_2=\text{CHCH}_2\text{Br} \xrightarrow[\text{2. CH}\equiv\text{CH}]{\text{1. }[\text{Ni}(\text{CO})_4]}$$

$$[\text{Ni}(\eta^1\text{-CH}_2\text{CH}=\text{CH}_2)(\eta^2\text{-CH}\equiv\text{CH})\text{BrL}_n] \longrightarrow [\text{Ni}(\eta^1\text{-CH}=\text{CHCH}_2\text{CH}=\text{CH}_2)\text{BrL}_n]$$

$$\downarrow \text{CO}$$

$$\xleftarrow{\text{CO}} [\text{Ni}(\eta^1\text{-COCH}=\text{CHCH}_2\text{CH}=\text{CH}_2)\text{BrL}_n]$$

$$\downarrow \text{ROH, alcoholysis}$$

$$\text{CH}_2=\text{CHCH}_2\text{CH}=\text{CHCOOR} \quad (57)$$

intramolecular insertion but before the second carbonyl insertion via sterically promoted β-hydrogen elimination (equation 58)[46]. The carbonylation of allylic halides has been reviewed in detail previously[104].

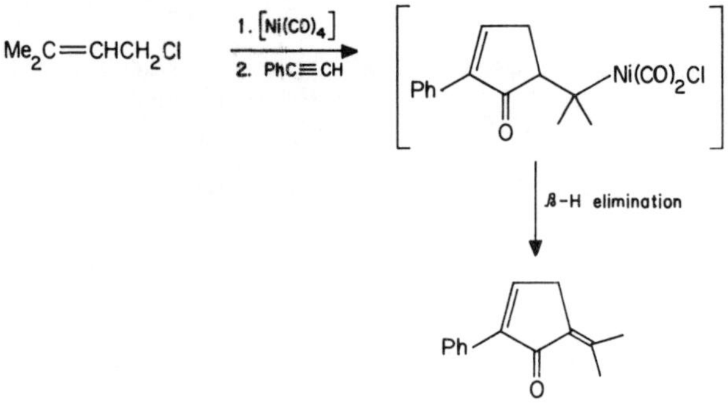

$$Me_2C=CHCH_2Cl \xrightarrow[2.\ PhC\equiv CH]{1.\ [Ni(CO)_4]} \left[\begin{array}{c} \text{Ph} \diagup\!\!\!\diagdown\!\!\!\diagdown\text{—Ni(CO)}_2\text{Cl} \\ \text{O} \end{array} \right]$$

$$\downarrow \beta\text{-H elimination} \quad (58)$$

$$\text{Ph} \diagup\!\!\!\diagdown\!\!\!=\!\!\diagdown$$

C. Acyl Halides

1. Decarbonylation[105]

Aroyl halides undergo catalytic decarbonylation in the presence of $[RhCl(CO)(PPh_3)_2]$ at high temperature to yield the corresponding aryl halide (equation 59). This route

$$PhCOX \xrightarrow{[RhCl(CO)(PPh_3)_2]} PhX + CO \qquad (59)$$

provides a convenient means of introducing a halogen substituent on to an aromatic ring system. Aroyl cyanides yield aryl nitriles similarly[106].

With aliphatic acyl halides, the tendency of the alkylrhodium intermediate to undergo β-hydrogen elimination leads to olefin formation. Isomerization of the initially formed olefin usually occurs, unless it is removed by continuous distillation (e.g. equation 60)[107].

$$Me(CH_2)_6COBr \longrightarrow CH_2=CH(CH_2)_4Me + (E)\text{-}MeCH=CH(CH_2)_3Me + (Z)\text{-}MeCH=CH(CH_2)_3Me$$

$$\qquad\qquad\qquad\qquad (30) \qquad\qquad\qquad (31) \qquad\qquad\qquad\qquad (32)$$

$$\qquad\qquad\qquad\qquad\qquad\qquad\qquad\qquad\qquad\qquad\qquad\qquad\qquad\qquad\qquad (60)$$

The products **30**, **31**, and **32** were formed in the ratio 14:5:1. The decarbonylation of acyl halides by homoleptic metal carbonyls is uncommon and occasionally leads to unusual products. For example[108], α-chlorodiphenylacetyl chloride reacts with $[Co_2(CO)_8]$ to yield products resulting from the coupling of diphenylcarbene. The reaction is believed to occur via initial ketene formation and decarbonylation.

VI. ALCOHOLS AS SUBSTRATES

A. Carbonylation

The carbonylation of alcohols is a reaction of prime industrial importance and is discussed fully in Chapter 7.

B. Hydrosilylation (Silane Alcoholysis)

The silane alcoholysis reaction (equation 61) is catalysed by a variety of heterogeneous and homogeneous transition metal systems[109]. $[Co_2(CO)_8]$ is an efficient catalyst, the mechanism of action being shown in equations 62 and 63[110,111]. Equation 63 has been shown to occur with inversion at silicon when optically active silylcobalt complexes are employed.

$$R_3SiH + R'OH \rightarrow R_3SiOR' + H_2 \quad (61)$$

$$[Co_2(CO)_8] + HSiR_3 \rightarrow [CoH(CO)_4] + [Co(SiR_3)(CO)_4] \quad (62)$$

$$[Co(SiR_3)(CO)_4] + R'OH \rightarrow R_3SiOR' + [CoH(CO)_4] \quad (63)$$

The alcoholysis reaction has been described for a variety of alcoholic substrates, among which phenols generally show the lower reactivity[111], presumably a factor related to their acidic nature.

VII. KETONES AND KETENES AS SUBSTRATES

A. Decarbonylation

The stoichiometric or catalytic decarbonylation of ketones is an uncommon reaction in comparison with the many synthetically useful decarbonylations of aldehydes and acyl halides which are known. The mechanism of decarbonylation for these latter substrates seems to involve oxidative addition of a C—X (X = H, halide) bond to the metal centre at an early stage in the sequence, and obviously no analogous process is possible for a simple ketone.

One of the commonest reagents for stoichiometric decarbonylation reactions is [RhCl(PPh$_3$)$_3$], which is converted into [RhCl(CO)(PPh$_3$)$_2$] during the reaction. Frequently such systems can be made catalytic by raising the operating temperature until [RhCl(CO)(PPh$_3$)$_2$] effectively liberates carbon monoxide and can participate in further decarbonylation cycles. Only stoichiometric decarbonylations of ketones have been reported. Thus, it seems necessary for a further functionality within the substrate to be present, possibly indicating that coordination to the metal centre by an external group is necessary. An example is the decarbonylation of acetylenic ketones[112,113] (equation 64).

$$(PhC \equiv C)_2CO + [RhCl(PPh_3)_3] \rightarrow PhC \equiv CC \equiv CPh + [RhCl(CO)(PPh_3)_2] + PPh_3 \quad (64)$$

Other decarbonylations of ketones by [RhCl(PPh$_3$)$_3$] are known[105,114], but few involving other metal complexes have been reported. Examples include the use of nickel, cobalt, and iron carbonyls[71] in the decarbonylation of diphenylcyclopropenone (equation 65).

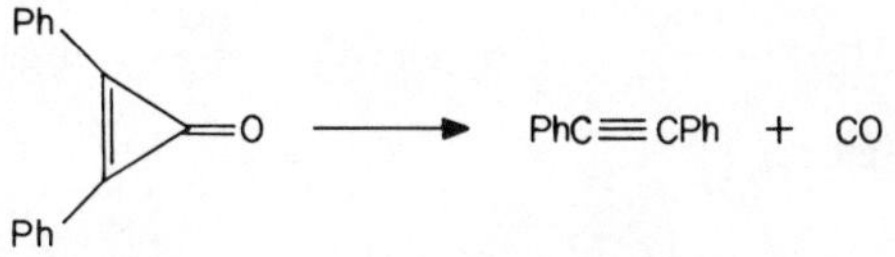

$$\text{Ph}_2\text{C}_3\text{=O} \longrightarrow \text{PhC} \equiv \text{CPh} + \text{CO} \quad (65)$$

Clearly, the synthetic utility of ketone decarbonylation is extremely limited in comparison with that of aldehydes and acyl halides. Decarbonylation reactions of ketenes have also been investigated, employing various complexes of Rh,[115] V, and Ti[116] as stoichiometric reagents or using carbonyls such as [Co$_2$(CO)$_8$], [Co$_4$(CO)$_{12}$], and [Co(C$_5$H$_5$)(CO)$_2$] as catalysts[115]. The decarbonylation of diphenylketene to tetraphenylethylene occurs smoothly at 110 °C (equation 66). Syntheses of these types, based on the

$$2\ Ph_2C=C=O \longrightarrow Ph_2C=CPh_2 + 2CO \quad (66)$$

TABLE 2. Reductive amination of ketones

Ketone	Amine	Carbonyl complex	Product	Conditions and Yield
Me_2CO	Pr^iNH_2	$[Fe(CO)_5]$	Pr^i_2NH	20 °C, 1 atm CO; 93% yield
Cyclohexanone	$c\text{-}C_6H_{11}NH_2$	$[Fe(CO)_5]$	$(c\text{-}C_6H_{11})_2NH$	20 °C, 1 atm CO; 100% yield
Cyclohexanone	Pr^iNH_2	$[Rh_6(CO)_{16}]$	$Pr^i(c\text{-}C_6H_{11})NH$	110–160 °C, 100–300 atm CO; 100% yield
BzMeCO	NH_3	$[Co_2(CO)_8]$ + PBu_3	BzMeNH	150–180 °C, 100–300 atm CO + H_2; 69% yield

metal-catalysed generation of carbenes, may prove to be useful in studies directed towards the preparation of hindered tetrasubstituted olefins, such as the elusive tetra-*tert*-butylethylene.

B. Reducive Amination

The reductive amination of carbonyl compounds is a reaction traditionally catalysed heterogeneously[117]. The need to develop efficient homogeneous systems for the reductive amination of carbonyl compounds has arisen since the production of such carbonyl compounds via homogeneously catalysed processes (e.g. aldehydes via hydroformylation) may result in a sufficient price reduction for these compounds to become desirable as precursors in large-scale amine production.

In the case of ketones, various carbonyl complexes have been employed to effect reductive amination (e.g. Table 2). Under mild conditions of temperature and pressure, $[HFe(CO)_4]^-$ {generated *in situ* from $[Fe(CO)_5]$ + alkali} effects stoichiometric reductive amination[118–113]. Under more extreme conditions, rhodium and cobalt carbonyls are catalytically active[124]. Watanabe *et al.*[122] proposed that reductive amination occurs via intermediate Schiff base and iminium ion formation followed by N=C bond reduction (equation 67). The mechanism gains some support from the fact that Schiff bases and

$$—NH_2 + HCHO \rightarrow [—NHCH_2OH] \xrightarrow{-H_2O} [—N=CH_2]$$
$$\downarrow [HFe(CO)_4]^-$$
$$[—N(Me)=CH_2]^+ \xleftarrow{-H_2O} [—N(Me)CH_2OH] \xleftarrow{HCHO} —NHCH_3 \quad (67)$$
$$\downarrow [HFe(CO)_4]^-$$
$$—NMe_2$$

iminium ions are known to be subject to reduction by $[FeH(CO)_4]^{-}$ [119,125]. In addition to carbonyl complexes, various other cobalt and rhodium compounds are reported to effect reductive amination of ketones[127].

C. Reductive Alkylation and Arylation

Ketones are subject to reductive alkylation or arylation by aldehydes in the presence of $[FeH(CO)_4]^-$. The reaction is described in Section VIII.E.

VIII. ALDEHYDES AS SUBSTRATES

A. Carbonylation

Typical catalysts for the carbonylation of saturated aldehydes are the halides of nickel, cobalt, and iron[94,126], which presumably react to form carbonyls under operating conditions. In the presence of water, hydrocarboxylation occurs (e.g. equation 68).

$$RCHO + CO + H_2O \rightarrow RCH(OH)COOH \tag{68}$$

B. Decarbonylation

The stoichiometric decarbonylation of aldehydes to alkanes by $[RhCl(PPh_3)_3]$ has found synthetic utility in many areas of organic chemistry. The red $[RhCl(PPh_3)_3]$ is converted into the yellow $[RhCl(CO)(PPh_3)_2]$ during the reaction. Stoichiometric decarbonylation occurs under very mild conditions (room temperature or mild warming) in such solvents as dichloromethane, benzene, toluene, or xylene. For secondary aldehydes, harsher conditions must be employed. However, in boiling toluene or xylene[105-127], a brick red complex (33) forms and terminates the decarbonylation[105].

$$[Rh(\mu - Cl)(PPh_3)_2]_2$$
$$(33)$$

The process can be made catalytic by operating at ca. 200 °C, employing either $[RhCl(PPh_3)_3]$ or $[RhCl(CO)(PPh_3)_2]$ as catalyst precursors. At such elevated temperatures, $[RhCl(CO)(PPh_3)_2]$ is able to eliminate CO at some stage in the decarbonylation sequence and hence enter into the catalytic cycle. Disadvantages of such high operating temperatures include the propensity of aliphatic aldehydes to undergo the aldol condensation reaction.

The stoichiometric decarbonylation of aldehydes to olefins is catalysed by $[Ru_2Cl_3(PEt_2Ph)_6]Cl$, where the catalyst is converted to $[RuCl_2(CO)(PEt_2Ph)]$[128]. Hydrogen is not released during the reaction, but is consumed by the reduction of the aldehyde to the alcohol (e.g. equation 69)[105]. The reaction occurs at 80-90 °C. Studies using deuterated aldehydes indicate that the mechanisms for the rhodium- and ruthenium-catalysed decarbonylations are different[105]. The scope of stoichiometric and catalytic aldehyde decarbonylation has been reviewed in detail previously[105,129].

$$2RCHO \rightarrow RH + RCH_2OH \tag{69}$$

C. Reductive Amination

Aldehydes are subject to reductive amination[117] under similar conditions to those described for ketones (see Section VII.B). $[FeH(CO)_4]^-$ effects the stoichiometric reductive amination of butanal with aniline at 20 °C under 1 atm of CO to produce N-n-butylaniline in 100% yield[119]. The corresponding reaction of α-ethylhexanal with isopropylamine at 110–160 °C under 100–300 atm of CO/H_2 yields 2-ethylhexylisopropylamine catalytically in the presence of $[Rh_6(CO)_{16}]$[124].

Mechanistically, the reactions are probably related to those of ketones, as described previously. In addition to carbonyl complexes, the reductive amination of aldehydes is promoted by other rhodium and cobalt species, a notable example being the production of N-methylaniline from formaldehyde and aniline in the presence of vitamin B_{12r} and related model cobalt compounds[130].

D. N-Acylamino Acid Synthesis

$[Co_2(CO)_8]$ is reported[131] to catalyse the reaction of an aldehyde with an amide in the presence of carbon monoxide to yield the N-acylamino acid (equation 70). The reaction

$$R^1CHO + R^2CONH_2 + CO \rightarrow R^1CH(COOH)NHCOR^2 \qquad (70)$$

proceeds smoothly under typical hydroformylation conditions, thus allowing the two processes to be effectively combined (equation 71). With the product N-acetylalanine, for

$$R^1CHCH_2 + R^2CONH_2 + 2CO + H_2 \rightarrow R^1CH_2CH_2CH(COOH)NHCOR^2 \quad (71)$$

example, the $\alpha:\beta$ ratio was 550:1. In the presence of stoichiometric amounts of $[Co_2(CO)_8]$, the reaction shown in equation 70 proceeds even at room temperature and atmospheric pressure. The mechanism of this process is not clear, since certain combinations of aldehydes and amides give anomalous results. Benzaldehyde and acetamide, for example, yield N-acetylbenzylamine.

Clearly, the chemistry described by equations 70 and 71 could be of considerable synthetic utility.

E. Reductive Alkylation and Arylation

Aldehydes may be employed to alkylate or arylate a variety of carbonyl or active methylene compounds in the presence of $[FeH(CO)_4]^-$[132]. The carbonyl compound itself may be an aldehyde or a ketone; equation 72 represents the reaction, which is believed to

$$R'COCH_2R + R''CHO \underset{+H_2O}{\overset{-H_2O}{\rightleftharpoons}} R'COC(R)=CHR'' \xrightarrow{[HFe(CO)_4]^-} R'COCHRCH_2R'' \qquad (72)$$

R', R'' = H, alkyl, aryl, part of ring; R = alkyl, aryl, heteroaryl

occur via condensation followed by irreversible reduction. For example, acetone undergoes alkylation in the presence of benzaldehyde, using this method, yielding 1-phenylbutan-3-one in 70% yield (equation 73).

$$Me_2CO + PhCHO \rightarrow PhCH_2CH_2COMe \qquad (73)$$

IX. EPOXIDES AND ETHERS AS SUBSTRATES

A. Rearrangement.[133]

1. Epoxides

$[Co_2(CO)_8]$ catalyses the rearrangement of epoxides to ketones in alcoholic solution[134] (e.g. equation 74). The proposed mechanism[135] involves attack of the epoxide by

$$\text{Methyloxirane} \xrightarrow{[Co_2(CO)_8]} Me_2CO \qquad (74)$$

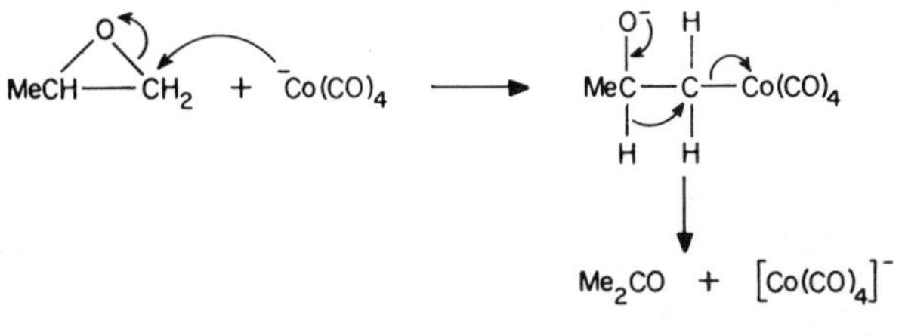

$$\text{MeCH}-\text{CH}_2 \text{ (epoxide)} + {}^-Co(CO)_4 \longrightarrow \underset{\underset{H}{|}}{\overset{\overset{O^-}{|}}{Me C}}-\underset{\underset{H}{|}}{\overset{\overset{H}{|}}{C}}-Co(CO)_4 \qquad (75)$$

$$\downarrow$$

$$Me_2CO + [Co(CO)_4]^-$$

[Co(CO)$_4$]$^-$, formed by disproportionation (equation 75). In the presence of CO, however, methanolic solutions of epoxides yield largely hydroxylic esters under similar conditions (see Section IX.C.2).

B. Reduction

1. Epoxides

Terminal epoxides are reduced to olefins by anions such as [HFe(CO)$_4$]$^-$ [136]. The reaction provides the basis for the synthesis of Z- or E-olefins from epoxides utilizing the related nucleophile [Fe(Cp)(CO)$_2$]$^-$, and hence a simple route to olefin isomerization via epoxidation[137] (equation 76). The thermal elimination yields the Z-olefin while acidific-

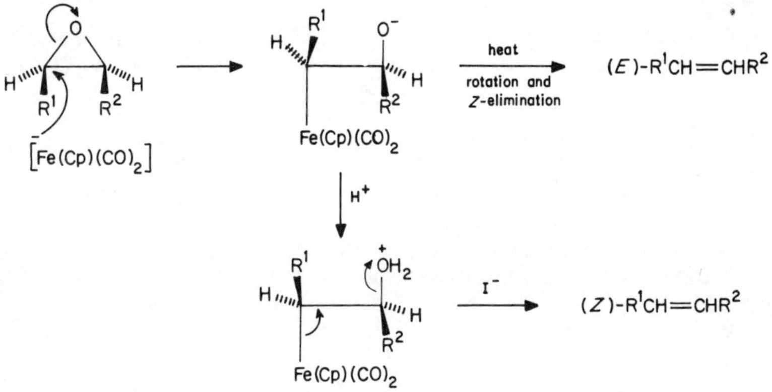

$$\begin{array}{c}\text{[reaction scheme 76]}\end{array}$$

ation followed by elimination produces a complex of the E-olefin, which releases the organic product on treatment with iodide ion.

C. Carbonylation[133]

1. Simple ethers

Simple ethers may be carbonylated using a variety of heterogeneous or homogeneous catalysts, including nickel, cobalt, and other metal carbonyls, under fairly extreme conditions of temperature and pressure[138,139]. Generally, the major products are the ester and acid, the latter presumably originating from hydrolysis of the ester by adventitious water (equation 77)[133].

$$\text{ROR} \xrightarrow{\text{CO}} \text{RCOOR} \underset{-\text{H}_2\text{O} + \text{ROH}}{\overset{+\text{H}_2\text{O}, -\text{ROH}}{\rightleftharpoons}} \text{RCOOH} \qquad (77)$$

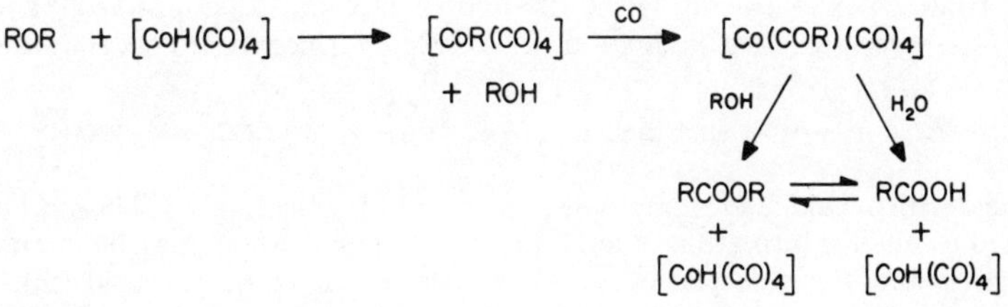

$$\text{ROR} + [\text{CoH(CO)}_4] \longrightarrow [\text{CoR(CO)}_4] \xrightarrow{\text{CO}} [\text{Co(COR)(CO)}_4]$$
$$+ \text{ROH}$$

with ROH / H$_2$O branches giving:

$$\text{RCOOR} \rightleftharpoons \text{RCOOH} \qquad (78)$$
$$+ \qquad +$$
$$[\text{CoH(CO)}_4] \quad [\text{CoH(CO)}_4]$$

The probable mechanism involves cleavage of the ether C—O bond by the catalyst, followed by the standard carbonylation sequence (e.g. equation 78)[12]. No synthetically useful applications of the carbonylation of simple ethers on a laboratory scale appear to have been reported.

2. Epoxides

The products of the carbonylation of epoxides depend on the nature of the solvent employed. Benzene does not participate in the carbonylation reaction, so α,β-unsaturated acids are produced by employing $[Co_2(CO)_8]$ as the catalyst[140] (equation 79). In

$$\text{oxirane} + CO \xrightarrow{[Co_2(CO)_8]} CH_2=CHCOOH \tag{79}$$

hydroxylic solvents, however, the corresponding hydroxyesters are produced[141] (e.g. equation 80). Thus, oxirane reacts with carbon monoxide in the presence of $Na[Co(CO)_4]$

$$\text{R-oxirane} + CO \xrightarrow[\text{MeOH}]{[Co_2(CO)_8]} RCH(OH)CH_2COOMe \tag{80}$$

in methanolic solution employing mild conditions of temperature and pressure to produce methyl 3-hydroxypropionate in ca. 50% yield[142]

Under hydroformylation conditions, epoxides yield hydroxyaldehydes, which may be subject to dehydration and subsequent hydrogenation under the reaction conditions employed (equation 81). The reaction is catalysed by cobalt carbonyls with $[Co_2(CO)_8]^{143-145}$ and $[Co_2(CO)_6(PR_3)_2]^{146}$ being the major catalysts employed. Both $[CoH(CO)_4]^{142}$ and hydridoiron carbonyls[147] may be used as stoichiometric reagents.

$$(RCH_2)_2\text{-oxirane} \xrightarrow{CO/H_2} RCH_2CH(CHO)CH(OH)CH_2R$$
$$\downarrow \text{dehydration} \tag{81}$$
$$RCH_2CH(CHO)CH_2CH_2R \xleftarrow{\text{hydrogenation}} RCH_2CH(CHO)CH=CHR$$
$$+ RCH_2C(CHO)=CHCH_2R$$
$$\downarrow \text{hydrogenation}$$
$$RCH_2CH(CH_2OH)CH_2CH_2R$$

3. Cyclic ethers

Larger ring cyclic ethers, such as tetrahydrofuran[148], are readily carbonylated in the presence of a variety of catalysts, including nickel and cobalt carbonyls[143,148]. The carbonylation of tetrahydrofuran has been studied in particular, since it represents a possible route to adipic acid. The formation of lactones as intermediates in diacid synthesis has been suggested (equation 82). Under hydroformylation conditions, cyclic

$$\text{thf} \xrightarrow{CO} \text{tetrahydrofuran-2-one} \xrightarrow{CO} HOOC(CH_2)_4COOH \tag{82}$$

ethers yield hydroxyaldehydes employing $[Co_2(CO)_8]$ as the catalyst. The mechanism[142] is believed to involve formation of an acylcobalt complex, which may be intercepted by hydrogen to yield the aldehyde or by water to yield the carboxylic acid (equation 83).

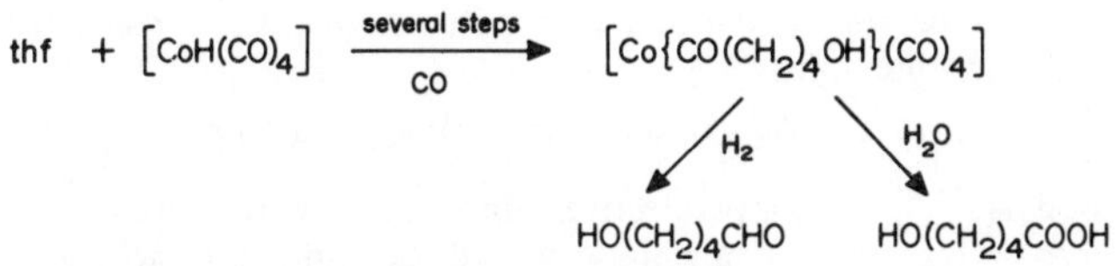

6. Transition metal carbonyls in organic synthesis

$$thf + [CoH(CO)_4] \xrightarrow[CO]{\text{several steps}} [Co\{CO(CH_2)_4OH\}(CO)_4]$$

$$\swarrow H_2 \qquad \searrow H_2O \qquad (83)$$

$$HO(CH_2)_4CHO \qquad HO(CH_2)_4COOH$$

D. Polymerization Under Hydrosilylation Conditions

Although polymerization reactions are not generally within the scope of this chapter, it is noteworthy that tetrahydrofuran is polymerized by $Co_2(CO)_8$ under hydrosilylation conditions[111]. Thus, in the presence of Et_3SiH, polymerization, postulated to occur via a cationic mechanism,[109] results (equation 84).

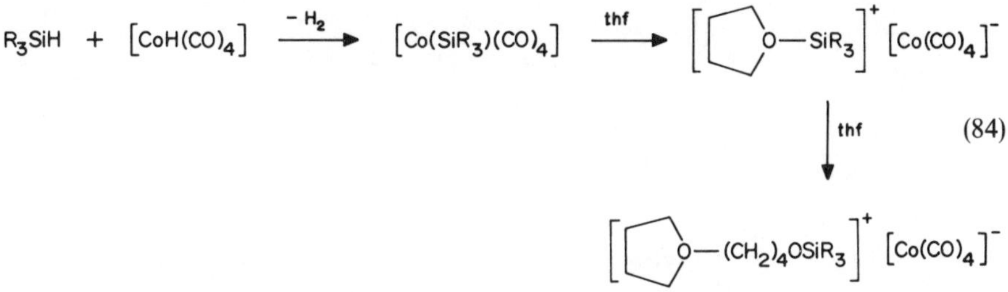

$$R_3SiH + [CoH(CO)_4] \xrightarrow{-H_2} [Co(SiR_3)(CO)_4] \xrightarrow{thf} \left[\begin{array}{c}\square\end{array}O-SiR_3\right]^+ [Co(CO)_4]^-$$

$$\downarrow thf \qquad (84)$$

$$\left[\begin{array}{c}\square\end{array}O-(CH_2)_4OSiR_3\right]^+ [Co(CO)_4]^-$$

X. CARBOXYLIC ACIDS AS SUBSTRATES

A. Decarbonylation

Decarbonylation of carboxylic acids is expected to be difficult to accomplish owing to the lack of a reactive C—H (aldehyde) or C—Cl (acyl chloride) bond to facilitate oxidative addition. It is reported[149] that decarbonylation to the olefin, e.g. equation 85, is accomplished by treatment with $[RhCl_3(PPhEt_2)_3]$, which is converted to trans-$[RhCl(CO)(PPhEt_2)_2]$.

$$Me(CH_2)_4COOH \rightarrow EtCH=CHMe \qquad (85)$$

No successful attempts to employ trans-$[RhCl(CO)(PR_3)_2]$ complexes at high temperature for catalytic decarbonylation appear to have been reported.

B. Hydrosilylation[109]

Carboxylic acids are susceptible to the $[Co_2(CO)_8]$-catalysed hydrosilylation reaction, as are a variety of other protic compounds.[109] The corresponding silyl ester is produced, with evolution of hydrogen (equation 86).

$$R_3SiH + R'COOH \rightarrow R'COOSiR_3 + H_2 \qquad (86)$$

XI. ACID ANHYDRIDES AS SUBSTRATES

A. Decarbonylation

In certain cases, the decarbonylation of aromatic acid anhydrides may be effected by $[RhCl(CO)(PPh_3)_2]$ or $[RhCl(PPh_3)_3]$[150,151]. The reactions are not selective and tend to give a number of products (e.g. equation 87). Possible mechanisms for such reactions have been elaborated upon previously.[105]

$$(PhCO)_2O \rightarrow \text{fluoren-9-one} + Ph_2 + Ph_2CO + PhCOOH \quad (87)$$

Cobalt carbonyls[152,153] are more selective, removing one carbonyl from a pair of *ortho*-substituted carbonyl groups on aromatic anhydrides and polycarboxylic acids (e.g. equation 88). The proposed mechanism[105] follows the standard decarbonylation pathway (equation 89). The corresponding reaction with $[Fe(CO)_4]^{2-}$ may be utilized to produce aldehydes from carboxylic acid anhydrides according to equation 90[154].

$$\text{phthalic anhydride} \xrightarrow[{[Co_2(CO)_8]}]{CO/H_2} PhCOOH \quad (88)$$

$$\text{phthalic anhydride} \xrightarrow{[HCo(CO)_4]} [Co\{CO(o\text{-}HO_2CC_6H_4)\}(CO)_4]$$

$$\xrightarrow{-CO} [Co(o\text{-}HO_2CC_6H_4)(CO)_4] \quad (89)$$

$$\downarrow H_2$$

$$PhCOOH + [CoH(CO)_4]$$

$$\text{phthalic anhydride} \xrightarrow{[Fe(CO)_4]^{2-}} [Fe\{CO(o\text{-}OOCC_6H_4)\}(CO)_4]^{2-} \xrightarrow{H^+} o\text{-}HO_2CC_6H_4CHO \quad (90)$$

With $[Ni(CO)_2(PR_3)_2]$, the anhydrides are found to yield the corresponding olefin via total decarbonylation[155] (equation 91). Thioanhydrides similarly undergo decarbonylation to olefins in the presence of $[Fe_2(CO)_9]$[156].

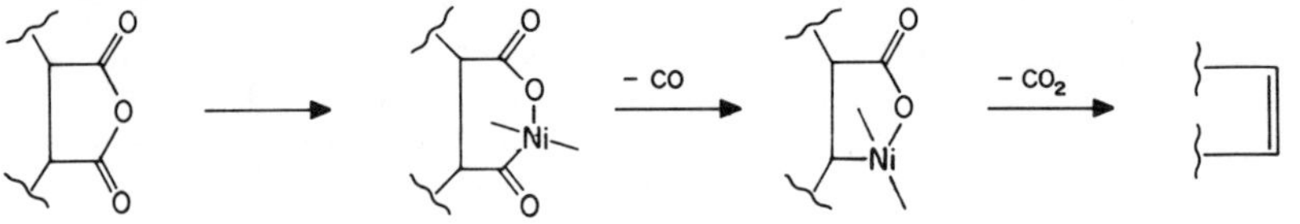

(91)

XII. ESTERS AS SUBSTRATES

A. Carbonylation of Esters and Lactones[133]

Several reports[157,158] indicate that anhydrides may be prepared by the carbonylation of esters in the presence of cobalt and nickel catalysts (equation 92). Conditions are extreme and the reaction appears to have little synthetic utility.

$$RCOOR' \xrightarrow{CO} RCOOC(O)R' \quad (92)$$

Simple lactones may be catalytically carbonylated to the carboxylic acid containing an additional carbon atom using cobalt carbonyl[159] or nickel carbonyl[148] catalysts. Equation 93 illustrates the reaction for γ-butyrolactone. While not of prime importance in themselves, such reactions are of interest in understanding the involvement of lactones in the carbonylation of cyclic ethers, with the formation of adipic acid by carbonylation of tetrahydrofuran being an important example (see Section IX.C).

$$\gamma\text{-butyrolactone} + H_2O + CO \rightarrow CH_2(CH_2COOH)_2 \quad (93)$$

B. Decarboxylation of Pyrones

The α-pyrone **34** undergoes photochemical extrusion of CO_2 in the presence of $[Fe(CO)_5]$ to yield the $[Fe(diene)(CO)_3]$ complex **35** (equation 94). The decomplexation of the substituted cyclobutene with iron(III) chloride or cerium(IV) ion may then generate the free diene.

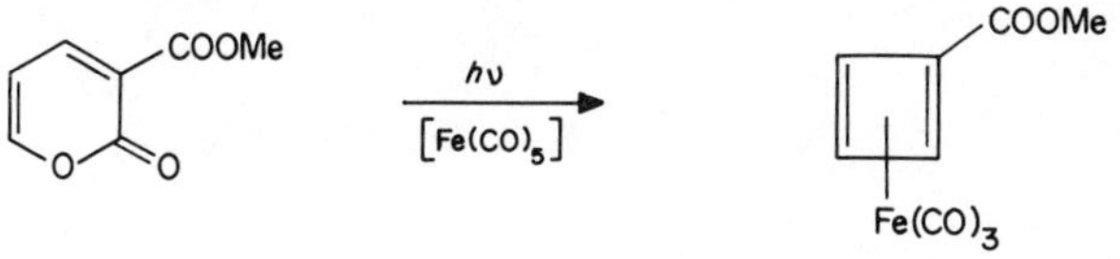

$$\text{(34)} \xrightarrow[{[Fe(CO)_5]}]{h\nu} \text{(35)} \quad (94)$$

XIII. DIALKYLACETALS AS SUBSTRATES[133]

A. Carbonylation

Under extreme conditions, dialkylacetals undergo carbonylation in the presence of cobalt catalysts, ultimately yielding the dialkylacetal of the α-alkoxyaldehyde[148,160]. The proposed mechanism[148] is shown in equation 95.

$$CH_2(OR)_2 + [CoH(CO)_4] \xrightarrow{-ROH} [Co(CH_2OR)(CO)_4] \xrightarrow{CO} [Co(COCH_2OR)(CO)_4]$$
$$\downarrow H_2 \quad (95)$$
$$ROCH_2CH(OR)_2 \xleftarrow{2ROH} ROCH_2CHO + [CoH(CO)_4]$$

XIV. *ORTHO*-ESTERS AS SUBSTRATES

A. Carbonylation[133]

The $[Co_2(CO)_8]$-catalysed carbonylation of *ortho*-esters[161] is a synthetically useful route to aldehydes containing one more carbon atom than the alkoxy group of the *ortho*-ester. The reaction occurs under mild conditions of temperature and pressure, with

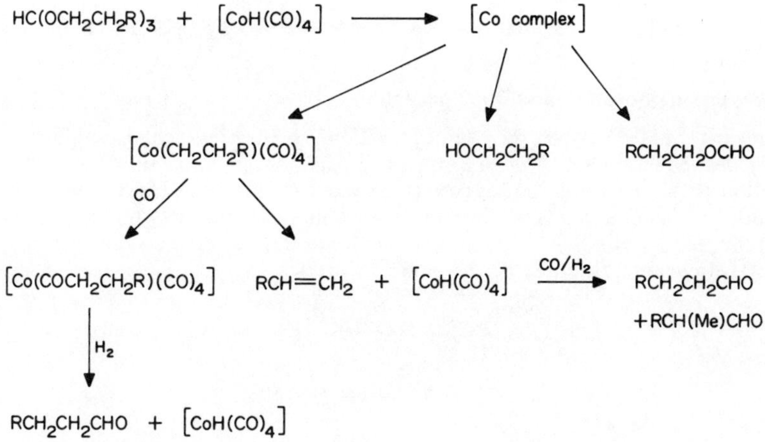

$$HC(OCH_2CH_2R)_3 + [CoH(CO)_4] \longrightarrow [Co\ complex]$$

$$\swarrow \quad \downarrow \quad \searrow$$

$$[Co(CH_2CH_2R)(CO)_4] \quad HOCH_2CH_2R \quad RCH_2CH_2OCHO$$

$$\downarrow CO \qquad \qquad \qquad \qquad \qquad (96)$$

$$[Co(COCH_2CH_2R)(CO)_4] \quad RCH=CH_2 + [CoH(CO)_4] \xrightarrow{CO/H_2} RCH_2CH_2CHO + RCH(Me)CHO$$

$$\downarrow H_2$$

$$RCH_2CH_2CHO + [CoH(CO)_4]$$

generally only small amounts of isomeric aldehydes as secondary products. The scope of the reaction has previously been discussed in some detail[133], with reviewers proposing the mechanism shown in equation 96.
The overall process may thus be summarized by equation 97.

$$RC(OR')_3 \xrightarrow{CO/H_2} RCOOR' + R'OH + R'CHO \qquad (97)$$

XV. SULPHOXIDES AS SUBSTRATES

A. Deoxygenation

The removal of a sulphoxide moiety from an organic substrate is frequently performed in two stages; initially the sulphoxide is reduced to the thioether and then the thioether is reduced to the alkane. The latter step is usually performed using traditional reductants (Raney nickel, lithium in liquid ammonia, etc.), but the former step, sulphoxide deoxygenation, may be effected by a number of transition metal complexes. Sulphoxide deoxygenation reactions have previously been discussed in detail[162,163].

[Fe(CO)$_5$] has been utilized in sulphoxide deoxygenation reactions[164], the proposed mechanism[35] being shown in equation 98. Alternative mechanisms involving iron—

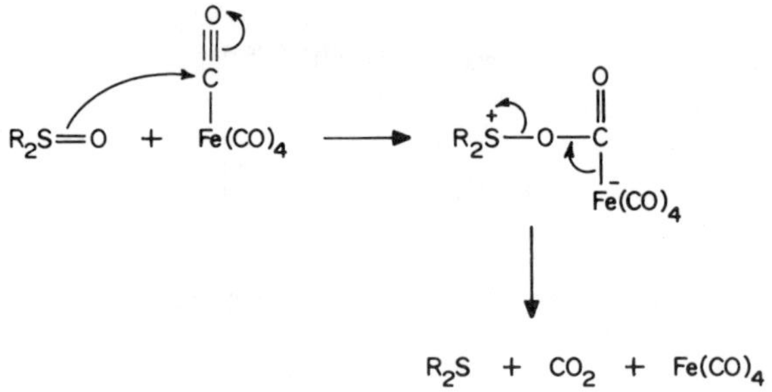

$$R_2S{=}O + Fe(CO)_4 \longrightarrow R_2\overset{+}{S}{-}O{-}\overset{O}{\underset{Fe(CO)_4}{C^-}} \qquad (98)$$

$$\downarrow$$

$$R_2S + CO_2 + Fe(CO)_4$$

sulphur bonded intermediates or disproportionation products have been mentioned in a review[35] of organic syntheses effected by [Fe(CO)$_5$]. The method is reported to be synthetically useful for dialkyl, diaryl, and heterocyclic mono- and di-sulphoxides[35].

XVI. ⟩C=S AND —N=S COMPOUNDS AS SUBSTRATES

A. Dehydrosulphuration and Desulphuration

A review[35] has described the use of [Fe(CO)$_5$] in effecting dehydrosulphuration of thioamides (equation 99)[165], N-substituted thioamides (equation 100)[35] and thioureas (equation 101)[35]. Reaction 100 has been employed in the synthesis of cyclic Schiff bases[35]. Reaction 101 is of limited versatility since only a limited number of thioureas are stable under the reaction conditions. Diphenylamine, for example, can be produced in 72% yield from 1,1-diphenyl-2-thiourea by this route[35].

$$RCSNH_2 \xrightarrow{Bu_2O} RC{\equiv}N \qquad (99)$$

$$RCSNHR' \xrightarrow{Bu_2O} RCH{=}NR' \qquad (100)$$

$$R_2NCSNH_2 \xrightarrow{Bu_2O} R_2NH \qquad (101)$$

6. Transition metal carbonyls in organic synthesis 325

Cyclic thionocarbonates are classically desulphurized by treatment with a phosphite (equation 102). The same reaction is induced by $[Fe(CO)_5]^{166}$, although the system is not

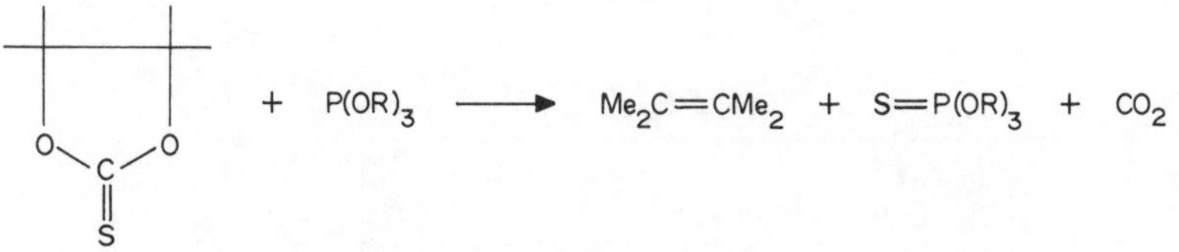

$$\underset{\underset{S}{\overset{\|}{C}}}{O\diagdown \diagup O} + P(OR)_3 \longrightarrow Me_2C=CMe_2 + S=P(OR)_3 + CO_2 \qquad (102)$$

stereospecific, e.g. compound **36** is formed from the (Z)-thionocarbonate derivative in 79.1% yield whereas (Z)-cycloheptane is produced in only 35% yield from its (E)-thionocarbonate derivative.

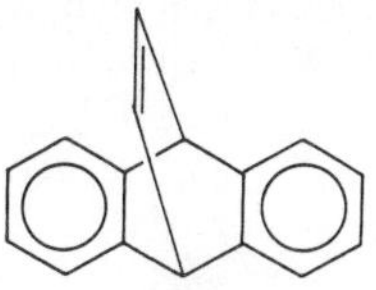

(**36**)

Both aliphatic and aromatic thioketones are desulphurized by $[HFe(CO)_4]^{-167}$. Utilization of the corresponding deuterioiron tetracarbonyl allows the isolation of specifically deuterated alkanes. For example, adamantanethione is desulphurized to 2,2-

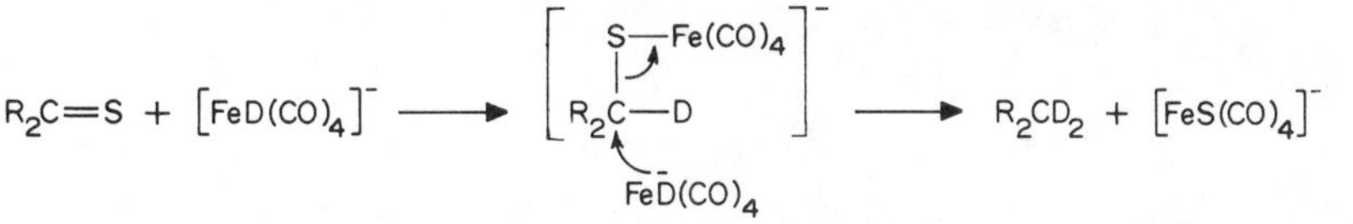

$$R_2C=S + [FeD(CO)_4]^- \longrightarrow \left[\begin{array}{c} S-Fe(CO)_4 \\ | \\ R_2C-D \\ | \\ FeD(CO)_4 \end{array} \right]^- \longrightarrow R_2CD_2 + [FeS(CO)_4]^- \qquad (103)$$

dideuterioadamantane in 78% yield. The proposed mechanism[167] is shown in equation 103. Thioamides[167] are similarly desulphurized (see equation 99) by $[FeH(CO)_4]^-$; the reaction shown in equation 104 proceeds in 51% yield.

$$CH_3CSNHPh \xrightarrow{[HFe(CO)_4]^-} C_2H_5NHPh \qquad (104)$$

The reaction of thioketones with $[Fe_2(CO)_9]$ can be modified[168,169] to produce lactones, or thiolactones, provided a site for metallation is available, according to equation 105.

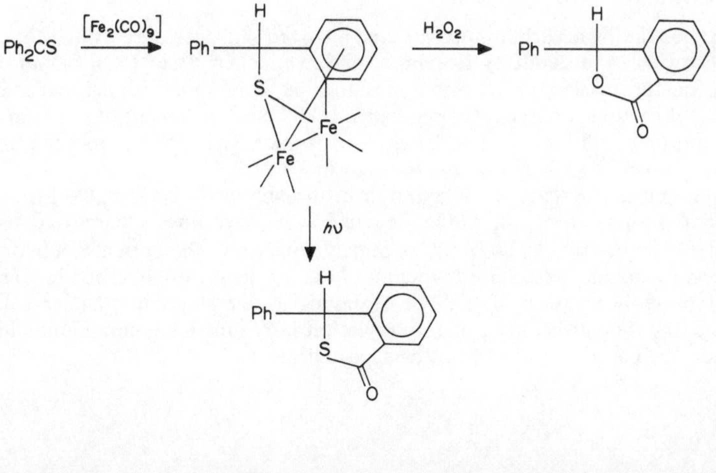

(105)

TABLE 3. (Z)- and (E)-but-2-ene episulphide desulphurization by iron carbonyl complexes

Episulphide	Carbonyl	Total yield (%)	But-2-ene (%)	
			(E)-	(Z)-
(Z)-	$[Fe_2(CO)_9]$	80.5	6.4	93.6
(E)-	$[Fe_2(CO)_9]$	81.9	97.5	2.5
(Z)-	$[Fe_3(CO)_{12}]$	Not determined	5.0	95.0
(E)-	$[Fe_3(CO)_{12}]$	Not determined	97.3	2.7

Desulphurization of sulphur diimides has also been reported[170] (equation 106). When R = Ph, azobenzene is produced in 19% yield after refluxing for 20 h in cyclohexene. A trace amount of phenylamine was also formed[170].

$$RN{=}S{=}NR + [Fe(CO)_5] \longrightarrow RN{=}NR + RNH_2 \quad (106)$$

The desulphurization of episulphides[171] provides a convenient and stereospecific olefin synthesis. The reaction is effected by $[Fe_2(CO)_9]$ and $[Fe_3(CO)_{12}]$. The proposed mechanism[172] is shown in equation 107. Some examples of this desulphurization reaction are given in Table 3.

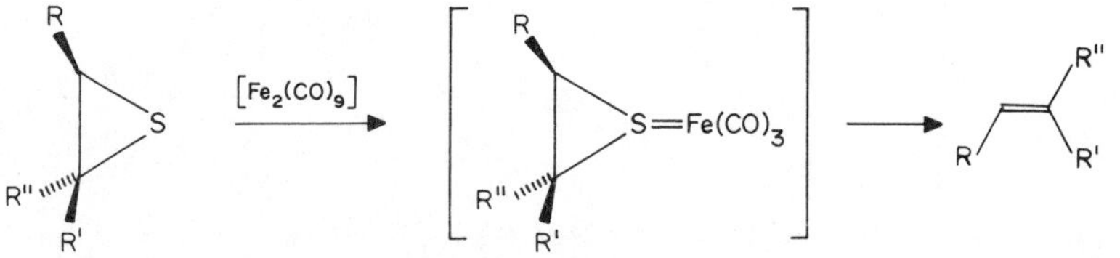

$$\text{(107)}$$

Clearly, iron carbonyls may be used to good advantage in certain desulphurization processes as a viable alternative to the more traditional reagents.

XVII. $>$C=N—, —N=N— AND —C≡N COMPOUNDS AS SUBSTRATES

A. Carbonylation

The carbonylation of carbon–nitrogen and nitrogen–nitrogen unsaturated compounds has been reviewed in depth by Rosenthal and Wender[173]. $[Co_2(CO)_8]$ is an effective catalyst for the cyclization of these unsaturated compounds under carbonylation conditions. Reactions are typically performed under elevated conditions of temperature and carbon monoxide pressure and may lead to high yields of heterocyclic products. Typical examples of such syntheses are listed in Table 4.

Aromatic nitriles undergo carbonylation in the presence of $[Co_2(CO)_8]$ to yield N-substituted phthalimidines in moderate yield when pyridine is employed as a co-catalyst[174]. The reaction is, however, extremely sensitive to the presence of hydrogen in the carbon monoxide used, since reduction of the nitrile to the corresponding amine[175] followed by carbonylation of the N—H bonds in the standard manner can occur (equation 108). The mechanism of carbonylation of N-functional compounds has been discussed previously[173] and will not be repeated here.

6. Transition metal carbonyls in organic synthesis

TABLE 4. Reactions of $>$C=N— and —N=N— compounds with CO

Class of substrate and example	Typical carbonyl complexes	Class of major product and example
Schiff bases PhCH=NPh	$[Co_2(CO)_8]$; also $[Fe(CO)_5]$	Substituted phthalimidines 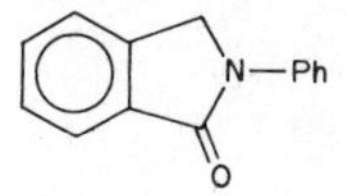 (38)
Aromatic ketoximes Ph_2C=NOH	$[Co_2(CO)_8]$	Substituted phthalimidines 38
Ketonic phenylhydrazones Ph_2C=NNHPh	$[Co_2(CO)_8]$	Substituted phthalimidine-N-carboxyanilides 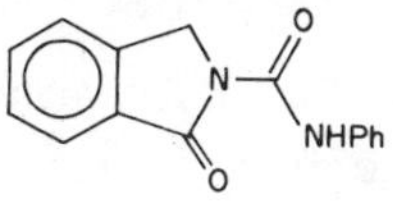
Aldehydic phenyl hydrazones PhCH=NNHPh	$[Co_2(CO)_8]$	Substituted phthalimidines 38
Azo compounds PhN=NPh	$[Co_2(CO)_8]$	Indazolones and quinazolones 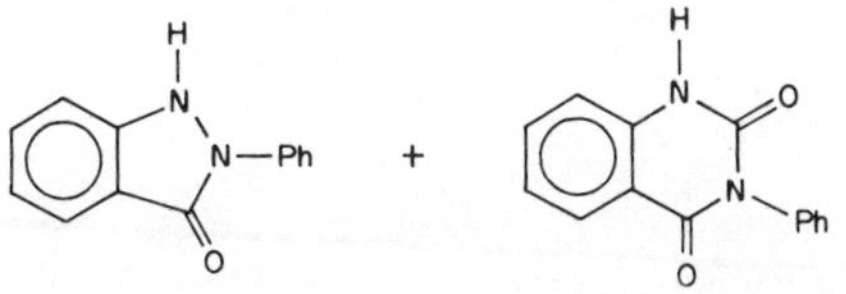

$$RC \equiv N + H_2 \xrightarrow{[Co_2(CO)_8]} RCH_2NH_2 \rightarrow [Co(NHCH_2R)(CO)_4] \xrightarrow{CO}$$

$$[Co(CONHCH_2R)(CO)_4]$$
$$\downarrow + 37$$
$$(RCH_2NH)_2CO + [Co_2(CO)_8] \quad (108)$$
$$(37)$$

XVIII. MAIN-GROUP ORGANOMETALLICS AS SUBSTRATES

A. Alkyllithium Reagents

Alkyllithium reagents react with many homoleptic metal carbonyls to yield the acyl carbonyl metallate anions (e.g. equations 109 and 110)[176,177]. The acyl carbonyl metallate

$$[Ni(CO)_4] + RLi \rightarrow Li[Ni(COR)(CO)_3] \quad (109)$$

$$[Fe(CO)_5] + RLi \rightarrow Li[Fe(COR)(CO)_4] \quad (110)$$

anions are highly nucleophilic and may be decomposed to aldehydes, ketones, or carboxylic acid derivatives (equations 111–116)[178–180]. The overall syntheses thus represent a number of useful alkyllithium carbonylation reactions. The corresponding

$$[Fe(COR)(CO)_4]^- \xrightarrow{H^+} RCHO \qquad (111)$$
$$(39)$$

$$39 \xrightarrow{R'Br} RCOR' \qquad (112)$$

$$39 \xrightarrow{O_2/R'OH} RCOOR' \qquad (113)$$

$$39 \xrightarrow{O_2/H_2O} RCOOH \qquad (114)$$

$$39 \xrightarrow{O_2/R'NH_2} RCONHR' \qquad (115)$$

$$39 \xrightarrow{R'NO_2/H^+} RCONHR' \qquad (116)$$

reactions of the acyl carbonyl nickelates can occur differently, however, yielding coupled products (e.g. equations 117–119)[178,181].

$$[Ni(COR)(CO)_3]^- \xrightarrow{H^+} RCOCOR \text{ or } R_2CO \qquad (117)$$
$$(40) \qquad (R = aryl) \quad (R = alkyl)$$

$$40 \xrightarrow{R'Br} RCOC(OH)RR' \qquad (118)$$

$$40 \xrightarrow{R'COX} (E)\text{-}R'OCOC(R)=C(R)OCOR' \qquad (119)$$

Similar chemistry occurs with lithium amides, which react with metal carbonyls to yield carbamoyl complexes (equation 120). With alkyl halides, the nickel carbamoyl yields the corresponding amide (equation 121)[182]. Related reactions of carbamoyl carbonyl metallates have also been described[5].

$$[Ni(CO)_4] + LiNR_2 \rightarrow Li[Ni(CONR_2)(CO)_3] \qquad (120)$$

$$[Ni(CONR_2)(CO)_3]^- \xrightarrow{R'X} R_2NC(O)R' \qquad (121)$$

B. Grignard Reagents and Organomercury(II) Halides

Carbamoyl carbonyl metallates are also available via reaction of suitable Grignard reagents with certain metal carbonyls (e.g. equation 122)[183,184]. Reactions with nitro compounds may then be utilized to produce ureas (equation 123)[183,184].

$$R_2NMgBr + [Fe(CO)_5] \rightarrow [Fe(CONR_2)(CO)_4]MgBr \qquad (122)$$

$$[Fe(CONR_2)(CO)_4]^- + R'NO_2 \xrightarrow{H^+} R_2NCONHR' \qquad (123)$$

The carbonylation of organometallics has also been applied to organomercury(II) halides, which yield ketones on treatment with $[Co_2(CO)_8]$ (equation 124). Similar chemistry is reported to occur with nickel carbonyl (e.g. equation 125)[185].

$$\text{RHgBr} \xrightarrow{[Co_2(CO)_8]} R_2CO \tag{124}$$

$$\text{RHgBr} \xrightarrow{[Ni(CO)_4]} R_2CO \tag{125}$$

XIX. TRANSITION METAL CARBENE COMPLEXES AS SUBSTRATES

Transition metal carbene complexes are intimately involved in the mechanism of olefin metathesis. Although they are generally poor sources of free carbenes, such complexes have found application in substituted cyclopropane synthesis (Section III.H) and other areas, as described below.

A. Thermolysis Reactions[66,186]

Thermolysis of carbene complexes may be used to generate olefins, probably via a bimolecular elimination (e.g. equation 126). In this case, both E- and Z-isomers are produced at 150 °C.

$$[(CO)_5Cr=C(Me)OMe] \rightarrow (Z)\text{- and }(E)\text{-MeOC(Me)=C(Me)OMe} \tag{126}$$

B. Carbene Cleavage[66,186]

The carbene fragment is cleaved and captured by reaction of the metal carbene complex with suitable electrophiles. For example, esters can be produced by reaction with molecular oxygen (e.g. equation 127). Corresponding products are similarly formed with sulphur and selenium.

$$[(CO)_5Cr = C(Ph)OMe] + O_2 \rightarrow PhCOOMe \tag{127}$$

Reaction with a carbene source generates the olefin from the metal carbene complex (e.g. equation 128).

$$[(CO)_5Cr=C(Ph)OMe] + PhHgCCl_3 \rightarrow Cl_2C=C(OMe)Ph$$
$$+ [Cr(CO)_5] + PhHgCl \tag{128}$$

XX. CONCLUDING REMARKS

In this chapter we have described some of the areas in which transition metal carbonyls have been used in organic synthesis. There remain many challenges in the development of laboratory- and industrial-scale syntheses with such reagents, the alkane activation problem being one example. Synthetically, transition metal reagents are of the greatest use in syntheses where a higher degree of regio- and stereo-selectivity is possible than for the more traditional reagents. Development of the N-acylamino acid synthesis, for example, may prove to be of importance.

There is little doubt that metal carbonyls will become of increasing importance in synthesis. The creation and cleavage of carbon—carbon bonds selectively and under mild conditions will undoubtedly be developed to a fine art in the future.

XXI. REFERENCES

1. W. H. Jones (Ed.), *Catalysis in Organic Syntheses*, Academic Press, New York, 1980.
2. C. W. Bird, *Transition Metal Intermediates in Organic Synthesis*, Logos Press, London, 1967.
3. G. Parshall, *Homogeneous Catalysis*, Wiley–Interscience, New Yoek, 1980.

4. K. Weissermel and H. J. Arpe, *Industrial Organic Chemistry*, Verlag Chemie, Berlin, 1978.
5. D. St. C. Black, W. R. Jackson, and J. M. Swan, in *Comprehensive Organic Chemistry* (Eds. D. Barton and W. D. Ollis), Vol. 3, Pergamon Press, Oxford, 1979, p. 1127.
6. B. R. James, *Homogeneous Hydrogenation*, Wiley–Interscience, New York 1974.
7. B. R. James, *Adv. Organomet. Chem.*, **17**, 319 (1978).
8. J. L. Speier, *Adv. Organomet. Chem.*, **17**, 407 (1978).
9. E. S. Brown, *Aspects of Homogenous Catalysis*, Vol. 2, Reidel, Berlin; p. 57.
10. T. Keii, *Kinetics of Ziegler–Natta Polymerization*, Chapman and Hall, London, 1972.
11. P. Cossee, in *The Stereochemistry of Macromolecules* (Ed. A. D. Kettey), Vol. 1, Marcel Dekker, New York, 1967.
12. J. Falbe, *Carbon Monoxide in Organic Synthesis* (translated by C. R. Adams), Springer-Verlag, New York, 1970.
13. J. Falbe, *New Syntheses with Carbon Monoxide*, Springer-Verlag, Berlin, 1980.
14. I. Wender and P. Pino, *Organic Syntheses via Metal Carbonyls*, Wiley–Interscience, 1968 (Vol 1) and 1977 (Vol. 2).
15. D. E. Webster, *Adv. Organmet. Chem.*, **15**, 147 (1977).
16. A. E. Shilov and A. A. Shteinman, *Coord. Chem. Rev.*, **24**, 97 (1977).
17. G. W. Parshall, *Catalysis*, **1**, 335 (1977).
18. A. E. Shilov, *Pure Appl. Chem.*, **50**, 725 (1978).
19. J. K. Hoyano and W. A. G. Graham, *J. Am. Chem. Soc.*, **104**, 3723 (1982).
20. A. H. Janowicz and R. G. Bergman, *J. Am. Chem. Soc.*, **104**, 352 (1982).
21. F. D. Mango, *Coord. Chem. Rev.*, **15**, 109 (1975).
22. J. Halpern, in *Organic Syntheses via Metal Carbonyls* (Eds. I. Wender and P. Pino), Vol. 2, Wiley, New York, 1977, p. 705.
23. F. D. Mango, *Top. Curr. Chem.*, **45**, 39 (1974).
24. R. B. Woodward and R. Hoffmann, *Angew. Chem., Int. Ed. Engl.*, **8**, 781 (1969).
25. L. A. Paquette, *Acc. Chem. Res.*, **4**, 280 (1971).
26. L. Cassar, P. E. Eaton, and J. Halpern, *J. Am. Chem. Soc.*, **92**, 3515 (1970).
27. V. I. Labunskaya, A. D. Shebaldova, and M. L. Khidekel', *Russ. Chem. Rev.*, **43**, 1 (1974).
28. L. Cassar and J. Halpern, *Chem. Commun.*, 1082 (1970).
29. W. G. Dauben and A. J. Kielbania, *J. Am. Chem. Soc.*, **93**, 7345 (1971).
30. D. M. Roundhill, D. N. Lawson, and G. Wilkinson, *J. Chem. Soc A*, 845 (1968).
31. S. Sarel, R. Ben-Shoshan, and B. Kisan, *Isr. J. Chem.*, **10**, 787 (1922).
32. R. M. Tuggle and D. L. Weaver, *J. Am. Chem. Soc.*, **92**, 5523 (1970).
33. K. W. Barnett, D. L. Beach, D. L. Garin, and L. A. Kaempfe, *J. Am. Chem. Soc.*, **96**, 7127 (1974).
34. A. J. Birch and H. Fitton, *Aust. J. Chem.*, **22**, 971 (1969).
35. H. Alper, in *Organic Syntheses via Metal Carbonyls* (Eds. I. Wender and P. Pino), Vol. 2, Wiley, New York, 1977, p. 545.
36. B. F. Hallam and P. L. Pauson, *J. Chem. Soc.*, 642 (1958).
37. A. J. Birch, P. E. Cross, J. Lewis, D. A. White, and S. B. Wild, *J. Chem. Soc A*, 332 (1968).
38. H. Alper and J. T. Edward, *J. Organomet. Chem.*, **14**, 411 (1968).
39. R. K. Kochhar and R. Petit, *J. Organomet. Chem.*, **6**, 272 (1966).
40. J. D. Holmes and R. Petit, *J. Am. Chem. Soc.*, **85**, 2531 (1963).
41. L. Watts, J. D. Fitzpatrick, and R. Petit, *J. Am. Chem. Soc.*, **87**, 3253 (1965).
42. P. Heimbach, *Angew. Chem., Int. Ed. Engl.*, **12**, 975 (1973).
43. W. Reppe, N. Kutepow, and A. Magin, *Angew. Chem., Int. Ed. Engl.*, **8**, 727 (1969).
44. G. Henrici-Olive and S. Olive, *Angew. Chem., Int. Ed. Engl.*, **10**, 105 (1971).
45. H. Muller, D. Wittenberg, H. Seibt, and E. Scharf, *Angew. Chem., Int. Ed. Engl.*, **4**, 327 (1969).
46. R. Baker, *Chem. Rev.*, **73**, 487 (1973).
47. H. W. B. Reed, *J. Chem. soc.*, 1931 (1954).
48. C. V. Pittman, L. R. Smith, and R. M. Hanes, *J. Am. Chem. Soc.*, **97**, 1742 (1975).
49. D. Wittenberg and E. Scharf, *Belg. Pat.*, 641 663, 1962–63.
50. G. Wilke, *Angew. Chem., Int. Ed. Engl.*, **2**, 105 (1963).
51. B. Weinstein and A. H. Fenselau, *J. Org. Chem.*, **32**, 2278 (1967), and references cited therein.
52. R. E. Benson and R. V. Lindsey, *J. Am. Chem. Soc.*, **81**, 4247 (1959).
53. F. W. Hoover and R. V. Lindsey, *J. Org. Chem.*, **34**, 3051 (1969).

54. M. Rosenblum, *Acc. Chem. Res.*, **7**, 122 (1974).
55. A. M. Rosan, M. Rosenblum, and J. Tancrede, *J. Am. Chem. Soc.*, **95**, 3062 (1973).
56. A. J. Pearson, *Tetrahedon Lett.*, 3617 (1975).
57. T. H. Whitesides, R. W. Archart, and R. W. Slaven, *J. Am. Chem. Soc.*, **95**, 5792 (1973).
58. A. J. Birch and I. D. Jenkins, in *Transition Metal Organometallics in Organic Synthesis* (Ed. H. Alper), Academic Press, New York, 1976 p. 1.
59. H. J. Dauben and D. Bertelli, *J. Am. Chem. Soc.*, **83**, 497 (1961).
61. A. J. Pearson, *Acc. Chem. Res.*, **13**, 463 (1980).
61. E. O. Greaves, G. R. Knox, P. L. Pauson, S. Toma, G. A. Sim, and D. I. Woodhouse, *J. Chem. Soc., Chem. Commun.*, 257 (1974).
62. A. J. Pearson, *Aust. J. Chem.*, **29**, 1841 (1976).
63. R. H. Grubbs and R. A. Grey, *J. Am. Chem. Soc.*, **95**, 5765 (1973).
64. J. C. Barborak and R. Pettit, *J. Am. Chem. Soc.*, **89**, 3080 (1967).
65. K. H. Dötz and E. O. Fisher, *Chem. Ber.*, **105**, 1356 (1972).
66. J. P. Collman and L. S. Hegedos, *Principles and Applications of Organotransition Metal Chemistry*, University Science Books, Mill Valley, USA, 1980.
67. M. J. Bennett, W. A. G. Graham, R. P. Stewart, and R. M. Tuggle, *Inorg. Chem.*, **12**, 2944 (1973).
68. D. M. Roe and A. G. Massey, *J. Organomet. Chem.*, **23**, 547 (1970).
69. P. Pino and G. Braca, in *Organic Syntheses via Metal Carbonyls* (Eds. I. Wender and P. Pino), Vol. 2, Wiley, New York, 1977, p. 419.
70. W. Reppe and H. Vetter, *Justus Liebigs Ann. Chem.*, **582**, 143 (1953).
71. C. W. Bird and J. Gudec, *Chem. Ind.* (*London*), 570 (1959).
72. D. P. Tate, J. M. Augl, W. M. Ritchey, B. L. Ross, and J. G. Grasselli, *J. Am. Chem. Soc.*, **86**, 3261 (1964).
73. P. M. Maitlis and S. McVey, *J. Organomet. Chem.*, **4**, 254 (1965).
74. P. M. Maitlis and S. McVey, *J. Organomet. Chem.*, **19**, 169 (1969).
75. P. M. Maitlis, J. M. Kang, and S. McVey, *Can. J. Chem.*, **46**, 3189 (1968).
76. W. reppe and A. Magin, *Ger. Pat.*, 1 071 077, 1965.
77. P. Pino, G. Braca, G. Sbrana, and A. Cuccura, *Chem. Ind.* (*London*), 1732 (1968).
78. B. F. Crowe, *Chem. Ind.* (*London*), 1000 (1960).
79. B. F. Crowe, *Chem. Ind.* (*London*), 1506 (1960).
80. W. Reppe, *Justus Liebigs Ann. Chem.*, **582**, 1 (1953).
81. W. Reppe, O. Schlichting, K. Khager, and T. Toepel, *Justus Liebigs Ann. Chem.*, **560**, 1 (1948).
82. K. P. C. Vollhardt, *Acc. Chem. Res.*, **10**, 1 (1977).
83. R. L. Funk and K. P. C. Vollhardt, *J. Chem. Soc., Chem. Commun.*, 833 (1976).
84. Y. Wataksaki and H. Yamazaki, *Synthesis*, 26 (1976).
85. H. Bonnemann and R. Brinkman, *Synthesis*, 600 (1975).
86. R. F. Heck and D. S. Breslow, *J. Am. Chem. Soc.*, **85**, 2779 (1963).
87. J. Y. Merour, J. L. Roustan, C. Charrier, J. Collin, and J. Benain, *J. Organomet. Chem.*, **51**, C24 (1973).
88. M. P. Cooke and R. M. Parlman, *J. Am. Chem. Soc.*, **97**, 6863 (1975).
89. J. K. Crandall and W. J. Michaely, *J. Organomet. Chem.*, **51**, 375 (1973).
90. E. Yoshisato and S. Tsutsumi, *J. Org. Chem.*, **33**, 869 (1978).
91. I. Rhee, N. Mizuta, M. Ryang, and S. Tsutsumi, *Bull. Chem. Soc. Jap.*, **41**, 1417 (1968).
92. E. J. Corey and L. S. Hegedus, *J. Am. Chem. Soc.*, **91**, 1233 (1969).
93. L. Cassar and M. Foa, *J. Organomet. Chem.*, **51**, 381 (1973).
94. N. L. Bauld, *Tetrahedron Lett*, 1841 (1963).
95. E. J. Corey and E. Hamanaka, *J. Am. Chem. Soc.*, **89**, 2758 (1967).
96. I. D. Webb and G. T. Borcherdt, *J. Am. Chem. Soc.*, **73**, 2654 (1951).
97. E. J. Corey, M. F. Sommelhack, and L. S. Hegedus, *J. Am. Chem. Soc.*, **90**, 2416 (1968).
98. E. J. Corey, M. F. Semmelhack, and L. S. Hegedus, *J. Am. Chem. Soc.*, **90**, 2417 (1968).
99. E. J. Corey and M. F. Semmelhack, *Tetrahedron Lett*, 6237 (1966).
100. R. F. Heck, *Acc. Chem. Res.*, **2**, 10 (1969).
101. G. P. Chiusoli, *Acc. Chem. Res.*, **6**, 422 (1973).
102. F. Guerrieri and G. P. Chiusoli, *J. Organomet. Chem.*, **15**, 209 (1968).
103. G. P. Chiusoli and L. Cassar, *Angew. Chem., Int. Ed. Engl.*, **6**, 124 (1967).

104. G. P. Chiusoli and L. Cassar, in *Organic Syntheses via Metal Carbonyls* (Eds. I. Wender and P. Pino), Vol. 2, Wiley, New York, 1977, p. 297.
105. J. Tsuji, in *Organic Syntheses via Metal Carbonyls* (Eds. I. Wender and P. Pino), Vol. 2, Wiley, New York, 1977, p. 595.
106. J. Blum, E. Oppenheimer, and E. D. Bergmann, *J. Am. Chem. Soc.*, **89**, 2338 (1967).
107. K. Ohno and J. Tsuji, *J. Am. Chem. Soc.*, **90**, 99 (1968).
108. P. Hong and N. Hagihara, *Abstr. Annu. Meet. Jap. Chem. Soc., 1968*, 1971; cited in ref. 105.
109. J. F. Harrod and A. J. Chalk, in *Organic Syntheses via Metal Carbonyls* (Eds. I. Wender and P. Pino), Vol. 2, Wiley, New York, 1977, p. 673.
110. B. J. Aylett and J. M. Campbell, *J. Chem. Soc. A*, 1910 (1969).
111. A. J. Chalk, *Chem. Commun.*, 847 (1970).
112. E. Muller and A. Segnitz, *Synthesis*, 147 (1970).
113. E. Muller and A. Segnitz, *Justus Liebigs Ann. Chem.*, **9**, 1583 (1973).
114. K. Kaneda, H. Azuma, M. Wayaku, and S. Teranishi, *Chem. Lett.*, 215 (1974).
115. P. Hong, K. Sonogashira, and H. Hagihara, *Nippon Kagaku Zasshi*, **89**, 74 (1968).
116. P. Hong, K. Sonogashira, and H. Hagihara, *Bull. Chem. Soc. Jap.*, **39**, 1821 (1966).
117. M. V. Klyuev and M. L. Khidekel', *Russ. Chem. Rev.*, **49**, 14 (1980).
118. Y. Watanabe, T. Mitsudo, M. Yamashita, S. C. Shim, and Y. Takegami, *Chem. Lett.*, 1265 (1974).
119. Y. Watanabe, M. Yamashita, T. Mitsudo, M. Tanaka, and Y. Takegami, *Tetrahedron Lett.*, 1879 (1974).
120. Y. Watanabe, S. C. Shim, T. Mitsudo, M. Yamashita, and Y. Takegami, *Chem. Lett.*, 699 (1975).
121. Y. Watanabe, S. C. Shim, T. Mitsudo, M. Yamashita, and Y. Takegami, *Chem. Lett.*, 995 (1975).
122. Y. Watanabe, S. C. shim, T. Mitsudo, M. Yamashita, and Y. Takegami, *Bull. Chem. Soc. Jap.*, **49**, 1378 (1976).
123. G. P. Boldrini, M. Panunzio, and A. Umani-Ronchi, *Synthesis*, 733 (1974).
124. L. Marko and J. Bakos, *J. Organomet. Chem.*, **81**, 411 (1974).
125. T. Mitsudo, Y. Watanabe, M. Yamashita, and Y. Takegami, *Chem. Lett.*, 1385 (1974).
126. J. Tsuji and T. Nogi, *Tetrahedron Lett.*, 1801 (1966).
127. J. Tsuji and K. Ohono, *Synthesis*, 157 (1969).
128. R. H. Prince and K. A. Raspin, *J. Chem. Soc. A*, 612 (1969).
129. F. Jardine, *Prog. Inorg. Chem.*, **27**, 1 (1981).
130. G. N. Schrauzer and R. H. Wihdgassen, *Nature (London)*, **214**, 492 (1967).
131. H. Wakamutsu, J. Uda, and N. Yamakami, *Chem. Commun.*, 1540 (1971).
132. G. F. Cainelli, M. Panunzio, and A. Ulmani-Ronchi, *J. Chem. Soc., Perkin Trans. 1*, 1273 (1975).
133. F. Piacenti and M. Bianchi, in *Organic Syntheses via Metal Carbonyls* (Eds. I. Wender and P. Pino), Vol. 2, Wiley, New York, 1977, p. 1.
134. J. L. Eisenmann, *J. Org. Chem.*, **27**, 2706 (1962).
135. R. F. Heck, in *Organic Syntheses via Metal Carbonyls* (Eds. I. Wender and P. Pino), Vol. 1, Wiley, New York, 1968, p. 373.
136. Y. Takegami, Y. Watanabe, T. Mitsudo, and H. Masada, *Bull. Chem. Soc. Jap.*, **42**, 202 (1969).
137. W. P. Giering, M. Rosenblum, and J. Tancrede, *J. Am. Chem. Soc.*, **94**, 7170 (1972).
138. S. K. Battacharya and S. K. Palit, *J. Appl. Chem.*, **12**, 174 (1962).
139. S. K. Battacharya and S. K. Palit, *Brennst.-Chem.*, **43**, 169 (1962).
140. W. A. McRae and J. L. Eisenmann, *U.S. Pat.*, 3 024 275, 1962.
141. J. L. Eisenmann, R. L. Yamortino, and J. F. Howard, *J. Org. Chem.*, **26**, 2102 (1961).
142. R. F. Heck, *J. Am. Chem. Soc.*, **85**, 1460 (1963).
143. L. Roos, R. W. Goetz, and M. Orchin, *J. Org. Chem.*, **30**, 3023 (1965).
144. A. Rosenthal and G. Kan, *Tetrahedron Lett.*, 477 (1967).
145. C. Yokohawa, Y. Watanabe, and Y. Takegami, *Bull. Chem. Soc. Jap.*, **37**, 677 (1964).
146. C. W. Smith, G. N. Schrauzer, R. J. Windgassen, and K. F. Koetitz, *U.S. Pat.*, 3 43 819, 1969.
147. Y. Takegami, Y. Watanabe, H. Masada, and I. Kanaya, *Bull. Chem. Soc. Jap.*, **40**, 1456 (1967).
148. W. Reppe, H. Kröper, H. J. Pistor, and O. Weissbarth, *Justus Liebigs Ann. Chem.*, **582**, 87 (1953).

149. R. H. Prince and K. A. Raspin, *Chem. Commun.*, 156 (1966).
150. J. Blum and Z. Lipshes, *J. Org. Chem.*, **34**, 3076 (1969).
151. J. Blum, D. Milstein, and Y. Sasson, *J. Org. Chem.*, **35**, 3233 (1970).
152. I. Wender, S. Friedman, W. A. Steiner, and R. B. Anderson, *Chem. Ind.* (*London*), 1958 (1964).
153. S. Friedman, S. R. Harris, and I. Wender, *Ind. Eng. Chem., Prod. Res. Dev.*, **9**, 347 (1970).
154. Y. Watanabe, M. Yamashita, T. Mitsudo, M. Igami, K. Tomi, and Y. Takegami, *Tetrahedron Lett.*, 1063 (1975).
155. B. M. Trost and F. Chen, *Tetrahedron Lett.*, 2603 (1971).
156. G. P. Chiusoli, S. Merzoni, and G. Mondelli, *Tetrahedron Lett.*, 2777 (1964).
157. S. K. Bhattacharyya and S. K. Palit, *J. Appl. Chem.*, **12**, 174 (1962).
158. W. Reppe and H. Friederich, *U.S. Pat.*, 2 730 546, 1956.
159. Y. Mori and J. Tsuji, *Bull. Chem. Soc. Jap.*, **42**, 777 (1969).
160. J. D. C. Wilson, II, *U.S. Pat.*, 2 555 950, 1951.
161. F. Piacenti, C. Cioni, and P. Pino, *Chem. Ind.* (*London*), 1240 (1960).
162. J. A. Davis, *Adv. Inorg. Chem. Radiochem.*, **24**, 115 (1981).
163. J. Drabowicz, T. Numata, and S. Oae, *Org. Prep. Proced. Int.*, **9**, 63 (1977).
164. H. Alper and E. C. H. Keung, *Tetrahedron Lett.*, 53 (1970).
165. H. Alper and J. T. Edwards, *Can. J. Chem.*, **46**, 3112 (1968).
166. J. Daub, V. Trautz, and U. Erhardt, *Tetrahedron Lett.*, 4435 (1972).
167. H. Alper, *J. Org. Chem.*, **40**, 2694 (1975).
168. H. Alper and A. S. K. Chan., *J. Am. Chem. Soc.*, **95**, 4905 (1973).
169. H. Alper and W. G. Roct, *J. Chem. Soc., Chem. Commun.*, 956 (1974).
170. S. Otsuka, T. Yoshida, and A. Kanamura, *Inorg. Chem.*, **7**, 1833 (1968).
171. R. B. King, *Inorg. Chem.*, **2**, 326 (1963).
172. B. M. Trost and S. D. Ziman, *J. Org. Chem.*, **38**, 932 (1973).
173. A. Rosenthal and I. Wender, in *Organic Syntheses via Metal Carbonyls* (Eds. I. Wender and P. Pino), Vol. I, Wiley, New York, 1968, p. 405.
174. A. Rosenthal and J. Gervay, *Chem. Ind.* (*London*), 1623 (1963).
175. S. Murahashi and S. Horiie, *Bull. Chem. Soc. Jap.*, **33**, 78 (1960).
176. M. Ryang and S. Tsutsumi, *Synthesis*, 55 (1971).
177. J. P. Collman, *Acc. Chem. Res.*, **8**, 342 (1975).
178. S. K. Myeong, Y. Sawa, M. Ryang, and S. Tsutsumi, *Bull. Chem. Soc. Jap.*, **38**, 330 (1965).
179. Y. Sawa, M. Ryang, and S. Tsutsumi, *J. Org. Chem.*, **35**, 4183 (1970).
180. M. P. Cooke, *J. Am. Chem. Soc.*, **92**, 6080 (1970).
181. M. Ryang, *Organomet. Chem. Rev., Sect. A*, **5**, 67 (1970).
182. S. Fukuoka, M. Ryang, and S. Tsutsumi, *J. Org. Chem.*, **36**, 2721 (1971).
183. M. Yamashita, K. Mizushina, Y. Watanabe, T. Mitsudo, and Y. Takegami, *J. Chem. Soc., Chem. Commun.*, 670 (1976).
184. M. Yamashita, Y. Watanabe, T. Mitsudo, Y. Takegami, *Tetrahedron Lett.*, 1585 (1976).
185. Y. Hirota, M. Ryang, S. Tsutsumi, *Tetrahedron Lett.*, 1531 (1971).
186. E. O. Fischer, *Adv. Organomet. Chem.*, **14**, 1 (1976).

CHAPTER 7

Olefin and alcohol carbonylation

GORDON K. ANDERSON

Department of Chemistry, University of Missouri–St. Louis, St. Louis, Missouri 63121, USA

JULIAN A. DAVIES

Department of Chemistry, College of Arts and Sciences, University of Toledo, Toledo, Ohio 43606, USA

I. INTRODUCTION	336
II. OLEFIN CARBONYLATION	336
III. MECHANISTIC DISCUSSION OF OLEFIN CARBONYLATION	337
A. Cobalt Systems	337
B. Nickel Systems	341
C. Palladium Systems	342
IV. OLEFIN CARBONYLATION REACTIONS	343
A. Monoenes as Substrates	343
1. Substrate effects	343
2. Effects of active hydrogen compounds	344
3. Effects of metal complex	345
B. Dienes as Substrates	345
C. Cumulenes as Substrates	346
D. Unsaturated Amines as Substrates	346
E. Unsaturated Amides as Substrates	347
F. Unsaturated Alcohols as Substrates	347
G. Unsaturated Organic Halides as Substrates	348
V. CONCLUDING REMARKS ON OLEFIN CARBONYLATION	348
VI. ALCOHOL CARBONYLATION	348
A. Cobalt-catalysed Methanol Carbonylation	349
B. Rhodium-catalysed Methanol Carbonylation	350
C. Iridium-catalysed Methanol Carbonylation	352
D. Other Catalytic Systems	354
VII. RECENT ADVANCES IN ALCOHOL CARBONYLATION	355
VIII. CONCLUDING REMARKS ON ALCOHOL CARBONYLATION	356
IX. REFERENCES	356

I. INTRODUCTION

Carbonylation is a very general term, usually employed in a non-specific manner to denote a reaction in which carbon monoxide is introduced into an organic molecule. This may be achieved by a conceptually simple 'direct' process, such as the conversion of an ether into a carboxylic acid ester via the 'insertion' of carbon monoxide into the ethereal carbon—oxygen bond. Such reactions are typically catalysed by transition metal carbonyls, such as $[Co_2(CO)_8]$ or $[Ni(CO)_4]$.

In this chapter, the carbonylations of olefins and alcohols are discussed. In the case of unsaturated substrates, carbon monoxide reacts in the presence of active hydrogen-containing nucleophiles to yield derivatives of carboxylic acids; when the hydrogen-containing nucleophile is H_2, the product is an aldehyde and the reaction is hydroformylation, which is discussed separately in Chapter 8. The application of this type of chemistry to the synthesis of functionalized molecules from olefinic substrates is the subject of the first part of this chapter.

II. OLEFIN CARBONYLATION

The general reaction, without regard to mechanism, may be described as in equation 1. Typical examples of the active hydrogen-containing compounds employed (HX), and the corresponding carboxylic acid derivatives produced, are shown in Table 1. Inspection of Table 1 reveals a number of problems which may be encountered during olefin carbonylation. For example, when $HX = H_2O$ the product is RCOOH. The carboxylic acid itself may also function as an active hydrogen compound, however, $HX = RCOOH$, leading to the formation of an anhydride, $(RCO)_2O$. Accordingly, mixtures of products may be expected in certain cases, even when isomerization of the olefinic substrate (see below) is not encountered.

$$CH_2{=}CH_2 + CO + HX \xrightarrow{[M]} EtCOX \quad (1)$$

With unsymmetrical olefins, it is apparent that functionality may be introduced at two possible sites, leading to isomeric products (equation 2). In this respect, olefin carbonylation is similar to the thermodynamically less favourable hydroformylation reaction.

$$RCH{=}CH_2 + CO + HX \xrightarrow{[M]} RCH_2CH_2COX + RCH(Me)COX \quad (2)$$

When the active hydrogen-containing nucleophile and the olefinic moiety are part of the same molecule, then carbonylation can lead to the formation of useful cyclic

TABLE 1. Active hydrogen-containing compounds and carboxylic acid derivatives[a]

HX	RCOX	Reference for examples
H_2O	RCOOH	1,2
R'OH	RCOOR'	1,2
R'SH	RCOSR'	1,2,3
R'_2NH	$RCONR'_2$	1,2,4
RCOOH	$(RCO)_2O$	1,2,3
HCl	RCOCl	1,2,5

[a] See equation 1.

compounds (equation 3). Typical examples include X= —O—, —NH—, and —NHCO—.

$$CH_2=CHCH_2XH + CO \xrightarrow{[M]} \underset{O}{\bigcirc}\!\!\!\!\!\diagdown_X \quad (3)$$

Although the carbonylation of olefins in the presence of water (hydrocarboxylation) and acid catalysts has been known for many years (the Koch reaction[6-8]; ca. 1931), the development of homogeneous catalytic systems employing transition metal catalysts came several years later (ca. 1940) and derives largely from the work of Reppe and associates at BASF[9,10]. For this reason, such carbonylation reactions are often referred to as Reppe reactions, particularly in the older literature. The following sections describe such reactions classified according to the type of olefinic substrate (monoene, diene, etc.) employed. Previous reviews[1,2,5,11-20] have made the compilation of comprehensive listings of reactions unnecessary and accordingly this section is intended to describe the practical features of each reaction type and to provide examples of applications. Prior to discussing the transformations possible with the various substrate types, a section outlining the basic mechanistic features of the more commonly encountered catalyst systems $\{[Co_2(CO)_8], [Ni(CO)_4]$ and $[PdCl_2(PR_3)_2]\}$ is included.

A very large body of work exists concerning the carbonylation of η^3-allyl complexes of the transition metals, either independently synthesized or generated in situ. This class of reaction differs mechanistically from the chemistry under consideration in this section and accordingly the carbonylation of η^3-allyl species is generally excluded. The chemistry of η^3-allylmetal complexes is discussed in Chapter 3 and carbonylation reactions have been exhaustively reviewed in the past[21].

III. MECHANISTIC DISCUSSION OF OLEFIN CARBONYLATION

In contrast to olefin hydroformylation, the carbonylation of olefins in the presence of active hydrogen compounds has not been the subject of extensive mechanistic studies. Early ideas on reaction mechanisms invoked initial ketene[3] or cyclopropanone[3,22] formation, via reaction of the olefinic substrate with CO, followed by hydrolysis or alcoholysis to the corresponding acid or ester. Since then, kinetic and spectroscopic data have allowed mechanisms to be proposed which are more in keeping with present ideas on the fundamental processes occurring in homogeneously catalysed reactions. The three major classes of catalyst (cobalt, nickel, and palladium) are considered in this section; for the cobalt systems, at least, the situation is such that entirely new pathways are still being proposed in the current literature. There is no doubt that this area could profit from definitive mechanistic study.

A. Cobalt Systems

In a similar fashion to cobalt-catalysed olefin hydroformylation, the use of cobalt salts or cobalt metal in carbonylation is believed to lead to in situ formation of cobalt carbonyls. Since reactions are frequently performed in the presence of a Lewis base accelerator (e.g. pyridine), the possible carbonyls span the range from $[Co_2(CO)_8]$ to $[CoL_6][Co(CO)_4]_2$[23]. Natta et al.[24] proposed, by analogy with hydroformylation, that the primary species in the catalytic cycle is $[CoH(CO)_4]$, formed via cleavage of $[Co_2(CO)_8]$. Several routes may be envisaged for formation of $[CoH(CO)_4]$, including hydrogenation by adventitious hydrogen in the CO feed or by hydrogen formed via the water gas shift reaction. Direct cleavage by alcoholysis (equation 4) has also been proposed[23].

$$[Co_2(CO)_8] \underset{+CO}{\overset{-CO}{\rightleftarrows}} [Co_2(CO)_7] \overset{ROH}{\longrightarrow} [CoH(CO)_4] + [Co(OR)(CO)_3] \quad (4)$$

The similarities between hydroformylation and olefin carbonylation have led to considerable mechanistic speculation concerning the formation of carboxylic acids from $[CoH(CO)_4]$. In essence, all such proposals involve addition of the Co—H bond across the olefin (i.e. olefin insertion or hydride transfer processes). In fact, another possibility exists, involving initial formation of a carboalkoxy–cobalt complex, Co—COOR, and addition of the Co—C bond across the olefin. Mechanisms based on the hydride route were developed in the 1950s and refined by Heck and Breslow[25] in the mid-1960s, whereas evidence for the carboalkoxy route has only recently been presented[26].

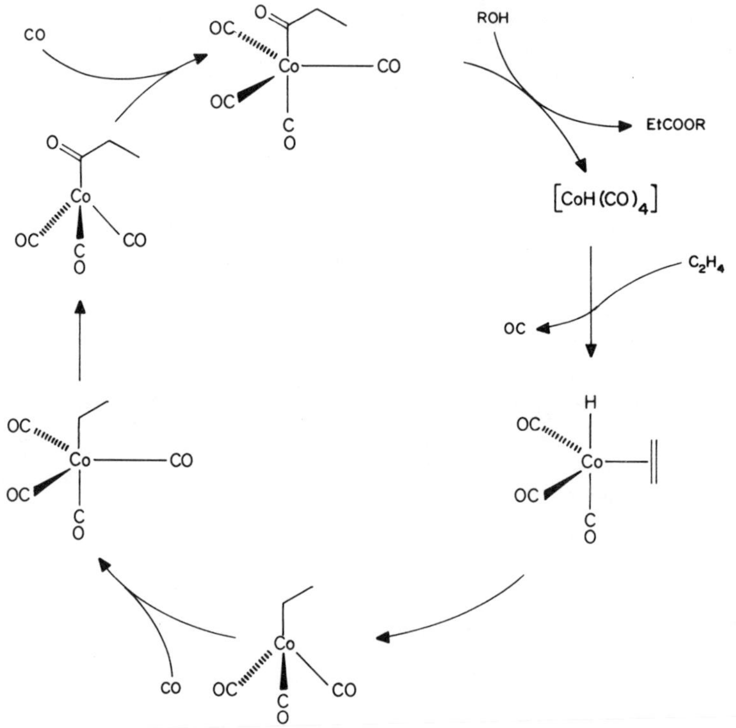

FIGURE 1. Hydride route for cobalt-catalysed olefin carbonylation.

The essential features of the hydride route are shown in Figure 1. The model has been refined by Heck and Breslow's studies[25] of the reactivity of acylcobalt complexes, since early proposals[24,27] involved protonation of the olefin by $[CoH(CO)_4]$ to yield a carbocation which is subsequently attacked by a coordinated carbonyl to yield an acylcobalt species (equation 5). The pK_a of $[CoH(CO)_4]$ is, in fact, slightly higher than that of hydrochloric acid. Evidence against a carbocation-based mechanism through protonation of the olefin by $[CoH(CO)_4]$[24,27] or other acids[28] comes from experiments performed in deuterated solvents (see Section III.A.1, equation 16, for example), where specific deuteration across the double bond occurs.

$$CH_2{=}CH_2 + [CoH(CO)_4] \longrightarrow CH_3CH_2^+ + [Co(CO)_4]^- \rightleftharpoons [Et\text{---}CO\text{---}Co(CO)_3]$$

$$\downarrow CO \mid ROH \qquad (5)$$

$$EtCOOR + [CoH(CO)_4]$$

Kinetic studies[23] led to a proposal that cleavage of the acylcobalt complex (Figure 1) occurs via reaction with an alkoxycobalt intermediate (see equation 5) according to equation 6. Heck and Breslow,[25] however, detected [CoH(CO)$_4$] formation after

$$[Co(COCH_2CH_2R)(CO)_3] + [Co(OR)(CO)_3] \longrightarrow RCH_2CH_2COOR + \tfrac{1}{2}[Co_4(CO)_{12}] \qquad (6)$$

elimination of the acid derivative from the acylcobalt complex, as shown in Figure 1. Reviewers[1] have suggested that the concentration of acylcobalt species is likely to be small and hence cleavage is the slow step in the cycle. This assumption reconciles the kinetic data[23] with the cycle proposed by Heck and Breslow[25].

The use of pyridine as an accelerator may simply reflect an enhanced ability to cleave the acylcobalt intermediate[1,29] (equations 7 and 8) or may be interpreted in terms of the

$$[Co(COCH_2CH_2R)(CO)_4] + C_5H_5N \longrightarrow RCH_2CH_2CO\overset{+}{N}C_5H_5 \;\; [Co(CO)_4]^-$$

$$\downarrow R'OH \qquad (7)$$

$$RCH_2CH_2COOR' + C_5H_5\overset{+}{N}H \;[Co(CO)_4]^-$$

$$C_5H_5\overset{+}{N}H \;[Co(CO)_4]^- \rightleftharpoons C_5H_5N + [CoH(CO)_4] \qquad (8)$$

formation of pyridine-substituted carbonyls[30] which exhibit higher reactivity than [CoH(CO)$_4$] alone.

For many years, the hydride route seemed established for the cobalt systems, although other routes have been postulated for palladium (see Section II.C). Recent work has now demonstrated that a carboalkoxy route must also be considered. Milstein and Huckaby[26] prepared the complex [Co(COOMe)(CO)$_4$] (equation 9) and its triphenylphosphine

$$MeOC(O)COCl + Na[Co(CO)_4] \longrightarrow [Co(COCOOMe)(CO)_4]$$

$$\downarrow -CO \qquad (9)$$

$$[Co(COOMe)(CO)_3(PPh_3)] \xleftarrow{+PPh_3} [Co(COOMe)(CO)_4]$$

$$\qquad (2) \qquad\qquad\qquad (1)$$

derivative (2). Complex 2 was characterized by X-ray crystallography. The complex 1 adds butadiene to yield a π-allyl complex, 3 (equation 10). Reaction of [CoH(CO)$_4$] with methyl penta-2,4-dienoate also yields 3 (equation 11). Clearly, these results show that the

$$[Co(COOMe)(CO)_4] + CH_2=CHCH=CH_2 \xrightarrow{-CO}$$
$$[Co(\eta^3\text{-}CH_2CHCHCH_2COOMe)(CO)_3] \quad (10)$$
$$(3)$$

$$[CoH(CO)_4] + CH_2=CHCH=CHCOOMe \longrightarrow 3 \quad (11)$$

carboxylate functionality may arise via addition of Co—COOR across the olefinic substrate. The catalytic process (equation 12) was studied and a catalytic cycle proposed (Figure 2).

$$CH_2=CHCH=CH_2 + MeOH \xrightarrow[4000 \text{ psi CO}]{[Co_2(CO)_8]/C_5H_5N} MeCH=CHCH_2COOMe \quad (12)$$

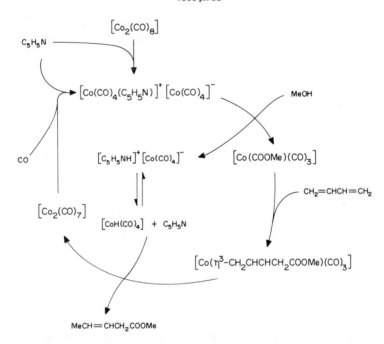

FIGURE 2. Carboalkoxy route for cobalt-catalysed olefin carbonylation.

Additionally, the π-allyl complex 3 was shown to be cleaved by $[CoH(CO)_4]$ and $[C_5H_5NH]^+[Co(CO)_4]^-$ to yield methyl pent-3-enoate together with minor amounts of the pent-4-enoate isomer, in agreement with the cleavage steps proposed in Figure 2. The unproved step in the catalytic cycle is the methanolysis of the ion pair $[Co(CO)_4(C_5H_5N)]^+[Co(CO)_4]^-$ to yield the carbomethoxycobalt complex. The formation of the ion pair has previously been discussed by Wender et al.,[31] in terms of the pyridine effect on $[Co_2(CO)_8]$-catalysed reactions, but no studies of its alcoholysis have yet been described. Competition experiments (equation 13 vs. 14) indicate that the Co—COOR addition route will predominate in the catalytic cycle.

$$[Co(COOBu)(CO)_3] \xrightarrow[-40\,°C \text{ (instantaneous)}]{CH_2=CHCH=CH_2} [Co(\eta^3\text{-}CH_2CHCHCH_2COOBu)(CO)_3] \quad (13)$$

$$[C_5H_5NH]^+[Co(CO)_4]^- \xrightarrow[25\,°C\,(>2\,h)]{CH_2=CHCH=CH_2} [Co(\pi^3\text{-}CH_2CHCHMe)(CO)_3] \quad (14)$$

In view of the importance of these reaction types (e.g. methyl pent-3-enoate is a potential precursor for dimethyl adipate and thus nylon 66)[26,32], it is necessary for further mechanistic studies to be carried out in this area to elucidate the relative contributions of the possible hydride and carboalkoxy routes.

B. Nickel Systems

The carbonylation of allylic halides in the presence of $[Ni(CO)_4]$ has been studied in detail by Heck[33], who obtained infrared spectroscopic data that indicated that acylnickel dicarbonyl halide complexes are formed. Figure 3 shows the essential features of the

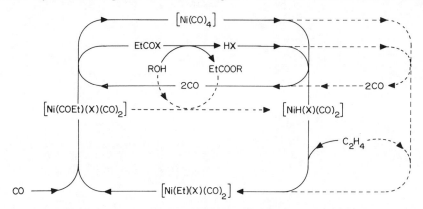

FIGURE 3. Catalytic cycle for $[Ni(CO)_4]$-catalysed olefin carbonylation.

proposed mechanism. Previous workers[22] had proposed a direct alcoholysis of $[Ni(CO)_4]$, followed by olefin insertion and reductive elimination steps (equation 15).

$$[Ni(CO)_4] \xrightarrow{ROH} [NiH(COOR)(CO)_3] \xrightarrow{C_2H_4} [Ni(Et)(COOR)(CO)_3]$$
$$\downarrow CO \quad (15)$$
$$[Ni(CO)_4] + EtCOOR$$

This early attempt at generating a workable mechanism did not account satisfactorily for the accelerating effects of hydrogen halides added to the catalytic system. A mechanism involving carbocation formation[34], similar to that once proposed for the cobalt system, has also been described, but is unsatisfactory because of the evidence from deuteration experiments, mentioned previously.

The initial step in the carbonylation sequence, i.e. reaction of $[Ni(CO)_4]$ with HX to yield $[NiH(X)(CO)_2]$, is likely to proceed via a dissociative mechanism (S_N1 type) involving an unsaturated 16-electron species such as $[Ni(CO)_3]$. This would explain the rate acceleration by u.v. irradiation[35] and the retardation by use of high CO pressures[14]. An alternative pathway can also be envisaged by considering the reaction of the olefin with the hydrogen halide to form an alkyl halide, capable of oxidative addition to nickel(0) to generate $[NiR(X)(CO)_2]$. This alternative pathway is also illustrated in Figure 3, and bears similarities to the mechanisms of the cobalt- and rhodium-catalysed carbonylation of methanol (see below).

C. Palladium Systems

Two major classes of mechanism have been proposed for the palladium-catalysed carbonylation of olefins. A hydride route[36] (Figure 4) may involve four- or five-coordinate hydridopalladium olefin intermediates and bears similarities to the hydride routes proposed for cobalt and nickel catalysts. An alternative route involves a carboalkoxy intermediate[37] (Figure 5) similar to that recently proposed for cobalt catalysts. Both

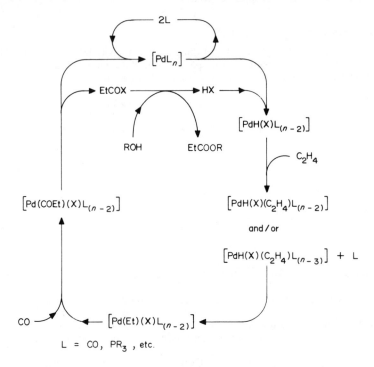

FIGURE 4. Hydride route for palladium-catalysed olefin carbonylation.

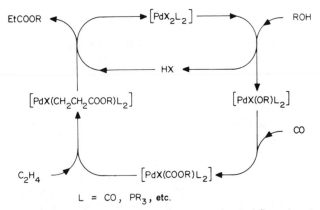

FIGURE 5. Carboalkoxy route for palladium-catalysed olefin carbonylation.

7. Olefin and alcohol carbonylation

routes have been substantially accepted[1,38] as a working basis for discussing the palladium-catalysed carbonylation of olefins. There appears to be no definitive evidence to exclude either pathway totally and, indeed, it is entirely possible that both routes may be operative under given conditions. Most qualitative observations are readily rationalized by either pathway. For example, the enhanced rate of carbonylation on addition of acids[39] may reflect increased metal hydride formation (Figure 4) or facile cleavage of a metal—carbon bond (Figure 5).

The attraction of the hydride route, with its many similarities to the proposed cobalt and nickel cycles, can also be attributed to a lack of chemical evidence for certain of the intermediates in the carboalkoxy cycle. Although several types of carboalkoxy complex have been known for some years, the parent carbohydroxy species, MCOOH, remained elusive until fairly recently. Interest in identifying such compounds arises not only because of their proposed involvement in olefin carbonylation, but also because of their intermediacy in catalytic carbon monoxide oxidation by water[40]. Carbohydroxy complexes of platinum[41], iridium[42], rhenium[43], and iron[44] have now been identified, with the platinum complex[41,45] $trans$-[PtCl(COOH)(PEt$_3$)$_2$] being directly analogous to the palladium species proposed in Figure 5.

IV. OLEFIN CARBONYLATION REACTIONS

A. Monoenes as Substrates

1. Substrate effects

Carbonylation of monoenes in the presence of water generally results in the formation of mixtures of isomeric carboxylic acids. Substrate reactivity patterns are shown in Table 2 for the [Co$_2$(CO)$_8$]-catalysed hydrocarboxylation reaction. For terminal monoenes, a dependence of reactivity upon molecular weight is observed. Thus, entries 1–3 and 5 (Table 2) illustrate a reactivity sequence $C_2 \approx C_3 > C_4 > C_5$. This seems to be a general trend, with high molecular weight α-olefins almost invariably exhibiting lower reactivity. Comparison of data for isomeric olefins, e.g. entries 3 $vs.$ 4 and 5 $vs.$ 6, shows that internal monoenes are less susceptible to hydrocarboxylation than the corresponding terminal olefins. It has also been noted[1,51] that cyclic monoenes generally show an even lower

TABLE 2. Substrate reactivity in the [Co$_2$(CO)$_8$]-catalysed hydrocarboxylation of monoenes

No.	Monoene	CO pressure (atm.)	Temperature (°C)	Products	Yield (%)	Ref.
1	C$_2$H$_4$	200	285	EtCOOH	85	46
2	MeCH=CH$_2$	123	130	Pr^{n}COOH	64	47
				Me$_2$CHCOOH	20	
3	EtCH=CH$_2$	150	180	Bu^{n}COOH	52	48
				Bu^{s}COOH	23	
4	MeCH=CHMe	250	210	Bu^{n}COOH	24	49
				Bu^{s}COOH	13	
5	Pr^{n}CH=CH$_2$	180	145	Pen^{n}COOH	52	50
				Pr^{n}CH(Me)COOH	17	
				Et$_2$CHCOOH	5	
6	EtCH=CHMe	180	145	Pen^{n}COOH	49	50
				Pr^{n}CH(Me)COOH	17	
				Et$_2$CHCOOH	6	
7	Cyclohexene	197	165	c-HexCOOH	89	51

reactivity than linear olefins of similar molecular weight, but entry 7 in Table 2 shows that high reactivity can be observed in some cases. Similar reactivity patterns to those described above emerge using [Ni(CO)$_4$]/HI-promoted hydrocarboxylation systems[52].

Strained cycloalkenes exhibit a greater reactivity than linear alkenes. Thus, bicyclo[2.2.1]heptene undergoes hydrocarboxylation in the presence of [Ni(CO)$_4$] under very mild conditions (1 atm CO, 50 °C)[53]. In the presence of D$_2$O, the *exo*-product is formed (equation 16), corresponding to an *exo-Z* addition[1].

$$\text{(16)}$$

2. Effects of active hydrogen compounds

The effect of employing homologous active hydrogen compounds in monoene carbonylation is illustrated by a comparison of the [Co$_2$(CO)$_8$]-catalysed formation of esters from cyclohexene[54]. In experiments performed at 165 °C, the percentage of ester produced was found to decrease on changing the alcohol from MeOH (81%; 246 atm CO, 6 h) to Bu^{n}OH (45%; 230 atm CO, 7 h) or Pr^{i}OH (27%; 226 atm CO, 6 h). Increasing molecular weight and/or branching thus appears to decrease selectivity. In comparison, hydrocarboxylation of cyclohexene at 165 °C yielded 89% of cyclohexanecarboxylic acid (197 atm CO, 3 h)[51].

Similar trends have been described for carbonylation of hex-1-ene in the presence of various alcohols at 190 °C[55], with selectivity towards ester formation decreasing in the order MeOH > EtOH > Bu^{i}OH. The rate of consumption of olefin, however, does not appear to be clearly related to the structure of the alcohol[55], with reactivity and selectivity following different orders.

An interesting effect was observed[56] using optically active 2-methylbutanol, (S)-(−)-EtCH(Me)CH$_2$OH, as a solvent for the carbonylation of α-methylstyrene using the achiral catalyst precursor [PdCl$_2$(PPh$_3$)$_2$]. An optical yield of 0.2% of the corresponding S-esters was obtained (390 atm Co, 100 °C, 46 h; 70% yield of α- and β-esters). More impressive results in asymmetric carbonylation have been obtained using palladium complexes of chiral phosphines and these are discussed in Section III.A.3.

In addition to the synthesis of acids and esters, employing water and alcohols, respectively, as the active hydrogen-containing compounds, a variety of syntheses have been described which utilize other active hydrogen-containing components (see Table 1).

The synthesis of amides, by olefin carbonylation in the presence of ammonia, has been known for some years[4,57] and has been reviewed extensively[1]. The general reaction is shown in equation 17. Primary and secondary amines may be employed in place of

$$RCH=CH_2 + NH_3 + CO \longrightarrow RCH_2CH_2CONH_2 + RCH(Me)CONH_2 \quad (17)$$

ammonia to yield *N*-substituted amides. Nickel[58], cobalt[59], iron[60], ruthenium[61], and rhodium[62] catalysts have been employed for such syntheses, with nickel and cobalt systems being utilized most extensively.

Thioesters have been prepared by olefin carbonylation in the presence of thiols employing [Ni(CO)$_4$][3] (equation 18). Similarly[3], the synthesis of acid anhydrides is possible, employing carboxylic acids as the active hydrogen-containing components (equation 19). Acyl chlorides may also be synthesized employing dry HCl (equation 20). Palladium[63], rhodium[64], and ruthenium[64] catalysts have been described for the acyl halide synthesis. Aspects of the syntheses described by equations 18–20 have been reviewed previously[5].

$$CH_2 = CH_2 + RSH + CO \longrightarrow EtCOSR \qquad (18)$$
$$CH_2 = CH_2 + RCOOH + CO \longrightarrow (RCO)_2O \qquad (19)$$
$$CH_2 = CH_2 + HCl + CO \longrightarrow EtCOCl \qquad (20)$$

3. Effects of metal complex

Hydrocarboxylation of monoenes may be effected stoichiometrically using $[Ni(CO)_4][65]$ or $[Co_2(CO)_8]/CO[1]$. The $[Ni(CO)_4]$-promoted carbonylation reaction is accelerated by dissociation of the photolabile carbonyl groups by irradiation[66] (see mechanistic discussion, Section II.B). The addition of hydrogen halides[53] to $[Ni(CO)_4]$ also enhances reactivity, in keeping with the mechanistic picture developed by Heck[33]. Catalytic carbonylation of monoenes is, however, a relatively simple procedure and is generally superior to stoichiometric processes, even on a small scale.

Catalytic reactions employing a variety of catalyst precursors, including iron[1], ruthenium[61], osmium[67], cobalt[68], rhodium[69], iridium[70], nickel[65], palladium[71], and platinum[67,72] complexes, have been investigated. Of these, $[Co_2(CO)_8]$, or cobalt carbonyls generated *in situ*, have been most widely used, although palladium complexes of chiral phosphines have been of most interest in asymmetric syntheses, and $[Ni(CO)_4]$ in situations where high selectivity is demanded and ketone formation must be minimized.

The $[Co_2(CO)_8]$ catalyst precursor is activated by pyridine[73], much as in hydroformylation, and is typically operated at 150–180 °C and 200–250 atm $CO[1]$. Use of pyridine complexes of cobalt carbonyls[1] as precursors has now led to the synthesis of heterogenized analogues based on poly(4-vinylpyridine) and related functionalized polymers[74]. These supported systems may be of considerable utility in assisting catalyst handling, separation, etc. The $[Ni(CO)_4]$ catalyst is generally much less active than $[Co_2(CO)_8]$. For example, a comparison of $[Ni(CO)_4]$ and $[Co_2(CO)_8]$ as precursors for the hydrocarboxylation of cyclohexene showed[70] the cobalt system to be 10^7 times as active as the nickel catalyst. This value appears to reflect the lower limit of the activity of the nickel system and the use of halide accelerators generally gives more acceptable rates. The $[Ni(CO)_4]$ catalyst usually leads to lower levels of by-products, in the form of aldehydes and ketones, than $[Co_2(CO)_8]$.

Palladium chloride catalysts, modified with chelating chiral phosphines such as diop, are effective in causing asymmetric induction in acid and ester synthesis[75–77]. In one study, optical yields of 3–20% of various esters were obtained from α-methylstyrene. Palladium and platinum chloro complexes are activated by addition of $SnCl_2 \cdot 2H_2O$ in olefin carbonylation[78], hydroformylation[79], and hydrogenation[80]. Recent results on precatalytic ligand rearrangement reactions occurring in some such systems[81–83] suggest that the widely accepted role[84] of these tin(II) promoters is in some doubt.

B. Dienes as Substrates

A variety of products can be envisaged via carbonylation of a non-conjugated diene. Carbonylation to yield an unsaturated monocarboxylic acid and/or a saturated dicarboxylic acid, or their derivatives (e.g. esters)[85], is the expected reaction. In addition, saturated monocarboxylic acids have been obtained[86], presumably by hydrogenation due to hydrogen gas present in the CO feed or formed via the water gas shift reaction. Cyclization products are also possible[87], envisaged as forming via an intramolecular olefin insertion into a metal acyl intermediate.

Conjugated dienes generally yield even more complex product mixtures. For example, buta-1,3-diene[32,88] may yield dicarboxylic acids where functionality is introduced at $C_1/C_4, C_2/C_4$, etc. Additionally, the reaction conditions promote the Diels–Alder [4 + 2] cycloaddition reaction[32], yielding 4-vinylcyclohexene, which itself may undergo carbonylation[89].

The product distribution in diene carbonylation is very sensitive to changes in reaction conditions, e.g. solvent[32] and temperature[39], allowing some control in synthetic procedures. $[Co_2(CO)_8]$[32] and palladium chloro complexes[87] are the preferred catalysts, with iron[90], rhodium[61,91], iridium[61,91], and nickel[90] complexes generally giving poorer results.

C. Cumulenes as Substrates

Although allene is the only cumulene to have been studied in any depth, the carbonylation of cumulenes is here considered separately from dienes, since the chemistry involved may well prove to be applicable to higher cumulated hydrocarbons.

The thermal oligomerization of allenes[92] is complex, yielding dimers, trimers, and tetramers, while catalysis by nickel(0) leads to a slight selectivity for trimer formation[93]. Under carbonylation conditions, catalysis by iron and ruthenium complexes yields products resulting from the carbonylation of dimers and trimers[61], along with resinous materials (presumably from higher oligomers) and small amounts of the simple carbonylation products, methacrylate esters. At lower temperatures, where the oligomerization is retarded, selectivity towards the monomeric esters is possible with platinum, nickel, and ruthenium catalysts[61,67,94].

The direct carbonylation of higher cumulenes (i.e. without addition of an active hydrogen-containing compound) is, of course, well known (e.g. equation 21)[95]. In the

$$Ph_2C=C=C=CPh_2 \xrightarrow[\text{[CO]}]{[Co_2(CO)_8]} \text{(indanone with Ph and } CH=CPh_2 \text{ substituents)} \qquad (21)$$

presence of water, the same substrate yields only the corresponding product where hydrogenation of the conjugated double bonds has occurred.

D. Unsaturated Amines as substrates

Lactams may be prepared by carbonylation of suitable unsaturated amines using $[Co_2(CO)_8]$[14,19]. Reaction conditions typically involve temperatures of ca. 300 °C and operating presssures of ca. 700 atm CO, although milder conditions have been successful in some cases. Allylamine[96–98] yields 4-butanelactam (γ-butyrolactam) according to equation 22. Where olefin isomerization is possible[97], both five- and six-membered

$$CH_2=CHCH_2NH_2 \xrightarrow{\text{[CO]}} \text{(γ-butyrolactam)} + \text{small amounts of substituted pyridines} \qquad (22)$$

lactams may be produced (equation 23). By consideration of equation 23 it is not

$$CH_3CH=CHCH_2NH_2 \xrightarrow{\text{[CO]}} \text{(lactone)} + \text{(six-membered lactam)} \qquad (23)$$

surprising that carbonylation of amine derivatives of cyclic monoenes[97] yields both fused and bridged bicyclic ring systems (equation 24). Using catalysts other than $[Co_2(CO)_8]$, such as rhodium or iron carbonyl complexes[98], the carbonylation of allylamine yields pyrollidone, according to equation 22, and also more substantial amounts of the substituted pyridines (equation 25).

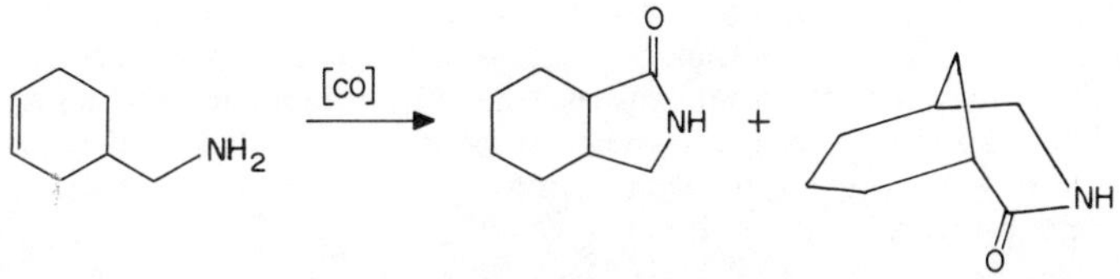

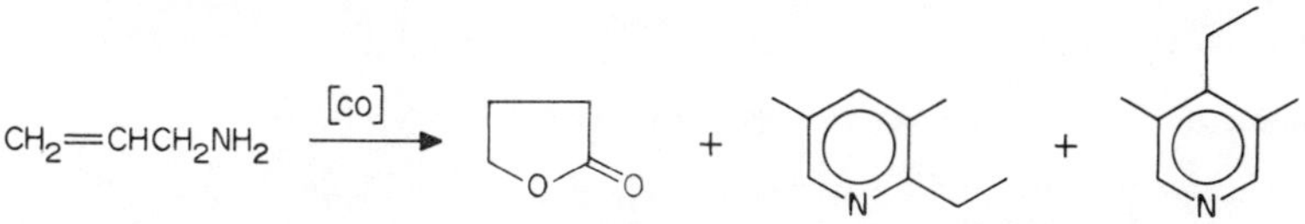

A study of the carbonylation of allylamine by [Fe(CO)$_5$] allowed the identification of **4** (equation 26) by ^{1}H n.m.r. spectroscopy and elemental analysis as an intermediate, although the overall mechanism remains unclear[98].

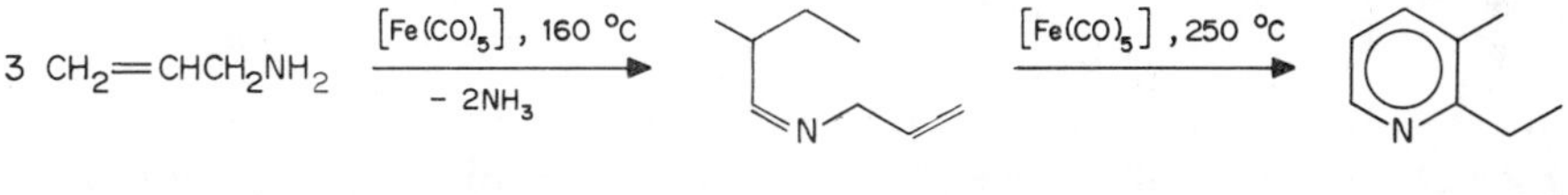

E. Unsaturated Amides as Substrates

The synthesis of cyclic imides via the cobalt carbonyl promoted carbonylation of unsaturated amides has been reported[99] (equation 27). Unsaturated amides capable of olefin isomerization may yield both possible imides, differing in ring size by one carbon atom (equation 28). Generally only five- and six-membered rings may be synthesized by this route, with five-membered ring products predominating unless steric effects of substituents dictate otherwise.

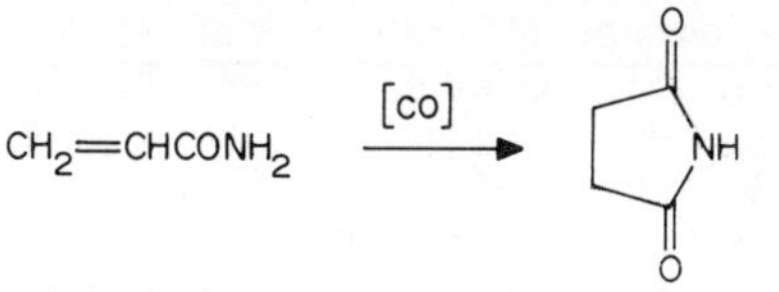

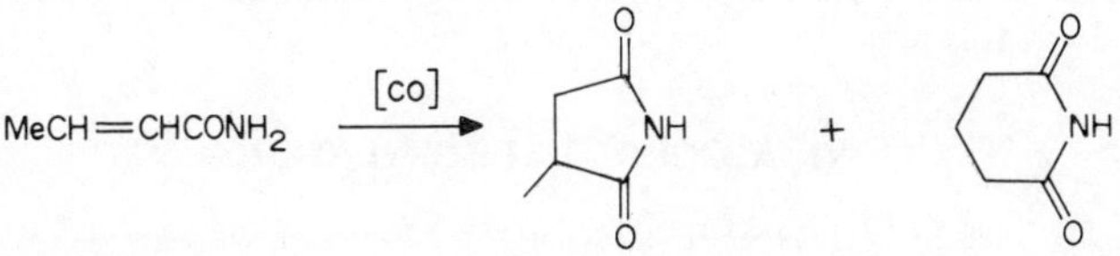

F. Unsaturated Alcohols as Substrates

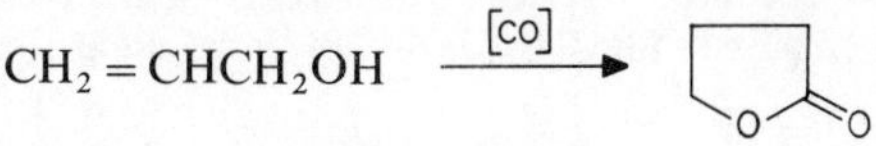

Carbonylation of unsaturated alcohols yields lactones, according to equation 29. [Co$_2$(CO)$_8$] is an effective catalyst for this cyclization[100]. As with unsaturated amines and amides, the formation of five- and six-membered rings is most favoured, with five-membered ring products predominating unless steric effects of substituents dictate otherwise[100,101]. Results of typical syntheses have been tabulated previously[1]. Where isomerization of an allylic alcohol to the corresponding saturated aldehyde is possible, yields tend to be low. Blocking the isomerization process, e.g. equation 30, allows good yields of cyclized products to be obtained[100].

$$CH_2 = CHC(Me)_2CH_2OH \xrightarrow{[CO]}$$

3,4,4-trimethyl-γ-butyrolactone + 5,5-dimethyl-δ-valerolactone   (30)
51%   14%

G. Unsaturated Organic Halides as Substrates

A vast amount of work on the carbonylation of allylic halides has been reported and recently reviewed[21]. With the exception of this body of work, which is outside the scope of this chapter, a few results are noteworthy. The carbonylation of vinyl chloride, in the presence of alcohols, is catalysed by $[PdCl_2(PPh_3)_2]^{39}$. The products are α- and β-chloro esters (equation 31), formed in ca. 85% total yield. The reaction conditions are severe,

$$CH_2 = CHCl \xrightarrow{EtOH/[CO]} MeCHClCOOEt + CH_2ClCH_2COOEt \quad (31)$$

however, requiring ca. 700 atm CO and temperatures of ca. 100 °C. Acrylonitrile functions similarly, with the cyano group acting as a pseudo-halide. Mixtures of α- and β-cyano esters are formed by carbonylation in the presence of alcohols, employing $[Co_2(CO)_8]$ as the catalyst[30].

V. CONCLUDING REMARKS ON OLEFIN CARBONYLATION

A variety of simple and cyclic carboxylic acid derivatives may be produced by carbonylation of an olefinic substrate in the presence of an active hydrogen-containing compound. The major drawback to stoichiometric and catalytic syntheses based on this chemistry is the tendency for more than one product to be formed, even in relatively simple cases. Work on olefin hydroformylation has shown that this problem can be largely overcome, provided mechanistic data are available to assist catalyst design and tailoring. In the present case, the available mechanistic information is not sufficiently detailed to permit the application of tailoring methods in a truly meaningful manner. Future applications in regiospecific carbonylation and in asymmetric synthesis will require detailed study in this area.

VI. ALCOHOL CARBONYLATION

The general reaction of this type involves the conversion of an alchohol to a carboxylic acid, as shown in equation 32.

$$ROH + CO \xrightarrow{[M]} RCOOH \quad (32)$$

The carbonylation of alcohols, and of methanol in particular, perhaps provides the most significant industrial development in the area of homogeneous catalysis.

Acetic acid, the product of methanol carbonylation, has been synthesized by a number of catalytic processes involving metal ions or complexes[102]. The hydrolysis of acetylene to acetaldehyde, catalysed by mercury(II) ion, was used as a basis for acetic acid production until the late 1950s, when a free radical oxidation of short-chain alkanes using manganese or cobalt salts as catalysts was developed. In 1965 BASF outlined a synthesis of acetic acid from methanol catalysed by cobalt ions in the presence of iodide, and an iodide-promoted rhodium or iridium catalysis for the above reaction was discovered by Monsanto in 1968.

The significance of these processes may be appreciated with it is realised that in 1977 the world-wide production of acetic acid was approximately 2.5 million tons. The cobalt- and rhodium-catalysed reactions have been developed industrially, and each of these will be discussed in the following sections.

A. Cobalt-catalysed Methanol Carbonylation

The cobalt-catalysed reaction, involving $[Co_2(CO)_8]$ or a range of cobalt(II) salts in the presence of an iodide promoter, was first reported in the mid-1960s[103,104]. The hydridocobalt species, $[HCo(CO)_4]$, in the presence of hydrogen iodide catalyses methanol carbonylation with a high selectivity for acetic acid, but very high pressures of carbon monoxide are necessary to achieve a reasonable reaction rate. When the cobalt is introduced as CoI_2, it is converted into $[CoH(CO)_4]$ under the high pressure reaction conditions[105] (equations 33 and 34). The hydridocobalt complex reacts with methyl iodide,

$$2CoI_2 + 2H_2O + 10CO \longrightarrow [Co_2(CO)_8] + 4HI + 2CO_2 \qquad (33)$$

$$[Co_2(CO)_8] + H_2O + CO \longrightarrow 2CoH(CO)_4 + CO_2 \qquad (34)$$

formed by interaction of methanol with HI, to form a methylcobalt species which subsequently undergoes carbonyl insertion to yield an acetylcobalt moiety. Hydrolysis of $[Co(COMe)(CO)_4]$ liberates acetic acid and regenerates the catalytic species[105,106]. The catalytic cycle is depicted in Figure 6.

The major side-product in the carbonylation of methanol is methyl acetate, formed by

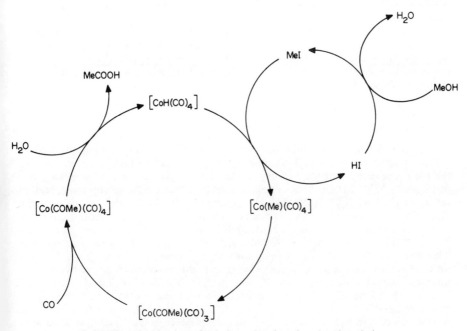

FIGURE 6. Mechanism of cobalt-catalysed methanol carbonylation.

methanolysis, instead of hydrolysis, of $[(Co(COMe)(CO)_4]$. In the reaction catalysed by $[CoH(CO)_4]$ a selectivity for acetic acid production of 90% is exhibited[102], but where traces of hydrogen are present in the carbon monoxide feedstock methane, acetaldehyde and ethanol are formed as by-products.

More recently, cobalt(II) salts with tertiary phosphines and an iodide source have been studied[107], where the catalyst stability may be maintained at lower CO pressures, but large amounts of methyl acetate are formed in addition to acetic acid.

B. Rhodium-catalysed methanol carbonylation

The discovery, by Paulik and Roth of Monsanto[108], that in the presence of an iodide promoter certain complexes of rhodium and iridium catalyse the carbonylation of methanol under relatively low pressure conditions represents a significant advance on the BASF cobalt-catalysed process. The advantages of the system based on rhodium include operating at lower carbon monoxide pressures (30–40 atm, compared with 500–700 atm for the cobalt system) and lower temperatures (ca. 180 °C, compared with ca. 230 °C), as well as much lower catalyst concentrations being necessary and > 99% selectivity toward acetic acid production[102]. Despite the greater expense involved in using rhodium, the much higher activity of this catalyst system makes its use worthwhile and the Monsanto process is expected to produce 1 million tons of acetic acid per annum by the mid-1980s[84].

The rhodium and iridium catalyst systems have been extensively studied and the carbonylation mechanisms are well understood. In fact, the iridium-catalysed reaction is more complex than that of rhodium, and will be considered separately in the next section.

Studies have been performed with a variety of rhodium compounds and iodide promoters and, with the exception of alkali metal iodides[102], most combinations of rhodium compounds and iodide species give rise to identical catalytic activity, indicating that one, common catalytic species is involved. Only for rhodium compounds of bidentate ligands is a lower initial rate of reaction observed[109].

The single catalytic species is the $[RhI_2(CO)_2]^-$ anion, and the initial dependences on the form of the rhodium (where such dependences are observed) simply reflect the rate at which the catalytic species is formed. Thus, the rhodium may be added in the form of the rhodium(III) halide or a tertiary phosphine complex, but the same active species is formed (equations 35 and 36).

$$RhCl_3 + 3CO + H_2O + 2I^- \rightarrow [RhI_2(CO)_2]^- + CO_2 + 2H^+ + 3Cl^- \qquad (35)$$

$$[RhCl(CO)(PPh_3)_2] + CO + 2MeI + 2I^- \rightarrow [RhI_2(CO)_2]^- + 2Ph_3\overset{+}{P}MeI^- + Cl^- (36)$$

A number of studies have indicated that the carbonylation of methanol, catalysed by soluble rhodium complexes, exhibits first-order kinetics with respect to the catalyst and the iodide promoter and zero-order kinetics with respect to carbon monoxide and methanol[102,109,110]. It has been proposed, therefore, that oxidative addition of methyl iodide to $[RhI_2(CO)_2]^-$ is the rate-determining step in the catalytic cycle. Since the $[RhI_2(CO)_2]^-$ ion is stable under ambient conditions, its reaction with methyl iodide has been investigated[111]. When $[RhI_2(CO)_2]^-$, which exhibits carbonyl stretching bands at 2064 and 1989 cm^{-1}, is treated with MeI, new infrared absorptions appear at 2062 and 1711 cm^{-1}, the latter being assigned to an acetyl moiety. Thus, the first observable intermediate is the product of oxidative addition of MeI followed by carbonyl insertion, and is formulated as $[RhI_3(COMe)(CO)]^-$. The X-ray crystal structure of this material, as its $[Me_3(C_6H_5)N]^+$ salt, indicates that it is dimeric, with weak rhodium—iodide bridges[112].

Addition of carbon monoxide to this complex produces a new species, with infrared bands at 2141, 2084, and 1708 cm^{-1}, which slowly decomposes to regenerate the

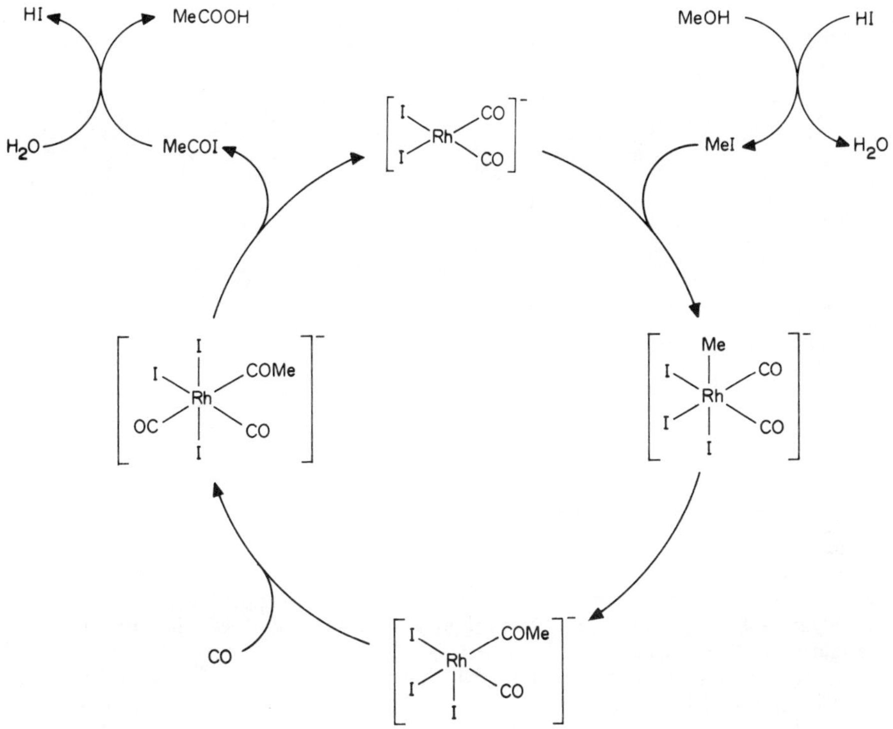

FIGURE 7. Mechanism of rhodium-catalysed methanol carbonylation.

$[RhI_2(CO)_2]^-$ anion[111]. This is assigned the structure $[RhI_3(COMe)(CO)_2]^-$, which further reacts by reductive elimination of acetyl iodide. Thus, the catalytic cycle is as shown in Figure 7. Under the reaction conditions the released acetyl iodide would react rapidly with water or methanol to produce acetic acid or methyl acetate, respectively. Infrared studies at high temperature (100 °C) and high pressure (6 atm) indicate that $[RhI_2(CO)_2]^-$ is the major rhodium-containing species under the operating conditions[111], providing further evidence that oxidative addition of methyl iodide is the rate-determining step.

Chloride and bromide sources do not act as promoters for the carbonylation reaction[102], since it is generally observed that the rate of oxidative addition of organic halides to d^8 metal complexes decreases in the order $I > Br > Cl$[113]. Other promoters have been found and the 'pseudo-halide' pentachlorobenzenethiol, although considerably less effective than methyl iodide, is suggested to be a suitable promoter in an industrial setting since it will be less corrosive[114].

Other adaptations of the rhodium-catalysed process include the use of supported catalysts. These are of two types: rhodium complexes deposited on inert supports and polymer-bound catalysts. The former approach involves vapour-phase reactions, where the kinetics appear to be the same as in the homogeneous case[115-117]. In these studies little could be deduced about the form of the metal, but in a more detailed study[118] of $RhCl_3$ deposited on a number of oxide supports it was concluded that the acid-base properties of the support are important in determining the form in which the rhodium exists on the surface. Thus Al_2O_3 or $Al_2O_3-SiO_2$ favour the formation of anionic rhodium

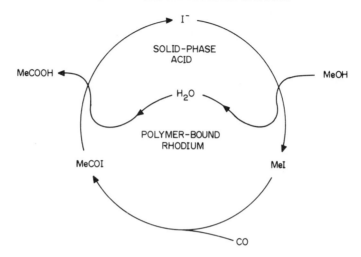

FIGURE 8. Catalytic cycle for methanol carbonylation using $[RhI_2(CO)_2]^-$ heterogenized by ionic attachment.

species, $[-O-RhCl(CO)_2]^-$, silica itself forms only neutral oxygen adducts, whereas TiO_2 forms both complex types; to act as a suitable support, a material must have an adequate basicity to complex to the rhodium, but must also be sufficiently acidic to produce the anionic form[118].

Polymer-bound rhodium complexes have been obtained by exchanging $[RhCl(CO)(PPh_3)_2]$ with a phosphinated polystyrene[119]. An infrared absorption at $1978\,cm^{-1}$ was suggested to be due to the polymer-supported analogue of $[RhCl(CO)(PPh_3)_2]$, but later studies[102] showed that, under the carbonylation conditions, the tertiary phosphine moieties become quaternized and the rhodium on the surface is in an anionic form and, hence, activity similar to the homogenous case should be expected.

A novel approach involving immobilizing the metal on a polymer functionalized with suitable anionic groups has recently been followed. Polystyrene functionalized with pentachlorobenzenethiol moieties has been used to prepare heterogenized rhodium catalysts[120], but they are less active than conventional homogeneous systems and rhodium is gradually lost from the polymer during reaction. A related heterogenized catalyst design involves ionic attachment of the $[RhI_2(CO)_2]^-$ anion to a resin, coupled with a solid-phase acid catalyst for conversion of inactive NaI to the active promoter MeI[121]. Leaching of rhodium could be minimized by a suitable choice of solvent and rhodium concentration. A simplified catalytic cycle is shown in Figure 8. This supported catalyst exhibits activity identical with that of its homogeneous counterpart, but has the inherent advantage of ready separation from the reaction medium.

C. Iridium-catalysed Methanol Carbonylation

Iridium complexes, in the presence of an iodide promoter, are also excellent catalysts for the carbonylation of methanol[108], but the catalytic system is more complex and, hence, it is more difficult to balance the various factors to obtain a satisfactory yield and selectivity. It has been suggested that this system is more sensitive to the nature of the iridium compound used[122], and that maximum catalytic activity is attained under conditions which differ from those employed in the rhodium system. It has been pointed out that the

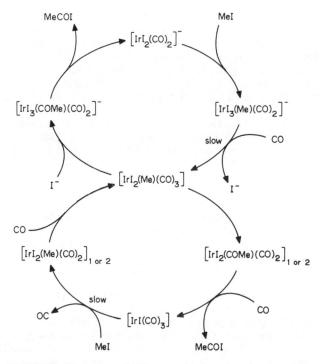

FIGURE 9. Mechanism of iridium-catalysed methanol carbonylation.

iridium-catalysed reaction is zero order with respect to methyl iodide and CO, but first order in methanol[123], in contrast to the rhodium system, and that NaI retards catalysis by iridium[124], whereas it has no effect on the rhodium-catalysed reaction.

A more detailed study of the iridium-catalysed carbonylation of methanol has recently appeared[125]. Two catalytic cycles emerge, as shown in Figure 9, involving neutral or anionic complexes, as well as a competitive water gas shift reaction under many conditions. The neutral and anionic catalyst precursors are related according to equation 37. The catalytic cycle which predominates, therefore, is dependent on the iodide concentration.

$$[IrI(CO)_3] \underset{-I^+}{\overset{+I^-}{\rightleftharpoons}} [IrI_2(CO)_2]^- + CO \qquad (37)$$

The anionic iridium cycle, which is favoured by high iodide concentrations, appears identical with the rhodium-catalysed reaction. The oxidative addition of methyl iodide is rapid in this case, however, owing to the readier formation of iridium(III) complexes, whereas conversion to the acetyl species is slow. Thus, $[IrI_3(Me)(CO)_2]^-$ is the major species present under the reaction conditions and, in the absence of a significant CO pressure, it does not undergo carbonyl insertion. Since iodide is released in the formation of $[IrI_2(Me)(CO)_3]$, this step is inhibited by iodide, and the conditions of high iodide concentration which favour the anionic complex cycle in fact reduce the rate of reaction.

The neutral complex $[IrI(CO)_3]$ is the predominant species under the reaction conditions at low iodide concentrations. In this cycle, oxidative addition of methyl iodide is the rate-determining step, as it is in the rhodium-catalysed reaction, and it has been shown that $[IrI(CO)_3]$ reacts with MeI to produce the dimeric complex

$[IrI_2(Me)(CO)_2]_2{}^{102}$. Under the carbonylation conditions it is not clear whether a monomeric or dimeric complex is formed at this stage in the cycle. A similar situation prevails for the corresponding acetyl complex, which rapidly loses acetyl iodide by reductive elimination to yield $[IrI(CO)_3]$ (Figure 9).

The rate of methanol carbonylation catalysed by iridium is very similar to that catalysed by rhodium under comparable conditions[102,122], but the complex nature of the iridium system suggests that the activity might be difficult to maintain on an industrial scale.

A number of iridium and rhodium complexes containing bidentate tertiary phosphine and arsine ligands have recently been studied as methanol carbonylation catalysts[126,127] although, in view of earlier comments on the use of bidentate ligand complexes[109], it is not clear how successful these will prove to be.

D. Other Catalytic Systems

A number of other metal complexes will catalyse the carbonylation of alcohols, but none is as successful as the systems discussed above. Thus, early studies involved the use of mercury[102,128,129] and nickel[130] compounds, and more recently nickel complexes with tertiary phosphines[131–133] and organometallic promoters[134] have been employed. Most of these are not selective, however, producing significant yields of methyl acetate and dimethyl carbonate, as well as acetic acid, from methanol.

In addition to isolated reports of alcohol carbonylation catalysed by rhenium[135] and ruthenium[136], a number of studies involving copper and silver have been made. Copper(I) salts have been used as catalyst precursors for alcohol carbonylation[137], and the catalytic species under CO pressurization and acidic conditions has been suggested to be the $[Cu(CO)_3]^+$ cation[138]. In $BF_3 \cdot H_2O$, the catalytic species derived from silver(I) oxide promoted the carbonylation of *tert*-butanol to give C_5- and C_9- carboxylic acids under mild conditions[139], and the copper and silver species formed under these conditions were suggested to be the $[Cu(CO)_n]^+$ ($n = 3$ or 4) and $[Ag(CO)_2]^+$ cations[140].

The best documented catalysis by metal complexes, excluding those of the cobalt triad, is that by palladium. However, although the carbonylation may lead to acetic acid it more commonly, in the case of methanol, produces dimethyl carbonate and dimethyl oxalate. Palladium(II) compounds are usually introduced, although $[Pd(CO)(PPh_3)_3]$ has been employed[141] in the carbonylation of methyl, ethyl, or benzyl alcohol to the corresponding dialkyl oxalates. Alkoxycarbonyl complexes of palladium are believed to be formed as reaction intermediates[142,143], and the complexes $[Pd(COOCH_3)(OCOCH_3)(PPh_3)_2]$ and $[Pd(COOCH_3)_2(PPh_3)_2]$ have been isolated[143]. The latter undergoes reductive elimination to yield dimethyl oxalate[144] (equation 38). The carbonylation of methanol,

$$[Pd(COOCH_3)_2(PPh_3)_2] \longrightarrow MeOCOCOOMe + \text{`}[Pd(PPh_3)_2]\text{'} \qquad (38)$$

catalysed by palladium(II) acetate in the presence of excess of tertiary phosphine, yields predominantly dimethyl oxalate and dimethyl carbonate[145]. The nature of the tertiary phosphine is critical, indicating that it is involved in the catalytic cycle, and its influence is mainly electronic: trialkylphosphines inhibit methanol carbonylation almost completely. The mechanism proposed for a stoichiometric reaction is shown in Figure 10, but it is easy to see how this might be made catalytic. Low pressures of carbon monoxide and the presence of tertiary amines favour the formation of dimethyl carbonate[145].

Alcohol carbonylation catalysed by palladium commonly involves the generation of products containing more than one CO or RO unit. An unusual example is the carbonylation of propargyl alcohols, which gives rise to a cyclic product containing two carbon monoxide units[146] (equation 39).

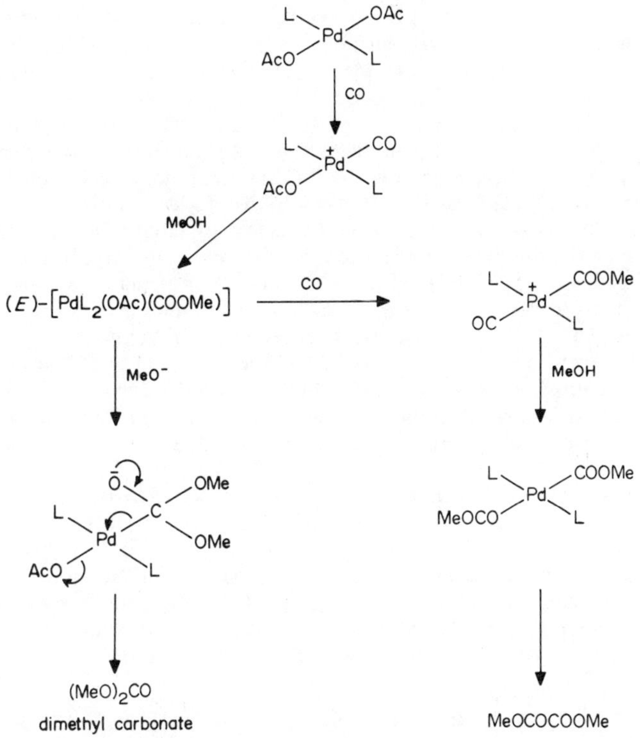

FIGURE 10. Palladium-catalysed formation of dimethyl carbonate and dimethyl oxalate.

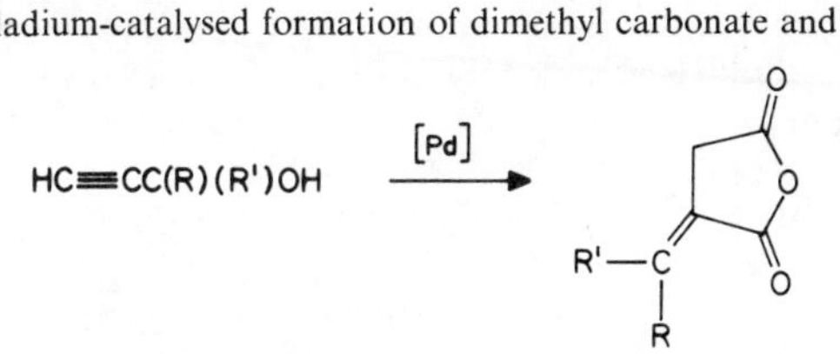

$$HC\equiv CC(R)(R')OH \xrightarrow{[Pd]} \text{(structure)} \tag{39}$$

VII. RECENT ADVANCES IN ALCOHOL CARBONYLATION

One of the major areas of interest in catalysis at present is the development of heterogenized analogues of homogeneous catalysts which, it is hoped, will combine the advantages of a homogeneous system, such as selectivity and catalyst tailoring, with ready product separation. This approach has already been discussed for metals supported on carbon and various oxides in Section VI.B. An area which has recently come to the fore is the trapping of catalytic species in zeolites, and this has also been applied to rhodium and iridium catalysts for the carbonylation of alcohols.

A general finding appears to be that zeolite-supported rhodium catalysts exhibit a higher activity than any other type of supported catalyst. Rhodium-X zeolite has been used to study the carbonylation of various alcohols[147,148]; with methanol a $>90\%$ selectivity for carbonylation is found, whereas for isopropanol the major product is propene. With ethanol intermediate behaviour is observed, carbonylation being favoured at low temperatures. These results were explained in terms of the relative ease of dehydration of the reactants on the surface of a polar catalyst[147]. The selectivity for carbonylation is increased by raising the concentration of the alkyl iodide promoter, and it has been suggested that the carbonylation mechanism is very similar to that of the homogeneous system. Infrared and ESCA studies have indicated that a dicarbonylrhodium species exists on the surface in the presence of carbon monoxide, whereas when both methyl iodide and CO are present an acetylrhodium moiety is detected[149]. It has been suggested that the zeolite lattice may assist in the production of catalytically active rhodium centres[149]. The same catalytic species are again apparently formed irrespective of the form in which the rhodium is introduced; hence, a catalyst derived from Linde 13X zeolite exchanged with $[RhCl(NH_3)_5]Cl_2$ is also active for methanol carbonylation[150].

The vapour-phase carbonylation of methanol over rhodium-Y zeolite produces methyl acetate initially, but acetic acid as the major product during the later stages of the reaction[151,152]. The reaction occurs at 150–200 °C and 1 atm CO, and shows a first-order dependence on methyl iodide concentration, but a zero-order dependence on methanol and carbon monoxide concentrations. The energy of activation was determined to be 56.5 kJ mol^{-1} [152], and the mechanism is thought to be analogous to the homogeneous one.

Infrared studies of RhNaX zeolite wafers in the presence of carbon monoxide reveal bands at 2085 and 2014 cm^{-1}; these are higher than those in $[RhI_2(CO)_2]^-$, but are believed to be due to analogous species within the zeolite lattice[153]. Kinetic studies of zeolite entrapped rhodium and iridium species, with a methyl iodide promoter, indicate[154] that oxidative addition of MeI is the rate-determining step in the rhodium case (as it is in the homogeneous rhodium system), whereas for iridium the addition of methanol and subsequent methyl group migration are the slowest steps.

VIII. CONCLUDING REMARKS ON ALCOHOL CARBONYLATION

The carbonylation of methanol is one of the success stories of homogeneous catalysis from an industrial standpoint. The rhodium-catalysed reaction is well understood and provides the cheapest route to acetic acid currently available[155]. It seems unlikely that a viable, alternative process will be developed in the near future, except perhaps by modification of the present system. Studies of supported rhodium and iridium catalysts are directed toward this end, and it is in this area that further advances appear most likely to be made.

IX. REFERENCES

1. P. Pino, F. Piacenti and M. Bianchi, in *Organic Syntheses via Metal Carbonyls* (Eds. I. Wender and P. Pino), Vol. 2, Wiley, New York, 1977, p. 233.
2. C. W. Bird, *Transition Metal Intermediates in Organic Synthesis*, Logos Press, London, 1967, p. 149.
3. W. Reppe and H. Kröper, *Justus Liebigs Ann. Chem.*, **582**, 38 (1953).
4. J. F. Olin and T. E. Deger, *U.S. Pat.*, 2422631, 1947.
5. Ya. T. Eidus and K. V. Puzitskii, *Russ. Chem. Rev.*, **33**, 438 (1964).
6. J. C. Woodhouse, *U.S. Pat.*, 2003477 (1931).
7. G. B. Carpenter, *U.S. Pat.* 1981801, 1931.
8. H. Koch and W. Hagg, *Angew. Chem.*, **70**, 311 (1958).
9. W. Reppe and H. Kröper, *Ger. Pat.*, 765969, 1953.
10. W. Reppe and H. Kröper, *Ger. Pat.*, 879987, 1953.
11. A. Mullen, in *New Syntheses with Carbon Monoxide* (Ed. J. Falbe), Springer-Verlag, New York, 1980, p. 243.
12. J. K. Stille and D. E. James, in *The Chemistry of Double-Bonded Functional Groups* (Ed. S. Patai), Wiley, New York, 1977, p. 1099.

13. J. Tsuji, *Organic Synthesis with Palladium Compounds*, Springer-Verlag, New York, 1980, p. 81.
14. J. Falbe, *Carbon Monoxide in Organic Synthesis*, Springer-Verlag, New York, 1970, p. 78.
15. C. W. Bird, *Chem. Rev.*, **62**, 283 (1962).
16. G. W. Parshall, *Homogeneous Catalysis*, Wiley, New York, 1980, p. 82.
17. Ya. T. Eidus, K. V. Puzitskii, A. L. Lapidus, and B. K. Nefedov, *Russ. Chem. Rev.*, **40**, 429 (1971).
18. Ya. T. Eidus, A. L. Lapidus, K. V. Puzitskii, and B. K. Nefedov, *Russ. Chem. Rev.*, **42**, 199 (1973).
19. J. Falbe, *Angew. Chem.*, **78**, 532 (1966).
20. H. M. Colquhoun, *Chem. Ind. (London)*, 747 (1982).
21. G. P. Chiusoli and L. Cassar, in *Organic Syntheses via Metal Carbonyls* (Eds. I. Wender and P. Pino), Vol. 2, Wiley, New York, 1977, p. 297.
22. G. Dupont, P. Piganiol, and J. Vialle, *Bull. Soc. Chim. Fr.*, 529 (1948).
23. R. Ercoli, M. Avanzi, and G. Moretti, *Chim. Ind. (Milan)*, **37**, 865 (1955).
24. G. Natta, P. Pino, and E. Mantica, *Gazz. Chim. Ital.*, **80**, 680 (1950).
25. R. F. Heck and D. S. Breslow, *J. Am. Chem. Soc.*, **85**, 2779 (1963).
26. D. Milstein and J. L. Huckaby, *J. Am. Chem. Soc.*, **104**, 6150 (1982).
27. N. S. Imyanitov, in *Hydroformylation* (Ed. N. S. Imyanitov), Khimiya, Leningrad, 1972, p. 13.
28. M. Almasy, L. Szabo, J. Farkas, and T. Bota, *Acad. Repub. Pop. Rom. Stud. Cercet. Chim.*, **8**, 495 (1960); *Chem. Abstr.*, **55**, 19427 (1961).
29. N. S. Imyanitov and D. M. Rudkovskii, *Zh. Org. Khim.*, **2**, 231 (1966).
30. A. Matsuda, *Bull. Chem. Soc. Jap.*, **40**, 135 (1967).
31. I. Wender, H. Sternberg, and M. Orchin, *J. Am. Chem. Soc.*, **74**, 1216 (1952).
32. A. Matsuda, *Bull. Chem. Soc. Jap.*, **46**, 524 (1973).
33. R. F. Heck, *J. Am. Chem. Soc.*, **85**, 2013 (1963).
34. H. Kröper, *Anlagerung von Kohlenmonoxide and Verbindungen mit Aciden Wasserstoff (Carbonylierung)*, Vol. IV, Part 2, Houber-Hewyl, Stuttgart, 1955, p. 385; cited in ref. 1.
35. A. Davison, N. McFartane, and L. Pratt, *J. Chem. Soc.*, 3652 (1962).
36. J. Tsuji, *Acc. Chem. Res.*, **2**, 144 (1969).
37. O. L. Kaliya, O. N. Temkin, N. G. Mekhryakova, and R. M. Flid, *Dokl. Akad. Nauk SSSR*, **199**, 1321 (1971).
38. D. M. Fenton, *J. Org. Chem.*, **38**, 3192 (1973).
39. K. Bittler, N. von Kutepow, D. Neubauer, and H. Reis, *Angew. Chem.*, **80**, 352 (1968).
40. J. Halpern, *Comments Inorg. Chem.*, **1**, 1 (1981).
41. H. C. Clark and W. J. Jacobs, *Inorg. Chem.*, **9**, 1229 (1970).
42. A. J. Deeming and B. L. Shaw, *J. Chem. Soc. A*, 443 (1969).
43. C. P. Casey, M. A. Andrews, and J. E. Rinz, *J. Am. Chem. Soc.*, **101**, 741 (1979).
44. N. Grice, S. C. Kao, and R. Pettit, *J. Am. Chem. Soc.*, **101**, 1627 (1979).
45. M. Catellani and J. Halpern, *Inorg. Chem.*, **19**, 566 (1980).
46. W. Reppe and H. Kröper, *Ger. Pat.*, 863 194, 1953.
47. F. Piacenti, P. P. Negiani, and F. Calderazzo, *Atti Soc. Toscana Sci. Nat., B*, 47 (1962).
48. F. Piacenti and M. Bianchi, unpublished results, cited in ref. 1.
49. N. S. Imyanitov, B. E. Kuvaev, and D. M. Rudkovskii, in *Carbonylation of Unsaturated Hydrocarbons* (Ed. D. M. Rudkovskii), Khimiya, Leningrad, 1968, p. 176.
50. F. Piacenti, M. Bianchi and R. Lazzaroni, *Chim. Ind. (Milan)*, **50**, 318 (1968).
51. R. Ercoli, *Chim. Ind. (Milan)*, **37**, 1029 (1955).
52. D. R. Levering and H. Kröper, *Justus Liebigs Ann. Chem.*, **582**, 38 (1953).
53. C. W. Bird, R. C. Cookson, J. Hudec, and R. O. Williams, *J. Chem. Soc.*, 410 (1963).
54. P. Pino and R. Ercoli, *Chim. Ind. (Milan)*, **36**, 536 (1954).
55. V. Yu. Gangin, M. G. Katsnel'son and D. M. Rudkovskii, in *Carbonylation of Unsaturated Hydrocarbons* (Ed. D. M. Rudkovskii), Khimiya, Leningrad, 1968, p. 178.
56. G. Consiglio and P. Pino, unpublished results, cited in ref. 1.
57. H. J. Nienburg and E. Keunecke, *Ger. Pat.*, 863 799, 1953.
58. D. M. Newitt and S. A. Momen, *J. Chem. Soc.*, 2945 (1949).
59. B. F. Crowe and O. C. Elmer, *U. S. Pat.*, 2 742 502, 1956.
60. A. Striegler and J. Weber, *J. Prakt. Chem.*, **29**, 281 (1965).
61. J. J. Kealy and R. E. Benson, *J. Org. Chem.*, **26**, 3126 (1961).

62. Y. Iwashita and M. Sakuraba, *J. Org. Chem.*, **36**, 3927 (1971).
63. J. F. Knifton, *U.S. Pat.*, 3 880 898, 1975.
64. T. Alderson and V. A. Engelhardt, *U.S. Pat.*, 3 065 242, 1962.
65. W. Reppe, *Justus Liebigs Ann Chem.*, **582**, 1 (1953); *Experientia*, **5**, 93 (1949); *Ger. Pat.*, 65 361, 1939.
66. B. Fell and J. M. J. Tetterboo, *Angew. Chem.*, **77**, 813 (1965).
67. E. L. Jenner and R. V. Lindsey, *U.S. Pat.*, 2 876 254, 1959.
68. G. Natta and P. Pino, *Chim. Ind. (Milan)*, **31**, 109 (1949).
69. T. Alderson and C. L. Aldridge, *U.S. Pat.*, 3 161 672, 1963.
70. N. S. Imyanitov and D. M. Rudkovskii, *Kinet. Katal.*, **8**, 1240 (1967).
71. S. Brewis and P. R. Hughes, *Chem. Commun.*, 157 (1965).
72. L. J. Kehoe and R. A. Shell, *J. Org. Chem.*, **35**, 2846 (1970).
73. A. Matsuda and H. Uchida, *Bull. Chem. Soc. Jap.*, **38**, 710 (1965).
74. A. J. Moffat, *J. Catal.*, **18**, 193 (1970); **19**, 322 (1970).
75. G. Consiglio and P. Pino, *Chimia*, **30**, 193 (1976).
76. G. Consiglio, *Helv. Chim. Acta*, **59**, 124 (1976).
77. C. Botteghi, G. Consiglio, and P. Pino, *Chimia*, **27**, 477 (1973).
78. J. J. Mrowca, *U.S. Pat.*, 3 859 319, 1975.
79. H.C. Clark and J. A. Davies, *J. Organomet. Chem.*, **213**, 503 (1981).
80. B. R. James, *Homogenous Hydrogenation*, Wiley, New York, 1973.
81. G. K. Anderson, H. C. Clark, and J. A. Davies, *Inorg. Chem.*, **22**, 427 (1983).
82. G. K. Anderson, H. C. Clark and J. A. Davies, *Inorg. Chem.*, **22**, 434 (1983).
83. G. K. Anderson, C. Billard, H. C. Clark, J. A. Davies and C. S. Wong, *Inorg. Chem.*, **22**, 439 (1983).
84. C. Masters, *Homogenous Catalysis—A Gentle Art*, Chapman and Hall, London, 1981.
85. J. Tsuji, S. Hosaka, J. Kiji, and T. Susuki, *Bull. Chem. Soc. Jap.*, **39**, 141 (1966).
86. N. S. Imyanitov and D. M. Rudkovskii, *Zh. Prikl. Khim.*, **40**, 2825 (1967).
87. S. Brewis and P. R. Hughes, *Chem. Commun.*, 489 (1965).
88. A. M. Hyson, *U.S. Pat.*, 2 586 341, 1952.
89. W. E. Billups, W. E. Walker, and T. C. Shields, *Chem. Commun.*, 1067 (1971).
90. N. S. Imyanitov and D. M. Rudkovskii, *Zh. Prikl. Khim.*, **39**, 2811 (1966).
91. D. G. Kuper and W. B. Hughes, *U.S. Pat.*, 3 746 747, 1973.
92. B. Weinstein and A. H. Fenslau, *J. Org. Chem.*, **32**, 2278 (1967).
93. R. E. Benson and R. V. Lindsey, *J. Am. Chem. Soc.*, **81**, 4247 (1959).
94. S. Kunichika, Y. Sakakibara, and T. Okamoto, *Bull. Chem. Soc. Jap.*, **40**, 885 (1967).
95. P. Kim and N. Hagihara, *Bull. Chem. Soc. Jap.*, **38**, 2022 (1965).
96. W. F. Gresham, *Bri. Pat.*, 628 659, 1949.
97. J. Falbe and F. Korte, *Chem. Ber.*, **98**, 886 (1965).
98. J. Falbe, H. Weitkamp, and F. Corte, *Tetrahedron Lett.*, 2677 (1965).
99. J. Falbe and F. Corte, *Chem. Ber.*, **95**, 2680 (1962); **98**, 1928 (1965); *Angew. Chem., Int. Ed. Engl.*, **1**, 266 (1962).
100. J. Falbe, H.-J. Schulze-Steinen and F. Korte, *Chem. Ber.*, **98**, 886 (1965).
101. A. Matsuda, *Bull. Chem. Soc. Jap.*, **41**, 1876 (1968).
102. D. Forster, *Adv. Organomet. Chem.*, **17**, 255 (1979).
103. N. von Kutepow, W. Himmele, and H. Hohenschutz, *Chem.-Ing.-Tech.*, **37**, 383 (1965).
104. H. Hohenschutz, N. von Kutepow, and W. Himmele, *Hydrocarbon Process.*, **45**, 141 (1966).
105. T. Mizoroki and M. Nakayama, *Bull. Chem. Soc. Jap.*, **39**, 1477 (1966).
106. J. Falbe, *Carbon Monoxide in Organic Synthesis*, Springer-Verlag, New York, 1970, p. 82.
107. J. D. Holmes, *U.S. Pat.*, 4 133 963, 1979.
108. F. E. Paulik and J. F. Roth, *Chem. Commun.*, 1578 (1968).
109. D. Brodzki, C. Leclere, B. Denise, and G. Pannetier, *Bull. Soc. Chim. Fr.*, 61 (1976).
110. J. Hjortkjaer and V. W. Jensen, *Ind. Eng. Chem., Prod. Res. Dev.*, **15**, 46 (1976).
111. D. Forster, *J. Am. Chem. Soc.*, **98**, 846 (1976).
112. G. W. Adamson, J. J. Daly, and D. Forster, *J. Organomet. Chem.*, **71**, C17 (1974).
113. R. Cramer, *Acc. Chem. Res.*, **1**, 186 (1968).
114. K. M. Webber, B. C. Gates, and W. Drenth, *J. Catal.*, **47**, 269 (1977).
115. R. G. Schultz and P. D. Montgomery, *J. Catal.*, **13**, 105 (1969).

116. K. K. Robinson, A. Hershman, J. H. Craddock, and J. F. Roth, *J. Catal.*, **27**, 389 (1972).
117. A. Krzywicki and G. Pannetier, *Bull. Soc. Chim. Fr.*, **1–2**, 64 (1977).
118. A. Krzywicki and M. Marczewski, *J. Mol. Catal.*, **6**, 431 (1979).
119. M. S. Jarrell and B. C. Gates, *J. Catal.*, **40**, 255 (1975).
120. K. M. Webber, B. C. Gates, and W. Drenth, *J. Mol. Catal.*, **3**, 1 (1977).
121. R. S. Drago, E. D. Nyberg, A. El A'mma, and A. Zombeck, *Inorg. Chem.*, **20**, 641 (1981).
122. D. Brodzki, B. Denise, and G. Pannetier, *J. Mol. Catal.*, **2**, 149 (1977).
123. T. Matsumoto, T. Mizoroki, and A. Ozaki, *J. Catal.*, **51**, 96 (1978).
124. T. Mizoroki, T. Matsumoto, and A. Ozaki, *Bull. Chem. Soc. Jap.*, **52**, 479 (1979).
125. D. Forster, *J. Chem. Soc. Dalton Trans.*, 1639 (1979).
126. C. M. Bartish, *U.S. Pat.*, 4 102 920, 1978; 4 102 921, 1978.
127. C. M. Bartish, *Ger. Pat.*, 2 800 986, 1978.
128. B. K. Nefedov, N. S. Sergeeva, and Ya. T. Eidus, *Izv. Akad. Nauk SSSR, Ser. Khim.*, 2733 (1972).
129. B. K. Nefedov and Ya. T. Eidus, *Kinet. Katal.*, **16**, 443 (1975).
130. W. Reppe, H. Kröper, N. von Kutepow, and H. J. Pistor, *Justus Liebigs Ann. Chem.*, **582**, 72 (1953).
131. N. Rizkalla and A. N. Naglieri, *Ger. Pat.*, 2 749 955 1978.
132. T. Isshiki, Y. Kijima, and Y. Mujauchi, *Ger. Pat.*, 2 842 267, 1979.
133. J. D. Holmes, *U.S. Pat.*, 4 133 963, 1979.
134. A. N. Naglieri and N. Rizkalla, *Ger. Pat.*, 2 749 954, 1978.
135. K. Teranishi, T. Shimizu, and T. Nakamata, *Jap. Pat.*, 80 813, 1976.
136. G. Braca, G. Sbrana, G. Valentini, G. Andrich, and G. Gregorio, *Fundam. Res. Homogeneous Catal.*, **3**, 221 (1979).
137. K. V. Puzitskii, S. D. Pirozhkov, T. N. Myshenkova, K. G. Ryabova, and Ya. T. Eidus, *Izv. Akad. Nauk SSSR, Ser. Khim.*, 443 (1975).
138. Y. Souma and H. Sano, *Bull. Chem. Soc. Jap.*, **46**, 3237 (1973).
139. Y. Matsushima, T. Koyano, and K. Shiozawa, *Jap. Pat.*, 123 613, 1975.
140. Y. Souma, *Osaka Kogyo Gijutsu Shikensho Hokoku*, **356**, 1 (1979); *Chem. Abstr.*, **92**, 221517 (1980).
141. U. Romano and F. Rivetti, *Ger. Pat.*, 2 814 708, 1978.
142. L. N. Zhir-Lebed, N. G. Mekhryakova, V. A. Golodov, and O. N. Temkin, *Zh. Org. Khim.*, **11**, 2297 (1975).
143. F. Rivetti and U. Romano, *J. Organomet. Chem.*, **154**, 323 (1978).
144. F. Rivetti and U. Romano, *Chem. Ind. (Milan)*, **62**, 7 (1980).
145. F. Rivetti and U. Romano, *J. Organomet. Chem.*, **174**, 221 (1979).
146. J. Tsuji and T. Nogi, *Jap. Pat.*, 09 046, 1968.
147. B. Christensen and M. S. Scurrell, *J. Chem. Soc., Faraday Trans. 1*, **73**, 2036 (1977).
148. B. Christensen and M. S. Scurrell, *J. Chem. Soc., Faraday Trans. 1*, **74**, 2313 (1978).
149. S. L. T. Andersson and M. S. Scurrell, *J. Catal.*, **59**, 340 (1979).
150. M. S. Scurrell and R. F. Howe, *J. Mol. Catal.*, **7**, 535 (1980).
151. T. Yashima, Y. Orikasa, N. Takahashi, and N. Hara, *J. Catal.*, **59**, 53 (1979).
152. N. Takahashi, Y. Orikasa, and T. Yashima, *J. Catal.*, **59**, 61 (1979).
153. J. Yamanis and K.-C. Yang, *J. Catal.*, **69**, 498 (1981).
154. P. Gelin, Y. Ben Taarit, and C. Naccache, *Stud. Surf. Sci. Catal.*, **7B**, 898 (1981).
155. H. D. Grove, *Hydrocarbon Process.*, **51**, 76 (1972).

CHAPTER **8**

Olefin hydroformylation

JULIAN A. DAVIES

Department of Chemistry, College of Arts and Sciences, University of Toledo, Toledo, Ohio 43606, USA

I. INTRODUCTION	362
II. HISTORICAL ASPECTS	362
III. INDUSTRIAL ASPECTS	363
A. Unmodified Cobalt Catalysts	364
B. Modified Cobalt Catalysts	364
C. Rhodium Catalysts	364
IV. MECHANISTIC ASPECTS	365
A. Cobalt-Catalysed Systems	365
1. Unmodified catalysts	365
a. Hydridocobalt carbonyls	365
b. Olefin complexation	369
c. Metal alkyl formation	369
d. Metal acyl formation	371
e. Metal acyl cleavage	372
f. Cluster formation	374
g. By-product formation	374
h. Free-radical reactions	375
i. Conclusions	377
2. Modified catalysts	377
a. Catalyst stability	378
b. Selectivity	379
c. Hydroformylation activity	379
d. Hydrogenation activity	380
e. Conclusions	380
B. Rhodium-Catalysed Systems	380
1. Unmodified catalysts	380
a. Hydridorhodium carbonyls	380
b. Olefin complexation	381
c. Metal alkyl formation	382
d. Metal acyl formation and cleavage	383
e. Cluster formation	383
f. Conclusions	383

    2. Modified catalysts 384
        a. Reaction mechanism 384
        b. Conclusions 386
V. CONCLUDING REMARKS 387
VI. REFERENCES. 387

I. INTRODUCTION

The olefin hydroformylation, or oxo, reaction converts C_n olefins into C_{n+1} aldehydes via a catalytic reaction with hydrogen and carbon monoxide (equation 1).

$$RCH=CH_2 + CO + H_2 \rightarrow RCH_2CH_2CHO + RCH(CHO)Me \qquad (1)$$

The hydroformylation reaction, which does not proceed in the absence of a catalyst, is an important example of an industrially significant process which utilizes homogeneous catalysts. Whilst a great many organic syntheses are catalysed by transition metal complexes, very few such processes meet the economic and technological demands necessary for industrial use. Olefin hydroformylation and the carbonylation of methanol (Chapter 7) are two examples of homogeneously catalysed reactions which have been developed successfully on an industrial scale. Such classes of reaction are thus important as examples of how work on homogeneous catalysis can be developed from an initial observation up to an industrial-scale process. In the case of the olefin hydroformylation reaction, it is illustrative to note that new ideas concerning the mechanisms of both cobalt- and rhodium-catalysed systems are still being reported in the current literature from academic research groups (Sections IV.A and IV.B), although the reaction has been industrially profitable for many years.

In this chapter, some of the background to the olefin hydroformylation reaction is described and work on the mechanistic aspects of the process are discussed.

Only the industrially significant cobalt- and rhodium-catalysed reactions are described here, since catalysis by other metals has not been widely studied in recent years and Pruett reviewed such chemistry in 1979[1]. Catalysis by 'other metals' is largely a neglected area which certainly could profit from more detailed studies in the future. As a result of its importance to organometallic chemistry, the olefin hydroformylation reaction has been the subject of a number of reviews[1-10], to which the reader is referred for details of applications of olefin hydroformylation. In particular, the paper by Pruett[1] provides much detail in areas such as secondary reactions and by-product formation, effects of various classes of olefinic substrate, and catalyst recycling.

II. HISTORICAL ASPECTS

The hydroformylation reaction developed from industrial research on heterogeneous Fischer–Tropsch catalysts for the reductive polymerization of carbon monoxide. A historical survey[11] of Fischer–Tropsch chemistry reported that the first hydrogenation of carbon monoxide was discovered in 1902[12] and developed for the production of higher hydrocarbons during the 1920s[13,14]. By 1936, Ruhrchemie AG was operating the first commercial Fischer–Tropsch plant, with the typical heterogeneous catalyst consisting of kieselguhr, cobalt, thorium oxide, and magnesium oxide (*ca.* 66, 30, 2 and 2%, respectively). In 1938, Otto Roelen[15,16] of Ruhrchemie AG was investigating the effects of olefins on the Fischer–Tropsch reaction, using the 'heterogeneous' catalyst system, and discovered that aldehydes were formed. The hydroformylation reaction, or oxo process (from oxonation) as it was then known, was thus discovered. The christening of the process as hydroformylation occurred in 1949[17] and it was then soon discovered that the actual catalytic process was in fact homogeneous[18] with the heterogeneous Fischer–Tropsch catalyst acting as a source of soluble cobalt species which homogeneously catalyse hydroformylation. The Fischer–Tropsch process is discussed in detail in Chapter 9.

III. INDUSTRIAL ASPECTS

Olefin hydroformylation, based on cobalt- or rhodium-catalysed systems, produces about 4 million tonnes[6] of aldehydes and their derivatives annually (1981 figures). This represents a major utilization of olefinic feedstocks, accounting for about one fifth[19] of the total propene consumed, for example, by the chemical industry (1976 figure). The growth of hydroformylation technology is exemplified by comparison of the 1981 production with the production reported for 1965[20] of 0.5–0.6 million tonnes (worldwide capacity, excluding communist block countries).

Although production totals have increased, the range of end-uses for aldehydes has remained fairly constant over the last 20 years. Aldehydes themselves are not normally the desired end-product from olefin hydroformylation, but rather are used as intermediates, largely for alcohol production. A major product is butanal, from propene, which is hydrogenated to butan-1-ol for use as a solvent or used to form 2-ethylhexanal via an aldol condensation. 2-Ethylhexanal may be reduced to the alcohol and esterified (e.g. with phthalic anhydride) to produce a plasticizer used in PVC production. The condensation and reduction steps may be combined by slight process modifications (e.g. using zinc acetylacetonate to catalyse the dimerization), allowing conversion without separation of the initially formed aldehydes, as in the aldox process[6,19]. Although butanal is the major aldehyde produced by olefin hydroformylation, there is significant production of long-chain aldehydes, with end-uses in lubricant, plasticizer, and detergent production. A few examples of commercial uses of aldehydes are shown in Table 1. Details of plant capacities, production figures, etc., are available elsewhere[19,20].

Commercially, only cobalt- and rhodium-catalysed processes are of significance. While unmodified cobalt systems are still of major importance, recent years have seen the adoption of modified cobalt and rhodium catalysts on an industrial scale. Key factors in these changes are the search for systems with higher selectivities and activities which operate under mild (i.e. energy-efficient) conditions of temperature and pressure. The development of better catalytic processes, particularly those which will operate using

TABLE 1. Some commercial uses of aldehydes

Aldehyde	Usage	Product Use
Butanal	Hydrogenation to butan-1-ol	Industrial solvent
Butanal	2-Ethylhexanal produced via aldol condensation for:	
	(i) reduction to 2-ethylhexanol[a] followed by esterification with phthalic anhydride;	Plasticizer for PVC
	(ii) oxidation to 2-ethylhexanoic acid.	Plasticizer
Long-chain aldehydes	Hydrogenation to long-chain alcohols for:	
	(i) adipate ester synthesis;	High-temperature lubricants and plasticizers
	(ii) sulphonation, etc.	Biodegradable detergents

[a]The condensation/hydrogenation steps are known as the aldox process.

existing plant technology, is an active area of research in both academic and industrial laboratories. Less chemically oriented factors, such as process automation and environmental control, are also likely to become of increasing importance.

The question of selectivity towards the formation of a single, useful product is clearly a major area of concern. Considering the hydroformylation of propene, to yield n- and iso-aldehydes, only butanal is a major industrial chemical and, since cobalt catalysts can catalyse hydrogenation to butanol, and further steps (e.g. the aldox process) can be combined, butanal from propene yields useful products in a direct manner. The isoaldehyde, however, is not a major industrial chemical and the patent literature[21] reports methods for recycling unwanted isoaldehyde via decomposition back to propene, carbon monoxide, and hydrogen. A system to produce only butanal from propene in an efficient way would obviously be very useful. A similar situation also exists for other important olefinic substrates.

The following sections briefly outline the industrial aspects of the cobalt and rhodium catalyst systems.

A. Unmodified Cobalt Catalysts[6,10,19,20]

Cobalt, in the form of metal, salts, or dicobalt octacarbonyl, is the catalyst precursor in industrial plants, typically operating under 200–350 atm of synthesis gas at 110–180 °C. The alkene feed is usually in 100–1000-fold excess over the catalyst and is converted to a mixture consisting largely of aldehydes, alcohols, and alkanes. Under normal conditions, about 10% of the aldehyde is hydrogenated to the alcohol and perhaps 1% of the olefin hydrogenated to the alkane. The cobalt catalyst is only moderately effective for reduction under hydroformylation conditions. The next step, decobaltation, typically involves treatment of the organic products with aqueous base to remove the catalyst as a water-soluble $[Co(CO)_4]^-$ salt. $[CoH(CO)_4]$ is regenerated by treatment of the salt with sulphuric acid. The organic phase is then distilled and the separated aldehydes may be transferred for hydrogenation over a heterogeneous catalyst, if alcohols are required. The distillation by-products, such as the aldol condensation products, are cracked to aldehydes thermally or catalytically.

B. Modified Cobalt Catalysts[6,10,19,20]

Phosphine-modified cobalt catalysts operate under lower synthesis gas pressures than unmodified catalysts, give higher n- to iso-ratios and are more efficient hydrogenation catalysts, should alcohols be required. Activity in hydroformylation is lower, however, and the increase in hydrogenation activity results in substantial conversion of the olefin to alkane.

Hydroformylation and hydrogenation of the aldehyde generally occurs in a single stage of the process at 160–200 °C under 50–100 atm of synthesis gas using $[Co_2(CO)_8]/PBu^n_3$ as the catalyst precursor. Separation may be achieved via distillation, since the modified catalysts are thermally stable under such conditions, and the catalyst directly recycled. The formation of substantial amounts, often ca. 15%, of alkane is a major disadvantage of the modified catalysts, although plant requirements are less substantial since the conversion to alcohols and catalyst recycling steps are simpler.

C. Rhodium Catalysts[6,10,19,20]

Unmodified rhodium catalysts are not industrially attractive, since the high cost of rhodium is augmented by very low n- to iso-ratios. With ligand-modified rhodium catalysts, the high catalyst cost is balanced by operation under low temperatures and pressures, typically 80–120 °C and 15–25 atm of synthesis gas, and high n- to iso-ratios. Although rhodium catalysts are efficient for hydrogenation, high ligand concentrations and high partial pressures of carbon monoxide minimize the reduction of both alkene (ca. 2%) and aldehyde (< 1%). The aldehydes may be obtained in ca. 95% yield from terminal alkenes, although the alkene feed must normally be cleaned to remove sulphur compounds which inhibit the activity of the expensive rhodium catalyst, typically used at metal concentrations of 10^{-2}–10^{-3}% with respect to the olefin.

The cost of rhodium should, in theory, be balanced by increased activity and selectivity, resulting in lower manufacturing costs. Masters,[6] however, has pointed out that a 3–5 ppm loss of rhodium for a 100 000 tonne plant producing butanal from propene at an 85% loading corresponds to an $85 000–$127 000 p.a. capital loss (for rhodium at $30 000 per kg). The price penalty for butanal here rises to 10–15 cents per kg. Masters[6] also observed the restrictions which may result simply from the availability of rhodium in the earth's crust and its accessibility to the western world, since a high proportion of rhodium comes from Soviet Block countries. If the problems with cobalt catalysts could be overcome, its availability and 3500-fold price advantage over rhodium would clearly become major factors in determining industrial trends.

IV. MECHANISTIC ASPECTS

The following sections describe mechanistic aspects of olefin hydroformylation for cobalt- and rhodium-catalysed systems. Stoichiometric reaction chemistry is also discussed where appropriate. Initially, each step of the cobalt-catalysed reaction is considered with a view to differentiating experimental observation and hypothesis. The differences between unmodified cobalt catalysts and ligand-modified cobalt systems are then discussed in terms of the mechanistic effects of ligand substitution. Finally, the rhodium systems are described.

A. Cobalt-Catalysed Systems

1. Unmodified catalysts

The following sections discuss the mechanism of hydroformylation using unmodified cobalt catalysts. The initial sections are based on the concepts put forward by Heck and Breslow[22] in the early 1960s. Although the mechanistic picture has been refined considerably since that time, the Heck and Breslow mechanism serves as a useful basis for the discussion. The catalytic cycle shown in Scheme 1 describes the basics of the model.

a. Hydridocobalt carbonyls

In the absence of added ligands, cobalt metal, cobalt salts, or dicobalt octacarbonyl, any of which may be employed as catalyst precursors, are readily converted into $[CoH(CO)_4]$ under typical hydroformylation conditions. $[CoH(CO)_4]$, the key intermediate in olefin hydroformylation and many other carbonylation reactions catalysed by soluble cobalt salts, has been described by Orchin[23] as 'the quintessential catalyst'. Some properties of $[CoH(CO)_4]$ are given in Table 2.

The involvement of $[CoH(CO)_4]$ in olefin hydroformylation is demonstrated both by the isolation of this complex from reactions performed under typical hydroformylation conditions[27] and by the stoichiometric reaction chemistry of $[CoH(CO)_4]$. The latter point is illustrated by comparison of the reaction products obtained in a typical cobalt-catalysed olefin hydroformylation with those of stoichiometric reactions of $[CoH(CO)_4]$ with olefins. For a range of olefinic substrates, both stoichiometric and catalytic syntheses

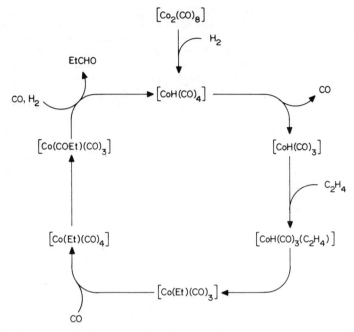

SCHEME 1. Cobalt-catalysed olefin hydroformylation.

TABLE 2. Properties of $[CoH(CO)_4]$

Melting point (°C)	-30^a
Boiling point (°C)	ca. 47^b
Solubility in water (mol l^{-1})	5.6×10^{-2c}
Acid dissociation constant, K_a	$< 2^d$

[a]Ref. 23.
[b]Ref. 24.
[c]Ref. 25.
[d]Acid dissociation constant similar to that of typical mineral acids[25,26]. Compare with values of 1.1×10^{-5}, 1.1×10^{-7}, and 8×10^{-8} for $[CoH(CO)_3\{P(OPh)_3\}]$, $[CoH(CO)_3(PPh_3)]$ and $[MnH(CO)_5]$ to observe effects of ligand substitution and metal on acidity[26].

yield the same products.[28] An example is the hydroformylation of hex-1-ene, reported to proceed under catalytic conditions with complete consumption of the substrate, producing hex-2-ene, hex-3-ene, 2-methylhexanal, and n-heptanal. Comparing this catalytic reaction, reported by Adkins and Krsek in 1949[29], with the stoichiometric reaction of hex-1-ene with $[CoH(CO)_4]$, Wender et al.[28] demonstrated that both C$_7$ aldehydes and the same isomeric C$_6$ olefins were produced and that the substrate was completely consumed. Indirect evidence also supported the involvement of $[CoH(CO)_4]$ in hydroformylation. For example, it was recognized at an early date that a strongly acidic species must be formed under hydroformylation conditions, since suitable substrates were found to undergo typical acid-mediated chemistry during reactions with synthesis gas in the presence of cobalt salts.[30] Thus, products formed from the reaction of pinacol may be

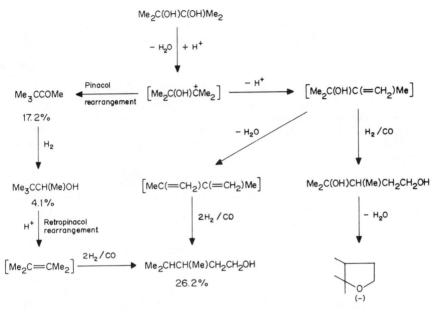

SCHEME 2. Reactions of pinacol under hydroformylation conditons.

explained using traditional carbocation chemistry, as shown in Scheme 2, in operation with the expected hydroformylation and hydrogenation reactions.

Further evidence that the acidic species is $[CoH(CO)_4]$ is shown by comparing the product obtained by addition of pyridine to $[Co_2(CO)_8]$ under hydroformylation conditions (230 atm CO/H_2, 120 °C) with that obtained by treating independently prepared $[CoH(CO)_4]$ with pyridine[28]. Both reactions yield the salt $[C_5H_5NH]^+[Co(CO)_4]^-$. Finally, high-pressure infrared spectroscopic studies have shown that under typical hydroformylation conditions (150 °C, 290 atm synthesis gas), $[Co_2(CO)_8]$ undergoes nearly complete hydrogenation to $[CoH(CO)_4]$[31].

The mechanism of formation of $[CoH(CO)_4]$ from $[Co_2(CO)_8]$ is one of the few examples of a dinuclear oxidative addition reaction which is well understood. Pino et al.[32] demonstrated in 1955 that the hydrogenolysis of $[Co_2(CO)_8]$ is inhibited by carbon monoxide, implying a dissociative process. The cleavage reaction is first order in both $[Co_2(CO)_8]$ and H_2 and, at low CO pressures, inverse first order in CO. A rate-determining (RD) cleavage preceded by a dissociative equilibrium in thus suggested (equation 2), involving a coordinatively unsaturated intermediate.

$$[Co_2(CO)_8] \rightleftharpoons [Co_2(CO)_7] + CO \xrightarrow[RD]{H_2} 2[CoH(CO)_4] \qquad (2)$$

A catalytic cycle for olefin hydroformylation, based on reaction chemistry of $[CoH(CO)_4]$, is shown in Scheme 1. The very first step in the sequence, dissociation of carbon monoxide from $[CoH(CO)_4]$, an 18-electron, coordinatively saturated species, to yield $[CoH(CO)_3]$, a 16-electron, coordinatively unsaturated species, remained difficult to prove without ambiguity for many years. Qualitatively, observations on the effects of carbon monoxide pressure on the rate of hydroformylation suggest that dissociation of a carbonyl ligand is involved. Thus, under typical catalytic conditions, the rate is inverse first order in carbon monoxide partial pressure[1], while for stoichiometric hydroformylation with [CoH(CO)$_4$], the presence of even 1 atm CO pressure grossly retards the rate of aldehyde formation[33,34]. As an example of this latter point, it is reported[34] that the stoichiometric hydroformylation of cyclopentene with [CoH(CO)$_4$] is more than 150 times faster under a nitrogen atmosphere than it is under an atmosphere of carbon monoxide.

It has been pointed out, however, that no conclusive rate expression for the catalytic reaction has yet been reported and that at least 15 separate rate constants are likely to be involved in describing the overall scheme[23]. Clearly, making detailed mechanistic arguments based on observed rate effects without knowledge of the complete rate expression is pointless. Acceptance of [CoH(CO)$_3$] as an intermediate in hydroformylation was aided by the successful observation and characterization of this species using matrix isolation methods in 1978[35], although the exact assignment of the i.r. data is problematic[35–37]. Based on a consideration of the thermal decomposition of [CoH(CO)$_4$] to [Co$_2$(CO)$_8$] and the effects of carbon monoxide pressure on stoichiometric olefin hydroformylation, Orchin[23] suggested that the equilibrium constant for the formation of [CoH(CO)$_3$] is probably small and Ungvary and Marko[38] have estimated that [CoH(CO)$_3$] is present to the extent of a few tenths of a percent in equilibrium with [CoH(CO)$_4$] at 20 °C under carbon monoxide and that the equilibrium amount of [CoH(CO)$_3$] is much higher under a nitrogen atmosphere[39].

The equilibrium between [CoH(CO)$_4$] and [CoH(CO)$_3$] may well be involved in some of the unusual effects observed when nucleophiles are added to hydroformylation systems. Of the many reported examples, it is illustrative to consider the effects of added nitriles on stoichiometric hydroformylation with [CoH(CO)$_4$][34,40]. In the hydroformylation of

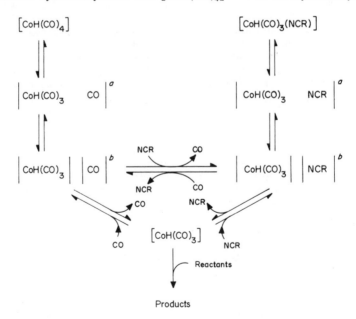

SCHEME 3. Effects of nucleophiles on the formation of [CoH(CO)$_3$].

[a]Caged pairs
[b]Solvent-separated pairs

cyclopentene, addition of appropriate amounts of nitriles to the system has beneficial effects on the product distribution, decreasing the amount of hydrogenation from 52% to 3%. Similarly, with pent-1-ene as the substrate, the extent of olefin isomerization is decreased and the *n*- to iso-ratio increased by addition of the nucleophile. While these effects are explicable in several ways, the unusual rate effects are worthy of comment. Thus, while addition of nucleophiles decreases the rate of stoichiometric hydroformylation of cyclopentene under a nitrogen atmosphere, it actually increases the rate under a carbon monoxide atmosphere. Considering the situation in the absence of a carbon monoxide atmosphere, it is apparent that addition of a nucleophile must reduce the concentration of any coordinatively unsaturated species in solution, i.e. that the amount of $[CoH(CO)_3]$ present will be reduced and the rate inhibited. In the presence of a carbon monoxide atmosphere, the situation has been depicted as shown in Scheme 3, where the amount of $[CoH(CO)_3]$ is determined by dissociation of a ligand (CO or NCR) and subsequent diffusion of $[CoH(CO)_3]$ from a cage to form a solvent-separated pair. Equilibration of the two possible solvent-separated pairs causes CO to be lost to the solution and hence the effective concentration of $[CoH(CO)_3]$ is increased, which is reflected by an enhanced rate. A definitive interpretation of these observations is not possible, however, since addition of NCR to the reaction medium alters the polarity, and other related parameters, which may be reflected by a change in equilibrium constant for one or more of the subsequent steps. Migratory insertion, for example, is known to be affected by the nature of the solvent medium[41,42].

b. Olefin complexation

Reaction of the 16-electron complex $[CoH(CO)_3]$ with the olefinic substrate to form an 18-electron η^2-olefin complex, as shown in Scheme 2, is an unconfirmed step in the catalytic cycle. Thus, reaction of olefins with $[CoH(CO)_4]$, believed to be in equilibrium with $[CoH(CO)_3]$, proceeds rapidly to produce aldehydes, acylcobalt, or alkylcobalt carbonyls, depending on the conditions. No experimental conditions have yet been found to allow interception of a π-complex in the reaction.

By high-pressure infrared studies of the reaction of the π-allyl complex $[Co(C_3H_5)(CO)_3]$ with H_2 (137 atm) in *n*-tetradecane solution at *ca.* 50 °C, King and Tanaka[43] observed formation of $[Co_4(CO)_{12}]$. Heating to 100 °C resulted in further reaction of the cluster and observation of new $\nu(CO)$ bonds tentatively assigned to a species $[CoH(CH_2{=}CHMe)(CO)_3]$. This complex persisted at H_2 pressures down to atmospheric, but could not be isolated or further characterized.

Although π-complexes have not been observed directly under definitive conditions, substantial indirect evidence confirms that they are implicated in the reaction mechanism. Thus, under catalytic conditions, (*R*)-3-ethylhexanal is produced in 3% yield with 70% optical purity by hydroformylation of (+)-(*S*)-3-methylhex-1-ene[44], whilst hydroformylation of 3-methylhex-1-ene-3-*d* yields 3-ethylhexanal-2-*d*[45,46]. These results indicate that isomerization occurs without dissociation of the organic group from the metal, since deuterium appears in the 2-position in the product and yet the stereochemistry of the chiral centre is maintained. An isomerization mechanism involving a series of σ-alkyl/π-olefin complex interconversions explains these results[23] (Scheme 4).

c. Metal alkyl formation

The proposed η^2-olefin complex $[CoH(CO)_3(\text{olefin})]$ is believed to undergo migratory insertion to yield a 16-electron alkylcobalt intermediate, $[CoR(CO)_3]$, which coordinates carbon monoxide to produce the coordinatively saturated alkyl $[CoR(CO)_4]$. Equation 3 illustrates these steps for a substituted olefinic substrate.

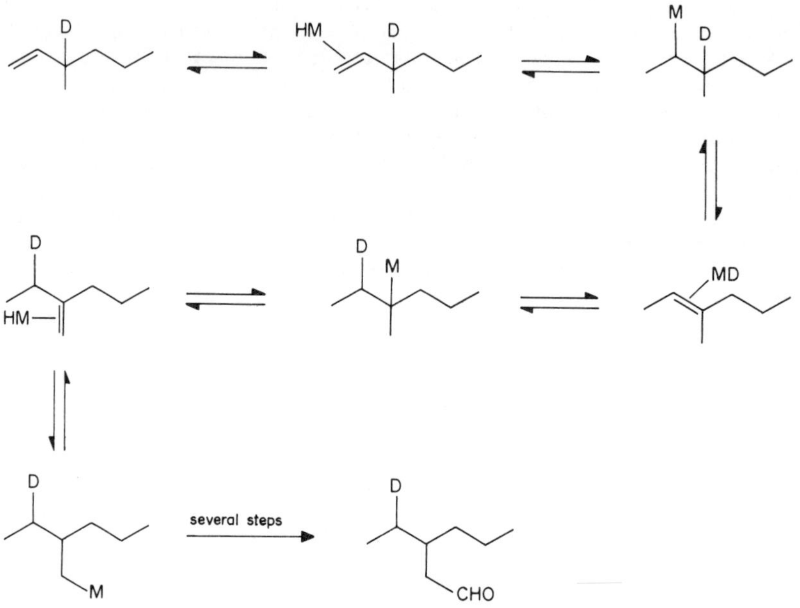

SCHEME 4. Isomerization mechanism for cobalt-catalysed olefin hydroformylation.

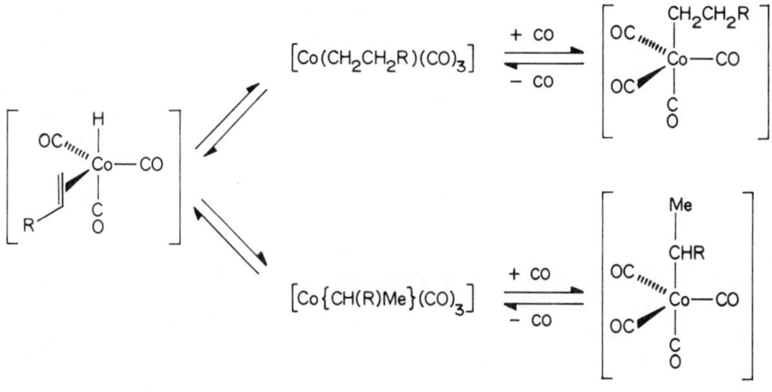

Neither the η^2-olefin complex nor the unsaturated alkyl have been observed as reaction intermediates. Saturated alkyls have not yet been observed under catalytic conditions either, since high-pressure infrared spectroscopy is unable to detect these intermediates owing to interfering absorptions from the other cobalt carbonyls present. The intermediacy of the saturated alkyls is confirmed via stoichiometric reaction chemistry. Alkyls of the type [CoR(CO)$_4$] are well known[22,47,48], but generally exhibit low thermal stability and are difficult to isolate. Their phosphine-substituted analogues[49], discussed in Section IV.A.2, are more amenable to study.

It has been stated that the mode of addition of the Co—H bond across the olefin determines the n- to iso-ratio of aldehydes produced[10] via further reactions of the alkyls shown in equation 3. That this is in fact not the case has been demonstrated by Taylor and Orchin[50], who showed that [CoH(CO)$_4$] reacts with CD_2=$CDCD_3$ in the gas phase, in the absence of carbon monoxide, to give *ca.* 70% Markownikoff addition. The aldehydes produced, 70% iso- and 30% n-C_4 aldehydes, correspond to the ratio dictated by the initial direction of Co—H addition. However, with CO present the gas-phase reaction gives *ca.* 60% normal product[51] whilst typical catalytic reactions produce about 70% linear aldehydes[1]. Clearly, the reversibility of the olefin insertion step allows for product control to occur at a later stage in the catalytic cycle.

Since the addition of the Co—H bond across the olefin does not control the product distribution and olefin isomerization via π-olefin/σ-alkyl interconversions also evidently occurs prior to the product-controlling step, it is not surprising that hydroformylation reactions of pent-1-ene and pent-2-ene yield almost the same ratios of n- and iso-C_6 aldehydes[52,53].

d. Metal acyl formation

When [CoH(CO)$_4$] reacts with a terminal monoene in the presence of carbon monoxide, it has been shown that more than 80% of the theoretical amount of CO is absorbed for formation of the acyl complex, [Co(COR)(CO)$_4$][47]. Such acyl complexes may be observed *in situ* by infrared spectroscopy, but often are too unstable for isolation from such a preparative route[22]. Phosphine-substituted analogues are of course well known[54].

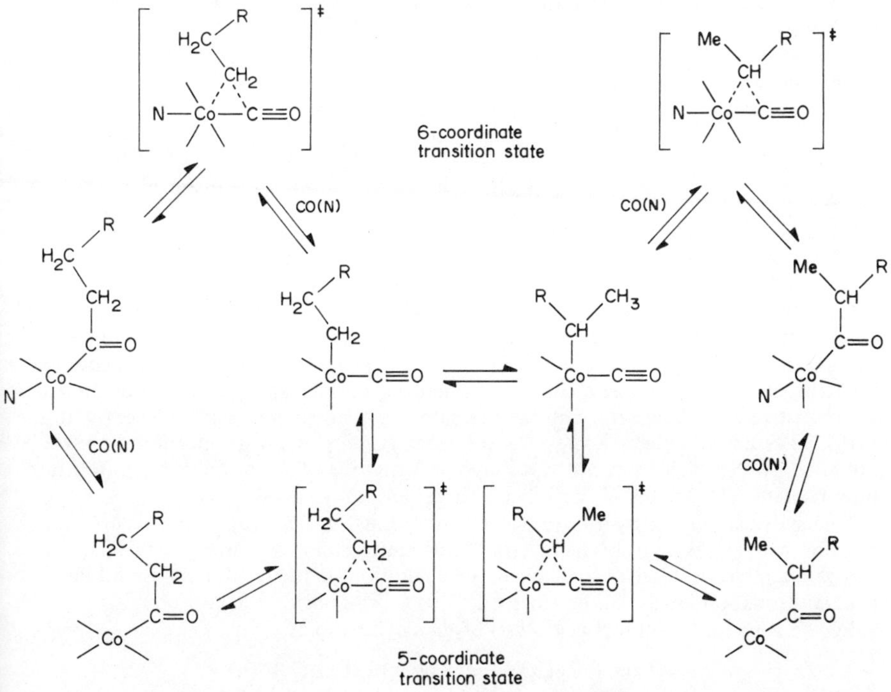

SCHEME 5. Five- and six-coordinate transition states in carbonyl insertion. Reprinted with permission from ref. 23. Copyright 1981 American Chemical Society.

The extensive literature on the carbonyl insertion reaction has been well documented in the past[42] and will not be covered in this chapter. Most important to the hydroformylation reaction is the influence of carbonyl insertion on product distribution. The n- to iso-ratio obtained in catalytic hydroformylation is controlled largely by the partial pressure of carbon monoxide. For simple olefinic substrates, high partial pressures of CO result in a predominance of the normal aldehyde[52]. These results are rationalized by considering the mechanism of carbonyl insertion. Orchin has suggested[23,51] that, depending on the availability of additional ligands, either five- or six-coordinate transition states are possible for the alkyl migration. The additional steric crowding of the branched alkyl results in a higher energy six-coordinate transition state, but has little effect on the energy of the less crowded five-coordinate transition state, in comparison with the n-alkyl isomer. Clearly, under reaction conditions where additional ligands are present, such as where high partial pressures of CO are employed or where tertiary phosphines are added, the route involving formation of a six-coordinate transition state is predominant and more favoured for the n-alkyl isomer. The ready isomerization of the n- and iso-alkyl complexes, via π-olefin/σ-alkyl complex conversion, allows for a predominant formation of the n-acyl complex, resulting in a high n- to iso-ratio. The situation is shown schematically in Scheme 5.

e. Metal acyl cleavage

Several possibilities exist for cleavage routes of the acyls to form the aldehydic products. This step is shown as a hydrogenolysis in Scheme 1, as originally proposed by Heck and Breslow[22], but evidence is mounting that this is not the case, even under catalytic conditions.

It is important to note that no second reactant is necessary to achieve cleavage of an acylcobalt carbonyl to yield aldehydes, since a disproportionation is possible (equation 4)[55,56]. Several factors are important in this disproportionation: the reaction is

$$2[Co(COCH_2CH_2R)(CO)_4] \rightarrow RCH_2CH_2CHO$$
$$+ RCH(Me)CHO + RCH=CH_2 + Co_2(CO)_8 + CO \qquad (4)$$

irreversible, produces two isomeric aldehydes from a single isomer of the acylcobalt complex, and is inhibited by carbon monoxide. Both pentanoyl- and 2-methylbutanoyl-cobalt carbonyls, for example, yield the same butenes on disproportionation[56]. The disporportionation reaction is favoured using pentane as a solvent, whereas using diethyl ether only isomerization of the acylcobalt complexes is favoured. The proposed mechanism[23] is shown in Scheme 6. The acyls are proposed to isomerize via equilibria involving σ-alkyl and π-olefin complex formation, whilst disproportionation can occur via reductive elimination through the reaction of either isomeric acyl complex with the hydride [CoH(CO)_3(olefin)]. Clearly, both isomerization and disproportionation will be inhibited by CO, since these processes involve formation of a coordinatively unsaturated intermediate. The origins of the solvent effects are not clear.

Under stoichiometric hydroformylation conditions, acyl cleavage must occur through reaction with [CoH(CO)_4]. The reaction is evidently rapid, since free [CoH(CO)_4] must still be available for this to occur. An oxidative addition–reductive elimination mechanism is likely (equations 5–8). Since [CoH(CO)_4] is a protic acid, it is also possible that the aldehyde is formed by direct acidolysis of the acyl complex.

$$[Co(COR)(CO)_4] \rightleftharpoons [Co(COR)(CO)_3] + CO \qquad (5)$$
$$[Co(COR)(CO)_3] + [CoH(CO)_4] \rightleftharpoons [Co_2(COR)(H)(CO)_7] \qquad (6)$$
$$[Co_2(COR)(H)(CO)_7] \rightarrow RCHO + [Co_2(CO)_7] \qquad (7)$$

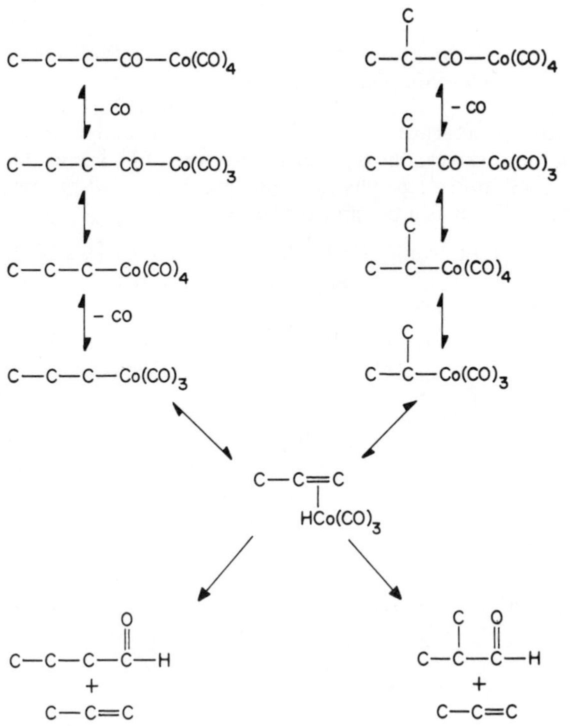

SCHEME 6. Disproportionation of acylcobalt carbonyls. Reprinted with permission from ref. 23. Copyright 1981 American Chemical Society.

$$[Co_2(CO)_7] + CO \rightleftharpoons [Co_2(CO)_8] \qquad (8)$$

In the case of catalytic reactions, cleavage of the acyl complex by $[CoH(CO)_4]$ is still possible, but the possibility of direct hydrogenolysis is also apparent (equations 9–12).

$$[Co(COR)(CO)_4] \rightleftharpoons [Co(COR)(CO)_3] + CO \qquad (9)$$

$$[Co(COR)(CO)_3] + H_2 \rightleftharpoons [Co(COR)(H)_2(CO)_3] \qquad (10)$$

$$[Co(COR)(H)_2(CO)_3] \rightarrow [CoH(CO)_3] + RCHO \qquad (11)$$

$$[CoH(CO)_3] + CO \rightleftharpoons [CoH(CO)_4] \qquad (12)$$

The relative contributions of the cleavage routes by H_2 and $[CoH(CO)_4]$ under operating conditions are not clear. High-pressure infrared studies[31,57] have been interpreted[23] to imply a possible dominance of $[CoH(CO)_4]$ cleavage, whereas kinetic data[58] have been interpreted[6] in terms of a dominant cleavage by H_2. No definitive statement is possible with the data presently available.

The acyl complexes $[Co(COR)(CO)_4]$ appear to form a major part of the species present under catalytic conditions[31]. This may imply that the cleavage step is rate-limiting, although since several steps in the cycle involve a dissociative loss of CO and the reaction is run under high partial pressures of CO, it is possible that an earlier step determines the rate[6]. Orchin[23] has suggested that loss of CO from $[CoH(CO)_4]$ may be rate-limiting.

f. Cluster formation

Polynuclear clusters of cobalt are less prevalent than those of rhodium, yet their possible involvement in olefin hydroformylation should not be ignored. So far, only dinuclear species have been mentioned in this section, but as early as 1955, Natta et al.[59] postulated equilibria involving tetranuclear species. Thus, although catalytic hydroformylation is inverse first order in carbon monoxide partial pressure at high pressures of CO, at low partial pressures a half-order dependence is observed. Natta et al. proposed the equilibria shown in equations 13–16 as a possible explanation.

$$[CoH(CO)_3(olefin)] + [CoH(CO)_4] \rightarrow RCHO + [Co_2(CO)_6] \quad (13)$$

$$[Co_2(CO)_6] \rightleftharpoons \tfrac{1}{2}[Co_4(CO)_{12}] \quad (14)$$

$$\tfrac{1}{2}[Co_4(CO)_{12}] + 2CO \rightleftharpoons [Co_2(CO)_8] \quad (15)$$

$$[Co_2(CO)_8] + H_2 \rightleftharpoons 2[CoH(CO)_4] \quad (16)$$

More recently, it has been shown that $[Co_2(CO)_8]$ and $[CoH(CO)_4]$ react to form $[Co_3(H)(CO)_9]$[60,61], whose possible involvement in hydroformylation is not yet known. In the presence of base (e.g. NEt_3), $[CoH(CO)_4]$ and $[Co_2(CO)_8]$ react to form a carbyne-bridged trinuclear cluster, $[Co_3(CO)_9(COH)]$[62]. A sequence inter-relating the various species is shown in equation 17. Under hydroformylation conditions, it is not known whether such species are formed.

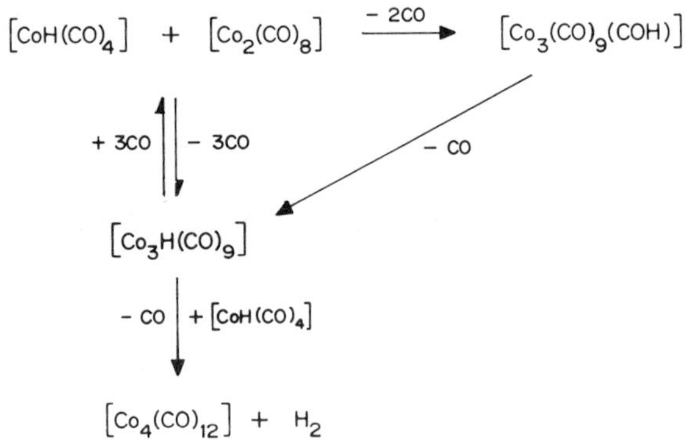

$$[CoH(CO)_4] + [Co_2(CO)_8] \xrightarrow{-2CO} [Co_3(CO)_9(COH)]$$

$$\Big\updownarrow {\scriptstyle +3CO \,\,/\,\, -3CO} \qquad \xrightarrow{-CO}$$

$$[Co_3H(CO)_9]$$

$$\Big\downarrow {\scriptstyle -CO \,\,/\,\, +[CoH(CO)_4]}$$

$$[Co_4(CO)_{12}] + H_2 \quad (17)$$

g. By-product formation

In addition to aldehydes, olefin hydroformylation produces varying amounts of alkanes, alcohols, ketones, and condensation products, depending on the exact reaction conditions. Catalytic hydrogenation of the olefinic substrate is envisaged as occurring via interception of the alkylcobalt species by H_2 or $[CoH(CO)_4]$, prior to the migratory insertion step (equations 18 and 19). Alternatives to these postulated two-electron transformations are described in the next section, where free-radical involvement in hydroformylation systems is considered.

$$[CoR(CO)_3] + H_2 \rightarrow RH + [CoH(CO)_3] \quad (18)$$

$$[CoR(CO)_3] + [CoH(CO)_3] \rightarrow RH + [Co_2(CO)_7] \quad (19)$$

Alcohol formation, via catalytic aldehyde reduction, follows kinetics which show an inverse dependence on the square of the carbon monoxide partial pressure[63], thus explaining why the cobalt system is relatively ineffective for this reduction under the high

CO pressures encountered in typical operating systems. The addition of a Co—H bond across the C=O of an aldehyde could occur in two possible directions as shown in the structures **1** and **2**. Clearly, **1** leads to alkoxycobalt complex formation, whereas **2** leads to

$$\begin{array}{cc}
\begin{array}{c} H \\ \diagdown \\ R \diagup \end{array} \!\!\! \begin{array}{c} C\!=\!=\!=\!O \\ | \quad\quad | \\ H\!-\!-\!-\!-\!Co(CO)_n \end{array} & \begin{array}{c} H \\ \diagdown \\ R \diagup \end{array} \!\!\! \begin{array}{c} C\!=\!=\!=\!O \\ | \quad\quad | \\ (OC)_nCo\!-\!-\!-\!-H \end{array} \\
(\mathbf{1}) & (\mathbf{2})
\end{array}$$

hydroxyalkylcobalt complex formation. Both routes have been postulated[63,64] to explain alcohol formation.

An intermediate formed from **1** is attractive in explaining formate ester formation during hydroformylation, since CO insertion into the Co—O bond of an alkoxycobalt intermediate, followed by reduction, provides an attractive pathway. Formate ester yields of ca. 4% have been reported[65]. An intermediate formed from **2**, however, seems necessary to explain the stoichiometric reaction of $[CoH(CO)_4]$ with methanal, yielding 2-hydroxyethanal[66], presumably via CO insertion into the Co—C bond of the hydroxymethylcobalt intermediate. This dichotomy remains unresolved.

Ketone formation is not a problem under typical hydroformylation conditions and only becomes substantial under low pressures with high concentrations of unhindered olefins[67]. A mechanism involving a binuclear reductive elimination has been proposed[68] (equation 20), but remains largely untested.

$$[Co(COR)(CO)_3] + [RCo(CO)_4] \rightarrow R_2CO + [Co_2(CO)_7] \quad (20)$$

Aldol condensation products, particularly 2-ethylhexanal, have been mentioned previously as being often desirable products from hydroformylation; their formation will not be considered further here.

h. Free radical reactions

In the preceding sections, the mechanism of olefin hydroformylation has been considered in terms of conventional two-electron reactions. There is mounting evidence, however, that radical pathways are intimately involved in the chemistry of cobalt carbonyls for the reaction types under consideration. The first definitive evidence[69] comes from a study of the reaction between $[CoH(CO)_4]$ and 1,1-diphenylethylene, where ^{1}H n.m.r. studies showed a CIDNP effect on raising the temperature of the reaction system from −78 to 31 °C. A signal attributable to a methyl group was observed as a doublet emission at 1.6 ppm, rapidly changing to a doublet absorption with an upfield shift as the temperature rose. These observations are clearly evidence for intermediate formation of a germinate radical pair, $Ph_2\dot{C}Me\dot{C}o(CO)_4$, via hydrogen abstraction from $[CoH(CO)_4]$ by $Ph_2C=CH_2$. The hydrogen abstraction reaction has precedent in Sweany and Halpern's observation[70] of a CIDNP effect in the reaction of $[MnH(CO)_5]$ with α-methylstyrene. Both reactions[69,70] show second-order kinetics, inverse deuterium isotope effects, and deuterium exchange between reactants and products. Modelled on Sweany and Halpern's description, the mechanism shown in equations 20–23 for free-radical hydrogenation of $Ph_2C=CH_2$ has been proposed[69].

$$Ph_2C=CH_2 + [CoH(CO)_4] \rightleftharpoons \overline{Ph_2\dot{C}Me\dot{C}o(CO)_4} \quad (20)$$

$$\overline{Ph_2\dot{C}Me\dot{C}o(CO)_4} \rightarrow \dot{C}o(CO)_4 + Ph_2\dot{C}Me \quad (21)$$

$$Ph_2\dot{C}Me + [CoH(CO)_4] \rightarrow Ph_2CHCH_3 + \dot{C}o(CO)_4 \quad (22)$$

$$2\dot{C}o(CO)_4 \rightarrow [Co_2(CO)_8] \quad (23)$$

It is possible, therefore, to envisage a situation for a particular olefinic substrate where diffusion of the germinate pair from the solvent cage (i.e. equation 21) leads to hydrogenation (i.e. equations 22 and 23), whereas recombination within the cage leads to the formation of an alkylcobalt complex which undergoes migratory insertion to yield aldehydes via the normal sequence[71]. Partitioning between hydrogenation and hydroformylation may thus be determined by the recombination or escape steps. This has been shown to be the case for stoichiometric hydroformylation with $[MnH(CO)_5]$, another reaction which shows all the characteristics (e.g. CIDNP effects) of a radical process[72].

Ungvary and Marko[73] utilized the concept of partitioning between recombination and escape to explain the kinetics of the reaction between $[CoH(CO)_4]$ and styrene in the presence of carbon monoxide. The rates of ethylbenzene and of α-phenylpropionylcobalt tetracarbonyl formation are both first order in $[CoH(CO)_4]$ and styrene and independent of CO and $[Co_2(CO)_8]$ concentrations. Large inverse isotope effects were observed and both $[Co_2(CO)_8]$ and some isomeric acyl complex were also found as by-products. The mechanism show in Scheme 7, essentially outlined in the preceding paragraphs, was proposed. It seems very likely that further contributions in this area will soon delineate the extent of radical involvement in this area of chemistry.

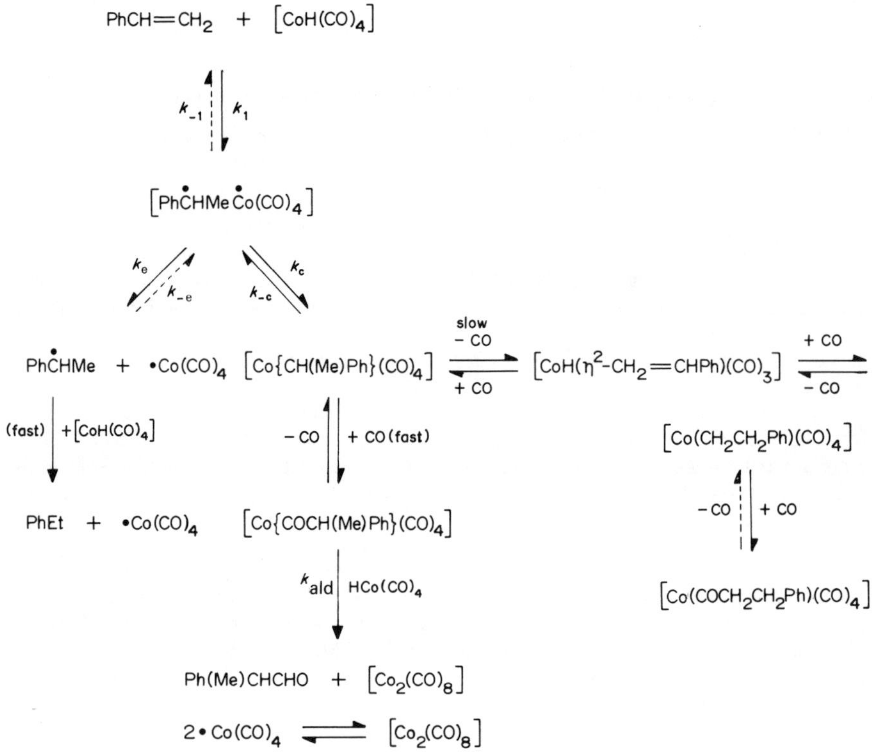

SCHEME 7. Radical involvement in the reaction of $[CoH(CO)_4]$ with styrene. Reprinted with permission from ref.73. Copyright 1982 American Chemical Society.

i. Conclusions

Although mechanistic studies of olefin hydroformylation using unmodified cobalt catalysts have answered many questions, there are clearly still areas worthy of further investigation. The recent demonstration of free radical involvement in partitioning hydrogenation and hydroformylation, for example, is possibly a more widespread occurrence than at present thought. The stereochemistry of olefin hydroformylation has not been described in the preceding sections, since the very small body of work in this area has only recently been discussed[23]. This topic would also profit from detailed study in the future. Many of the individual steps proposed for the catalytic hydroformylation cycle remain unproved to date; it is to be hoped that the future will see the conflicting evidence for some of these steps resolved.

2. Modified catalysts

In comparison with data for unmodified cobalt catalysts, any mechanistic discussion of modified systems must explain the major differences between the two cases. Principle among these differences are the following:
(i) modified catalysts can operate under lower partial pressures of carbon monoxide than their unmodified analogues[10];
(ii) higher n- to iso-rates are obtained with modified catalysts[74];

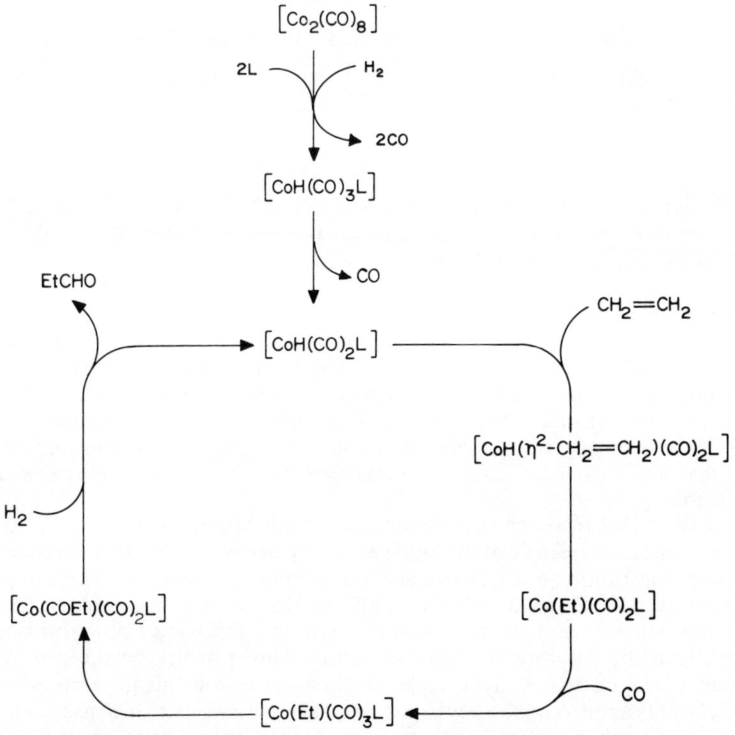

SCHEME 8. Olefin hydroformylation by modified cobalt catalysts.

(iii) the activity in hydroformylation is generally lower for modified systems[75];
(iv) the activity in hydrogenation is generally higher for modified systems[74].

The mechanism of the modified cobalt system in olefin hydroformylation is generally assumed to follow a cycle similar to that described in the preceding sections for unmodified catalysts. A generalized cycle is shown in Scheme 8, which forms the basis for the discussion of modified catalysts which follows. Much of the validity of this discussion is based on the correctness of the proposed mechanism for the unmodified system, new aspects of which are still being discovered.

a. Catalyst stability

In ligand-modified catalyst systems, $[Co_2(CO)_8]$ is the usual cobalt source and the modifying ligand is most typically a trialkylphosphine. It is proposed that under operating conditions, equilibria are established leading to the formation of mononuclear hydridocobalt species (equations 24 and 25).

$$[Co_2(CO)_8] + 2L \rightleftharpoons [Co_2(CO)_6L_2] + 2CO \tag{24}$$

$$[Co_2(CO)_6L_2] + H_2 \rightleftharpoons 2[CoH(CO)_3L] \tag{25}$$

Under mild reaction conditions it has been observed[76] that the conversion of $[Co_2(CO)_8]$ to $[CoH(CO)_3(PR_3)]$ is incomplete, even with Co to PR_3 ratios as high as 1:5. Under other conditions, it is known that the simple substitution reaction in equation 24 may be suppressed in favour of disproportionation (equation 26)[77-79].

$$[Co_2(CO)_8] + 2L \rightleftharpoons [Co(CO)_3(L)_2]^+[Co(CO)_4]^- + CO \tag{26}$$

The formation of ionic products is generally favoured at lower temperatures and, indeed, the ionic species may be converted into the dinuclear complex by heating[80] in some cases (e.g. $L = AsPh_3$, $SbPh_3$) (equation 27).

$$[Co(CO)_3L_2]^+[Co(CO)_4]^- \rightleftharpoons [Co_2(CO)_6L_2] + CO \tag{27}$$

The formation of substituted hydrides, e.g. $[CoH(CO)_3L]$ and $[CoH(CO)_2L_2]$, is also well known through preparative chemistry. It is generally observed that the thermal stability of substituted hydrides exceeds that of the parent carbonyl, $[CoH(CO)_4]$. Using melting points as an indication of this is illustrative: $[CoH(CO)_4]$, $-30\,°C$[23]; $[CoH(CO)_3\{P(OPh)_3\}]$, $ca.\ 0\,°C$[81]; $[CoH(CO)_3(PPh_3)]$, $ca.\ 20\,°C$[81]. Thermal stability is a key factor in determining the operating conditions for catalytic experiments. Thus, in the case of unmodified catalysts, a high reaction temperature is needed to generate $[CoH(CO)_4]$ *in situ* and, at such high temperatures, the partial pressure of CO must be kept high to prevent decomposition to metallic cobalt[82]. Since phosphine-substituted carbonyls are thermally more stable with respect to decomposition to metallic cobalt, it follows that lower partial pressures of carbon monoxide may be employed at a given temperature.

It is possible that the enhanced stability of the substituted carbonyls is electronic in nature. The metal centre in $[CoH(CO)_4]$ is evidently electron-poor, since the complex is a strong acid. Substitution of CO, particularly by a trialkylphosphine, is likely to promote electroneutrality at the metal. Whether trialkylphosphines are considered to be pure σ-donors or σ-donors/π-acceptors, it is still apparent that replacing a powerful π-acceptor carbonyl ligand by PR_3 will increase the electron density at the metal centre. This has important effects on the catalytic cycle, as discussed below, but in terms of stability, electroneutrality appears to be a key factor, since an increased electron density at the metal centre allows for enhanced Co—CO π-bonding. This additional bond strength is reflected in the greater thermal stability of the substituted complexes.

b. Selectivity

The origins of enhanced selectivity for n-aldehyde formation with phosphine-modified cobalt catalysts appear to be both steric and electronic in nature. Thus, replacement of a small carbonyl ligand by a bulkier tertiary phosphine is expected to result in crowding at the metal centre. In terms of the catalytic reaction, this is likely to result in the preferential formation of n-alkyl- and n-acyl-cobalt intermediates, since the transition states involved in the migratory insertion steps will be of lower energy than for the sterically more demanding iso- species. Clearly, for these steric effects to be operative, the tertiary phosphine must remain bound to the metal throughout the reaction. Electronic effects thus become important, since displacement of PR_3 by $CO^{83,84}$ (equation 28)

$$[CoH(CO)_3(PR_3)] + CO \rightleftharpoons [CoH(CO)_4] + PR_3 \qquad (28)$$

would result in subsequent catalysis by the unmodified route with the corresponding decrease in selectivity.

Experiments using $[Co_2(CO)_8]/2PR_3$ catalyst precursors have confirmed the dependence of selectivity on both the electronic and steric properties of the PR_3 ligands[83]. Masters[6] has previously discussed the effects of PR_3 ligands on hydroformylation in terms of Tolman's electronic and steric ligand parameters[85].

c. Hydroformylation activity

Modified cobalt catalysts generally exhibit lower activity in olefin hydroformylation than their unmodified counterparts, assuming conditions where phosphine displacement is minimized for the former case. Masters[6] suggested that this is due to a difference in the rate-determining steps of the two cycles. If it is assumed that hydrogenolysis of a cobalt acyl is the rate-determining step for the unmodified catalyst, it can be rationalized that a different rate-determining step will occur with the modified system. Thus, as discussed above, the phosphine is likely to render the metal centre more electron-rich, thereby promoting oxidative addition and hydride transfer in the modified catalyst. Conversely, it is argued that the steric hindrance of the phosphine will inhibit olefin coordination and that this may be the rate-determining step for the modified catalyst.

Orchin[23] has suggested that dissociation of CO from $[CoH(CO)_4]$ may be rate-limiting in the unmodified systems. Based on discussion given earlier, it is expected that phosphine substitution will result in enhanced Co—CO π-bonding in $[CoH(CO)_3(PR_3)]$ and thus it would be expected that a slower rate-determining step would result for the modified catalyst.

With the data currently available it does not seem to be possible to differentiate between these two possibilities. Evidently the kinetics of the two systems differ, but whether this is due to a change in the rate-determining step is not yet clear. With relatively unreactive olefins, such as cyclohexene, infrared studies[31] indicate that $[CoH(CO)_3(PR_3)]$ (Co:PR_3 = 2:1, R = n-Bu, 190 °C, 80 atm synthesis gas) is the major species in solution. Pruett[1] has emphasized that such results probably should not be extrapolated to typical fast systems (e.g. with ethylene).

The possibility that loss of CO from $[CoH(CO)_3(PR_3)]$ is rate-limiting has been mentioned above and the fact that such a step is likely to be slow has led to a proposal of an associative mechanism for ligand-modified systems[4]. Later discussion of rhodium-catalysed systems (Section IV.B) will show that the concept of dual catalytic pathways (associative and dissociative) has recently been examined and may not be mechanistically valid.

d. Hydrogenation activity

Both olefin and aldehyde hydrogenation are more prevalent with phosphine-modified catalysts than with the ligand-free system. This would seem to result from the electronic effects of phosphine substitution. As mentioned above, substitution of CO by PR_3 will tend to increase the electron density at the metal centre. This will clearly favour the oxidative addition of H_2 and, further, increase the hydride character of any Co—H intermediates. An increase in hydridic nature will favour migratory insertion via hydride transfer and thus lead to enhanced alkane formation from metal alkyls and alcohol formation from alkoxymetal species.

The possibility that olefin hydrogenation results from a radical-based process, as described previously for unmodified catalysts, does not appear to have been examined as yet. This is evidently important, since olefin hydrogenation is the major disadvantage of modified cobalt catalysts on an industrial scale.

e. Conclusions

The effects of modifying ligands may be understood by considering both the electronic and steric effects that these impose at the metal centre. The separation of steric effects from electronic effects is difficult to achieve and Tolman's work[85] to put these concepts on a semi-quantitative basis has helped a clearer picture to evolve. Once again, the problem encountered in studying such complex systems is the lack of detailed kinetic data for the individual steps in the reaction sequence. Even the rate-determining step is not known with any certainty. As with the unmodified systems, much still remains to be accomplished.

B. Rhodium-Catalysed Systems

1. Unmodified catalysts

Unmodified rhodium catalysts are not industrially attractive because of the problem of low n- to iso-ratios alluded to earlier. The study of such systems is of importance principally because of the light shed on modified catalysts through understanding of the simpler system. The proposed catalytic cycle for unmodified rhodium catalysts is shown in Scheme 9 and it is immediately apparent that the steps involved parallel those for unmodified cobalt systems. The cycle shown forms the basis for the following discussion.

a. Hydridorhodium carbonyls

The rhodium catalyst precursor may be introduced into the reaction system in any one of several forms. On a laboratory scale, use of $[Rh_2(CO)_4Cl_2]$ or $[Rh_4(CO)_{12}]$ is common while on a larger scale, rhodium salts or dispersed rhodium metal may be employed conveniently. It is now known that these precursors lead to $in\ situ$ formation of $[RhH(CO)_4]$ under hydroformylation conditions. Although the existence of $[RhH(CO)_4]$ was postulated as early as 1943[86], its detection and characterization proved to be exceedingly difficult. For the unmodified cobalt systems, the effects of equilibria involving tetra-, di-, and mono-nuclear species have been discussed previously (Section IV.A.1) and may be represented by equation 29.

$$\tfrac{1}{4}[Co_4(CO)_{12}] \underset{-CO}{\overset{+CO}{\rightleftharpoons}} \tfrac{1}{2}[Co_2(CO)_8] \underset{-\tfrac{1}{2}H_2}{\overset{+\tfrac{1}{2}H_2}{\rightleftharpoons}} [CoH(CO)_4] \quad (29)$$

In the case of rhodium, it is now known that $[Rh_4(CO)_{12}]$ reacts to form $[Rh_2(CO)_8]$ in a solid carbon monoxide matix[87], although the very existence of $[Rh_2(CO)_8]$ was

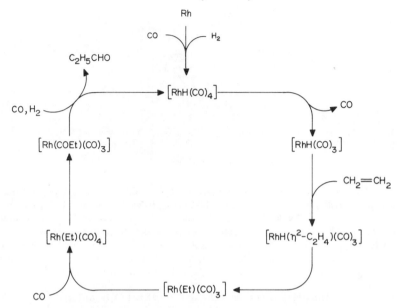

SCHEME 9. Rhodium-catalysed olefin hydroformylation.

questioned[88] until fairly recently; similar reactivity is indicated by infrared studies on hydrocarbon solutions[89]. Attempts to observe [RhH(CO)$_4$] spectroscopically proved more difficult[89,90] than for the cobalt system[43,91] or for the corresponding iridium system[31] (not discussed in this chapter). King et al.[91] tentatively assigned a ν(CO) mode at 2019 cm^{-1} to [RhH(CO)$_4$] during a high-pressure infrared study of ethylene carbonylation, but complete identification of this complex was achieved by Vidal and Walker[92]. Using Fourier transform infrared techniques to study the fragmentation of [Rh$_4$(CO)$_{12}$] in dodecane solution under 1241–1370 atm of CO at 5–12 °C, the formation of [Rh$_2$(CO)$_8$] was detected. Addition of hydrogen (CO:H$_2$ = 9:2; 1542 atm) led to the formation of new species and, by subtraction of bands due to the tetra- and di-nuclear compounds, the spectrum of [RhH(CO)$_4$] was observed (2070m, 2039vs, 2008w cm^{-1}) and assigned by analogy to the cobalt and iridium analogues.

The first step in the catalytic cycle, dissociation of CO from [RhH(CO)$_4$] to form a coordinatively unsaturated 16-electron species, [RhH(CO)$_3$], is unproved. The detection and characterization of coordinatively unsaturated species of this type are generally difficult and, in this specific case, complicated by an apparent series of equilibria with polynuclear complexes (Scheme 10). Cluster involvement is further discussed in Section B.1.e.

The intimate details of the degradation of rhodium clusters to [RhH(CO)$_3$] are evidently not as well understood as the corresponding cobalt reactions, largely owing to the complexity of the infrared spectra obtained under hydroformylation conditions. Clearly this area is sufficiently fundamental to olefin hydroformylation to warrant detailed study.

b. Olefin complexation

High-pressure infrared spectroscopy has provided tentative evidence for the formation of [RhH(CO)$_3$(olefin)] complexes. For example, King and Tanaka[43] reported that

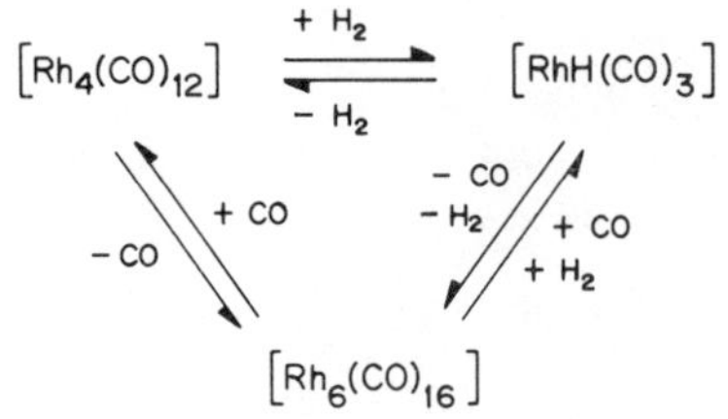

$$[Rh_4(CO)_{12}] \underset{-H_2}{\overset{+H_2}{\rightleftharpoons}} [RhH(CO)_3]$$

$$[Rh_4(CO)_{12}] \xrightarrow[-CO]{+CO} [Rh_6(CO)_{16}] \xleftarrow[+H_2]{-CO,\,-H_2,\,+CO} [RhH(CO)_3]$$

SCHEME 10. Equilibria involving the postulated [RhH(CO)$_3$]. Reproduced from ref. 1 by permission of Academic Press.

[Rh$_4$(CO)$_{12}$] reacts with a 1:1 C$_2$H$_4$–H$_2$ mixture (6 atm, 37 °C), producing a species which was tentatively identified by its i.r. spectrum [v(CO) = 2075w, 2026s cm^{-1}] as a trigonal bipyramidal species with local C_{3v} symmetry, such as [RhH(CO)$_3$(olefin)], with three equatorial carbonyls. This species was found to have a very different regime of stability from [RhH(CO)$_4$], reported by Vidal and Walker[92], but could not be isolated or identified unambiguously. King and Tanaka[43] also presented i.r. data interpreted in terms of reaction between [Rh(CO)$_2$(π-allyl)] and hydrogen to yield the propene complex [RhH(CO)$_3$(η^2-MeCH=CH$_2$)].

The reaction of [RhH(CO)$_3$] with the olefin to form [RhH(CO)$_3$(olefin)] may well be rate-limiting, in cases where sterically demanding olefins are involved. Thus, many i.r. studies[43,92–94] indicate that rhodium acyls build up during hydroformylation of simple α-olefins (maybe indicating a rate-limiting hydrogenolysis or other cleavage step in these cases), but for cyclic olefins, [Rh$_4$(CO)$_{12}$] build-up occurs[93,94]. Scheme 10 indicates that clustering of [RhH(CO)$_3$] would cause this to occur, indicating that the reaction of [RhH(CO)$_3$] with cyclic olefins must be slow. Under high-temperature conditions, [Rh$_6$(CO)$_{16}$], the thermodynamically favoured cluster, is observed spectroscopically when cyclic olefins are the substrate.

King and Tanaka's work indicates that it may be possible to prepare and study [RhH(CO)$_3$(olefin)] complexes under conditions similar to those used in hydroformylation. This is clearly worthwhile, since unambiguous characterization of the hydridoolefin complex and an understanding of its reactivity would put the mechanism of hydroformylation on a much firmer basis.

c. Metal alkyl formation

Much as in the unmodified cobalt system, the rhodium catalyst cycle is believed to proceed from [RhH(CO)$_3$(olefin)] via migratory insertion to produce the 16-electron species [RhR(CO)$_3$], which coordinates CO, forming [RhR(CO)$_4$]. The coordinatively unsaturated species has never been observed and identified, but [RhR(CO)$_4$] complexes have been implicated by observation under high-pressure conditions by infrared spectroscopy. Thus, treatment of various rhodium(I) catalyst precursors with CO–C$_2$H$_4$–H$_2$ mixtures allowed observation of a complex by i.r. methods, identified as [Rh(Et)(CO)$_4$][95]. The ethylrhodium species was found to be unstable except under CO pressures and, if correctly identified, represents the first [RhR(CO)$_4$] complex. Although [CoR(CO)$_4$] species of limited thermal stability have been known for some time, the ethylrhodium complex represents the observation of a proposed hydroformylation intermediate, which is stable only under high-pressure hydroformylation conditions. Other rhodium alkyls have now been observed under similar experimental conditions[43].

d. Metal acyl formation and cleavage

Rhodium acyls have been observed under high-pressure conditions by infrared spectroscopy[93,94]. As noted previously, with α-olefins, the coordinatively saturated rhodium acyl is the species which tends to build up in solution[93,94], indicating that its cleavage may be rate-limiting. King and coworkers[43,95] noted, however, that with ethylene as a substrate it is the ethylrhodium species and not the propionylrhodium complex which is predominant in solution. Possibly, then, the formation of the acyl complex is also a slow step under certain conditions.

The cleavage step is represented by a hydrogenolysis in Scheme 9 but, as in the cobalt system, there is really no evidence that this is correct, since a binuclear reductive elimination of aldehyde from $[Rh(COR)(CO)_n]$ and $[RhH(CO)_n]$ is also possible. Indeed, since $[RhH(CO)_4]$ is known to be a strong protic acid[92] {more acidic than even $[CoH(CO)_4]$}, the possibility of a direct acidolysis of a rhodium acyl becomes likely.

The intimate details of acyl formation and cleavage have not been studied in such depth as for cobalt and much remains to be learned here.

e. Cluster formation

Aggregation to form polynuclear clusters is more commonly observed for rhodium than for cobalt. Johnson and Lewis[96] pointed out that aggregation processes are often initiated by cluster degradation involving the reversible ejection of a coordinatively saturated fragment and oligomerization of the remaining coordinatively unsaturated fragments (e.g. equations 29 and 30).

$$[Rh_4(CO)_{12}] \rightleftharpoons \text{`}[Rh_3(CO)_8]\text{'} + [Rh(CO)_4] \qquad (29)$$

$$2\text{`}[Rh_3(CO)_8]\text{'} \rightarrow [Rh_6(CO)_{16}] \qquad (30)$$

The nature of the clustering processes are important because these can lead to an effective decrease in the available rhodium present in solution, since there is no evidence that polynuclear clusters actively participate in hydroformylation. Further, clustering seems to result from many expected side-reactions in hydroformylation systems. For example, when $[RhH(CO)_4]$ acts as an acid[92], the $[Rh(CO)_4]^-$ produced is known to be trapped by $[Rh_4(CO)_{12}]^{92,97}$ (equations 31 and 32).

$$[RhH(CO)_4] + B^- \rightarrow [Rh(CO)_4]^- + HB \qquad (31)$$

$$[Rh(CO)_4]^- + [Rh_4(CO)_{12}] \rightarrow [Rh_5(CO)_{15}]^- \qquad (32)$$

The $[Rh_5(CO)_{15}]^-$ anion is evidently a stable species under high-pressure conditions, since Heaton et al.[98] observed by high-pressure ^{13}C n.m.r. that degradation of $[Rh_{12}(CO)_{30}]^{2-}$ stops at the pentarhodium cluster under high pressures of synthesis gas. The real involvement of polynuclear clusters in hydroformylation, other than as a source of mononuclear species, still remains a mystery. There do not appear to be any reports of hydroformylation catalysis by polynuclear rhodium species in which the involvement of mononuclear complexes has been rigorously excluded.

f. Conclusions

Clearly, our knowledge of unmodified rhodium catalysts does not match that of the unmodified cobalt catalysts. Much of the catalytic cycle described for the rhodium system is taken to hold 'by analogy' with cobalt. This is evidently undesirable, since there are a number of fundamental areas (e.g. radical involvement and cluster formation) which are likely to differ mechanistically. High-pressure infrared studies have been particularly useful in studying rhodium catalysts and it is expected that much mechanistic information could be acquired through use of the newly developed high-pressure ^{13}C n.m.r. probe[1], in conjuction with i.r., to study systems under operating conditions.

2. Mofified catalysts

The use of modified rhodium catalysts for olefin hydroformylation can be attributed to the early work of Slaugh and Mullineaux[90,100]. Much of the work necessary to understand the fundamentals of the reaction mechanism came from Wilkinson's group[101–106], although Brown and coworkers[107–109] have now questioned some aspects of the mechanistic interpretation. The key issue in discussing ligand-modified rhodium catalysts is the effect of the rhodium to ligand ratio on the n- to iso-ratio of the aldehydes produced. As a general rule, increasing the ligand concentration increases the n- to iso-ratio. With this in mind, the differences between the modified and unmodified rhodium systems are discussed in the following paragraphs.

a. Reaction mechanism

Several different rhodium(I) complexes act as catalyst precursors for olefin hydroformylation[100] and all of them appear to form $[RhH(CO)_2(PR_3)_2]$ under reaction conditions. This complex is also presumed to be formed when the source of rhodium is rhodium metal or rhodium salts. When chloro complexes are used as precursors, e.g. $[RhCl(CO)(PR_3)_2]$, then addition of amines to scavenge HCl will often reduce the induction period[102] during which $[RhH(CO)_2(PR_3)_2]$ is being formed (equation 33).

$$[RhCl(CO)(PR_3)_2] \xrightarrow{+H_2} [RhCl(H)_2(CO)(PR_3)_2] \xrightarrow[-HNR_3Cl]{+NR_3} [RhH(CO)(PR_3)_2]$$

$$\downarrow +CO \qquad (33)$$

$$[RhH(CO)_2(PR_3)_2]$$

Pruett[1] and Pruett and Smith[110] noted that $[RhH(CO)_2(PR_3)_2]$ is unlikely to exist in isolation, but rather in equilibrium with a variety of closely related species (Scheme 11). Brown et al.[107] similarly demonstrated by ^{1}H n.m.r. that the tetraphosphole complex, **3**, reacts with CO to yield **4** and **5** (equation 34), whose equilibration is slow on the n.m.r. time scale.

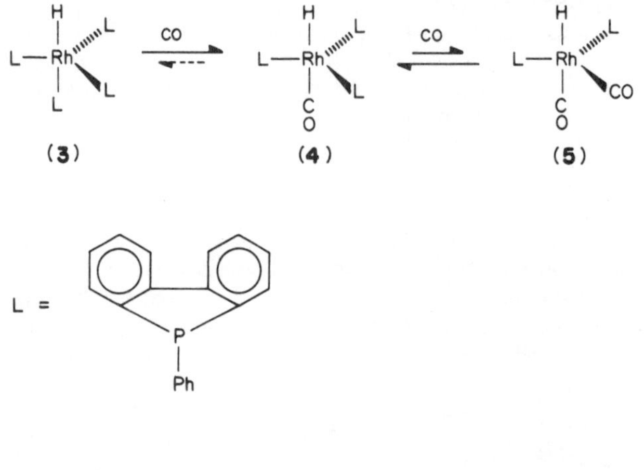

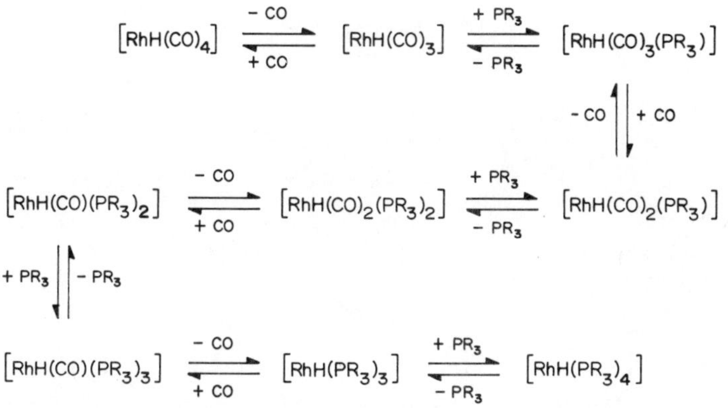

8. Olefin hydroformylation

$$[RhH(CO)_4] \underset{+CO}{\overset{-CO}{\rightleftarrows}} [RhH(CO)_3] \underset{-PR_3}{\overset{+PR_3}{\rightleftarrows}} [RhH(CO)_3(PR_3)]$$

$$\updownarrow \begin{array}{c} -CO \\ +CO \end{array}$$

$$[RhH(CO)(PR_3)_2] \underset{+CO}{\overset{-CO}{\rightleftarrows}} [RhH(CO)_2(PR_3)_2] \underset{-PR_3}{\overset{+PR_3}{\rightleftarrows}} [RhH(CO)_2(PR_3)]$$

$$\updownarrow \begin{array}{c} +PR_3 \\ -PR_3 \end{array}$$

$$[RhH(CO)(PR_3)_3] \underset{+CO}{\overset{-CO}{\rightleftarrows}} [RhH(PR_3)_3] \underset{-PR_3}{\overset{+PR_3}{\rightleftarrows}} [RhH(PR_3)_4]$$

SCHEME 11. Equilibria involving [RhH(CO)$_2$(PR$_3$)$_2$]. Adapted with permission from refs. 1 and 110. Copyright 1969, 1979 American Chemical Society.

The selection of [RhH(CO)$_2$(PR$_3$)$_2$] as the key intermediate is based on the observation[111] that C$_2$H$_4$ (1 atm, 25 °C) reacts only with this complex (R = Ph) when present as a mixture with [RhH(CO)(PR$_3$)$_2$]. Similarly, methylenecyclopropane reacts only with **5** when added to a solution of both **4** and **5** at 25 °C[107].

The problem arises in defining the number of PR$_3$ ligands bound to rhodium throughout the catalytic cycle. Wilkinson and coworkers[102-105] proposed both associative and dissociative routes, shown in Scheme 12, where the former cycle involves more sterically hindered intermediates. The increased *n*- to iso-ratio with increasing PR$_3$ concentration could thus be ascribed to a dominance of the associative route in the presence of excess ligand, which favours *n*-aldehyde formation because the transition states associated with the key insertion steps will be less hindered than for iso-aldehyde formation.

The controversial step in the associative cycle is the initial interaction of [RhH(CO)$_2$(PR$_3$)$_2$] with the olefinic substrate. A genuine coordination of the olefin would represent an 18- to 20-electron conversion which, whilst not totally unknown, does not fit well with accepted ideas on the mechanisms of catalytic reactions. Possibly a concerted reaction of Rh—H with an 'uncoordinated' olefin to yield an alkyl directly may occur[6]. The fact that the reaction rate does decrease with increasing PR$_3$ concentration is in accord with a lower reactivity for the saturated species (associative route) than for the unsaturated species (dissociative route).

The newest ideas on this problem come from Brown and coworkers' n.m.r. studies[107-109]. Terminal olefins were shown to react with [RhH(CO)$_2$(PRh$_3$)$_2$] to give linear and branched acylrhodium complexes, of the type known to accumulate under hydroformylation conditions. For example, the complexes **6** and **7** were formed in a 91:9 ratio from styrene. Over a 2-h period, the isomer ratio changed to favour **7**, with both *n*-

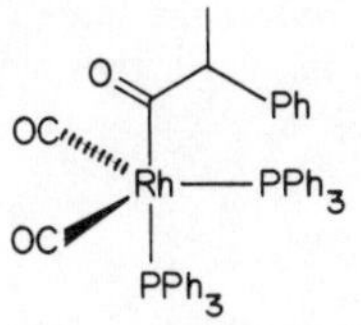

(6) 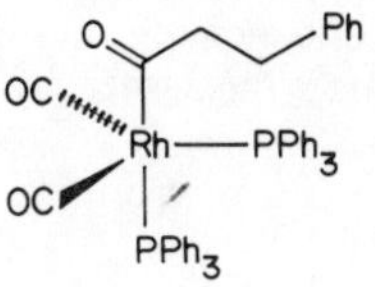   (7)

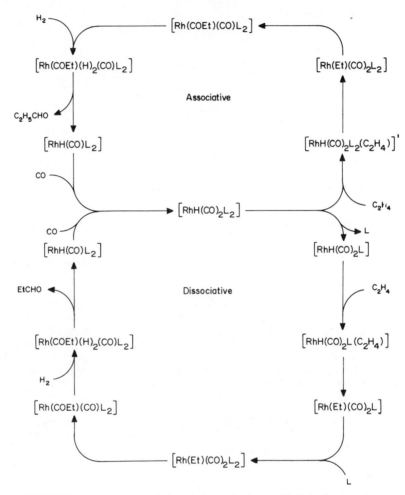

SCHEME 12. Associative and dissociative cycles for modified rhodium catalysts.

and iso-aldehydes being formed (e.g. by hydride donation from a rhodium hydride). The isomerization was inhibited by excess of PPh_3 and aldehyde formation was suppressed.

Brown and coworkers[107-109] have observed that the isomerization of the acyl complexes is likely to be important at higher temperatures in catalytic reactions where such complexes are likely to be trapped by H_2, after a dissociative ligand loss. Accordingly, the *n*- to iso-ratios of such catalytic reactions may reflect the kinetic liability of the acyl complexes rather than a change in the catalytic pathway. Further work will no doubt be forthcoming to clarify this issue.

b. Conclusions

A major criticism[107] of the mechanistic discussion of modified rhodium catalysts is that almost all of the available information dates to pre-1970. Whilst there is clearly nothing wrong with this *per se*, the developments in instrumentation since that time mean that we are now in a position to investigate more aspects of the systems than was previously possible. In particular, multinuclear magnetic resonance methods[112] and the newly developed high-pressure n.m.r. techniques[98] are beginning to be applied in such areas and much should be accomplished in the future.

V. CONCLUDING REMARKS

In this chapter, some aspects of the mechanisms of catalytic olefin hydroformylation with cobalt and rhodium catalysts have been discussed. It is important to realize that this is only one area of hydroformylation chemistry that is currently developing at a rapid pace. Other areas of interest, in which our level of fundamental understanding is perhaps more restricted, include stereochemistry and asymmetric synthesis, the use of supported catalysts, and catalysis by 'other metals'. There is little doubt that olefin hydroformylation is one of the success stories of organometallic chemistry and has an important place in both the history and the future of the chemistry of the metal—carbon bond.

VI. REFERENCES

1. R. L. Pruett, *Adv. Organomet. Chem.*, **7**, 1 (1979).
2. J. Falbe, *Carbon Monoxide in Organic Synthesis*, Springer-Verlag, Berlin, 1970.
3. P. Pino, F. Piacenti and M. Bianchi, in *Organic Syntheses via Metal Carbonyls* (Eds. I. Wender and P. Pino), Vol. 2, Wiley, New York, 1977, p. 43.
4. F. E. Paulik, *Catal. Rev.*, **6**, 49 (1972).
5. L. Marko, in *Aspects of Homogeneous Catalysis* (Ed. R. Ugo), Vol. 2, Riedel, Dordrecht, 1973.
6. C. Masters, *Homogeneous Catalysis—A Gentle Art*, Chapman and Hall, London, 1981, p. 102.
7. M. Orchin and W. Rupilius, *Catal. Rev.*, **6**, 85 (1972).
8. P. J. Davidson, R. R. Hignett, and D. T. Thompson, *Catalysis*, Vol. 1, Chemical Society, London, 1977, p. 369.
9. C. W. Bird, *Transition Metal Intermediates in Organic Synthesis*, Academic Press, New York, 1967, p. 117.
10. G. W. Parshall, *Homogeneous Catalysis*, Wiley–Interscience, New York, 1980, p. 85.
11. C. Masters, *Adv. Organometal. Chem.*, **17**, 61 (1979).
12. P. Sabatier and J. B. Sendersens, *C. R. Acad. Sci.*, **134**, 514 (1902).
13. F. Fischer and H. Tropsch, *Brennst.-Chem.*, **4**, 276 (1923).
14. F. Fischer and H. Tropsch, *Ger. Pat.*, 484 337, 1925.
15. O. Roelen, *Ger. Pat.*, 849 548, 1938.
16. O. Roelen, *U.S. Pat.*, 2 317 066, 1943.
17. H. Adkins and G. Krsek, *J. Am. Chem. Soc.*, **71**, 3051 (1949).
18. I. Wender, M. Orchin and H. Storch, *J. Am. Chem. Soc.*, **72**, 4842 (1950).
19. A. L. Waddams, *Chemicals from Petroleum*, 4th ed., John Murray, London, 1978.
20. A. L. Waddams, *Chemicals from Petroleum*, 2nd ed., John Murray, London, 1968.
21. J. Falbe, H. Tummes, and H. Hahn, *U.S. Pat.*, 4 039 584, 1977.
22. R. F. Heck and D. S. Breslow, *J. Am. Chem. Soc.*, **83**, 4023 (1961).
23. M. Orchin, *Acc. Chem. Res.*, **14**, 259 (1981).
24. J. A. Roth and M. Orchin, *J. Organomet. Chem.*, **187**, 103 (1980).
25. H. W. Sternberg, I. Wender, R. A. Friedel, and M. Orchin, *J. Am. Chem. Soc.*, **75**, 2718 (1953).
26. J. P. Collman and L. S. Hegedus, *Principles and Applications of Organotransition Metal Chemistry*, University Science Books, Mill Valley, CA, 1980, p. 68.
27. L. Kirch, I. J. Goldfarb, and M. Orchin, *J. Am. Chem. Soc.*, **78**, 5450 (1956).
28. I. Wender, H. W. Sternberg, and M. Orchin, *J. Am. Chem. Soc.*, **75**, 3041 (1953).
29. H. Adkins and G. Krsek, *J. Am. Chem. Soc.*, **71**, 3051 (1949).
30. I. Wender, S. Metlin and M. Orchin, *J. Am. Chem. Soc.*, **73**, 5704 (1951).
31. R. Whyman, *J. Organomet. Chem.*, **81**, 97 (1974).
32. P. Pino, R. Ercoli, and F. Calderazzo, *Chim. Ind. (Milan)*, **37**, 783 (1955).

33. G. L. Karapinka and M. Orchin, *J. Org. Chem.*, **26**, 4187 (1961).
34. G. L. Karapinka and M. Orchin, *Abstr. 137th Am. Chem. Soc. Meet.*, Cleveland, OH, April 5–14, 1960, p. 92–0.
35. P. Wermer, B. A. Ault, and M. Orchin, *J. Organomet. Chem.*, **162**, 189 (1978).
36. R. L. Sweany, *Inorg. Chem.*, **19**, 3512 (1980).
37. R. L. Sweany, *J. Am. Chem. Soc.*, **104**, 3739 (1982).
38. F. Ungvary and L. Marko, *J. Organomet. Chem.*, **20**, 205 (1969).
39. J. Terapone, *PhD Thesis*, University of Cincinnati, 1969; cited in ref. 33.
40. L. Roos and M. Orchin, *J. Org. Chem.*, **31**, 3015 (1966).
41. R. J. Mowby, F. Basolo, and R. G. Pearson, *J. Am. Chem. Soc.*, **86**, 3994 (1964).
42. A. Wojcicki, *Adv. Organomet. Chem.*, **11**, 87 (1973).
43. R. B. King and K. Tanaka, *J. Indian Chem. Soc.*, **59**, 124 (1982).
44. F. Piacenti, S. Pucci, M. Bianchi, R. Lazzaroni, and P. Pino, *J. Am. Chem. Soc.*, **90**, 6847 (1968).
45. C. P. Casey and C. R. Cyr, *J. Am. Chem. Soc.*, **93**, 1280 (1971).
46. C. P. Casey and C. R. Cyr, *J. Am. Chem. Soc.*, **95**, 2240 (1973).
47. R. F. Heck and D. S. Breslow, *Chem. Ind.* (*London*), 467 (1960).
48. W. Beck and R. E. Nitzschmana, *Chem. Ber.*, **97**, 2098 (1964).
49. Z. Nagy-Magos, G. Bor, and L. Marko, *J. Organomet. Chem.*, **14**, 205 (1968).
50. P. Taylor and M. Orchin, *J. Am. Chem. Soc.*, **93**, 6504 (1971).
51. M. Orchin and W. Rupilus, *Catal. Rev.*, **6**, 85 (1972).
52. F. Piacenti, P. Pino, R. Lazzaroni, and M. Bianchi, *J. Chem. Soc. C*, 488 (1966).
53. I. J. Goldfarb and M. Orchin, *Adv. Catal.*, **14**, 1 (1957).
54. R. F. Heck and D. S. Breslow, *J. Am. Chem. Soc.*, **82**, 4438 (1960).
55. W. Rupilus and M. Orchin, *J. Org. Chem.*, **37**, 936 (1972).
56. W. Rupilus and M. Orchin, *Symp. Chem. Hydroformylation Relat. Reaction, Proc., Veszprem, Hungary, 1972*, p. 59; *Chem. Abstr.*, **77**, 87513 (1972).
57. N. H. Alemdarogly, J. L. M. Penninger, and E. Oltay, *Monatsh. Chem.*, **44**, 184 (1963).
58. G. Natta, *Brennst.-Chem.*, **36**, 176 (1955).
59. G. Natta, R. Ercoli, and S. Castellano, *Chim. Ind.* (*Milan*), **37**, 6 (1955).
60. G. Fachinetti, personal communication to M. Orchin, cited in ref. 23.
61. G. Fachinetti, L. Balochi, F. Secco, and M. Venturini, *Angew. Chem., Int. Ed. Engl.*, **20**, 204 (1981).
62. H. N. Adams, G. Fachinetti, and J. Strahle, *Angew. Chem., Int. Ed. Engl.*, **20**, 125 (1981).
63. L. Marko, *Proc. Chem. Soc., London*, 67 (1962).
64. C. L. Aldridge and H. B. Jonassen, *J. Am. Chem. Soc.*, **85**, 886 (1963).
65. R. Krummer, H. J. Nienburg, H. Hohenschutz, and M. Strohmeyer, *Adv. Chem. Ser.*, No. 132, 19 (1974).
66. J. A. Roth and M. Orchin, *J. Organomet. Chem.*, **127**, C27 (1979).
67. E. Naragon, A. Millendorf and J. Vergillio, *U.S. Pat.*, 2 699 453, 1955.
68. J. Bertrand, C. Aldridge, S. Husebeye, and H. Jonassen, *J. Org. Chem.*, **29**, 790 (1964).
69. T. E. Nalesnik and M. Orchin, *Organometallics*, **1**, 222 (1982).
70. R. L. Sweany and J. Halpern, *J. Am. Chem. Soc.*, **99**, 8335 (1977).
71. M. Orchin, personal communication, 1982.
72. T. E. Nalesnik and M. Orchin, *J. Organomet. Chem.*, **222**, C5 (1981).
73. F. Ungvary and L. Marko, *Organometallics*, **1**, 1120 (1982).
74. L. H. Slaugh and R. D. Mullineaux, *J. Organomet. Chem.*, **13**, 469 (1968).
75. E. R. Tucci, *Ind. Eng. Chem., Prod. Res. Dev.*, **8**, 215 (1969).
76. M. Van Boven, N. H. Alemdaroglu, and J. M. L. Penninger, *Ind. Eng. Chem., Prod. Res. Dev.*, **14**, 259 (1975).
77. W. Reppe and W. J. Schweckendieck, *Justus Liebigs Ann. Chem.*, **560**, 104 (1948).
78. A. Sacco, *J. Inorg. Nucl. Chem.*, **8**, 566 (1958).
79. F. A. Cotton and G. Wilkinson, *Advanced Inorganic Chemistry*, 3rd ed., Wiley, New York, 1972, p. 886.
80. J. A. McCleverty, A. Davison, and G. Wilkinson, *J. Chem. Soc.*, 3890 (1965).
81. W. Hieber and E. Lindner, *Chem. Ber.*, **94**, 1417 (1961).
82. J. Berty, E. Oltay, and L. Marko, *Chem. Tech.* (*Liepzig*), **9**, 283 (1957).
83. E. R. Ticci, *Ind. Eng. Chem., Prod. Res. Dev.*, **9**, 516 (1970).

84. W. Rupilus, J. McCoy, and M. Orchin, *Ind. Eng. Chem., Prod. Res. Dev.*, **10**, 142 (1971).
85. C. A. Tolman, *Chem. Rev.*, **77**, 313 (1977).
86. W. Hieber and H. Lagally, *Z. Anorg. Allg. Chem.*, **251**, 96 (1943).
87. L. A. Hanlan and G. A. Ozin, *J. Am. Chem. Soc.*, **96**, 6324 (1974).
88. F. Calderazzo, R. Ercoli, and G. Natta, in *Organic Syntheses via Metal Carbonyls* (Eds. I. Wender and P. Pino), Vol. 1, Wiley, New York, 1968, pp. 4 and 16.
89. R. Whyman, *J. Chem. Soc., Chem. Commun.*, 1194 (1970).
90. R. Whyman, *J. Chem. Soc., Chem. Commun.*, 1381 (1960).
91. R. B. King, M. Z. Iqbal, and A. D. King, *Proc. Symp. Rhodium Homogeneous Catal.*, 85 (1978).
92. J. L. Vidal and W. E. Walker, *Inorg. Chem.*, **20**, 249 (1981).
93. G. Csontos, B. Heil, and L. Marko, *Ann. N.Y. Acad. Sci.*, **239**, 47 (1974).
94. B. Heil and L. Marko, *Chem. Ber.*, **104**, 3418 (1971).
95. R. B. King, A. D. King, and M. Z. Iqbal, *J. Am. Chem. Soc.*, **101**, 4893 (1979).
96. B. F. G. Johnson and J. Lewis, *Adv. Inorg. Chem. Radiochem.*, **24**, 225 (1981).
97. A. Fumagalli, T. F. Koetzle, F. Takusagawa, P. Chini, S. Martinego, and B. T. Heaton, *J. Am. Chem. Soc.*, **102**, 1740 (1980).
98. B. T. Heaton, J. Jonas, T. Eguchi, and G. A. Hoffman, *J. Chem. Soc., Chem. Commun.*, 331 (1981).
99. L. H. Slaugh and R. D. Mullineaux, *Ger. Pat. Appl.*, 1 186 455, 1961.
100. L. H. Slaugh and R. D. Mullineaux, *U.S. Pat.*, 3 239 566, 3 239 569 and 3 239 570, 1966.
101. J. A. Osborn, J. F. Young, and G. Wilkinson, *J. Chem. Soc., Chem. Commun.*, 17 (1965).
102. D. Evans, J. A. Osborn, and G. Wilkinson, *J. Chem. Soc. A*, 3133 (1968).
103. C. K. Brown and G. Wilkinson, *J. Chem. Soc. A*, 2753 (1970).
104. G. Yagupsky, C. K. Brown, and G. Wilkinson, *J. Chem. Soc. A*, 1392 (1970).
105. G. Wilkinson, *Bull. Soc. Chim. Fr.*, 5055 (1968).
106. J. A. Osborn, F. H. Jardine, J. F. Young, and G. Wilkinson, *J. Chem. Soc. A*, 1711 (1966).
107. J. M. Brown, L. R. Canning, A. G. Kent, and P. J. Sidebottom, *J. Chem. Soc.*, 721 (1982).
108. J. M. Brown and A. G. Kent, *J. Chem. Soc., Chem. Commun.*, 723 (1982).
109. J. M. Brown, *Chem. Ind.* (*London*), 737 (1982).
110. R. L. Pruett and J. A. Smith, *J. Org. Chem.*, **34**, 327 (1969).
111. D. Evans, G. Yagupsky, and G. Wilkinson, *J. Chem. Soc. A*, 2660 (1968).
112. J. A. Davies, in *The Chemistry of the Metal—Carbon Bond* (Eds. F. R. Hartley and S. Patai), Wiley, Chichester, 1982, Chapter 21, p. 813.

The Chemistry of the Metal—Carbon Bond, Vol. 3
Edited by F. R. Hartley and S. Patai
© 1985 John Wiley & Sons Ltd.

CHAPTER **9**

The Fischer–Tropsch synthesis

G. HENRICI-OLIVÉ and S. OLIVÉ

Department of Chemistry, University of California at San Diego, La Jolla, California 92093, USA

I. INTRODUCTION	392
II. HISTORY	392
III. PRESENT STATUS OF TECHNICAL DEVELOPMENT	393
IV. THE PRODUCTS OF THE FISCHER–TROPSCH SYNTHESIS	395
A. Primary Products	395
B. Secondary Reactions of the α-Olefins	396
C. Modified Fischer–Tropsch Synthesis	397
V. DISTRIBUTION OF MOLECULAR WEIGHTS	398
A. The Schulz–Flory Distribution Function	398
B. Experimental Molecular Weight Distributions	402
VI. KINETICS AND THERMODYNAMICS OF THE FISCHER–TROPSCH REACTION	405
VII. REACTION MECHANISM	408
A. Suggested Mechanisms	408
B. The Carbide Theory in the Light of Homogeneous Coordination Chemistry	409
C. Details and Support for the CO Insertion Mechanism	411
1. The reaction scheme	411
2. Evidence for individual steps	414
3. Further mechanistic evidence for the CO insertion mechanism	418
4. Secondary reactions	419
5. Comparison with hydroformylation	420
6. Influence of the dispersity of the metal centres	421
7. The role of alkali metal promoters	423
VIII. PRODUCT SELECTIVITY	425
A. Consequences of the Schulz–Flory Molecular Weight Distribution	425
B. Deviations from the Schulz–Flory Distribution	428
IX. OUTLOOK	430
X. REFERENCES	430

I. INTRODUCTION

The Fischer–Tropsch synthesis is essentially a polymerization reaction—or perhaps better an oligomerization, since in most cases the average molecular weight of the product is not very high—where carbon—carbon bonds are formed between carbon atoms derived from carbon monoxide, under the influence of hydrogen and a metal catalyst, and with elimination of water. Without anticipating the detailed reaction mechanism, the main reaction of the Fischer–Tropsch (FT) synthesis may be formulated as:

$$n(CO + 2H_2) \xrightarrow{catalyst} -(CH_2)_n- + nH_2O \qquad (1)$$

Depending on the catalyst and reaction conditions, the products are linear hydrocarbons, oxygenated derivatives thereof, or mixtures of both. Usually a wide range of molecules with different chain lengths and with a molecular weight distribution characteristic of polymerization reactions is formed.

In this chapter we shall concentrate mainly on the chemistry and mechanism of the FT synthesis, with particular emphasis on the importance of intermediate metal—carbon bonds. However, the story of the FT synthesis would be incomplete without at least a brief account of its historical development and the present status of technical progress.

The major part of the chapter will then be dedicated to the application of principles of polymer chemistry, as well as concepts originating in the field of molecular catalysis with defined transition metal compounds, to the experimental facts of the FT synthesis. The aim is the discussion of a reaction mechanism which is in accord with the mentioned principles and concepts. Although literature reports have been amply consulted, and are duly referenced, the mechanistic implications represent, in many instances, the authors' present opinion.

II. HISTORY

Great discoveries in chemistry are often the consequence of accidental observations. Thus, the invention of the Ziegler polymerization of ethylene at low pressure was triggered when residual nickel catalyst in an autoclave led to an unexpected dimerization of ethylene to but-1-ene, whereas actually higher olefins were expected on an aluminium alkyl catalyst. In the search for an elucidation of the 'nickel effect', other metal compounds were tested, until the classical Ziegler catalyst, $TiCl_4$–AlR_3, emerged[1]. The oxo (hydroformylation) reaction was discovered by Roelen[2] while searching for the synthesis of hydrocarbons in homogeneous media.

This was not so with the FT synthesis. When the Kaiser Wilhelm Institut für Kohleforschung in Mülheim, FRG, was founded in 1914, the general idea of producing oil from coal was already taken into consideration. However, World War I delayed this long-term research goal. Roelen, who was one of the coworkers of Fischer and Tropsch, described the development of these early days[3]. Since 1913, the direct liquefaction of coal according to Bergius was known. On the other hand, BASF had obtained patents for the formation of hydrocarbons and, mainly, oxygenated derivatives thereof, from carbon monoxide and hydrogen under high pressure, with alkali-activated cobalt and osmium catalysts. In 1919, Fischer, then Director of the Kaiser Wilhelm Institut für Kohleforschung, decided that the indirect liquefaction of coal, via synthesis gas, was the preferred route. Repeating and developing the BASF patents, Fischer and Tropsch obtained a first patent on 'Synthol', a mixture of oxygen-containing derivatives of hydrocarbons, produced from CO and H_2 at over 100 atm and 400 °C, with alkali-treated iron shavings as catalysts. Combining this knowledge intelligently with that of the methanol synthesis with ZnO catalyst reported by Patard[4] in 1924, Fischer, Tropsch, and their coworkers in 1925 synthesized for the first time small amounts of ethane and higher hydrocarbons under normal pressure, at 370 °C, on an Fe_2O_3–ZnO catalyst. The first patent was applied for, only 2 months after the first laboratory observations, still in 1925. The following years saw the laborious development of improved catalysts until, in 1934, the FT synthesis was ready for transfer to industry. Ruhrchemie AG was the first company to be licensed to build a commercial plant, working at normal pressure and with a cobalt–thorium oxide catalyst.

The further development has been described in detail by Pichler[5], another coworker of Fischer and Tropsch at that time and only a few highlights are given here. In 1936, Fischer and Pichler[6] patented the synthesis at medium pressure (5–30 atm), with cobalt catalysts, which resulted in higher yields, higher molecular weight products, and a longer life of the catalyst. During World War II, nine FT plants were operated in Germany, based on the medium-pressure cobalt-catalysed process. From 1937 on, the research shifted back to iron catalysts. Fischer and Pichler[7] had discovered that alkalized iron catalysts, at medium pressure, resulted in further improvements of product yields and catalyst lifetime; moreover, the iron catalysts permitted a broader variation of the process variables and hence of the reaction products. However, these important discoveries did not mature to technical utilization before the end of the war.

After 1945, there was general access to the German research and development work through the various FIAT and BIOS Reports and other publications. The Bureau of Mines, in the USA, obtained, together with plans and documents, an entire FT pilot plant[8], and in the late 1940s several test facilities were built by the Bureau of Mines as well as by the private industry[8,9]. However, in the early 1950s, the world price of oil began to fall, and the great euphoria over cheap energy from oil exploded, not only in USA but also in Europe. Coal mines were shut down. Apart from price, the ease of automated transport systems and feed lines for refineries, production plants, industrial and domestic heating, etc., helped oil to oust the dirty and inconvenient raw material coal.

When the big shake-up came in the early 1970s, with the oil embargo, and the escalation of crude oil prices in its aftermath, scientific and industrial investigation reverted to coal as a source of carbon, and to the FT synthesis as one of the most potent ways to transform it into gasoline, oil, and other useful chemicals. Although, even with today's crude oil prices, coal-based routes are not yet completely competitive economically, a sound effect of the embargo shock has been to create a feeling of awareness and responsibility for the future. The eventual exhaustion of petroleum sources has been predicted for a considerable period of time. Commercial processes are now available to enable a coal-to-fuel industry to be started, although advanced technologies, improving efficiency and reducing costs, need to be developed.

III. PRESENT STATUS OF TECHNICAL DEVELOPMENT

All technically important developments since 1945 are based on the work of Fischer and Pichler[7] concerning the FT synthesis at medium pressure, with alkalized iron catalysts.

The only large commercial facility for the production of gasoline, gas oil, and paraffins from coal, by the FT process, is located in Sasolburg, in South Africa. This country possesses large coal resources, but has no oil. In 1955, the South African Coal, Oil, and Gas Corporation started the production of synthesis gas ($CO + H_2$) from coal using a Lurgi coal gasification process, and the synthesis of hydrocarbons by the FT reaction[10]. Two parallel process designs for the FT synthesis are used in this plant (Sasol I), the fixed-bed reactor and the circulating fluid-bed reactor. Among other variables, the two processes differ in the way they eliminate the heat of reaction. The fixed-bed reactor was originally developed by Ruhrchemie and Lurgi, in FRG. The precipitated, extruded iron catalyst is located in a large number of parallel tubes, surrounded by boiling water, the temperature of which is regulated by the steam pressure. A typical reactor of 3 m diameter and 17 m length contains over 2000 such tubes, each of 5 cm diameter, with a total of 40 m^3 of catalyst. The working pressure is *ca.* 25 atm at 220–240 °C; the synthesis gas is fed at an H$_2$ to CO ratio of 1.8[5]. The fluid-bed reactor, developed by the Kellogg Company in USA in pilot-plant operations, was industrialized to a large scale for the first time in Sasolburg. After years of scaling-up problems and improvements it is now completely reliable and technically mature[5b]. It works at a higher temperatue (310–340 °C), at similar pressure (24 atm), and at an H$_2$ to CO ratio of 6. The low particle size iron catalyst is moved together with the synthesis gas through the reactor in an upward gas stream and, after separation of the reaction products, is recycled. The heat of reaction is removed by sets of coolers built into the reactor.

Tables 1 and 2 compare the reaction products obtained by the two processes. The fixed-bed process leads to a higher average molecular weight of the product, and to 95% linear molecules; the fluid-bed process results in a higher olefin content, but has 40% of branched material (mainly methyl branches). Oxygen-containing compounds and aromatics are insignificant in the former, but not in the latter process. The straight-chain material from the fixed-bed process is a useful feedstock for the production of chemicals such as plasticizers, detergents with high biodegradability, and synthetic lubricants, whereas the fluid-bed process is more useful for the production of gasoline. In fact, Sasol I operates an intimate combination of both processes, permitting a flexible adaptation to market requirements.

The overall yield of gasoline and/or diesel fuel can be improved by a further work-up of the primary FT products, using known technology such as oligomerization of light olefins, hydrocracking, and reforming. Sasol I has provided the South Africans with a unique experience which has encouraged them to build two much larger facilities for the FT process (Sasol II and III) which, by 1984, were expected to provide half of that country's liquid fuel needs. Both of these two plants will operate according to the fluid-bed process[8].

TABLE 1. Comparison of the products obtained with fixed-bed and circulating catalysts (in wt.-%)[5b]

Fraction	Fixed-bed catalyst (220–240 °C)	Circulating catalyst (310–340 °C)
C_3–C_4	5.6	7.7
C_5–C_{11} (gasoline)	33.4	72.3
Gas oil	16.6	3.4
Alkanes, m.p. < 60 °C	22.1	3.0
Alkanes, m.p. 95–97 °C	18.0	—
Alcohols, ketones	4.3	12.6
Acids	Traces	1.0

TABLE 2. Comparison of the composition of the liquid fraction for fixed-bed and circulating catalysts (in vol.-%)[5b]

Components	Fixed-bed catalyst		Circulating catalyst	
	C_5–C_{10}	C_{11}–C_{18}	C_5–C_{10}	C_{11}–C_{18}
Olefins	50	40	70	60
Alkanes	45	55	13	15
Oxygen-containing compounds	5	5	12	10
Aromatics	—	—	5	15

Apart from these large industrial plants, there appears to be scientific and industrial research going on at many places. The motivation is evident. The price of the crude oil has increased by a factor of over ten from pre-embargo 1973 to date. It is likely to increase further, and at a faster pace than that of coal, because increasingly more costly drilling and recovery procedures will be necessary, in particular in such new oil fields as in Alaska or in the North Sea. Moreover, coal is estimated to represent over 65% of the world's recoverable fossile fuel resources, whereas petroleum is estimated as less than 10%[11]. Although the FT synthesis is only one of the various processes for the transformation of coal to a useful feedstock, it will certainly continue to attract much scientific and industrial interest io the future.

IV. THE PRODUCTS OF THE FISCHER–TROPSCH SYNTHESIS

A. Primary Products

The product mixture contains mainly olefins and alkanes, but also variable amounts of alcohols, aldehydes, acids, esters, and aromatic compounds, depending on the reaction conditions. In the early days of the process[12,13] it was already found that the main products obtained with cobalt catalysts were linear olefins and alkanes, and that the small amounts of non-linear products consisted predominantly of monomethyl-branched compounds. Since olefins and alkanes were found to have the same skeleton, it was assumed that the olefins were the primary products of the synthesis, and that the alkanes were formed therefrom by subsequent hydrogenation.

The olefin component of the hydrocarbons consists mainly of α- and β-olefins (alk-1- and -2-enes). The fact that the concentration of α-olefins is generally much higher than that corresponding to thermodynamic equilibrium led Friedel and Anderson[14] to the conclusion that α-olefins should be the primary products of the synthesis. This conclusion was unequivocally confirmed by Pichler et al.[15], who investigated the product composition as a function of the residence time of the synthesis gas on the catalyst, varying the space velocity. The synthesis was carried out at normal pressure and 200 °C, with a precipitated cobalt–thorium oxide–Kieselguhr catalyst, at an H_2 to CO ratio of 2. Under these conditions the main product consists of hydrocarbons in the gasoline and gas oil range. The result of the gas chromatographic analysis of the products is represented in Figure 1. The clear answer is that the shorter the residence time, the larger is the α-olefin fraction of the product. At longer residence times (low space velocities) the primarily formed α-olefins are transformed in secondary reactions into β-olefins, linear alkanes and methyl-branched products. The mechanistic implications of these findings cannot be

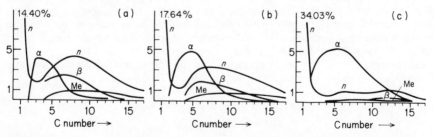

FIGURE 1. Composition of the FT mixture of hydrocarbons as a function of the space velocity: (a) $78\,h^{-1}$; (b) $337\,h^{-1}$; (c) $2380\,h^{-1}$. $n = n$-Alkanes; $\alpha = \alpha$-olefins; $\beta = \beta$-olefins; Me = methyl-branched alkanes. Catalyst = Co–ThO$_2$. (The space velocity is defined as the volume of synthesis gas per volume of catalyst × hour.) After pichler et al.[15].

overestimated. Whatever reaction mechanism might be suggested, it has to take into account that the main primary reaction product consists of linear molecules with one, and only one, double bond in the α-position. The alkane products are formed by subsequent hydrogenation of the olefins, and not by hydrogenolysis of a metal—carbon bond.

In contrast to the primary and secondary products mentioned, the small alcohol fraction found under the given reaction conditions was independent of the space velocity. Pichler et al. concluded convincingly that the alcohols are also primary products of the process. They indicated that there might be a common precursor for linear α-olefins and linear alcohols.

B. Secondary Reactions of the α-Olefins

The development of very efficient gas chromatographic methods (capillary columns, radio gas chromatography) has permitted a detailed determination of the many isomers contained in the synthesis products, and hence provided an insight into the course of the secondary reactions. Pichler et al.[16] reported a complete analysis of the C_7 hydrocarbon fraction of Sasol's fixed-bed catalysis product. They showed that this fraction is $> 91\%$ linear, and that the non-linear part is $> 90\%$ monomethyl-, 6% dimethyl-, and 2.5% ethyl-branched. The approximate composition of the linear part is 51% heptane, 31% hept-1-ene, 17% (Z)- and (E) hept-2-ene, and only 1% (Z)- and (E) hept-3-ene. This study shows that the main secondary reaction of the primary olefins is hydrogenation, with some double bond migration also being present.

Another important contribution towards the mechanism of the secondary reactions of α-olefins was made by Schulz et al.[17]; ^{14}C-labelled ethylene, propylene, but-1-ene, or hexadec-1-ene were added to the synthesis gas, and the fate of these α-olefins was monitored by radio gas chromatography. Both the normal-pressure cobalt catalyst system and the medium-pressure iron catalyst system were investigated. With cobalt catalysts, over 90% of the added olefin reacts, whereas with iron systems the reaction is slower. Most of the olefin is converted into alkane by hydrogenation.

The fraction of the labelled olefin transformed into alkane is comparable to the fraction of the corresponding FT product with the same carbon number found as alkane. This is considered to be additional proof for the FT synthesis to result primarily in α-olefins, which are then hydrogenated in a secondary reaction. However, the tagged α-olefins are also incorporated into growing chains. With increasing carbon number of the olefin, its incorporation decreases. About 30% of the ethylene and propylene, but only 6% of hexadec-1-ene, are incorporated into the FT synthesis products (this is in accord with the relative polymerization activity of these olefins with transition metal catalyst). The incorporation of ethylene does not lead to chain branching, whereas that of propylene gives methyl branches. Simultaneously, α-olefins are split, with breaking of the double bond. Since this splitting occurs in a hydrogen atmosphere, additional methane is formed by this secondary process. In the cobalt system, and with about 0.3–0.8 vol.-% of the labelled olefin in the gas feed, 4-% of the ^{14}C was found as methane.

Interestingly, $[2\text{-}^{14}C]$-propene also gave a considerable amount of labelled methane. Hence, the following decomposition reactions have to be considered:

$$CH_2 = CH_2 \longrightarrow 2\ CH_4 \qquad (2)$$

$$MeCH = CH_2 \longrightarrow \begin{array}{l} C_2H_4 + CH_4 \\ \\ 3\ CH_4 \end{array} \qquad (3)$$

$$RCH=CH_2 \longrightarrow RMe + CH_4 \qquad (4)$$

The splitting of the olefin is more important with cobalt catalysts than with iron systems, and at least part of the high methane content (Figure 1) is probably due to this kind of reaction. The incorporation reaction is also more frequent with the cobalt system; with the iron system hydrogenation predominates. Neither Co nor Fe produces significant cracking of alkanes. ^{14}C-labelled butane added to the synthesis gases could be recovered unchanged in over 99% yield. Even added 2-methylpentadecane, which certainly has a prolonged residence time in the reactor, remained essentially unaltered.

C. Modified Fischer–Tropsch Synthesis

The synthesis originally aimed at gasoline and motor oil, but certain modifications of the catalyst and/or the reaction conditions led to a broad variety of products, the formation of which is closely related to the FT synthesis proper. Using a ruthenium catalyst, and working at high pressure (1000–2000 atm) and temperature (140–200 °C), Pichler and Buffleb[18] were able to orient the synthesis to high-melting linear alkanes having molecular weights up to 10^6; the 'polymethylenes' obtained are essentially identical with Ziegler polyethylene.

As mentioned above, alcohols and other oxygenated compounds are always found among the FT products in greater or lesser amounts, and sometimes unwanted. From the old synthol work of Fischer and coworkers it was known that high pressure and high temperature favour the formation of oxygenated compounds. More recently, interest appears to have shifted back towards the original Synthol process. High-pressure and high-temperature processes have been reported to give good yields of aliphatic alcohols in the C_1–C_{18} range, important raw materials for detergents, plasticizers, etc.[19a–e]. In Russian patents[19d], iron catalysts have been used at temperatures in the range 160–220 °C, pressures of 50–300 atm, and H_2 to CO ratios of 5–20, to produce liquid oxygen-containing compounds in the range C_1–C_{20}, mostly alcohols. Subsequent hydrogenation over Cu–Cr catalysts increases the yield of alcohols. At high space velocity, selectivities of up to 86% (expressed as percentage of alcohols in total liquid product boiling below 160 °C) are claimed[19e].

Another modification involves the addition of ammonia to the synthesis gas. Ruhrchemie[20] patented in 1949 the preparation of primary amines, in the presence of 0.5–5% of NH_3 in the gas feed. Iron catalysts were used under conditions which, in the absence of ammonia, would result in hydrocarbons. W. R. Grace and Co., more recently, also claimed iron as the main component of the catalyst for a process of preparing linear aliphatic primary amines from mixtures of CO, H_2, and NH_3[21]. In this patent it was noted that an increasing amount of ammonia reduces the average molecular weight of the product amine, and that no other nitrogen-containing compounds are formed in the process. Evidently, the ammonia interferes with the FT chain-growth reaction, terminating the molecules. An extreme case of molecular weight reduction by NH_3 is given by the formation of acetonitrile from CO, H_2, and NH_3 at 500 °C and normal pressure on SiO_2-supported iron or molybdenum catalysts[22].

Related to the FT reaction is the synthesis of ethylene glycol from CO + H_2, with up to 70% selectivity, on a rhodium catalyst[23]. This high selectivity to a C_2 compound, as well as the use of a soluble catalyst system, however, differentiates this interesting reaction from the classic FT systems and is reminiscent of the hydroformylation reaction.

Other catalytic reactions of CO + H_2, such as the formation of methane (actually the oldest hydrocarbon synthesis, found in 1902 by Sabatier and Senderens[24]), or the technically very important production of methanol (known since 1924[4]), although closely related to the FT synthesis, will not be treated here explicitly, because they do not involve carbon—carbon bond formation.

V. DISTRIBUTION OF MOLECULAR WEIGHTS

A. The Schulz–Flory Distribution Function

Schulz[25], in 1935, derived an equation for the distribution of molecular weights of polymers obtained by a free-radical polymerization process, i.e. through a one-by-one addition of monomer to a growing chain. This distribution function is generally applicable if there is a constant probability of chain growth, α, and $\alpha < 1$ (i.e. any reaction delimiting the chain growth is present).

These conditions can safely be assumed to be present during FT synthesis. The reaction can be operated over long periods of time at a constant rate[5b]. While this constant rate, obtained on a given number of catalytic sites, indicates the absence of any chain termination in the kinetic sense (i.e. annihilation of active sites), the fact that a large number of molecules are formed per active site indicates that a chain transfer reaction takes place, in the course of which a product molecule leaves the active site, and a new chain is started at the same centre. Since α-olefins, as well as alcohols, form the primary products of the FT synthesis, we have to consider more than one particular chain transfer reaction. The probability of chain growth is then defind as

$$\alpha = \frac{r_p}{r_p + \sum r_{tr}} \quad (5)$$

where r_p and r_{tr} are the rates of chain propagation and chain transfer, respectively; α is assumed to be independent of the chain length.

The statistical derivation of the distribution function according to Schulz is lucid and straightforward. The probability of a chain growth step is α. The probability for the growth step taking place P times without interruption is

$$p_P = \alpha_1 \alpha_2 \alpha_3 \alpha_4 \ldots \alpha_P = \alpha^P \quad (6)$$

The number of molecules of degree of polymerization P, n_P, is proportional to the probability of their formation:

$$n_P = \text{constant} \times \alpha^P \quad (7)$$

The mass (or weight) fraction, m_P, is proportional to n_P, as well as to the molecular weight of the molecules under consideration, $M_P = M_M P$ (M_M = molecular weight of the monomer). Hence

$$m_P = A P \alpha^P \quad (8)$$

where A contains the constant M_M. The mass fraction is defined in such a way that the sum of all m_P is unity. Moreover, the mass fraction is considered to be a continuous function of P (this last statement is perfectly admissible for a large average molecular weight; its consequences for small average molecular weight will be discussed below). It then follows that

$$\int_0^\infty m_P dP = A \int_0^\infty P\alpha^P dP = 1 \quad (9)$$

and

$$A = 1 \bigg/ \int_0^\infty P\alpha^P dP \quad (10)$$

Solving the integral (taking into account that $\alpha < 1$, $\alpha^\infty = 0$), and combining equations (8) and (10), leads to the mass distribution function:

$$m_P = (\ln^2 \alpha) P \alpha^P \quad (11)$$

For practical purposes this is conveniently written in logarithmic form:

$$\log\left(\frac{m_P}{P}\right) = \log(\ln^2\alpha) + (\log \alpha)P \tag{12}$$

If a molecular weight distribution follows this law, a plot of $\log(m_P/P)$ versus P should result in a straight line. Logarithmic representations of experimental data are often regarded as less accurate, smoothing away errors and deviations. However, this one has an internal control, since α is contained in the slope ($\log \alpha$), as well as in the intercept with the ordinate [$\log(\ln^2 \alpha)$] of the straight line, and the values of α from both sources have to agree. Figure 2 shows a series of theoretical straight lines, in the range of α and P interesting for our present purpose. It should be mentioned, however, that the anatomy of equations 12 and 13 is such that the slope will result in the correct value of α (and hence of the average molecular weight; see below), even if only a few single points (mass fractions

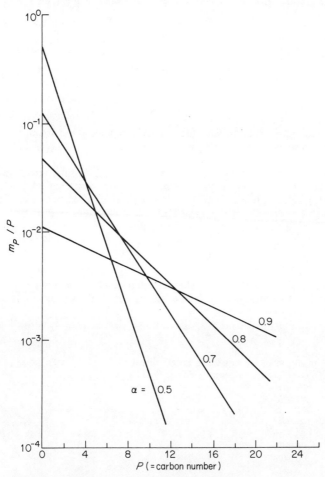

FIGURE 2. Theoretical Schulz–Flory plots (according to equation 12) for α-values and P range relevant for the FT synthesis.

m_P) are known with sufficient accuracy. The value derived from the intercept, however, will be in error if the experimental distribution is incomplete.

According to Schulz, the average degree of polymerization (number-average, P_n) is related to the mass distribution function as follows:

$$P_n = 1 \Big/ \int_0^\infty (m_P/P) \, dP \tag{13}$$

With the application of equation 11, this leads to

$$P_n = 1/\ln^2\alpha \int_0^\infty \alpha^P \, dP = -1/\ln\alpha \tag{14}$$

Flory[26] published in 1936 the theoretical distribution function for a different type of macromolecule formation, the polycondensation of bifunctional monomers, which takes place via the gradual growing together of x-mers (dimers, trimers, etc.):

$$x\text{-mer} + y\text{-mer} \rightarrow (x+y)\text{-mer} \tag{15}$$

with $x, y \geq 1$. Flory stated that, although based on an entirely different set of conditions, his equation is essentially equivalent to equation 11 of Schulz. Following Flory's argument, but using, for the purpose of easy comparison, similar symbols to those in the derivation of equation 11, α' is defined as the fraction of functional groups that have reacted at a given time:

$$\alpha' = (N_0 - N_t)/N_0 \tag{16}$$

(N_0 and N_t = the number of molecules present at the beginning and at time t, respectively). This defines α' also as the probability that a condensation reaction has taken place at a given end group.

For the degree of polymerization P to be realized, the condensation reaction must have taken place $(P-1)$ times. The probability that no condensation has taken place at both ends is $(1-\alpha')^2$. Hence, the probability of existence of each particular configuration is $\alpha'^{P-1}(1-\alpha')^2$. The probability that any of the P configurations exists is $P\alpha'^{P-1}(1-\alpha')^2$, and this is equal to the 'weight fraction distribution':

$$m_P = \frac{(1-\alpha')^2}{\alpha'} P\alpha'^P \tag{17}$$

In this form, $(1-\alpha')^2/\alpha'$ corresponds to $\ln^2\alpha$ in equation 11. For α (or $\alpha') \rightarrow 1$, equations 11 and 17 are essentially equivalent; for smaller values of α (or α'), there is a slight discrepancy (see Table 3).

According to Flory's derivations, the average degree of polymerization (number

TABLE 3. Comparison of the molecular weight distribution, $m_P = f(\alpha)P\alpha^P$, and of the average degree of polymerization, P_n, as given by Schulz[25] and Flory[26]

$\alpha(\alpha')$	Schulz: $f(\alpha) = \ln^2\alpha$	Flory: $f(\alpha') = (1-\alpha')^2/\alpha'$	Schulz: $P_n = -1/\ln\alpha$	Flory: $P_n = 1/(1-\alpha')$
0.99	1.01×10^{-4}	1.01×10^{-4}	99.4	100
0.9	0.0111	0.0111	9.5	10
0.8	0.0498	0.0500	4.5	5
0.7	0.1272	0.1286	2.8	3.3
0.5	0.4805	0.500	1.4	2
0.3	1.449	1.633	0.8	1.4

average) is given by

$$P_n = \frac{1}{1-\alpha'} \tag{18}$$

For α (or α') → 1, equations 14 and 18 are, again, essentially equivalent. For smaller α (or α'), P_n according to equation 18 is generally higher; calculated values for P_n are included in Table 3. In the lower range of α, equation 18 may be more appropriate than equation 14, since the method of integration over the range from 0 to ∞, although an excellent approximation for large degrees of polymerization, is less appropriate when applied to relatively small molecules, where the step-by-step growth mechanism has a greater bearing. Because of the close overall resemblance of the two mass distribution functions, it has become customary to call them, in either form, the 'Schulz–Flory distribution function'.

About 15 years later, Friedel and Anderson[27], based on earlier work of Herington[28], developed an equation for the products of the FT synthesis. Since again the same statistics are involved (as long as branching can be neglected), the resulting equation is equivalent to the former two (although apparently the authors were not aware of the work of Schulz and Flory).

For the sake of completeness, we want to mention briefly what kind of a molecular weight distribution is to be expected if $\alpha = 1$, i.e. if there is neither chain termination nor transfer. (Under certain conditions this type of distribution appears to be approximated in FT synthesis; see Section VIII.) For the ideal case that all molecules have started the growth at the same time, Flory[29] has shown that the molecular weight distribution is given by the Poisson function:

$$m_P = \frac{e^{-v}v^{(P-1)}P}{(P-1)!(v+1)} \tag{19}$$

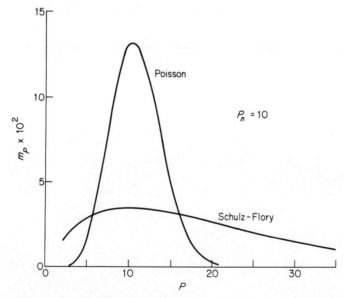

FIGURE 3. Poisson distribution, according to equation 19, and Schulz–Flory distribution, according to equation 11, for the same average degree of polymerization, $P_n = 10$.

where v is the average number of growth steps per molecule, which is related to the average degree of polymerization P_n by

$$v = P_n - 1 \tag{20}$$

For comparison, the theoretical Schulz–Flory and Poisson distributions, corresponding to the same average degree of polymerization, $P_n = 10$, are given in Figure 3 (equation 11 with $\alpha = 0.905$ and equation 19 with $v = 9$, respectively). Evidently, the Poisson distribution is considerably narrower.

B. Experimental Molecular Weight Distributions

As mentioned in Section III, the products of the FT synthesis vary considerably with the catalyst, reaction conditions (pressure, temperature), and process design. Figures 4–7 show that despite such variations, the Schulz–Flory molecular weight distribution function generally holds.

Figures 4 and 5 refer to FT syntheses oriented towards the production of hydrocarbons. Since the primary α-olefins are partly transformed into inner olefins (isomerization) and alkanes (hydrogenation), the weight fraction, m_P, for each degree of polymerization is the sum of olefins and alkanes with the same number of carbon atoms. Figure 4 is a plot of the data in Figure 1 according to equation 12. In the range C_4–C_{12}, the experimental points lie on a straight line. C_1 is too high and C_2 and C_3 are too low, as are the values beyond C_{13}. The value of α is 0.81 from the slope and 0.80 from the intercept.

Figure 5 shows the same plot for data of Storch *et al.*[9]. The range C_3–C_{13} gives a reasonable straight line. Again, the value of C_1 is too high and the values of C_2 and beyond C_{13} are too low. From the slope a value of $\alpha = 0.87$ is obtained and from the intercept $\alpha = 0.86$.

The reasons for the deviations in the range C_1–C_3 have already been mentioned: the α-olefins with the greatest ability to be coordinatively bonded to a transition metal (ethylene

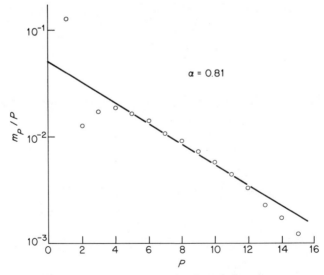

FIGURE 4. Hydrocarbon data of Figure 1a, represented according to equation 12.

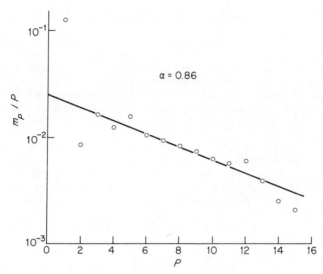

FIGURE 5. Molecular weight distribution of hydrocarbons[9], represented according to equation 12. Cobalt catalyst.

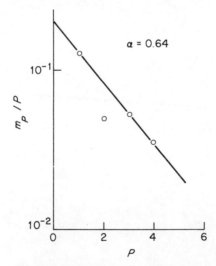

FIGURE 6. Data from the Sasol fluid-bed process, represented according to equation 12.

> propylene) are partly lost by decomposition to methane and, to a lesser extent, by incorporation into growing chains. At the upper end of the distribution, the hydrocarbons formed tend to have longer residence times on the catalyst owing to their lower mobility. Presumably, they are partly polymerized, partly cracked, or even carbonized. An additional uncertainty in this range may be caused by the fact that the high-boiling

fractions give very broad peaks on gas chromatography; their values tend to be underestimated.

Data from the Sasol fluid bed process, as reported by Pichler and Krüger[30], are represented according to equation 12 in Figure 6. Only the weight fractions of the C_1–C_4 hydrocarbons are given explicitly in the reference (C_1, 0.131; C_2, 0.101; C_3, 0.162; C_4, 0.132); the hydrocarbons above C_5 are grouped together (39 wt.-%). The accuracy of the weight fractions of the lower hydrocarbons is reflected in the fact that α from the slope as well as from the intercept is 0.64. As mentioned in Section III, the Sasol fluid bed process with iron catalysts leads to a relatively low molecular weight product. From the α value, an average degree of polymerization of ca. 2.6 can be estimated.

Yang et al.[31] have reported on a FT system leading to an even lower average degree of polymerization. The catalyst is coprecipitated Co–Cu–Al_2O_3, and H_2 and CO are applied at medium pressure, at a ratio ranging from 3:1 to 1:1, the temperature is 225–275 °C. Figure 7 shows data for one particular set of conditions (not identified in the reference). The value of α is 0.55 from the slope and 0.52 from the intercept, resulting in an estimated average degree of polymerization of $P_n \approx 2.2$.

As mentioned in Section IV.C, the FT synthesis can be oriented to give predominantly alcohols by suitable selection of the catalyst and reaction conditions. Using data of Kagan et al.[19e], we applied equation 12 also to alcohols. Figure 8 shows two sets of data, obtained with different catalysts. For the Fe_3O_4 catalyst modified with Al_2O_3 and K_2O (Figure 8a), an average value of $\alpha = 0.37$ has been calculated from the slope (0.38) and from the intercept (0.36). Data for the pure Fe_3O_4 system (Figure 8b) give the same value, $\alpha = 0.54$, from both the slope and intercept.

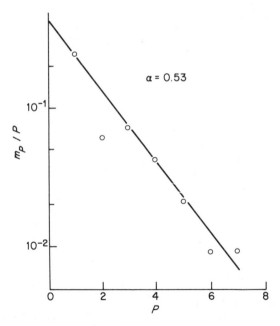

FIGURE 7. Low molecular weight hydrocarbons, as obtained on a Co–Cu–Al_2O_3 catalyst[31], represented according to equation 12.

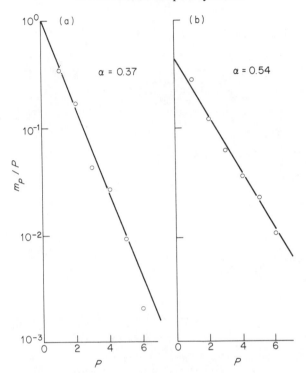

FIGURE 8. Molecular weight distribution of alcohols[19e], represented according to equation 12: (a) $Fe_3O_4 + Al_2O_3 + K_2O$; (b) Fe_3O_4.

From a mechanistic point of view it is of particular interest that amines produced by a modified FT process (cf. Section IV.C) present the same type of molecular weight distribution. This is shown if Figure 9. In the range C_4–C_{15} the points fit the expected straight line reasonably well. The values of α obtained from the slope ($\alpha = 0.68$) and from the intercept ($\alpha = 0.70$) are in satisfactory agreement; the estimated average degree of polymerization is $P_n \approx 3.2$.

VI. KINETICS AND THERMODYNAMICS OF THE FISCHER–TROPSCH REACTION

Most kinetic studies publised over the years have been based on actual process conditions. As Dry et al.[32] pointed out, this makes interpretations difficult, mainly owing to continuous changes of partial gas pressures along the catalyst bed. Using a specially constructed small differential reactor, Dry et al.[32] determined the basic rate law for the FT reaction on an iron catalyst at 225–265 °C. Maintaining the CO partial pressure constant, the H_2 partial pressure was varied, correcting at the same time the space velocity for constant contact time with the catalyst (i.e. working at a constant linear velocity). A clear first-order dependence of the overall rate on the H_2 partial pressure was found. The corresponding measurements at constant H_2 and varying CO partial pressure resulted in a zero-order dependence of the rate on the latter. Hence,

$$\text{rate} = K p_{H_2} \tag{21}$$

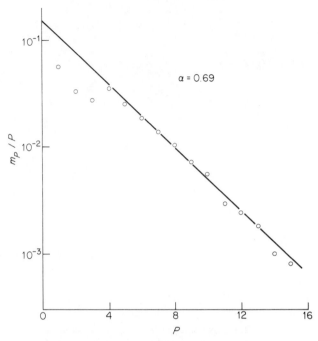

FIGURE 9. Modified FT process for the production of amines[21]; molecular weight distribution of amines according to equation 12.

where p_{H_2} is the partial pressure of hydrogen and K is a constant. This simple rate law has been found to be valid in the pressure range 1–2 MPa (10–20 bar), and with a synthesis gas ratio (H_2 to CO) varying from 1 to 7. From the temperature dependence of the reaction rate at a given set of conditions, an overall activation energy of ca. 70 kJ mol^{-1} (ca. 17 kcal mol^{-1}) was determined.

Under actual process conditions the macrokinetics are more complicated. A large body of data has been recollected by Storch et al.[33]. Rate equations featuring a reciprocal proportionality with p_{CO} have been reported[34,35]. At temperatures above 300 °C, the water gas shift reaction becomes important; water vapour has been found to depress the rate[32]. Overall activation energies in the range 84–113 kJ mol^{-1} (20–27 kcal mol^{-1}) have been reported for cobalt catalysts[5b].

There is a general consensus, that the reaction rate is amazingly low compared with those of other reactions in the field of heterogeneous catalysis[5b,36,37]. Dautzenberg et al.[37] investigated whether the low rate is due to the fact that only very few exposed surface metal atoms are active, or the active sites themselves have a very low intrinsic activity. They used a pulse technique, in the course of which a catalyst (ruthenium in their case) was repeatedly exposed to a CO–H_2 mixture at 210 °C during a variable time τ. Between exposures, the system was quenched by flushing with hydrogen alone and heated to 350 °C to force chain termination and product release. It was found that the average chain length, as well as the product molecular weight distribution, depend sensitively on τ. For instance, for $\tau = 8$ min, the C_{12}/C_6 molar ratio was 0.12, whereas under stationary conditions the same system would have resulted in a value of 0.74. On increasing the pulse time, the production of long-chain hydrocarbons was enhanced, and the relative distribution of the lighter hydrocarbons approached that obtained under steady-state conditions. Based on an elaborate kinetic model, the authors were able to estimate a formal rate constant of actual chain growth, $k_{propagation} \approx 1.5 \times 10^{-2} s^{-1}$, corresponding to a growth rate of about one —CH_2— group per minute per growing chain. This work appears to relegate the cause of the low overall rates to slow chain growth rather than to a small number of active sites, at least for the case of ruthenium catalysts. Moreover, the fact that the lighter hydrocarbons approach their steady-state situation at τ values where the long-chain hydrocarbons are still in considerable deficiency indicates that the chain initiation is not rate determining. This work does not answer the question of which partial step in the course of the complex transformation from CO to a —CH_2— group is actually rate determining. However, the experimental rate law (equation 21) indicates that the rate-determining step involves hydrogen.

The synthesis of hydrocarbons from CO and H_2 is a strongly exothermic reaction, evolving 146–176 kJ (35–40 kcal) per mole of CO converted to hydrocarbons, under the usual reaction conditions[5b,9]. Since the product distribution depends sensitively on the reaction temperature, heat removal is a very important factor in process design (cf. Section III).

The equilibrium constants for the formation of hydrocarbons of varying chain length depend not only on the temperature, but also on the chain length. Figure 10 shows that the yield of higher hydrocarbons can be expected to decrease with increasing temperature. Below 400 °C, the formation of the higher hydrocarbons is favoured over the lower hydrocarbons, since the equilibrium constant increases with increasing carbon number. However, a graph such as that in Figure 10 can only indicate general tendencies. The actual FT synthesis is governed by a large number of parallel primary reactions (e.g. formation of the hydrocarbons and alcohols), secondary reactions (e.g. hydrogenation and

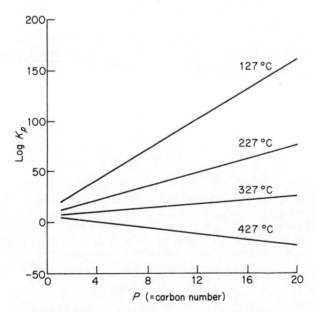

FIGURE 10. Equilibrium constant (K_p) as a function of the chain length of the product molecules and temperature (according to Pichler and Krüger 5b).

isomerization) and side reactions (e.g. water gas shift equilibrium, disproportionation of CO to C and CO_2), resulting in a complex system of simultaneous equilibria[38] which resists exact evaluation.

VII. REACTION MECHANISM

A. Suggested Mechanisms

The mechanism of the FT synthesis has attracted much interest in the past decade. Some authors favour the original proposal made by Fischer and Tropsch in 1926[39], according to which the synthesis proceeds via the formation of carbides. They assumed that the finely divided, carbon-rich carbides are decomposed by hydrogen, with regeneration of the catalyst metal and formation of the hydrocarbons. Evidently, this would imply the formation of CH_2 entities arranged in a row on the surface of the catalyst, unless one assumes that the CH_2, or in general CH_x, fragments move more or less freely along the catalyst surface. (For a recent review, see ref. 40.) The essential features of this scheme are given in equation 22. Storch et al.[9] suggested the formation of hydroxymethylene groups,

$$\overset{O}{\underset{|}{C}} \xrightarrow{H_2} \overset{}{\underset{|}{C}} \xrightarrow{H_2} \overset{H_2}{\underset{\triangle}{C}} \longrightarrow \cdots \overset{H_2}{\underset{|}{C}} \cdots \overset{H_2}{\underset{|}{C}} \cdots \overset{H_2}{\underset{|}{C}} \cdots \longrightarrow$$

$$-CH_2-CH_2-CH_2- \longrightarrow \text{etc.}$$

(22)

from carbon monoxide chemisorbed at the metal surface, and hydrogen chemisorbed in atomic form; C—C bonds are then established through a condensation reaction between the hydroxymethylene groups, with loss of H_2O. Here again, a perfect alignment of the hydroxymethylene groups in a row is implicit, unless one assumes a free movement of carbenes on the surface. The basics of this mechanism are summarized in equation 23. On

$$\overset{O}{\underset{|}{C}} \overset{O}{\underset{|}{C}} \xrightarrow{H_2} \overset{H}{\underset{\|}{C}} \overset{OH}{\underset{\|}{C}} \overset{H}{\underset{\|}{C}} \overset{OH}{\underset{\|}{C}} \xrightarrow{-H_2O} \overset{H}{\underset{\|}{C-C}} \overset{OH}{\underset{\|}{C}} \xrightarrow{H_2}$$

$$\overset{CH_3}{\underset{\|}{C}} \overset{OH}{\underset{}{}} \longrightarrow \text{etc.}$$

(23)

the other hand, Pichler and Schulz[41] and also the present authors[42,140] suggested mechanisms involving CO insertion into metal—H (initiation) and metal—alkyl (chain growth) bonds, analogous to the well known CO insertion into such bonds in homogeneous catalysis, and subsequent reduction of the acyl groups. The two mechanisms diverge in some intermediate steps, which will be discussed in Section VII.C, but the general growth pattern can be summarized as shown in equation 24, with R = H or

$$\underset{M(CO)_n}{\overset{R}{|}} \longrightarrow \underset{M(CO)_{n-1}}{\overset{\overset{R}{|}}{\overset{CO}{|}}} \xrightarrow[+H_2/CO]{-H_2O} \underset{M(CO)_n}{\overset{\overset{R}{|}}{\overset{CH_2}{|}}} + H_2O$$

(24)

alkyl). The important difference with equations 22 and 23 resides in the fact that the entire chain growth takes place at the same metal centre, at the surface of a catalyst. (Pichler and Schulz's scheme requires a second neighbouring metal centre to take care of the transformation of the oxygen to water, see below.)

B. The Carbide Theory in the Light of Homogeneous Coordination Chemistry

Ever since the carbide theory for the FT synthesis appeared, there has been controversy about the validity of this theory. After years of arguments and counterarguments, the problem appeared to be temporarily settled by Kummer et al.[43]. Using ^{14}CO or ^{12}CO for the formation of carbide on the surface of iron and cobalt catalysts, and running the synthesis with $^{12}CO + H_2$ or $^{14}CO + H_2$, respectively, they studied the distribution of radioactivity in the initially formed hydrocarbons. They concluded that the carbide route could not account for more than 10% of the product.

After this work, the carbide theory was more or less shelved until the 1970s when, motivated by the oil embargo, work on the FT synthesis was actively resumed. At that time, the concepts of coordination chemistry had made their way into the interpretation of catalysis, in particular in homogeneous phases, and reaction paths more related to those established for homogeneous catalysis were suggested for the FT synthesis. A clear example was to consider the growth step as the consequence of the insertion of CO into metal—alkyl bonds, as mentioned in the previous section.

In the meantime, Takeuchi and Katzer[44] investigated the methanol formed from labelled CO on an $Rh-TiO_2$ catalyst. A mixture of $^{13}C^{16}O$ and $^{12}C^{18}O$ was hydrogenated, and the resulting methanol had an isotope distribution very close to that of the CO feed. This clearly shows that the methanol synthesis had taken place without any separation of CO into its atomic components. It should be recalled that olefins and alcohols are the primary products of the FT synthesis (cf. Section IV.A) and that it has been suggested and supported that they have a common precursor[42a]. Hence, these findings appear to corroborate the CO insertion mechanism.

Other workers, however, favoured the carbide theory. Brady and Pettit[45] studied the behaviour of methylene fragments, as produced by the thermal decomposition of diazomethane, on various transition metals. They concluded that their observations are in agreement only with the mechanism depicted by equation 22. Their main findings are (a) in the absence of H_2 and CO only ethylene is formed, (b) in the presence of H_2 hydrocarbons up to C_{18} featuring a Schulz–Flory molecular weight distribution are observed, and (c) in the presence of H_2–CO the same type of distribution, but with an average degree of polymerization higher than that for H_2–CO alone, is observed. However, the findings of Brady and Pettit do not in fact contradict the assumption that the FT synthesis takes place by a CO insertion mechanism (equation 24), as discussed recently by the present authors[46]. The polymerization of CH_2 fragments originating from the decomposition of CH_2N_2 by a soluble metal hydride catalyst has been described[47]. Presumably, the CH_2 fragments are inserted into metal—H (initiation) and metal—alkyl (chain growth) bonds. In fact, convincing evidence for the insertion of a CH_2 fragment coordinated to a transition metal centre into a metal—alkyl bond located at the same centre has been reported recently[48].

Hence, the data of Brady and Pettit are more consistently interpreted by a step-by-step insertion polymerization of CH_2 units on a metal hydride catalyst which, of course, results in a Schulz–Flory distribution. This interpretation is corroborated by the fact that nickel and palladium catalysts, which under the reaction conditions used by Brady and Pettit would not be FT active, do form higher hydrocarbons from CH_2N_2–H_2. In the absence of H_2, the polymerization of CH_2 fragments cannot take place because no metal hydride catalyst is formed. Actually, Brady and Pettit reported that only ethylene (the product of the uncatalysed dimerization of CH_2 fragments) is formed.

The increases in the molecular weight, within the framework of a Schulz–Flory distribution, on introducing CH_2N_2 into an FT system appears then as a consequence of a copolymerization with different 'monomers', $(CO + H_2)$ and CH_2N_2 which, however, lead to the same incorporated unit, $-CH_2-$. Adapting equation 5 to this situation, one obtains

$$\alpha'' = \frac{\sum r_p}{\sum r_p + \sum r_{tr}} \quad (25)$$

where $\sum r_p$ is the sum of the rates of the two growth steps. If $\alpha'' > \alpha$, a Schulz–Flory distribution with a larger average molecular weight results.

Biloen et al.[49] also support the carbide theory of the FT synthesis. They suggest that CO dissociates in a fast step to give carbidic intermediates, from which methane and the higher hydrocarbons are produced. The experimental basis is similar to that of Kummer et al.[43] mentioned above, except for working with ^{13}C instead of ^{14}C and using mass analysis. The main evidence is the presence of ^{13}C in a fraction of some of the lower hydrocarbons (mainly CH_4, with some C_3H_8), if ^{13}C was deposited on to the surface of a cobalt catalyst by decomposition of ^{13}CO, and the synthesis was carried out with $^{12}CO-H_2$, at low conversion.

Although it was pointed out recently[50] that these data do not provide solid evidence for the carbide mechanism to account for a large fraction of the hydrocarbons, the fact remains that a considerable amount of the methane, and small amounts of lower hydrocarbons, do actually contain one, two, or even three ^{13}C atoms. Evidently, part of the methane and small amounts of lower hydrocarbons can be formed in a way different from CO insertion. However, do we really have to consider 'CH_x particles'[49] moving freely along the catalyst surface until they meet and form hydrocarbons?

It appears that modern coordination and organometallic chemistry offers more probable reaction paths[50]. The insertion of terminal (methylidene) or bridging (μ-methylene) carbene groups into metal—alkyl bonds, within the same complex or cluster, has repeatedly been reported[48,51,52]. In view of these interesting reactions, an occasional insertion of a carbene group into a growing FT chain appears probable. On the other hand, the dimerization of carbene groups is also known[53,54], but whenever it has been observed, it took place within the confines of a binuclear complex or a cluster (i.e. never 'through space'). Mobility of carbene ligands within the confines of a cluster or a bimetallic intermediate, involving an exchange between terminal and bridging positions, has also been reported[55,56].

It should also be remembered that the transfer of alkyl ligands from one metal to another is common practice in preparative organometallic chemistry[56]; here again the pathway is assumed through intermediate binuclear complexes[57]. Such internuclear alkyl group exchange is best documented for the Group III metal trialkyl compounds, but there is also increasing evidence that it takes place with transition metal compounds[57,58]. Also, rapid intranuclear motion of hydride ligands within metal clusters has amply been demonstrated by n.m.r. data; the activation energy for such motion may be as low as 12–20 kJ mol^{-1} (3–5 kcal mol^{-1})[57].

In the light of this knowledge from homogeneous coordination and organometallic chemistry, the formation of methane and some lower hydrocarbons from carbidic carbon, as observed by Biloen et al.[49] in the FT synthesis, may be visualized as follows. Carbide-carbon is generally located in the interstices between metal atoms, equidistant from several metal atoms[59]. If an oxidative addition of hydrogen takes place at any of the surrounding metal atoms, carbene ligands can be formed. These carbene ligands have a certain mobility, the movement taking place by changes from bridging to terminal positions or vice versa. Certainly, the carbene ligand remains coordinatively bonded to at least one metal atom at all times. The driving force for such movement is, presumably, an energy gain through more favourable ligand environment at the acceptor metal centre. The same restricted mobility may also be assumed for the alkyl ligands (growing chains in the FT synthesis), albeit only on the surface.

The most probable fate of the carbene ligand is the capture of two more hydrogen atoms to give methane. The small hydrogen, either as a molecule, or dissociated, can be assumed to have the easiest access to carbene ligands, even to those located in the inner part of the catalyst particles. In fact, the major part of the labelled carbide carbon ends up as methane[49]. It should also be noted that methane is frequently found in large excess among the reaction products of the FT synthesis, compared with the otherwise 'normal' (Schulz–Flory) distribution of molecular weights (see, e.g., Figures 4 and 5). Presumably, there is always some methane formation via the carbide–carbene route, parallel to the FT synthesis proper.

If a carbene ligand and an alkyl group (growing chain) happen to be simultaneously coordinated to the same metal centre, the first may be inserted into the metal—alkyl bond. For steric reasons, this may be possible only at the catalytically active surface metal centres.

An encounter of two methylene groups (at two neighbouring metal centres, which need not necessarily be FT active) evidently may lead to ethylene formation. However, ethylene has a large tendency to coordinate to a metal centre instead of escaping into the gas phase; as a consequence, it may become incorporated into growing chains, as has been shown by Schulz et al.[17] with the aid of tagged ethylene. Hence, this is another pathway for carbide carbon to enter the hydrocarbons.

The formation of C_3 hydrocarbons from carbide carbon only should have a very low probability in an FT system. In principle, C_3 formation may proceed from the insertion of carbide-based ethylene into a likewise carbide-based metal—methyl or metal—methylidene bond (in the latter case, cyclopropane would probably be formed first[60], but in the presence of hydrogen it would end up as propane[61]). However, in the inner parts of a catalyst particle, the steric conditions are probably not satisfactory for such chain prolongation. At the catalyst surface, on the other hand, the reaction of any carbide-based intermediate with a growing chain appears more probable. Actually, only very small amounts of molecules with three carbide-based carbon atoms (and none with more than three) have been claimed by Biloen et al.[49].

To summarize, the presence of some labelled, carbide-based carbon in the FT products does not seem to be in contradiction with the CO insertion mechanism suggested for the FT synthesis.

C. Details and Support for the CO Insertion Mechanism

1. The reaction scheme

Any plausible reaction scheme for the FT synthesis has to take into account the experimental fact that both α-olefins and alcohols are primary products, thus precluding one as the precursor of the other. It has also to be considered that the primary products have a Schulz–Flory molecular weight distribution. We suggested in 1976 the reaction mechanism shown in Scheme 1. At that time, it was based more on chemical intuition than on experimental proof of every step. In formulating the scheme, we followed a contemporary trend, probably first expressed by Nyholm[62], of looking at heterogeneous catalysis with transition metals more from the point of view of individual active centres and of their coordination chemistry than that of 'active surfaces'. Therefore, the scheme was based, as far as possible, on individual steps well established in the homogeneous catalysis with soluble transition metal complexes. Such individual steps are, e.g.,

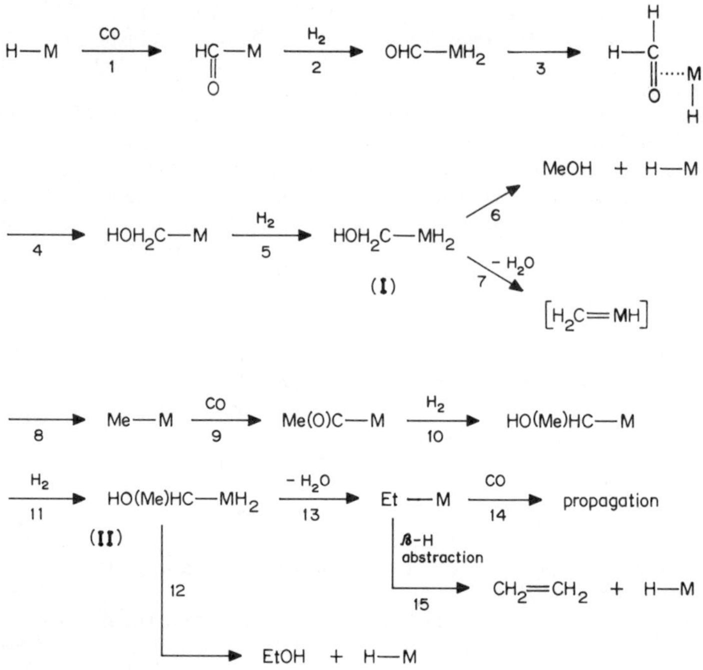

SCHEME 1. Suggested mechanism of the FT synthesis[42a,b]. All other ligands omitted for clarity.

coordination of CO and of olefins, oxidative addition and reductive elimination, insertion, and β-H abstraction (β-H transfer)[63]. In the meantime, many papers by various authors have appeared, confirming most of the intermediates assumed in the scheme, either directly or by analogous reactions. These supporting data will be discussed in the next section.

Scheme 1 starts with a metal hydride, which may be assumed to arise during the activation of the catalyst. Carbon monoxide is coordinated to the metal centre and inserted into the M—H bond (step 1). A hydrogen molecule is oxidatively added to the metal centre (step 2). Reductive elimination of the acyl ligand and one of the hydride ligands gives formaldehyde plus a metal hydride (step 3); the former, however, does not leave the metal centre, but remains coordinatively bonded through its C=O group (aldehydes are not primary products of the FT reaction; cf. Section IV). Addition of the metal hydride to the aldehyde is the next step (step 4), followed by another oxidative addition of H_2 (step 5). Intermediate I can react in one of two possible ways: reductive elimination (step 6) to give methanol and a metal hydride, which can continue the kinetic chain, or elimination of water with the intermediate formation of a carbenoid ligand (step 7), which rearranges to give a σ-bonded methyl ligand (step 8). Now the next CO molecule can be inserted (step 9). The next step (step 10) actually summarizes the results of three steps, corresponding to steps 2-4, with the difference that the carbon chain is now increased by one unit. Oxidative addition of H_2 (step 11) leads to the intermediate II, which can again undergo two alternative reactions, giving either the alcohol by reductive elimination (step 12), or the alkyl group by H_2O elimination (step 13, which summarizes the two steps corresponding to steps 7 and 8). The alkyl metal compound can either add

CO and thus contribute to the chain propagation (step 14) or, by β-H transfer, give an α-olefin (ethylene at this stage) and a metal hydride, which continues the kinetic chain (step 15).

This scheme explains, at least qualitatively, most of the experimental findings without violating known principles concerning the reaction patterns of transition metal centres. It is interesting to consider the coordination site requirements for an active metal centre. Assuming that the dissociation of the hydrogen molecule takes place by oxidative addition of both hydrogen atoms to the same metal centre (as is well established in homogeneous catalysis[63]), several of the steps in Scheme 1 require three empty coordination sites. Hence, active metal centres may be expected at edges or at surface defects, where incomplete coordination is the rule. (Compare similar considerations and observations by Rodriguez and van Looy[64] concerning the location of active sites in $TiCl_3$-based heterogeneous Ziegler–Natta catalysts.) In the case of a Phillips catalyst (CrO_3–SiO_2), the availability of at least three sites on certain surface chromium centres has actually been demonstrated[65].

The oxidative addition of hydrogen to the catalyst centre appears to be a slow reaction. This is indicated by the experimental rate law, $r_p \approx p_{H_2}$ (cf. equation 21), which has its parallel in the hydroformylation reaction[66]. In the particular case of hydroformylation with a rhodium catalyst, the oxidative hydrogen addition has been suggested to be the rate-determining step[67].

The metal hydride is assumed not only to be the carrier of the kinetic chain (cf. Scheme 1), but also to be responsible for the secondary reactions such as hydrogenation and incorporation of the α-olefins. Coordination and insertion of an α-olefin into an M—H bond can occur in either of two ways (equation 26). Step 16 and further growth will lead

$$RCH=CH_2 + HFe \xrightarrow[17]{16} \begin{array}{l} RCH(CH_3)Fe \\ \\ RCH_2CH_2Fe \end{array} \qquad (26)$$

to monomethyl-branched compounds, whereas step 17 gives a linear prolongation of the previously olefinic chain. This explains why methyl branching is the only significant branching experimentally observed (for further discussions, see Section VII.C.4).

The reaction mechanism suggested by Pichler and Schulz[41] is in agreement with Scheme 1 in the most important feature, i.e. the chain growth proceeding by successive CO insertions at the same metal centre. It diverges in several of the other steps. Thus, it assumes the aid of a neighbouring metal hydride, to take care of the oxygen (equation 27).

$$RC(O)M + HM' \rightarrow RCH(M)OM' \xrightarrow[-H_2O]{+H_2} RCH=M \qquad (27)$$

From the carbene ligand on, the sequence is similar to that in Scheme 1, i.e. transformation to an alkyl ligand and further CO insertion. However, the formation of product α-olefins is not assumed to take place by β-H abstraction and re-establishment of the chain carrier (step 15 in Scheme 1), but by a rearrangement of the carbene ligand (equation 28).

$$RCH_2CH=M \rightarrow RCH\!\!\!\overset{\vdots}{=}\!\!\!CH_2 \rightarrow RCH=CH_2 + M \qquad (28)$$
$$M$$

Since none of the suggested reaction mechanisms is definitively proved, we want to consider them, at the present, as 'working hypotheses'. However, in the next Section, we shall discuss experimental data that have accumulated in support of Scheme 1.

2. Evidence for individual steps

Surface science teaches that hydrogen is 'dissociatively adsorbed' on transition metal surfaces, whereas CO is 'molecularly adsorbed', and only partly dissociated (see, e.g., ref. 68). So far, we have assumed that, on a molecular basis, the reaction of hydrogen with surface metal atoms takes place by oxidative addition of a hydrogen molecule on to the same surface metal atom, by analogy with homogeneous systems[63]. However, we cannot exclude the possibility that the two hydrogen atoms may coordinate to two different metal centres, as is known, for instance, in the case of the $[Co_2(CO)_8]$ cluster compound (equation 29)[63].

$$[Co_2(CO)_8] + H_2 \rightarrow 2[CoH(CO)_4] \tag{29}$$

Moreover, we have to admit a certain mobility for hydride lingands, as long as the movement can take place by exchanges from end-on to bridge positions and *vice versa* (cf. Section VII.B).

When both H_2 and CO were applied together, under ultra-high vacuum conditions, on to a transition metal surface, the irreversible formation of a new entity was discovered by temperature-programmed desorption experiments[69]. The new surface complex desorbs at a temperature considerably higher than that of H_2 and CO desorption; its chemical composition is close to H:CO = 1[68-70]. Deluzarche et al.[71] presented chemical evidence for the surface entity to be a formyl complex. Instead of desorbing it, they trapped it with Me_2SO_4, Et_2SO_4, or MeI. Gas chromatographic detection of MeCHO, EtCHO, and MeCHO, respectively, was taken as evidence for the presence of formyl (CHO) groups on the metal surface.

We then can assume that the surface formyl complexes are formed by CO insertion into metal—hydrogen bonds (step 1 in Scheme 1). This step has long resisted direct experimental evidence, although it had been calculated as thermodynamically feasible and even favourable, by Goddard et al.[72] and Berke and Hoffman[73]. Numerous formyl complexes were actually prepared in the late 1970s, and in several instances considered as evidence for step 1, but they were generally synthesized in a different way, by treating carbonyl complexes with formic–acetic anhydride[74] or with borohydrides[74,75]. Recently the formation of a stable formyl complex from a rhodium hydride in the presence of CO was reported (equation 30)[76]

$$[RhH(oep)] + CO \rightarrow [Rh(CHO)(oep)] \tag{30}$$

(oep = octaethylporphyrin). The reaction was carried out in benzene solution. In another case[77], the insertion of CO into a metal—H bond appears as an obvious intermediate step, although it has not specifically been pointed out. The reaction of CO with Cp_2TiCl_2–Me_2AlCl in benzene solution (1 atm CO, 20–50 °C) leads to small (sub-stoichiometric) amounts of methane, formaldehyde, and diketene; on hydrolysis, acetaldehyde is found. This appears to be interpreted best by the sequence in equations 31 and 32. Thus, although the evidence is not yet abundant, step 1 in Scheme 1 appears probable.

$$MeTiCO \rightarrow TiC(O)Me \rightarrow TiH + CH_2{=}C{=}O \tag{31}$$

$$TiH + CO \rightarrow HTiCO \rightarrow TiCHO \tag{32}$$

Next in Scheme 1 comes the hydrogenation of the formyl ligand to a formaldehyde ligand, with the aid of a new, oxidatively added hydrogen molecule. The formaldehyde (and higher aldehydes further on) is assumed to stay π-bonded to the metal centre, because aldehydes do not count among the primary products of the FT synthesis (cf. Section IV.A).

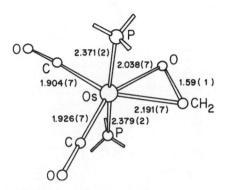

FIGURE 11. Molecular structure of the osmium–formaldehyde complex [Os(π-CH$_2$O)(CO)$_2$(PPh$_3$)$_2$][78]. Reproduced with permission. Copyright 1979 American Chemical Society.

A stable, π-bonded formaldehyde–transition metal complex has been prepared and characterized by Brown et al.[78]. The reaction (equation 33) resulted in a complex featuring, among others, an i.r. band at 1017 cm^{-1} which was assigned to ν_{CO} of the π-bonded formaldehyde. The structure, as determined by X-ray single crystal analysis, is shown in Figure 11. The strength of the π-bond between the metal and the aldehyde C=O is indicated by the considerable weakening of the C=O bond: the bond length is 1.59 Å, compared with 1.209 Å for free formaldehyde; concomitantly, the ν_{CO} stretching frequency is lowered from 1744 cm^{-1} for the free aldehyde to 1017 cm^{-1} for the π-complexed one. A similar arrangement has been found in a molybdenum complex with π-bonded benzaldehyde[79] and in a nickel complex with π-bonded benzophenone[80].

$$[Os(CO)_2(PPh_3)_3] + CH_2O \rightarrow [Os(\eta^2\text{-}CH_2O)(CO)_2(PPh_3)_2] + PPh_3 \quad (33)$$

The next step in Scheme 1 is the addition of the metal hydride to the coordinated formaldehyde, resulting in a hydroxymethyl complex (step 4). Although this step appears probable, it should be noted that the suggested mechanism can be simplified by eliminating the intermediate aldehyde complex postulated to form in step 3 (and corresponding steps later in the scheme). Instead, one may assume that the formyl ligand (and the corresponding subsequent higher acyl ligands) can change from their σ-bonded to a π-bonded state. Hydrogenation of the π-bonded acyl group would lead directly to the hydroxyalkyl ligand (equation 34).

$$RC(O)M \xrightarrow{2'} \underset{O}{RC{\parallel}\cdots M} \xrightarrow{H_2,\ 3'} \underset{O}{\overset{}{RC{\parallel}\cdots M{\vert}H}} \xrightarrow{4'} RCH(OH)M \quad (34)$$

This suggestion is based on the discovery, isolation, and characterization of a compound having an acyl ligand π-bonded to a transition metal by Fachinetti et al.[81]. It was obtained by reacting dimethylbiscyclopentadienylzirconium(IV), in toluene at 20 °C and atmospheric pressure, with CO (equation 35).

$$[Zr(Cp)_2(Me)_2] + CO \rightarrow [Zr(\pi\text{-}COMe)(Cp)_2(Me)] \quad (35)$$

The structure of the π-acetyl complex was determined by X-ray single crystal analysis (Figure 12). C(8), C(7), C(6), O, and Zr are coplanar; the CO interatomic distance is fairly

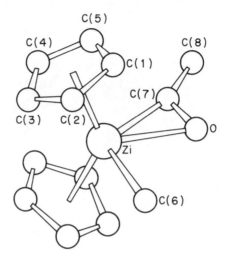

FIGURE 12. Molecular structure of [Zr(π-COMe)(Cp)$_2$(Me)][81]. Reproduced with permission. Copyright 1976 The Royal Society of Chemistry.

long (1.211 Å), indicative of a weakening of the bond; the corresponding ν_{CO} frequency is low (1545 cm^{-1}). A similar structure has been found by X-ray structural analysis for the product of reaction 36[82].

$$[\text{Ti}(\text{Cp})_2(\text{CO})_2] + \text{RCl} \xrightarrow[\text{toluene}]{-\text{CO}} [\text{Ti}(\text{Cl})(\pi\text{-COCH}_3)(\text{Cp})_2] \quad (36)$$

Based on i.r. and ^{1}H n.m.r. data, a π-acyl ligand has also been claimed for ruthenium complexes obtained according to equation 37[82].

$$[\text{RuHCl}(\text{CO})(\text{PPh}_3)] + 2\ \text{RCHO} \longrightarrow [\text{Ru}(\text{Cl})(\pi\text{-COR})(\text{CO})(\text{PPh}_3)_2] + \text{RCH}_2\text{OH} + \text{PPh}_3$$

R = Me or Et. (37)

From the hydroxymethyl complex (cf. equation 34) on, the hydrogenolysis to give methanol (step 6 in Scheme 1) is straightforward. The intermediate formation of a carbene ligand en route to the alkyl ligand is suggested as the most plausible way to eliminate water (cf. equation 1) from a mononuclear metal centre. The suggestion is based on a large body of evidence for the existence of transition metal carbene complexes[60,84], as well as their importance in catalysis, in particular in the metathesis of olefins[63]. The equilibrium between carbene–transition metal hydrides and metal alkyls is also well documented for several transition metals[85], which makes this reaction path very attractive.

A very instructive demonstration of the stepwise reduction of coordinated carbon monoxide, via formyl and hydroxymethyl ligands to a methyl ligand, on a mononuclear model rhenium compound has been reported by Sweet and Graham (equation 38)[86]. The

$$[\text{ReCp}(\text{CO})_2(\text{NO})]^+ \text{BF}_4^- \rightarrow [\text{ReCp}(\text{CO})(\text{CHO})(\text{NO})]$$
$$(1) \qquad\qquad (2)$$
$$\rightarrow [\text{ReCp}(\text{CO})(\text{CH}_2\text{OH})(\text{NO})] \rightarrow [\text{ReCp}(\text{CO})(\text{Me})(\text{NO})] \quad (38)$$
$$(3) \qquad\qquad (4)$$

reduction was carried out with $NaBH_4$, in a mixed H_2O–thf solvent; each step was achieved with one equivalent of the reducing agent. The first two steps required 15 min at 0 °C each, giving > 90% yield; the last step required 5 h at 25 °C, with 88% yield. Complexes 1–4 are stable, crystalline substances that have been characterized by i.r., ^1H n.m.r. and elemental analysis.

With step 9 in Scheme 1, the reaction cycle begins again, the difference being that the carbon monoxide is now inserted into a metal—alkyl bond instead of into a metal—H bond. The insertion of coordinated CO into a metal—alkyl bond present at the same transition metal centre is one of the key reactions in the hydroformylation and carbonylation of unsaturated hydrocarbons[63]. The most frequently studied model reaction is the carbonylation of the methyl ligand in pentacarbonylmethylmanganese, $[Mn(CO)_5(Me)]$, under the influence of excess of CO. A number of pertinent features have been clearly established. Using ^{13}CO, Noak and Calderazzo[87] have shown that the methyl group reacts with one of the four coordinated ^{12}CO ligands Z to the Me ligand and not with an incoming ^{13}CO, which only replenishes the octahedral coordination (equation 39).

$$\text{Me–Mn(CO)}_4\text{(OC)} + {}^{13}CO \longrightarrow {}^{13}CO\text{–Mn(CO)}_4\text{(COMe)} \quad (39)$$

The rate constant (9×10^{-3} l mol^{-1} s^{-1} at 30 °C) and the activation parameters ($E_a = 61.9$ kJ mol^{-1}, $\Delta H^* = 59.4$ kJ mol^{-1}, $\Delta F^* = 86.2$ kJ mol^{-1}, $\Delta S^* = -88$ J mol^{-1} k^{-1}) of the process have been determined[88]. Based on a study of the stereochemical changes in the carbonylation of (Z)-$[Mn(CO_4)(^{13}CO)(Me)]$, Noak and Calderazzo[87] were able to show that the reaction of the methyl group with CO is not an insertion of the CO into the M—C bond at the coordination site of the methyl group, but rather a migration of the methyl group to the site of the CO. It is by now generally assumed that all insertion reactions in the coordination sphere of a transition metal follow this pattern (e.g. also the chain growth in ethylene polymerization on a Ziegler–Natta catalyst[89]). Theoretical calculations corroborate this view[73,90]. The expression 'migratory insertion' is frequently used to describe this mechanistic detail.

The only new feature in the further course of Scheme 1 is the β-H abstraction (step 15), which can take place as soon as a β-H becomes available, and which leads to the primary products of the FT synthesis, the α-olefins. This reaction is, again, one of the key reactions in homogeneous catalysis[63]. For instance, in the dimerization, oligomerization, and polymerization of ethylene, β-H abstraction is the molecular weight-determining step (equation 40).

$$RCH_2CH_2M \rightarrow RCH=CH_2 + MH \quad (40)$$

From a kinetic point of view it is a chain transfer (not termination) reaction, since the resulting metal hydride is a chain carrier (i.e. it initiates a new chain), and as such it is suggested in Scheme 1. The analogy goes further than that. It has been shown for the ethylene polymerization that electron acceptor ligands on the metal favour β-H abstraction, resulting in a smaller average molecular weight of the products[91]. On the other hand it has been known since the early days of the FT synthesis that decreasing the ratio of metal to Kieselguhr in the catalyst leads to a decrease in the average molecular weight of the products[92]. Kieselguhr may be considered as a highly acidic (electron acceptor) 'ligand' of the surface metal atoms.

3. Further mechanistic evidence for the CO insertion mechanism

Any suggested reaction mechanism has to accommodate the fact that linear aliphatic primary amines are the only nitrogenated products formed when the FT synthesis is carried out in the presence of ammonia, and under conditions that otherwise would lead to linear hydrocarbons in the absence of NH_3. The amines present a Schulz–Flory molecular weight distribution, and the average chain length decreases with increasing amount of NH_3 in the synthesis gas (cf. Sections IV.C and V.B).

If the hydrocarbons were to be built up from 'carbidic intermediates CH_x'[49], α, ω-diamines would be expected among the reaction products. The CO insertion mechanism, on the other hand, can explain this selectivity. The Schulz–Flory distribution could, in principle, be expected if the primary products of the FT synthesis, the α-olefins, were to react with ammonia in a subsequent step, either outside the coordination sphere of the catalyst or after coordination to a metal centre. However, such subsequent amination could not explain the dependence of the molecular weight of the amines on the ammonia concentration.

The facts are best interpreted[93] by assuming that NH_3 acts as a chain transfer agent, at any stage of the growth cycle, giving the amine and a metal hydride. In the course of the growth cycle (see Scheme 1) there are alkyl, acyl, hydroxyalkyl, and carbene groups bound to the metal centre.

The alkyl ligand offers the most straightforward route (equation 41).

$$RM + NH_3 \rightarrow RNH_2 + MH \tag{41}$$

(M = metal centre). Amazingly, no reference to this simple type of reaction could be found in the literature. The acyl ligand might be visualized as reacting with ammonia to give an imine ligand which, on subsequent hydrogenation, results in a primary amine and a metal hydride (equation 42).

$$RC(O)M + NH_3 \xrightarrow{-H_2O} RC(NH)M \xrightarrow{H_2} RCH_2NH_2 + MH \tag{42}$$

Markó and Bakos[94] have shown that aldehydes and ketones are transformed into amines under similar reaction conditions, although primary and secondary amines are formed in comparable amounts if ammonia is used as the amination agent; in the modified FT system, on the other hand, only primary amines are formed. Moreover, it is felt that the mechanism depicted in equation 42 should lead to nitrogen-containing byproducts such as amides and nitriles, which are not found experimentally[21].

The OH groups of the α-hydroxyalkyl ligands could also react with ammonia (equation 43).

$$RCH(OH)M + NH_3 \xrightarrow{-H_2O} RCH(NH_2)M \xrightarrow{+H_2} RCH_2NH_2 + MH \tag{43}$$

The catalysed production of amines from alcohols and ammonia is well known[95]; it requires, however, temperatures of 300–400 °C, whereas the modified FT synthesis operates at 200 °C.

Finally, one might consider the carbene ligands, formulated as the result of step 7 (and corresponding later steps; see Scheme 1), to react with ammonia. Fischer and coworkers[96] have actually shown that ammonia is able to react with transition metal—carbene complexes (in particular complexes of chromium), but the result is an aminocarbene complex which is so stable that the reaction can be used to protect amino groups of amino acids during peptide synthesis. Similar results have been reported for platinum—carbene complexes[97]. This brief discussion indicates the simple, one-step amination of the alkyl—metal bond (equation 41) as the most probable reaction. Evidently, such a process would

$$RM \begin{matrix} \xrightarrow{CO} RC(O)M \\ \xrightarrow{NH_3} RNH_2 + MH \end{matrix} \qquad (44)$$

be competitive with the CO insertion (growth step) (equation 44). This readily explains the molecular weight reduction produced by increasing the amount of NH_3 in the feed gas mixture.

4. Secondary reactions

In Section IV it was noted that linear α-olefins are the predominant reaction products if the synthesis gas has a short residence time in the reactor. With longer residence time, other products appear in the series linear alkanes > β-olefins > methyl-branched alkanes (cf. Figure 1). We assumed in Section VII.C.1 that the chain carriers, metal—H species, are responsible for these secondary reactions. Presumably, the α-olefins can be coordinated to such metal centres, and undergo hydrogenation, as well as isomerization, as is known from homogeneous catalysts (equation 45)[63].

$$MH + \alpha\text{-alkene} \longrightarrow M\text{—alkyl} \begin{matrix} \xrightarrow{H_2} \text{alkane} + MH \\ \searrow \beta\text{-alkene} \end{matrix} \qquad (45)$$

At a first glance, there seems to be an enigma: why do the metal—alkyl species occurring in the course of the chain growth (cf. Scheme 1) resist hydrogenation and isomerization, whereas those proceeding from the coordination of α-olefins to metal hydride species (equation 26) do undergo these reactions? The explanation resides in the preferred orientation of an α-olefin when inserted into the metal—H (or metal—R) bond. With Group VIII metals, there is a remarkable tendency for the 'anti-Markownikoff' mode of insertion to occur; regioselectivities of 70% to almost 100% are the rule (equation 46)[98].

$$MH + CH_2 = CHR \rightarrow MCH(Me)R \qquad (46)$$

The migratory CO insertion is much slower for an isoalkyl ligand than for a linear alkyl ligand[63]. Thus, whereas in normal FT growth the CO insertion takes place before any other reaction can occur, the isoalkyl groups proceeding from coordination of α-olefins are prone to hydrogenation and isomerization.

The rare cases of CO insertion involving an isoalkyl group lead to methyl-branched hydrocarbons. However, the reaction of a growing chain with a coordinated propylene may also give methyl branches (equation 47).

$$MR + CH_2 = CHMe \rightarrow RCH_2(Me)CHM \qquad (47)$$

The corresponding reaction with coordinated ethylene gives a mere chain prolongation by two carbons; with but-1-ene, an ethyl branch will result. However, the 'polymerizability' of α-olefins decreases rapidly with increasing chain length[63]. In fact, ethyl branches are rare in the FT products, and longer branches are essentially absent[5b]. Internal olefins can safely be assumed not to coordinate under the conditions of the FT synthesis. As one may expect, the rate of consumption of CO, in moles per volume unit (cm^3) of catalyst bed and unit time (h), is highest when the residence time of the synthesis gas on the catalyst is lowerst (see Table 4). At longer residence times, coordination of α-olefins and secondary reactions block the active centres.

TABLE 4. Influence of space velocity (residence time) on the rate of CO consumption.

Space velocity[a] (h^{-1})	Residence time (s)	CO conversion[a] (%)	CO consumption ($10^3\,\text{mol cm}^{-3}\,\text{h}^{-1}$)
78	46.1	84.6	0.9
337	10.7	42.1	2.0
2830	1.5	10.1	3.4

[a]Data of Pichler et al.[15]; 32 vol.-% CO; normal pressure; ca. 200 °C.

5. Comparison with hydroformylation

The hydroformylation of olefins, detected in 1938 by Roelen[2], is one of the most important commercial examples of homogeneous catalysis with transition metal compounds. Under carbon monoxide and hydrogen pressures (mostly $CO:H_2 = 1:1$) of over 100 atm, at 150–180 °C, olefins are transformed to aldehydes having one carbon atom more than the starting olefins, according to the overall reaction shown in equation 48.

$$RCH=CH_2 + CO + H_2 \xrightarrow{\text{catalyst}} RCH_2CH_2CHO \qquad (48)$$

Oxides and complexes of several transition metals, in particular those of cobalt and rhodium, have been applied as catalyst precursors. The soluble, active species formed under the reaction conditions may be formulated generally as $[MH(CO)_m L_n]$, with M = transition metal ion, L = neutral ligand such as phosphite, phosphine, or amine, $m \geqslant 1$, $n \geqslant 0$, and $(m+n) = 3$ or 4^{63}.

It has been suggested that the mechanisms of the hydroformylation and the FT synthesis are closely related[42a,b]. Scheme 2 shows, in a simplified manner, the generally agreed course of the hydroformylation of a terminal olefin[99]. [For the sake of clarity, $(CO)_m$ and L_n ligands are omitted; it is understood that the insertions of olefins and CO are preceded by the coordination of these molecules to the metal centre.] Two possible routes are discussed in the literature for the last step, the transformation of the acyl metal species to the aldehyde: hydrogenolysis after oxidative addition of a hydrogen molecule (steps 3 and 4[99]), or reaction with a second catalyst species (step 5). Spectroscopic evidence for the importance of step 5, under certain reaction conditions, has been reported for the particular case of $MH = [CoH(CO)_4]^{100}$.

Evidently, the insertion of a CO molecule into a metal—carbon bond, with the

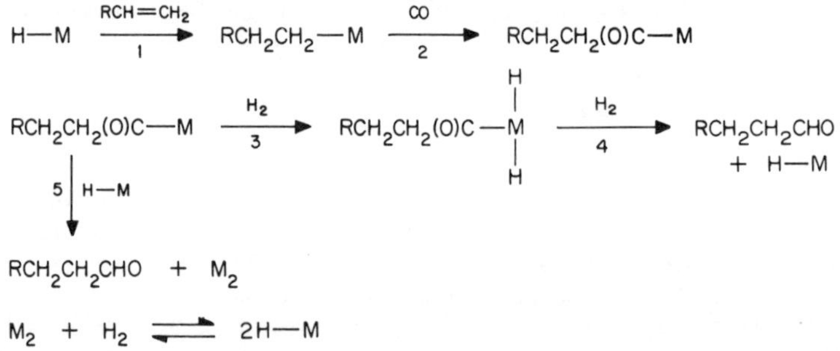

$$H-M \xrightarrow[1]{RCH=CH_2} RCH_2CH_2-M \xrightarrow[2]{CO} RCH_2CH_2(O)C-M$$

$$RCH_2CH_2(O)C-M \xrightarrow[3]{H_2} RCH_2CH_2(O)C-M(H)(H) \xrightarrow[4]{H_2} RCH_2CH_2CHO + H-M$$

$$\downarrow 5 \; H-M$$

$$RCH_2CH_2CHO + M_2$$

$$M_2 + H_2 \rightleftharpoons 2H-M$$

SCHEME 2. Mechanism of the hydroformylation[99,100]. All other ligands ommitted for clarity.

formation of an acyl species, is a common step in both reactions, hydroformylation (step 2 in Scheme 2), and FT synthesis (steps 9, 14,..., in Scheme 1). It appears enigmatic that, under closely related conditions and with similar catalysts, the further course of the reaction is thus different. In hydroformylation, the aldehyde is formed and immediately leaves the complex; no further chain growth takes place. In the FT reaction, on the other hand, the aldehyde (assumed as a plausible intermediate) does not leave the coordination sphere of the metal, but subsequent reaction with the remaining hydrogen ligand (step 4) opens the way to further chain growth via steps 6, 7, 8, etc.

Based on the findings of van Boven et al.[100], we suggested[42b] that the decisive difference between the homogeneous and the heterogeneous process is the availability of free, mobile, very reactive hydrido-metal species in the solution, which makes step 5 (Scheme 2) the more important mode of reaction of the acyl metal species in the homogeneous system. The simultaneous formation of the dinuclear metal complex, $M_2\{[Co_2(CO)_8]$ in the case of $[CoH(CO)_4]$ as active species}, makes further reactions in the coordination sphere impossible and causes the aldehyde to leave the metal centre. In the heterogeneous system, on the other hand, the MH species are fixed at their surface sites and can not encounter any acyl metal species, also fixed at the surface. Thus, the oxidative addition of molecular hydrogen to the latter is the only way of reaction. Actually, no convincing experimental evidence supporting reactions 3 and 4 in Scheme 2 under hydroformylation conditions is available[141], whereas step 5 has been made probable by high-pressure i.r. spectroscopy on the reacting system[100] (p_{H_2} = 50–60 atm, p_{CO} = 10–20 atom, $T \approx 100\,°C$). Moreover, the stoichiometric formation of aldehyde from olefin and $[CoH(CO)_4]$ under nitrogen, where CO as well as hydrogen must be provided by the hydrido-metal species, is known[101].

In one case, where aldehydes were obtained on a supposedly heterogeneous catalyst (cobalt carbonyl on a polymeric carrier), under the conditions of the hydroformylation, it was observed that some of the cobalt had leached into the liquid phase as $[CoH(CO)_4]$ and $[Co_2(CO)_8]$[102]. Hence, the hydroformylation conditions as required by the suggested mechanism (availability of $[CoH(CO)_4]$ for the reduction of the acyl species) are given. The phenomenon is not surprising, taking into account that most cobalt hydroformylation catalysts are formed from insoluble cobalt compounds under the influence of a CO–H_2 atmosphere and at temperatures above 50 °C[103].

6. Influence of the dispersity of the metal centres

The metallic state has frequently been assumed to be essential for the catalysts of the FT synthesis, and transition metal single crystals have been used as models for catalysts with regard to bonding and structure of adsorbed carbon monoxide, hydrogen, and their reaction products[68–70,104]. Highly sophisticated physical methods have been used in these studies (for a review of these methods see ref. 105). Much work has been carried out to correlate catalytic activity (or properties believed to be associated with activity) with the peculiarities of different crystallographic planes. Steps, kinks, holes, etc., in otherwise 'flat' surfaces have been found to increase activity[106]. Evidently, metal atoms situated at such irregularities have less metal—metal coordination, and hence more sites available to coordinate reactants.

Yokohama et al.[107] were able to show that crystallinity is not a prerequisite for catalytic activity. They prepared ribbons of amorphous Fe–Ni alloys containing inclusions of phosphorus or boron atoms to shift crystallization to relatively high temperatures. At temperatures well below the crystallization temperature, it was found that the activity of the amorphous state was up to several hundred times higher than that of the crystalline state, for the same composition. Again, the amorphous state, like steps or kinks on crystalline surfaces, can be expected to have a higher number of exposed atoms with an incomplete coordination sphere.

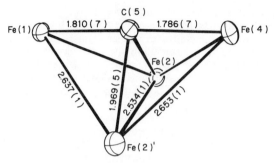

FIGURE 13. Fe$_4$C core cluster structure for [Fe$_4$(μ_4-C)(CO)$_{12}$]$^{2-}$, 'butterfly cluster'[109]. Reproduced with permission. Copyright 1976 American Chemical Society.

So far, however, no effective homogeneous catalyst for the hydrocarbon formation according the FT synthesis has been found. Whenever reasonable amounts of hydrocarbons have been detected in soluble systems (based on ruthenium complexes) it has been found that deposited metal was responsible, rather than the soluble complex[108]. The absence of mononuclear, soluble FT catalysts led Thomas et al.[109] to the proposal that multinuclear metal centres might be necessary for activity. Cluster compounds have been studied as models, in particular the iron 'butterfly' clusters with carbidic carbon atoms[110] (see Figure 13). In fact, it was found that some soluble clusters complexes were active, whereas related mononuclear complexes were not, but the activity was in general extremely low[109].

There is now some evidence that mononuclear species may be active for the hydrogenation of CO to hydrocarbons, as long as they are anchored on a solid support. Brenner and Hucul[111] deposited many mono-, di-, and poly-nuclear carbonyl complexes on to γ-Al$_2$O$_3$, by adsorption from pentane solution or by sublimation, at very low concentration. During temperature-programmed decomposition experiments under H$_2$, they first observed desorption of part of the carbonyl ligand, followed by CH$_4$ formation on the remaining 'subcarbonyl' species. The CH$_4$ evolution per metal centre was essentially the same for mono-, di-, and poly-nuclear species. Even more important, they could exclude the sintering of mononuclear to polynuclear species by observing the CH$_4$ formation as a function of the catalyst loading, in the case of [Mo(CO)$_6$]. For a nuclearity n being required for activity, the amount of CH$_4$ formed, m_{CH_4}, is proportional to the number of sites of this nuclearity which would increase with the power of n of the loading w (equations 49 and 50).

$$m_{CH_4} = aw^n \tag{49}$$

$$\log m_{CH_4} = \log a + n \log w \tag{50}$$

A graphical evaluation of experimental data (plot of log m_{CH_4} versus log w), for w varying over several orders of magnitude resulted in a straight line and a value of n close to unity. This is consistent with the activity residing in mononuclear species.

Although the activity of these interesting mononuclear catalysts is low, their discovery certainly helps to narrow the gap between the classical view of heterogeneous catalysis as a surface science, and the more recent trend of considering it from the point of view of individual transition metal centres and their coordination chemistry. Moreover, these findings appear to indicate that multicoordinated species such as **5**[41] and **6**[112,113] may not be involved in the hydrocarbon chain growth in the FT synthesis.

$$
\begin{array}{cc}
\text{R—C—O} & \text{C—O} \\
\text{(5)} & \text{(6)}
\end{array}
$$

7. The role of alkali metal promoters

The addition of small amounts of alkali metal compounds, in particular K_2O or K_2CO_3, to an iron catalyst for the FT synthesis is known to produce substantial changes in the catalytic performance[9,114]. The synthetic activity and the average molecular weight of the hydrocarbons formed are increased; moreover, the olefinic fraction in the hydrocarbon effluent is improved. Interpretations for these phenomena have been proposed[115,116], based on the electron donor capacity of K_2O. Electrons donated to the metal would enhance the adsorption of CO, while reducing that of H_2; weakening of the CO bond would facilitate the attack of this bond by hydrogen. Reduced hydrogenation ability, on the other hand, would lead to higher molecular weight (whereby the chain-terminating step is assumed to be hydrogenolysis of an alkyl—metal bond) and to a higher olefinic fraction.

As we showed recently, this type of interpretation is inconsistent with the views of molecular catalysis. In particular, the fact that the primary products of the synthesis are α-olefins excludes hydrogenolysis of alkyl—metal bonds as the molecular weight-determining factor. Based on well documented phenomena with well defined, soluble transition metal carbonyl complexes, a different interpretation of the alkali metal promotion was suggested.

Transition metal carbonyl complexes are bases, and tend to form complexes with Lewis acids (electron acceptors) rather than with electron donors[117]. For a considerable number of transition metal carbonylates, as well as donor (e.g. alkyl) substituted, neutral carbonyl complexes, it has been shown by X-ray structural analysis, i.r., and other measurements that the interaction with the electron acceptor (Mg^{2+}, K^+, Na^+, Li^+, AlX_3, etc.) occurs via the electron-rich oxygen of a carbonyl group[117,118], for instance in the particular case of $Na^+[FeR(CO)_4]^-$ [118]:

$$Na^+ \cdots\cdots\cdots OC—Fe(CO)_3R$$
(7)

The various X-ray and i.r. studies have demonstrated that such interaction strengthens the metal—carbon bond and weakens the carbon—oxygen bond of the involved carbonyl ligand.

Collman and coworkers[118] have shown that the close interaction of the sodium ion with the carbonyl ligand in complex 7 greatly favours the migratory insertion of the CO ligand into the metal—alkyl bond. They found that in thf at 25 °C, Li^+ or Na^+ cations cause the insertion reaction to occur two to three orders of magnitude faster than the rate observed if the cation was trapped in a crown ether, or if the bulky cation $[Ph_3P]_2N^+$ was used instead. In the resulting acyl complex, the cation is associated with the oxygen of the acyl group. Collman and coworkers assume that the rate of alkyl migration is so dramatically increased by the presence of the electron acceptors Li$^+$ or Na$^+$, because the latter stabilize the coordinatively unsaturated intermediate:

$$\begin{array}{c} Na^+ \\ O^- \\ \| \\ R-C-Fe(CO)_3 \end{array}$$

(8)

The presence of the acceptor cation may help to dissipate temporarily, from the metal, electron density which is released on to it through the loss of one π-acceptor ligand (CO), in the course of the formation of the acyl ligand (cf. Berke and Hoffmann[73], who indicated that π-acceptor ligands on the metal in the migratory plane would have a similar effect). It should, however, also be taken into account that the interaction of an alkali metal cation with the oxygen of a carbonyl group leads to a lowering of the energy of the lowest unoccupied molecular orbital (frontier orbital) of the carbonyl group[119]. This certainly contributes to facilitating the insertion reaction.

With all this excellent previous work in mind, we suggest that the effect of alkali metal promoters in the FT synthesis is similar to that in the above-mentioned carbonyl complexes. One has to consider that under the conditions of the FT synthesis (presence of CO, H_2, CO_2, and H_2O, high temperature) the surface iron atoms are in an environment different from that in the original (reduced) catalyst. In particular, there will be surface iron species featuring one or several CO ligands. The potassium promoter, on the other hand, in whatever form it might have been added to the catalyst, is certainly present as K^+, even if it would have been transformed to metallic potassium in the reductive activation of the catalyst. Thus, if a surface alkyliron carbonyl species has a suitably positioned neighbouring K^+ ion, the migratory insertion reaction should be greatly accelerated. Along the same lines, the further reaction of the acyl group (still in interaction with the cation) with hydrogen should also be promoted, owing to the lowering of the energy of the CO frontier orbital. Assuming that the β-hydrogen transfer from the growing chain to the transition metal, as the growth-ending step, is not affected by the alkali metal promoter, an increase in the average molecular weight follows cogently from the increased growth rate, since the average degree of polymerization (i.e. the number of C atoms in the hydrocarbon chain) is given by the ratio of the rate of chain propagation to the rate of chain transfer (equation 51).

$$P_n = \frac{r_p}{r_{tr}} = \frac{d[CH_2]/dt}{d[olefin]/dt} \tag{51}$$

The amount of alkali metal ions in the promoted catalysts is generally low (g K/g Fe < 0.01)[115]. Hence, there should always be present surface iron carbonyl complexes that have no suitable neighbouring K^+ ion and which, consequently, should produce chain growth and concomitant chain lengths according to unpromoted catalysts. In other words, a bimodal molecular weight distribution may result from promoted catalysts. Relevant molecular weight distribution data are scarce; however, there is some evidence for bimodal distribution produced by alkali metal-promoted iron catalysts compared with a normal distribution produced by unpromoted cobalt catalysts[120]. Clear examples of bimodal molecular weight distributions have been reported recently for the reaction products obtained with alkali metal-promoted cobalt catalysts[121].

Finally, the observed increase in the olefin fraction of the effluent also fits into the picture. With the reasonable assumption that a metal hydride is responsible for the

9. The Fischer–Tropsch synthesis

hydrogenation of the olefins, and that the latter must be coordinated to the metal centre prior to reaction, the following trivial explanation can be visualized. The reasons given above for the acceleration of the insertion of CO into the metal—alkyl bond should apply equally well to the chain initiation (insertion of CO into a metal—hydrogen bond) (equation 52). Hence, in the case of the promoted catalyst, chain initiation can compete more favourably with hydrogenation than in the non-promoted catalyst.

$$
\begin{array}{c}
(L_x)MH \xrightarrow{CO} (L_x)MH\!\!\begin{array}{c}CO\\\vdots\end{array} \longrightarrow \text{chain initiation}\\
\xrightarrow{RC\equiv C} \begin{array}{c}RC\!=\!C\\(L_x)MH\end{array} \longrightarrow \text{hydrogenation}
\end{array}
\qquad (52)
$$

L_x = all other ligands

In conclusion, the remarkable effect of potassium promoters in the FT synthesis may be satisfactorily interpreted based on well documented phenomena, and fitting smoothly within the framework of molecular catalysis[122].

VIII. PRODUCT SELECTIVITY

A. Consequences of the Schulz–Flory Molecular Weight Distribution

The mechanism of the FT synthesis, as a step-by-step polymerization with chain growth and chain transfer, imposes severe restrictions on the selectivity with regard to the molecular weight of the product. The validity of the Schulz–Flory distribution function (cf. Section V) determines the selectivity for a given chain length as

$$m_P = (\ln^2\alpha)P\alpha^P \qquad (52)$$

where m_P = mass fraction of product having a degree of polymerization P and α = probability of chain growth. Figure 14 shows this dependence for a few selected values of P, and indicates that C_2 peaks at $\alpha \approx 0.4$, C_4 at $\alpha \approx 0.6$, C_6 at $\alpha \approx 0.7$, etc. However, whatever the value of α, hydrocarbons of other degrees of polymerization are simultaneously present in amounts determined by equation 52.

At high CO conversion (long residence time), this relatively simple relationship tends to be complicated by the products of secondary reactions, in particular those leading to branched molecules. As discussed in Section III, certain catalysts (Fe > Co) and process designs (fluid bed > fixed bed) have a higher propensity to give branched material. For the cases of such branched products, Friedel and Anderson[27] introduced a correction factor with an adjustable parameter into their molecular distribution function (cf. Section V.A); the resulting equation proved to be very useful in organizing a large amount of experimental data.

To a certain degree it is possible to vary the parameter α, and hence the molecular weight, by process variables. Thus, alkali metal promoters shift the chain length to higher values (Section VII.C.7). The ratio of H_2 to CO in the gas feed also has some influence (the higher this ratio the lower the average molecular weight; cf. Section III). The lower the ratio of metal to carrier, for Kieselguhr, the smaller is the average molecular weight[92]. This principle has been used by Commereuc et al.[123] to shift the molecular weight distribution to the low side. Values of α in the range of 0.2–0.4 can be determined from their data, obtained with noble metal (Rh, Ir, Ru, Os) catalysts on alumina or silica. Slightly higher values (0.5–0.6) can be estimated from the scattered data with several iron catalysts.

In a system in which the β-hydrogen abstraction rate is extremely high compared with

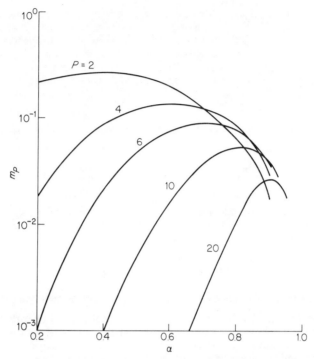

FIGURE 14. Consequences of a Schulz–Flory molecular weight distribution: molar fraction of selected chain lengths as a function of the growth probability α, calculated according to equation 52.

the rate of chain growth, one might expect that the former takes place as soon as a β-H becomes available, i.e. when an ethyl ligand is formed (cf. Scheme 1, step 13). This situation is well known in homogeneous catalysis, in the case of the selective dimerization of olefins[63]. In the FT synthesis this would lead preferentially to ethylene, apart from methane, which has several formation routes (e.g. the carbide route, cf. Section VII.B; decomposition of low olefins, Section IV.B). One such case has actually been claimed[124], for a reaction at a sub-atmospheric pressure of 18 kPa (140 Torr), with acidic catalysts ($\leqslant 11\%$ of metal on Al_2O_3), and at a ratio of CO to H_2 of 2. Methane and ethylene are the only hydrocarbons produced and the selectivity for ethylene is amazingly high (see Table 5).

There are several reports indicating that the average molecular weight can also be

TABLE 5. Synthesis of ethylene on Al_2O_3-supported catalysts[124]. $CO:H_2 = 2$; space velocity, $2500\,h^{-1}$; $p = 140\,Torr$

Metal	Temperature (°C)	CO conversion (%)	CH_4 (mol-%)	C_2H_4 (mol-%)
Co	350	14.4	8.9	91.1
Ni	300	16.9	20.9	79.1
Pt	300	17.5	35.1	64.9

TABLE 6. Probability of chain growth, α, as a function of metal dispersion[125a]. Catalyst: Ru on γ-Al$_2$O$_3$; $T = 250\,°C$; H$_2$:CO = 2; p, 12 atm; conversion = 22%

Dispersion (%)	Crystallite size (nm)	α
78	1.3	0.65
70	1.5	0.66
47	2.2	0.67
16	6.4	0.73

influenced by the particle size of the metal (dispersion of the metal on the carrier)[125]. It was found that α decreases with increasing particle size (see Table 6) and that this is accompanied by a decrease in specific activity, although the Schulz–Flory distribution remains valid[125c].

The influence of the particle size on the specific activity is a controversial issue. Both increases[126] and decreases[125] with diminishing particle size have been claimed, and a variety of interpretations have been suggested. Takasu et al.[127] have cautioned that in some cases the preparation of the samples (impregnation of, or precipitation on to, the support) may have made it difficult to obtain clean metal surfaces and to characterize them correctly. They used a model palladium catalyst prepared by evaporating metal on to a carbon film. The mean diameter of the metal particles was controlled by the amount of metal deposited, and determined by electron microscopy. The electron energy levels within the metal particles were characterized by X-ray photoelectron spectroscopy. the bonding energy of the $3d_{5/2}$ electron was found to increase, from the bulk metal value, as the particle size decreased, the major change (1.5 eV) taking place in the diameter range from 1.8 to 1.0 nm. The energy of the valence band in the palladium particles is assumed to shift with that of the $3d_{5/2}$ band[128]. Presumably, as a consequence, the activation energy of a catalytic test reaction (H$_2$/D$_2$ exchange) increases with decreasing particle size. These interesting findings appear to indicate that the observed variations of activity with particle size are due to activity changes of the individual metal centres[127], rather than to a decrease in the number of active sites[125c,129]. Alterations in the electronic structure of small metal particles that are attributable to strong metal–support interactions have in fact repeatedly been described[130,131] and, amazingly, this interaction has been found to increase in the series Al$_2$O$_3$ < SiO$_2$ < C, in the case of nickel particles[131]. It appears plausible that such changes translate into variations of the chain growth probability (cf. equation 5, Section V.A) as shown in equation 52. However, it should be taken into account that the changes in specific activity (r_p per metal centre) and in α cannot be expected to be proportional since r_p and r_{tr} are probably influenced to different degrees and possibly in opposite directions. In fact, a decrease in α by 10% has been reported to be accompanied by a decrease in specific activity of about an order of magnitude[125c].

$$\alpha = r_p/(r_p + r_{tr}) \qquad (52)$$

Finally, it may be noted that a certain increase in α can be obtained by the addition of ethylene to the feed gas[132]; this procedure may be considered as a copolymerization of two 'monomers', (CO + H$_2$) and CH$_2$=CH$_2$. Adapting equation 52 to this situation, one obtains

$$\alpha = \frac{\sum r_p}{\sum r_p + r_{tr}} \qquad (53)$$

where $\sum r_p$ is the sum of the rates of the different growth steps.

B. Deviations from the Schulz–Flory Distribution

Selectivity is a crucial problem is FT synthesis. Not always is the whole range of products, as dictated by the Schulz–Flory equation, really wanted. Thus, for gasoline, a branched product in the range C_5–C_{11} would be desirable; for heating purposes, a high content of low molecular weight hydrocarbons, particularly methane, ethane, and propane is necessary; for the production of plasticizers, detergents of high biodegradability, synthetic lubricants, etc., linear α-olefins in the range C_6–C_{16} would be the most favourable. Thus, approaches to selectivity, circumventing the Schulz–Flory distribution law, are the most recent branch of FT chemistry.

In Section V.A. it was mentioned that a very narrow molecular weight distribution should be obtained for $\alpha = 1$ (Poisson distribution). In the particular case of the FT synthesis this would require the β-H transfer to be completely suppressed. From the work of Pichler et al.[133] it is known that alkylruthenium species, as intermediates in the FT synthesis with unsupported ruthenium at high pressures ($\geqslant 1000$ atm) and relatively low temperatures (120–132 °C), tend to resist β-H abstraction. Unless quenched with alkali, or with hydrogen at high temperature and pressure, the growing chains remain 'living'. It was also shown that the lower alkylruthenium carbonyls are volatile in that temperature range.

A FT system resulting in a molecular weight distribution remarkably close to a Poisson distribution was reported by Madon[134] (see Figure 15). The reaction was carried out with 1% Ru on Al_2O_3 at 240 °C, 3.1 MPa (ca. 30 atm), and H_2:CO = 2. Apparently, the β-H abstraction was in fact effectively suppressed (note that at 1 atm, and otherwise comparable conditions, the same system produces hydrocarbons with a Schulz–Flory molecular weight distribution[125c]). This pronounced pressure influence appears to be restricted to ruthenium, since other catalysts (e.g. Fe, Co) would give products with a Schulz–Flory distribution under the conditions described by Madon.

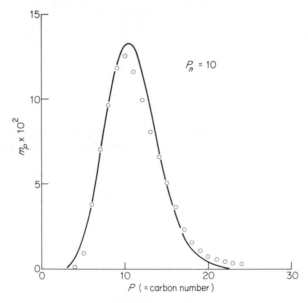

FIGURE 15. Poisson-type molecular weight distribution obtained on a ruthenium catalyst[134]. Solid line: theoretical Poisson distribution for $v = 9$, $P_n = 10$ (cf. equations 19, 20 and Figure 3).

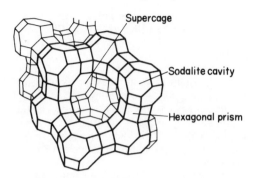

FIGURE 16. Framework of a faujastite-type zeolite[138]. Supercage, 1.25 nm diameter, 0.75 nm entrance opening; sodalite cavity, 0.65 nm diameter, 0.22 nm entrance opening; hexagonal prism, 0.22 nm diameter. Reproduced with permission. Copyright 1979 Plenum Publishing Corporation.

A promising approach to selectivity in the FT synthesis appears to be the trapping of catalytic transition metal species in the cavities of zeolites. Zeolites are crystalline aluminosilicates, the primary building blocks of which are tetrahedra consisting of either silicon or aluminium ions surrounded by four oxygen anions. These tetrahedra combine, linking silicon and aluminium ions through oxygen bridges, to yield highly ordered three-dimensional frameworks. There are 34 known natural and about 120 synthetic zeolite structures[135], all having a three-dimensional network of channels or linked cavities. Figure 16 shows the framework of one of them, the faujasite-type zeolite. In this particular case, the 'supercage' has a diameter of 1.25 nm (12.5 Å); the opening, consisting of a 12-membered oxygen ring, has a diameter of 0.75 nm (7.5 Å).

Whereas the silicon—oxygen tetrahedra building blocks are electrically neutral, each aluminium-based tetrahedron has a negative charge, balanced by a cation. Formally this may be written as

(9)

The zeolites have been used as 'shape-selective' catalysts since the early 1960s for a variety of reactions, whereby the product distribution depends on the ease with which reactant molecules diffuse through the zeolite channels, and on the residence time during which they are held in the vicinity of a catalytically active site. Ion exchange, cracking, hydrocracking, and 'selectoforming' (cracking of n-alkanes mainly to propane) are among the most important industrial uses[135].

In the particular case of the formation of hydrocarbons from methanol, a special zeolite developed by Mobil, ZSM-5, limits the maximum size of the product molecules to the C_9–C_{10} range. The reaction products are mainly branched-chain and aromatic hydrocarbons, which are ideal for high-octane fuel. The reaction mechanism is complicated, and involves both ethers and alkenes as intermediates[136].

The first application of zeolites as a chain-limiting catalyst in the Ft synthesis was reported by Nijs et al.[137]. A RuNa zeolite was prepared by partial exchange of the Na^+ ions (cf. 9) in a faujasite-type zeolite with $[Ru(NH_3)_6]Cl_3$. The ruthenium was then reduced with hydrogen at 300 °C. The FT synthesis was run under conditons where a usual Ru–SiO$_2$ catalyst would give a product with $60\% > C_{12}$ and a Schulz–Flory distribution. The zeolite catalyst generated a product with less than $1\% > C_{12}$[137a]. In the case of a Ru–La zeolite, a chain length limitation at C_5 was observed[137b]. Ballivet-Tkatchenko et al.[138] prepared zeolite catalysts by thermally decomposing metal carbonyls (of Fe, Co, and Ru) within the cavities of zeolites; they also observed selective formation of hydrocarbons in the C_1–C_9 range. A Schulz–Flory plot shows the usual picture up to C_9, and thereafter a sharp drop, indicating that higher hydrocarbons are only formed in traces.

Fraenkel and Gates[139] cautioned that the high-temperature reduction of the transition metal species within the zeolite cavities might lead to a partial destruction of the original aluminosilicate structure. They used cadmium vapour for the reduction of Co^{2+} ions exchanged into different types of zeolites. The intactness of the zeolite framework structures was checked by X-ray diffraction. On a zeolite characterized by relatively small cavities, propylene was formed exclusively (151 °C; ca. 6 atm; $CO:H_2 = 1$). Cadmium species remaining in the zeolite cavities, either alone or combined with cobalt species, are assumed to participate in the catalytic action. On a faujasite-type zeolite, the main product was n-butane, the C_4–C_7 mixture constituting ca. 70% of the hydrocarbon product and the distribution deviating considerably from the Schulz–Flory equation.

The size limiting effect of the zeolite cavities may be visualized as the consequence of a steric hindrance of the migratory insertion of CO into the growing chain. As the cavity becomes 'crowded', this step may become increasingly difficult, its probability eventually dropping to zero. The abstraction of a β-H will then free the molecule from its catalytic 'birth place', the metal centre.

IX. OUTLOOK

The revival of the Fischer–Tropsch synthesis, in the aftermath of the oil emgargo in the early 1970s, has brought about interesting chemistry[140]. However, considerable research and development work still needs to be done, in order to have this and other coal-based hydrocarbon synthesis routes available when the need appears. Today, the commodity of oil still appears to override growing scruples regarding dwindling world supplies of petroleum. However, price may become a powerful driving force in a foreseeable future, since oil prices are likely to increase faster than coal prices. Hopefully, the forceful research impetus developed during the past few years in many laboratories, in different countries, will continue to produce innovation, and prepare a smooth transition to an, at least partly, coal-based fuel and chemicals industry.

X. REFERENCES

1. K. Ziegler, E. Holzkamp, H. Breil, and H. Martin, *Angew. Chem.*, **67**, 541 (1955).
2. O. Roelen, *Ger. Pat.*, 849 548, 1938; *Angew. Chem.*, **60**, 524 (1948).
3. O. Roelen, *Erdöl Kohle*, **31**, 524 (1978).
4. G. Patard, *C. R. Acad. Sci.*, **179**, 1330 (1924).
5. (a) H. Pichler, *Adv. Catal.*, **4**, 271 (1952); (b) H. Pichler and G. Krüger, *Herstellung Flüssiger Kraftstoffe aus Kohle*, Gersbach & Sohn, Munich, 1973.
6. F. Fischer and H. Pichler, *Brennst.-Chem.*, **20**, 41 and 221 (1939); *Ger. Pat.*, 731 295, 1936.
7. F. Fischer and H. Pichler, *Ges. Abh. Kenntn. Kohle*, **13**, 407 (1937/51).
8. B. S. Lee, Synthesis of *Fuels from Coal, AIChE Monogr. Ser.*, No. 14, Vol. 78, 1982.
9. H. H. Storch, N. Golumbic, and R. B. Anderson, *The Fischer–Tropsch and Related Syntheses*, Wiley, New York, 1951.
10. M. E. Dry and J. C. Hoogendoorn, *Catal. Rev., Sci. Eng.*, **23**, 265 (1981).

11. H. R. Linden, J. D. Parent, and J. G. Seay, *Perspectives on U.S. and World Energy Problems*, Gas Research Institute, Chicago, Ill., 1979.
12. H. Tropsch and H. Koch, *Brennst.-Chem.*, **10**, 337 (1929).
13. H. Koch and H. Hilberath, *Brennst.-Chem.*, **22**, 135 and 145 (1941).
14. R. A. Friedel and R. B. Anderson, *J. Am. Chem. Soc.*, **72**, 1212 (1950).
15. H. Pichler, H. Schulz, and M. Elstner, *Brennst.-Chem.*, **48**, 78 (1967).
16. H. Pichler, H. Schulz, and D. Kühne, *Brennst.-Chem.*, **49**, 344 (1968).
17. H. Schulz, B. R. Rao, and M. Elstner, *Erdöl Kohle*, **23**, 651 (1970), and literature therein.
18. H. Pichler and H. Buffleb, *Brennst.-Chem.*, **21**, 257, 273 and 285 (1940).
19. (a) W. Wenzel, *Angew. Chem., Ausg. B*, **21**, 225 (1948); (b) M. D. Schlesinger, H. E. Benson, E. Murphy, and H. H. Storch, *Ind. Eng. Chem.*, **46**, 1322 (1954); (c) Ruhrchemie, *Ger. Pat.*, 939 385, 1951 and 974 811, 1961; (d) *Russ. Pat.*, 386 899, 1973 and 386 900, 1973; (e) Yu. Kagan, A. N. Bashkirov, L. A. Morozov, Yu. B. Kryukov, and N. A. Orlova, *Neftekhimiya*, **6**, 262 (1966).
20. Ruhrchemie, *Ger. Pat.*, 904 891, 1949.
21. W. R. Grace and Co., *U.S. Pat.*, 3 726 926, 1973.
22. Monsanto Co., *U.S. Pat.*, 4 179 462, 1979.
23. Union Carbide Corp., *U.S. Pat.*, 3 833 634, 1974.
24. P. Sabatier and J. B. Senderens, *C. R. Acad. Sci.*, **134**, 514 and 689 (1902).
25. G. V. Schulz, *Z. Phys. Chem., Abt. B*, **29**, 299 (1935); **30**, 375 (1935); **32**, 27 (1936).
26. P. J. Flory, *J. Am. Chem. Soc.*, **58**, 1877 (1936).
27. R. A. Friedel and R. B. Anderson, *J. Am. Chem. Soc.*, **72**, 1212 and 2307 (1950).
28. E. F. G. Herington, *Chem. Ind. (London)*, 347 (1946).
29. P. J. Flory, *J. Am. Chem. Soc.*, **62**, 1561 (1940).
30. H. Pichler and G. Krüger, *Herstellung Flüssiger Kraftstoffe aus Kohle*, Gersbach & Sohn, Munich, 1973, p. 224.
31. C. H. Yang, F. E. Massoth, and A. G. Oblad, *Adv. Chem. Ser.*, No. 178, 35 (1979).
32. M. E. Dry, T. Shingles, and L. J. Boshoff, *J. Catal.*, **25**, 99 (1972).
33. H. H. Storch, N. Golumbic and R. B. Anderson, *The Fischer–Tropsch and Related Syntheses*, Wiley, New York, 1951, Chapter 6.
34. W. Brötz, *Z. Elektrochem.*, **5**, 301 (1949).
35. W. Brötz and W. Rottig, *Z. Elektrochem.*, **56**, 896 (1952).
36. M. A. Vannice, *J. Catal.*, **37**, 449 and 462 (1975); **40**, 129 (1975).
37. F. M. Dautzenberg, J. N. Helle, R. A. van Santen, and H. Verbeek, *J. Catal.*, **50**, 8 (1977).
38. R. B. Anderson and P. H. Emmett, *Catalysis*, Vol. 4, Reinhold, New York, 1956.
39. F. Fischer and H. Tropsch, *Brennst.-Chem.*, **7**, 97 (1926).
40. E. L. Muetterties and J. Stein, *Chem. Rev.*, **79**, 479 (1979).
41. H. Pichler and H. Schulz, *Chem.-Ing.-Tech.*, **42**, 1162 (1970).
42. G. Henrici-Olivé and S. Olivé, (a) *Angew. Chem.*, **88**, 144 (1976); *Angew. Chem., Int. Ed. Engl.*, **15**, 136 (1976); (b) *J. Mol. Catal.*, **3**, 443 (1977/78); (c) *J. Mol. Catal.*, **4**, 379 (1978); (d) *J. Catal.*, **60**, 481 (1979).
43. J. T. Kummer, T. W. DeWitt, and P. H. Emmett, *J. Am. Chem. Soc.*, **70**, 3632 (1948).
44. A. Takeuchi and J. R. Katzer, *J. Phys. Chem.*, **85**, 937 (1981).
45. R. C. Brady and R. Pettit, (a) *J. Am. Chem. Soc.*, **102**, 6181 (1980); (b) *J. Am. Chem. Soc.*, **103**, 1287 (1981).
46. G. Henrici-Olivé and S. Olivé, *J. Mol. Catal.*, **16**, 111 (1982).
47. U. Mazzi, A. A. Orio, M. Nicolini, and A. Marzotto, *Atti Accad. Peloritana Pericolanti, Cl. Sci. Fis. Mat. Nat. (Engl.)*, **50**, 95 (1970); *Chem. Abstr.*, **76**, 154161j (1972).
48. D. L. Thorn and T. H. Tulip, *J. Am. Chem. Soc.*, **103**, 5984 (1981).
49. P. Biloen, H. N. Helle, and W. M. H. Sachtler, *J. Catal.*, **58**, 95 (1979).
50. G. Henrici-Olivé and S. Olivé, *J. Mol. Catal., J. Molec. Cat.*, **18**, 367 (1983).
51. K. Isobe, D. G. Andrews, B. E. Mann, and P. M. Maitlis, *J. Chem. Soc., Chem. Commun.*, 809 (1981).
52. J. C. Hayes, G. D. N. Pearson, and N. J. Cooper, *J. Am. Chem. Soc.*, **103**, 4648 (1981).
53. E. O. Fischer, B. Heckl, K. H. Doetz, J. Müller, and H. Werner, *J. Organomet. Chem.*, **16**, P29 (1969).
54. M. Cooke, D. L. Davis, J. E. Guerchais, S. A. R. Knox, K. A. Mead, J. Roné, and P. Woodward, *J. Chem. Soc., Chem. Commun.*, 862 (1981).

55. A. F. Dyke, S. A. R. Knox, K. A. Mead, and P. Woodward, *J. Chem. Soc., Chem. Commun.*, 861 (1981).
56. R. Aumann and E. O. Fischer, *Chem. Ber.*, **114**, 1853 (1981).
57. E. Band and E. L. Muetterties, *Chem. Rev.*, **78**, 639 (1978).
58. B. Olgemöller and W. Beck, *Angew. Chem.*, **92**, 863 (1980); *Angew. Chem. Int. Ed. Engl.*, **19**, 834 (1980); W. Beck and B. Olgemöller, *J. Organomet. Chem.*, **127**, C45 (1977).
59. S. Nagakura, *J. Phys. Soc. Jap.*, **12**, 482 (1957).
60. C. P. Casey, *Chemtech*, 378 (June 1979), and references cited therein.
61. R. Merta and V. Ponec, *J. Catal.*, **17**, 79 (1969).
62. R. S. Nyholm in W. M. H. Sachtler, G. C. A. Schuitt, and P. Zwietering (eds Sachtler, Schuitt and Zwietering), in *Proceedings of 3rd International Congress on Catalysis*, Vol. 1, North Holland, Amsterdam, 1965, p. 25.
63. E. g. G. Henrici-Olivé and S. Olivé, *Coordination and Catalysis*. Verlag Chemie, Weinheim, New York, 1977.
64. L. A. M. Rodriguez and H. M. van Looy, *J. Polym. Sci., A-1*, **4**, 1951 and 1971 (1961).
65. J. P. Hogan, *J. Polym. Sci., A-1*, **8**, 2637 (1970).
66. G. Natta, *Brennst.-Chem.*, **36**, 176 (1955).
67. G. Csoritos, B. Heil, and L. Markó, *Trans. N.Y. Acad. Sci.*, **239**, 47 (1974).
68. J. B. Benziger and R. J. Madix, *Surf. Sci.*, **115**, 279 (1982), and references cited therein.
69. D. W. Goodman, J. T. Yates, and T. E. Madey, *Surf. Sci.*, **93**, L135 (1980).
70. J. H. Craig, *Appl. Surf. Sci.*, **10**, 315 (1982).
71. A. Deluzarche, J. P. Hindermann, and R. Kieffer, *Tetrahedron Lett.*, **31**, 2787 (1978).
72. W. A. Goddard, S. P. Walch, A. K. Rappé, T. H. Upton, and C. F. Melius, *J. Vac. Sci. Technol.*, **14**, 416 (1977).
73. H. Berke and R. Hoffmann, *J. Am. Chem. Soc.*, **100**, 7224 (1978).
74. J. P. Collman and S. R. Winter, *J. Am. Chem. Soc.*, **95**, 4089 (1973).
75. C. P. Casey and S. M. Neumann, *J. Am. Chem. Soc.*, **99**, 1651 (1977); **100**, 2544 (1978); J. A. Gladysz and W. Tam, *J. Am. Chem. Soc.*, **100**, 2545 (1978); J. A. Gladysz and J. C. Selover, *Tetrahedron Lett.*, 319 (1978).
76. B. B. Wayland and B. A. Woods, *J. Chem. Soc., Chem. Commun.*, 700 (1981).
77. E. A. Grigoryan, Kh. R. Gyulumyan, F. S. D'yachovskii, and N. S. Enikolopyan, *Dokl. Akad. Nauk SSSR*, **250**, 649 (1980).
78. K. L. Brown, G. R. Clark, C. E. L. Headford, K. Marsden, and W. R. Roper, *J. Am. Chem. Soc.*, **101**, 503 (1979).
79. H. Brunner, J. Wachter, I. Bernal, and M. Creswick, *Angew. Chem.*, **91**, 920 (1979); *Angew. Chem., Int. Ed. Engl.*, **18**, 861 (1979).
80. T. T. Tsou, J. C. Huffman, and J. K. Kochi, *Inorg. Chem.*, **18**, 2311 (1979).
81. G. Fachinetti, C. Floriani, F. Marchetti, and S. Merlino, *J. Chem. Soc., Chem. Commun.*, 522 (1976).
82. G. Fachinetti, C. Floriani, and H. Stoeckli-Evans, *J. Chem. Soc., Dalton Trans.*, 2297 (1977).
83. R. R. Hitch, S. K. Gondal, and C. T. Sears, *Chem. Commun.*, 777 (1971).
84. E.g., E. O. Fischer, *Adv. Organomet. Chem.*, **14**, 1 (1976) (Nobel Address), and references cited therein; M. F. Lappert and P. L. Pye, *J. Chem. Soc., Dalton Trans.*, 837 (1978), and references cited therein.
85. L. S. Pu and A. Yamamoto, *J. Chem. Soc., Chem. Commun.*, 9 (1974); N. J. Cooper and M. L. H. Green, *J. Chem. Soc., Chem. Commun.*, 761 (1974); M. L. H. Green, *Pure Appl. Chem.*, **50**, 27 (1978); R. R. Schrock, *J. Am. Chem. Soc.*, **97**, 6577 (1975); G. Henrici-Olivé and S. Olivé, *Angew. Chem.*, **90**, 918 (1978); *Angew. Chem., Int. Ed. Engl.*, **17**, 862 (1978).
86. J. R. Sweet and W. A. G. Graham, *J. Am. Chem. Soc.*, **104**, 2811 (1982); *J. Organomet. Chem.*, **173**, C9 (1979).
87. K. Noak and F. Calderazzo, *J. Organomet. Chem.*, **10**, 101 (1967).
88. F. Calderazzo and F. A. Cotton, *Inorg. Chem.*, **1**, 30 (1962).
89. G. Henrici-Olivé and S. Olivé, *Coordination and Catalysis*, Verlag Chemie, Weinheim, New York, 1977, Section 7.4.1.
90. D. R. Armstrong, P. G. Perkins, and J. J. P. Stewart, *J. Chem. Soc., Dalton Trans.*, 1972 (1972); M. E. Ruiz, A. Flores-Riveros, and O. Novaro, *J. Catal.*, **64**, 1 (1980).
91. G. Henrici-Olivé and S. Olivé, *Adv. Polym. Sci.*, **15**, 1 (1974); *Chemtech*, 745 Dec. 1981.

92. *Ullmanns Enzyklopädie der Technischen Chemie*, Vol. 9, Urban undberg Verlag, Munich, Berlin, 1957, p. 684.
93. G. Henrici-Olivé and S. Olivé, *J. Mol. Catal.*, **4**, 379 (1978).
94. L. Markó and J. Bakos, *J. Organomet. Chem.*, **81**, 411 (1974).
95. E.g., Ref. 92, Vol. 3.
96. U. Klabunde and E. O. Fischer, *J. Am. Chem. Soc.*, **89**, 7142 (1967); K. Weiss and E. O. Fischer, *Chem. Ber.*, **109**, 1868 (1976).
97. M. H. Chisholm, H. C. Clark, W. S. Johns, J. E. H. Ward, and K. Yasufuku, *Inorg. Chem.*, **14**, 900 (1975).
98. G. Henrici-Olivé and S. Olivé, *Top. Curr. Chem.*, **67**, 107 (1976).
99. R. F. Heck and D. S. Breslow, *J. Am. Chem. Soc.*, **83**, 4023 (1961).
100. M. van Boven, N. H. Alemdaroglu and J. M. L. Penninger, *Ind. Eng. Chem., Prod. Res. Dev.*, **14**, 259 (1975).
101. M. Orchin and W. Rupilius, *Catal. Rev.*, **6**, 85 (1972).
102. A. J. Moffat, *J. Catal.*, **19**, 322 (1970).
103. J. Falbe, *Synthesen mit Kohlenmonoxyd*, Springer, Berlin, New York, 1967.
104. E.g., G. Ertl, *J. Vac. Sci. Technol.*, **14**, 435 (1077); F. P. Netzer and T. E. Madey, *J. Chem. Phys.*, **76**, 710 (1982).
105. E. L. Muetterties, *Angew. Chem.*, **90**, 577 (1978); *Angew. Chem., Int. Ed. Engl.*, **17**, 545 (1978).
106. E.g., K. Christmann and G. Ertl., *Surf. Sci.*, **60**, 635 (1976); G. A. Somorjai, cited in ref. 105.
107. A. Yokohama, H. Komiyama, H. Inoue, T. Matsumoto, and H. M. Kimura, *J. Catal.*, **68**, 355 (1981).
108. J. S. Bradley, *J. Am. Chem. Soc.*, **101**, 7419 (1979); M. J. Doyle, A. P. Kouwenhoven, C. A. Schaap, and B. van Oort, *J. Organomet. Chem.*, **174**, C55 (1979).
109. M. G. Thomas, B. F. Beier, and E. L. Muetterties, *J. Am. Chem. Soc.*, **98**, 1296, (1976).
110. J. H. Davis, M. A. Beno, J. M. Williams, J. Zimmie, M. Tachikawa, and E. L. Muetterties, *Proc. Natl. Acad. Sci. USA*, **78**, 668 (1981), and references cited therein.
111. A. Brenner and D. A. Hucul, *J. Am. Chem. Soc.*, **102**, 2484 (1980).
112. E. L. Muetterties and J. Stein, *Chem. Rev.*, **79**, 479 (1979).
113. A. Deluzarche, J. P. Hindermann, R. Kieffer, J. Cressely, R. Stupfler, and A. Kiennemann, *Spectra 2000*, **10**, 27 (1982); *Bull. Soc. Chim. Fr.*, II-329 (1982).
114. M. E. Dry, *Brennst.-Chem.*, **50**, 193 (1969).
115. M. E. Dry, T. Shingles, L. J. Boshoff, and G. J. Oosthuizen, *J. Catal.*, **15**, 190 (1969).
116. M. E. Dry, T. Shingles, and L. J. Boshoff, *J. Catal.*, **25**, 99 (1972); W. Rähse and D. Schneidt, *Ber. Bunsenges. Phys. Chem.*, **77**, 727 (1972); Yu. B. Kagan, A. N. Bashkirov, L. A. Morozov, Yu. B. Kryukov, and N. A. Orlava, *Neftekhimiya*, **6**, 262 (1966).
117. S. W. Ulmer, P. M. Skarstad, J. M. Burlich and R. E. Hughes, *J. Am. Chem. Soc.*, **95**, 4469 (1973); H. B. Chin and R. Rau, *J. Am. Chem. Soc.*, **98**, 2434 (1976); M. Y. Darensbourg, D. J. Darensbourg, D. Burns, and D. A. Drew, *J. Am. Chem. Soc.*, **98**, 3127 (1976); S. B. Butts, S. H. Strauss, E. M. Holt, R. E. Stimson, N. W. Alcock, and D. F. Shriver, *J. Am. Chem. Soc.*, **102**, 5093 (1980); **101**, 5864 (1979); R. B. Petersen, J. J. Stezowski, Ch. Wan, J. M. Burlich, and R. E. Hughes, *J. Am. Chem. Soc.*, **93**, 3532 (1971).
118. J. P. Collman, J. N. Cawse, and J. I. Bramman, *J. Am. Chem. Soc.*, **94**, 5905 (1972); J. P. Collman, R. G. Finke, J. N. Cawse, and J. I. Bramman, *J. Am. Chem. Soc.*, **100**, 4766 (1978).
119. Nguyen Trong Anh, *Tetrahedron Lett.*, 155 (1976).
120. R. B. Anderson, 'Catalysts for the Fischer–Tropsch Synthesis', in *Catalysis, Vol. IV, Hydrocarbon Synthesis, Hydrogenation and Cyclization*, (Ed. P. H. Emmett), Reinhold, New York, 1956 (cf. Fig. 76, p. 359, and Fig. 27, p. 110).
121. R. J. Madon and W. F. Taylor, 'Effect of Sulfur on the Fischer–Tropsch Synthesis; Alkali-Promoted Precipitated Cobalt-Based Catalysts', in *Hydrocarbon Synthesis from Carbon Monoxide and Hydrogen* (Eds. E. L. Kugler and F. W. Steffgen), *Adv. Chem. Ser.*, No. 178 (1979).
122. G. Henrici-Olivé and S. Olivé, *J. Mol. Catal*, **16**, 187 (1982).
123. D. Commereuc, Y. Chauvin, F. Hugues, J. M. Basset, and D. Oliver, *J. Chem. Soc., Chem. Commun.*, 154 (1980).

124. Ger. Pat. Auslegeschrift, 1 271 098, 1968.
125. (a) R. C. Everson, K. J. Smith, and E. T. Woodburn, *Proc. Natl. Meet. S. Afr. Inst. Chem. Eng., 3rd*, 2D-1 (1980); (b) H. H. Nijs and P. A. Jacobs, *J. Catal.*, **65**, 328 (1980); **66**, 401 (1980); (c) C. S. Kellner and A. T. Bell, *J. Catal.*, **75**, 251 (1982).
126. M. A. Vannice, *J. Catal.*, **40**, 129 (1975); **44**, 152 (1976).
127. Y. Takasu, T. Akimaru, K. Kasahara, Y. Matsuda, H. Miura, and I. Toyoshima, *J. Am. Chem. Soc.*, **104**, 5249 (1982).
128. Y. Takasu, R. Unwin, B. Tesche, A. M. Bradshaw and M. Grunze, *Surf. Sci.*, **77**, 219 (1978).
129. H. Topsøe, N. Topsøe, H. Bohlbro, and J. A. Dumesic, *Stud. Surf. Sci. Catal.*, 247 (1981).
130. S. J. Tauster, S. C. Fung, R. T. K Baker, and J. A. Horsley, *Science*, **211**, 1121 (1981); D. A. Hucul and A. Brenner, *J. Phys. Chem.*, **85**, 496 (1981).
131. M. Aral, T. Ishikawa, and Y. Nishiyama, *J. Phys. Chem.*, **86**, 577 (1982).
132. D. J. Dwyer and G. A. Somorjai, as discussed in ref. 46.
133. H. Pichler, B. Firnhaber, D. Kioussis, and A. Dawallu, *Makromol. Chem.*, **70**, 12 (1964); F. Bellstedt, *PhD Thesis*, University of Karlsruhe, 1971.
134. R. J. Madon, *J. Catal.*, **57**, 183 (1979).
135. D. A. Whan, *Chem. Ind.*, 532 (1981).
136. S. E. Voltz and J. J. Wise (Eds.), *Development Studies on Conversion of Methanol and Related Oxygenates to Gasoline*, Orn 1/FE-1, NTIS, Springfield, 1977.
137. H. H. Nijs, P. A. Jacobs, and J. B. Uytterhoeven, *J. Chem. Soc., Chem. Commun.*, (a) 180 (1979); (b) 1095 (1979).
138. D. Ballivet-Tkatchenko, G. Coudurier, H. Mozzanega, and I. Tkatchenko, in *Fundamental Research in Homogeneous Catalysis* (Ed. M. Tsutsui), Vol. 3, Plenum Press, New York, 1979, p. 257; D. Ballivet-Tkatchenko and I. Tkatchenko, *J. Mol. Catal.*, **13**, 1 (1981).
139. D. Fraenkel and B. C. Gates, *J. Am. Chem. Soc.*, **102**, 2478 (1980).
140. G. Henrici-Olivé and S. Olivé, *The Chemistry of the Catalyzed Hydrogenation of Carbon Monoxide*, Springer Verlag, Heidelberg, New York, (1984).
141. Note added in proof: For model cobalt alkyl and acyl complexes it has been found recently that the rate constant is one to two orders of magnitude higher for reaction with $HCo(CO)_4$ than for reaction with H_2[142,143].
142. F. Ungváry and L. Markó, *Organomettallics*, **2**, 1608 (1983).
143. C. Hoff, F. Ungváry, R. B. King, and L. Markó, *J. Amer. Chem. Soc.*, in press.

The Chemistry of the Metal—Carbon Bond, Vol. 3
Edited by F. R. Hartley and S. Patai
© 1985 John Wiley & Sons Ltd.

CHAPTER **10**

Olefin carbonylation

DONALD M. FENTON and ERIC L. MOOREHEAD

Science and Technology Division, Union Oil Company of California, P.O. Box 76, Brea, California 92621, USA

I. INTRODUCTION	435
II. NON-REDOX CARBONYLATION OF OLEFINS	436
A. Olefins	437
III. OXIDATIVE CARBONYLATION	439
IV. CARBONYLATION INTERMEDIATES	441
IV. MECHANISM OF NON-REDOX CARBONYLYATION	441
VI. MECHANISM OF OXIDATIVE CARBOXYLATION	442
VII. REFERENCES	443

I. INTRODUCTION

Reppe[1] was the first to show that olefins could be attacked by carbon monoxide using a nickel carbonyl catalyst in the presence of alcohols to give saturated esters, for example reaction 1. This process has been called carbonylation, hydrocarbalkoxylation[2], hydroalkoxycarbonylation[3], and hydratocarbonylation[4]. Here it will be called the non-redox carbonylation reaction. If there is also a substrate oxidation, the reactions are called oxidative carbonylations, for example reaction 2[5].

$$CH_2=CH_2 + CO + ROH \rightarrow EtCO_2R \qquad (1)$$

$$CH_2=CH_2 + CO + ROH + (O) \xrightarrow{\text{catalyst}} CH_2=CHCO_2R$$
$$+ RO_2CCH_2CH_2CO_2R + H_2O \qquad (2)$$

The Oxo reaction is a reductive carbonylation and will not be discussed here. There is considerable interest, as documented by recent reviews[6], in this chemistry because of the wide variety of synthetic possibilities, the significant changes in rate and product distribution caused by subtle changes in conditions and catalysts, and because of the potential for industrial applications. The wide variety of products potentially available by these carbonylation reactions is illustrated in Table 1, using ethylene as the olefin.

In some cases there is also the possibility of the loss of HX, if there was an appropriate hydrogen on the starting olefin, to yield an acrylate (equation 2). Many conbinations have

TABLE 1. Formal addition of HX, Y, and CO to olefins

$$H_2-\underset{\underset{X}{|}}{C}-\underset{\underset{H}{|}}{\overset{\overset{H}{|}}{C}}-\overset{\overset{O}{\|}}{C}-Y$$

X	Y
H	OR
Cl	NR_2^7
RO	OH
RO_2C	
RCO_2	O_2CR^8
$MeCH=CHCH_2^9$	Cl
$(RO_2C)_2(CH_3)CH^{10}$	Br
Cl_3C^{11}	R = alkyl, aryl, H
Lactone[12]	

not yet been attempted, but considering the divergent nature of the groups already tested, the possibilities for inclusion of additional groups and combinations in the future are certain. This chapter will address the key facets of the oxidative and non-redox carbonylation reaction. Particular attention will be paid to the palladium-catalysed system. Cobalt, rhodium, palladium, and platinum catalysts are equal to or better than nickel, which is the usual catalyst for acetylene carbonylation[13].

II. NON-REDOX CARBONYLATION OF OLEFINS

The non-redox carbonylation of olefins formally involves the addition of X and O=CY across the olefin double bond to give an acid or acid derivative. In this case there is no formal change in the oxidation state of the metal catalyst. Four different reacting groups are involved: olefin, carbon monoxide, X, and Y. Only metal-catalysed reactions are considered. For information on strong acid catalysis, see reviews on the Koch reaction[14].

$$RHC=CH_2 + CO + R'OH \rightarrow RCH_2CH_2CO_2R' + CH_3CHRCO_2R' \qquad (3)$$

Early work demonstrated that the addition of alcohols and carbon monoxide to symmetrical olefins yielded saturated esters. With unsymmetrical olefins, two regioisomers were usually observed (reaction 3). The catalysts for these reactions are mainly Group VIII metal complexes, usually homogeneous, but supported metals have also been used[15]. The most widely studied metal is palladium, followed by platinum and cobalt. Cobalt has greater industrial potential owing to its lower cost. Recently rhodium and iridium have been considered as catalysts for proponic acid synthesis. Hydrogen donor ligands such as hydrohalic acids, carboxylic acids, phenols, alcohols, and water are frequently used together with donor ligands such as phosphines and pyridines. The latter are used almost exclusively with the cobalt system. Some systems respond favourably to the presence of complexing salts such as tin(II) chloride (platinum system) and iron(II) chloride and related iron-containing compounds in addition to ionic salts such as lithium chloride. The solvent for this reaction is frequently the alcohol being reacted. There have been reports that ketones may provide some advantages in this system and are probably exerting some solvent polarity control[16]. Mild temperatures (around 100 °C) are frequently used for the palladium system whereas temperatures of 150–220 °C are used for the rhodium and cobalt systems. Super-atmospheric pressures of carbon monoxide are desirable.

A. Olefins

A wide variety of olefins (and some acetylenes) have been used for the non-redox carbonylation reaction. These include branched, internal, alpha, cyclic, conjugated, and allenic olefins[17]. Olefins with functional groups such as hydroxyl, halides, carbonyl, and imides have also been employed. In some cases the functional group can participate in the reaction; some examples are given in reactions 4–7[18–21].

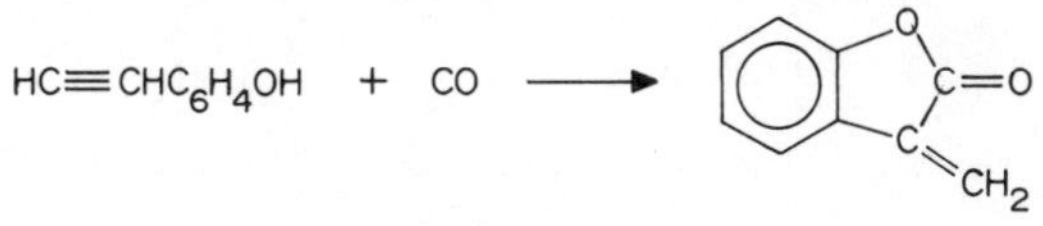

$$HC{\equiv}CHC_6H_4OH + CO \longrightarrow \text{(benzofuranone structure)} \quad (4)$$

$$CH_2{=}CHCH_2X \xrightarrow[X\,=\,OH,Cl,OAc]{CO/ROH} CH_2{=}CHCH_2CO_2R$$

or tetrahydrofuran-2-one (5)

$$CH{=}CHX \xrightarrow{CO/ROH} CH_2{=}CHCO_2R \quad (6)$$

$$\xrightarrow{CO/ROH} MeCHXCO_2R \quad (7)$$

3-vinylcyclohexene $\xrightarrow{CO/ROH}$ cyclohex-3-enyl-$CH_2CH(Me)CO_2R$

$$\xrightarrow{CO/ROH} RO_2CC_6H_{10}CH_2CH(Me)CO_2R \quad (8)$$

1,5-cod $\xrightarrow{CO/ROH}$ unsaturated monoester → saturated diester (9)

(z)-buta-1,4-diene $\xrightarrow{CO/ROH}$ (E)-MeCH=CHCH$_2$CO$_2$R

→ adiapates + α-methyl glutarate + α,α-ethyl succinate (10)

In the case of dienes one double bond or two can react, depending on the conditions, as in reactions 8–10[22]. The olefin reactivity decreases in the following order[23]: ethylene and strained olefins > α-olefins > cyclic > z-internal > E-internal > unsaturated esters. In general, the addition is E but in at least one case it was shown to be z[24]. The concentration of the olefin has little to do with percentage conversion. With unsymmetrical olefins two regioisomers are possible. Table 2 shows that under a variety of experimental conditions a large change in position of attachment of the carboxyl group is possible. The ratio of straight-chain product to the total branched-chain product is as high as 19 and as low as

TABLE 2. Range of normal to branched products in olefin carbonylation reactions

Olefin	Straight chain/branched chain ratio			
	Max.	Ref.	Min.	Ref.
Propylene	6.0	25	0.09	26
Hept-1-ene	19	27	1.4	27
Oct-1-ene	5.5	28	0.5	28
Styrene	6.6	29	0.1	22

TABLE 3. Effects of process variables on carbonylation of 1-octene and N-vinylphthalimide

	Conversion[a]		Straight/branched ratio[a]	
Increased in	1-Octene	N-Vinyl-phthalimide	1-Octene	N-Vinyl-phthahimide
Solvent polarity	I	No effect	I	I
CO pressure	I	I	D	D
H_2 pressure	I to max.	I	I	D
PPh_3 concentration	I to max.	D	I	D

[a] I = increase; D = decrease.

0.09. Obviously, the choice of experimental conditions is very important and allows considerable regioselective control. On the other hand, to illustrate the insensitivity of changes in olefin structure on regio-control on going from oct-1-ene to N-vinylphthalimide, a list of variables and their effects on conversion and ratio of straight- to branched-chain products is shown in Table 3[30]. The catalyst used was $[PdCl_2(PPh_3)_2]$.

The good correlation would probably be even better if the variables were checked over the same ranges. However, for other catalyst systems these same variables may be in a different order, e.g. with nickel and cobalt increases in carbon monoxide pressure can lead to conversion decreases.

Hydrohalic acids are often added to increase the reaction rate and keep the catalyst components soluble[31]. Hydrohalic acids have a pronounced beneficial effect on increasing the branched-chain isomer yield, and this and high temperature are the most influential variables for increasing branched-chain isomer yields. As expected, the addition of bases such as lithium acetate increases the normal to iso chain ratio. Chloride ion decreases the ratio[1] and chloride acceptors such as iron carbonyls increase the ratio.

The effect of reducing agents is important, particularly with the cobalt and rhodium systems where it is important to maintain the catalyst in the hydride form. For the palladium system, the addition of these reducing agents can also be very beneficial. Hydrogen, hydroquinone, and hydrazine have also been used. Oxidizing agents such as oxygen, copper(II) chloride, and benzoquinone impede the reaction[32]. Similarly, the addition of too much hydrogen can lead to the formation of reduced products such as aldehydes[33]. The hydrogen partial pressure is particularly important in some cobalt and rhodium work[34].

Knifton[27] showed that in the conversion of heptene to octanoates using a triphenylphosphinepalladium chloride–tin(II) chloride catalyst, the conversion of olefin and selectivity for the straight-chain isomer did not vary regardless of which alcohol was used from the group methanol, hexan-1-ol, 2-chloroethanol, and water. Propan-2-ol and phenol gave equivalent selectivities but decreased rates, possibly indicating a steric effect.

For platinum and palladium catalysts the triarylphosphines have been generally used[35]. Arylarsine complexes are also useful but trialkylphosphines are not as active. However, with cobalt and nickel the alkylphosphines have utility. Pyridine has a marked inhibition for the palladium catalyst but is highly beneficial for the cobalt catalyst.

Temperature and pressure are very important variables with the original catalysts of $[Ni(CO)_4]$ and $[Co_2(CO)_8]$. Originally, temperatures of around 200 °C and pressures of 2000 psig carbon monoxide were necessary, but when the triphenylphosphine was added the temperature required was reduced to 120 °C and the pressure was reduced to 500 psig. It was then discovered that the reaction can be made reversible with control of regioisomers. For example, with palladium catalysts, anhydrides having a β-hydrogen can be converted back to an olefin and carbon monoxide at 150 °C[36]. If the temperature is further increased to 200 °C, then the carboxylic acids which are also formed are converted back to the starting olefin, carbon monoxide, and water[37]. Similarly, esters and acid chlorides are converted back to olefins, carbon monoxide, and the corresponding alcohol or acid at temperatures around 220 °C[37a,38,39]. These reactions are illustrated in equations 11–14.

$$RCH_2CH_2C(O)OC(O)CH_2CH_2R \rightarrow 2RCH{=}CH_2 + 2CO + H_2O \quad (11)$$

$$RCH_2CH_2COOH \rightarrow RCH{=}CH_2 + CO + H_2O \quad (12)$$

$$RCH_2CH_2COOR \rightarrow RCH{=}CH_2 + CO + ROH \quad (13)$$

$$RCH_2CH_2COCl \rightarrow RCH{=}CH_2 + CO + HCl \quad (14)$$

If the volatiles formed in the decomposition reaction are kept under the reaction conditions, then true equilibrium can be attained. Therefore, data obtained on acid and anhydride formation using palladium catalysts above 150 °C are at least partially based on thermodynamic rather than kinetic control[40].

If the addition of an alcohol or an other acid to the olefin–carbon monoxide system is formally considered as the addition of a proton and base, then it should be possible to substitute other possible species for the proton. Butenyl chloride[9] has been used successfully according to reaction 15. Although this general reaction sequence has not been adequately investigated, there are many possibilities for synthetic utilization.

$$MeCH{=}CHCH_2Cl + C_2H_4 + CO + ROH$$
$$\xrightarrow{[Ni(CO)_4]} MeCH{=}CH(CH_2)_3CO_2R \quad (15)$$

Carbon dioxide has been used as a carbon monoxide substitute in cases where the unsaturated system is particularly active, e.g. strained rings and dienes according to reactions 16 and 17[41,42].

$$\triangle\!\!\!\!| + CO_2 \longrightarrow \text{(methylene lactone)} \quad (16)$$

$$CH_2{=}CHCH{=}CH_2 + CO_2 \xrightarrow[Pd]{Ni} CH_2{=}CHCH{=}CHCO_2H \quad (17)$$

III. OXIDATIVE CARBONYLATION

Blackham[43] was the first to demonstrate the oxidative carbonylation of olefins. In this system ethylene was reacted with carbon monoxide in the presence of a palladium chloride–benzonitrile adduct to yield β-chloropropionyl chloride according to reaction 18. Shortly thereafter, Tsuji et al.[44] improved the yield of the β-chloropropionyl chloride by using the [PdCl$_2$(PhCN)$_2$] complex. The scope of the reaction was widened when it was reported that mercury(II) ion could be used as an oxidizing agent and that alkoxides could be substituted for halides according to reaction 19[45]. More important, the reaction could be made catalytic in palladium by the use of co-oxidants such as those used in the palladium-catalysed oxidation of ethylene to acetaldehyde. For example, the reaction could be made stoichiometric in copper(II) ion[46] or ultimately catalytic in both palladium and copper but stoichiometric in oxygen[47]. The addition of these co-oxidants also provides for the synthesis of a variety of products such as acrylic acid from ethylene (reaction 20)[48].

$$PdCl_2(PhCN)_2 + C_2H_4 + CO \rightarrow ClCH_2CH_2C(O)Cl + Pd^0 \quad (18)$$

$$2ROH + C_2H_4 + CO + Hg^{2+} \rightarrow ROCH_2CH_2COOR + Hg^0 \quad (19)$$

$$CH_2{=}CH_2 + CO + \tfrac{1}{2}O_2 \xrightarrow[CuCl_2]{PdCl_2} CH_2{=}CHCO_2H \quad (20)$$

In these more complicated systems the solvent of choice is usually an alcohol and in addition to the expected β-alkoxypropionates there were also formed acrylates and succinates (reaction 21).

$$CH_2=CH_2 + CO + ROH + CuCl_2 \xrightarrow{catalyst} ROCH_2CH_2CO_2R$$
$$+ CH_2=CHCO_2R + RO_2CCH_2CH_2CO_2R + HCl + CuCl \quad (21)$$

Both α- and internal straight-chain olefins react well. Branched-chain olefins, such as isobutene, are also reactive but in this case hydride rearrangements are generally observed. In general, strong acids, such as the HCl liberated from the copper(II) chloride system, cause double bond migration. On the other hand, acetate ion promotes π-allyl formation, which can then lead to product rearrangement. So-called activated olefins, such as acrylates and cinnamates, fail to react. Conjugated dienes generally yield diesters with one non-conjugated double bond remaining. When α-olefins are used the major product contains the forming carboxy group ends attached to the terminal carbon of the former olefin. However, when the RO group is large as in the case of 2,6-dichlorophenoxy[49], then the carbonyl of the carboxyl groups tends to be attached to the internal carbon of the double bond. Styrene gives the same regiospecificity as do α-olefins again establishing that steric effects can be more controlling than electronic[50]. A large halide ion concentration tends to favour acrylate formation and the presence of a base such as acetate ion[51] or a *tert*-amine is desirable for decreased succinate formation. A high ratio of carbon monoxide to olefin tends to increase the succinate yield[52]. High concentrations of both the olefin and carbon monoxide are advantageous in that side reactions such as aldehyde and ketone formation from the olefin or carbon dioxide from the CO as well as various other oxidations such as alcohol oxidation are minimized[53]. It was also shown that water markedly increases the yield of carbon dioxide via the water gas shift reaction. As a result, drying agents have been used to minimize this effect. If copper(II) chloride is used as the co-oxidant with palladium, then the HCl liberated can attack the alcohol to yield an alkyl chloride, ether or an olefin. For this reason it is advantageous to use copper(II) alkoxy chloride or copper(I) chloride if oxygen is going to be the ultimate oxidant.

The preferred catalyst is palladium chloride. Co-oxidants such as $FeCl_3^{5a}$, $CuCl_2$, $EtONO_2^{15}$, benzoquinone[54], Hg^{2+} [45], and NO_x[55] and electrochemical oxidations[56] have been used. One advantage of using oxygen as the ultimate oxidant is that only small amounts of halide are necessary so that corrosion can be minimized. The principle disadvantage is the formation of water[5a]. It is important to keep the palladium in an oxidized form at all times during the course of the reaction, as in its reduced form it will catalyse non-redox carbonylations if the olefin and carbon monoxide still remain. In general the temperatures employed are mild, ranging from below room temperature to 150 °C. At low temperatures the formation of lactones has been reported.

$$C_2H_4 + CO + PdCl_2 \rightarrow ClCH_2CH_2COCl + Pd^0 \quad (21)$$

$$C_2H_4 + Cl_2 \rightarrow CH_2=CHCl \xrightarrow{catalyst} CH_2=CHCOCl \quad (22)$$

$$\begin{array}{cc} H_2CCH_2C(O)Pd \rightarrow CH_2=CHCO_2R & H_2CCH_2C(O)OOR \\ | & | \\ OOR & Pd \\ (1) & (2) \end{array} \quad (23)$$

The oxidation capacity can be controlled by either the Pd^{2+}–Pd^0 couple or by first oxiding the olefin and then performing a non-redox carbonylation reaction as in reactions 21 and 22[57]. In both cases the products formed are interconvertible by standard methods.

10. Olefin carbonylation

The alternative pathways to achieve the same goal makes these carbonylation reactions doubly useful. A third alternative pathway not yet elucidated could involve the intermediate formation of a hydroperoxide according to reaction 23. This alternative pathway would require that the intermediate be **1** not **2**. However, the only example of a carbonylation involving alkyl hydroperoxide did not give oxidation[58].

If the oxidative carbonylation is conducted with insufficient oxidizing strength then both oxidative and non-redox carbonylation will occur. In the case of propylene the products shown in equation 24 were observed in one reaction. The ratio of β-acetoxy acids to each other was the same as the ratio of the corresponding unsaturated acids, indicating a common intermediate. However, the ratio of saturated straight- to branched-chain products was much lower than with the corresponding unsaturated acids. These data are consistent with an oxidized palladium intermediate leading to unsaturated acid and β-acetoxybutyrate formation while reduced palladium catalyst leads to saturated acid formation.

$$\text{MeCH=CH}_2 + \text{CO} + \text{O}_2 \xrightarrow[\text{CuCl}_2]{\text{PdCl /HOAc}} \text{MeCH=CHCOOH}$$

$$+ \text{Pr}^n\text{CO}_2\text{H} + \text{CH}_2\text{=C(Me)CO}_2\text{H} + \text{AcOCH}_2\text{CH(Me)CO}_2\text{H}$$

$$+ \text{AcOCH(Me)CH}_2\text{CO}_2\text{H} + \text{Me}_2\text{CHCO}_2\text{H} \tag{24}$$

IV. CARBONYLATION INTERMEDIATES

Structure **3** shows that as X and Y groups react with the olefin and carbon monoxide to form C—C—C—COY three new bonds are formed, designated as a, b and c. Since the oxidized metal undergoes a reduction during the course of this reaction, it must also be involved with at least one of these groups too, possibly all four. The metal—carbon bonding is labelled 1–4.

$$\begin{array}{c} \text{O} \\ \text{C}\!=\!\!\overset{b}{\text{C}}\!-\!\!\overset{\|}{\text{C}} \\ {}^{a}\diagup{}^{2}\ \ {}^{3}\diagdown{}^{c} \\ \text{X}---\text{M}\!\!\leftarrow\!\!-\!\!-\!\!-\!\!-\!\!-\!\!\rightarrow\!\text{Y} \\ {}^{1}\qquad\quad {}^{4} \end{array}$$

(3)

Which combinations of these bond-forming reactions lead to products? Which intermediates have been discovered? Of all the combinations possible for palladium, only two are not yet known. These two possibilities are (1) the two-group combination of olefin and carbon monoxide and (2) the three-group combination of olefin, carbon monoxide, and Y. Although the two-group combination of olefin and carbon monoxide is not known, the corresponding acetylene complex is known, prepared from cyclopropanone. These complexes have been formed under a variety of conditions. They are all candidates for intermediates in the carbonylation reaction.

V. MECHANISM OF NON-REDOX CARBONYLATION

For cobalt, rhodium, iridium, and non-acidic palladium, the mechanism of the non-redox carbonylation is probably very similar to the well known Oxo reaction, where in an early stage the hydride metal carbonyl complex reacts with the olefin to form an akyl metal compound. This intermediate complex undergoes an internal insertion of the carbonyl to form an acyl metal complex. The final step is the attack of the solvent which in non-redox reactions is usually an alcohol, to yield the product and regenerate the hydride metal complex[25]. In one case (reaction 25) it was shown that the hydrogen and carbonyl group are added z- to each other[59].

$$\underset{\underset{\text{ROH}}{|}}{H_2C=CH_2} + HM(CO)_n \rightarrow MeCH_2M(CO)_n \rightarrow MeCH_2(O)CM(CO)_{n-1}$$
$$\rightarrow MeCH_2COOR + HM(CO)_{n-1} \qquad (25)$$

For the cobalt–pyridine system, the role of the pyridine is to aid in the solvolysis in the intermediate acyl metal complex. For the case where M is platinum, intermediate **4** has been isolated[60]. Since the cobalt carbonyl hydride complex is such a strong acid, it is not surprising that with α-olefins the proton adds according to Markovnikov's rule, thus yielding the most stable carbenium ion. As a result, the observed product distribution favours branched-chain acids and esters. However, when used with pyridine, straight-chain acids predominate[61]. Conversely, in the rhodium and iridium systems, the product distribution shows a much higher degree of linearity. This is presumably due to the greater steric requirements. The mechanism for palladium- and platinum-catalysed solvolytic carbonylation is probably more complex than for the cobalt, rhodium, or iridium systems. In this case under acidic conditions, e.g. in the presence of free HX, a Markovnikoff addition of the proton to yield the secondary carbenium ion could occur. This is most likely since palladium tends not to form stable hydrides like cobalt, rhodium, or iridium. As expected, the stable carbenium intermediate tends to yield branched acids and esters. The predominance of branched products in the palladium system also occurs when the halide concentration is kept high. In this case the excess halide presumably impedes the palladium–olefin reactions but not the reaction with carbenium ions. However, when the major products formed are straight-chain isomers, then it is likely that the first step is the attack of the palladium complex on the carbon-carbon double bond, followed by hydrogen addition.

Heck[50] has shown in a related system that the organic portion attaches to the less sterically hindered position of the olefin and is relatively insensitive to olefin polarity. The formation of the carbonyl group can be either an addition of carboalkoxy (one step) or carbon monoxide followed by alkoxy addition (two step).

A carboalkoxy intermediate complex has recently been isolated from a solvolytic reaction medium. Consistent with this mechanism is the fact that straight-chain products can be isomerized to the corresponding branched-chain products with no loss of carbon monoxide, thus indicating that a carbonyl chloride was migrating and adding back to the olefin[62]. In addition, it has been shown that deuterium exchange on esters occurs to greater extent at the β-carbon than at the α- or γ-position[63]. Also, where there is a choice (e.g. with allyl alcohol), the product reversibly formed was in one case [Pt(CO$_2$CH$_2$CR'= CH$_2$)Cl(PR$_3$)], indicating preferential reaction with the hydroxyl group over the olefin[64]. Heck et al.[65] also showed that carbon monoxide reaction would only occur with aryl halides if alcohols were also present[65].

VI. MECHANISM OF OXIDATIVE CARBOXYLATION

The mechanism of the oxidative carboxylation of olefins must take into accout the formation of the major products formed. These are α,β-unsaturated acids and esters, β-alkoxyesters (β-acyloxy acids), and succinates. These products are probably not interconvertable under the reaction conditions. It has been previously demonstrated that the β-acetoxypropionate was not in equilibrium with acrylic acid under the conditions of the reaction. Similarly, alkoxypropionates are not easily converted to succinates and the reverse reaction, succinates to alkoxypropionates, does not occur at the lower temperatures generally employed for oxidative carboxylation. Stille et al.[66] elegantly showed that (z)- and (E)-but-2-ene are converted mainly to 3-methyl-2-methoxybutane carboxylates by an E-addition of methoxy and carbomethoxy across the double bond of the olefin. In the presence of acetate, the major product was shown to be predominantly the 2,3-dimethylbutane dicarboxyate formed by the z-addition of two carbomethoxy groups. The key intermediates were thought to be **5** and **6**, shown here for (z)-but-2-ene. In each case the palladium was later replaced by carboalkoxy groups with retention of configuration[67]. The effect of the acetate was to remove the proton from the incoming carboalkoxy group and thus increase the probability of formation of **6** according to reaction 26[68]. An

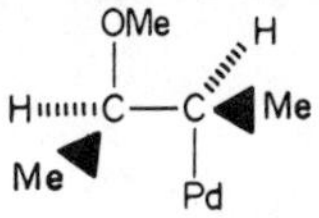

(5) 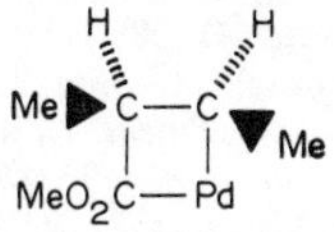   (6)

alternative choice is to have **6** be the sole intermediate either reacting with methanol in an E-fashion or with carbomethoxy in a z-fashion. Complex **6** also has the useful feature that acrylates or acrylic acid could easily be formed by the loss of HPd.

$$Pd—C(O)\overset{+}{O}(H)Me + AcO^- \rightarrow Pd—C(O)OMe + HOAc \quad (26)$$

In a different system, for oxidative carboxylation of propylene to give methacrylic, crotonic, β-acetoxybutyric, and β-acetoxyisobutyric acids, it was found that the ratio of methacrylic to crotonic in six cases out of seven was the same as the ratio of β-acetoxyisobutyric to β-acetoxybutyric[69]. Assuming that there was no equilibrium as was the case in the ethylene system, then intermediate **6** would explain the results. If intermediate **5** was also involved, this would mean that the E-addition of methoxy and the x-addition of carbomethoxy would both have to proceed with similar regioselectivity.

VII. REFERENCES

1. W. Reppe, *Justus Liebigs Ann. Chem.*, **582**, 1 (1953).
2. G. Cavinato and L. Toniolo, *J. Mol. Catal.*, **10**, 161 (1981).
3. V. Yu. Gankin, M. G. Katsnel'son and D. M. Rudkovskii, *Zh. Prikl. Khim.*, **41**, 2582 (1968).
4. D. M. Fenton, *J. Org. Chem.*, **38**, 1928 (1973).
5. (a) D. M. Fenton and P. J. Steinwand, *J. Org. Chem.*, **37**, 2034 (1972); (b) R. F. Heck, *J. Am. Chem. Soc.*, **94**, 2712 (1972).
6. (a) J. Falbe, *Carbon Monoxide in Organic Synthesis*, Springer Verlag, New York, 1920; (b) Ya. T. Eidus and A. L. Lapious, *Russ. Chem. Rev.*, **42**, 199 (1973); (c) J. Tsuji, Adv. Org. Chem., **6**, 150 (1977); (d) I. Wender and P. Pino, *Organic Synthesis via Metal Carbonyls*, Vol. 2, Wiley, New York, 1977, pp. 297–516.
7. P. Pino, F. Piacenti, M. Bianchi, and R. Lazzaroni, *Chim. Ind.* (*Milan*), **50**, 106 (1968).
8. D. M. Fenton, *US Pat.*, 3 641 073 (1972).
9. G. P. Chiusoli and G. Cometti, *J. Chem. Soc., Chem. Commun.*, 1015 (1972).
10. L. S. Hegedus and W. H. Darlington, *J. Am. Chem. Soc.*, **102**, 4981 (1980).
11. T. Susuki and J. Tsuji, *J. Org. Chem.*, **35**, 2982 (1970).
12. J. K. Stille and R. Divakaruni, *J. Am. Chem. Soc.*, **100**, 1303 (1978).
13. J. Falbe, *J. Organomet. Chem.*, **94**, 213 (1975).
14. G. Olah and J. Olah, *Friedel–Crafts and Related Reactions*, Vol. 3, Interscience, New York, 1964, pp. 1272–1296.
15. (a) S. Umemura, K. Matsui, Y. Ikeda, K. Masunaja, T. Kadota, K. Fujii, K. Nishihara, and M. Matsuda, *Br. Pat.*, 2024821 (1980); (b) E. N. Frankel and F. L. Thomas, *J. Am. Oil Chem. Soc.*, **50**, 39 (1973).
16. E. N. Frankel, F. L. Thomas, and W. F. Knolek, *J. Am. Oil Chem. Soc.*, **51**, 393 (1974).

17. J. Tsuji, T. Susuki, and A. Mullen, *Tetrahedron Lett.*, 3027 (1965).
18. J. R. Norton, K. E. Shenron, and J. Schwartz, *Tetrahedron Lett.*, **1**, 51 (1975).
19. (a) L. J. Kehoe and R. A. Schell, *J. Org. Chem.*, **35**, 2846 (1970); (b) D. M. Fenton, *US Pat.*, 3 692 849 (1972).
20. R. D. Closson, I. Casey, and V. Ihrman, *US Pat.*, 3 457 299 (1968).
21. A. Schoenberg, I. Bartoletti, and R. F. Heck, *J. Org. Chem.*, **39**, 3318 (1974).
22. K. Bittler, N. V. Kutepow, D. Neubauer, and H. Reis, *Angew. Chem., Int. Ed. Engl.*, **7**, 329 (1968).
23. (a) G. P. Chiusoli, G. Cometti, and V. Bellotti, *Gazz. Chim. Ital.*, **104**, 259 (1974); (b) A. Mullen, in *New Syntheses with Carbon Monoxide* (Ed. J. Falbe), Springer Verlag, New York, p. 243.
24. J. Knifton, *J. Org. Chem.*, **41**, 2885 (1970).
25. G. Cavinato and L. Toniolo, *Chemia*, **33**, 286 (1978).
26. DuPont, *Ger. Pat., German Patent*, 2 739 096 (1978).
27. J. F. Knifton, *J. Org. Chem.*, **41**, 793 (1976).
28. D. M. Fenton, *J. Org. Chem.*, **38**, 3192 (1973).
29. Y. Sugi, K. Bando, and S. Shin, *Chem. Ind. (London)*, 397 (1975).
30. G. Cavinato and L. Toniolo, *J. Organomet. Chem.*, **229**, 93 (1982).
31. J. R. Tsuji, M. Morikawa, and J. Kiji, *Tetrahedron Lett.*, **22**, 1437 (1963).
32. A. P. Budennyi, V. A. Rybakov, G. N. Gvozdovskii, B. P. Tarsov, V. M. Gavrilova, and T. A. Semenova, *Zh. Prikl. Khim.*, **54**, 2156 (1981).
33. A. Mullen, in *New Syntheses with Carlon Monoxide* (Ed. J. Falbe), Springer Verlag, New York, 1980, p. 1278.
34. G. T. Chen, *US Pat.*, 4 303 589 (1981).
35. N. V. Kutepow, K. Bittler, and D. Neubauer, *US Pat.*, 3 501 518 (1970).
36. D. M. Fenton, *US Pat.*, 3 668 249 (1972).
37. (a) D. M. Fenton, *US Pat.*, 3 530 198 (1970); (b) T. Sakakibara, *Chem. Commun.*, 1563 (1970).
38. J. Tsuji, *Synthesis*, **1**, 157 (1969).
39. J. Tsuji and K. Ohno, *J. Am. Chem. Soc.*, **90**, 94 (1968).
40. (a) D. M. Fenton, *US Pat.*, 3 578 688 (1971); (b) T. A. Foglia, I. Schmeltz, and P. A. Barr, *Tetrahedron Letters*, **30**, 11 (1974).
41. Y. Inoue, T. Hibi, M. Satake, and H. Hashimoto, *J. Chem. Soc., Chem. Commun.*, 982 (1979).
42. (a) Y. Sasaki, Y. Inoue, and H. Hashimoto, *J. Chem. Soc., Chem. Commun.*, 605 (1976); (b) H. Hoberg and D. Schaeffer, *J. Organomet. Chem.*, **255**, C15 (1983).
43. A. Blackham, *US Pat.*, 3 119 861 (1964).
44. J. Tsuji, M. Morikawa, and J. Kiji, *Tetrahedron lett.*, 1061 (1963).
45. D. M. Fenton, *US Pat.*, 3 316 280 (1967).
46. D. M. Fenton, *US Pat.*, 3 397 225 (1968).
47. D. M. Fenton and P. Steinwand, *J. Org. Chem.*, **37**, 2034 (1972).
48. (a) D. M. Fenton, K. L. Olivier, and J. Biale, *Am. Chem. Soc. Div. Pet. Chem. Prepr.*, **14** (3), C77; (b) D. M. Fenton, K. L. Olivier, and J. Biale, *Hydrocarbon Process.*, **95** (1972).
49. J. Hallgren and R. Matthews, *J. Organomet. Chem.*, **192**, C12 (1980).
50. R. F. Heck, *J. Am. Chem. Soc.*, **93**, 6896 (1971).
51. G. Cometti and G. P. Chiusoli, *J. Organomet. Chem.*, **181**, C14 (1979).
52. (a) R. F. Heck, *J. Am. Chem. Soc.*, **91**, 6707 (1969); (b) D. E. James and T. K. Stille, *J. Am. Chem. Soc.*, **98**, 1810 (1976).
53. (a) J. F. Blackburn and J. Schwartz, *J. Chem. Soc., Chem. Commun.*, 157 (1977); (b) A. Nikiforova, I. Noiseev, and Ya. Syrkin, *J. Gen. Chem. USSR*, **33**, 3237 (1963).
54. D. M. Fenton and J. Biale, *US Pat.*, 3 755 421 (1973).
55. J. Biale, D. M. Fenton, K. L. Olivier, and W. Schaffer, *US Pat.*, 3 381 030 (1968).
56. (a) T. Yukawa and S. Tsutsumi, *J. Org. Chem.*, **34**, 738 (1969); (b) D. M. Fenton, *US Pat.*, 3 481 845 (1969).
57. J. A. Scheben and I. L. Mador, *US Pat.*, 3 468 947 (1969).
58. J. Collin, D. Commerenc, and Y. Chalivin, *J. Mol. Catal.*, **14**, 113 (1982).
59. C. W. Bird, R. C. Cookson, J. Hudel, and R. O. Williams, *J. Chem. Soc.*, 410 (1963).
60. R. Bardi, A. M. Piazzesi, G. Cavinato, P. Cavoli, and L. Toniolo, *J. Organomet. Chem.*, **224**, 407 (1982).
61. N. S. Imyanitov and D. M. Rudkovskii, *J. Appl. Chem. USSR*, **41**, 157 (1968).

62. T. A. Foglia, P. A. Barr, and M. J. Idacavage, *J. Org. Chem.*, **41**, 3452 (1976).
63. P. A. Colfer, T. A. Foglia and P. E. Pfeffer, *J. Org. Chem.*, **44**, 2573 (1979).
64. H. Kurosawa and R. Okawara, *J. Organomet. Chem.*, **71**, C35 (1974).
65. A. Schoenberg, I. Bartoletti, and R. F. Heck, *J. Org. Chem.*, **39**, 3318 (1974).
66. J. K. Stille and R. Divakaruni, *J. Organomet. Chem.*, **169**, 239 (1979).
67. L. F. Hines and J. K. Stille, *J. Am. Chem. Soc.*, **94**, 485 (1972).
68. M. Hidai, M. Kokura, and Y. Uchida, *J. Organomet. Chem.*, **52**, 431 (1973).
69. K. L. Oliver, *US Pat.*, 3 621 054 (1971).

Author Index

This author index is designed to enable the reader to locate an author's name and work with the aid of the reference numbers appearing in the text. The page numbers are printed in normal type in ascending numerical order, followed by the reference numbers in brackets. The numbers in *italics* refer to the pages on which the references are actually listed.

If reference is made to the work of the same author in different chapters, the above arrangement is repeated separately for each chapter.

Abashev, G. G., 116 (231), 118 (263, 266), 119 (266), *136, 137*
Abel, E. W., (163), *93*
Abenhaim, D., 101 (13), 102 (31–35), 108 (99), 119, 120 (13, 99, 275), *132, 133, 137*
Abraham, M. H., 2, 24, 37 (22), *90*
Adams, H. N., 374 (62), *388*
Adamson, G. W., 350 (112), *358*
Adkins, H., 363 (17), 366 (29), *387*
Agawa, T., 190 (206), *198*
Agnes, G., 154 (76), *161*
Ahmad, M. S., 114 (165), *135*
Akermark, B., *167*, 172 (101), 173 (106), *196*, 223, 224 (60), *256*
Akhrem, A. H., 116 (223), *136*
Akimaru, T., 427 (127), *434*
Akita, M., 202 (5), *204*
Akutagawa, S., 154 (81), *161*
Albizati, K. F., 33 (135), *92*
Albrecht, H. P., 114 (166), *135*
Alcock, N. W., 423 (117), *433*
Alderson, T., 220 (43), *256*, 344 (64), 345 (69), *358*
Aldridge, C. L. 354 (69), *358*, 375 (64), *388*
Aleksandorozicz, P., 190 (211), *198*
Alekseeva, Z. D., 118 (256), *136*
Alemdaroglu, N. H., 378 (76), *388*, 420 (100), *433*
Alemdarogly, N. H., 373 (57), *388*
Alexakis, A., 103, 104, 108, 109 (61), *133*
Almdöf, J., 223, 224 (60), *256*
Almemark, M., 223, 224 (60), *256*
Almsay, M., 338 (28), *357*

Alper, H., 302 (35, 38), 303 (35), 324 (35, 164, 165), 325 (167–169), *330, 333*
Aldridge, C., 375 (68), *388*
Altpeter, B., 8 (58), *91*
Avernhe, G., 116 (245), *136*
Amice, P., 114 (151), *134*
Amiet, R. G., 267 (40), *293*
Amin, N. V., 190 (209), *198*
Andell, O. S., 159 (100), 162
Andersen, N. H., 37 (152), *93*
Anderson, G. K., 345 (81, 83), *358*
Anderson, J. K., 345 (82), *358*
Anderson, R. B., 322 (152), *333*, 393 (9), 395 (14), 401 (27), 402, 403 (9), 406 (33), 407 (9), 408 (9, 38), 423 (9), 424 (120), 425 (27), *430, 431, 433*
Andersson, S. L. T., 356 (149), *359*
Anding, C. R., 116 (228), *136*
Andrac, M., 110 (113), *134*
Andreev, A. A., 28 (120), *92*
Andrews, D. G., 410 (51), *431*
Andrews, L. C., 223 (57), *256*
Andrews, M. A., 343 (43), *357*
Andrews, P. S., 270 (55), *293*
Andrews, S. B., 8 (56), *91*, 129 (436), 131 (483), *140, 141*
Andrich, G., 354 (136), *359*
Anisimov, K. N., 262 (16–21), *292*
Anselme, J. P., 102 (42), *137*
Ansheles, V. R., 212 (19), *255*
Antonova, N. D., 127 (409), *140*
Aoki, S., 2 (8), 37, 47 (153), 70 (8), 77 (248), *89, 93, 95*

Author index

Arai, H., 49 (181a), *94*
Arai, M., 125 (371), *139*
Aral, M., 427 (131), *434*
Aratani, T., 78 (259), *95*
Archart, R. W., 305 (57), *331*
Armbrecht, F. M., 3 (27, 28, 33), 4 (33), 8 (55), *90*
Armstrong, D. R., 417 (90), *432*
Arnaud, P., 114 (152), *134*
Arora, P., 116 (239), *136*
Arpe, H. J., (4), *294*, 297 (4), *330*
Arsenijevic, L., 116 (230, 248), *136*
Arsenijevic, V., 116 (230, 248), *136*
Artamkina, G. A., 126 (397), *139*
Artamkina, Y. A., 127 (403), *139*
Asahi Chem. Ind. Co. Ltd., 84 (291), *96*
Asamiya, K., 165 (47), *195*
Ashley-Smith, J., 255 (114), *257*
Ast, W., 86 (306), 87 (311, 317), 318 (88), *97*
Astakhova, A. S., 176 (124), *197*
Aswraf, M. C., 174 (113), *197*
Atkins, K. E., 181 (153), 184 (173), *197*, *198*
Attridge, C. J., 18 (88), *91*, 131 (480), *141*
Auerbach, R. A., 128 (411), *140*
Auger, J., 109, 110 (109), *134*
Augl, J. M., 308 (72), *331*
Ault, B. A., 368 (35), *388*
Aumann, R., 410 (56), *432*
Avanzi, M., 337–339 (23), *357*
Avram, E., 275 (76), *296*
Avram, M., 168 (76), *196*, 275 (76), *294*
Aylett, B. J., 2, 68 (4), *89*, 315, 317 (110), *332*
Azuma, H., 315, 317 (114), *332*

Baarschers, W. H., 11 (63), *91*
Baba, S., 168 (77, 79), *196*
Bachi, M. D., 75 (246), *95*
Bäckval, J. E., 223, 224 (60), *256*
Backvall, J., 185 (181), *198*
Bäckvall, J. E., 159 (100), *162*
Baibulatova, N. Z., 179 (139), *197*
Bailey, J. V., 121 (304), *137*
Bailey, P. M., 288 (96), *294*
Bailey, W. I., 260 (5), 283 (92), *292*, *294*
Baillargeon, D. J., 120 (277), *137*
Baillargeon, V. P., 70 (228), *95*
Baker, D. J., 150 (50), *160*
Baker, R., 86 (305), *97*, 144 (5), 148 (29, 30, 34), 149 (37), 151 (5), 152 (73), 154 (80, 82, 83), 158 (96), 159 (34), *159–162*, 163 (3), 178 (134), 179 (140, 141), *194*, *197*, 303, 304, 312, 314 (46), *330*
Baker, R. T. K., 427 (130), *434*
Bakos, J., 316, 317 (124), *332*, 418 (94), *433*
Balabane, M., 188 (197), *198*
Baldwin, J. E., 114 (176), *135*
Ballivet-Tkatchenko, D., 430 (138), *434*

Balochi, L., 374 (61), *388*
Balsamo, A., 116 (220, 238), *136*
Ban, R., 223 (57), *256*
Band, E., 410 (57), *432*
Bando, K., 437 (29), *444*
Barborak, J. C., 306 (64), *331*
Barbot, F., 102 (21, 25, 27, 30), *132*
Bardi, R., 442 (60), *444*
Barefield, E. K., 150 (46), *160*
Baret, P., 115 (208), *135*
Barili, P. L., 116 (220), *136*
Barnes, C. L., 45 (173), *93*
Barnett, K. W., 301 (33), *330*
Barnette, W. E., 61 (205), *94*
Barr, P. A., 439 (40b), 442 (62), *444*, *445*
Bartish, C. M., 354 (126, 127), *359*
Bartoletti, I., 437 (21), 442 (65), *444*, *445*
Barton, D., 297, 301, 302, 304–306, 312, 328 (5), *330*
Barton, D. H. R., 2, 70, (11), *89*
Bashkirov, A. N., 397, 404, 405 (19e), 423 (116), *431*, *433*
Basnett, B. L., 267 (40), *293*
Basolo, F., 369 (41), *388*
Basset, J. M., 85 (296), *96*, 425 (123), *433*
Battacharya, S. K., 319 (138, 139), *332*
Battiste, M. A., 167 (70–72), *196*
Bau, R., 260 (7), 279 (86), *292*, *294*
Bauer, L. N., 122 (319), *138*
Baukov, Y. I., 8 (53), *90*
Bauld, N. L., 151 (55), *161*, 312, 317 (94), *331*
Bauman, B. A., 128 (413), *140*
Baykut, F., 113 (129), *134*
Beach, D. L., 301 (33), *330*
Bechter, M., 164 (7), 165 (25), *195*
Beck, J. A., 272 (61), *293*
Beck, W., 370 (48), *388*, 410 (58), *432*
Beckwith, A. L., 71, 74 (236), *95*
Beckwith, A. L. J., 71 (234, 237), 74 (237), *95*
Beetz, J., 123 (336), *138*
Behr, A., *162*
Beier, B. F., 422 (109), *433*
Bekker, R. A., 27 (117), *92*
Beletskaya, I. P., 25 (109), 31 (131), 53 (109, 131, 190–194), 55 (190, 191), *92*, *94*, 126 (380, 382, 384, 385, 387, 390, 392, 393, 397, 399), 127 (403, 407), 128 (380), *139*, *140*, 191 (221), *199*
Bell, A. T., 427, 428 (125c), *434*
Bell, H. C., 79, 81 (266–268), *96*
Bellassoued, M., 101 (20), 104 (71), 107 (89), 108 (71, 97), 116 (229), *132*, *133*, *136*
Bellegarde, D., 7, 61 (49), *90*
Bellstedt, F., 428 (133), *434*
Belotti, V., 437 (23a), *444*
Beltskaya, I. P., 126 (383), *139*
Belykh, Z. D., 118 (258), 119 (271), *136*, *137*

Benain, J., 311 (87), *331*
Bender, D. D., 179 (145), *197*
Benezra, C., 116 (241), *136*
Benfield, F. W. S., 227 (70), *256*
Benn, R., 145 (12, 13), 150 (47, 48), *160*
Benner, L. S., 422 (112), *433*
Bennett, 269, 270 (54), *293*
Bennett, M. J., 308 (67), *331*
Beno, M. A., 422 (110), *433*
Ben-Shoshan, R., 301 (31), *330*
Benson, H. E., 397 (19b), *431*
Benson, R. E., 304 (52), *330*, 344, 345 (61), 346 (61, 93), *357*, *358*
Ben Taarit, Y., 356 (154), *359*
Benziger, J. B., 414, 421 (68), *432*
Bercaw, J. E., 275 (71), *294*
Berenblyum, A. S., 176 (124), *197*
Bergamaschi, E., 157 (92), *162*
Bergbreiter, D. E., 127 (405), *140*
Bergman, R. G., 266 (38), 275 (71), 278 (80), *293*, *294*, 298 (20), *330*
Bergmann, E. D., 314, 317 (106), *332*
Bergstrom, D. E., 125 (370), *139*, 194 (243), *199*
Berke, H., 414, 417, 424 (73), *432*
Bernardon, C., 120 (280), 121 (303), *137*
Bernadou, F., 103 (62, 64), 104 (62, 64, 66, 70), 105 (62, 64, 66), *133*
Bernal, I., 415 (79), *432*
Bernhardt, J. C., 126 (401), 128 (418, 419, 427), *139*, *140*
Bernstein, P. B., 185 (177), *198*
Bersellini, U., 152 (70), 154 (77, 78), *161*
Bertelli, D., 305 (59), *331*
Bertrand, J., 375 (68), *388*
Bertrand, M. T., 117, 118 (250–252), *136*
Berty, J., 378 (82), *388*
Bestian, H., 220, 229 (41), 253 (41, 110), *256*, *257*
Bevan, P. C., 152 (73), *161*
Bhattacharyya, S. K., 322 (157), *333*
Biale, J., 439 (48a, 48b), 440 (54, 55), *444*
Bianchi, M., 318, 319, 322–324 (133), *332*, 336, 337, 339 (1), 343 (1, 48, 50), 344, 345, 348 (1), *356*, *357*, 362 (3), 369 (44), 371, 372 (52), *387*, *388*, 436 (7), *443*
Bilhou, J. L., 85 (296), *96*
Billard, C., 345 (83), *358*
Billups, W. E., 180 (149), *197*, 346 (89), *358*
Biloen, P., 410, 411, 418 (49), *431*
Binger, P., 184 (168), *198*
Birch, A. J., 122 (314), *138*, 302, (34, 37), 305 (37, 58), *330*, *331*
Bird, C. W., 272, 289 (65), *293*, 297 (2), 308, 315 (71), *329*, *331*, 336 (2), 337 (2, 15), 344, 345 (53), *356*, *357*, 362 (9), *387*, 442 (59), *444*

Biresaw, G., 114 (173), *135*
Birkenstock, U., 233 (93), *257*
Birkhahn, M., 5 (39), *90*
Bittler, K., 343, 346, 348 (39), *357*, 437 (22), 438 (35), *444*
Bittner, C. W., 211 (5), *255*
Bjorkman, E. E., 185 (181), *198*
Black, D. St. C., 297, 301, 302, 304–306, 312, 328 (5), *330*
Black, M., 223, 224 (56), *256*
Black, R. J., 21 (100a), *92*
Blackburn, J. F., 440 (53a), *444*
Blackett, B. N., 154 (80), *161*
Blackham, A., 439 (43), *444*
Blagoer, B., 116 (221), *136*
Blanchard, E. P., 113, 114 (144–146), *134*
Blaser, H. U., 154 (83a), *161*
Blaukat, U., 125 (353), *138*
Blindheim, U., 219, 229 (39), 231, 233 (39, 88), 234, 236 (39), 237, 239 (88), 242 (39, 88), 244–246 (88), 247, 248 (39), 250 (39, 88), *256*, *257*
Bloch, A., 122 (324), *138*
Bloch, H. S., 88 (321), *97*
Blom, J. E., 164 (13), *195*
Blount, J. F., 172 (100), *196*
Blum, J., 314, 317 (106), 321 (150, 151), *332*, *333*
Bobek, M., 122 (324), *138*
Boelhouwer, C., 86 (303, 304, 309), *97*
Bogadanovic, B., 244 (102), *257*
Bogavac, M., 116 (230), *136*
Bogdanovic, B., 145 (17, 18), 150 (49, 52), 152 (64, 71), 156 (87), *160*, *161*, 217 (32), 219 (32, 36), 233 (93), 239 (96), *255*, *257*, 290 (108), *294*
Bogdanvic, B., 242, 244 (98), *257*
Bohlbro, H., 427 (129), *434*
Böhm, H., 83 (280), *96*
Boileau, A. M., 284 (93), *294*
Boireau, G., 102 (33–35), *132*
Bokii, N. G., 166 (53), *195*
Boldrini, G. P., 316, 317 (123), *332*
Bonhomme, J., 123 (335), *138*
Bönnemann, H., 233 (93), 246 (97), *257*, 310 (85), *331*
Bor, G., 370 (49), *388*
Borcherdt, G. T., 151 (54), *161*, 312 (96), *331*
Bortolin, R., 182 (156), *197*
Bos, K. D., 77 (253), *95*
Bosc, J. J., 111 (119), 112 (122), 113 (140), *134*
Boschi, T., 169 (88), *196*
Boshoff, L. J., 405, 406 (32), 423 (115, 116), 424 (115), *431*, *433*
Bota, T., 338 (28), *357*
Botteghi, C., 345 (77), *358*
Bouchoule, C., 102 (22, 49), 110 (49), *132*

Bourhis, M., 111 (119), 112 (122), 113 (140), *134*
Boven, M. van, 420 (100), *433*
Braatz, J., 165 (32), *195*, *196*
Braatz, J. A., 167 (65, 69), *196*
Braca, G., 308 (69, 77), 309 (69), *331*, 354 (136), *359*
Bradley, J. S., 269 (50), *293*, 422 (108), *433*
Bradshaw, A. M., 427 (128), *434*
Brady, R. C., 409 (45), *431*
Bramman, J. I., 423 (118), *433*
Bray, L. S., 268, 284 (45), *293*
Breil, H., 158 (95), *162*
Brenner, A., 422 (111), 427 (130), *433*, *434*
Brenner, W., 148, 149 (36), 152 (72), *160*, *161*
Breslow, D. S., 311 (86), *331*, 338, 339, (25), *357*, 365 (22), 370 (22, 47), 371 (22, 54), 372 (22), *387*, *388*, 420 (99), *433*
Breusch, F. L., 113 (129), *134*
Brewis, S., 180 (146), *197*, 345 (71, 87), *358*
Brewster, R. Q., 126 (388), *139*
Brickman, S. J., 158 (93a), *162*
Briel, H., 392 (1), *430*
Brinkman, R., 310 (85), *331*
British Petroleum Co. Ltd., 212 (13), *255*
Brocas, J. M., 68, 70 (221), *95*
Brodzki, D., 350 (109), 352 (122), 354 (109, 122), *358*, *359*
Broline, B. M., 114 (176), *135*
Brötz, W., 406 (34, 35), *431*
Brower, K. R., 27 (116), *92*
Brown, C. K., 384, 385 (103, 104), *389*
Brown, E. S., 159 (98), *162*, 297 (9), *330*
Brown, J. M., 384–386 (107–109), *389*
Brown, K. L., 415 (78), *432*
Browning, J., 272 (63, 64), *293*
Brunner, H., 415 (79), *432*
Buchachenko, A. L., 126 (380, 383, 384), 128 (380), *139*
Bucourt, R., 116 (225), *136*
Budding, H. A., 7 (50), *90*
Budennyi, A. P., 438 (32), *444*
Buffet, H., 115 (208), *135*
Buffleb, H., 397 (18), *431*
Bühler, R., 264 (24), *293*
Bulten, E. J., 77 (253), *95*
Bumagin, N. A., 25 (109), 31 (131), 53 (109, 190–193), 54 (194), 55 (190, 191), *92*, *94*
Bumagina, I. G., 25 (109), 53 (109, 191–193), 55 (191), *92*, *94*
Bunce, S. C., 114 (153), *134*
Bundel, Yu. G., 125 (350), 127 (409), *138*, *140*
Bürger, G., 151 (56), *161*, 165 (20), *195*
Burger, U., 114 (184), *135*
Burkhard, J., 114 (190), *135*
Burkhart, J. P., 164 (14), 175 (117), *195*, *197*
Burlich, J. M., 423 (117), *433*

Burlitch, J. M., 3 (26), *90*, 129 (446, 461, 464, 465), 130 (446), 131 (446, 477, 480), *140*, *141*
Burns, D., 423 (117), *433*
Burns, T. P., 114, 116 (170), *135*
Burns, W., 288 (97), *294*
Burrowes, T. G., 74 (243), *95*, 174 (113), *197*
Bursics, A. R. L., 51 (187), *94*
Burstinghaus, R., 9 (59), *91*
Burtlich, M., 129 (460), *141*
Buse, C. T., 64 (211), *94*
Bush, W. V., 211 (5), 213 (23), *255*
Bussiere, H., 107, 113 (83), *133*
Butts, S. B., 423 (117), *433*
Buzas, H., 120 (294), *137*

Cadiot, P., 37 (154), *93*, 115 (204), *135*
Cainelli, G. F., 318 (132), *332*
Cais, M., 123 (329), *138*
Calderazzo, F., 252 (106), *257*, 343 (47), *357*, 368 (33), 381 (88), *387*, *389*, 417 (87, 88), *432*
Calderon, N., 84 (292), *96*
California Research Corp., 211 (6), 212 (18), *255*
Campari, G., 154, 158 (74), *161*
Campbell, H. C., 289 (102), *294*
Campbell, J. M., 315, 317 (110), *332*
Campistron, I., 86 (307), *97*
Canceill, J., 119 (268), *137*
Canning, L. R., 384–386 (107), *389*
Caporusso, A. M., 261 (11), *292*
Carlson, R. M., 12 (72), *91*
Carlton, L., 261 (12), *292*
Carpenter, G. B., 337 (7), *356*
Carrick, W. L., 84 (283), *96*
Carson, G. L. B., 123 (330), *138*
Cartier, G. E., 114 (153), *134*
Casey, C. P., 343 (43), *357*, 369 (45, 46), *388*, 411 (60), 414 (75), 416 (60), *432*
Casey, I., 437 (20), *444*
Cason, J., 122 (308), 123 (308, 327, 337), *137*, *138*
Cassar, L., 144 (8), 145 (20), 146, 151, 152 (8), 154 (20), 156 (8), 157 (8, 89a, 90), *159–161*, 298 (26), 299 (26, 28), 312 (93), 313 (103), *330*, *331*, 337, 348 (21), *357*
Cast, J. R., 131 (487), *141*
Castaing, M., 114 (163), *135*
Castaing, M. D., 34 (136), *92*
Castellano, S., 374 (59), *388*
Castellucci, N. T., 115 (206), *135*
Castro, B., 118 (253), *136*
Catellani, M., 147 (26), 152 (66, 70), 158 (93), *160–162*
Cavinato, G., 438 (30), 442 (60), *444*
Cavoli, P., 442 (60), *444*

Cawse, J. N., 423 (118), *433*
Cekovic, Z., 74 (240), *95*
Cesarotti, E., 202 (8), *204*
Cetkovic, S., 116 (248), *136*
Chalivin, Y., 441 (58), *444*
Chalk, A. J., 314 (109), 315 (111), 317, 321 (109, 111), *332*
Chambers, R. L., 21, 23 (100), *92*
Chan, A. S. K., 325 (168), *333*
Chan, D. M. T., 189 (202–204), 190 (205), *198*
Chan, T. H., 37 (151), *93*
Chang, C. C., 291 (116), *294*
Chang, C. T., 291 (116), *294*
Charpentier, J. P., 106 (81), *133*
Charrier, C., 311 (87), *331*
Chatt, J., 220 (51), 227 (72), *256*
Chatterjee, H., 191 (223), *199*
Chauvin, J., 219 (37), *256*
Chauvin, M., 244, 245 (100), *257*
Chauvin, Y., 84, 87 (293), *96*, 177 (130), *197*, 206, 249 (2), *255*, 425 (123), *433*
Chen, F., 322 (155), *333*
Chen, G. T., 438 (34), *444*
Chen, S. L., 32, 33 (133), *92*
Chenault, J., 120 (288), *137*
Cheng, Y. M., 131 (479), *141*
Chenicek, J. A., 88 (321), *97*
Chernova, A. D., 122 (312), *137*
Chin, H. B., 279 (86), *294*, 423 (117), *433*
Chini, P., 383 (97), *388*
Chino, K., 72 (235), 76 (247), 77 (235), *95*
Chiraleu, F., 168 (76), *196*
Chirico, R., 262 (23), *292*
Chisholm, H., 268 (53), 283 (92), *293*, *294*, 418 (97), *433*
Chiusoli, G. P., 144 (8, 9), 145 (16, 20), 146 (8, 24, 25), 147 (26), 151 (8, 9, 24, 25, 57, 58), 152 (8, 66, 67, 69, 70), 154 (9, 20, 74–78), 156 (8, 85, 86), 157 (8, 85, 90–92), 158 (74, 93), *159–162*, 291 (116), *294*, 313 (101–103), 314 (104), 322 (156), *331–333*, 337, 348 (21), *357*, 436 (9), 437 (23a), 439 (9), 440 (51), *443*, *444*
Choffield, T. H., 288 (97), *294*
Chong, J. M., 33 (134), *92*
Christ, H., 165 (24, 26, 28), *195*
Christensen, B., 355 (147, 148), 356 (147), *359*
Christmann, K., 421 (106), *433*
Chuang, Y. D., 291 (116), *294*
Chujo, Y., 156 (88), *161*, 191 (216), *199*
Chung, H., 176 (125), *197*
Cioffari, A., 71 (232, 233), *95*
Cioni, C., 323 (161), *333*
Clardy, J., 16 (80a), *91*
Claremon, D. A., 61 (205), *94*
Clark, G. R., 415 (78), *432*
Clark, H. C., 3 (25), *90*, 227 (73), *256*, 343 (41), 345 (79, 81–83), *357*, *358*, 418 (97), *433*
Clauss, K., 220, 229 (41), 253 (41, 110), *256*, *257*
Closson, R. D., 437 (20), *444*
Coates, G. E., 222 (53), *256*
Cochran, J. C., 37 (143), *93*
Colburn, R. E., 290 (110), *294*
Colette, J. W., 120 (293), *137*
Colfer, P. A., 442 (63), *445*
Colin, G., 5 (42), *90*
Colleuille, Y., 86 (308), *97*
Collin, J., 311 (87), *331*, 441 (58), *444*
Collins, D. J., 165 (51), 173 (103, 105), *195*, *196*
Collins, D. M., 260 (5), *292*
Collins, P. A., 151 (62a), *161*
Collman, J. P., 222 (52), *256*, 306 (66), 327 (177), 329 (66), *331*, *333*, 366 (26), *387*, 414 (74), 423 (118), *432*, *433*
Collum, D. B., 16 (79), *91*
Colquhoun, H. M., 337 (20), *357*
Colvin, W. E., 37 (151), *93*
Cometti, G., 151 (58), *161*, 436 (9), 437 (23a), 439 (9), 440 (51), *443*, *444*
Commereuc, D., 177 (130), *197*, 425 (123), *433*, 441 (58), *444*
Conia, J. M., 114 (151, 174, 175), *134*, *135*
Considine, J. L., 36 (140), *93*
Consiglio G., 159 (102), *162*, 344 (56), 345 (75–77), *357*, *358*
Conti, F., 165 (33–35), 166 (55), *195*
Conway, W. P., 170, 172 (94), *196*
Cook, A. H., 148 (29, 34), 159 (34), *160*
Cooke, M., 410 (54), *431*
Cooke, M. P., 311 (88), 328 (180), *331*, *333*
Cookson, R. C., 152, (73), 154 (80), 158 (96), *161*, *162*, 344, 345 (53), *357*, 442 (59), *444*
Cooley, J. H., 71 (231), *95*
Cooper, N. J., 410 (52), 416 (85), *431*, *432*
Cope, A. C., 289 (101, 102, 104), *294*
Copeland, A. H., 152 (73), *161*
Corbin, T. F., 112 (125), 114 (167), *134*, *135*
Corcoran, D. E., 16 (80), *91*
Corey, E. J., 19 (92, 93), 31 (132), *91*, *92*, 119, 120 (276), *137*, 144 (11), 145 (15), 151, 154 (59), *160*, *161*, 312 (95, 97, 98), 313 (99), *331*
Corey, J. S., 312 (92), *331*
Corrigan, P. A., 270 (57), *293*
Corte, F., 346 (98), 347 (98, 99), *358*
Cortese, N. A., 179 (143), *197*
Cossee, P., 147 (28), *160*, 226 (69), *256*, 297 (11), *330*
Costa, M., *162*
Costanzo, S. J., 122, 123 (309), *137*
Costerousse, G., 116 (224), *136*

Cotton, F. A., 252 (106), *257*, 260 (5), 283 (92), *292*, *294*, 378 (79), *388*, 417 (88), *432*
Cotton, J. D., 77 (250), *95*
Coulson, D. E., 183 (165), *198*
Coulson, D. R., 183 (167), *198*
Courtois, G., 101 (5), 102 (5, 50), 103 (5, 50, 55, 56), 104 (5, 67), 105 (75, 78, 79), 106 (75, 79), 109 (109–111), 110 (5, 109, 111), 111 (55, 116, 118), 112 (118, 121), 117, 118 (250–252), 119 (267), *132–134*, *136*, *137*
Craddock, J. H., 351 (116), *359*
Craig, J., 167 (69), *196*
Craig, J. H., 414, 421 (70), *432*
Craig, Z. E., 289 (103), *294*
Cram, D. J., 46 (177), *94*
Cramer, R., 220 (44), 223 (59), 224 (59, 64), 251 (104, 105), 252 (104), *256*, *257*, 351 (113), *358*
Crandall, J. K., 311 (89), *331*
Creemers, H. M. J. C., 5 (41), 7 (41, 51), *90*
Cressley, J., 423 (113), *433*
Crimmin, M. J., 148 (29), 154 (82), *160*, *161*
Crociani, B., 169 (88), *196*
Cross, B., 101 (10), *132*
Cross, P. E., 302, 305 (37), *330*
Cross, R. C., 154 (80), *161*
Cross, R. J., 3 (27), *90*
Crossandeau, M. C., 86 (307), *97*
Crotti, P., 116 (220, 238), *136*
Crowe, B. F., 308 (78, 79), *331*, 344 (59), *337*
Csontos, G., 382, 383 (93), *389*
Csoritos, G., 413 (67), *432*
Cullen, N. R., 28 (121, 122), *92*
Curdy, R. M., 120 (292), *137*
Curran, D. P., 189 (200), *198*
Curtain, D. Y., 126 (369), *139*
Cuvigny, T., *162*
Cyr, C. R., 369 (45, 46), *388*

Dale, J., 278 (79), *294*
Dalin, M. A., 212 (19), *255*
Dall'Asta, G., 85 (297), *96*
Dallatomasina, F., 152 (66, 69), 154 (77), *161*
Daly, J. J., 350 (112), *358*
Damasewitz, G. A., 152 (68), *161*
Damiel, A., 84 (285), *96*
Damrauer, R., 129 (462, 463), 130 (470), 131 (483), *141*
D'Aniello, M. J., 150 (46), *160*
Dardoize, F., 116 (242), *136*
Darensbourg, D. J., 423 (117), *433*
Darensbourg, M. Y., 423 (117), *433*
Darlington, W. H., 174 (107), *196*, 436 (10), *443*
Darragh, K. V., 129 (442–444), 130, (443, 444), *140*

Daub, J., 325 (166), *333*
Daub, H. J., 305 (59), *331*
Dauben, W. G., 120 (293), *137*, 300 (29), *330*
Daude, G., 43 (168), *93*
Dautzenberg, F. M., 406 (37), *431*
Daviaud, G., 103 (54), 107 (90), 116 (240), *133*, *136*
Daviaurd, G., 101 (18, 19), *132*
David, S., 2, 70 (14), *89*
Davidsohn, W., 67 (218), *95*
Davidson, J. L., 265 (31), 266 (32), (59), *293*
Davidson, P. J., 77 (250), *95*, 362 (8), *387*
Davies, A. G., 2 (23), 18 (87), 24 (23), 25, 28 (108), 37, 43, 45 (23), 67 (215), 70 (23, 229), *90–92*, *94*, *95*
Davies, J. A., 387 (112), *389*
Davies, R. E., 267 (40), *293*
Davies, S., 159 (101), *162*
Davis, D. D., 17 (83), 21 (100, 100a), 23 (100), *91*, *92*
Davis, D. L., 410 (54), *431*
Davis, J. A., 324 (162), *333*, 345 (79, 81–83), *358*
Davis, J. H., 422 (110), *433*
Davis, R. E., 268, 284 (45), *293*
Davison, A., 341 (35), *357*, 378 (80), *388*
Dawallu, A., 428 (133), *434*
Day, V. W., 265 (28), *293*
De Benneville, P. L., 121 (305), *137*
Deeming, A. J., 227 (71), *256*
Deeming, H. J., 343 (42), *357*
Deger, T. E., 336, 344 (4), *356*
Degreve, Y., 270 (58), *293*
Dehmolow, E. V., 130 (471), *141*
De Jeso, B., 68, 70 (221), *95*
Deluzarche, A., 414 (71), *432*
Deniau, J. P., 108 (102), 121 (301), *133*, *137*
Denis, J. M., 114 (174, 175), *135*
Denise, B., 350 (109), 352 (122), 354 (109, 122), *358*, *359*
Dent, W. T., 164 (8), 193 (233), *195*, *199*
Denzel, T., 123 (332), *138*
Deshpande, K. G., 123 (342), *138*
Desio, P. J., 119–124 (274), *137*
Dettlaf, G., 268 (46), *293*
Deubauer, D., 343, 346, 348 (39), *357*
Devos, D., 79 (265), *96*
Dewar, M. J. S., 220 (50), *256*
De Witt, T. W., 409, 410 (43), *431*
Dickinson, R. S., 268, 270 (44), 271 (44, 60), *293*
Dickson, R. S., 260 (8), 270 (56, 57), *292*, *293*
Didier, P., 2, 66 (17), *90*
Dietl, H., 165 (27, 28), *195*, 275, (77), *294*
Dietsche, T. J., 165 (38), 170 (38, 92–94), 172 (93, 94, 99), *195*, *196*
Dinulescu, I. G., 168 (76), *196*, 275 (76), *294*

Divakarumi, R., 169 (83), *196*
Divakaruni, R., 436 (12), 442 (66), *443*, *445*
Doering, E. W., 113, 114 (143), *134*
Doetz, K. H., 410 (53), *431*
Dohlaine, H., 114 (159), *134*
Dohring, A., 148, 149, 159 (33), *160*, 184 (169), *198*
Domaille, P. J., *162*
Domiano, P., 116 (238), *136*
Donaldson, P. B., 269, 270 (54), *293*
Donati, M., 165 (33–35), 166 (55), *195*
Döring, I., 103 (58), *133*
Döts, K. H., 306 (65), *331*
Dötz, K. H., 131 (488), *141*
Doucoureau, A., 36 (142), *93*
Dowd, S. R., 129–131 (446), *140*
Doyle, M. J., 422 (108), *433*
Drabowicz, J., 324 (163), *333*
Dradi, E., 154 (74), 158 (74, 93), *161*, *162*
Drago, R. S., 352 (121), *359*
Drenth, W., 7 (50), *90*, 351 (114), 352 (120), *358*, *359*
Drew, D. A., 423 (117), *433*
Driggs, R. J., 126 (401), *139*
Druliner, J. D., 159 (97), *162*
Dry, M. E., 393 (10), 405, 406 (32), 423 (114–116), 424 (115), *430*, *431*, *433*
Dubac, J., 35 (137), *92*, 131 (484, 485), *141*
Dubini, M., 151 (57), *161*
Dubois, J. E., 64 (211), *94*
Duchek, I., 6 (45), *90*
Dumarten, G., 34 (136), *92*
Dumesic, J. A., 427 (129), *434*
Du Mont, W.-W., 88 (320), *97*
Duncanson, L. A., 220 (51), *256*
Dunkerton, L. V., 189 (201), *198*
Dunne, K., 167 (58), *195*
Dupont, G., 337, 341 (22), *357*
DuPont de Nemours and Co., E. I., 220 (45), *256*, 437 (26), *444*
Dwyer, D. J., 427 (132), *434*
D'yachovskii, F. S., 414 (77), *432*
Dyatkin, B. L., 129 (441), *140*
Dyckman, W. J., 67 (220), *95*
Dyke, A. F., 410 (55), *432*
Dzhemilev, U. M., 179 (138, 139), *197*

Eaborn, C., 2, 52 (21), *90*
Eaton, P. E., 298, 299 (26), *330*
Eavenson, C. W., 278 (84), 279 (85), *294*
Echsler, K. J., 10 (61), 15 (76), *91*
Edidus, Ya., T., 336, 337, 345 (5), *356*
Edward, J. T., 302 (38), *330*
Edwards, J. T., 324 (165), *333*
Efraty, A., 262 (15, 23), 278 (82, 83), *292*, *294*
Eguchi, T., 383, 387 (98), *389*
Ehmann, W. J., 273, 277 (68), *294*

Eicher, S., 289 (105), *294*
Eichler, S., 289 (106), *294*
Eidus, Ya. T., 337 (17, 18), 354 (128, 129, 137), *357*, *359*, 435 (6b), *443*
Eisch, J. J., 152 (68), *161*
Eisenmann, J. L., 318 (134), 320 (140, 141), *332*
Eisert, M. A., 129 (435, 466), *140*, *141*
Ekanayake, N., *162*
El A'mma, A., 352 (121), *359*
Elhafez, F. A. Abd, 46 (177), *94*
Elissondo, B., 4 (38), *90*
Elmer, O. C., 344 (59), *357*
El'perina, E. A., 128 (420), *140*
Elsevier, C. J., 113 (138), *134*
Elstner, M., 395 (15), 396, 411 (17), 420 (15), *431*
Emmett, P. H., 408 (38), 409, 410 (43), *431*
Emptoz, G., 115 (205), 122 (311), *135*, *137*
Enda, J., 190 (206), *198*
Endo, M., 48 (180), *94*
Engelhardt, V, A., 344 (164), *358*
English, A. D., 159 (97), *162*
Enikolopyan, N. S., 414 (77), *432*
Ercoli, R., 337–339 (23), 343 (51), 344 (51, 54), *357*, 367 (32), 374 (59), 381 (88), *387*–*389*
Erhardt, U., 325 (166), *333*
Erthl, G., 421 (104), *433*
Ertl, G., 421 (106), *433*
Ethyl Corp., 212 (9, 10, 15), *255*
Evans, B. R., 154, 155 (84), *161*
Evans, D., 384 (102), 385 (102, 111), *389*
Evans, D. A., 120 (277), *137*
Everson, R. C., 427 (125a), *434*
Evikk, E. R., 266 (38), *293*
Evnin, A. B., 29, 30 (125, 126), 55 (198), *92*, *94*
Ewers, J., 219 (35), 231 (78), 234 (35), *255*, *256*
Eylander, C., 12 (71), *91*

Fachinetti, G., 374 (60–62), *388*, 415 (81), 416 (81, 82), *432*
Fajkos, J., 114 (162, 164), *134*, *135*
Falbe, J., 297 (12, 13), 320 (12), *330*, 337 (14, 19), 341 (14), 346 (14, 19, 97, 98), 347 (97–100), 348 (100), 349 (106), *357*, *358*, 362 (2), 364 (21), *387*, 421 (103), *433*, 435 (6a), 436 (13), *443*
Fallon, G. D., 270 (57), *293*
Faraone, F., 164 (17), *195*
Farbenindustrie, I. G., 151 (53), *161*
Farbwerke Hoechst, 231 (85), *257*
Farcasiu, D., 23 (103), *92*
Farina, M., 253 (108), *257*
Farkas, J., 338 (28), *357*

Fauvarque, J. F., 118 (264, 265), *137*
Favre, E., 108 (96), *133*
Feast, W. J., 87 (315), *97*
Fel'dblyum, V. Sh., 206 (1), 219 (38), 229 (77), 231 (38), 233, 234 (77), *255, 256*
Feldkamp, J., 117 (249), *136*
Felkin, H., 159 (101), *162*
Fell, B., 345 (66), *358*
Fellmann, P., 64 (211), *94*
Fenselau, A. H., 304 (51), *330*
Fenslau, A. H., 346 (92), *358*
Fenton, D. M., 343 (38), *357*, 435 (4, 5a), 436 (8), 437 (19b, 28), 438 (36, 37a), 439 (37a, 40a, 45–47, 48a, 48b), 440 (5a, 45, 54, 55, 56b), *443, 444*
Fenton, S. W., 289 (104), *294*
Ferretti, M., 116 (220), *136*
Fesev, R., 228 (76), *256*
Fessenden, R. J., 122, 123 (308), *137*
Fettel, H., 122 (310), *137*
Fiato, R. A., 167 (71, 72), *196*
Fiaud, J. C., 185 (182), 191 (218), 192 (226), *198, 199*
Ficini, J., 188 (195), *198*
Fiene, M. L., 288 (100), *294*
Finar, J., 16 (80a), *91*
Finke, R. G., 428 (118), *433*
Fiorenza, J. C., 68 (223a), *95*
Firnhaber, B., 428 (133), *434*
Fischer, E. O., 165 (20), *195*, 329 (186), *333*, 410 (53, 56), 416 (84), 418 (96), *431–433*
Fischer, F., 362 (13, 14), *387*, 393 (6, 7), 408 (39), *430, 431*
Fisher, E. O., 131 (488), *141*, 151 (56), *161*, 306 (65), *331*
Fitton, H., 302 (34), *330*
Fitzpatrick, J. D., 303 (41), *330*
Fleming, I., 21, 22 (101), 37 (151), *92, 93*
Fletterick, R., 130 (467), *141*
Flid, R. M., 342 (37), *357*
Flores-Riveros, A., 417 (90), *432*
Floriani, C., 415 (81), 416 (81, 82), *432*
Flory, P. J., 400 (26), 401 (29), *431*
Foa, M., 157 (89a), *161, 162*, 312 (93), *331*
Foglia, T. A., 439 (40b), 442 (62, 63), *444, 445*
Ford, G. P., 220 (50), *256*
Forchester, D., 348 (102), 350 (102, 111, 112), 351 (102, 111), 352 (102), 353 (125), 354 (102), *358, 359*
Fortunak, J. M., 185 (183), *198*
Fortunak, J. M. D., 54 (196), *94*
Foss, V. L., 128 (416), *140*
Fotin, V. V., 116 (231), 118 (262), 119 (270, 271), *136, 137*
Four, P., 2, 70 (12), *89*
Fox, J. J., 2, 66 (18), *90*
Fraenkel, D., 430 (139), *434*

Franchimont, A., 126 (386), *139*
Frangin, Y., 101 (20), 103 (63, 65), 104 (68, 69, 71), 107 (88, 89), 108 (71, 96, 97), *132, 133*
Frankel, E. N., 436 (15b, 16), 440 (15b), *443*
Fraser, D. J. J., 165 (22), *195*
Fraser, P. J., 268, 270 (44), 271 (44, 60), *293*
Freedman, H. H., 36 (139), *93*, 291 (112), *294*
Freiberg, J., 122 (318), *138*
Freidman, S., 322 (152, 153), *333*
Freidrich, E. C., 114 (148, 173), *134, 135*
Freon, P., 101 (13), 102 (31, 32, 34), 108 (98, 102), 109 (106–108), 113 (127, 128), 119 (13, 275), 120 (13, 275, 283, 285, 287, 289, 291, 294, 295), 121 (106–108, 283, 295, 297, 298, 301, 303), 123 (127, 128, 283, 285, 295, 297, 338), *132–134, 137, 138*
Frey, F. W., 2 (24), *90*
Fried, J. H., 19 (96), *91*, 115 (212, 213), *135*
Friedel, R. A., 366 (25), *387*, 395 (14), 401, 425 (27), *431*
Friederich, H., 322 (158), *333*
Friedrich, L. E., 167 (71), *196*
Fritch, J. R., 265 (28), *293*
Fritz, H. P., 77 (255), *95*
Fröhlich, C., 291 (114), *294*
Fuelbier, H., 152 (68b), *161*
Fujii, H., 114 (200), *135*
Fujii, K., 436, 440 (15a), *443*
Fujimoto, H., 169 (84), *196*
Fujinara, M., 261 (13), *292*
Fujita, E., 11 (66), 24 (106a), *91, 92*
Fujita, J., 151 (61), *161*
Fujita, T., 114 (201), 124 (201, 343), *135, 138*, 178 (133), *197*
Fujiwara, Y., 194 (244), *199*
Fukui, K., 226 (67), *256*
Fukui, M., 164 (16), *195*
Fukuoka, S., 328 (182), *333*
Fullerton, T. J., 165 (31, 38), 170 (31, 38, 92, 93), 172 (93, 99), *195, 196*
Fumagalli, A., 383 (97), *389*
Funakoshi, W., 169 (84), *196*
Funfschilling, P. C., 185 (177), *198*
Fung, S. C., 427 (130), *434*
Funk, R. L., 281 (88), 282 (90), *294*, 309 (83), *331*
Furakawa, J., 114 (185–187, 194), 124 (343), *135, 138*
Furet, C., 5, 6 (43), *90*
Furukawa, J., 101, 107 (3), 108 (3, 103), 110, 112, 113 (3), 114 (3, 189, 191–193, 201), 115 (3), 124 (201), *132, 133, 135*, 150 (43), *160*, 182 (158), *197*, 226 (68), *256*
Furukawa, M., 116 (243), *136*
Furusato, M., 11 (64), 84 (286), *91, 96*

Author index

Gaasch, J. F., 228 (75), *256*
Gabhe, S. Y., 37 (152), *93*
Gall, M., 128 (411), *140*
Gallazzi, M. C., 146 (23), *160*
Galliulina, R. F., 122 (312), *137*
Gambaro, A., 41 (159a), 43 (166, 169, 170), 44 (169), *93*
Gangin, V. Yu., 344 (55), *357*
Ganis, P., 43 (170), *93*
Gankin, V. Yu., 435 (3), *443*
Garavaglia, F., *162*
Gardener, H. C., 125 (351), *138*
Gardner, S. A., 270 (55), *293*
Garg, B. K., 67 (217), *95*
Garin, D. L., 301 (33), *330*
Gasc, J. C., 116 (225), *136*
Gassman, P. G., 86 (310), *97*
Gastambide, B., 119 (269), *137*
Gastinger, R. G., 266 (37), *293*
Gatehouse, B. M., 271 (60), *293*
Gatellani, M., 343 (45), *357*
Gates, B. C., 351 (114), 352 (119, 120), *358, 359*, 430 (139), *434*
Gathouse, B. M. K., 165 (51), *195*
Gatti, G., 182 (156), *197*
Gaube, W., 152 (68a, 68b), *161*
Gaudemar, F., 110 (113), *134*
Gaudemar, J. L., 107, 120 (85), *133*
Gaudemar, M., 101 (4, 8, 20), 102 (52), 103 (52, 63, 65), 104 (52, 68, 69), 107 (4, 8, 84, 86, 88, 89, 92, 93), 108 (93–97), 110 (4, 113), 115 (216), 116 (229, 242, 247), *132–134, 136*
Gauemar, M., 116 (222), *136*
Gavinato, G., 435 (2), 437, 442 (25), *443, 444*
Gavrilova, G., 53 (189), *94*, 125 (350, 352), *138*
Gavrilova, V. M., 438 (32), *444*
Gedye, R. N., 116 (239), *136*
Geibel, W., 287 (95), *294*
Geiseler, G., 113 (130), *134*
Geist, R., 264 (24), *293*
Gelin, P., 356 (154), *359*
Gellert, H. G., 215 (26a 26b), *255*
Gemert, J. F. van, *256*
Genchard, N., 219 (37), *256*
Genet, J. P., 188 (195–197), *198*
Genetti, R. A., 265 (27), *293*
Georgian, V., 11 (63), *91*
Gerard, F., 101 (11, 12), 102 (11, 12, 24, 26, 29), *132*
Germer, A., 184 (168), *198*
Gervay, J., 326 (174), *333*
Giacomelli, G., 261 (11), *292*
Gibson, H. H., 131 (487), *141*
Giering, W. P., 319 (137), *332*

Giese, B., 125 (354–357, 359–363, 365), *138, 139*
Girard, C. 114 (174), *135*
Giroldini, W., 152 (70), *161*
Gladysz, J. A., 414 (75), *432*
Glanvill, J. O., 260 (4), *292*
Gleize, P. A., 2, 70 (9), *89*
Glockner, P., 298 (106), *294*
Glowinski, R., 128 (431), *140*
Goasdoue, N., 116 (247), *136*
Gocmen, M., 121 (289), *137*
Goddard, R., 145 (13), 156 (89), *160, 161*
Goddard, W. A., 224 (63), *256*, 414 (72), *432*
Godet, J. Y., 114 (163), *135*
Godleski, S. A., 188 (193, 194), *198*
Godschalx, J., 194 (238), *199*
Godschalx, J. P., 39 (156a), *93*
Godschlax, J., 38 (156), *93*
Goetz, R. W., 320 (143), *332*
Gold, V., 2 (1), *89*
Golden, H. J., 150 (50), *160*
Goldfarb, I. J., 365 (27), 371 (53), *387, 388*
Goller, E. J., 108, 120 (100, 101), *133*
Golodov, V. A., 354 (142), *359*
Golse, R., 107 (83), 111 (119), 112 (122), 113 (83, 140), 120 (290), *133, 134, 137*
Golubev, V. B., 126 (390), *139*
Golumbic, N., 393, 402, 403 (9), 406 (33), 407, 408, 423 (9), *430, 431*
Gondal, S. K., *432*
Gonzalez-Gomez, J. A., 125 (365), *139*
Goodard, R., 287 (95), *294*
Goodman, D. W., 414, 421 (69), *432*
Goodyear Tire and Rubber Co., 212 (7, 16, 17), *255*
Gordon, M. E., 3 (26), *90*, 129 (460, 464), 131 (481), *141*
Goswami, R., 16 (80), 19 (97), *91*
Goudurier, G., 430 (138), *434*
Gowland, F. W., 189 (199), *198*
Grace and Co., W., R., 397, 406, 418 (21), *431*
Graffe, B., 130 (469), *141*
Graham, W. A. G., 298 (19), 308 (67), *330, 331*, 416 (86), *432*
Grandall, J. K., 71, 74 (238), *95*
Grard, C., 246 (97), *257*
Grasseli, J. G., 308 (72), *331*
Gray, C. E., 17 (83), *91*
Gream, G. E., 71, 74 (236), *95*
Greaves, E. O., 305 (61), *331*
Green, M., 255 (114), *257*, 260 (6), 270 (59), 272 (63, 64), 284 (94), *292–294*
Green, M. L. H., 222 (53), 227 (70), *256*, 416 (85), *432*
Gregorio, G., 354 (136), *359*
Gremaud, D., 29 (127), *92*
Gresham, W. F., 346 (96), *358*

Greswick, M., 415 (79), *432*
Grey, R. A., 306 (63), *331*
Gribble, A. D., 152 (73), *161*
Grice, N., 343 (44), *357*
Griffin, C. E., 115 (206), *135*
Grignon, J., 37 (145, 147), 38 (145), *93*
Grigoryan, E. A., 414 (77), *432*
Grim, S. O., 131 (489), *141*, 260 (4), *292*
Grimmin, M. J., 86 (305), *97*
Grobel, B. T., 10 (60), 12 (68), *91*
Gros, E. G. 116 (228), *136*
Gross, B., 110 (113), *134*
Gross, H., 122 (318), *138*
Grotjahn, D. B., 37 (152), *93*
Grove, D. M., 260 (6), *292*
Grove, H. D., 356 (155), *359*
Grubbs, R. H., 84 (295), *96*, 306 (63), *331*
Grubisch, N., 11 (63), *91*
Grudzinskas, C. V., 32, 33 (133), *92*
Grugel, C., 35, 36 (138), 78 (256), *92*, *95*
Gruning, R., 5 (40), *90*
Grunze, M., 427 (128), *434*
Grützmacher, H. F., 125 (358), *138*
Guccura, A., 308 (77), *331*
Gudec, J., 308, 315 (71), *331*
Guerchais, J. E., 410 (54), *431*
Guerrieri, F., 146, 151 (25), 152 (63, 67), 156 (86), 157 (90), *160*, *161*, 313 (102), *331*
Guetté, J. P., 116 (232–237), *136*
Guibe, F., 2, 70 (12), *89*
Guillerm, G., 27 (116), *92*
Guillern, G., 27 (114), *92*
Gusev, A. I., 262 (20, 21), *292*
Gusev, B. P., 128 (420), *140*
Gvozdovskii, G. N., 438 (32), *444*
Gyulumyan, Kh. R., 414 (77), *432*

Haas, C. K., 129 (437, 447, 449, 452), *140*, *141*
Haegle, G., 114 (159), *134*
Hafner, W., 164 (6), *195*
Hagg, W., 337 (8), *356*
Hagihara, H., 315, 317 (115, 116), *332*
Hagihara, N., 175, 176 (120, 121), 182 (159), *197*, 264 (25), 266 (36), *293*, 346 (95), *358*
Hagiwara, I., 66 (214a), *94*
Hahn, 364 (21), *387*
Hahn, R. C., 112 (125), 114 (167), *134*, *135*
Haiduc, I., 279 (85), *294*
Hallam, B. F., 302 (36), *330*
Hallgren, J., 440 (49), *444*
Halliday, D. E., 149 (37), *160*
Halpern, H., *256*, 298 (22), 299 (28), 300, 301 (22), *330*, 343 (40, 45), *357*, 375 (70), *388*
Hamanaka, E., 312 (95), *331*
Hambling, J. K., 212 (20), 213 (20, 21), *255*
Hammack, E. S., 181 (153), *197*
Hammamoto, I., 192 (229), 199

Hamsen, A., 10 (61), 11 (65), *91*
Hanes, R. M., 303 (48), *330*
Hanlan, L. A., 380 (87), *389*
Hannon, S. J., 19 (91), *91*
Hansen, E. M., 8 (55), *90*
Hansen, H.-J., 45 (172), *93*
Hansen, J. R., 23, 60 (104), *92*
Hanson, E. M., 129, 130 (457, 458), *141*
Hara, M., 178 (135), 180 (148), *197*
Hara, N., 356 (151), *359*
Harada, T., 43 (167), *93*
Harama, M., 111 (116, 118), 112 (118, 121), *134*
Hardinger, S. A., 190 (208–210), *198*
Hardt, P., 145 (17), *160*, 219 (36), 244 (102), *255*, *257*
Harris, S. R., 322 (153), *333*
Harrison, I. T., 114 (172), 115 (213), *135*
Harrison, P. G., 75 (245), *95*
Harrison, R., 11 (63), *91*
Harrod, J. F., 314, 317, 321 (109), *332*
Hartley, F. R., 164 (11), *195*, 223 (61a), *256*
Hartman, G. D., 23 (102), 50 (186), *92*, *94*
Hartmann, H., 27 (115), *92*
Hartwig, W., 2, 70 (11), *89*
Haruta, J., 128 (415, 417), *140*
Harvey, D. R., 79, 81 (261), *95*
Harvie, I. J., 165 (41), *195*
Hasegawa, K., 86 (300), *96*
Hasegawa, S., 165 (18, 19), *195*
Hashimoto, H., 114 (195, 196, 200), 115 (209–211), *135*, 183 (160, 161), 186 (189), *198*, 267 (42, 43), *293*, 439 (41, 42a), *444*
Hashimoto, K., 183 (164), *198*
Hashimoto, S., 69, 70 (227), *95*
Hata, G., 177 (127–129), 178 (132), 184 (170–172), *197*, *198*, 220 (42), 231 (86), *256*, *257*
Hatcher, A. S., 12 (72), *91*
Hatta, T., 2, 70 (10), *89*
Haudegond, J. P., 177 (130), *197*
Hauser, C. R., 122 (321), *138*
Hayama, N., 127 (408), 128 (428), *140*
Hayashi, T., 159 (103), *162*, 191 (222), *199*
Hayashi, Y., 194 (239), *199*
Hayes, J. C., 410 (52), *431*
Haynes, P., 181 (154), *197*
Headford, C. E. L., 415 (78), *432*
Heathcock, C. H., 64 (211), *94*
Heaton, B. T., 383 (97, 98), 387 (98), *389*
Heck, R. F., 57 (199a), *94*, 125 (366–369, 372), 126 (373, 374, 400), 128 (430), *139*, *140*, 179 (143–145), 194 (241, 242), *197*, *199*, 260 (10), *292*, 311 (86), 313 (100), 318 (135), 320 (142), *331*, *332*, 338, 339 (25), 341, 345 (33), *357*, 365 (22), 370 (22, 47), 371 (22, 54), 372 (22), *387*, *388*, 420 (99),

433, 435 (5b), 437 (21), 440 (50, 52a), 442 (50, 65), *443–445*
Heckl, B., 410 (53), *431*
Heeren, J. K., 131 (489), *141*
Hegedus, L., 144 (7, 11), 145 (15), 151 (7, 60), 152 (65), 154 (76c, 84), 155 (84), *159–162*, 174 (107), 192 (228), *196*, *199*, 222 (52), *256*, 306 (66), 312 (92, 97, 98), 329 (66), *331*, 366 (26), *387*, 436 (10), *443*
Heil, B., 382, 383, (93, 94), *389*, 413 (67), *432*
Heimbach, P., 144 (3), 145 (17), 148 (36), 149 (36, 38, 39, 41), 150 (41, 42, 44, 45), 151 (3), 152 (71, 72), *159–161*, 219 (36), 244 (102), *255*, *257*, 303 (42), *330*
Held, J., 88 (320), *97*
Helden, R. van, 176 (122, 123), 193 (234), *197*, *199*
Helle, H. N., 410, 411, 418 (49), *431*
Helle, J. N., 406 (37), *431*
Helling, J. F., 265 (26), *293*
Helm, D. van der, 45 (173), *93*
Henc, B., 145 (12, 13), *160*, 239 (96), *257*
Henderson, J., 131 (487), *141*
Henderson, K. H., 165 (49), *195*
Hénin-Vichard, F., 119 (269), *137*
Henn, L., 25 (111, 112), *92*
Henrici-Olive, G., 253 (109), 254 (109, 111), *257*, 303 (44), *330*, 408 (42), 409 (42, 46), 410 (50), 412 (42, 63), 413, 414 (63), 416 (63, 85), 417 (63, 89, 91), 418 (93), 419 (63, 98), 420 (42, 63), 421 (42), 425 (122), 426 (63), 430, (140), *431–434*
Henry, M. C., 67 (218), *95*
Henry, P. M., 224 (65), *256*
Henry-Basch, E., 101 (13), 102 (31, 32, 34), 108 (98, 102), 109 (106, 108), 113 (127, 128), 119 (13, 275), 120 (13, 275, 283, 285, 287, 291), 121 (106, 108, 283, 299, 301, 303), 123 (127, 128, 283, 285), *132–134*, *137*
Herbeck, R., 83 (280), *96*
Herington, E. F. G., 401 (208), *431*
Herisson, J. L., 84, 87 (293), *96*
Hersh, W. H., 85 (298), *96*
Hersham, A., 351 (116), *359*
Hershberger, S., 126 (377), 127 (402), *139*
Herwig, W., 129 (432), *140*
Heuck, K., 125 (359, 360), *138*
Heulbier, H., 152 (68a), *161*
Hey, H., 152 (71), *161*
Hibi, T., 439 (41), *444*
Hida, M., 115 (209–211), *135*
Hidai, M., 443 (68), *445*
Hieber, W., 378 (81), 380 (86), *388*, *389*
Higashimura, T., 86 (299, 301), *96*
Hignett, R. R., 362 (8), *387*
Hilberath, H., 395 (13), *431*
Hillard, R. L., 280 (87), *294*
Himbert, G., 25 (110–112), *92*
Himmele, W., 349 (103, 104), *358*
Hindermann, J. P., 414 (71), 423 (113), *432*, *433*
Hinenoya, M., 86 (302), *96*
Hines, L. F., 443 (67), *445*
Hinney, H. R., 190 (208, 209), *198*
Hirao, T., 170 (89), 190 (206), *196*, *198*
Hirota, Y., 328 (185), *333*
Hirsch, H. P., 260 (8), *292*
Hitch, R. R., *432*
Hjortkjaer, J., 350 (110), *358*
Ho, B. Y. K., 43 (171), *93*
Hoberg, H., 114 (177), *135*, *162*, 291 (113, 114), *294*, 439 (42b), *444*
Hoehn, W. M., 121 (302), *137*
Hoffman, E. G., 145 (12), *160*
Hoffman, G. A., 383, 387 (98), *389*
Hoffmann, R., 223, 225–227, 233 (58), *256*, 298 (24), *330*, 414, 417, 424 (73), *432*
Hoffmann, R. W., 41, 43 (160), *93*
Hogan, J. P., 413 (65), *432*
Hoge, R., 25 (111), *92*
Hohenschutz, H., 349 (103, 104), *358*, 375 (65), *388*
Hohlfeld, R., 5 (39), *90*
Höhm, H., 123 (332), *138*
Holle, S., 145, 148 (19), *160*
Holliday, A. K., 77 (254), *95*
Holmes, J. D., 303 (40), *330*, 350 (107), 354 (133), *358*, *359*
Holmstead, R. L., 114 (148), *134*
Holt, E. M., 423 (117), *433*
Holysz, R. P., 122 (320), *138*
Holzkamp, E., 215 (26b), *255*, 392 (1), *430*
Holzman, G., 211 (5), 213 (23), *255*
Hong, P., 314 (108), 315 (115, 116), 317 (108, 115, 116), *332*
Hoogendoorn, J. C., 393 (10), *430*
Hoogzand, C., 272 (66), 273 (69), 289 (66), *293*, *294*
Hoornhaert, C., 75 (246), *95*
Hoover, F. W., 304 (53), *330*
Hopf, H., 184 (168), *198*
Hopper, S. P., 129 (439, 440, 442, 452), 130 (452), 131 (478), *140*, *141*
Hoppin, C. R., 84 (295), *96*
Horie, S., 89 (322), *97*
Horii, S., 326 (175), *333*
Horino, H., 125 (371), *139*
Horiuchi, C. A., 165 (42, 50), *195*
Horn, C. F., 83, 84 (278), *96*
Horsley, J. A., 427 (130), *434*
Hosaka, S., 169 (82, 85), 180 147), *196*, *197*, 345 (85), *358*
Hoshino, I., 63 (208a), *94*

Hoskin, D. H., 4 (35), *90*
Hosokawa, T., 275 (74, 75), *294*
Hosomi, A., 48 (180), *94*
Hossain, M. B., 45 (173), *93*
Hostettler, F., 83, 84 (278), *96*
Hotta, Y., 115 (214), *135*
Houlihan, F., 204 (12), *204*
House, H. O., 114 (149, 150), 128 (411), *134, 140*
Howard, J. A. K., 260 (6), *292*
Howard, J. F., 320 (141), *332*
Howe, R. F., 356 (150), *359*
Howsam, R. W., 165 (48), *195*
Hoyano, J. K., 298 (19), *330*
Hübel, W., 262 (22), 272 (66), 273 (69), 278 (81), 289 (66), *292–294*
Hubert, A. J., 278 (79), *294*
Huckaby, J. L., 338, 339, 341 (26), *357*
Hucul, D. A., 422 (111), 427 (130), *433, 434*
Hudec, J., 344, 345 (53), *357*
Hudel, J., 442 (59), *444*
Hudrlik, P. F., 202 (6), *204*
Huet, F., 120 (283, 285), 121 (283), 122 (311), 123 (283, 285), *137*
Huet, J., 107 (91), *133*
Huffman, J. C., 268 (53), *293*, 415 (80), *432*
Hugel, H., 6 (44), *90*
Hughes, P. R., 180 (146), *197*, 345 (71, 87), *358*
Hughes, R. E., 130 (467), *141*, 423 (117), *433*
Hughes, R. P., 147 (27), *160* 168 (73), *196*
Hughes, W. B., 131 (489), *141*, 346 (91), *358*
Hugues, F., 425 (123), *433*
Huisgen, R., 114 (184), *135*
Hung, T., 175 (115), *197*
Hunt, J. B., 131 (487), *141*
Hunton, D. E., 168 (73), *196*
Hurd, L. D., 122 (320), *138*
Hurk, J. W. G. van den, 83 (276), *96*
Hurwitz, M. J., 126 (396), *139*
Husebeye, S., 375 (68), *388*
Hutchinson, J. J., 115 (207), *135*
Hüttel, R., 164 (7), 165 (24–28, 37), 166 (52), *195*
Hyashi, T., 113 (134), *134*
Hyson, A. M., 346 (88), *358*

ICI, 231 (81), *256*
Idacavage, M. J., 442 (62), *445*
Igami, M., 322 (154), *333*
Iguchi, H., 48 (180), *94*
Ihrman, K. G., 288 (97), *294*
Ihrman, V., 437 (20), *444*
Iida, K., 203 (9), *204*
Iijima, S., 201 (3), 202 (7), 203 (9), *204*
Ikawa, T., 167 (59), *196*
Ikeda, Y., 436, 440 (15a), *443*
Ikegami, S., 15 (78), *91*
Ikegami, T., 84 (286), *96*
Ilin, V. I., 166 (53), *195*
Imamura, S., 165 (45, 46), 169 (46), 192, 193 (230), *195, 199*
Imanaka, T., 194 (244, 245), *199*
Imoto, M., 83 (277), *96*
Imyanitov, N. S., 338 (27), 339 (29), 343 (49), 345 (70, 86), 346 (90), *357, 358*, 442 (61), *444*
Inagaki, S., 226 (67), *256*
Inoue, H., 421 (107), *433*
Inoue, N., 125 (371), *139*
Inoue, S., 151 (61, 62, 62b), *161*
Inoue, Y., 183 (160, 161, 164), *198*, 267 (42, 43), *293*, 439 (41, 42a), *444*
Inouye, Y., 114 (202), *135*
Ioffe, I. A., 77 (251), *95*
Ippolito, R. M., 37 (152), *93*
Iqbal, M. Z., 381 (91), 382 383 (95), *389*
Irvine, J. L., 123 (331), *138*
Ishida, N., 202 (5), *204*
Ishihara, Y., 46 (178), *94*
Ishii, Y., 69 (225), *95*, 164 (12), 165 (18, 19), 167 (64), *195, 166*
Ishikawa, H., 204 (11), *204*
Ishikawa, N., 194 (240), *199*
Ishikawa, T., 427 (131), *434*
Ishizu, J., 151 (56a), *161*
Ishzu, J., *162*
Isobe, K., 410 (51), *431*
Israeli, Z. H., 116 (227), *136*
Isshiki, T., 354 (132), *359*
Ito, K., *195*
Ito, T., 175 (114), *197*
Itoh, A., 40 (158), *93*, 186 (189), *198*
Itoh, K., 69 (225), *95*, 164 (16), *195*
Itoh, Y., 170 (89), *196*, 267 (42, 43), *293*
Itoi, K., 83 (281, 282), 84 (288), *96*
Ivanova, N. P., 25 (107), *92*
Iwashita, Y., 344 (62), *358*
Izumi, T., 126 (375), 128 (429), *139, 140*
Izumi, Y., 2 (1), *89*, 114 (200), *135*

Jablonski, C., *256*
Jackson, W. R., 74 (243), *95*, 165 (51), 170 (91), 173 (102–105), 174 (113), *195–197*, 297, 301, 302, 304–306, 312, 328 (5), *330*
Jacobs, P. A., 427 (125b), 430 (137), *434*
Jacobs, W. J., 343 (41), *357*
Jacobus, J., 114 (188), *135*
Jacques, J., 119 (268), *137*
James, B. R., 297 (6, 7), *330*, 345 (80), *358*
James, D. E., 337 (12), *356*, 440 (52b), *444*
Janowicz, A. H., 298 (20), *330*
Japan Synthetic Rubber Co. Ltd., 83 (275), *96*
Jarboe, C. J., 122 (325), *138*

Jardine, F., 317 (129), *332*
Jardine, F. H., 384 (106), *389*
Jarrell, M. S., 352 (119), *359*
Jarvie, A. W. P., 17 (84), *91*
Jarvis, J. A. J., 223 (55), *256*
Jautelat, M., 114 (183), *135*
Jean, A., 118 (255), *136*
Jellinek, F., 201 (1, 2), *204*
Jenkins, I. D., 305 (58), *331*
Jenner, E. L., 220 (43), *256*, 345, 346 (67), *358*
Jensen, H., 253 (110), *257*
Jensen, V. W., 350 (110), *358*
Jerkunica, J. M., 55, 60 (199), *94*
Jesson, J. P., 159 (97), *162*
Job, A., 219 (33), *255*
Johanson, K. H., 224 (66), *256*
Johns, W. S., 418 (97), *433*
Johnson, B. F. G., 227 (71), *256*, 383 (96), *389*
Johnson H. T., 21, 23 (100), *92*
Johnson, P. J., 118 (261), *136*
Johnson, R. J., 262 (23), *292*
Johnson, S. A., *162*
Johnson, T. H., 86 (310), *97*
Jolly, P. W., 144 (1, 3), 145 (1, 12, 13, 19), 148 (1, 19, 33), 149 (1, 33), 150 (1), 151 (1, 3), 152, 154 (1), 156 (1, 89), 159 (1, 33), *159–161*
Jolly-Goudget, M., 159 (101), *162*
Joly, P. W., 175 (115), 184 (169), *197*, *198*
Jonas, J., 383, 387 (98), *389*
Jonassen, H., 375 (68), *388*
Jonassen, H. B., 375 (64), *388*
Jones, C. W., 131 (487), *141*
Jones, D. N., 165 (40), *195*
Jones, F. N., 122 (321), *138*
Jones, F. R., 120 (279), *137*
Jones, P. R., 108 (100, 101), 119 (274), 120 (100, 101, 274), 121 (274, 306), 122 (274, 306, 315, 325), 123 (274, 315), 124 (274), *133*, *137*, *138*
Jones, R., 122, 123 (309), *137*
Jones, R. W., 12 (72), *91*
Jones, S. R., 80 (271), *96*, 164 (11), *195*
Jones, W. H., 297 (1), *329*
Joska, J., 114 (162, 164), *134*, *135*
Jula, T. F., 129 (463), 131 (478, 485), *141*
Julémont, M., 145 (14), *160*
Julia, M., *162*
Jung, M. E., 2, 70 (13), *89*
Jungheim, L. N., 187 (192), *198*
Justand, A., 173, (106), *196*
Jutand, A., 118 (264, 265), *137*

Kad, G. L., 114 (160, 161), *135*
Kadota, K., 436, 440 (15a), *443*
Kadyrmatova, T. P., 118, 119 (266), *137*
Kaempfe, L. A., 301 (33), *330*
Kagan, H. B., 202 (8), *204*
Kagan, Yu., 397, 404, 405 (19e), *431*
Kagan, Yu. B., 423 (116), *433*
Kagotani, M., 177, 178 (131), 179 (137), *197*
Kahle, G. R., 88 (319), *97*
Kaji, A., 2, 70 (7), *89*, 192 (229), *199*
Kajimoto, T., 169 (87), *196*
Kakos, A. G., 165 (51), *195*
Kakui, T., 194 (236), *199*
Kaliya, O., 342 (37), *357*
Kalman, J. R., 79, 81 (266, 267, 270), *96*
Kamernitskii, A. V., 116 (223, 226), *136*
Kamimura, A., 2, 70 (7), *89*
Kan, G., 320 (144), *332*
Kanatani, R., 202 (5), *204*
Kanaya, I., 320 (147), *332*
Kaneda, K., 194 (244, 245), *199*, 315, 317 (114), *332*
Kang, J. M., 308 (75), *331*
Kano, K., 69, 70 (227), *95*
Kanumura, A., 326 (170), *333*
Kao, L. C., 179 (143), *197*
Kao, S. C., 343 (44), *357*
Kapteiju, F., 86 (309), *97*
Karapinka, G. L., 368 (33, 34), *388*
Karapinko, G. L., 84 (283), *96*
Karmann, H. G., 239 (96), *257*
Karol, J., 84 (283), *96*
Kasahara, A., 126 (375), 128 (429), *139*, *140*, 165 (47), *195*
Kasahara, K., 427 (127), *434*
Kasatkin, A. N., 31 (131), 54 (194), *92*, *94*
Kashin, A. N., 25 (109), 53 (109, 190–193), 55 (190, 191), *92*, *94*
Kashireninov, O. E., 83 (272), *96*
Kasugi, M., 49 (181a), *94*
Katano, K., 2, 12, 66 (16), *90*
Kataoka, H., 187 (190), 190 (213), *198*, *199*
Katayama, Y., 84 (286), *96*
Kato, A., 108 (103), *133*
Kato, K., 151 (62) *161*
Katsnel'son, M. G., 344 (55), *357*, 435 (3), *443*
Katz, T. J., 85 (298), *96*
Katzenellenbogen, J. A., 101 (15), *132*
Katzer, J. R., 409 (44), *431*
Kauffman, T., 15 (76), *91*
Kauffman, W. J., 108, 120 (100, 101), *133*
Kauffmann, T., 8 (58), 10 (61), 11 (65), *91*
Kavaliunas, A. V., 165 (21, 22), *195*
Kawabata, N., 101, 107 (3), 108 (3, 103), 110, 112, 113 (3), 114 (3, 185–*187*, 189, 191–193, 198, 201), 115 (3), 124 (201, 343), *132*, *135*, *138*
Kawaguchi, S., 168 (77–79), *196*
Kawasaki, A., 84 (289), *96*
Kealy, J. J., 344–346 (61), *357*
Keck, G. E., 37 (150a), *93*

Kehoe, L. J., 345 (72), *358*, 437 (19a), *444*
Keii, T., 297 (10), *330*
Keim, W., 144 (4), 145 (17), 151 (4), *159*, *160*, *162*, 176 (125), *197*, 219 (36, 40) 232 (92), 244 (102), *255–257*
Keinan, E., 2 (9), 38, 39, 52 (157), 63 (208), 70 (9), *89*, *93*, *94*, 191 (219, 220), *199*
Keith, A. N., 266, 270 (32), *293*
Kellay, E. A., 265 (30), *293*
Keller, K. P., 124–129, 131 (347), *138*
Kellner, C. S., 427, 428 (125c), *434*
Kelulé, A., 126 (386), *139*
Kennedy, J. D., 36 (140), *93*
Kent, A. G., 384–386 (107, 108), *389*
Kerber, R., 86 (306), 87 (311, 317), 88 (318), *97*
Kerk, G. J. M. van der, 3 (30), 7 (51), 17 (82), 18 (82, 86), 83 (276), *90*, *91*, *96*
Kessenikh, A. V., 126 (380, 383), 128 (380), *139*
Ketley, A. D., 165 (32), 167 (65, 67, 69), *195*, *196*
Keuk, B. P., 111 (120), *134*
Keunecke, E., 344 (57), *357*
Keung, E. C. H., 324 (164), *333*
Keyton, D. J., 71, 74 (238), *95*
Khabil, H., 116 (239), *136*
Khager, K., 309 (81), *331*
Khangazheev, S. Kh., 37 (150), *93*
Khidekel, M. L., 176 (124), *197*, 298 (27), 316, 317 (117), *330*, *332*
Khutoryanskii, V. A., 53 (192), *94*
Kieeneman, A., 87 (312), *97*
Kieffer, R., 87 (312), *97*, 414 (71), 423 (113), *432*, *433*
Kielbania, A. J., 300 (29), *330*
Kiennemann, A., 423 (113), *433*
Kiji, J., 150 (43), *160*, 165 (46), 168 (80), 169 (46, 82), 182 (158), 192, 193 (230), *195–197*, *199*, 438 (31), 439 (44), *444*
Kiji, K., 345 (85), *358*
Kijima, Y., 354 (132), *359*
Kilbourn, B. T., 223 (55), *256*
Killian, L., 29 (124), *92*
Kim, C. U., 19 (92), *91*
Kim, P. 346 (95), *358*
Kimura, H. M., 421 (107), *433*
Kimura, S., 165 (19), *195*
Kimura, T., 194 (245), *199*
Kin, A. O., 113 (133), *134*
Kindaichi, Y., 175 (114), *197*
King, A. D., 381 (91), 382, 383 (95), *389*
King, A. O., 113 (132), *134*
King, J.C., 114 (197), *135*
King, R. B., 278 (82–84), 279 (85), *294*, 326 (171), *333* 369 (43), 381 (43, 91), 382, 383 (43, 95), *388*, *389*

Kioussis, D., 428 (133), *434*
Kirch, L., 365 (27), *387*
Kirmse, W., 115 (203), *135*
Kirrmann, A., 123 (341), *138*
Kirsch, H. P., 270 (56), *293*
Kisan, B., 301 (31), *330*
Kishi, Y., 43 (165), *93*
Kita, Y., 128 (415, 417), *140*
Kitagawa, Y., 186 (189), *198*
Kitayama, M., 114 (189), *135*
Kitazume, T., 194 (240), *199*
Klabunde, U., 418 (96), *433*
Klager, K., 259,. 289 (1), *292*
Klei, B., 201–203 (4), *204*
Kleijn, H., 113 (138, 139), *134*
Klein, H., 113 (131), *134*
Klein, R. S., 2, 66 (18), *90*
Kleiner, F. G., 66 (213), *94*
Kleschik, W. A., 64 (211), *94*
Kliegman, J. M., 164 (15), *195*
Kline, J. B., 223, 224 (59), *256*
Klun, T. P., 186 (184, 185), *198*
Kluth, J., 149 (39), 150 (42), *160*
Klyuev, M. V., 316, 317 (117), *332*
Knapp, S., 16 (80a), *91*
Knifton, J., 437 (24), *444*
Knifton, J. F., 192, 193 (231), *199*, 344 (63), *358*, 437, 438 (27), *444*
Knolek, W. F., 436 (16), *443*
Knox, G. R., 305 (61), *331*
Knox, S. A. R., 269 (48, 49), 271 (49), 272 (61), *293*, 410 (54, 55), *431*, *432*
Knox, S. D., 165 (40), *195*
Knunyants, I. L., 129 (441), *140*
Kobayashi, M., 113 (136, 137), *134*
Kobayashi, Y., 187 (190), 190 (207, 213), *198*, *199*
Kock, H., 337 (8), *356*, 395 (12, 13), *431*
Kocheshkov, K. A., 101, 107, 108, 110, 112, 113, 115 (1), 119–121, 123 (272), *132*, *137*, 395 (13), *431*
Kochi, J. K., 78 (257), *95*, 125 (351), *138*, 415 (80), *432*
Kochkin, D. A., 67 (219), *95*
Kochlar, R. K., 302 (39), *330*
Koelsch, C. F., 122 (322), *138*
Koetitz, K. F., 320 (146), *332*
Koetzle, T. F., 383 (97), *389*
Koetzle, T. K., 223 (57), *256*
Koga, G., 102 (42), *132*
Koga, N., 102 (42), *132*
Kohl, F., 77 (252), *95*
Köhler, J., 101 (14), 120 (14, 278), *132*, *137*
Kohll, C. F., 193 (234), *199*
Kokura, M., 443 (68), *445*
Kolesnikov, S. P., 77 (251), *95*
Kollonitsch, J., 120, 122, 123 (282), *137*

Kolobova, N. E., 262 (16–21), *292*
Komarov, N. V., 28 (120), *92*
Komissarov, Y. F., 129 (441), *140*
Komiyama, H., 421 (107), *433*
Komorniczyk, K., 27 (115), *92*
Kondratenko, N. V., 127 (404), *140*
Konietzny, A., 288 (96), *294*
Konig, K., 6, 27, 41 (47), *90*
Konishi, H., 158 (94), *162*
Konishi, M., 159 (103), *162*, 191 (222), *199*
Kopp, W., 246 (97), *257*
Korableva, L. G., 176, (124), *197*
Koreeda, M., 46 (176), *93*
Korte, D. E., 154, 155 (84), *161*
Korte, F., 346 (97), 347 (97, 100), 348 (100), *358*
Kostyuk, A. S., 8 (53), *90*
Kosugi, M., 37 (148), 38 (155), 40, 53 (159), 63 (208a), 66 (214a), *93*, *94*
Koton, M. M., 128 (421, 423), *140*
Kouwenhoven, A. P., 422 (108), *433*
Kovano, T., 354 (139), *359*
Kovar, R. F., 129 (434), *140*
Kovtun, E. A., 28 (120), *92*
Kowaldt, F. H., 219 (40), 232 (91c), *256*, *257*
Kozlova, L. S., 118 (260), *136*
Kozyrod, R. P., 79, 81, 82 (269a), *96*
Kramer, A. V., 77 (251), *95*
Kratzer, I., 165 (25), *195*
Kratzer, J., 164 (7), *195*
Krause, L. J., 124 (345), *138*
Kretchmer, R. A., 128 (431), *140*
Kretzchmar, G., 125 (363), *139*
Kriegesmann, R., 4 (36), 8 (58), 10 (61), 11 (65), *90*, *91*
Krivoruchko, V. A., 116 (226), *136*
Kroll, W. R., 215 (26b), *255*
Kröner, M., 145 (17), 152 (64), 156 (87), *160*, *161*, 219 (36), 244 (102), *255*, *277*, *290* (108), *294*
Kröper, H., 320, 323 (148), *332*, 336 (3), 337 (3, 9, 10), 341, (34), 343 (46), 344 (3, 52), 354 (130), *356*, *357*, *359*
Krsek, G., 363 (17), 366, (29), *387*
Krüerke, U., 273 (69), *294*
Krüger, C., 145 (12, 13), 156 (89), *160*, *161*, 287 (95), *294*
Krüger, G., 393, 394, 398 (5b), 404 (30), 406, 407, 419 (5b), *430*, *431*
Krummer, R., 375 (65), *388*
Kryczka, B., 116 (244), *136*
Kryukov, Yu. B., 397, 404, 405 (19e), 423 (116), *431*, *433*
Krzywicki, A., 351 (117), 352 (118), *359*
Kucherov, V. F., 128 (420), *140*
Kuehne, M. E., 114 (197), *135*
Kuhlein, K., 36 (141), 69 (226), *93*, *95*

Kühlhorn, H., 215 (26a), *253*
Kühne, D., 396 (16), *431*
Kuivila, A. G., 74 (239), *95*
Kuivila, H. G., 23 (105, 106), 36 (140), 37 (143), 60 (204), 70 (229, 230), 74 (241), *92–95*
Kulkarni, S. M., 123 (342), *138*
Kumada, M., 159 (103), *162*, 191 (222), 194 (235, 236), *199*, 202 (5), *204*
Kumar Das, V. G., 45 (173), *93*
Kummer, J. T., 409, 410 (43), *431*
Kumuda, M., 113 (134), *134*
Kunakova, P. V., 179 (138), *197*
Kunakova, R. V., 179 (139), *197*
Kune, A., 2, 70 (10), *89*
Kunichika, S., 346 (94), *358*
Kuper, D. G., 346 (91), *358*
Kurachi, Y., 164 (16), *195*
Kuraev, B. E., 343 (49), *357*
Kuramitsu, T., 274 (70), *294*
Kurino, K., 37 (148), *93*
Kurosawa, H., 227 (73), *256*, 442 (64), *445*
Kurtev, B., 116 (221), *136*
Kurts, A. L., 127 (407, 409), *140*
Kutepow, N., von, 303, 308 (43), *330*, 343, 346, 348, (39), 349 (103, 104), 354 (130), *357–359*, 437 (22), 438 (35), *444*
Kuwajima, I., 127 (410), *140*

Labadie, J. W., 25, 40, 55, 56 (108a), *57*, *92*
Labunskaya, V. I., 298 (27), *330*
Lacombe, S., 116 (245), *136*
Laflamme, P., 113, 114 (143), *134*
Lagally, H., 380 (86), *389*
Lagowski, J. J., 291 (115), *294*
Lahournere, J. C., 12 (73), *91*
Laine, R. M., 260 (7), *292*
Lambert, R. L., 3 (29, 31), *90*, 129, 130 (453, 454), *141*
Landa, S., 114 (190), *135*
Landgrebe, J. A., 130 (475), *141*
Landis, M. E., 29 (127), *92*
Lange, G., 121 (296), *137*
Lange, R., 152 (68a, 68b), *161*
Langlais, M., 120 (294), 121, 123 (297), *137*
Lankelma, H. P., 115 (208), *135*
Lanoiselée, M., 101 (17), *132*
Lantseva, L. T., 129 (441), *140*
Lapidus, A. L., 337 (17, 18), *357*
Lapious, A. L., 435 (6b), *443*
Lapkin, I. I., 116 (231), 118 (256–260, 262, 263, 266), 119 (266, 270, 271), *136*, *137*
Laporterie, A., 35 (137), *92*
Lappert, M. F., 68 (222), 77 (250), *95*, 416 (84), *432*
Lardicci, L., 261 (11), *292*
Larock, R. C., 124, 125 (348, 349), 126 (348,

349, 376, 377, 401), 127 (348, 349, 402, 496), 128 (348. 349, 418, 419, 425–427), 129, 131 (348, 349), *138–140*, 164 (14), 168 (74, 75), 175 (117), *195–197*
Larrabee, C. E., 289 (103), *294*
Laurent, A., 116 (244, 245), *136*
Lavigne, A. A., 121, 122 (306), *137*
Lavrent'ev, I. P., 176 (124), *197*
Lawrence, J. P., 84 (292), *96*
Lawrence, T., 71 (234), *95*
Lawson, D. N., 301 (30), *330*
Lazzaroni, R., 369 (44), 371, 372 (52), *388*, 436 (7), *443*
Leach, D. R., 127 (406), *140*
Lebedev, S. A., 31 (131), 54 (194), *92*, *94*
Leclere, C., 350, 354 (109), *358*
Lee, B. S., 393, 394 (8), *430*
Lefebvre, G., 206 (2), 219 (37), 220 (47), 244, 245 (100), 249 (2), *255–257*
Legoff, E., 114 (168), *135*
Legras, Y., 188 (197), *198*
Le Guilly, L., 120 (284, 288), *137*
Lehmkuhl, H., 103 (57–60), *133*, 150 (47, 48), *160*
Lehnert, G., 87 (316), *97*
Lenox, R. S., 101 (15), *132*
Lequan, M., 27 (114, 116), *92*, 118 (255), *136*
Lerch, A., 114 (178), *135*
Lescheva, A. I., 219, 231 (38), *256*
Leto, J. R., 288 (99, 100), *294*
Leto, M. F., 288 (99), *294*
Leusink, A. J., 7 (50), *90*
Levering, D. R., 344 (52), *357*
Levy, M., 84 (285), *96*
Lewis, J., 227 (71), *256*, 302, 305 (37), *330*, 383 (96), *389*
Libman, J., *66*
Liermain, A., 107, 113 (83), 120 (290), *133*, *137*
Light, L. A., 2, 70 (13), *89*
Lilienfield, W. M., 120 (281), *137*
Limasset, J. C., 114 (151), *134*
Limburg, W. W., 66 (214), *94*
Linden, H. R., 395 (11), *431*
Lindner, E. 378 (81), *388*
Lindsey, R. V., 220 (43), *256*, 304 (52, 53), *330*, 345 (67), 346 (67, 93), *358*
Lipscomb, W. N., 291 (112), *294*
Lipshles, Z., 321 (150), *333*
Litvinovskaya, R. P., 116 (223, 226), *136*
Livantsova, L. I., 27 (117), *92*
Ljungvist, A., 172 (101), *196*
Lloyd, D. J., 260 (8), *292*
Loeb, W. L., 83 (247), *96*
Loffler, H. P., 74 (242), *95*
Logue, M. W., 25 (108b), *92*
Loh, T. L., 11 (63), *91*

Long, R., 164 (8), 193 (233), *195*, *199*
Long, W. P., 84 (284), *96*
Look, B. F., 131 (487), *141*
Looy, H. M. van, 413 (64), *432*
Lopez, M. I., 114 (155), *134*
Lorberth, J., 5 (39, 40), 68 (223), *90*, *95*
Lotts, K. D., 4 (35), *90*
Love, R. A., 223 (57), *256*
Loven, R., 74, 75 (244), *95*
Lucas, D., 69 (224), *95*
Lucas, M., 116 (232–237), *136*
Luehr, H., *162*
Luijten, J. G. A., 17, 18 (82), *91*
Lukes, J. H., *195*
Lundberg, C., 215, 216 (29), *253*
Lupin, M. S., 167 (61, 63), *196*
Lutsenko, I. F., 6 (46), 8 (52, 53), 27 (117), *90*, *92*, 126 (395), 128 (416), *139*, *140*

Macchia, B., 116 (220, 238), *136*
Macchia, F., 116 (220, 238), *136*
MacDonald, T. L., 58 (201), *94*
Machigin, E. V., 8 (52), *90*
Madden, P. D., 154 (80), *161*
Madey, T. E., 414 (69), 421 (69, 104), *432*, *433*
Madix, R. J., 414, 421 (68), *432*
Madon, R. J., 424 (121), 428 (134), *433*, *434*
Mador, I. L., 440 (57), *444*
Maekawa, M., 11 (64), *91*
Maertens, D., 87 (316), *97*
Maganem, F., 27 (116), *92*
Magin, A., 303 (43), 308 (43, 76), *330*, *331*
Magorskaya, O. I., 53 (192), *94*
Mahalingam, S., 58 (201), *94*
Mai, V. A., 131 (481), *141*
Mais, R. H. B., 223, 224 (56), *256*
Maitlis, P. M., 163 (4), *194*, 265 (29, 30), 275 (72, 73, 77), 277 (78), 278 (72), 288 (96), 291 (111), *293*, *294*, 308 (73–75), *331*, 410 (51), *431*
Maitte, P., 130 (469), *141*
Majerski, Z., 114 (188), *135*
Makarova, L. G., 124–129 (346), *138*
Makhaev, V. D., 126 (378, 379), 128 (379), *139*
Makin, P. H., 77 (254), *95*
Maksimenko, O. A., 126 (382, 387, 392, 393), *139*
Maleeva, A. I., 28 (119), *92*
Malleron, J. L., 185 (182), 191 (218), 192 (226), *198*, *199*
Mal'tsev, V. V., 84 (290), *96*
Mammarella, R. E., 3 (31), 19 (89), *90*, *91*
Manas, M. M., 184 (174), *198*
Manchard, P. S., 172 (100), *196*
Mandelbaum, A., 123 (329), *138*

Author index

Mango, F. D., 266 (33), *293*, 298, 301 (21, 23), 330
Mann, B. E., 410 (51), *431*
Manojlovic-Muir, L., 266, 270 (32), *293*
Mantica, E., 337, 338 (24), *357*
Mantovani, J., *256*
Manuel, G., 37 (144), *93*
Manyik, R. M., 181 (153), 184 (173), *197*, *198*
Manzocchi, A., 116 (219), *136*
Maracci, F. 261 (11), *292*
Marchetti, F., 415, 416 (81), *432*
Marczewski, M., 352 (118), *359*
Mark, V., 211 (3), *255*
Mark, W., 267 (40), *293*
Marko, L., 316, 317 (124), *332*, 362 (5), 368 (38), 370 (49), 374, 375 (63), 376 (73), 378 (82), 382, 383 (93, 94), *387–389*, 413 (67), 418 (94), *432*, *433*
Marmor, R. T., 131 (486), *141*
Marquet, B., 116 (244, 245), *136*
Marraccini, A., 154 (76), *161*
Marsden, K., 415 (78), *432*
Martin, C. W., 130 (475), *141*
Martin, G., 122 (316), *138*
Martin, H., 215 (26a), *255*, 392 (1), *430*
Martin, H. A., 201 (1, 2), *204*
Martinego, S., 383 (97), *389*
Marton, D., 41 (159a), 43 (166, 169, 170), 44 (169), *93*
Maruoka, K., 116 (218), *136*
Maruyama, K., 41 (161), 45 (174), 46 (161, 175, 178), 47 (161, 179), 49 (174, 182, 183), 50 (184, 185), 64 (210), 66 (210, 212), *93*, *94*
Maruyama, T., 84 (289), *96*
Marx, B., 108 (98), *133*
Marzotto, A., 409 (47), *431*
Masada, H., 319 (136), 320 (147), *332*
Massa, W., 5 (39), *90*
Massey, A. G., 308 (68), *331*
Masson, A., 104 (67), *133*
Massoth, F. E. 404 (31), *431*
Massy-Barbot, M., 101 (19), *132*
Masters, C., 345, 350 (84), *358*, 362 (6, 11), 363–365, 373, 379, 385 (6), *387*
Masuda, T., 86 (299. 301), *96*
Masunaja, K., 436, 440 (15a), *443*
Matarasso-Tchiroukhine, E., 37 (154), *93*
Mateescu, G. D., 275 (76), *294*
Mathews, F. E., 215 (27), *255*
Matsuda, A., 339 (30), 341 (32), 345 (73), 346 (32), 348 (30, 101), *357*, *358*
Matsuda, I., 69 (225), *95*
Matsuda, M., 436, 440 (15a), *443*
Matsuda, Y., 427 (127), *434*
Matsui, K., 436, 440 (15a), *443*
Matsui, M., 2, 12, 66 (16), *90*

Matsumoto, K., *41*
Matsumoto, T., 353 (123, 124), *359*, 421 (107), *433*
Matsumura, T., 185 (176), *198*
Matsumura, Y., 167 (66), *196*
Matsushima, Y., 354 (139), *359*
Matsushita, H., 191 (223, 224), *199*
Matthews, R., 440 (49), *444*
Matz, J. R., 174 (112), *197*
Mauzé, B., 36 (142), *93*, 102 (36–40, 50, 51, 53), 103 (50, 51, 62), 104, (53, 62, 72), 105 (53, 62, 75, 77, 78), 106 (75, 77, 80), 111 (53, 117), *132–134*
Mawby, R. J., 369 (41), *388*
May, G. L., 79, 81 (267), *96*
Mays M. J., 272, 286 (62), *293*
Mazerolles, P., 131 (484, 485), *141*
Mazzi, U., 409 (47), *431*
McAlister, D. R., 275 (71), *294*
McCleverty, J. A., 378 (80), *388*
McCoy, J., 379 (84), *389*
McCrae, D. A., 37 (152), *93*
McDonald, J. H., 16 (79), *91*
McDonnell-Bushnell, L. P., 266 (38), *293*
McFarlane, N., 341 (35), *357*
McKennis, J. S., 267 (40), *293*
McLain, S. J., 220 (48), 254 (112), *256*, *257*
McLane, R., 103 (60), *133*
McMurry, J. E., 174 (112), *197*
McNiff, M., 165 (37), *195*
McQuillin, F. J., 165 (41, 48, 49), 167 (58), *195*
McRae, W. A., 320 (140), *332*
McVey, S., 308 (73–75), *331*
Mead, K. A., 410 (54, 55), *431*, *432*
Meakin, P., 159 (97), *162*
Medema, D., 176 (122, 123), 193 (234), *197*, *199*
Mehler, K., 150 (48), *160*
Meijer, H. J. 83 (276), *96*
Meijer, H. J. L., 201–203 (4), *204*
Meijer, J., 12 (71), *91*, 113 (138), *134*
Meinhart, J. D., 188 (193), *198*
Meinwald, J., 16 (80a), *91*, 130 (467), *141*
Meister, J., 125 (354, 356), *138*
Meixner, J., 125 (355, 357, 364), *138*, *139*
Mekhryakova, N. G., 342 (37), 354 (142), *357*, *359*
Melius, C. F., 414 (72), *432*
Mellor, J. M., 80 (271), *96*
Menapace, L. W., 74 (239), *95*
Mendelsohn, J., 7, 61 (49), *90*
Meremyi, R., 262 (22), 278 (81), *292*, *294*
Merijan, A., 265 (26), *293*
Merlino, S., 415, 416 (81), *432*
Merour, J. Y., 204 (12), *204*, 311 (87), *331*
Merta, R., 411 (61), *432*
Merzoni, S., 145 (16), 146, 151 (25), *160*, 322

(156), *333*
Mesnard, D., 105 (76), 106 (76, 81), *133*
Messmer, R. P., 224 (66), *256*
Metlin, S., 366 (30), *387*
Metzner, P. J., 165 (39), *195*
Meuniet-Pieret, J., 270 (58), *293*
Meyer, F. J., 121 (296), *137*
Meyer, K., 215 (26a), *255*
Meyer, N., 11 (67), *91*
Michaelv, W. J., 311 (89), *331*
Michel, J., 113 (127, 128), 120 (287, 291), 121 (299), 123 (127, 128), *134*, *137*
Michel, U., 129 (438), *140*
Miginiac, L., 36 (142), *93*, 101 (5–7, 17), 102 (5, 36–41, 43–48, 50, 51, 53), 103 (5, 50, 51, 55, 56, 62, 64), 104 (5, 53, 62, 64, 66, 67, 70, 72, 73), 105 (53, 62, 64, 66–74, 76–78, 79), 106 (75, 76, 79, 81), 107 (43, 73), 108 (73), 109 (104, 105, 109–111), 110 (5, 104, 109, 111–113), 111 (53, 55, 73, 117, 118, 122), 112 (73, 118), 117, 118 (250–252), *132–134*, *136*
Miginiac, Ph., 101 (6, 7, 11, 12, 18, 19), 102 (11, 12, 21–26, 28–30, 49), 103 (54), 107 (90), 110 (49, 112–114), 111 (115, 116), 112 (121), 116 (240), 119 (267), *132–134*, *136*, *137*
Migita, T., 37 (148), 38 (155), 40 (159), 49 (181a), 53 (159), 63 (208a), 66 (214a), *93*, *94*
Millendorf, A., 375 (67), *388*
Miller, D. J., 188 (193), *198*
Miller, M. J., 192 (227), *199*
Miller, R. G., 150 (50, 51), *160*
Milstein, D., 30 (128), 31 (130), 53 (128), 55 (128, 130), 56, 57 (130), 92, 168 (81), *196*, 321 (151), *333*, 338, 339, 341 (26), *357*
Minami, I., 66 (214b), *94*, 190 (212), 191 (217), *199*
Minasz, R. J., 129–131 (446), *141*
Minato, A., 113 (134), *134*
Minematsu, H., 182 (159), *197*
Minkiewicz, J. V., 179 (143), *197*
Minot, C., 35 (137), *92*
Mirskov, R., 25 (107), *92*
Mirskov, R. G., 27 (113), 37 (150), *92*, *93*
Mislow, K., 114 (188), *135*
Mitchell, M., 126 (377), *139*
Mitchell, M. A., 126 (376), *139*, 168 (74, 75), *196*
Mitchell, T. N., 54 (197), *94*
Mitra, A., 16 (79), *91*
Mitsudo, T., 316, 317 (118–122, 125), 319 (136), 322 (154), 328 (183, 184), *332*, *333*
Mitsuyasu, T., 178 (135, 136), 181 (151, 152), *197*
Mittelmeijer, M. C., 86 (304), *97*

Miura, H., 427 (127), *434*
Miyake, A., 177 (127–129), 184 (170–172), *197*, *198*, 231 (86), *257*
Miyake, H., 2, 70 (7), *89*
Miyano, S., 114 (195, 196, 200), 115 (209–211), *135*
Miyaura, N., 192 (225), 194 (237), *199*
Mizoroki, T., 349 (105), 353 (123, 124), *358*, *359*
Mizushina, K., 328 (183), *333*
Mizuta, N., 311, 312 (91), *331*
Mladenova, M., 116 (221), *136*
Moberg, C., 148 (32), *160*, *162*, 179 (142), 185 (181), *197*, *198*
Moffat, A. J., 345 (74), *358*, 421 (102), *433*
Moffat, J., 275 (77), 277 (78), *294*
Moffett, R. B., 121 (302), *137*
Moiseev, I. I., 223 (61a, 61c), *256*
Mol, J. C., 86 (309), *97*
Moland, J. C., 86 (305), *97*
Molander, G. A., 190 (214), *199*
Momen, S. A., 344 (58), *357*
Mondelli, G., 322 (156), *333*
Monsanto Co., 397 (22), *431*
Montani, I., 275 (74, 75), *294*
Montemarano, J., 67 (220), *95*
Montgomery, P. D., 351 (115), *358*
Montino, F., 151 (57), *161*
Moore, G. L., 128 (414), *140*
Morandini, F., 159 (102), *162*
Moreau, J. L., (82, 85–88, 92, 93), 108 (87, 93–95), 111 (82), 120 (85), *133*
Morelli, D., 165 (33), *195*
Moreo-Monas, M., 174 (108), *196*
Moretti, G., 337–339 (23), *357*
Mori, Y., 180 (148), *197*, 322 (159), *333*
Moriarty, R. E., 260 (7), *292*
Moriarty, R. M., 78 (258), *95*
Morikawa, M., 164 (5), 168 (80), 170, 173 (5, 90), 192, 193 (230), *195*, *196*, *199*, 438 (31), 439 (44), *444*
Morikawa, S., 20 (99), *91*
Moritani, I., 158 (94), *162*, 184 (175), *198*
Moriya, H., 203 (9), *204*
Moriya, O., 72, 77 (235), *95*
Morizawa, Y., 188 (198), *198*
Morooka, Y., 167 (59), *196*
Morozov, L. A., 397, 404, 405 (19e), 423 (116), *431*, *433*
Morrison, J. A., 124 (345), *138*
Morrow, S. D., 190 (208–210), *198*
Motherwell, R. S. H., 2, 70 (11), *89*
Motherwell, W. B., 2, 70 (11), *89*
Mozzanega, H., 430 (138), *434*
Mrowca, J. J., 345 (78), *358*
Mueller, D. C., 3 (34), *90*, 129, 130 (453–456), 131 (455), *141*

Muetterties, E., 269 (51, 52), *293*
Muetterties, E. L., 408 (40), 410 (57), 421 (105), 422 (109, 110), 423 (113), *431–433*
Mui, J. Y. P., 3 (26), *90*, 129 (446, 460–464), 130, 131 (446), *141*
Muir, K. W., 266, 270 (32), *293*
Mujuachi, Y., 354 (132), *359*
Mukaiyama, T., 11 (64), 43 (167), *91*, *93*, 127 (410), *140*
Mukayama, T., 42 (164), *93*
Mullen, A., 337 (11), *356*, 437 (17, 23b), 438 (33), *444*
Müller, E., 122 (310), *137*, 152 (71), *161*, 315, 317 (112, 113), *332*
Müller, F. J., 246 (103), *257*
Müller, F. W., 156 (87), *161*
Müller, H., 148 (35), *160*, 303 (45), *330*
Müller, J., 410 (53), *431*
Mullineaux, R. D., 377, 378 (74), 384 (99, 100), *388*, *389*
Münclnich, R., 264 (24), *293*
Murahashi, S., 184 (175), *198*, 326 (175), *333*
Murai, S., 114 (156, 199), 126 (398), *134*, *135*, *139*
Murphy, E. 397 (19b), *431*
Murphy, G. J., 3 (31), *90*, 129 (440, 445), 130 (445), *140*
Murray, M., 51 (187), *94*
Musco, A., 182 (155, 156), 183 (162, 163), *197*, *198*
Mushak, P., 167 (70, 72), *196*
Musikhina, V. N., 118 (257), *136*
Mutin, R., 85 (296), *96*
Muzart, J., 175 (116), *197*
Myeong, S. K., 328 (178), *333*
Mynott, R., 145 (12, 13, 19), 148 (19, 33), 149 (33), 150 (47, 48), 159 (33), *160*, 287 (95), *294*
Myshenkora, T. N., 354 (137), *359*

Naccache, C., 356 (154), *359*
Nadeau, R., 122 (325), *138*
Nagagawa, T., 114 (198), *135*
Nagakura, S., 410 (59), *432*
Nagaoka, H., 43 (165), *93*
Nagel, K., 215 (26a), *255*
Nagihara, H., 314, 317 (108), *332*
Nagihara, N., 261 (13), *292*
Naglieri, A. N., 354 (131, 134), *359*
Nagy-Magos, Z., 370 (49), *388*
Naiman, A., 282 (91), *294*
Nakamata, T., 354 (135), *359*
Nakamura, A., 260 (2), 264 (25), *292*, *293*
Nakamura, Y., 83 (277), *96*
Nakanisi, Y., 78 (259), *95*
Nakao, T., 114 (198), *135*
Nakayama, M., 349 (105), *358*

Nalesnik, T. E., 375 (69), 376 (72), *388*
Nametkin, N. S., 83 (273), *96*
Namy, J. L., 102 (33–35), *137*
Nannini, G., 116 (238), *136*
Naragon, E., 375 (67), *388*
Naragund, K. S., 123 (342), *138*
Narasaka, K., 11 (64), *91*, 127 (410), *140*
Naruta, Y., 45 (174), 46 (175), 47 (179), 49 (174, 182, 183), 50 (184, 185), *93*, *94*
Nashed, M. A., 2, 12, 66 (16), *90*
Natalie, Jr., K. J., 52 (188), *94*
Natta, G., 253 (107), *257*, 337, 338 (24), 345 (68), *357*, *358*, 373 (58), 374 (59), 381 (88), *388*, *389*, 413 (66), *432*
Nedelec, L., 116 (224, 225), *136*
Nefedov, B. K., 337 (17, 18), 354 (128, 129), *357*, *359*
Nefedow, O. M., 77 (251), *95*
Neff, H., 113 (131), *134*
Negiani, P. P., 343 (47), *357*
Negishi, E., 113 (132, 133, 136, 137), *134*, 191 (223, 224), *199*
Negishi, E. E., 2, 18, 24, 37, 52, 64 (2), *89*
Nehl, H., 103 (57–59), *133*
Nelson, J. V., 120 (277), *137*
Nenitzescu, C. D., 275 (76), *294*
Nesmeyanov, A. N., 124, 125 (346), 126 (346, 389, 391, 395), 127 (346), 128 (346, 412, 416), 129 (346), *138–140*, 262 (16–21), *292*
Nesmeyanova, O. A., 126 (389), *139*
Nestle, M. O., 260, 272 (3), *292*
Netzer, F. P., 421 (104), *433*
Neubauer, D., 437 (22), 438 (35), *444*
Neumann, S. M., 414 (75), *432*
Neumann, W. P., 6, 27 (47), 35 (138), 36 (138, 141), 37 (149), 41 (47), 66 (213), 69 (226), 77 (249), 78 (256), *90*, *92–95*, 125 (333), *138*
Newitt, D. M., 344 (58), *357*
Ng, S.-W., 45 (173), *93*
Nguen Trong Anh, 424 (119), *433*
Nicholas, K., 268, 284 (45), *293*
Nicholas, K. M., 260, 272 (3), *292*
Nicholson, J. K., 164 (9), *195*
Nicolau, K. C., 61 (205), *94*
Nicolini, M., 409 (47), *431*
Nienburg, H. J., 83 (280), *96*, 344 (57), *357*, 375 (65), *388*
Nijs, H. H., 427 (125b), 430 (137), *434*
Nikanorov, V. A., 53 (189), *94*, 125 (350, 352), *138*
Nikiforova, A., 440 (53b), *444*
Nishi, S., 191 (216), *199*
Nishida, T., 83 (282), *96*
Nishihara, K., 436, 440 (15a), *443*
Nishimura, J., 114 (185–187, 189, 191–194, 201), 124 (201), *135*

Nishioka, S., 275 (74), *294*
Nishiyama, Y., 427 (131), *434*
Nitzschmana, R. E., 370 (48), *388*
Nivert, C., 102 (53), 104 (53, 72, 73), 105 (53), 107, 108 (73), 111 (53, 73), 112 (73), *133*
Noak, K., 417 (87), *432*
Nobbs, M. S., 179 (140, 141), *197*
Nobbs, S., 154 (83), *161*
Nogi, T., 317 (126), *332*, 354 (146), *359*
Noguchi, Y., 116 (243), *136*
Noiseev, I., 440 (53b), *444*
Noller, C. R., 112 (123), *134*
Noltes, J. G., 3 (32), 5 (41), 7 (41, 51), 17 (82), 18 (82, 86), 77 (253), *90*, *91*, *95*
Nordberg, R. E., 185 (181), *198*
Norman, N. C., 284 (94), *294*
Norman, R. O. C., 79 (261–264), 80 (264), 81 (261–263), *95*, *96*
Normant, J. F., 103, 104, 108, 109 (61), *133*
Norton, J. R., 437 (18), *444*
Noth, H., 28 (123), *92*
Novaro, O., 417 (90), *432*
Noyori, R., 167 (68), *196*
Nozaki, H., 12 (69), 40 (158), 78 (259), *91*, *93*, *95*, 115 (214), 116 (218), *135*, *136*, 186 (189), 188 (198), *198*
Nozaki, S., 151 (62), *161*
Numata, T., 324 (163), *333*
Nurkeeva, Z. S., 84 (290), *96*
Nüssel, H. G., 239 (96), *257*
Nützel, K., 101, 107, 108, 110–115, 118 (2), 119 (124), *132*, *137*
Nyberg, E. D., 352 (121), *359*
Nyholm, R. S., 411 (62), *432*

Oberkirch, W., 145 (17), *160*, 219 (36), 244 (102), *255*, *257*
Obeschalova, N. V., 206 (1), 219 (38), 229 (77), 231 (38), 233, 234 (77), *255*, *256*
Oblad, A. G., 404 (31), *431*
Ochiai, M., 11 (66), 24 (106a), *91*, *92*
Ochsler, B., 6 (45), *90*
Oda, R., 167 (66), *196*
Odaira, Y., 129 (459), *141*
Odic, Y., 62 (206), *94*
Oertle, K., 175 (117), *197*
Ofstead, E. A., 84 (292), *96*
Ogawa, M. K., 125 (370), *139*
Ogawa, T., 2, 12, 66 (16), *90*
Ogura, T., 168 (77–79), *196*
Ohashi, K., 190 (212), *199*
Öhler, E., 101 (16), *132*
Ohno, A., 127 (408), *140*
Ohno, K., 181 (151, 152), 182 (157), *197*, 314 (107), 316 (127), 317 (107, 127), *332*, 439 (39), *444*
Ohshiro, Y., 190 (206), *198*

Ohta, M., 21 (94), 60 (203), *91*, *94*
Ohta, N., 170 (89), *196*
Oka, S., 127 (408), *140*
Okada, I., 67 (216), *94*
Okamota, H., 69, 70 (227), *95*
Okamoto, T., 127 (408), 128 (428), *140*, 346 (94), *358*
Okawara, M., 2 (8), 19, 21 (94), 37, 47 (153), 60 (203), 70 (89), 72 (235), 76 (247), 77 (235, 248), *89*, *91*, *93–95*
Okawara, R., 67 (216), *94*, 442 (64), *445*
Okawara, T., 116 (243), *136*
Okukado, N., 113 (133), *134*
Okumoto, H., 190 (207), 193 (232), *198*, *199*
Okuyama, S., 128 (415), *140*
Olah, G., 436 (14), *443*
Olah, J., 436 (14), *443*
Ol'dekop, Yu. A., 128 (422), *140*
Oleneva, G. I., 27 (117), *92*
Olgemöller, B., 410 (58), *432*
Olin, J. F., 336, 344 (4), *356*
Olivé, S., 253 (109), 254 (109, 111), *257*, 303 (44), *330*, 408 (42), 409 (42, 46), 410 (50), 412 (42, 63), 413, 414 (63), 416 (63, 85), 417 (63, 89, 91), 418 (93), 419 (63, 98), 420 (42, 63), 421 (42), 425 (122), 426 (63), 430 (140), *431–434*
Oliver, D., 425 (123), *433*
Oliver, J. P., 217 (30), *255*
Oliver, K. L., 443 (69), *445*
Olivier, K. L., 439 (48a, 48b), 440 (55), *444*
Olofson, R. A., 4 (35), *90*, 128 (413), *140*
Oltay, E., 373 (57), 378 (82), *388*
Omura, H., 126 (398), *139*
Onions, A., 148 (30), *160*
Ono, N., 2, 70 (7), *89*, 192 (229), *199*
Onoue, H., 184 (175), *198*
Onsager, O. T., 215 (28, 29), 216 (29), 219, 229 (39), 231, 233 (39, 88), 234, 236 (39), 237, 239 (88), 242 (39, 88), 244–246 (88), 247, 248 (39), 250 (39, 88), *255–257*
Oort, B. van, 422 (108), *433*
Oosthuizen, G. J., 423, 424 (115), *433*
Openheimer, E., 314, 317 (106), *332*
Orchin, M., 320 (143), *332*, 340 (31), *357*, 362 (7), 363 (18), 365 (23, 27), 366 (23–25, 28, 30), 367 (28), 368 (23, 33–35, 40), 369 (23), 371 (23, 50, 51, 53), 372 (23, 51, 55, 56), 373 (23), 375 (66, 69), 376 (71, 72), 377, 378 (23), 379 (23, 84), *387–389*, 421 (101), *433*
Orikasa, Y., 356 (151, 152), *359*
Orio, A. A., 409 (47), *431*
Orlava, N. A., 423 (116), *433*
Orlova, L. D., 119 (271), *137*
Orlova, N. A., 397, 404, 405 (19e), *431*
Orlova, Zh. I., 166 (53), *195*

Orpen, A. G., 284 (93, 94), *294*
Osborn, J. A., 384 (101, 102, 106), 385 (102), *389*
Oshima, K., 12 (69), 40 (158), *91*, *93*, 115 (214), *135*, 188 (198), *198*
Osipov, O. A., 83 (272), *96*
Osman, S. M., 114 (165), *135*
Oster, B. W., *162*
Ota, S., 151 (61), *161*
Otsuka, S., 164 (10), *195*, 203 (10), *204*, 260 (2), *292*, 326 (170), *333*
Otton, J., 86 (308), *97*
Ouchi, T., 83 (277), *96*
Owston, P. G., 223 (55, 56), 224 (56), *256*
Ozaki, A., 353 (123, 124), *359*
Ozin, G. A., 380 (87), *389*

Paetsch, J. D. H., 131 (486), *141*
Pale, P., 175 (116), *197*
Palenik, G. J., 266 (37), *293*
Palit, S. K., 319 (138, 139), 322 (157), *332*, *333*
Pallini, L., *162*
Palmer, M. H., 123 (326), *138*
Pampus, G., 87 (316), *97*
Panasenko, A. A., 179 (138), *197*
Pankratova, V. N., 122 (312), *137*
Pannetier, G., 350 (109), 351 (117), 352 (122), 354 (109, 122), *358*, *359*
Pansard, J. L., 107 (84), *133*
Panuzio, M., 172 (101), *196*, 316, 317 (123), 318 (132), *332*
Paquette, L. A., 288 (98), *294*, 298 (25), *330*
Parent, J. D., 395 (11), *431*
Parlman, R. M., 311 (88), *331*
Parshall, G., 297 (3), *329*
Parshall, G. W., 87 (314), *97*, 159 (99), *162*, 165 (43, 44), *195*, 297 (17), *330*, 337 (16), *357*, 362, 364, 370, 377 (10), *387*
Pashchenko, N. M., 229, 233, 234 (77), *256*
Pasynskii, A. A., 262 (16–21), *292*
Patard, G., 392, 397 (4), *430*
Patel, B. A., 179 (143), *197*
Patrick, T. B., 29 (127), *92*
Patterson, Jr., J. W., 19 (96), *91*
Pauley, G. S., 291 (112), *294*
Paulik, F. E., 350, 352 (108), *358*, 362, 379 (4), *387*
Pauson, P. L., 302 (36), 305 (61), *330*, *331*
Pearlman, P. S., 48, 63 (181), *94*
Pearson, A. J., 122 (313, 314), *137*, *138*, 305 (56, 61), 306 (62), *331*
Pearson, G. D. N., 410 (52), *431*
Pearson, R. G., 266 (34), *293*, 369 (41), *388*
Peiffer, G., 101 (9), *132*
Pelczar, F. L., 36 (140), *93*
Pellacani, L., 114 (157), *134*
Penfold, B. R., 272 (64), *293*

Penninger, J. M. L., 373 (57, 76), *388*, 420 (100), *433*
Perchenko, V. N., 83 (273), *96*
Perego, C., 183 (162), *198*
Perevalova, E. G., 126 (389, 391), 128 (412), *139*, *140*
Pereyre, M., 2 (5, 6), 4 (38), 5 (42, 43), 6 (43), 7 (48, 49), 31 (129), 34 (136), 37 (145–147), 38 (145, 146), 41 (146, 162), 42 (162), 43 (168), 61 (48, 49), 62 (48, 206), 66 (5, 6, 48), 68 (5, 48), 70 (5, 6), *89*, *90*, *92–94*, 114 (163), *135*
Peries, R., 33 (135), *92*
Perkins, P., 422 (112), *433*
Perkins, P. G., 417 (90), *432*
Perraud, R., 114 (152), *134*
Peruzzo, V., 41 (159a), 43 (166, 169, 170), 44 (169), *93*
Pestrikov, S. V., 223 (61a), *256*
Pete, J. P., 175 (116), *197*
Peter, R., 112 (124), *134*
Petersen, D. R., 36 (139), *93*
Petersen, R. B., 423 (117), *433*
Peterson, D., 202 (6), *204*
Peterson, D. J., 8 (54), 23, 60 (104), *90*, *92*
Petit, G. R., 11 (62), *91*
Petit, R., 302 (39), 303 (40, 41), 306 (64), *330*, *331*
Petitt, R., 268, 284 (45), *293*
Petrosyan, V. S., 126, 128 (379), *139*
Petrov, A. A., 27 (118), 28 (119), *92*
Pett, N. P., 128 (411), *140*
Pettit, R., 267 (40), *293*, 343 (44), *357*, 409 (45), *431*
Pfeffer, P. E., 442 (63), *445*
Philipsborn, G., von, 114 (166), *135*
Phillipou, G., 71, 74 (237), *95*
Phillips, L. R., 52 (188), *94*
Phillips Petroleum Co., 219 (34), *255*
Phung, N. H., 219 (37), 220 (47), *256*
Piacenti, F., 318, 319, 322 (133), 323 (133, 161), 324 (133), *332*, *333*, 336, 337, 339 (1), 343 (1, 47, 48, 50), 344, 345, 348 (1), 356, 357, 362 (3), 369 (44), 371, 372 (52), *388*, 436 (7), *443*
Piau, F., 188 (195, 196), *198*
Piazzesi, A. M., 442 (60), *444*
Piccolo, O., 159 (102), *162*
Pichler, H., 393 (5a, 5b, 6, 7), 394 (5a, 5b), 395 (15), 396 (16), 397 (18), 398 (5b), 404 (30), 406, 407 (5b), 408, 413 (41), 419 (5b), 420 (15), 423 (41), 428 (133), *430*, *431*, *434*
Piers, E., 33 (134), *92*
Pietropaolo, R., 164 (17), *195*
Piganiol, P., 337, 341 (22), *357*
Pigmolet, H., 239 (94), *257*

Pignolet, L. H., 222, 223 (54), *256*
Pinazzi, C. P., 86 (307), *97*
Pines, H., 211 (3, 4), *255*
Pinhey, J. T., 79, 81 (266–270), 82 (269, 269a), 83 (269b), *96*
Pinke, P. A., 150 (50, 51), *160*
Pinkerton, F. H., 54 (195), *94*
Pino, P., 297 (14), 303 (69, 77), 309 (69), 323 (161), *330*, *331*, 336 (1), 337 (1, 24), 338 (24), 339, 343 (1), 344 (1, 54, 56), 345 (1, 68, 75, 77), 348 (1), *356–358*, 362 (3), 367 (32), 369 (44), 371, 372 (52), *387*, *388*, 435 (6d), 436 (7), *443*
Piret, P., 270 (58), *293*
Piron, J., 270 (58), *293*
Pirozhkov, S. D., 354 (137), *359*
Pis'man, I. I., 212 (19), *255*
Pistor, H. J, 320, 323 (148), *332*, 354 (130), *359*
Pittman, C. V., 303 (48), *330*
Plate, N. A., 84 (290), *96*
Plazzogna, G., 43 (166), *93*
Plieninger, H., 264 (24), *293*
Poller, R. C., 2, 43, 45, 70 (20), *90*
Pommier, J. C., 2 (5, 17), 7 (48), 23 (106), 60 (204), 61, 62 (48), 66 (5, 17, 48), 68 (5, 48, 221), 69 (224), 70 (5, 221), *89*, *90*, *92*, *94*, *95*
Ponamarev, S. V., 8 (52), *90*
Ponaras, A. A., 71 (231), *95*
Ponec, V., 411 (61), *432*
Ponomarev, S. V., 6 (46), *90*
Popplestone, R. J., 148 (30), *160*, 178 (134), *197*
Pornet, J., 102 (41, 43–48), 107 (43), 109 (104, 105), 110 (104), 111 (120), *132–134*
Porri, L., 146 (23), *160*
Post, H. W., 66 (214), *94*
Potenza, J. A., 262 (23), *292*
Powell, J., 147 (27), *160*, 164 (9), 167 (61), *195*, *196*
Pozdnyakova, M. V., 83 (273), *96*
Praat, A. P., 147 (28), *160*
Pradere, J. P., 122 (317), *138*
Pratt, A. J., 12 (73a), *91*
Pratt, J. R., 54 (195), *94*
Pratt, L., 341 (35), *357*
Pregaglia, G. F., 165 (34), *195*
Preseglio, G., 154 (77), *161*
Prest, D. W., 272 (63), *293*
Prevost, Ch., 101 (6, 7, 10), 110 (112–114), *132*, *134*
Pri-Bar, I., 48, 63 (181), *94*
Price, J. R., 121 (307), *137*
Prince, R. H., 317 (128), 321 (149), *332*, *333*
Prince, S. R., 268 (47), *293*
Prinz, E., 253 (110), *257*

Prokai, B., 3 (27), 68 (222), *90*, *95*, 131 (482), *141*
Prout, F. S., 121 (300), *137*
Pruett, R. L., 362, 368, 371, 379, 382 (1), 384, 385 (1, 110), *387*, *389*
Pu, L. S., 416 (85), *432*
Pucci, S., 369 (44), *388*
Puddaphatt, R. J., 77 (254), *95*
Puddephatt, R. J., 25, 28 (108), 67 (215), *92*, *94*
Pullukat, T., 83 (279), *96*
Puzitskii, K. V., 336 (5), 337 (5, 17, 18), 345 (5), 354 (137), *356*, *357*, *359*
Pye, P. L., 416 (84), *432*

Que, L., 239 (94), *257*
Que, L. Fr., 222, 223 (54), *256*
Quina, F. H., 123 (330), *138*
Quiniou, H., 122 (317), *138*
Quintard, J. C., 2, 66, 70 (6), *89*
Quintard, J. P., 4 (38), 34 (136), *90*, *92*

Racah, E. J., 71 (231), *95*
Ragazzini, M., 253 (108), *257*
Rahm, A., 34 (136), *92*
Rähse, W., 423 (116), *433*
Raithby, P. R., 272 (62), *293*
Rakhlin, V. I., 37 (150), *93*
Rakhmankulov, D. L., 179 (139), *197*
Randrianoelina, B., 109 (104, 105), 110 (104), *133*
Rankel, L. A., 283 (92), *294*
Rao, B. R., 396, 411 (17), *431*
Rappe, A. K., 414 (72), *432*
Raschack, M., 114 (166), *135*
Raspin, K. A., 317 (128), 321 (149), *332*, *333*
Ratier, M., 114 (163), *135*
Rau, R., 423 (117), *433*
Rausch, M. D., 129 (433, 434), *140*, 265 (27), 266 (37), 270 (55), *293*
Rawson, R. J., 114 (172), 115 (213), *135*
Razuvaev, G. A., 128 (421–423), *140*
Read, G., 261 (12), *292*
Reed, H. W. B., 303 (47), *330*
Reetz, M. T., 112 (124), *134*
Reich, R., 219 (33), *255*
Reid, J. A., 123 (326), *138*
Reinheimer, H., 275 (77), 277 (78), *294*
Reinher, D., 154 (83a), *161*
Reininger, K., 101 (16), *132*
Reis, H., 343, 346, 348 (39), *357*, 437 (22), *444*
Reist, E. J., 123 (337), *138*
Renger, B., 6 (44), *90*
Rennison, S. C., 265 (26), *293*
Rensing, A., 15 (76), *91*
Renson, M., 122 (323), 123 (335, 336, 340), *138*

Repic, O., 114 (171), *135*
Reppe, W., 259, 289 (1), *292*, 303 (43), 308 (43, 70, 76), 309 (80, 81), 320 (148), 322 (158), 323 (148), *330–333*, 336 (3), 337 (3, 9, 10), 343 (46), 344 (3), 345 (65), 354 (130), *356–359*, 378 (77), *388*, 435, 438 (1), *443*
Reshetova, I. G., 116 (223, 226), *136*
Reuter, J. M., 2, 66 (15), *89*
Reutov, O. A., 53 (189, 190, 192), 55 (190), *94*, 125 (350, 352), 126 (378, 379, 382, 385, 386, 390, 392, 393, 397, 399), 127 (403, 407, 409), 128 (379), *138–140*
Reyx, D., 86 (307), *97*
Rhee, I., 126 (398), *139*, 311, 312 (91), *331*
Rheinwald, G., 86 (306), 87 (311, 317), 88 (318), *97*
Rhodes, Y. E., 114 (158), *134*
Rhum, D., 128 (414), *140*
Ricci, A., 68 (223a), *95*
Richter, P., 113 (130), *134*
Richter, W., 291 (113), *294*
Rieber, N., 4 (37), *90*, 126 (394), *139*
Riecke, R. D., 165 (21, 22), *195*
Riediker, M., 174 (110, 111), 194 (239), *196*, *199*
Riefling, B., 128 (426), *140*
Riegel, B., 120 (281), *137*
Riegel, H. J., *162*
Rieke, R. D., 114 (170), 116 (170, 217), *135*, *136*
Rinz, J. E., 343 (43), *357*
Ritchey, W. M., 308 (72), *331*
Rivetti, F., 354 (141, 143–145), *359*
Rizkalla, N., 354 (131, 134), *359*
Robbins, M. D., 23, 60 (104), *92*
Roberts, J. D., 223, 224 (59), *256*
Roberts, R. M., 123 (331), *138*
Robinson, D. T., 179 (140), *197*
Robinson, K. K., 351 (116), *359*
Robinson, S. D., 166 (56), *195*
Roct, W. G., 325 (169), *333*
Rodewald, G., 88 (320), *97*
Rodriguez, L. A. M., 413 (64), *432*
Roe, D. M., 308 (68), *331*
Roelen, O., 362 (15, 16), *387*, 392 (2, 3), 420 (2), *430*
Rogachev, B. G., 176 (124), *197*
Roloff, A., 150 (44), *160*
Romano, U., 354 (141, 143–145), *359*
Romanova, T. N., 223 (61c), *256*
Rone, J., 410 (54), *431*
Roos, B., 223, 224 (60), *256*
Roos, L., 320 (143), *332*, 368 (40), *388*
Roper, W. R., 415 (78), *432*
Rosai, A., 116 (238), *136*
Rosan, A. M., 304 (55), *331*

Rösch, N. 224 (66), *256*
Rosenberg, E. 130 (468), *141*
Rosenblum, M., 304 (54), 319 (137), *331*, *332*
Rosenthal, A., 320 (144), 326 (173, 174), *332*, *333*
Ross, B. L., 308 (72), *331*
Roth, J. A., 366 (24), 375 (66), *387*, *388*
Roth, J. F., 350 (108), 351 (116), 352 (108), *358*, *359*
Rothwell, I. P., 268 (53), *293*
Rottig, W., 406 (35), *431*
Roundhill, D. M., 301 (30), *330*
Roustan, J. L., 204 (12), *204*, 311 (87), *331*
Rowe, B. A., 79, 81 (269, 269b), 82 (269), 83 (269b), *96*
Rowe, J. M., 166 (57), *195*
Rowley, R. J., *93*
Rozenberg, V. I., 53 (189), *94*, 125 (350, 352), 127 (409), *138*, *140*
Rubotton, G. M., 114 (155), *134*
Rücker, G., 117 (249), *136*
Rudkovski, 339 (29), 343 (49), 344 (55), 345 (70, 86), 346 (90), *357*, *358* 435 (3), 442 (61), *443*, *444*
Rufinska, H., 150 (47, 48), *160*
Ruhrchemie, 397 (19c, 20), *431*
Ruitenberg, K., 113 (138, 139), *134*
Ruiz, M. E., 417 (90), *432*
Runge, T. A., 187 (191), *198*
Rupilius, W., 362 (8), *387*, 421 (101), *433*
Rupilus, W., 371 (51), 372 (51, 55, 56), 379 (84), *388*, *389*
Russell, C. E., 174 (107), *196*
Ruth, J. L., 194 (243), *199*
Ruwett, A., 123 (340), *138*
Ryang, M., 144, 151 (10), *160*, 311, 312 (91), 327 (176), 328 (178, 179, 181, 182, 185), *331*, *333*, 354 (137), *359*
Rybakov, V. A., 438 (32), *444*
Rykov, S. V., 126 (380, 383, 384), 128 (380), *139*
Ryu, I., 114 (156, 199), 126 (398), *134*, *135*, *139*
Rzaev, Z. M., 67 (219), *95*

Sabatier, P., 362 (12), *387*, 397 (24), *431*
Sacco, A., 378 (78), *388*
Sachtler, W. M. H., 410 (49), 411 (49, 62), *418* (49), *431*, *432*
Sacquet, M. C., 130 (469), *141*
Saegusa, T., 156 (88), *161*, 170 (89), 191 (216), *196*, *199*
Saft, M. S., 190 (208, 209), *198*
Saihi, M. L., 31 (129), *92*
Saitkulova, F. G., 116 (231), 118 (259, 263, 266), 119 (266, 270), *136*, *137*
Saito, K., 151 (62), *161*

Saito, O., 165 (23), *195*
Saito, T., 40 (158), *93*
Sakai, S., 164 (12), 167 (64), *195*, *196*
Sakakibara, M., 164 (12), *195*
Sakakibara, T., 129 (459), *141*
Sakakibara, Y., 127 (408), 128 (428), *140*, 346 (94), *358*
Sakikabara, T., 438 (37b), *444*
Sakuraba, M., 344 (62), *358*
Sakurai, H., 37 (151), 48 (180), 84 (286), *93*, *94*, *96*
Sakutagawa, 203 (10), *204*
Salerno, G., 144 (9), 147 (26), 151 (9), 152 (63, 66, 69, 70), 154 (9, 74, 77, 78), 157 (92), 158 (74, 93), *159–162*
Saloman, R. G., 2, 66 (15), *89*
Salz, R., 145 (12, 13, 19), 148 (19), *160*
Samorjai, G. A., 421 (106), *433*
Sanchez, C., 87 (312), *97*
Sanchez-Delgado, R. A., 267 (41), *293*
Sandal, V. R., 36 (139), *93*
Sano, H., 2 (8), 37, 47 (153), 70 (8), *89*, *93*, 354 (138), *359*
Santen, R. A., van, 406 (37), *431*
Santeniello, E., 116 (219), *136*
Santi, R., *162*
Sarafidis, C., 55 (198), *94*
Sarel, S., 301 (31), *330*
Sarkar, T. K., 37 (152), *93*
Sas, W., 190 (211), *198*
Sasaki, N., 86 (299, 301), *96*
Sasaki, Y., 183 (160, 161, 164), *198*, 439 (42a), *444*
Sasazawa, K., 38 (155), *93*
Sasson, Y., 321 (151), *333*
Satake, M., 439 (41), *444*
Satao, K., 50 (184), *94*
Sato, F., 201 (3), 202 (6a, 7), 203 (9), 204 (11), *204*
Sato, K., 151 (61, 62, 62b), *161*, 193 (232), *199*
Sato, M., 201 (3), 202 (6a, 7), 203 (9), 204 (11), *204*
Satoch, J. Y., 165 (50), *195*
Satoh, J. Y., 165 (42), *195*
Sauer, H., 215 (26a), *155*
Sauer, J., 78 (256), *95*
Sauerbier, M., 122 (310), *137*
Savchenko, I. A., 127 (407), *140*
Savel'eva, N. I., 8 (53), *90*
Savost'yanova, I. A., 27 (117), *92*
Sawa, Y., 328 (178, 179), *333*
Sawada, S., 114 (202), *135*
Sawyer, A. K., 2 (19), *90*
Sbrana, G., 308 (77), *331*, 354 (136), *359*
Scarpa, N. M., 23 (105), *92*
Schaap, C. A., 422 (108), *433*
Schaefer, D., *162*

Schaeffer, D., 439 (42b), *444*
Schaffer, W., 440 (55), *440*
Schalk, D. E., 87 (313), *97*
Scharf, E., 148 (35), *160*, 303 (45, 49), *330*
Schaub, R. E., 32, 33 (133), *92*
Scheben, J. A., 440 (57), *444*
Scheffold, R., 129 (438), *140*
Schell, R. A., 437 (19a), *443*
Schenkluhn, H., 149 (39), 150 (42, 45), *160*
Scherbakov, V. I., 124 (344), 129, 130 (448), *138*, *140*
Schick, K. P., 148, 149, 159 (33), *160*
Schlesinger, M. D., 397 (19b), *431*
Schlewer, G., 116 (241), *136*
Schleyer, P., 114 (188), *135*
Schlichting, O., 259, 289 (1), *292*, 309 (81), *331*
Schmeisser, M., 123 (339), *138*
Schmeltz, I., 439 (40b), *444*
Schmuck, R., 125 (358), *138*
Schmid, H., 45 (172), *93*, 166 (52), *195*
Schmidt, A., 29 (124), *92*
Schmidt, R., 5 (39), *90*
Schmidt, V., 101 (16), *132*, 145, 146 (22), *160*
Schmuff, N. R., 185 (180), 192 (227), *198*, *199*
Schneider, B., 3, 4 (33), *90*
Schneider, J., 215 (26b), *253*
Schneidt, D., 423 (116), *433*
Schoenberg, A., 437 (21), 442 (65), *444*, *445*
Schöllkopf, U., 4 (37), *90*, 114 (178), 126 (394), *135*, *139*
Schrauzer, G. N., 289 (105, 106), *294*, 317 (130), 320 (146), *332*
Schriewer, M., 35, 36 (138), *92*
Schroch, R. R., 220 (48), *256*
Schrock, R. R., 84 (294), 87 (314), *96*, *97*, 254 (112), *257*
Schroeder, W. D., 126 (388), *139*
Schroer, U., 37 (149), *93*
Schroth, G., 145 (12), 150 (47, 48), *160*
Schuitt, G. C. A., 411 (62), *432*
Schülte, K. E., 117 (249), *136*
Schultz, R. G., 167 (60, 62), *196*, 351 (115), *358*
Schulz, G. V., 398, 400 (25), *431*
Schulz, H., 395 (15), 396 (16, 17), 408 (41), 411 (17), 413 (41), 420 (15), 423 (41), *431*
Schulze-Steinen, H. J., 347, 348 (100), *358*
Schumann, H., 70 (228), 88 (320), *95*, *97*
Schumann, I., 70 (228), *95*
Schumann, K., 168 (73), *196*
Schwartz, J., 174 (109–111), 194 (239), *196*, *199*, 437 (18), 440 (53a), *444*
Schwarzenbach, K., 114 (179, 180), *135*
Schwarzenberg, K., 120 (279), *137*
Schwarzhans, K. E., 77 (255), *95*
Schweckendieck, W. J., 378 (77), *388*

Schweig, A., 37 (144), *93*
Schwind, H., 145, 146 (22), *160*
Scurrell, M. S., 355 (147, 148), 356 (147, 149, 150), *359*
Sears, C. T., *432*
Seay, J. G., 395 (11), *431*
Secco, F., 374 (61), *388*
Seebach, D., 6 (44), 9 (59), 10 (60), 11 (67), 12 (68), *90*, *91*
Seevogel, K., 145 (12, 13), *160*
Segnitz, A., 315, 317 (112, 113), *332*
Seibt, H., 148 (35), *160*, 303 (45), *330*
Seidel, W. C., *162*
Seifert, P., 78 (256), *95*
Seikina, M., 2, 70 (10), *89*
Seitz, D. E., 15 (75), *91*
Seitz, S. P., 61 (205), *94*
Sekiya, S., 183 (164), *198*
Sekutowski, C., 156 (89), *161*
Sekutowski, J. C., 145 (12), *160*
Seldes, A. M., 116 (228), *136*
Selman, C. M., 84 (287), *96*
Selover, J. C., 414 (75), *432*
Semenova, T. A., 438 (32), *444*
Semmelhack, F., 151, 154 (59), *161*
Semmelhack, M. F., 144 (5, 11), 145 (15), 151 (5), 154 (76b), 158 (93a), *159–162*, 312 (98), 313 (99), *331*
Senderens, J. B., 397 (24), *431*
Sendersens, J. B., 362 (12), *387*
Senicher, V. S., 28 (120), *92*
Sentralinstituttet for Industriell Forskning, 231 (80, 87), 236 (87), 247 (80), 252 (87), *256*, *257*
Serelis, A. K., 71, 74 (237), *95*
Sergeeva, N. S., 354 (128), *359*
Sergi, S., 164 (17), *195*
Serino, A. J., 189 (201), *198*
Serres, B., 131 (484), *141*
Servens, C., 5, 6 (43), 37, 38 (145, 146), 41 (146, 162), 42 (162), *90*, *93*
Seyferth, D., 3 (26–29, 31, 33, 34), 4 (33), 8 (55, 56), 18 (88), 19 (89, 90), 29, 30 (125, 126), 55 (198), *90–92*, *94*, 113, 114 (141), 129 (435–437, 439, 440, 442–447, 449–458, 460–466), 130 (443–447, 449–458, 470, 472–474, 476), 131 (446, 451, 455, 476–486, 489), *134*, *140*, *141*, 260, 272 (3), *292*
Shanzer, A., *66*
Shapiro, H., 2 (24), *90*
Sharanina, L. G., 27 (118), *92*
Sharipova, F. V., 179 (138), *197*
Sharma, R. A., 122 (324), *138*
Shaumann, E., 12 (70), *91*
Shaw, A. W., 211 (5), 213 (23), *255*
Shaw, B. L., 164 (9), 166 (54, 56), 167 (61, 63), *195*, *196*, 227 (72), *256*, 343 (42), *357*

Shelbaldova, A. D., 298 (27), *330*
Shechter, H., 112 (125), 114 (167), *134*, *135*
Sheldon, R. A., 78 (257), *95*
Shell, R. A., 345 (72), *358*
Shell Development, 232 (90), *257*
Shell International Research Maatschappij, 231 (82), *257*
Shell Oil Co., 231 (79, 83), 239 (79), *256*, *257*
Shenkluhn, H., 150 (44), *160*
Shenron, K. E., 437 (18), *444*
Shenvi, S., 64 (209), *94*
Sherman, P. D., 120 (279), *137*
Sheverdina, N. I., 101, 107, 108, 110, 112, 113, 115 (1), 119–121, 123 (272), *132*, *137*
Shibano, T., 175, 176 (120, 121), *197*
Shibasaki, M., 15 (78), *91*
Shield, T. C., 180 (149), *197*
Shields, T. C., 346 (89), *358*
Shier, G. D., 183 (166), *198*
Shih, H. M., 129, 130 (450), 131 (484), *141*
Shilov, A. E., 297 (16, 18), *330*
Shim, S. C., 316, 317 (118, 120–122), *332*
Shimizu, I., 66 (214b), *94*, 190 (212), 191 (215, 217), *199*
Shimizu, T., 354 (135), *359*
Shimizu, Y., 38 (155), 40, 53 (159), *93*
Shimoji, K., 12 (69), *91*
Shin, S., 437 (29), *444*
Shingles, T., 405, 406 (32), 423 (115, 116), 424 (115), *431*, *433*
Shiozawa, K., 354 (139), *359*
Shirley, D. A., 123 (328), *138*
Shoda, S., 43 (167), *93*
Shono, T., 167 (66), *196*
Shostakovskii, M. F., 25 (107), *92*
Shriner, R. L., 115 (215), *136*
Shriver, D. F., 423 (117), *433*
Shteinman, A. A., 297 (16), *330*
Sidebottom, P. J., 384–386 (107), *389*
Siebeneick, H. U., 114 (166), *135*
Siegel, H., 184 (168), *198*
Siegel, S., 120 (281), *137*
Siirala-Hansen, K., 154 (76a), *161*
Silveira, A., 113 (132), *134*
Sim, G. A., 305 (61), *331*
Simmons, H. D., 129, 130 (446, 450, 451), 131 (446, 451), *140*, *141*
Simmons, H. E., 113 (142, 144–146), 114 (142, 144–146, 169), *134*, *135*
Simons, L. H., 291 (115), *294*
Sims, J. J., 114 (154), *134*
Singh, G., 129, 130 (451), 131 (451, 489), *141*
Sinn, H., 215, 216 (29), *255*
Sjöberg, K., 154, 155 (84), *161*
Sitting, M., 214 (24), *255*
Skarstad, P. M., 423 (117), *433*
Slater, S., 269 (51, 52), *293*

Slaugh, L. H., 377, 378 (74), 384 (99, 100), 388, 389
Slaven, R. W., 305 (57), 331
Smidt, J., 164 (6), 195
Smissman, E. E., 116 (227), 136
Smith, C. W., 320 (146), 332
Smith, D. S., 289 (101), 294
Smith, J. A., 384, 385 (110), 389
Smith, J. J., 84 (283), 96
Smith, K. J., 427 (125a), 434
Smith, L. R., 303 (48), 330
Smith, P. J., 2, 24, 37, 43, 45, 70 (23), 90
Smith, R. D., 113 (142, 144), 114 (142, 144, 169), 134, 135
Smith, T. N., 148 (30, 34), 149 (37), 159 (34), 160
Smith, W. E., 130, 131 (476), 141
Snoble, K. A. J., 114 (150), 134
Snyder, E. I., 15 (77), 91
Sobata, T., 168 (77, 79), 196
Soga, O., 50 (184), 94
Söll, M., 215 (26b), 255
Sommelhack, M. F., 312 (97), 331
Somorjai, G. A., 427 (132), 434
Sonoda, N., 114 (156, 199), 126 (398), 134, 135, 139
Sonogashira, K., 315, 317 (115, 116), 332
Sorokin, G. V., 83 (273), 96
Souma, Y., 354 (138, 140), 359
Soussan, G., 120 (286), 121 (298), 123 (338), 137, 138
Speckamp, W. N., 74, 75 (244), 95
Speier, J. L., 297 (8), 330
Spencer, J. L., 260 (6), 272 (63, 64), 292, 293
Spierenburg, J., 79 (265), 96
Spirikhin, L. V., 179 (138), 197
Sreekumar, C., 15 (74), 91
Stacy, G. W., 120 (292), 137
Staicu, S., 168 (76), 196
Stakem, F. G., 179 (144, 145), 197
Staley, S. W., 114 (147), 134
Stamm, H., 116 (246), 136
Stampf, J. L., 116 (241), 136
Stansfield, R. F. D., 269 (48, 49), 271 (49), 272 (61), 284 (93), 293, 294
Stauffer, R. D., 150 (50, 51), 160
Stedronsky, E. R., 228 (75), 256
Stein, J., 408 (40), 423 (113), 431, 433
Steinbach, R., 112 (124), 134
Steinrucke, E., 145 (17), 160, 219 (36), 244 (102), 255, 257
Steinseifer, F., 8 (58), 10 (61), 91
Steinwand, P., 439 (47), 444
Steinward, P. J., 435, 440 (5a), 443
Stepovik, L. P. 122 (312), 137
Sternberg, H., 340 (31), 357
Sternberg, H. W., 366 (25, 28), 367 (28), 387

Sternhell, S., 79, 81 (266, 268, 270), 96
Sternhey, S., 79, 81 (267), 96
Steudle, H., 116 (246), 136
Stewart, J. J. P., 417 (90), 432
Stewart, J. M., 260 (4), 292
Stewart, R. P., 308 (67), 331
Stezowski, J. J., 423 (117), 433
Still, W. C., 8, 12, 13 (57), 15 (74), 16 (79), 19 (95), 91
Stille, J. K., 25 (108a), 30 (128), 31 (130), 38 (156), 39 (156a), 40 (108a), 48 (181), 53 (128), 55 (108a, 128, 130), 56 (108a, 130), 57 (130), 63 (181), 64 (209), 70 (228), 92–95, 169 (83), 194 (238), 196, 199, 337 (12), 356, 436 (12), 442 (66), 443 (67), 443, 445
Stille, T. K., 440 (52b), 444
Stimson, R. E., 423 (117), 433
Stiverson, R. K., 152 (65), 161
Støard, A., 223, 224 (60), 256
Stobbe, S., 145 (13), 156 (89), 160, 161
Stock, L. M., 79, 81 (260), 95
Stoeckli-Evans, H., 416 (82), 432
Stone, F. G. A., 51 (187), 94, 255 (114), 257, 260 (6), 269 (48, 49), 270 (59), 271 (49), 272 (61, 64), 291 (111), 292–294
Storch, H., 363 (18), 387
Storch, H. H., 393 (9), 397 (19b), 402, 403 (9), 406 (33), 407, 408, 423 (9), 430, 431
Strahle, J., 374 (62), 388
Strange, A., 2, 70 (11), 89
Strange, E. H., 215 (27), 255
Straub, H., 124–129, 131 (347), 138
Strauss, J. U., 170 (91), 173 (104), 196
Strauss, J. U. G., 173 (102), 196
Strauss, S. H., 423 (117), 433
Strege, P. E., 165 (36, 38), 170 (36, 38, 92, 94, 97), 172 (94, 99), 195, 196
Striegler, A., 344 (60), 357
Strohmeier, W., 246 (103), 257
Strohmeyer, M., 375 (65), 388
Struble, D., 71, 74 (236), 95
Struchkov, Yu. T., 166 (53), 195, 262 (20, 21), 292
Strunk, R. J., 74 (241), 95
Stuckwisch, C. G., 121 (304), 137
Studiengesellschaft, Kohle, 231 (84), 257
Stupfler, R., 423 (113), 433
Su, T. L., 2, 66 (18), 90
Subramaniam, R. V., 67 (217), 95
Suga, K., 178 (133), 197
Suggs, J. W., 19 (93), 91
Sugi, Y., 437 (29), 444
Suginome, H., 192 (225), 199
Sullivan, D. A., 266 (37), 293
Sumi, K., 11 (66), 91
Sumiya, T., 66 (214a). 94

Summerbell, R. G., 122 (319), *138*
Sun Oil Co., 231 (89), *257*
Susuki, T., 169 (86), *196*, 345 (85), *358*, 436 (11), 437 (17), *443*, *444*
Suzuki, A., 192 (225), 194 (237), *199*
Suzuki, H., 69 (225), *95*, 167 (59), *196*
Suzuki, K., 113 (134), *134*
Suzuki, R., 177, 178 (131), 179 (137), *197*
Suzuki, Y., 202 (6a), *204*
Swan, J. M., 297, 301, 302, 304–306, 312, 328 (5), *330*
Sweany, R. L., 368 (36, 37), 375 (70), *388*
Sweet, J. R., 416 (86), *432*
Syhora, K., 115 (212), *135*
Syrkin, Ya., 440 (53b), *444*
Szabo, L., 338 (28), *357*

Tachikawa, M., 422 (110), *433*
Tada, S., 11 (66), *91*
Taft, R. W., 244 (101), *257*
Tagawa, H., 128 (417), *140*
Tagliavini, G., 43 (166, 169, 170), 44 (169), *93*
Takagi, K., 126 (377), 127 (408), 128 (428), *139*, *140*
Takahashi, H., 12 (69), *91*, 164 (5), 169 (87), 170, 173 (5, 90), *195*, *196*
Takahashi, K., 177 (127–129), 178 (132), 184 (170–172), *197*, *198*
Takahashi, M., 167 (59), 194 (235), *196*, *199*
Takahashi, N., 356 (151, 152), *359*
Takahashi, S., 175, 176 (120, 121), 182 (159), *197*
Takahashi, T., 187 (190), *198*
Takahashi, Y., 164 (12), 165 (18), 167 (64), *195*, *196*
Takai, K., 115 (214), *135*
Takakis, I. M., 114 (158), *134*
Takami, Y., 83 (272), *96*, 175 (114), *197*
Takano, I. 63 (208a), *94*
Takasu, Y., 427 (127, 128), *434*
Takaya, H., 167 (68), *196*
Takayama, K., 37 (148), *93*
Takegami, Y., 316, 317 (118–122, 125), 319 (136), 320 (145, 147), 322 (154), 328 (183, 184), *332*, *333*
Takeuchi, A., 409 (44), *431*
Takuwa, A., 50 (184), *94*
Tam, W., 414 (75), *432*
Tamao, H., 202 (5), *204*
Tamao, K., 194 (235, 236), *199*
Tamaru, Y., 177, 178 (131), 179 (137), *197*
Tamelen, E. E. van, 11 (62), *91*
Tams, K., 113 (134), *134*
Tamura, M., 261 (13), *292*
Tamura, R., 2, 70 (7), *89*, 192 (228), *199*
Tamura, Y., 128 (415, 417), *140*
Tanable, Y., 192 (225), *199*

Tanaka, K., 145 (17), *160*, 165 (47), *195*, 219 (36), 244 (102), *255*, *257*, 369 381–383 (43), *388*
Tanaka, M., 58 (200), *94*, 316, 317 (119), *332*
Tanaka, Y., 46 (176), *94*
Tancrede, J., 304 (55), 319 (137), *331*, *332*
Tarbell, D. S., 121 (307), *137*
Tardella, P. A., 62 (207), *94*, 114 (157), *134*
Tarnow, M., 152 (68a), *161*
Tarsov, B. F., 438 (32), *444*
Tartiari, V., 183 (162), *198*
Tate, D. P., 308 (72), *331*
Tates, J. T., 414, 421 (69), *432*
Tatibouet, F., 120 (284, 288, 289, 295), 121, 123 (295), *137*
Tatsuno, Y., 164 (10), *195*
Tatsuoka, T., 16 (80a), *91*
Taube, T., 145, 146 (22), *160*
Tauster, S. J., 427 (130), *434*
Tawara, K., 191 (216), *199*
Taylor, J. S., 113, 114 (146), *134*
Taylor, P., 371 (50), *388*
Taylor, R. D., 17 (81), *91*
Taylor, W. F., 424 (121), *433*
Teichmann, B., 112 (126), *134*
Temkin, O. N., 342 (37), 354 (142), *357*, *359*
Temple, J. S., 174 (109, 111), 194 (239), *196*, *199*
Teng, K., 25 (108b), *92*
Teranishi, K., 354 (135), *359*
Teranishi, S., 194 (244, 245), *199*, 315, 317 (114), *332*
Terapone, J., 368 (39), *388*
Ter-Asaturova, N. I., 83 (273), *96*
Teratake, S., 20 (98, 99), *91*
Terawaki, Y., 116 (243), *136*
Terenghi, M. G., *162*
Terres, E., 214 (25), *255*
Tesche, B., 427 (128), *434*
Tetterboo, J. M. J., 345 (66), *358*
Teuben, J. H., 201–203 (4), *204*
Teyssie, Ph, 145 (14), *160*
Tezuka, Y., 168 (78), *196*
Thames, S. F., 54 (195), *94*
Thatchenko, I., 430 (138), *434*
Theodore, L. J., 37 (152), *93*
Theolier, A., 85 (296), *96*
Thieffry, A., 2, 70 (14), *89*
Thiele, K. H., 101 (14), 120 (14, 278), *132*, *137*
Thomas, C. B., 79 (262–264), 80 (264), 81 (262, 263), *96*
Thomas, E. J., 12 (73a), *91*
Thomas, F. L., 436 (15b, 16), 440 (15b), *443*
Thomas, J., 109, 121 (106–108), *133*, *134*
Thomas, M. G., 422 (109), *433*
Thompson, D. T., 362 (8), *387*
Thompson, M. R., 265 (28), *293*

Thorn, D. L., 223, 225–227, 233 (58), *256*, 409, 410 (48), *431*
Thurmann, E. N., 126 (381), *139*
Ticci, E. R., 379 (83), *388*
Tien, R. Y., 36 (140), *93*
Tilney Bassett, J. F., 260 (9), *292*
Timms, R. N., 165 (51), 173 (103, 105), *195*, *196*
Todd, L. J., 129 (435, 466), *140*, *141*
Toepel, T., 259, 289 (1), *292*, 309 (81), *331*
Tojo, T., 150 (43), *160*
Tolman, C. A., 145 (21), 149, 150 (40), 159 (97), *160*, *162*, 223 (62), 239 (95), 242, 244 (99), 255 (113), *256*, *257*, 379, 380 (85), *389*
Tolstikov, G. A., 179 (139), *197*
Tolstikova, G. A., 179 (138), *197*
Toma, S., 305 (61), *331*
Tomi, K., 322 (154), *333*
Tomita, H., 182 (158), *197*
Toniolo, L., 435 (2), 437 (25), 438 (30), 442 (25, 60), *443*, *444*
Topsøe, N., 427 (129), *434*
Torelli, V., 116 (224), *136*
Torisawa, Y., 15 (78), *91*
Toyo Rayon Co. Ltd., 212 (8), *255*
Toyoshima, I., 427 (127), *434*
Trautz, V., 325 (166), *333*
Travers, S., 101, 107 (8), *132*
Traylor, T. G., 19 (91), 23 (102), 50 (186), 55, 60 (199), *91*, *92*, *94*
Trius, A., 174 (108), 184 (174), *196*, *198*
Tronich, W., 130 (472–474), *141*
Tropsch, H., 362 (13,14), *387*, 395 (12), 408 (39), *431*
Trost, B. M., 38, 39, 52 (157), 54 (196), 63 (208), *93*, *94*, 148 (31), *160*, 163 (2), 165 (31, 36, 38, 39), 170 (31, 36, 38, 92–197), 172 (93–95, 98, 99), 185 (176–180, 183), 186 (184–188), 187 (191, 192), 189 (199, 200, 202–204), 190 (205, 214), 191 (219, 220), 192 (227), *194–196*, *198*, *199*, 322 (155), 326 (172), *333*
Tsai, A., 2 (1), *89*
Tsay, Y., 156 (89), *161*
Tsou, T. T., 415 (80), *432*
Tsuda, T., 156 (88), *161*, 191 (216), *199*
Tsuji, J., 66 (214b), *94*, 163 (1), 164 (5), 165 (45, 46), 168 (80), 169 (46, 82, 85–87), 170, 173 (5, 90), 175 (118, 119), 176 (126), 178 (135, 136), 180 (147, 148), 181 (151, 152), 182 (157), 187 (190, 207, 212, 213), 191 (215, 217), 192 (199), 193 (230, 232), *194–199*, 314 (105, 107), 315 (105), 316 (127), 317 (105, 107, 126, 127), 322 (105, 159), *332*, *333*, 337 (13), 342 (36), 345 (85), 354 (146), *357–359*, 435 (6c), 436 (11), 437 (17), 439 (38, 39, 44), *443*, *444*

Tsuji, J. R., 438 (31), *444*
Tsuji, Y., 181 (150), *197*
Tsukiyama, K., 167 (64), *196*
Tsukui, N., 2, 70 (7), *89*
Tsutsumi, S., 129 (459), *141*, 311 (90, 91), 312 (91), 327 (176), 328 (178, 179, 182, 185), *331*, *333*, 440 (56a), *444*
Tucci, E. R., 378 (75), *388*
Tueting, D., 25, 40, 55, 56 (108a), *92*
Tuggle, R. M., 301 (32), 308 (67), *330*, *331*
Tukusagawa, F., 383 (97), *389*
Tulip, T. H., 409, 410 (48), *431*
Tumanova, Z. M., 126 (395), *139*
Tummes, H., 364 (21), *387*
Turkel, R. M., 129 (466), *141*
Turnbull, P., 115 (212), *135*
Turnbull, R. J., 115 (213), *135*
Tzschach, A., 6 (45), *90*
Tzu-Jung, P., 114, 116 (170), *135*

Uchida, A., 86 (302), *96*
Uchida, H., 345 (73), *358*
Uchida, T., 194 (236), *199*
Uchida, Y., 443 (68), *445*
Uchino, M., 244, 245 (100), *257*
Uchiyama, T., 194 (244), *199*
Uda, J., 318 (131), *332*
Ueno, H., 190 (207), *198*
Ueno, Y., 2 (8), 19, 21 (94), 37, 47 (153), 60 (203), 70 (8), 72 (235), 76 (247), 77 (235, 248), *89*, *91*, *93–95*
Uglova, E. V., 126 (378, 379), 128 (379), *139*
Ugo, R., 165 (33), *195*
Uhm, S. T., 114 (170), 116 (170, 217), *135*, *136*
Ukhim, L. Yu., 166 (53), *195*
Ukita, T., 24 (106a), *92*
Ulmani-Ronchi, A., 318 (132), *337*
Ulmer, S. W., 423 (117), *433*
Umani-Ronchi, A., 316, 317 (123), *332*
Umemura, S., 436, 440 (15a), *443*
Ungrary, F., 368 (38), 376 (73), *388*
Union Carbide Corp., 397 (23), *431*
Universal Oil Products, Co., 212 (11, 12), *255*
Uno, H., 49 (183), 50 (185), *94*
Unwin, R., 427 (128), *434*
Upton, T. H., 224 (63), *256*, 414 (72), *432*
Uraneck, C. A., 88 (319), *97*
Urasaki, T., 169 (84), *196*
Urch, C. J., 21, 22 (101), *92*
Ushida, S., 47 (179), *94*
Uytterhoeven, J. B., 430 (137), *434*

Vahrenhorst, A., 10 (61), *91*
Vail, P. D., 190 (209), *198*
Vais, J., 114 (190), *135*
Valade, J., 5 (42), 7 (49), 12 (73), 61 (49), *90*, *91*

Valente, L. F., 113 (137), *134*
Valentini, G., 354 (136), *359*
Valpey, R. S., 188 (194), *198*
van Boven, M., 378 (76), *388*, 420 (100), *433*
Van Dam, P. B., 86 (304), *97*
van den Hurk, J. W. G., 83 (276), *96*
van der Helm, D., 45 (173), *93*
van der Kerk, G. J. M., 3 (30), 7 (51), 17 (82), 18 (82, 86), 83 (276), *90*, *91*, *96*
van Gemert, J. T., 220 (46), *256*
van Helden, R., 176 (122, 123), 193 (234), *197*, *199*
van Looy, H. M., 413 (64), *432*
Vannice, M. A., 406 (36), 427 (126), *431*, *434*
van Oort, B., 422 (108), *433*
van Santen, R. A., 406 (37), *431*
van Tamelen, E. E., 11 (62), *91*
Van Vuuren, P. J., 130 (467), *141*
Varagnat, J., 86 (308), *97*
Varapath, S., 151 (60), *161*, *162*
Vargaftik, M. N., 223 (61a), *256*
Vasil'kovskaya, G. V., 212 (19), *255*
Vechirko, E. P., 127 (404), *140*
Vedejs, E., 128 (424), *140*
Venturini, M., 374 (61), *388*
Verbeek, F., 5, 7 (41), *90*
Verbeek, H., 406 (37), *431*
Vereschagin, L. I., 123 (333, 334), *138*
Vergillio, J., 375 (67), *388*
Verhoeven, T. R., 148 (31), *160*, 185 (178, 179, 183), 186 (186–188), *198*
Verkuijlen, E., 86 (303, 309), *97*
Vermeer, P., 12 (71), *91*,113 (138, 139), *134*
Vetter, H., 308 (70), *331*
Vevioalles, J., 87 (311), *97*
Vialle, J., 337, 341 (22), *357*
Vidal, J. L., 381–383 (92), *389*
Vig, O. P., 114 (160, 161), *135*
Vigneau, M., 116 (225), *136*
Villani, F. J., 113 (132), *134*
Ville, G., 266 (39), *293*
Villemin, D., 87 (311), *97*
Villieras, J., 118 (253, 254), *136*
Vinson, J. R., 158 (96), *162*
Vittorelli, P., 45 (172), *93*
Vitulli, G., 146 (23), *160*
Vlasov, V. M., 27 (113), *92*
Vogt, S., 114 (171), *135*
Vol'eva, V. B., 126 (380, 383, 385, 390, 399), 128 (380), *139*
Volger, H., 165 (29), *195*
Volger, H. V., 165 (30), *195*
Vollhardt, K. P. C., 265 (28), 266 (39), 272, 274 (67), 278 (80), 280 (67, 87), 281 (88), 282 (90, 91), 290 (110), *293*, *294*, 309 (82, 83) 310 (82), *331*, 422 (112), *433*
Voltz, S. E., 429 (136), *434*

von Kutepow, N., 349 (103, 104), 354 (130), *358*, *359*
von Philipsborn, G., 114 (166), *135*
Vo-Quang, L., 115 (204, 205), *135*
Vo-Quang, Y., 115 (205), *135*
Voronkov, M. G., 37 (150), *93*
Vovsi, D., 84 (285), *96*
Vrieze, K., 147 (28), *160*

Wachter, J., 415 (79), *432*
Waddams, A. L., 363, 364 (19, 20), *387*
Wade, K., 222 (53), *256*
Wade, P. A., 190 (208), *198*
Wade, P. W., 190 (210), *198*
Wagner, S. D., 154 (76a), *161*
Wakamustu, H., 318 (131), *332*
Wakatsuki, Y., 266 (35), 274 (70), *293*, *294*
Wakselman, C., 123 (341), *138*
Walch, S. P., 414 (72), *432*
Waldman, M. C., 28 (121, 122), *92*
Walker, W. E., 180 (149), 181 (153), 184 (173), *197*, *198*, 346 (89), *358*, 381–383 (92), *389*
Wallendael, S. V., 188 (193), *198*
Walling, C., 71 (231–233), *95*
Walter, D., 145 (17), *160*, 219 (36), 233 (93), 239 (96), 244 (102), *255*, *257*
Walter, W., 12 (70), *91*
Wan, Ch., 423 (117), *433*
Wancowicz, D. J., 128 (413), *140*
Wang, H., 219, 229 (39), 231, 233 (39, 88), 234, 236 (39), 237, 239 (88), 242 (39, 88), 244–246 (88), 247, 248 (39), 250 (39, 88), *256*, *257*
Ward, J. E. H., 418 (97), *433*
Wardell, J. L., 17 (81, 85), 18 (85), *91*
Warner, C. R., 74 (241), *95*
Warwick, P., 194 (243), *199*
Washburne, S. S., 18 (88), *91*, 131 (480, 483, 485), *141*
Wataksaki, Y., 310 (84), *331*
Watanabe, K., 151 (62b), *161*
Watanabe, M., 72, 77 (235), *95*
Watanabe, S., 178 (133), *197*
Watanabe, Y., 316, 317 (118–122, 125), 319 (136), 320 (145, 147), 322 (154), 328 (183, 184), *332*, *333*
Waterman, E. L., 154 (76a, 84), 155 (84), *161*
Watson, J. M., 123 (331), *138*
Watts, L., 303 (41), *330*
Wayaku, M., 315, 317 (114), *332*
Wayland, B. B., 414 (76), *432*
Weaver, D. L., 301 (32), *330*
Webb, I. D., 151 (54), *161*, 312 (96), *331*
Webber, K. M., 351 (114), 352 (120), *358*, *359*
Weber, J., 344 (60), *357*
Weber, L., 165 (38), 170 (38, 92, 95), 172 (95,

98, 99), *195*, *196*
Webster, D. E., 297 (15), *330*
Weeks, P. D., 128 (424), *140*
Wege, D., 151 (62a), *161*
Weichmann, H., 6 (45), *90*
Weidenbruch, M., 123 (339), *138*
Weidner, U., 37 (144), *93*
Weil, T. A., *256*
Weimann, B., 149 (39), 150 (42), *160*
Weinstein, B., 304 (51), *330*, 346 (92), *358*
Weiss, E., 268 (46), *293*
Weiss, K., 418 (96), *433*
Weiss, R. C., 15 (77), *91*
Weissbarth, O., 320, 323 (148), *332*
Weisser, J., *162*
Weissermel, K., 297 (4), *330*
Weitkamp, H., 346, 347 (98), *358*
Welch, A. J., 270 (59), *293*
Wen, T. C., 291 (116), *294*
Wender, I., 297 (14), 322 (152, 153), 326 (173), *330*, *333*, 340 (31), *357*, 363 (18), 366 (25, 28, 30), 367 (28), *387*, 435 (6d), *443*
Wenderoth, B., 112 (124), *134*
Wenzel, W., 397 (19a), *431*
Wermer, P., 368 (35), *388*
Werner, C. M., 3 (32), *90*
Werner, H., 228 (76), *256*, 410 (53), *431*
Westermann, J., 112 (124), *134*
Westmijze, H., 113 (139), *134*
Whan, D. A., 429 (135), *434*
White, D. A., 166 (57), *195*, 302, 305 (37), *330*
White, D. L., 29, 30 (126), *92*
Whitesides, G. M., 127 (405), *140*, 228 (75), *256*, 273, 277 (68), *294*
Whitesides, T. H., 305 (57), *331*
Whitfield, G. H., 193 (233), *199*
Whitmore, F. C., 126 (381), *139*
Whitten, D. G., 123 (330), *138*
Whyman, R., 367, 373, 379 (31), 381 (89, 90), *387*, *389*
Wihdgassen, R. H., 317 (130), *332*
Wild, S. B., 302, 305 (37), *330*
Wilke, G., 144 (1, 3), 145 (1, 12, 13, 17), 148, 149 (1, 33), 150 (1), 151 (1, 3), 152 (1, 64, 71, 72), 154 (1, 79), 156 (1, 87, 89), 158 (95), 159 (1, 33), *159–162*, 175 (115), *197*, 219 (32, 36), 233 (93), 239 (96), 244 (102), 246 (97), *255*, *257*, 287 (95), 290 (107–109), *294*, 303 (50), *330*
Wilkes, J., 213 (22), *255*
Wilkes, J. B., 212 (14), *255*
Wilkinson, G., 165 (43, 44), *195*, 267 (41), *293*, 301 (30), *330*, 378 (79, 80), 384 (101–106), 385 (102–105, 111), *388*, *389*
Wilkinson, J., 164 (8), *195*
Wilkinson, P. R., 220 (46), *256*

Willemsens, L. C., 3 (30), 79 (265), *90*, *96*
Williams, G. J. B., 223 (57), *256*
Williams, J. M., 422 (110), *433*
Williams, R. O., 344, 345 (53), *357*, 442 (59), *444*
Willis, C. J., 3 (25), *90*
Willson, J. S., 79 (263, 264), 80 (264), 81 (263), *96*
Wilson, B., 87 (315), *97*
Wilson, J. D. C., 323 (160), *333*
Wilson, S. R., 52 (188), 87 (313), *94*, *97*
Windgassen, R. J., 320 (146), *332*
Wingler, F., 114 (181, 182), *135*
Winter, M. J., 266 (39), 269 (48, 49), 271 (49), 272 (61), *293*
Winter, S. R., 414 (74), *432*
Winton, P. M., 154 (83), *161*, 179 (141), *197*
Wise, J. J., 429 (136), *434*
Wiseman, F. L., 114 (147), *134*
Witte, J., 87 (316), *97*
Wittenberg, D., 148 (35), *160*, 303 (45, 49), *330*
Wittig, G., 114 (179–181, 183), 121 (296), 129 (432), *135*, *137*, *140*
Witting, G., 114 (182), 115 (207), *135*
Wobke, B., 88 (320), *97*
Woerlee, E. F. G., 86 (305), *97*
Wojcicki, A., 369, 372 (42), *388*
Wollenberg, R. H., 31 (132), 33 (135), *92*
Wolters, J., 79 (265), *96*
Wong, C. S., 345 (83), *358*
Wong, H. S., 172 (100), *196*
Woodburn, E. T., 427 (125a), *434*
Woodhouse, D. I., 305 (61), *331*
Woodhouse, J. C., 337 (6), *356*
Woodruff, R. A., 129, 130 (454), *141*
Woods, B. A., 414 (76), *432*
Woodward, P., 269 (48, 49), 271 (49), 272 (61), 284 (93), *293*, *294*, 410 (54, 55), *431*, *432*
Woodward, R. B., 298 (24), *330*
Wrackmeyer, B., 28 (123), 29 (124), *92*
Wright, T. L., 79, 81 (260), *95*
Wu, E. S. C., 154 (76b), *161*
Wursthorn, K. R., 19 (89, 90), *91*
Wykpiel, W., 6 (44), *90*

Yagupolski, L. M., 127 (404), *140*
Yagupsky, G., 384 (104), 385 (104, 111), *389*
Yamada, T., 191 (215), *199*
Yamakami, N., 318 (131), *332*
Yamamoto, A., 151 (56a), *161*, *162*, 165 (23), *195*, 416 (85), *432*
Yamamoto, H., 12 (69), *91*, 186 (189), 194 (236), *198*, *199*
Yamamoto, K., 18 (88), *91*, 131 (480), *141*, 150 (43), *160*, 182, (158), *197*

Yamamoto, T., 86 (302), *96*, 151 (56a), *161*, *162*, 165 (23), *195*
Yamamoto, Y., 41 (161), 46 (161, 178), 47 (161), 64 (210), 66 (210, 212), *93*, *94*, 158 (94), *162*
Yamanis, J., 356 (153), *359*
Yamashita, M., 316, 317 (118–122, 125), 322 (154), 328 (183, 184), *332*, *333*
Yamashita, S., 114 (198), *135*
Yamazaki, H., 261 (13, 14), 266 (35, 36), 274 (70), 291 (14), *292*, *294*, 310 (84), *331*
Yamomoto, Y., 46 (175), *93*
Yamortino, R. L., 320 (141), *332*
Yang, C. H., 404 (31), *431*
Yang, K.-C., 356 (153), *359*
Yano, T., 194 (237), *199*
Yashima, T., 356 (151, 152), *359*
Yashina, O. G., 123 (333, 334), *138*
Yasuda, H., 181 (150), *197*
Yasufuku, K., 418 (97), *433*
Yatagai, H., 41 (161), 46 (161, 175), 47 (161), 64, 66 (210), *93*, *94*
Yates, J. B., 37 (150a), *93*
Yokohoma, A., 421 (107), *433*
Yokohawa, C., 320 (145), *332*
Yokota, K., 159 (103), *162*, 191 (222), *199*
Yoshida, J., 194 (235, 236), *199*
Yoshida, T., 164 (10), *195*, 326 (170), *333*
Yoshida, Z., 177, 178 (131), 179 (137), *197*
Yoshimo, A., 49 (181a), *94*
Yoshimura, T., 167 (66), *196*
Yoshisato, E., 311 (90), *331*

Young, D. M., 83, 84 (278), *96*
Young, J. F., 384 (101, 106), *389*
Young, J. R., 122, 123 (315), *138*
Yukawa, T., 440 (56a), *444*
Yus, M., 145 (18), *160*

Zamlouty, G., 111 (115), *134*
Zapata, A., 15 (75), *91*
Zarnegar, B. M., 123 (330), *138*
Zavara, T. V., 123 (333, 334), *138*
Zavgorodnii, V. S., 27 (118), 28 (119), *92*
Zazzaroni, R., 343 (50), *357*
Zdunneck, P., 101, 120 (14), *132*
Zeise, W. C., 220 (49), *256*
Zelenova, L. M., 179 (138), *197*
Zentgraf, R., 28 (123), *92*
Zhir-Lebed, L. N., 354 (142), *359*
Ziegler, K., 215 (26a, 26b), *255*, 392 (1), *430*
Ziman, S. D., 326 (172), *333*
Zimmermann, H., 145 (17), *160*, 219 (36), 244 (102), *255*, *257*
Zimmie, J., 422 (110), *433*
Zitzman, J., 118 (261), *136*
Zlotskii, S. S., 179 (139), *197*
Zombeck, A., 352 (121), *359*
Zorin, V. V., 179 (139), *197*
Zosel, K., 215 (26a, 26b), *255*
Zubov, P. I., 67 (219), *95*
Zuckerman, J. J., 2 (3), 43 (171), 45 (173), *89*, *93*, 130 (468), *141*
Zwick, W., 125 (361, 362), *138*, *139*
Zwietering, P., 411 (62), *432*

Subject Index

Acetamide, 318
Acetaminomalonate, 177
Acetic acid, 348, 350, 351, 354
Acetonyl-oxiranes, 63
Acetoxydihydropyran, 189
2-Acetoxymethyl-3-allytrimethylsilane, 189
π-Acetyl complex, 415
N-Acetylalanine, 318
N-Acetylbenzylamine, 318
Acetylene, 309
Acetylenic ketones, 315
Acetyl iodide, 351
Acid anhydrides, 321, 344
Acid chlorides, cross-coupling reactions with, 56
Acidolysis, 372
Acrylonitrile, 348
Active hydrogen compounds, 344
Active hydrogen-containing nucleophiles, 336
Actylene, 303
N-Acylamino acid, 318
Acyl bromides, 313
Acyl halides, decarbonylation, 314
Acylium ion, 305
Acyloins, 178
Acyloxation reaction, 79
Adamantanethione, 325
1,4-Addition
 of carbanionic nucleophile, 59
 with cyclic enones, 9
Adipic acid, 320
Alcohol carbonylation, 314, 348, 355, 405
Alcoholysis, 337
Aldehydes, 154
 carbonylation of, 317
 commercial uses of, 363
 decarbonylation of, 317
 exothermic reactions, 43
 hydroformylation reaction, 362
 reductive alkylation and arylation, 318
 reductive amination, 317
Aldehydic phenyl hydrazones, 327

Aliphatic alcohols, 397
Alkane activation, 297
Alkane products in Fischer–Tropsch synthesis, 395
Alkene elimination, 17
Alkene protection and storage, 301
Alkenonylquinones, allylation of, 50
Alkenylaluminium, 191
1-Alkenylboranes, 191
Alkenyl bromides, Bu_3SnH reductions, 71
Alkenylmercury(II) chloride, 168
(Z)-Alkenylpentafluorosilicates, 193
Alkenylstannanes, Diels–Alder reactions of, 34
Alkoxycarbonyl complexes, 354
Alkoxycobalt, 375
α-Alkoxyl organolithium reagents, 13
Alkylaluminium compounds, 218
O-Alkylation, 187
Alkylation
 of fused ring systems, 80
 regioselectivity of, 11, 63
Alkyl halides
 carbonylation reactions, 311
 coupling reactions, 311
Alkylisopropenylmalonic acids, 165
Alkyllithium, 327
Alkyne complexes, 268
Alkyne dimers, 268
Alkyne dimetal complexes, 268
Alkyne oligomerization, 259
 catalytic cyclotrimerization, 272
 cotrimerizations, 282
 cyclization pathways, 260
 cyclodimerization, 262, 267
 cyclopentadienylcobalt-induced cocylizations, 280
 cyclotetramerization cyclooctatraene, 288
 dimerization, 261
 linear dimerization, 261
 reactions of diynes, 277
 retrocyclization, 266

Alkyne protection and storage, 307
Alkyne sequential linking, 287
Alkyne tetramer complexes, 283, 284
Alkyne tetramer formations, 288
Alkyne trimers, 268
  dimetal complexes of, 270
Alkynyl bromides, Bu$_3$SnH reductions, 71
Alkynyl-lead and tin compounds, coupling reaction with acid chlorides, 25
Alkynylstannanes from cyclopropene, 28
Alkynyl-tin compounds
  coupling reaction with acid chlorides, 25
  Diels–Alder reactions of, 29
Alkynyl transfer, 25
Alkyoxyethynyltins, reaction with halogenated ketones, 27
Allene cyclooligomerization, 304
Allenyltins, 27
C-Allylation, 190
Allyl alcohol, 164, 174
Allylation of acyl halides, 40
Allyl carbanions, stability of, 11
Ally chloride, 164, 165
Allyl complexes, 303, 305, 337
Allyldiethylamine, 184
Allyl enol carbonates, 191
Allyl-Grignard reagents, 204
Allylic carbonates, 190, 193
Allylic compounds, 303
Allylic halides, 341
  coupling reactions, 312
Allylic rearrangement, 38, 41, 43
Allyl ketones, 66
Allyllead compounds, 36
Allylnickel complexes, 143
  addition to activated olefins, carbonyl compounds, epoxides, and quinones, 154
  addition to dienes, 159
  attack by nucleophiles, 148
  β-hydride elimination, 149
  bis(allyl)nickel, 152
  carbon dioxide reaction, 156
  C–C coupling, 151
  C–Mg bond reactions, 159
  carbonylation, 156
  carboxylation, 156
  coupling, 145
  electrocyclic reaction, 147
  electrophilic attack, 146
  HCN addition, 159
  insertion of double or triple bonds, 152
  insertion processes, 157
  insertion reactions, 158
  ligand effects, 148, 154
  non-electrocyclic reaction, 147
  oxidative addition, 152
  reactivity, 144
  regiochemistry, 147
  regioselectivity, 154
  stereochemistry, 146
  syn = anti isomerization, 146, 150
Allylpalladium complexes, 163
  allylic compound reactions, 184
  allylic halide reactions, 193
  carbonucleophile reactions, 170
  carbonylation reactions, 180
  catalytic reactions, 175
  cocyclization reactions, 181
  conjugated diene reactions, 175
  dimerization with carbonycleophiles, 176
  miscellaneous reactions, 174
  nucleophilic attack, 167
  preparation, 164
    from allylic compounds, 164
    from diolefins and PdCl$_2$, 166
    from miscellaneous methods, 167
    from olefinic compounds, 165
  reactions of allenes, 183
  reaction with carbon dioxide, 174, 183
  reaction with carbon monoxide, 164, 168, 180
  reaction with isocyanides, 169
  toichiometric carbon–carbon bond formation, 168
Allylsilanes, 164
Allyl-stannanes, 191
  coupling with halogenated carbonyl compounds, 48
Allyl-tin compounds
  hydroxymethylation of, 47
  reaction with aldehydes, 41
  reaction with chloral, 41
  thermal reactions of, 37
Allyltitanium, 201
Allyltitanium complexes, oxidative addition reaction, 203
Allyl transfer reactions with organic halides, acetates, and acid halides, 37
Allyl transition metal complexes, 201
Allyltrimethyltin, reactivity of, 37
α,β-unsaturated acids, 320
α,β-unsaturated epoxides, 190
α,β-unsaturated ketones
  1,2-additions to, 7
  conjugate addition, 48
α-elimination, 3
α-substituted nitriles, esters and ketones, additions to aldehydes and ketones, 6
α-substituted organotin compounds, 5
Ambrettolide, 87
Amide, 328
Amide synthesis, 344
Amines, 406
Angular ethynyl group, 31

Subject index

Anti-Cram product, 46
Anti-elimination, 17
Anti-7-trimethylstannylnorcarane, 4
Aromatic halides, 311
 carbonylation of, 312
Aromatic ketoximes, 327
Aromatic nitriles, 326
Aroyl cyanides, 314
α-Arylated ketones, route to, 83
Arylation
 of fused ring systems, 80
 of polymethylbenzenes, 81
α-Arylation of unsymmetrical ketones, 66
Aryl bromides, reactions with, 38
Aryl esters, 312
Aryl halides, coupling reactions, 56, 311
Aryl iodides, 312
Aryllead tricarboxylate, 78
 as arylating agent, 81
Arylmercury compounds, 194
Aryl nitrides, 314
Aryl-plumbanes, 52
 reactions with acid halides, alkyl and aryl halides tetracyanoethylene, 53
Aryl-stannanes, 52
 reactions with acid halides, alkyl and aryl halides tetracyanoethylene, 53
Azo compounds, 327
Azobenzene, 326

Benzaldehyde, 181, 318
Benzene, 309
Benzil, 312
Benzocyclobutenes, 280, 281, 309
Benzoquinone, 182
p-Benzoquinones, 308
Benzyl chloride, 311
Benzyl halides, cross-coupling reactions with, 56
Benzyl mercaptan, 309
Benzyl-plumbanes, 52
 reaction with acid halides, alkyl and aryl halides, tetracyanoethylene, 53
Benzyl-stannanes, 52
 reaction with acid halides, alkyl and aryl halides, tetracyanoethylene, 53
Benzylic halides, 311
β-substituted organo-tin and lead compounds, 1,2-elimination, 17
BF_3-catalysed addition of crotyl- and prenyl-tin compounds to quinones, 45
Biallyl, 312
Biaryls, formulation of, 79
Bicyclic β-lactams, 75
Bicyclo[4.3.1]decenone, 184
Bicyclo[3.1.0]hex-2-enes, 301
Bimodal molecular weight distributions, 424

Bis(alkynyl)stannanes, heterocyclic compounds from, 29
Bis(cyclopentadienyl)tin, 77
1,1′-Bishomocubane, 300
Bis(pentamethylcyclopentadienyl)tin(II), 77
(*E*)-1,2-Bis(tributylstannyl)ethene, 31
2,2-Bis(trifluoromethyl)-3,6-divinyltetrahydropyran, 181
Bromoacetal, homolytic cyclization of, 72
7-Bromo-7-norcaranyltriphenyltin, 4
5-Bromopent-1-ene, 311
Buta-1,3-diene, 303
2-(Buta-1,3-dienyl)cyclopropane-1,1-dicarboxylate, 188
4-Butanelactam, 346, 347
But-1-en-3-ol, 192
But-2-en-1-ol, 192
But-3-enoate, 168
4-(But-3-enyl)dodeca-1,6,8,11-tetraene, 176
γ-Butyrolactone, 72, 322

Cadmium organometallics, 119
 acylation, 123
 addition reactions, 119
 alkyl, allylic, and benzylic halides substitution reactions, 122
 alkylation, 122
 allylic reactions, 119
 α-bromoesters, 122
 carbene and carbenoid intermediates, 124
 phenylic reactions, 120
 substitution reactions, 122
Carbene, 410
 cleavage, 329
 dimerization of, 410
 insertion of, 410
 mobility of, 410
Carbene complexes, 306
Carbinyl carbanion equivalents, 12
Carboalkoxy–cobalt complex, 338
Carbocyclization, reactions of, 58
3-Carbomethoxy-5,6-dimethylhept-5-en-2-one, 177
Carbon–carbon coupling reactions, 55
Carbon dioxide fixation, 202
Carbon monoxide, 193
 insertion of, 336
Carbonyl addition, 41
Carbonylation, 126, 156, 311, 314, 317, 322, 323, 336
Carbonyl compounds
 hydroxymethylation of, 11
 reductive amination of, 316
Carbonyl insertion, 349, 350, 353
Carbonyl metallate, 311, 327
Carboxylic acids, 312, 336
 decarbonylation of, 321

hydrosilylation, 321
Carroll rearrangement, 191
Chain growth probability, 398
Chemoselectivity
  in aldol reactions, 47
  of reactions, 38
α-Chiral aldehydes, addition to, 6
Chloral, 27
γ-Chloroallyl anion equivalent, 36
γ-Chlorobutyrophenone, tributyltin hydride reductions, 74
α-Chlorodiphenylacetyl chloride, 314
5-Chloropent-3-enoate, 169
Cholest-4-en-3-one, 173
Chromocene, 287
CIDNP effect, 375
Citronelloyl chloride, tributyltin hydride reductions, 74
Cobaltacycloheptatriene, 272, 275
Cobaltacyclopentadiene, 272, 274
Cobalt carbonyls, 337
Cobal-catalysed methanol carbonylation, 349
Cobalt-catalysed olefin hydroformylation, 337
Cobalt catalysts, 364, 365, 377, 379
Cobalt cyclobutadiene complexes, 264, 267
Coenzyme Q_1 (75%), synthesis of, 49
Conjugated dienes, 302, 346
Copolymerization reactions, 84
Cross-coupling, 55–56
Crotyl-tin compounds
  reaction with aldehydes, 41
  reaction with chloral, 41
  reactivity of, 43
Cumulenes, 303
  carbonylation of, 346
N-Cyanothebaine, 302
Cyclic transition-state structure, 43
Cyclization catalysts, 282
Cycloalkanes, 298
Cyclobutadiene complexes, 262, 265, 266, 273, 278, 280, 306
Cyclobutadiene iron, 280
Cyclobutene, 323
Cyclodestannylation, 60, 61
Cyclododeca-1,7-diyne, 278
Cyclododeca-1,5,9-triene, 303
Cyclohexanone, 311
  pyrrolidine enamine of, 177
Cyclohexanone enamine, 184
Cyclohexenols, 58
Cyclohexenone, 165
Cycloocta-1,5-diene, 303
Cyclooctatetraene, 259, 287, 288, 299
  nickel-catalysed formation, 290
  ring-opening reaction of, 287
Cyclooctyne, 269
Cycloolefin polymerization, 87

Cyclooligomerization, 303
Cyclopentadiene, 288
Cyclopentadienones, 265, 273, 303
Cyclopentadienylcobalt-induced cocyclizations, 280
Cyclopentene, ring-opening polymerization of, 87
Cyclopropanation, 188
Cyclopropanes, 21, 60, 167, 306, 329
Cyclopropenium ion, 301
Cyclopropylidene intermediate, 4
Cyclotetradeca-1,7-diyne, 278
Cyclotetramerization, 259
Cymantrene derivative, 288

Deca-2,8-diene, metathesis of, 84
Decarbonylation, 317
Dehydrosulphuration, 324
Dendrolasin, 13
15-Deoxy-16-hydroxyprostaglandins, 32
Desulphuration, 324
Deuterium isotope effects, 375
Dialkoxymethyltributyltin, 12
Dialkylacetals, carbonylation of, 323
Diaroylnickel, 312
Diarylmethanes, formation of, 79
Dibenzyl ketone, 311
1,11-Dibromoundeca-2,5,9-triene, 312
Dicobalt octacarbonyl, 364
Dicyclopropane, 300
2,2-Dideuterioadamantane, 325
Diels–Alder [4 + 2]cycloaddition, 346
Diels–Alder reactions, 34, 36, 306
Diene carbonylation, 346
Diene cyclooligomerization, 303
Dienes, 301, 345
1,3-Dienes, 11, 202, 204
Dienyl complexes, 305
Diesel fuel, 394
gem-Difluorocyclopropanes, 3
Dihydrobenzofuran, 76
Dihydroindole, 76
Dihydrojasmone, 19
Dihydropentalene, 292
Dihydropyrazole, 68
Diimides, 326
3,6-Diisopropenyl-1-phenyl-2-piperidone, 183
Diketene, 168
β-Diketones, reactions with, 82
Dimetallacyclobutene, 268
Dimethyl adipate, 341
Dimethyl carbonate, 354
5,6-Dimethylhept-5-en-2-one, 178
Dimethyl oxalate, 354
Dimolybdenum systems, 271, 284
Dinuclear metallacycles, 290
Dinuclear oxidative addition reaction, 367

Subject index

Di(octa-2,7-dienyl)ketone, 179
Diphenylamine, 324
Diphenylbutadiyne, 278
Diphenylcarbene, 314
Diphenylcyclopropenone, 315
Diphenyl disulphide, 312
Diphenylketene, 315
1,1-Diphenyl-2-thiourea, 324
1,3-Dipolar cycloaddition reactions, 5
trans-3,4-Disubstituted 4-butanolides, synthesis of, 12
3,6-Divinylpiperidines, 182
3,6-Divinyltetrahydropyrans, 181
α,ω-Diyne, 309
Diyne reactions, 277
Diynes, macrocyclic, 278
n-Dodeca-1,3,6,10-tetraene, 176

Ecdysone, 184
Electrophilic olefin complexes, nucleophillic attack, 304
Electrophilic substitution reactions, 305
1,3-Elimination, 21
Enol stannanes, 191
   aldol reactions of, 64
   alkylation with alkyl halides, 61–62
   allylic alkylation, 63
   chemoselectivity, 63
   erythro-selectivity with triphenyltin enolates, 64
   metallotropic forms, 61
   reaction with allyl acetates, 63
   reaction with halo ketones and halo aldehydes, 63
   regio-selectivity, 61
   regiospecificity of reaction, 62
   stereo-selectivity, 61
   threo-selectivity with trialkyltin enolates, 64
Episulphides, 326
Epoxides
   carbonylation of, 320
   rearrangement of, 318
   reduction of, 319
Erythro-β-hydroxycarbonyl compounds, 47
Erythro-selectivity of homoallylic alcohols, 46
Esters, 329
   carbonylation of, 322
Ethane codimerization, 253
Ethene dimerization, 209
Ethene oligomerization, 248
Ether carbonylation, 319, 320
2-Ethoxy-3-acetozypropene, 189
2-Ethoxyallylic acetate, 189
Ethoxynorcaranes, 4
(E)-2-Ethoxyvinyllithium, 33
Ethylene, 311
   polymerization of, 83

Ethylene glycol, 397
2-Ethylidenehepta-4,6-dienoic acid, 183
2-Ethylidene-nopinane, 172
3-Ethylidene-1-phenyl-6-vinyl-2-piperidone, 182
Ethyl ketones, 311
Ethylpent-3-enoate, 192
Ethylpropiolate, 288

β-Farnesene, 39
Farnesoate, 172
Fatty acid esters, homo-metathesis of, 86
Ferracyclopentadienes, 268
Ferroles, 264, 267, 268, 279
Fischer–Tropsch reaction kinetics, 405
Fischer–Tropsch synthesis, 362, 391
   alkali metal promoters, 423
   carbide theory, 409
   CO insertion, 408, 411, 418
   commercial, 393
   copolymerization, 410
   dispersity of metal centres, 421
   history, 392
   mechanisms of, 408, 412
   migratory insertion, 417, 424
   modified, 397
   molecular catalysis, 392
   molecular weight distribution, 402
   polymerization degree, 400
   polymerization reaction, 392
   present status, 393
   primary products, 395
   product selectivity, 425
   products of, 395
   rate-determining step, 413
   reaction scheme, 411
   secondary reactions, 396, 419
   selectivity, 425, 428
'Flyover' complexes, 270
Formaldehyde, 181
Formaldehyde ligand, 414
Formate ester, 375
β-Formylcarboxylic acid esters, 308
Formyl complex, 414
Friedel–Crafts conditions, 305
Fused ring systems
   alkylation of, 80
   arylation of, 80

γ-substituted organotin compounds, 21
Gasoline, 394
Geranylacetone, 171
Geranylgeraniol, 172
Germinate radical pair, 375
Glutarimide, 347
Graft polymerization, 84
Grignard reagents, 328

Halogen–lithium exchange, 8
Halohydrins, 49
Heck reaction, 124
n-Hexadeca-1,5,7,10,15-pentaene, 176
Hexadiene, 304
Hexa-1,5-dienyl chloride, 194
Hexa-1,5-diynes, 309
Hex-2-ene, cross-metathesis with methyl oleate, 86
Hexafluoro (Dewar) benzene, 302
Hexaphenylbenzene, 262, 263, 273
[Hg{Co(CO)$_4$}$_2$] catalyst, 278
Homoallyl alcohols, 201
Homoallylstannanes
  oxyselenation of, 61
  route to, 9
Homoenolate dianion systhesis, 16
Humulene, 174, 186, 312
Hydride abstraction, 305
Hydride ligands, 410
Hydride transfer, 338, 379
β-Hydride transfer, spirocyclic compounds from, 59
Hydridocobalt, 378
Hydridocobalt carbonyls, 365
Hydridoiron carbonyls, 320
Hydridoolefin complex, 382
Hydridorhodium carbonyls, 380
Hydrocarbons, 404
  carbonylation, 417
  hydroformylation, 417
  methyl-branched, 419
Hydrocarboxylation, 303, 337, 343, 345
Hydroformylation, 336, 345, 420
Hydrogen, oxidative addition, 410, 412
β-Hydrogen elimination, 227
Hydrogen–lithium exchange, 8
Hydrogenation, 337, 345
Hydrogenolysis, 373
Hydrosilylation, polymerization under, 321
Hydrosilylation reaction, 321
Hydroxyalkylcobalt, 375
Hydroxyalkyl ligand, 415
β-Hydroxy-amines, 7
Hydroxydendrolasin, 13
(E)-11-Hydroxydodecenoate, 194
β-Hydroxy-esters, 7
β-Hydroxy-ketones, 7
Hydroxymethylation
  of allytins, 47
  of carbonyl compound, 11
β-Hydroxy-nitriles, 7
γ-Hydroxypropylstannanes, conversion to, 19
β-Hydroxystannanes, route to, 8
γ-Hydroxyvinylstannanes, conversion to α,β-unsaturated γ-lactones, 24
Hyperconjugation, 24

Iminium ion formation, 316
Indan, 308
Indazolones, 327
Internuclear alkyl group exchange, 410
Iridium-catalysed methanol carbonylation, 352
Iron–cyclobutadiene system, 267
Isoalkyl ligand, 419
Isobutene, dimerization of, 215
Isomerization, 336
Isophorone, 165
Isoprene, 177, 180, 182
Isoprene synthon, 51
Isoquinoline systems, 282
Itaconate, 169

Ketenimine, 170
Ketones, 154
  decarbonylation of, 315
  reductive alkylation and arylation, 316
  reductive amination of, 316
Ketonic phenylhydrazones, 327
Kinetic enolate, 61
Koch reaction, 337

Lactams, 346
Lactones, 320, 325, 347
Lactons, carbonylation of, 322
Lavandalol, 47
Lead tetraacetate, 78
Lewis acids, 233, 239
Lewis bases, 236, 337
Limonene synthesis, 40
Linde 13X zeolite, 356
Lithium amides, 328
LTA/CF$_3$COOH oxidant solution, 80

Manganese complexes, 288
Markownikoff addition, 371
Masked aldehyde anionic equivalent, 12
Mercury organometallics, 124
  acylation, 127
  addition reactions, 124, 129
  alkene additions, 124
  alkylation, 126
  alkyne additions, 124
  carbene and carbenoid intermediates, 129
    B–C bonds, 131
    carbon–carbon double bonds, 129
    carbon–carbon triple bonds, 130
    C–H bonds, 131
    Ge–C bonds, 131
    Si-C bonds, 131
    Wittig alkene synthesis, 131
  dimerization, 128
  Heck reaction, 124
  insertion reactions, 130
  substitution reactions, 126

Subject index

Mesitylene, reactions with, 81
Mesityl oxide, substitution reaction of, 165
Metal acyl cleavage, 372, 383
Metal acyl formation, 371, 383
Metal alkyl formation, 369, 382
Metal hydride, 343, 413, 415
Metallacycle formation, 267, 275
Metallacycloheptatriene complexes, 272
Metallacyclopentadiene, 266, 270
Methacrylate esters, 346
Methanal, 375
Methane, carbide–carbene route to, 411
Methane formation, 397, 411
Methanol carbonylation, 348, 356, 362
    cobalt-catalysed, 349
    iridium-catalysed, 352
    rhodium-catalysed, 350
Methanol synthesis, 409
Methyl acetate, 349, 351, 354
2-Methylbut-1-ene, 165
Methyl cyclohexanecarboxylate, 174
Methylenation of aldehydes, reagent for, 15
Methylenecyclopropane, 385
Methyl 3-hydroxpropionate, 320
Methyl iodide
    carbonylation of, 311
    oxidative addition of, 350, 351, 353
3-Methyl-2-methylenebut-3-enyl alcohol, 183
3-Methyl-2-methylenebut-3-enylamine, 183
cis-2-Methyl-3-methyleneocta-1,5,7-triene, 183
trans-2-Methyl-3-methyleneocta-1,5,7-triene, 183
Methyl oleate, cross-metathesis of hex-2-ene with, 86
Methyloxirane, 318
Methyl penta-2,4-dienoate, 339
Methyl pent-3-enoate, 340, 341
4-Methylpent-3-enoate, 181
3-Methylsuccinimide, 347
1-Methyl-2-(trimethylstannyl)pyrrole, 54
Molybdenum complexes, 283
Monarch butterfly, 172
Monoenes, 343
Monsanto acetic acid process, 311
Myrcene, 39, 167, 178

Naproxen, 82
Nickelanonatetraene, 289
Nickel aroyl, 312
Nickel carbamoyl, 328
Nickel-catalysed cyclooctatetraene formation, 290
Nickel-catalysed olefin carbonylation, 341
Nickel complex catalysis, 231
Nickel(0) complexes, 303
Nickel hydride formulation, 150

Nitriles, 310
Nitroalkanes, 178, 190
Nitromethane, 178
Nona-3,8-dienoate, 180
Non-cage fused ring systems, 300
Non-conjugated dienes, 302
Norbornadiene, 299
Norbornene, 23
N-substituted amides, 344
N-substituted thioamides, 324
Nucleophilic acetaldehyde equivalent, 33
Nylon-66, 341

Ocimene, 167
Octa-3,7-dienoate, 193
2-Octa-2,7-dienyl-cyclohexanone, 177
Octa-1,3,7-triene, 175, 176
1-Octene carbonylation, 438
Oesterone, 282
Olefin activation via complex formation, 220
Olefin carbonylation, 336, 435
    effects of process variables, 438
    experimental conditions, 437
    functional groups, 437
    intermediates, 441
    mechanistic discussion, 337
    nickel-catalysed, 341
    non-redox, 436
        mechanism of, 441
        olefins used, 436
    normal to branched products, 437
    olefin reactivity, 437
    oxidative, 439
        major products formed, 442
        mechanism of, 442
    palladium-catalysed, 342
    reactions, 343
α-Olefin codimerization, 253
Olefin complexation, 369, 381
Olefin coordination, 379
Olefin ethylation, 213
Olefin hydroformylation, 361
    by-product formation, 374
    cluster formation, 374, 383
    cobalt-catalysed, 337
    electronic effects, 380
    free radical reactions, 375
    historical aspects, 362
    industrial aspects, 363
    mechanistic aspects, 365
    migratory insertion, 369
    nucleophile addition, 368
    oxidative addition, 379
    steric effects, 380
Olefin insertion, 225, 338, 341, 345
Olefin isomerization, 346, 347

Olefin metathesis, 84, 329
  cross-metathesis reactions of aliphatic unsaturated acids, esters, and polymers, 86
  stereospecificity of metathesis, 85
Olefin oligomerization, 205
  acid catalysis, 214
  and polymerization, 206
  anionic, 211
  base catalysis, 211
  coordinative transition metal complex catalysis, 219
  dimerization, 206, 208
  organometallic synthesis, 215
Olefin products in Fischer–Tropsch synthesis, 395
Open-chain transition states, 45
Organoaluminium compounds, 215, 216
Organoboranes, reactions with, 28
Organolead alkoxides, 67
Organomercury(II) halides, 328
Organotin alkoxides, reactions of, 66
Organotin amines, 67
Organotin enamines, 70
  metallotropic forms, 68
Organotin hydrides, 70
  intramolecular cyclization reactions, 71
  Ph_3SnH reduction, 75
Organozirconium species, 174
Ortho-esters, carbonylation of, 323
Oxazolidines, 179
Oxetanes, 48
Oxidative addition, 77
Oxidative coupling, 8
Oxidative dimerization, 79
Oxiranes, 48, 320
Oxyselenation of homoallylstannanes, 61

Palladium(II) acetate, 354
Palladium-catalysed olefin carbonylation, 342
Palladium catalysis in coupling with allyl bromides and acetates, 38
Palladium(II) catalysis mechanism, 56
Palladium chloro complexes, 346
Palladium complexes, 275, 287
  with alkynes, 265
Palladium(II) compounds, 354
Palladium cyclobutadiene complex, 291
Pentachlorobenzenethiol, 351, 352
Pent-3-enonate, 180
Perfluoroacetone, 181
Perylene synthesis, 16
dl-PGE_2, 32
Phenols, 81, 315
Phenylamine, 326
1-Phenylbutan-3-one, 318
Phenylcyclopropene, 167

2-Phenyl-3,6-divinyltetrahydropyran, 181
Phenylhydrazones, 154, 179
1-Phenyl-1-phospha-3-methylcyclopent-3-ene, 177
Phenyl-stannanes, 191
(Phenylthio) styrene, 11
(Phenylthio)(triphenylplumbyl) methyllithium, 11
(Phenylthio)(triphenylstannyl) methyllithium, 11
1-Phenyl-2-vinylhepta-4,6-dien-1-ol, 181
Pheromone of Monarch butterfly, 172
Phosphine, 364
Phosphite, 325
Phosphono-methyl triorganostannanes, addition to aldehydes and isocyanates, 6
Phthalic anhydride, 322
Phthalimidines, 326
Pinacol, 366
Platinum complexes with alkynes, 265
Plumbylation reaction, 79
Plumbyl-diazoalkanes, 5
Poisson distribution, 401, 428
Polyester-modified polyalkenylenes, 88
Polyesters, 84, 87
Polymer-bound rhodium complexes, 352
Polymethylbenzenes, arylation of, 81
Polypropylene, 84
Poly(vinyl fluoride), 84
Poly(4-vinylpyridine), 345
Prelog–Djerassi lactone, 46
Prenyl acetate, 172
Primary amines, 312, 397, 418
Progesterone, 173
Propargyl alcohols, 27, 354
Propene
  dimerization of, 209–211, 235, 237–240, 242–247
  oligomerization of, 214
Propene–butene codimerization, 249
Pyrazoles, 5, 27
Pyridine, 442
Pyrilium salts, 166
Pyrollidone, 347
Pyrone decarboxylation, 323
Pyrrolidine enamine of cyclohexane, 177

Quadricyclane, 299
  reactions with α-pyrones, 29
Quinazolones, 327
Quinones, 49
Quintessential catalyst, 365

Recifeiolide, 186
Redox reactions, 55
Reductive elimination, 77, 341
Regioselectivity of alkylation, 11, 63

Subject index

Reppe reaction, 303, 337
Rhenium alkyne complexes, 272
Rhenium complexes, 286
Rhodium acyls, 382, 383
Rhodium-catalysed dimerization, 251
Rhodium-catalysed methanol carbonylation, 350
Rhodium catalysts, 40, 251, 364, 380, 384
Rhodium complexes, polymer-bound, 352
Rhodium-X zeolite, 355
Rhodium-Y zeolite, 356
Ring closure reactions, 22
Ring-opening reaction of cyclooctatetraene, 287
Ring synthesis, 303, 306

Sasol fluid-bed process, 403, 404
Schiff base, 316, 324, 327
Schulz–Flory distribution, 398, 425
  deviations from, 428
Secondary amines, 312
Silane alcoholysis, 314
Silylcobalt complexes, 315
Silyl ester, 321
Sodium tetracarbonylferrate, 311
Spirocycles, 188
α-Stannylalkyl bromide, olefinic coupling products from, 15
α-Stannylalkyl iodide, cross-coupling product from, 15
α-Stannylated nitrosamines, addition to aromatic aldehydes, 6
Stannylated ynamines, coupling reactions with aryl halides, acyl and imidoyl chlorides, and ketenes, 25
α-Stannylcarbinols, conversion to α-stannylalkyl halides, 15
Stannyl-diazoalkanes, 5
Stannyl dithianes, 1,2-additions, 10
Stannylenes, 78
β-Stannyl ketones, reactions with alkyl halides, 19
α-Stannyl-organolithium, route to, 8
γ-Stannyl tertiary and benzyl alcohols, 21
Stereochemistry of aldol addition, 41
Stereoconvergence reaction
  of tributylcrotyltin with aldehydes, 45
  without allylic rearrangement, 45
Stereoselective synthesis of (Z)-2-alkenyltins, 46
Stereoselectivity, 43
  of aldol additions, 47
Stereospecific cycloaddition, 35
Steric hindrance, 379
Succinimide, 347
Sulphoxide deoxygenation, 324
Synthesis gas, 364

Tantalum complex catalysis, 254
Ternary catalyst, 83
Terpenes, 203
δ-Terpinol, 52
Testosterone, 173
Tetraalkyltin, 84
Tetracyanoethylene, 36, 50, 306
Tetrafluorobenzyne, 308
Tetrahydrofuran, 320, 321
Tetrahydrofuran-2-one, 320
Tetramethylallene, 302
Tetramethyl-p-benzoquinone, 308
Tetraorganotin compounds, hydride abstraction from, 55
Tetraphenylcyclopentadienone, 262, 308
Tetraphenylethylene, 315
Tetraphosphole complex, 384
Tetrasubstituted cyclooctatetraenes, 288
Tetra-*tert*-butylethylene, 316
Thermodynamic enolate, 61
Thioamides, 324
Thioanhydrides, 322
Thioester, 309, 344
Thioether, 324
Thioketones, 325
Thiolactones, 325
α-Thiolated acetaldehyde, synthetic equivalent of enolate of, 12
Thiols, 344
Thionocarbonates, 325
Thioureas, 324
Threo-selectivity of enol stannanes, 64
Tin/lithium exchange, 13
Titanium complex catalysis, 252
Transannular 1,4-aryl shift, 74
Transition metal carbene complexes, 329, 416
Transition metal carbonyls, 295
  acid anhydrides as substrates, 321
  alcohols as substrates, 314
  aldehydes as substrates, 317
  alkenes as substrates, 297, 301, 307
  bimolecular elimination, 329
  carbon–nitrogen unsaturated compounds as substrates, 326
  carbon–sulphur compounds as substrates, 324
  carboxylic acids as substrates, 321
  dialkylacetals as substrates, 323
  epoxides as substrates, 318
  esters as substrates, 322
  ethers as substrates, 318
  ketones and ketenes as substrates, 315
  main-group organometallics as substrates, 327–329
  metal carbene complexes as substrates, 329
  nitrogen–nitrogen unsaturated compounds as substrates, 326

nitrogen–sulphur compounds as substrates, 324
organic halides as substrates, 311
ortho-esters as substrates, 323
rearrangement/isomerization, 302
sulphoxides as substrates, 324
thermolysis, 329
Transition metal complex catalysis, 219
Transmetallation, 12
Trialkylphosphine, 378
Trialkyl(trihalomethyl)tins, halocarbenes from, 3
Tributylcrotyltin, 49
Tributylstannyl hydrazines, 69
Tributylstannyl propanamide, 16
β-Tributylstannylpropionaldehyde, 19
Tributylstannylpropionaldehyde, reaction with Wittig reagents, 59
2-(Tributylstannyl)propionitrile, route to cyclopropane derivatives, 20
Tributylstannylpyrrole, 69
Tributyl(trihalomethyl)tin compounds, addition to aldehydes, 5
Trienes, 305
Trienyl acetate, 16
1,1,1-Trifluoro-2-trifluoromethylhexa-3,5-dien-2-ol, 181
Trihaloallyltin, 42
Trimethylenemethane, 189
1-Trimethylsilylallyl acetates, 190
α-Trimethylsilylstannane, 15
(Trimethylstannyl)alkyl-substituted cyclohexenones, 58
Trimethylstannyl carbene, 4
β-Trimethylstannylethylidenetriphenylphosphorane, 19
Trimethylstannylisoprene, 51
Trimethyl(trifluoromethyl)tin, 3
Triorganostannyl esters, polymerization reactions of, 67
Triphenylstannyl carbene, 4
Triphenyl(trihalomethyl)plumbanes, reactions with olefinic compounds, 3
Triple bond, addition across, 27
Tris(dibenzylideneactone)dipalladium, 165
1,2,4-Trisubstituted benzenes, 271
Z-Trisubstituted homoallylic alcohols, synthesis of, 16

Unsaturated alcohols, 347
Unsaturated amides, 347
Unsaturated amines, 346
Unsaturated organic halides, 348
Unsymmetrical biaryls, yield with ArPb(OAc)$_3$, 81
Uranium powder, 291
Ureas, 328

Vanadacyclopentadiene, 262
Vapour-phase reactions, 351
Vinylbicyclo[3.2.1]octanone, 184
Vinyl chloride, 348
Vinylcyclopropane, 301
1-Vinyl-3,3-dimethylcyclohex-1-ene, 306
N-Vinylphthalimide carbonylation, 438
Vinylstannanes, 30
 reactions with allyl acetates, 31
 reactions with benzyl and aromatic halides, 31
 stereoselective reactions with acyl halides and acyl anhydrides, 31
Vinyl transfer, 30
Vitamin A, 172
Vitamin B$_{12}$, 317
Vitamin D, 185, 302
Vitamin E, 186
Vitamin K$_1$, 50

Water gas shift reaction, 337, 345, 353, 406
Wittig alkene synthesis, 115, 131
Woodward–Hoffmann rules, 298

o-Xylylenes, 309

Zeise's salt, 220
Zeolites, 429–430
Ziegler–Natta catalysts, 83, 219, 225, 252, 277, 417
Zinc organometallics
 addition reactions, 113
 allenic addition reactions, 107
  carbonyl derivatives, 107
  imines, 107
  isolated acetylenic triple bonds, 108
 allenic substitution reactions, 111
  derivatives with mobile halogen atom, 112
  derivatives with reactive alkoxy group, 112
 allylic addition reactions, 101
  addition to functionalized alkynes, 104
  addition to non-functionalized alkynes, 103
  carbonyl derivatives and epoxides, 101
  conjugated enynes, 105
  C=N derivatives and nitriles, 102
  isolated acetylenic triple bonds, 103
  isolated olefinic double bonds, 102
 allylic substitution reactions, 110
  derivatives with mobile halogen atom, 110
  derivatives with reactive alkoxy group, 111
 benzylic addition reactions, 101
  addition to functionalized alkynes, 104
  addition to non-functionalized alkynes, 103

Subject index

carbonyl derivatives and epoxides, 101
C=N derivatives and nitriles, 102
conjugates enynes, 105
isolated acetylenic triple bonds, 103
isolated olefinic double bonds, 102
carbene and carbenoid intermediates from, 113
   carbon–carbon double bonds, 113
   carbon–carbon triple bonds, 115
   C=N derivatives, 115
   Wittig alkene synthesis, 115
insertion reactions, 113
propargylic addition reactions, 107
   carbonyl derivatives, 107
   imines, 107
   isolated acetylenic triple bonds, 108
propargylic substitution reactions, 111
   derivatives with mobile halogen atom, 112
   derivatives with reactive alkoxy group, 112
reactions of, 101
Reformatsky and related reactions, 115

addition reactions, 116
   carbonyl derivatives, 116
   C=N derivatives and nitriles, 116
   derivatives with alkoxy or acyloxy group, 119
   derivatives with mobile halogen atom, 118
   substitution reactions, 118
   terminal alkynes, 117
saturated and miscellaneous addition reactions, 108
   carbonyl derivatives and epoxides, 108
   conjugated enynes, 109
   imines, 109
   terminal alkynes, 109
saturated and miscellaneous substitution reactions, 112
   derivatives with mobile halogen atom, 112
   derivatives with reactive alkoxy group, 113
   palladium-catalysed reaction with aryl, alkenyl, and allenic halides, 113
substitution reactions, 110